Signals and Communication Technology

This series is devoted to fundamentals and applications of modern methods of signal processing and cutting-edge communication technologies. The main topics are information and signal theory, acoustical signal processing, image processing and multimedia systems, mobile and wireless communications, and computer and communication networks. Volumes in the series address researchers in academia and industrial R&D departments. The series is application-oriented. The level of presentation of each individual volume, however, depends on the subject and can range from practical to scientific.

Indexing: All books in "Signals and Communication Technology" are indexed by Scopus and zbMATH

For general information about this book series, comments or suggestions, please contact Mary James at mary.james@springer.com or Ramesh Nath Premnath at ramesh.premnath@springer.com.

Hossam Mahmoud Ahmad Fahmy

Concepts, Applications, Experimentation and Analysis of Wireless Sensor Networks

Third Edition

 Springer

Hossam Mahmoud Ahmad Fahmy
Faculty of Engineering, Computer Engineering
and Systems Department
Ain Shams University
Cairo, Egypt

ISSN 1860-4862 ISSN 1860-4870 (electronic)
Signals and Communication Technology
ISBN 978-3-031-20711-2 ISBN 978-3-031-20709-9 (eBook)
https://doi.org/10.1007/978-3-031-20709-9

This Springer imprint is published by the registered company Springer Nature Switzerland AG
The registered company address is: Gewerbestrasse 11, 6330 Cham, Switzerland

Dedicated to my family;
parents, brothers, and sister with whom I
grew up warmly...
wife and daughters who gave my life a caring
touch ...
Dedication is not only for those who are in our
world ...

Preface

Writing a book is tempting, many ideas and topics, idea after idea, and topic upon topic, what to elaborate, which to mention, the reader must find a satisfying answer, enough knowledge; overlooking or going-by are painful choices for the author, space is limited, a hard decision is to be made, without compromising what should be transferred to the audience. Authoring a scientific book is like navigating the oceans in boat or a glass submarine, looking and searching for known and unknown species, appreciating diversified colors and variety of sizes, and collecting for a near benefit or for the future. In a third navigation for this book, I explored night and day, when cold and hot, whether windy or breezy, without missing any chance to know and learn.

Networking is a field of integration, of hardware and software, protocols and standards, simulation and testbeds, wired and wireless, VLSI and communication; it is an orchestrated harmony that collaborates dependably, all for the good of a connected, well-performing network. That is the charm of networking, of life in a civilization that recognizes differences and goes on.

In introductory computer networking books, chapters sequencing follows the bottom up or top down architecture of the seven layers protocol. This book is more moves ahead, both horizontally and vertically, the view and understanding are getting clearer, chapters ordering is based on topics significance to the elaboration of wireless sensor networks (WSNs) concepts and issues.

An in-depth focus is accorded to the notions of WSNs, their applications, and their analysis tools, and meticulous care has been accorded to the definitions and terminology. To make WSNs felt and seen, the adopted technologies as well as their manufacturers are presented in detail. With such a depth, this book is intended for a wide audience, it is meant to be helper and motivator, for both senior undergraduates and postgraduates as well as researchers and practitioners; concepts and WSNs-related applications are laid out, research and practical issues are backed by appropriate literature, and new trends are put under focus. For senior undergraduate students, it familiarizes with conceptual foundations, applications, and practical projects implementations. For graduate students and researchers, energy-efficient routing protocols, transport layer protocols, and cross-layering protocols approach are presented. Testbeds and simulators provide a must-follow emphasis on the analysis methods and tools for WSNs. For practitioners, besides applications and deployment, the manufacturers and components of WSNs at several platforms and testbeds are fully explored.

The contents of this edition are distributed over five parts. Part I (WSNs Concepts and Applications) includes Chaps. 1, 2, and 3. Part II (Network and Transport Layers, Cross-Layering) comprises Chaps. 4, 5, and 6. Part III (WSNs Experimentation and Analysis) is composed of Chaps. 7 and 8. Part IV (WSNs Manufacturers and Datasheets) contains Chaps. 9 and 10.

Chapter 1 introduces the basics of sensors and WSNs, the types of WSNs, and the standards specifically innovated to bring WSNs to useful life. Chapter 2 presents the distinctive protocol stack in WSNs. Chapter 3 is updated to lay out the plentiful applications of WSNs in military, industry, environment, agriculture, health, daily life, multimedia, and robotic WSNs.

Chapter 4 is added to this edition with a focus on the energy and lifetime-aware routing protocols designed to maintain sustained WSNs functionality. Chapter 5 is devoted to exhibiting characterizing transport layer protocols in WSNs. Chapter 6 tackles the cross-layered approach for protocol design, a WSN necessity to cope with their limited resources and energy constraints.

Analysis techniques of WSNs are prime to study, understand, and implement WSNs; these are the goals of Chaps. 7 and 8. Chapter 7 presents the testbeds, as available in research institutes and projects, to investigate protocols and practical deployments. Chapter 8 takes care of exhaustively surveying and comparing the simulation tools existing in the WSN realm.

Chapters 9 and 10 are meant to provide the full spectrum of the WSN industry, from a wide diversity of manufacturers to a full variety of products and their specs. They are not to be left over; they must be checked whenever a product or a manufacturer is mentioned in the text.

Part V (Ignition) is a single concluding chapter. Chapter 11 motivates the takeoff in WSNs study, research, and implementation.

Exercises at the end of each chapter are not just questions and answers; they are not limited to recapitulate ideas. Their design objective is not bound to be a methodical review of the provided concepts, but rather as a motivator for lot more of searching, finding, and comparing beyond what has been presented in the book.

Talking numbers, this book extends over 11 chapters, and embodies 335 acronyms, 192 figures, 38 tables, and more than 1000 references.

With the advances of technology, authoring a book is becoming easier, as information is attainable; but it is certainly hard to ensure details and depth are not missed. A book is a step in a long path sought to be correct, precise as possible, nonetheless errors are non-escapable, they are avoided iteratively, with follow up and care.

The preface is the first get-together between the author and the audience, it is the last written words; it is laying on the ground after the end line, to restore taken breath, to enjoy relaxing after long painful efforts, mentally and physically, and last but not least to relax in preparation for a new challenge. Bringing a book to life consumes months and months, days and nights, events after events, familial, social, and at the wide world of technology, sports, and politics. This book has seen many events and recorded some.

Writing is an agitation that stops only when the manuscript is submitted to the publisher. Ideas popping at bedtime are pins that hurt unless instantly written, what a sleep that is sleepless! Writing is a selfish, non-shareable addiction, a tenacious obsession that insists on full devotion. Authors have the blessing of delivering something that lasts. Don't we have access to books that go back hundreds of years, even before the Internet? With the Internet magic, authors are seen and heard everywhere, and readers are reached wherever they are.

An author has their ups and downs, as everybody, but they are visible like nobody. Could they manage to hide some of their letdowns? They have to, for the sake of their book, their readership. If you find somebody talking to themselves, tumbling, wearing a differently colored pair of shoes, don't laugh at them, they are probably writing a book!

Writing with care and feelings can be a title for my books; let's go with this third augmented edition.

Cairo, Egypt   Hossam Mahmoud Ahmad Fahmy

About the Book

This book focuses on the notions of WSNs, their applications, and their analysis tools; meticulous care has been accorded to the definitions and terminology. To make WSNs felt and seen, the adopted technologies as well as their manufacturers are presented in detail. In introductory computer networking books, chapters sequencing follows the bottom up or top down architecture of the seven layers protocol. This book is some more steps along, both horizontally and vertically, the view and understanding are getting clearer, chapters ordering is based on topics significance to the elaboration of wireless sensor networks (WSNs) concepts and issues.

With such a depth, this book is intended for a wide audience, and it is meant to be helper and motivator, for both senior undergraduates and postgraduates as well as researchers and practitioners. Concepts and WSNs-related applications are laid out, research and practical issues are backed by appropriate literature, and new trends are put under focus. For senior undergraduate students, it familiarizes with conceptual foundations, applications, and practical projects implementations. For graduate students and researchers, energy-efficient routing protocols, transport layer protocols, and cross-layering protocols approach are presented. Testbeds and simulators provide a must-follow emphasis on the analysis methods and tools for WSNs. For practitioners, besides applications and deployment, the manufacturers and components of WSNs at several platforms and testbeds are fully explored.

Contents

1.8.14   EnOcean 44

Part IV   WSNs Manufacturers and Datasheets

About the Author

Hossam M. A. Fahmy is Professor of Computer Engineering and has served as chair of the Computer Engineering Systems Department, Faculty of Engineering, Ain Shams University, Cairo, Egypt, from 2006 to 2008, and from 2010 to 2012. He participates in many academic activities in Egypt and abroad. Prof. Fahmy is recipient of Ain Shams University Appreciation Award in Engineering Sciences. He has published and refereed extensively for Springer, Elsevier, and IEEE journals and for several refereed international conferences. His teaching and research areas are focused on computer networks, MANETs, WSNs, VANETs, fault tolerance, software, and Web engineering. In 2016, he authored the book *Wireless Sensor Networks: Concepts, Applications, Experimentation and Analysis*, published by Springer, and in 2021, he authored its second edition titled *Concepts, Applications, Experimentation and Analysis of Wireless Sensor Networks*; he also wrote the book *Wireless Sensor Networks: Energy Harvesting and Management Techniques for Research and Industry* published by Springer, 2020.

Professor Fahmy founded and chaired the IEEE International Conference on Computer Engineering and Systems (ICCES) from 2006 to 2008, and from 2010 to 2013. He is a senior member of IEEE, IEEE Region 8 Distinguished Visitor (2013–2015; 2015–2018), and chair of the Distinguished Visitor Committee of the IEEE Computer Society. He speaks Arabic, French, and English.

List of Acronyms

ACA	Autonomous Component Architecture
ACC	Active Congestion Control
ACK	Acknowledgement
ACQUIRE	ACtive QUery forwarding in sensoR nEtworks
ADC	Analog to Digital Conversion/ Analog to Digital Converter
ADCP	Aerial Data Collection Problem
AER	Address Event Representation
AFOSR	Air Force Office of Scientific Research
ALBA	Adaptive Load-Balanced Algorithm
AM API	Aggregate Manager API
AmI	Ambient Intelligence
AMR	Anisotropic Magneto-Resistive
AoA	Angle of Arrival
AODV	Ad hoc On-demand Distance Vector
AP	Access Point
API	Application Program Interface
APS	Ad hoc Positioning System
APTEEN	AdaPtive Threshold sensitive Energy Efficient sensor Network
ARQ	Automatic Repeat reQuest
ART	Asymmetric and Reliable Transport
ATM	Asynchronous Transfer Mode
BAN	Body Area Network
BEER	Balanced Energy Efficient Routing
BER	Bit Error Rate
BGP	Border Gateway Protocol
BIOSARP	Biologically Inspired self-Organized Secure Autonomous Routing Protocol
BLE	Bluetooth Low Energy
BSD	Berkeley Software Distribution
CBMPR	Cluster-Based Multipath Routing
CBR	Constant Bit Rate

CCA	Clear Channel Assessment
CCD	Charge-Coupled Device
ChSim	CHannel SIMulator
CIF	Common Intermediate Format
CINEMa	Cyborg Insect Networks for Exploration and Mapping
CKN	Connected k-Neighborhood
CLB	Cross-Layer Routing Protocol for Balancing Energy Consumption
CLDP	Cross-Link Detection Protocol
CLEEP	Cross-Layer Energy-Efficient Protocol
CLOD	Cross-Layer Optimal Design
CMOS	Complementary Metal-Oxide Semiconductor
CODA	COngestion Detection and Avoidance
COM	Component Object Model
COOJA	COntiki Os JAva
COP	Computer Operating Properly
CORBA	Common Object Request Broker Architecture
COST	COmponent-oriented Simulation Toolkit
COTS	Commercial Off-The-Shelves
COVID-19	COronaVirus Disease 2019
CPLD	Complex Programmable Logic Device
CPU	Central Processing Unit
CSI	Channel State Information
CSLRP	Coverage, Sink Location, and Routing Problem
CSMA/CA	Carrier Sense Multiple Access/ Collision Avoidance
CSS	Central Supervisory Station
DARPA	Defense Advanced Research Projects Agency
DCF	Distributed Coordination Function
DD	Directed Diffusion
DDR3	Double Data Rate Type Three
DHAC	Distributed Hierarchical Agglomerative Clustering
DiffServ	Differentiated Services
DIMM	Dual In-line Memory Module
DOA	Direction of Arrival
DoS	Denial-of-Service
DREAM	Distance Routing Effect Algorithm for Mobility
DSP	Digital Signal Processor
DSR	Dynamic Source Routing
DTC	Distributed TCP Cache
DTSN	Distributed Transport for Sensor Networks
DV-hop	Distance Vector-hop
DVR	Digital Video Recorder
E^2R	Energy Efficient Routing
EAR	Energy Aware Routing
EBCR	Energy-Balanced Cooperative Routing

EBGR	Energy-efficient Beaconless Geographic Routing
EC-CKN	Energy Consumed Uniformly-Connected k-Neighborhood
ECDC	Energy and Coverage-aware Distributed Clustering
ECG	Electro-Cardiogram
ECLP	Enhanced Cross-Layer Protocol
EDT	Eastern Daylight Time
EEG	Electro-Encephalogram
EEMHR	Energy-Efficient Multilevel Heterogeneous Routing
EEOR	Energy-Efficient Opportunistic Routing
EEPROM	Electrically Erasable Programmable Read-Only Memory
EIGRP	Enhanced Interior Gateway Routing Protocol
EKG	Electro-Cardiogram
EMI	Electromagnetic Interference
EPRB	Ethernet PRogramming Board
EQSR	Energy efficient and QoS aware multipath Routing
ERTP	Energy efficient and Reliable Transport Protocol
ESM	Experience Sampling Method
ESN	Environmental Sensor Network
EST	Eastern Standard Time
E-TORA	Energy-aware Temporarily Ordered Routing Algorithm
ExOR	Extremely Opportunistic Routing
FEAR	Fair Energy Aware Routing
FFD	Full Function Device
FIFO	First-In First-Out
FOV	Field of View
FPGA	Field Programmable Gate Array
FPS	Frames Per Sec
FPSLIC	Field Programmable System Level Integrated Circuit
FSK	Frequency-Shift Keying
GaAs	Gallium Arsenide
GCP	Global Control Processor
GDSTR	Greedy Distributed Spanning Tree Routing
GEAR	Geographic and Energy Aware Routing
GEM	Graph EMbedding for routing
GENI	Global Environment for Network Innovation
GFSK	Gaussian Frequency-Shift Keying
GloMoSim	Global Mobile Information System Simulator
GLRM	Grid-based Load-balanced Routing Method
GMT	Greenwich Mean Time
GPI	Geographic Priority Index
GPIO	General-Purpose Input/Output
GPO	GENI Project Office
GPRS	General Packet Radio Service
GPS	Global Positioning System

GPSR	Greedy Perimeter Stateless Routing
GRAB	GRAdient Broadcast
GSN	Global Sensor Network
GTS	Guaranteed Time Slots
GUI	Graphical User Interface
HARP	Hierarchical Adaptive and Reliable routing Protocol
HIL	Hardware-in-the-Loop
HMRP	Hierarchy-based Multipath Routing Protocol
HOL	Head-Of-Line
HSN	clustered Heterogeneous Sensor Network
IACK	Implicit Acknowledgement
IACUC	Institutional Animal Care and Use Committee
ID	Identification
IDE	Integrated Development Environment
IDL	Interface Definition Language
IEMA	Itinerary Energy Minimum Algorithm
IEMF	Itinerary Energy Minimum for First-source-selection
IETF	Internet Engineering Task Force
IGRP	Interior Gateway Routing Protocol
IID	Independent and Identically Distributed
INET	Internetworking Framework
IP	Ingress Protection/Internet Protocol
JPEG	Joint Photographic Experts Group
JRPRA	Joint Routing, Power control and Random Access
JSP	JavaServer Pages
JTAG	Joint Test Action Group
LAF	Location-Aided Flooding
LAN	Local Area Network
LCD	Liquid Crystal Display
LDO	Low Dropout Regulator
LEACH	Low-Energy Adaptive Clustering Hierarchy
LEACH-C	Low-Energy Adaptive Clustering Hierarchy Centralized
LE-MHR	Lifetime extended Multi-levels Heterogeneous Routing
LMCRTA	Lifetime Maximization Cooperative Routing with Truncated Automatic repeat request
LOS	Line of Sight
LQI	Link Quality Indication
LR-WPAN	Low Rate Wireless Personal Area Network
MANET	Mobile Ad hoc Network
MBD	Muzzle Blast Detection
MCC	Motor Control Center
MCSA	Motor Current Spectral Analysis
MCU	Micro-Controller Unit
M-EECP	Multihop Energy-Efficient Clustering Protocol

MEMS	Micro-Electro-Mechanical System
MF	Mobility Framework
MIMO	Multiple-Input and Multiple-Output
MiNT	Miniaturized Network Testbed for Mobile Wireless Research
MIP	Multiagent-based Itinerary Planning
MITM	Man-In-The-Middle
MiXiM	MiXed siMulator
MMC/SD	Multimedia Card/Secure Digital
MMSPEED	Multipath and Multi-SPEED
MORE	MAC-independent Opportunistic Routing and Encoding
MOVIE	Mint-m cOntrol and Visualization InterfacE
mp-MILP	multi-parametric Mixed-Integer Linear Program
MPR	Mote Processor Radio board
MTT	Multi-target Tracking
MURI	Multidisciplinary Research Program of the University Research Initiative
NACK	Negative Acknowledgement
NAM	Network Animator
NED	NEtwork Description
NFC	Near Field Communication
NFS	Network File System
NIST	National Institute of Standards and Technology
NPE	Network Processor Engine
NRT	Network Research Testbed
NSF	National Science Foundation
NTP	Network Time Protocol
NUI	Natural User Interface
OEM	Original Equipment Manufacturer
OGF	On-demand Geographic Forwarding
OLSR	Optimized Link State Routing
OMF	ORBIT Management Framework
OML	ORBIT Measurement Framework and Library
OMNeT++	Objective Modular Network Testbed in C++
OPNET	Optimized Network Engineering Tool
OSPF	Open Shortest Path First
OTAP	Over-The-Air Programming
PAN	Personal Area Network
PANEL	PAN Coordinator Election
PARSEC	PARallel Simulation Environment for Complex systems
PAWiS	Power Aware Wireless Sensors
PCB	Printed Circuit Board
PCCP	Priority-based Congestion Control Protocol
PCFG	Probabilistic Context Free Grammar
PCMCIA	Personal Computer Memory Card International Association

PCR	Peak Cell Rate
PDA	Personal Digital Assistant
PEGASIS	Power-Efficient GAthering in Sensor Information Systems
PF	Positif Framework
PHEIC	Public Health Emergency of International Concern
pHEMT	Pseudomorphic High Electron Mobility Transistor
PHP	Hypertext Preprocessor
PHY	Physical Layer
PIR	Passive Infra Red/ Pyroelectric Infrared
PLL	Phase Locked Loop
PNC	PicoNet Coordinators
PNNI	Private Network-to-Network Interface
POC	Proof of Concept
PRNG	Pseudo Random Number Generator
Prowler	Probabilistic Wireless Network Simulator
PSFQ	Pump Slowly Fetch Quickly
PSK	Phase-Shift Keying
PSRAM	Pseudostatic Random-Access Memory
QoS	QoS Quality of Service
RAHMoN	Routing Algorithm for Heterogeneous Mobile Network
RBC	Reliable Bursty Convergecast
RBCR	Relay selection Based Cooperative Routing
RF	Radio Frequency
RFC	Request for Comment
RFD	Reduced Function Device
RHE	Radio Harsh Environment
RIP	Routing Information Protocol
RISC	Reduced Instruction Set Computing
RMASE	Routing Modeling Application Simulation Environment
RMI	Remote Method Invocation
RMST	Reliable Multi-Segment Transport
ROAM	Routing On-demand Acyclic Multipath
RON	Resilient Overlay Networks
RPM	Revolutions Per Minute
RS	Reservation System
RSS	Really Simple Syndication
RSSI	Received Signal Strength Indicator
RTC	Real-Time Clock
RTS	Request To Send
RTT	Round Trip Time
RWS	Robotic Wireless Sensors
RWSN	Robotic Wireless Sensor Network
SAR	Sequential Assignment Routing
SAS	Safety and Automation System

SCU	Small Combat Unit
SDRAM	Synchronous Dynamic Random Access Memory
SDRT	Segmented Data Reliable Transport
S-EECP	Single-hop Energy-Efficient Clustering Protocol
SENS	Sensor, Environment, and Network Simulator
SENSE	SEnsor Network Simulator and Emulator
SET-IBOOS	Secure and Efficient data Transmission-Identity-Based Online/Offline digital signature
SET-IBS	Secure and Efficient data Transmission- Identity-Based digital Signature
SF	Serial Forwarder
SFA	Slice-based Federation Architecture
SIMD	Single Instruction Multiple Data
SIMM	Single In-line Memory Module
SINR	Signal to Interference plus Noise Ratio
SMA	Sub-Miniature version A
SMD	Surface-Mount Device
SMP	Sensor Management Protocol
SNAA	Sensor Network Authentication and Authorization
SNMP	Simple Network Management Protocol
SNMS	Simple Network Management System
SNR	Signal to Noise Ratio
SNS	Sensor Network Server
SOA	Service-Oriented Architecture
SO-DIMM	Small Outline Dual In-line Memory Module
SPDT	Single Pole Double Throw
SPI	Serial Peripheral Interface
SPIN	Sensor Protocols for Information via Negotiation
SPIN-BC	SPIN for Broadcast Networks
SPIN-EC	SPIN with Energy Conservation
SPIN-RL	SPIN with Reliability
SRAM	Static Random Access Memory
SS	Signal Strength
SSDoS	Security Service DoS
STCP	Sensor TCP
STD	State Transition Diagram
SUE	System Under Examination
SUNSHINE	Sensor Unified aNalyzer for Software and Hardware in Networked Environments
SVM	State Vector Machine
SWD	Shockwave Detection
SW-GDS	Soldier Wearable Gunfire Detection System
SXGA	Super eXtended Graphics Array
TARS	Trace-Announcing Routing Scheme
TBRPF	Topology dissemination Based on Reverse-Path Forwarding

TCP	Transport Control Protocol
TCP/IP	Transport Control Protocol/Internet Protocol
TDMA	Time Division Multiple Access
TDOA	Time Difference of Arrival
TOA	Time of Arrival
TORA	Temporally Ordered Routing Algorithm
TOSSF	TinyOS Scalable Simulation Framework
TPGFPlus	Two-Phase Geographic Greedy Forwarding
TS	Testbed Server
TSMP	Time Synchronized Mesh Protocol
TWIST	TKN Wireless Indoor Sensor network Testbed
UART	Universal Asynchronous Receiver/Transmitter
UASN	Underwater Acoustic Sensor Network
UAV	Unmanned Aerial Vehicles
UDG	Unit Disk Graph
UDP	User Datagram Protocol
USART	Universal Synchronous/Asynchronous Receiver/Transmitter
USB	Universal Serial Bus
USRP	Universal Software Radio Peripheral
UWB	Ultra Wide Band
VC	Virtual Circuit
VGA	Video Graphics Array
VHDL	Verilog Hardware Description Language
VS	Virtual Sink
VSG	Vector Signal Generator
VTOL	Vertical Take-Off and Landing
WAC	World Athletics Championships
WARP	Wireless Open Access Research Platform
WHO	World Health Organization
WINLAB	Wireless Information Network Laboratory
WLAN	Wireless LAN
WLCSP	Wafer Level Chip Scale Package
WMN	Wireless Mesh Network
WMSN	Wireless Multimedia Sensor Network
WPAN	Wireless Personal Area Network
WSDL	Web Service Definition Language
WSN	Wireless Sensor Network
WT	Watchdog Timer
XLP	Cross-Layer Protocol
XLS	Cross-Layer Simulator
XML	Extensible Markup Language
XSM	eXtreme Scale Mote
XSS	eXtreme Scale Stargate
XTC	Extended C
ZC	Zero-Crossing

List of Figures

List of Tables

Part I
WSNs Concepts and Applications

Chapter 1
Introduction

Good beginnings lead to happy endings, most of the time.

Beginnings are usually uneasy, sometimes stiff; the acquaintance with newness does not go without tensity, the first year in school, in college, at work, the first year of marriage, the early months of retirement, even when first time using a new gadget. Some fear change, a new TV, watch, mobile phone, software, color, and brand. Befriending own habits, as human nature, grows with years; juniors are usually more receptive and adaptive. The first 15 minutes of a movie are decisive, to stay or leave right away. The first book chapter is the hardest; it introduces the author, the book, and the topic. Writing is not dumping words and machinely composing sentences, it is a live dialog between the author and the audience, they see each other in their minds, while writing and while reading, issues, debates, controversies, questions and answers, noise, smiles, brainstorming, head-scratching. Chapter 1 of this book bears his task with willingness, enthusiasm, and goodwill.

Significant developments in scalable standards are now pacing adoption and presenting wireless sensor networks (WSNs) in applications welcomed at IT, industry, home, work, and everywhere. Wireless sensors can be deployed quickly in an ad hoc fashion and used to report environmental changes, ensure the efficiency of industrial processes in an oil refinery, determine how much power the blade servers in a data center are using, or tell if the refrigerator is still as energy-efficient as when it was purchased.

In the 15 years that WSNs have been around, improvements in their architecture and protocols have continued to push applications to the mainstream. Semiconductor technology continues to follow Moore's law, providing smaller, more powerful, and cheaper wireless devices. There are now established and reliable low-power standards supporting the multiplicity of WSNs applications. The Internet, the largest known network, has extended into the world of low-power embedded WSN devices.

Stepwise, WSNs are to be introduced with depth and focus.

H. M. A. Fahmy, *Concepts, Applications, Experimentation and Analysis of Wireless Sensor Networks*, Signals and Communication Technology, https://doi.org/10.1007/978-3-031-20709-9_1

1.1 Sensing, Senses, Sensors

Sensing is what distinguishes the living from stones and rocks. Alive creatures have several levels and ways of sensing, without sensing there is no communication with the outside word, and there is no life. Lecturing on zoology or botany is not an objective, but a quick reminder on senses of the living is recalled (Birds and Blooms 2013).

Many animals see the world completely differently to humans. Being able to see helps animals locate food, move around, find mates, and avoid predators, whether they live at the bottom of the ocean or soar high in the sky. Eyesight is important for most animals and nearly all animals can see, 95% of all species have eyes. Some animals live in complete darkness in caves or underground, where they cannot see anything, their eyes often no longer work, but they have developed an extra-sensitive sense of touch to feel their way around. However, only two animal groups have evolved the ability to hear, vertebrates like mammals, birds, and reptiles, and arthropods, such as insects, spiders, and crabs. No other animals can hear. Some animals have a remarkable sense of hearing, finely tuned to where and how they live, many animals hear sounds that humans cannot. Human senses of smell and taste are feeble compared to those of many other animals, a keen sense of smell allows animals to find food and mates, as well as to stay out of danger, it can stop an animal wandering into a rival's territory or help it find its way.

Animals communicate using visual signals, sounds, touch, smells, and taste. Vision, touch, and taste work well over short distances, but sounds travel much further and scent marks that can last long after the animal has moved on. Sometimes the aim is to deceive, blending into the background, pretending to be a twig or playing dead; animals give out all sorts of false information to avoid danger or help catch their next meal. Their tricks and deceptions vary from camouflage and mimicry to distracting, startling, scaring, and confusing others (National Museum Scotland 2013).

An insect's acute sense of smell enables it to find mates, locate food, avoid predators, and even gather in groups. Insects have sense organs for taste, touch, smell, hearing, and sight. Some insects have sense organs for temperature and humidity as well as stresses and movements of their body parts. Some insects rely on chemical cues to find their way to and from a nest, or to space themselves appropriately in a habitat with limited resources. Insects, you may have noticed, do not have noses. So how are they able to sense the faintest of scents in the wind? Antennae sometimes are called "feelers." However, antennae as primarily "smellers" are the insect's noses because they are covered with many organs of smell. These organs help the insect to find food, a mate, and places to lay eggs. Insects even can decide which direction to fly by using their sense of smell (O. Orkin Insect Zoo 2013).

How do fish sense movement? Fish have the five senses that people have, but have a sixth sense that is more than a sense of touch. Fish have a row of special cells inside a special canal along the surface of the fish's skin. This is called the "lateral line" which allows them to detect water vibrations. This sixth sense allows fish to detect movement around them and changes in water flow. Detecting movement helps fish find prey or escape from predators. Detecting changes in water flow help fish chose where to swim (Negron 2020).

What about birds? They depend less on the senses of smell and taste than people do.

The odors of food, prey, enemies, or mates quickly disperse in the wind. Birds possess olfactory glands, but they are not well developed in most species, including the songbirds in our backyards. The same is true for taste, which is related to smell. While humans have 9000 taste buds, songbirds have fewer than 50. That means the birds we feed around must locate their food by sight or touch, two senses that are highly developed in birds (Birds and Blooms 2013).

Plants, unlike animals, do not have ears, eyes, or tongues to help them feel and acquire information from their environment. But without being helpless, they do sense their environment in other ways and respond accordingly. Plants can detect various wavelengths and use colors to tell them what the environment is like. When a plant grows in the shadow of another, it will send a shoot straight up toward the light source; it has also been shown that plants know when it is day and when it is night. Leaf pores on plants open up to allow photosynthesis during the daytime and close at night to reduce water loss. Plants also respond to ultraviolet light by producing a substance that is essentially a sunscreen so that they do not get sunburned. Plants can sense weather changes and temperatures as well. Plants have specific regulators, plant hormones, minerals, and ions that are involved in cell signaling and are important in environmental sensing. In fact, without these, the plants will not grow properly (UCSB ScienceLine 2020)

Reminding of human senses is easy, the use of eye contacts, the eye attraction to what is beautiful, the love of perfumes, the appreciation of beautiful music, the relieving touch of softness, and the tantalizing taste of sweeties. It is all senses. Human interaction with the environment is an eternal task that grows and expands with expansion of ambitions, with technology. This book is interested in presenting wireless sensor networks (WSNs) in comprehensive details that are far beyond what birds, insects, mammals can.

As an opening start, the goal of this chapter is to present a thorough survey of WSNs.

1.2   Preliminaries of Wireless Sensor Networks

With the recent technological advances in wireless communications, processor, memory, radio, low power, highly integrated digital electronics, and micro-electromechanical systems (MEMS), it has become possible to significantly develop tiny and small-size, low-power, and low-cost multifunctional sensor nodes (Warneke and Pister 2002). A wireless sensor network (WSN) is a network that is made of tens to thousands of these sensor nodes which are densely deployed in an unattended environment with the capabilities of sensing, wireless communications, and computations (i.e., collecting and disseminating environmental data) (Akyildiz et al. 2002a). These nodes are capable of wireless communications, sensing, and computation (software, hardware, algorithms). So, it is obvious that a WSN is the result of the combination of sensor techniques, embedded techniques, distributed information processing, and communication mechanisms.

Functionally, smart sensor nodes are low-power devices equipped with one or more sensors, a processor, memory, power supply, a radio interface, and some additional components that will be detailed later. A variety of mechanical, thermal, biological, chemical, optical, and magnetic sensors may be attached to the sensor node to measure properties of the environment. Since the sensor nodes have limited memory and are typically deployed in difficult-to-access locations, a radio interface is implemented for wireless communication to transfer the data to a basestation (e.g., a laptop, a personal handheld device, or an access point to a fixed infra- structure). Battery is the main power source in a sensor node. Also, a secondary power supply that harvests power from the environment such as solar panels may be added to the node depending on the appropriateness for the environment where the sensor will be deployed (Yick et al. 2008).

Regarding their practicality and low cost, WSNs have great potential for many applications in scenarios such as military target tracking and surveillance (Yick et al. 2005), natural disaster relief (Castillo-Effen et al. 2004), biomedical health monitoring (Gao et al. 2005), and hazardous environment exploration and seismic sensing (Wener-Allen et al. 2006). In military target tracking and surveillance, a WSN can assist in intrusion detection and identification. Specific examples include spatially correlated and coordinated troop and tank movements. With natural disasters, sensor nodes can sense and detect the environment to forecast disasters before they occur. In biomedical applications, surgical implants of sensors can help monitor a patient's health. For seismic sensing, ad hoc deployment of sensors along the volcanic area can detect the development of earthquakes and eruptions. In detail, Chap. 2 of this book elaborates on WSNs applications.

Energy is the driver and concern of living beings that have the need to eat and drink, and of modern technologies that need gas, winds, and sun. Noteworthy, one of the most important WSN limitations is energy conservation; therefore, the main WSN focus is on power conservation through appropriate optimization of communication and operation management. Several analyses of energy-efficient use for sensor networks have been realized, and several algorithms that lead to energy-efficient and lifetime aware routing protocols for WSNs are developed as detailed in Chap. 4; also, efficient transport layer protocols have been proposed, as will be presented later in Chap. 5 of this book. Chapter 6 introduces cross-layering as an energy-saving approach typical to WSNs.

What is the size of a WSN and where to place nodes? The environment plays a key role in determining the size of the WSN network, the deployment scheme, and the network topology. The network size varies with the monitored environment. For indoor environments, fewer nodes are required to form a network in a limited space whereas outdoor environments may require more nodes to cover a larger area.

An ad hoc deployment is preferred over preplanned deployment when the environ-ment is inaccessible by humans or when the network is composed of hundreds to thousands of nodes. Obstructions can also limit communication between nodes, which in turn affects the network connectivity, or topology. The position of sensor nodes is not usually predetermined, although the application can provide some guidelines and insights that can lead to the construction of an optimal design that satisfies application requirements and meets wireless network limitations.

To go from here and there, a better route is to be selected, several routing, power management, and data dissemination protocols have been designed for WSNs, depending on both their architecture and the applications they are intended to support. WSN protocols support the proliferation of WSNs and efficiently make them an integral constituent of daily life. To make wireless sensor networks practi-cally useful and functioning, these protocols are designed to overcome the unique constraints of small memory, tiny size, limited energy, and to fulfill standards of scalability, adaptivity, fault tolerance, low latency, and robustness.

In the coming section, an overview of MANETs is provided as a step that leads to WSNs.

1.3 Mobile Ad Hoc Networks (MANETs)

At first, it is needed to strengthen up basics, a Mobile Ad hoc NETwork (MANET) is one that comes together as needed, not necessarily with the support of an existing Internet infrastructure or any fixed station, it is an autonomous system of mobile hosts serving as routers and connected by wireless links (Cordeiro and Agrawal 2002). This contrasts the single-hop cellular network that supports the need for wireless communication by installing basestations as access points, such that the communication between wireless nodes rely on the wired backbone and the fixed basestations. In a MANET, there is no infrastructure and the network topology changes unpredictably since nodes are free to move. As for the mode of operation, ad hoc networks are peer-to-peer multihop mobile wireless networks where infor-mation packets are transmitted in a store and forward manner from source to destination via intermediate nodes as shown in Fig. 1.1. Topology changes as the nodes move, for instance as node MH2 changes its point of attachment from MH3 to MH4 other nodes must follow the new route to forward packets to MH2. It is to be clear that not all nodes are within radio reach of each other; otherwise, there would not be any routing problem. Bidirectional links between nodes indicate that they are within radio range of each other, for instance MH1 and MH3. Unidirectional links indicates that a node may transmit while the other cannot, for instance, MH4 can send to MH7, while MH7 cannot.

The following sections go further in the WSN journey.

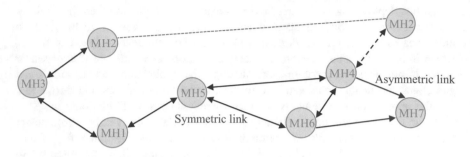

Fig. 1.1 Mobile ad hoc network. Cordeiro and Agrawal (2002)

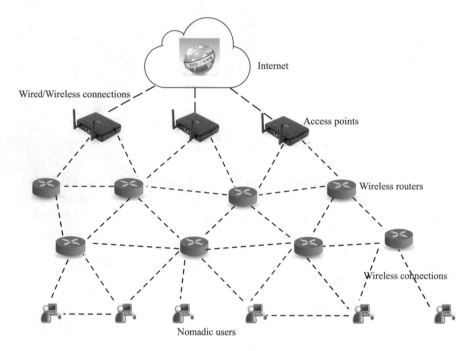

Fig. 1.2 A three-tier architecture for wireless mesh networks

1.4 Wireless Mesh Networks (WMNs)

Mesh network architectures have been conceived by both industry and academia. A wireless mesh network is a fully wireless network that employs multihop communications to forward traffic to and from wired Internet entry points. Different from flat ad hoc networks, a wireless mesh network (WMN) introduces a hierarchy in the network architecture by the implementation of dedicated nodes (wireless routers) communicating among each other and providing wireless transport services to data traveling from users to other users or to access points (access points are special wireless routers with a high-bandwidth wired connection to the Internet backbone). As shown in Fig. 1.2, the network of wireless routers forms a wireless backbone tightly integrated into the mesh network, which provides multihop connectivity between nomadic users and wired gateways. The meshing among wireless routers and access points creates a wireless backhaul communication system, which provides each mobile user with a low-cost, high-bandwidth, and easy multihop interconnection service with a limited number of Internet entry points, and with other wireless mobile users. Backhaul is used to indicate the service of forwarding traffic from the user originator node to an access point from which it can be distributed over the external network, the Internet in this case.

The mesh network architecture addresses the emerging market requirements for building wireless networks that are highly scalable and cost-effective, offering a solution for the easy deployment of high-speed ubiquitous wireless Internet.

Mesh networking has more than a benefit (Raffaele et al. 2005):

- Reduction of installation costs. Currently, one of the major efforts to provide wireless Internet, beyond the boundaries of indoor WLANs, is through the deployment of WiFi hot spots. Basically, a hot spot is an area that is served by a single WLAN or a network of WLANs, where wireless clients access the Internet through an 802.11-based access point. The downside of this solution is a tolerable increase in the infrastructure costs, because a cabled connection to the wired backbone is needed for every access point in the hot spot. As a consequence, the hot spot architecture is costly, unscalable, and slow to deploy. On the other hand, building a mesh wireless backbone enormously reduces the infrastructural costs because the mesh network needs only a few access points connected to the wired backbone.

- Large-scale deployment. In recently standardized WLAN technologies, i.e., 802.11a and 802.11g, increased data rates have been achieved by using more spectrally efficient modulation schemes. However, for a specific transmit power, shifting toward more efficient modulation techniques reduces coverage, i.e., the further from the access point the lower the data rate available. Moreover, for a fixed total coverage area, more access points should be installed to cover small-size cells. Obviously, this miniaturization of WLANs cells further hinders the scalability of this technology, especially in outdoor environments. On the other hand, multihop communications offer long-distance communications via hopping through intermediate nodes. Since intermediate links are short, these transmissions could be at high data rates, resulting in increased throughput compared to direct communications. The wireless backbone can realize a high degree of spatial reuse through wireless links covering longer distance at higher speed than conventional WLAN technologies.

- Reliability. The wireless backbone provides redundant paths between each pair of endpoints, significantly increasing communications reliability, eliminating single points of failure and potential bottleneck links within the mesh. Network fault tolerance is increased against potential problems such as node crash, path failure due to temporary obstacles or external radio interference, by the existence of multiple possible destinations, i.e., any of the exit points toward the wired Internet, and alternative routes to these destinations.

- Self-management. The adoption of peer-to-peer networking to build a wireless distribution system provides all the advantages of ad hoc networking, such as self-configuration and self-healing. Consequently, network setup is automatic and transparent to users. For instance, when adding additional nodes in the mesh, these nodes use their meshing functionalities to automatically discover all possible wireless routers and determine the optimal paths to the wired network. In addition, the existing wireless routers reorganize, taking into account the new available routes. Thus, the network can easily be expanded, because the network self-reconfigures to assimilate the new elements.

With the differences between WSN and WMN, many similarities coexist:

- The goal of any WSN and WMN is to create and maintain network connectivity as easy as possible, in order to get as many data, as fast, easy, secure as needed from source to destination node(s), while consuming the least possible number of resources, such as the wireless spectrum, node energy, node memory, node processing power, and financial budget.
- Multihop networks are created, which requires some form of node addressing and a routing protocol.

Many popular WSN and WMN technologies share the limited 2400−2500 MHz ISM band of the wireless spectrum.

Table 1.1 compares sensor and mesh nodes (Bouckaert et al. 2010).

Table 1.1 Sensor and mesh node characteristics

		Sensor nodes	Mesh nodes
General	Target form factor	Small or tiny $O(mm^3)$	Larger $O(cm^3)$
	Antenna	Integrated	External
	Power consumption	$O(mW)$	$O(W)$
	Power	Small battery or energy harvesting	Unlimited due to external power source
	Price	Relatively cheap (a few dollars or less)	Relatively expensive ($50–$500 and up)
	RAM/ROM	KBytes	MBytes
	Processing power	Very limited	Relatively high
Network	Bandwidth	Low (a few Mbps and frequently less)	Relatively high (several Mbps)
	Interface(s)	Single, often proprietary	Single or multiple, often standardized
	Max packet size	Small O(Bytes)	Larger O(KBytes)
	IP capabilities	Limited or none	IP capable
	Sleeping schemes	Often used	Rarely used
	Delay per hop	$O(ms)$ to several seconds	$O(ms)$
	Mobility	None to highly mobile	Most often limited or none

1.5   Closer Perspective to WSNs

1.5.1   Wireless Sensor Nodes

To get closer to how a wireless sensor network is built, an insight into a sensor node is to come first. Specifically, a sensor node is made up of basic components as shown in Fig. 1.3:

- Sensing units. Sensing units are usually composed of two subunits, sensors and analog to digital converters (ADCs). The analog signals produced by the sensors based on the observed phenomenon are converted to digital signals by the ADC and then fed into the processing unit.
- Processing unit. The processing unit is generally associated with a small storage unit and manages the procedures that make the sensor node collaborates with the other nodes to carry out the assigned sensing tasks.
- Transceiver unit. A transceiver unit connects the node to the network.
- Power unit. Power units may be supported by a power scavenging unit such as solar cells.
- Application-dependent additional components such as a location finding system, a power generator, and a mobilizer. Most of the sensor network routing techniques and sensing tasks require the knowledge of location with high accuracy; thus, it is common that a sensor node has a location finding system. A mobilizer may sometimes be needed to move sensor nodes when it is required to carry out the assigned tasks.

All of these subunits may need to fit into a matchbox sized module whose size may be smaller than even a cubic centimeter, which is light enough to remain suspended in the air. Added to size, there are also some other stringent specifications of sensor nodes (Khan et al. 1999):

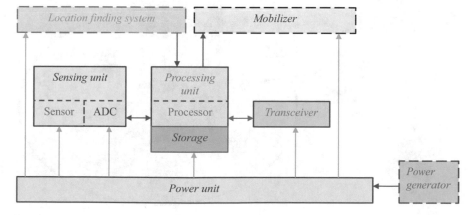

Fig. 1.3 Components of a sensor node. Akyildiz et al. (2002b)

- Consume extremely low power.
- Operate in high volumetric densities.
- Have low production cost, can be easily replaced, and the malfunction of any does not halt other sensors.
- Are autonomous and operate unattended.
- Are adaptive to the environment.

1.5.2 Architecture of WSNs

The term *architecture* has been adopted to describe the activity of designing any kind of system; it is the complex or carefully designed structure of something; one of its common uses is in describing information technology, such as computer architecture and network architecture. The architecture of WSNs is built up of main entities as shown in Fig. 1.4:

- The sensor nodes that form the sensor network. Their main objectives are making discrete, local measurement about phenomenon surrounding these sensors, forming a wireless network by communicating over a wireless medium, and collecting data and routing data back to the user via a sink (basestation).
- The sink (basestation) communicates with the user via Internet or satellite communication. It is located near the sensor field or well-equipped nodes of the sensor network. Collected data from the sensor field are routed back to the sink by a multihop infra-structureless architecture through the sink.

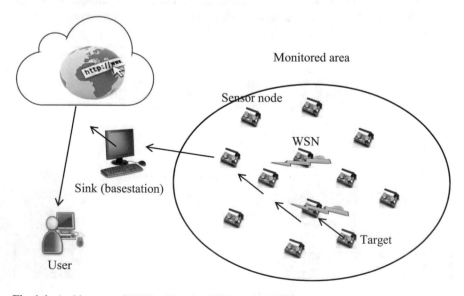

Fig. 1.4 Architecture of WSNs. (Based on Tilak et al. (2002))

- The phenomenon which is an entity of interest to the user, it is sensed and analyzed by the sensor nodes.
- The user who is interested in obtaining information about a specific phenomenon to measure/monitor its behavior.

Although many protocols and algorithms have been proposed for traditional wireless ad hoc networks, they are not well suited for the unique features and application requirements of sensor networks, as detailed in this section. For further illustration, the differences between WSNs and MANETs are outlined below (Akyildiz et al. 2002a):

- The number of sensor nodes in WSNs can be several orders of magnitude higher than the nodes in MANETs.
- Sensor nodes are densely deployed.
- Sensor nodes are prone to failures.
- The topology of a sensor network changes very frequently.
- Unlike a node in MANETs, a sensor node may not have a unique global IP address due to the numerous numbers of sensors and the resulting high overhead.
- Sensor nodes, as deployed in high numbers, are extremely cheap and considerably tiny, unlike MANET nodes (e.g., PDAs and laptops).
- The communication paradigm used in WSNs is broadcasting, whereas MANETs are based on point-to-point communications.
- The topology of a WSN changes very frequently.
- Limited energy and bandwidth conservation is the main concern in WSN protocol design, which is not really worrisome in MANETs.

1.6 Types of WSNs

WSNs can be deployed on ground, underground, and underwater. Five functional types can be distinguished, specifically terrestrial, underground, underwater, multimedia, and mobile WSNs (Yick et al. 2005). What follows provides the details of each type.

1.6.1 Terrestrial WSNs

Terrestrial WSNs deployed in a given area (Yick et al. 2008). There are two ways to deploy sensor nodes on WSNs:

- In unstructured WSN, which contains a dense collection of sensor nodes. Sensor nodes may be deployed in an ad hoc manner into the field, once deployed the network is left unattended to perform monitoring and reporting functions. In an unstructured WSN, network maintenance such as managing connectivity and detecting failures is difficult since there are so many nodes.

- In structured WSN, all or some of the sensor nodes are deployed in a preplanned manner. The advantage of a structured network is that fewer nodes can be deployed with lower network maintenance and management cost. Fewer nodes are beneficially deployed since they are placed at specific locations to provide coverage while ad hoc deployment can have uncovered regions.

Sensor nodes are deployed on the sensor field within reach of the transmission range of each other and at densities that may be as high as 20 nodes/m^3. Densely deploying hundreds or thousands of sensor nodes over a field requires maintenance of topology along three phases:

- Predeployment and deployment phase. Sensor nodes may either be thrown in the deployment field as a mass from an airplane or an artillery shell, or placed one by one by a human or a robot.
- Post-deployment phase. After deployment, topology changes due to change in sensor nodes position, reachability (that may be effected by jamming, noise, moving obstacles, etc.), remaining energy, malfunctioning, and task details.
- Redeployment of additional nodes. Additional sensor nodes can be redeployed to replace malfunctioning nodes or to account for changes in task dynamics.

In a terrestrial WSN, reliable communication in a dense environment is a must. Sensor nodes must be able to effectively communicate data back to the basestation. While battery power is limited and may not be rechargeable, terrestrial sensor nodes however can be equipped with a secondary power source such as solar cells, it is important for sensor nodes to conserve energy. For a terrestrial WSN, energy can be conserved with multihop optimal routing, short transmission range, in-network data aggregation, eliminating data redundancy, minimizing delays, and using low duty-cycle operations.

1.6.2 Underground WSNs

Underground WSNs consist of a number of sensor nodes buried underground or in a cave or mine used to monitor underground conditions (Li and Liu 2007, 2009; Li et al. 2007). Additional sink nodes are located aboveground to relay information from the sensor nodes to the basestation. An underground WSN is more expensive than a terrestrial WSN in terms of equipment, deployment, and maintenance. Underground sensor nodes are expensive because appropriate equipment parts must be selected to ensure reliable communication through soil, rocks, water, and other mineral contents. The underground environment makes wireless communication a challenge due to signal losses and high levels of attenuation. Unlike terrestrial WSNs, the deployment of an underground WSN requires careful planning and energy and cost considerations. Energy is an important concern in underground

WSNs. Like terrestrial WSN, underground sensor nodes are equipped with a limited battery power, and once deployed into the ground, it is difficult to recharge or replace a sensor node's battery. As usual, a key objective is to conserve energy in order to increase the network lifetime, which can be achieved by implementing efficient communication protocol.

1.6.3   Underwater Acoustic Sensor Networks (UASNs)

Underwater acoustic sensor networks (UASNs) technology provides new opportunities to explore the oceans, and consequently, it improves understanding of the environmental issues, such as the climate change, the life of ocean animals, and the variations in the population of coral reefs. Additionally, UASNs can enhance the underwater warfare capabilities of the naval forces since they can be used for surveillance, submarine detection, mine countermeasure missions, and unmanned operations in the enemy fields. Furthermore, monitoring the oil rigs with UASNs can help taking preventive actions for the disasters such as the rig explosion that took place in the Gulf of Mexico in 2010. Last but not least, earthquake and Tsunami forewarning systems can also benefit from the UASN technology (Erol-Kantarci et al. 2011).

Ocean monitoring systems have been used for the past several decades, where traditional oceanographic data collection systems utilize individual and disconnected underwater equipment. Generally, this equipment collects data from their surroundings and sends these data to an on-shore station or a vessel by means of satellite communications or underwater cables. In UASNs, this equipment is replaced by relatively small and less expensive underwater sensor nodes that house various sensors on board, e.g., salinity, temperature, pressure, current speed sensors. The underwater sensor nodes are networked, unlike the traditional equipment, and they communicate underwater via acoustics.

In underwater, radio signals attenuate rapidly; hence, they can only travel to short distances while optical signals scatter and cannot travel far in adverse conditions, as well. On the other hand, acoustic signals attenuate less, and they are able to travel further distances than radio signals and optical signals. Consequently, acoustic communication emerges as a convenient choice for underwater communications. However, it has several challenges (Heidemann et al. 2006):

- The bandwidth of the acoustic channel is low; hence, the data rates are lower than they are in terrestrial WSNs. Data rates can be increased by using short-range communications, which means more sensor nodes will be required to attain a certain level of connectivity and coverage. In this respect, large-scale UASN poses additional challenges for communication and networking protocols.
- The acoustic channel has low link quality, which is mostly due to the multipath propagation and the time variability of the medium.

- Furthermore, the speed of sound is slow (approximately 1500 m/s) yielding large propagation delay.
- In mobile UASNs, the relative motion of the transmitter or the receiver may create the Doppler effect.
- UASNs are also energy limited similar to other WSNs.

Due to the above challenges, UASN rooms research studies in novel medium access, network, transport, localization, synchronization protocols, and architectures (Jornet et al. 2008; Vuran and Akyildiz 2008; Lee et al. 2010; Ahna et al. 2011). The design of network and management protocols is closely related with the network architecture, and various UASN architectures have been proposed in the literature. Moreover, localization has been widely addressed since it is a fundamental task used in tagging the collected data, tracking underwater nodes, detecting the location of an underwater target, and coordinating the motion of a group of nodes, Furthermore, location information can be used to optimize the medium access and routing protocols (Chandrasekhar et al. 2006; Erol-Kantarci et al. 2011; Zhou et al. 2011).

Underwater sensor nodes must be able to self-configure and adapt to harsh ocean environment; they are equipped with a limited battery, which cannot be replaced or recharged. The issue of energy conservation for underwater WSNs involves developing efficient underwater communication and networking techniques.

1.6.4 Multimedia WSNs

Multimedia WSNs have been proposed to enable monitoring and tracking of events in the form of multimedia such as video, audio, and imaging (Akyildiz et al. 2007). Multimedia WSNs consist of a number of low-cost sensor nodes equipped with cameras and microphones. These sensor nodes interconnect with each other over a wireless connection for data retrieval, processing, correlation, and compression. They are deployed in a preplanned manner into the environment to guarantee coverage. Challenges in multimedia WSN include the following:

- High-bandwidth demand.
- High energy consumption.
- Quality of service (QoS) provisioning.
- Data processing and compressing techniques.
- Cross-layer design.

Multimedia content such as a video stream requires high bandwidth in order for the content to be delivered quickly; consequently, high data rate leads to high energy consumption. Thus, transmission techniques that support high bandwidth and low energy consumption have to be developed. QoS provisioning is a challenging task in a multimedia WSN due to the variable delay and variable channel capacity. It is important that a certain level of QoS must be achieved for reliable content delivery. In-network processing, filtering, and compression can significantly improve network performance in terms of filtering and extracting redundant information and merging contents. Similarly, cross-layer interaction among protocol layers can improve the processing and delivering of data.

1.6.5 Mobile WSNs

Mobile WSNs consist of a collection of sensor nodes that can move on their own and interact with the physical environment (Di Francesco et al. 2011). There are several comparative issues between mobile and static sensor nodes:

- Like static nodes, mobile nodes have the ability to sense, compute, and communicate.
- Contrarily, mobile nodes have the ability to reposition and organize themselves in the network. A mobile WSN can start off with some initial deployment and nodes can then spread out to gather information. Information gathered by a mobile node can be communicated to another mobile node when they are within range of each other.
- Another key difference is data distribution. In a static WSN, data can be distributed using fixed routing or flooding while dynamic routing is used in a mobile WSN.

Mobility in WSNs is useful for several reasons, as presented in what follows (Anastasi et al. 2009):

- Connectivity. As nodes are mobile, a dense WSN architecture is not a pressing requirement. Mobile elements can cope with isolated regions such that the constraints on network connectivity and on nodes (re)deployment can be relaxed. Hence, a sparse WSN architecture becomes a feasible option.
- Cost. Since fewer nodes can be deployed, the network cost is reduced in a mobile WSN. Although adding mobility features to the nodes might be expensive, it may be possible to exploit mobile elements, which are already present in the sensing area (e.g., trains, buses, shuttles, or cars) and attach sensors to them.
- Reliability. Since traditional (static) WSNs are dense and the communication paradigm is often multihop ad hoc, reliability is compromised by interference and collisions; moreover, message loss increases with the increase in number of hops. Mobile elements, instead, can visit nodes in the network and collect data directly through single-hop transmissions; this reduces not only contention and collisions, but also the message loss.
- Energy efficiency. The traffic pattern inherent to WSNs is converge cast; i.e., messages are generated from sensor nodes and are collected by the sink. As a consequence, nodes closer to the sink are more overloaded than others, and subject to premature energy depletion. This issue is known as the funneling effect, since the neighbors of the sink represent the bottleneck of traffic. Mobile elements can help reduce the funneling effect, as they can visit different regions in

the network and spread the energy consumption more uniformly, even in the case of a dense WSN architecture.

However, mobility in WSNs also introduces significant challenges, which do not arise in static WSNs, as illustrated below:

- Contact detection. Since communication is possible only when the nodes are in the transmission range of each other, it is necessary to detect the presence of a mobile node correctly and efficiently. This is especially true when the duration of contacts is short.
- Mobility-aware power management. In some cases, it is possible to exploit the knowledge on the mobility pattern to further optimize the detection of mobile elements. In fact, if visiting times are known or can be predicted with certain accuracy, sensor nodes can be awake only when they expect the mobile element to be in their transmission range.
- Reliable data transfer. As available contacts might be scarce and short, there is a need to maximize the number of messages correctly transferred to the sink. In addition, since nodes move during data transfer, message exchange must be mobility-aware.
- Mobility control. When the motion of mobile elements can be controlled, a policy for visiting nodes in the network has to be defined. That is, the path and the speed or sojourn time of mobile nodes have to be defined in order to improve (maximize) the network performance.
- Challenges also include deployment, localization, navigation and control, coverage, maintenance, and data processing.

Mobile WSN applications include environment monitoring, target tracking, search and rescue, and real-time monitoring of hazardous material. Mobile sensor nodes can move to areas of events after deployment to provide the required coverage. In military surveillance and tracking, they can collaborate and make decisions based on the target. Mobile sensor nodes can achieve a higher degree of coverage and connectivity compared to static sensor nodes. In the presence of obstacles in the field, mobile sensor nodes can plan ahead and move appropriately to unobstructed regions to increase target exposure.

1.7 Performance Metrics of WSNs

Metric is the standard of measurement; it varies with the measured environment. Time delay is a widely used metric, it is the time needed to obtain a response after applying certain input, its units are coarsely seconds, but specifically at which scale? In an electronic environment, time delay units are microseconds and less; in electro-mechanical environment, they are milliseconds or more; in pure mechanical systems, they are seconds and above. In athletic run sports, speed is the metric; its unit scale varies with distances, from the 100-m race till the marathon. Generally, speed varies

a)   b)

Fig. 1.5 Fastest runners with different metrics. (**a**) Usain Bolt hits 9.58 sec for 100 m. (**b**) Cheetah fastest runner on earth

with who is running and where, a professional human runner spends 2:15 hours in a 42.195-km marathon, while a cheetah that is three times faster just needs 25 minutes to reach the end point; the metric is time, the same, but for humans it measured in hours, while it is in minutes for the cheetah (Fig. 1.5). One of the metrics for goods is weight, its unit is kilograms or pounds, but for coal it is a multiplicity of kilograms for home use and tons for industry. On the other hand, gold weight is calculated in grams or ounces for personal use and in kilograms for gold traders. Lifetime a metric related to living being existence, it is left for the reader to have some thoughts about it, at least for mind relief.

Back to WSNs, for a WSN to perform satisfactorily, some metrics are also defined, measured, and interpreted far from confusion. Several metrics, close to WSNs characteristics as introduced in previous sections of this chapter, evaluate sensor network performance. Specifically:

- Network lifetime. It is measure of energy efficiency, as sensor nodes are battery operated, WSN protocols must be energy-efficient to maximize system lifetime. System lifetime can be measured by generic parameters such as the time until half of the nodes die, or by application directed metrics, such as when the network stops providing the application with the desired information about the environment, it is also calculated as the time until message loss rate exceeds a given threshold.
- Energy consumption. It is the sum of used energy by all WSN nodes. The consumed energy of a node is the sum of the energy used for communication, including transmitting, receiving, and idling. Assuming each transmission consumes an energy unit, the total energy consumption is equivalent to the total number of packets sent in the network.
- Latency. It is the end-to-end delay that implies the average time between sending a packet from the source, and the time for successfully receiving the message at the destination. Measurement takes into account the queuing and the propagation delay of the packets. The observer is interested in getting information about the environment within a given delay. The precise units of latency are application-dependent.

- Accuracy. It is the freedom from mistake or error, correctness, conformity to truth, exactness. Obtaining accurate information is the primary objective of the observer. There is a trade-off between accuracy, latency, and energy efficiency. A WSN should be adaptive such that its performance achieves the desired accuracy and delay with minimal energy expenditure. For example, the WSN task, the application, can either request more frequent data dissemination from the same sensor nodes, or it can direct data dissemination from more sensor nodes with the same frequency.
- Fault tolerance. Sensors may fail due to surrounding physical conditions or when their energy runs out. It may be impractical to replace existing sensors; in response, the WSN must be fault-tolerant such that non-serious failures are hidden from the application in a way that does not hinder it. Fault tolerance may be achieved through data replication, as in the SPIN protocol (Xiao et al. 2006). However, data replication itself requires energy; there is a trade-off between data replication and energy efficiency; generally, data replication should be application-specific; higher priority data according to the application might be replicated for fault tolerance.
- Scalability. As a prime factor, it is WSN adaptability to increased workload, which is to include more sensor nodes than what was anticipated during network design. A scalable network is one that can be expanded in terms of the number of sensors, complexity of the network topology, data quality; e.g., sampling rate, sensor sensitivity, and amount of data while the cost of the expansion installation and operational cost, communication time, processing time, power, and reliability is no worse than a linear, or nearly linear, function of the number of sensors (Pakzad et al. 2008). WSN scalability needs to consider an integrated view of the hardware and software. For hardware, scalability involves sensitivity and range of MEMS sensors, communication bandwidth of the radio, and power usage. The software issues include reliability of command dissemination and data transfer, management of large volume of data, and scalable algorithms for analyzing the data. The combined hardware–software issues include high-frequency sampling, and the tradeoffs between onboard computations compared with wireless communication between nodes.
- Network throughput. It is a common metric for all networks. The end-to-end throughput measures the number of packets per second received at the destination.
- Success rate. It is also a common metric. It is the total number of packets received at the destinations versus the total number of packets sent from the source.

1.8   WSNs Standards

A standard is a required or agreed level of quality or attainment. There are standards for health, industry, and education. The International Organization for Standardization known as ISO is an international standard-setting body composed of representatives from various national standards organizations. Founded on February 23, 1947, long time before WSNs were born, ISO promotes worldwide proprietary, industrial, and commercial standards. The WSN standards are tightly coupled to the ISM frequency bands that are recalled in the next paragraph and widely evoked in the coming subsections.

The Industrial, Scientific and Medical (ISM) radio bands were first established in 1947 by the International Telecommunications Union (ITU) in Atlantic City. The ISM bands are defined by the ITU-R in 5.138, 5.150, and 5.280 of the Radio Regulations (ITU 1947). Individual countries' use of the bands designated in these sections may differ due to variations in national radio regulations. ISM are radio bands (portions of the radio spectrum) reserved internationally for the use of radiofrequency (RF) energy for industrial, scientific, and medical purposes other than telecommunications (Table 1.2). Examples of applications in these bands include radiofrequency process heating, microwave ovens, and medical diathermy machines. The powerful emissions of these devices can create electromagnetic interference and disrupt radio communication using the same frequency, so these devices were limited to certain bands of frequencies.

Table 1.2 ISM bands defined by ITU-R

Frequency range		Bandwidth	Center frequency	Availability
00.000 kHz	150 kHz	150 kHz	75 kHz	Region 1 low power, narrow band
6.765 MHz	6.795 MHz	30 kHz	6.780 MHz	Subject to local acceptance
13.553 MHz	13.567 MHz	14 kHz	13.560 MHz	Radiofrequency identification
26.957 MHz	27.283 MHz	326 kHz	27.120 MHz	Citizen band (CB) radio models
40.660 MHz	40.700 MHz	40 kHz	40.680 MHz	Radio models
433.050 MHz	434.790 MHz	1.74 MHz	433.920 MHz	Region 1 and subject to local acceptance
866.000 MHz	868.000 MHz	2 MHz	867.000 MHz	Region 1. Very narrow band, few channels.
902.000 MHz	928.000 MHz	26 MHz	915.000 MHz	Region 2 only (with some exceptions)
2.400 GHz	2.4835 GHz	83.5 MHz	2.441 GHz	Region 1, region 2, region 3
5.725 GHz	5.875 GHz	150 MHz	5.800 GHz	Region 3 has extended the upper range, additional ~150 MHz.
24.000 GHz	24.250 GHz	250 MHz	24.125 GHz	
61.000 GHz	61.500 GHz	500 MHz	61.250 GHz	Subject to local acceptance
122.000 GHz	123.000 GHz	1 GHz	122.500 GHz	Subject to local acceptance
244.000 GHz	246.000 GHz	2 GHz	245.000 GHz	Subject to local acceptance

Region 1 comprises Europe, Africa, the Middle East west of the Arabian Gulf including Iraq, the former Soviet Union, and Mongolia
Region 2 covers the Americas, Greenland, and some of the eastern Pacific Islands
Region 3 contains most of non-former-soviet-union Asia, east of and including Iran, and most of Oceania

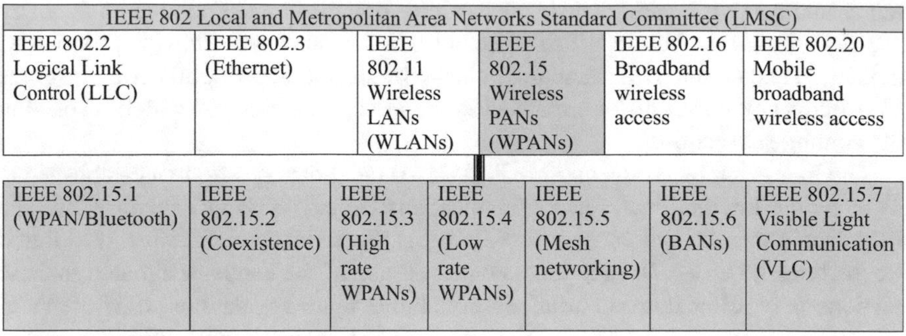

IEEE 802 Local and Metropolitan Area Networks Standard Committee (LMSC)					
IEEE 802.2 Logical Link Control (LLC)	IEEE 802.3 (Ethernet)	IEEE 802.11 Wireless LANs (WLANs)	IEEE 802.15 Wireless PANs (WPANs)	IEEE 802.16 Broadband wireless access	IEEE 802.20 Mobile broadband wireless access

IEEE 802.15.1 (WPAN/Bluetooth)	IEEE 802.15.2 (Coexistence)	IEEE 802.15.3 (High rate WPANs)	IEEE 802.15.4 (Low rate WPANs)	IEEE 802.15.5 (Mesh networking)	IEEE 802.15.6 (BANs)	IEEE 802.15.7 Visible Light Communication (VLC)

Fig. 1.6 IEEE 802 standards with focus on IEEE 802.15

Wireless sensor standards have been developed with a key design requirement for low-power consumption. The standards define the functions and protocols necessary for sensor nodes to interface with a variety of networks. A detailed description of such standards is enlightened in the Sections to follow. The IEEE 802.15 is a working group focusing on wireless personal area networks (WPANs); it has seven different approved standards and several ongoing standards discussions that are in different phases of the standardization process (IEEE 2019). All 802.15.x approved standards propose PHY and MAC layers; they do not provide network, transport, or application layers, implying that this task is left for other parties. ZigBee as will be illustrated in Sect. 1.8.2 is a company alliance that constructs network and application layers to 802.15.4 devices. Figure 1.6 lists the IEEE 802 standards with a focus on IEEE 802.15.

As instance, the 802.15.1 is a standard of the lower transport layers of the Bluetooth stack that contains a MAC and a PHY layer specifications. Task group 2 has delivered 802.15.2 as a recommended practice for coexistence of WPAN devices with other radio equipment in unlicensed frequency bands. Also, task group 3 of 802.15 presented a standard in 2003 that was intended for high-rate WPAN with application areas such as multimedia and digital imaging. High rate in this context is transfer rates of 11, 22, 33, 44, and 55 Mbps. Task group 3 had two sub-working groups called 802.15.3a and 802.15.3b, where the former was supposed to present a new PHY based on ultra-wide band (UWB) radio technique, and the latter came up in 2005 with an amendment to the MAC sublayer. In subgroup 3a, two different proposals of UWB techniques were discussed as a new PHY layer, but two industry alliances could not come to a consensus on which one to adopt. Consequently, IEEE decided to postpone further meetings in this subgroup and there is no UWB PHY standard yet to high-rate WPAN. Task group 4 of 802.15, as will be further elaborated in Sect. 1.8.1, has developed a standard intended for low data transfer rates of WPANs as opposite to the high transfer rates of 802.15.3. In Fig. 1.7, the wireless standards space, including IEEE 802.15 standards, is portrayed based on data rate and wireless equipment range.

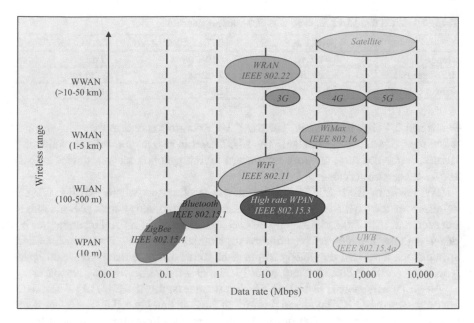

Fig. 1.7   Wireless standards space

Table 1.3   IEEE 802.15.4 High-level characteristics

			868 MHz	1 channel – 20 Kbps
		Low-band (BPSK)	915 MHz	10 channels – 40 Kbps
Frequency band	Two PHYs	High-band (O-QPSK)	2.4 GHz	16 channels – 250 Kbps
Channel access	CSMA/CA and slotted CSMA/CA			
Range	10–20 m			
Latency	15 msec			
Addressing	Short 8 bit or 64-bit IEEE			

Acronyms: *BPSK* binary-phase shift keying, *CSMA/CA* carrier sense multiple access with collision avoidance, *O-QPSK* offset quadrature phase shift keying

1.8.1   IEEE 802. 15.4 Low Rate WPANs

IEEE 802.15.4 is the proposed standard for low rate wireless personal area networks (LR-WPANs) with focus on enabling WSNs (Gutierrez et al. 2001; Callaway et al. 2002; Howitt and Gutierrez 2003). IEEE 802.15.4 focuses on low cost of deployment, low complexity, and low-power consumption; it is designed for wireless sensor applications that require short-range communication to maximize battery life. WSN applications using IEEE 802.15.4 include residential, industrial, and environment monitoring, control and automation.

IEEE 802.15.4 devices are designed to follow the physical and data link layer protocols. As illustrated in Table 1.3, the physical layer supports 868/915 MHz low

Table 1.4 IEEE 802.15.4 compared with 802 wireless standards

	802.11b WLAN	802.15.1 WPAN	802.15.4 LR-WPAN
Range	~100 m	~10–100 m	10 m
Raw data rate	11 Mbps	1 Mbps	≤0.25 Mbps
Power consumption	Medium	Low	Ultra-low

bands and 2.4 GHz high bands. The MAC layer controls access to the radio channel using the CSMA/CA mechanism. The MAC layer is also responsible for validating frames, frame delivery, network interface, network synchronization, device association, and secure services.

The intent of IEEE 802.15.4 is to address applications where existing WPAN solutions are too expensive and the performance of a technology such as Bluetooth is not required. IEEE 802.15.4 LR-WPANs complement other WPAN technologies by providing very low-power consumption capabilities at very low cost, thus enabling applications that were previously impractical. Table 1.4 illustrates a basic comparison between IEEE 802.15.4 and other IEEE 802 wireless networking standards.

As previously stated, like all IEEE 802 standards, the IEEE 802.15.4 standard encompasses only PHY layer and portions of the data link layer (DLL). Higher layer protocols are at the discretion of the individual applications utilized in an in-home network environment. In traditional wired networks, the network layer is responsible for topology construction and maintenance, as well as naming and binding services, which incorporate the necessary tasks of addressing, routing, and security. The same services exist for wireless in-home networks, but are far more challenging to implement because of the premium placed on energy conservation. In fact, it is important for any network layer implementation built on the already energy conscious IEEE 802.15.4 standard to be equally conservative. Network layers built on the standard must be self-organizing and self-maintaining, to minimize total cost to the consumer user.

The IEEE 802 standard splits the DLL into two sublayers, the MAC and logical link control (LLC) sublayers. The LLC is standardized in 802.2 and is common among the 802 standards such as 802.3, 802.11, and 802.15.1, while the MAC sublayer is closer to the hardware and may vary with the physical layer implementation. The features of the IEEE 802.15.4 MAC are association and disassociation, acknowledged frame delivery, channel access mechanism, frame validation, guaranteed time slot management, and beacon management.

Figure 1.8 shows how IEEE 802.15.4 fits into the International Organization for Standardization (ISO) open systems interconnection (OSI) reference model. The IEEE 802.15.4 MAC provides services to an IEEE 802.2 type I LLC through the service-specific convergence sublayer (SSCS), or a proprietary LLC can access the MAC services directly without going through the SSCS. The SSCS ensures compatibility between different LLC sublayers and allows the MAC to be accessed through a single set of access points. Using this model, the 802.15.4 MAC provides

Higher layers		
Network layer		
Data link layer	IEEE 802.2 LLC, type I SSCS	Other LLC
	IEEE 802.15.4 MAC	
Physical layer	IEEE 802.15.4 868/915 MHz	IEEE 802.15.4 2.4 GHZ

Acronyms:
SSCS (service-specific convergence sublayer)

Fig. 1.8   IEEE 802.15.4 follow up of the ISO OSI model

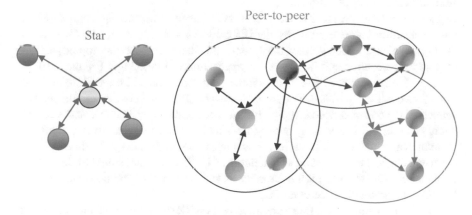

Fig. 1.9   Star and peer-to-peer topology organized as cluster network

features not utilized by 802.2 and therefore allows the more complex network topologies.

The IEEE 802.15.4 standard allows the formation of the star and peer-to-peer topology for communication between *network* devices (Fig. 1.9):

- In the star topology, the communication is performed between network devices and a single central controller, called the PAN coordinator. A network device is either the initiation point or the termination point for network communications. The PAN coordinator is in charge of managing all the star PAN functionality.
- In the peer-to-peer topology, every network device can communicate with any other within its range. This topology also contains a PAN coordinator, which acts as the root of the network. Peer-to-peer topology allows more complex network formations to be implemented, e.g., ad hoc and self-configuring networks. The routing mechanisms required for multihopping are part of the network layer and are therefore not in the scope of IEEE 802.15.4.

1.8.2   ZigBee

The ZigBee standard was publicly available as of June 2005 (ZigBee Alliance 2013); it defines the higher layer communication protocols built on the IEEE 802.15.4 standards for LR-PANs. ZigBee got its name from the way bees zig and zag while tracking between flowers and relaying information to other bees about where to find resources. ZigBee is a simple, low-cost, and low-power wireless communication technology used in embedded applications. ZigBee devices use very little power and can operate on a cell battery for many years. ZigBee has been introduced by IEEE with IEEE 802.15.4 standard and the ZigBee Alliance to provide the first general standard for such applications.

ZigBee is built on the robust radio (PHY) and medium access control (MAC) communication layers defined by the IEEE 802.15.4 standard for LR-WPANs. On the higher layer, ZigBee defines mesh, star, and cluster tree network topologies with data security features and interoperable application profiles (Figs. 1.9 and 1.10).

Table 1.5 compares ZigBee with wireless standards that address mid- to high data rates for voice, PC LANs, video, etc. However, ZigBee meets the unique needs of sensors and control devices, typically, low bandwidth, low latency, and very low energy consumption for long battery lives and for large device arrays. ZigBee is simpler than Bluetooth; it has a lower data rate and spends most of its time sleepy. It is accepted that standards such as Bluetooth and WLAN are not suited for low-power applications, due to their high node costs as well as complex and power-demanding RF-ICs and protocols (Lee et al. 2007).

As Fig. 1.11 illustrates, there are three types of devices that form mesh networks connecting hundreds to thousands of (Lee et al. 2007) devices together (Safaric and Malaric 2006):

- ZigBee coordinator, it initiates network formation, stores information, and can bridge networks together.
- ZigBee routers, they link groups of devices together and provide multihop communication across devices.
- ZigBee end device, it consists of the sensors, actuators, and controllers that collects data and communicates only with the router or the coordinator.

Fig. 1.10 ZigBee over IEEE 802.15.4 buildup

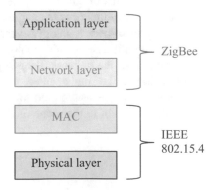

Table 1.5 ZigBee compared with wireless standards

	Bluetooth	UWB	ZigBee	WiFi
IEEE specification	802.15.1	802.15.3a	802.15.4	802.11a/b/g
ISM frequency band	2.4 GHz	3.1–10.6 GHz	868/915 MHz, 2.4 GHz	2.4 GHz, 5 GHz
Application	Wireless connectivity between devices such as phones, PDA, laptops, headsets	Real-time video and music, multimedia wireless network, WPAN	Industrial control and monitoring, sensor networks, building automation, home control and automation, toys, games	Wireless LAN connectivity, broadband internet access
Max signal rate	1 Mbps	110 Mbps	250 Kbps	54 Mbps
Nominal range	10 m	10 m	10–100 m	100 m
Transmission power	0–10 dBm	−41.3 dBm/ MHz	(−25) – 0 dBm	15–20 dBm
Channel bandwidth	1 Mbps	500 MHz– 7.5 GHz	0.3/0.6 MHz; 2 MHz	22 MHz
Modulation type	GFSK	BPSK, QPSK	BPSK (+ ASK), O-QPSK	BPSK, QPSK COFDM, CCK, M-QAM
Basic cell	Piconet	Piconet	Star	BSS
Extension of the basic cell	Scatternet	Peer-to-peer	Cluster tree, mesh	ESS
Max number of cell nodes	8 active devices, 255 in park mode	8	>65,000	Unlimited in ad hoc networks (IBSS), up to 2007 devices in infrastructure networks
Encryption	E0 stream cipher	AES block cipher (CTR, counter mode)	AES block cipher (CTR, counter mode)	RC4 stream cipher (WEP), AES block cipher
Authentication	Shared secret	CBC-MAC (CCM)	CBC-MAC (ext. of CCM)	WPA2 (802.11i)
Data protection	16-bit CRC	32-bit CRC	16-bit CRC	32-bit CRC
Properties	Cost, easy setup, low interference, device connection requires up to 10 seconds	Low power, high throughput, low interference, wall penetration	Reliability, very low power, low cost, security, devices can join an existing network in under 30 ms	Speed, flexibility, device connection requires 3–5 seconds

Acronyms: *AES* advanced encryption standard, *ASK* amplitude shift keying, *BPSK/QPSK* binary/ quadrature phase SK, *BSS/IBSS/ESS* basic/independent basic/extended service set, *CBC-MAC* cipher block chaining message authentication code, *CCK* complementary code keying, *CCM* CTR with CBC-MAC, *COFDM* coded OFDM, *CRC* cyclic redundancy check, *FHSS/DSSS* frequency hopping/direct sequence spread spectrum, *GFSK* Gaussian frequency SK, *M-QAM* M-ary quadrature amplitude modulation, *MB-OFDM* multiband OFDM, *O-QPSK* offset-QPSK, *OFDM* orthogonal frequency division multiplexing, *WEP* wired equivalent privacy, *WPA* WiFi-protected access.

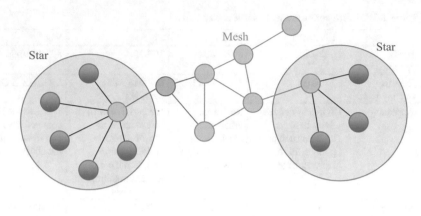

Fig. 1.11   ZigBee network model

1.8.3   WirelessHART

WirelessHART was released in September 2007 (Kim et al. 2008; Song et al. 2008). The WirelessHART standard provides a wireless network communication protocol for process measurement and control applications; it is based on IEEE 802.15.4 for low-power 2.4 GHz operation. WirelessHART is compatible with all existing devices, tools, and systems; it is reliable, secure, and energy-efficient and supports mesh networking, channel hopping, and time-synchronized messaging. Network communication is secure with encryption, verification, authentication, and key management. Power management options enable the wireless devices to be more energy-efficient. WirelessHART is designed to support mesh, star, and combined network topologies. As shown in Fig. 1.12, WirelessHART network consists of wireless field devices, gateways, process automation controller, host applications, and network manager:

- Wireless field devices are connected to process or plant equipment.
- Gateways enable the communication between the wireless field devices and the host applications.
- Handheld which is a portable WirelessHART enabled computer used to configure devices, run diagnostics, and perform calibrations.
- The network manager configures the network and schedule communication between devices; it also manages the routing and network traffic. The network manager can be integrated into the gateway, host application, or process automation controller.

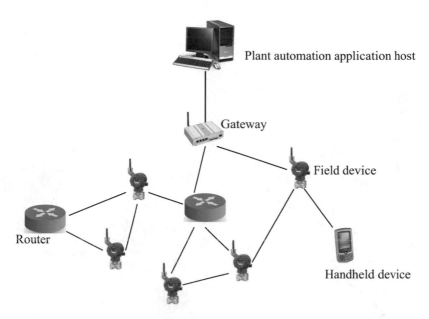

Fig. 1.12 WirelessHART mesh networking

OSI layer	WirelessHART	
Application	Command oriented. Predefined data types and application procedures	
Presentation		
Session		
Transport	Auto-segmented transfer of large data sets, reliable stream transport, negotiated segment sizes	
Network		Power-optimized redundant path, mesh to the edge network
Data link	A binary, Byte oriented, token passing, master/slave protocol	Secure, time synched TDMA/ CSMA, frequency agile with ARQ
Physical	Simultaneous analog & digital signaling 4-20mA copper wiring	2.4 GHz wireless, 802.15.4 based radios, 10dBm Tx Power
	Wired FSK/PSK & RS 485	Wireless 2.4 GHz

Acronyms:
ARQ (Automatic Repeat -reQuest), CSMA (carrier sense multiple access), FSK (frequency shift keying), PSK (phase shift keying), TDMA (time division multiple access).

Fig. 1.13 WirelessHART protocol stack

Figure 1.13 illustrates the architecture of the WirelessHART protocol stack in accordance with the OSI 7-layer communication model.

1.8.4 ISA100.11a

ISA100.11a was officially approved in September 2009 by ISA Standards & Practices Board. It is the first standard of ISA100 family with foundations for process

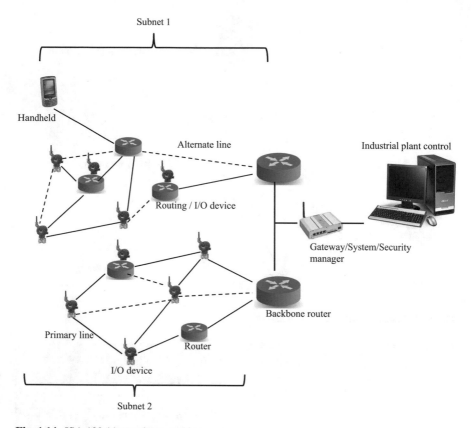

Fig. 1.14 ISA 100.11a mesh networking

automation, and provisions for secure, reliable, low data rate wireless monitoring. Specifications for the OSI layer, security, and system management are comprised (Costa and Amaral 2012). ISA100.11a focuses on low energy consumption, scalability, infrastructure, robustness, and interoperability with other wireless devices. ISA100.11a networks use only 2.4 GHz radio and channel hopping to increase reliability and minimize interference. It offers both meshing and star network topologies and delivers simple, flexible, and scalable security functionality.

ISA100.11a defines for devices different role profiles that represent various functions and capabilities, such as I/O devices, routers, provisioning devices, backbone routers, gateway, system manager, and security manager. Each device may resume more than a role, while its capabilities are reported to the system manager upon joining the network (Fig. 1.14):

- I/O device (sensor and actuator) provides and/or consumes data, which is the basic goal of the network.
- Handheld which is a portable computer used to configure devices, run diagnostics, and perform calibrations.

- Router is a job accorded to devices responsible for routing data packet from source to destination and propagating clock information. A router role also enables a device to act as a proxy that permits new devices to join the network.
- Device with provisioning role, for preconfiguring devices with necessary information to join a specific network.
- Backbone router, routes data packets from one subnet connected to the backbone network to a destination (e.g., another subnet connected to the backbone). The backbone router is implemented with both ISA100.11a wireless network interface and backbone interface.
- Gateway acts as an interface between ISA100.11a field network and the host applications in the control system.
- System manager is the administrator of the ISA100.11a wireless network. It monitors the network and is in charge of system management, device management, network run-time control, and communication configuration (resource scheduling), as well as time-related services.
- Security manager provides security services based on policies specified by this standard.

Both WirelessHART and ISA100.11a use a simplified version of the seven-layered open systems interconnection (OSI) basic reference model, as depicted in Fig. 1.15 (Petersen and Carlsen 2011). ISA100.11a divides the DLL into a MAC sublayer, a MAC extension, and an upper DLL. The MAC sublayer is a subset of IEEE Standard 802.15.4 MAC, with the main responsibility of sending and receiving individual data frames. The MAC extension includes additional features not supported by IEEE Standard 802.15.4 MAC, mainly concerning changes to the carrier sense multiple access with collision avoidance (CSMA/CA) mechanisms by including additional spatial, frequency, and time diversity. The upper DLL handles link and mesh aspects above the MAC level, and it is responsible for routing within a DL subnet.

OSI layers		
Application	Application layer	Upper application layer
		Application sublayer
Presentation	Not defined	Not defined
Session	Not defined	Not defined
Transport	Transport layer	Transport layer
Network	Network layer Services	Network layer
	Network layer	
Data link	Logical Link Control	Upper data link layer
	MAC sublayer	MAC extension
		MAC sublayer
Physical	Physical layer	Physical layer
	WirelessHART	ISA100.11a

Fig. 1.15 The WirelessHART and ISA100.11a protocol stack

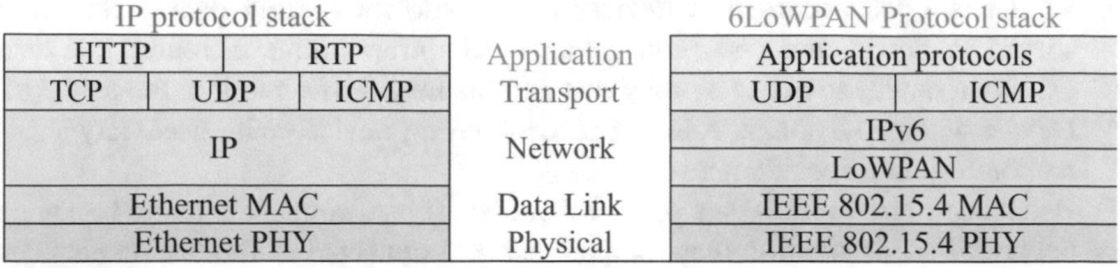

IP protocol stack				6LoWPAN Protocol stack	
HTTP		RTP	Application	Application protocols	
TCP	UDP	ICMP	Transport	UDP	ICMP
IP			Network	IPv6	
				LoWPAN	
Ethernet MAC			Data Link	IEEE 802.15.4 MAC	
Ethernet PHY			Physical	IEEE 802.15.4 PHY	

Acronyms:
HTTP (HyperText Transfer Protocol), ICMP (Internet Control), RTP (Real -time Transport Protocol), TCP (Transport Control Protocol), UDP (User Datagram Protocol).

Fig. 1.16 IP and 6LoWPAN protocol stacks

1.8.5   6LoWPAN

IPv6-based low-power wireless personal area networks (6LoWPAN) enable IPv6 packets communication over an IEEE 802.15.4-based network (Mulligan 2007; Montenegro et al. 2007; Shelby and Bormann 2011). Low-power devices can communicate directly with IP devices using IP-based protocols. Utilizing 6LoWPAN, low-power devices have all the benefits of IP communication and management. 6LoWPAN standard provides an adaptation layer, new packet format, and address management. Because IPv6 packet sizes are much larger than the frame size of IEEE 802.15.4, the adaptation layer is used. The adaptation layer accomplishes header compression, which creates smaller packets fitting into the IEEE 802.15.4 frame size. Address management mechanism handles the forming of device addresses for communication. 6LoWPAN is designed for applications with low data rate devices that require Internet communication (Fig. 1.16).

The Wireless Embedded Internet is created by connecting networks of wireless embedded devices; each network is a stub on the Internet. A stub network is a network where IP packets are sent from or destined to, but which does not act as a transit to other networks. The 6LoWPAN architecture is made up of low-power wireless area networks (LoWPANs), which are IPv6 stub networks. The overall 6LoWPAN architecture is presented in Fig. 1.16. A LoWPAN is the collection of 6LoWPAN nodes which share a common IPv6 address prefix (the first 64 bits of an IPv6 address), meaning that regardless of where a node is in a LoWPAN its IPv6 address remains the same. Three different kinds of LoWPANs have been defined (Fig. 1.17):

- Simple LoWPAN, connected through one LoWPAN Edge Router to another IP network. A backhaul link (point to point, e.g., GPRS) is shown in the figure, but it could also be a backbone link (shared).

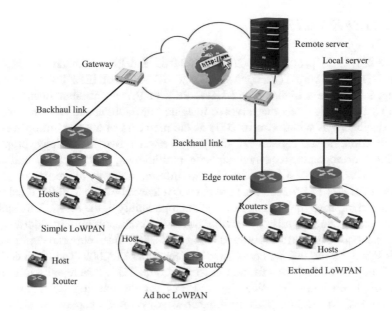

Fig. 1.17 The 6LoWPAN architecture. (Based on Shelby and Bormann (2011))

- Extended LoWPAN, that encompasses the LoWPANs of multiple edge routers via a backhaul link (e.g., Ethernet) interconnecting them. Edge routers share the same IPv6 prefix and the common backbone link.
- Ad hoc LoWPAN, that is not connected to the Internet, but instead operates without an infrastructure.

A LoWPAN consists of one or more edge routers along with nodes, which may function as host or router. The network interfaces of the nodes share the same IPv6 prefix distributed by the edge router and routers throughout the LoWPAN. Each node is identified by a unique IPv6 address and is capable of sending and receiving IPv6 packets. In order to facilitate efficient network operation, nodes register with an edge router. LoWPAN nodes may participate in more than one LoWPAN at the same time (called multihoming), and fault tolerance can be achieved between edge routers. Nodes are free to move throughout the LoWPAN, between edge routers, and even between LoWPANs. Topology change may also be caused by wireless channel conditions, without physical movement.

LoWPANs are connected to other IP networks through edge routers, as seen in Fig. 1.17. The edge router plays an important role as it routes traffic in and out of the LoWPAN, while handling 6LoWPAN compression and Neighbor Discovery for the LoWPAN. If the LoWPAN is to be connected to an IPv4 network, the edge router will also handle IPv4 interconnectivity. Edge routers have management features tied into overall IT management solutions. Multiple edge routers can be supported in the same LoWPAN if they share a common backbone link.

1.8.6   IEEE 80215.3

IEEE 802.15.3 as proposed in 2003 is a MAC and PHY standard for high-rate WPANs (11 to 55 Mbps) (Tseng et al. 2003; IEEE 2013a, b). IEEE 802.15.3a was an attempt to provide a higher speed UWB PHY enhancement amendment to IEEE 802.15.3 for applications that involve imaging and multimedia. But the proposed PHY standard was withdrawn in 2006 as the members of the task group were not able to come to an agreement choosing between two technology proposals, multiband orthogonal frequency division multiplexing (MB-OFDM) and direct sequence UWB (DS-UWB), backed by two different industry alliances.

The IEEE 802.15.3b-2005 amendment was released on May 5, 2006. It enhanced 802.15.3 to improve implementation and interoperability of the MAC. This includes minor optimizations while preserving backward compatibility. In addition, this amendment corrected errors, clarified ambiguities, and added editorial clarifications.

IEEE 802.15.3c-2009 was published on September 11, 2009. The IEEE 802.15.3 Task Group 3c (TG3c) was formed in March 2005. TG3c developed a millimeter-wave-based alternative physical layer (PHY) for the existing 802.15.3 WPAN Standard 802.15.3–2003. This millimeter-wave WPAN operates in clear band including 57–64 GHz unlicensed band defined by FCC 47 CFR 15.255. The mmWPAN permits high coexistence (close physical spacing) with all other micro-wave systems in the 802.15 family of WPANs. In addition, the mmWPAN allows very high data rate over 2 Gbps applications such as high-speed Internet access, streaming content download (video on demand, HDTV, home theater, etc.), real-time streaming, and wireless data bus for cable replacement. Also, there are optional data rates in excess of 3 Gbps.

At the MAC layer, WPAN high-rate technology (802.15.3) is based on central-ized signaling and peer-to-peer traffic structure; the nodes are classified as PicoNet Coordinators (PNC) and Devices (DEV). A PNC assigns guaranteed time slots to all nodes for communication. More precisely, there is a period for contention, followed by a contention free period, which contains guaranteed time slots being allocated by the PNC as shown in Fig. 1.18.

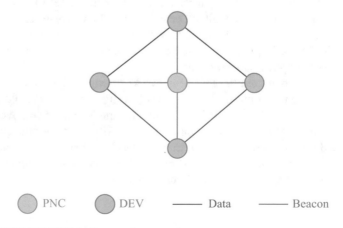

Fig. 1.18   IEEE 802.15.3 MAC network structure

1.8.7 Wibree, BLE

Released in 2006 by Nokia, it is a wireless communication technology designed for low-power consumption, short-range communication, and low-cost devices, it is called Baby-Bluetooth, and renamed Bluetooth Low Energy (BLE) technology (Pei et al. 2008). Wibree allows the communication between small battery-powered devices and Bluetooth devices. Small battery-powered devices include watches, wireless keyboard, and sports sensors which connect to host devices such as personal computer or cellular phones. This standard operates on 2.4 GHz and has a data rate of 1 Mbps, with 5–10 m as a linking distance between the devices.

Wibree may be deployed on a stand-alone chip or on a dual-mode chip along with conventional Bluetooth; it works with Bluetooth to make devices smaller and more energy-efficient. Bluetooth-Wibree utilizes the existing Bluetooth RF and enables ultra-low-power consumption.

A key point must be taken into consideration, BLE is incompatible with standard Bluetooth, and BLE devices do not interoperate with classical Bluetooth products. However, implementing a dual-mode device could achieve such interoperability. A dual-mode device is an integrated circuit that includes both a standard Bluetooth radio and a BLE radio, each mode operates separately, not at the same time, though they can share an antenna. Several vendors offer dual-mode chips, such as Broadcom, CSR, EM Microelectronics, Nordic Semiconductor, and Texas Instruments. Complete modules also are available from connectBlue (Frenzel 2012). Table 1.6 compares Bluetooth, Wibree, and ZigBee.

Table 1.6 Bluetooth, Wibree compared

	Bluetooth	Wibree
Band	2.4 GHz	2.4 GHz
Antenna/HW	Shared	
Power	100 mW	10 mW
Target battery life	Days – months	1–2 years
Peak current consumption	<30 mA	<15 mA
Range	10–30 m	10 m
Data rate	1–3 Mbps	1 Mbps
Application throughput	0.7–2.1 Mbps	0.27 Mbps
Active slaves	7	Unlimited
Component cost	$3	Bluetooth +20¢
Network topologies	Point to point, scatternet	Point to point, star
Security	56–128 bit encryption	128-bit AES
Time to wake and transmit	100+ msec	<6 msec

Acronyms: *AES* advanced encryption standard

1.8.8   Z-Wave

Z-Wave, a proprietary technology developed by Zensys A/S of Denmark, is focusing exclusively on the residential market (Reinisch et al. 2007; Z-Wave Alliance 2012). The two wireless networking standards, Z-Wave and ZigBee, are competing to become the standard for automated home control. ZigBee, an IEEE802.15.4-based standard proposed by a large group of worldwide manufacturers represented by the ZigBee Alliance, has a broader focus that includes both home and commercial control systems (Fig. 1.19).

Z-Wave uses a two-way RF system that operates in the 908 MHz ISM bands (868 MHz in Europe and 908 MHz in the United States). Z-Wave allows transmission at 9.6 and 40 Kbps data rates using binary frequency shift keying (BFSK) modulation.

The recent Z-Wave 400 series single chip supports the 2.4 GHz band and offers bit rates up to 200 Kbps. In contrast to other wireless networking technologies such as Bluetooth and wireless LAN, Z-Wave features lower power consumption and lower data rates. With very short transmit times and efficient design, Z-Wave nodes can easily be powered from a battery with long lifetime. Applications like residential lighting control take no more than 250 ms. Z-Wave relies on the fact that its targeted residential applications require the transmission of small amounts of data, and therefore, it uses a data rate of just 9.6 Kbps.

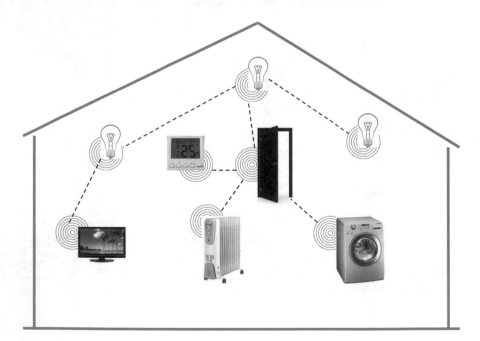

Fig. 1.19   Z-Wave WSNs home control

The second-generation Z-Wave chipset, the ZW0201, is used as a mixed signal chip, integrating an RF transceiver, Z-Wave protocol storage and handling, and capacity for application storage and handling. The ZW0201 as the core of Z-Wave features a low standby current of 0.1 uA. This current rises to 25 mA on transmission, but as the protocol has been designed to keep transmit and receive time to an absolute minimum it is possible to run a node from a battery. The device includes an 8-bit CPU core running at 8 MHz with up to 32 Kbits of flash memory; it has enough capacity to handle both an application and the wireless communication protocol.

On top of the link layer, a source-routing protocol gives designers the ability to set up a Z-Wave mesh network. Based on the network topology data in the initiator's memory, Z-Wave's source-routing protocol allows the initiator to generate a complete route from the initiator to the destination.

Z-Wave defines two types of devices, controllers and slaves. Controllers poll or send commands to the slaves, which reply to the controllers or execute the commands. The Z-Wave routing layer performs routing based on a source-routing approach. When a controller transmits a packet, it includes the path to be followed in the packet. A packet can be transmitted over up to four hops, which is sufficient in a residential scenario and hard limits the source-routing packet overhead. A controller maintains a table that represents the full topology of the network. A portable controller (e.g., a remote control) tries first to reach the destination via direct transmission; if that option fails, the controller estimates its location and calculates the best route to the destination. Slaves may act as routers; routing slaves store static routes (typically toward controllers) and are allowed to send messages to other nodes without being requested to do so. Slaves are suitable to be monitoring sensors where the delay contributed by polling is acceptable, as well as for actuators that perform actions in response to activation commands. Routing slaves are used for time critical and non-solicited transmission applications such as alarm activation.

1.8.9 Impulse Radio Ultra-Wide Bandwidth Technology, 802.15.4a

UWB is one of the enabling technologies for sensor network applications; in particular, impulse radio-based UWB (IR UWB) technology has a number of inherent properties that are well suited to sensor network applications. UWB systems have potentially low complexity and low cost; with noise like signal properties that create little interference to other systems, they are resistant to severe multipath and jamming and have very good time domain resolution allowing for precise locating and tracking. Various ultra-wide band wireless sensor network applications include locating and imaging of objects and environments, perimeter intrusion detection, video surveillance, in vehicle sensing, outdoor sports monitoring, monitoring of highways, bridges, and other civil infrastructure (Zhang et al. 2009).

Recognizing these interesting applications, a number of UWB-based sensor network concepts have been developed both in the industrial and in the government/military domain. Of particular importance are systems based on the UWB impulse radio IEEE 802.15.4a standard, which via well-defined flexible PHY and MAC layer is suitable for a wide variety of applications. Furthermore, it works together with the ZigBee networking standard, the dominant technology in WSN systems.

Design-wise, among two options within the 802.15.4a standard, the UWB LR-WPAN option is built to provide communications and high precision ranging (ranging is a process or method to determine the distance from one location or position to another location or position) / locating capability (1 meter accuracy and better), high aggregate throughput, and ultra-low power, as well as adding scalability to data rates, longer range, and lower power consumption and cost.

Several features are provided to satisfy the requirements for data communications:

- Extremely wide bandwidth characteristics that can ensure very robust performance under harsh multipath and interference conditions.
- Concatenated forward error correction coding to provide flexible and robust performance.
- Optional UWB pulse control features to provide improved performance under some channel conditions, while supporting reliable communications and precision ranging capabilities.

The 802.15.4a UWB PHY has its own attributes:

- In addition to the 850 Kbps mandatory data rate, variable data rates such as 110 Kbps, 1.70 Mbps, 6.81 Mbps, and 27.24 Mbps are also provided.
- Data can be communicated either between any UWB device and a coordinator, or between peer-to-peer coordinators.
- The UWB PHY design also enables heterogeneous networking, i.e., networks that consist of nodes with different capabilities and requirements. A network has at least one full-function device (FFD) and several reduced-function devices (RFDs). FFDs are typically more expensive (they are a minority of the network devices); they are often configured to handle higher processing complexity. For FFDs, higher energy consumption is not a real concern since they are usually connected to a permanent power supply. On the other hand, sensor nodes are usually RFDs with extremely stringent limits on complexity and energy consumption.
- The UWB PHY layer, which includes modulation, coding, and multiple-access schemes (MCM), has been designed in such a way that it allows both FFDs and RFDs to achieve optimum performance, such as allowing the FFD devices to employ coherent reception (enhanced performance at the cost of energy consumption and complexity), while RFDs use simple energy detectors (non-coherent receivers) for reduced current drain and design simplicity. Furthermore, such a flexible MCM scheme does not worsen the possible performance of

Table 1.7 UWB compared with ZigBee and WiFi

	2.4 GHz ZigBee	2.4 GHz WiFi	UWB[a]
Data rate	Low, 250 Kbps	High, 11 Mbps for 802.11b and 100+ Mbps for 802.11n	Medium, 1 Mbps mandatory, and up to 27 Mbps for 802.15.4a
Transmission distance	Short, <30 meters	Long, up to 100 meters	Short, <30 meters
Location accuracy	Low, several meters	Low, several meters	High, <50 cm
Power consumption	Low, 20 mW–40 mW	High, 500 mW–1 W	Low, 30 mW
Multipath performance	Poor	Poor	Good
Interference resilience	Low	Medium	High with high complexity receivers, low with simplest receivers
Interference with other systems	High	High	Low
Complexity and cost	Low	High	Low, medium, high are possible

[a]Frequency band: 3.1–10.6 GHz in US, 6–8.5 GHz in Europe, 3.4–4.8 GHz in Japan

the FFDs, i.e., the performance of FFDs with flexible MCM is almost as good as with an MCM that is designed for homogeneous coherent-receiver networks. Table 1.7 compares UWB with ZigBee and WiFi.

1.8.10 INSTEON

INSTEON is a solution developed for home automation by SmartLabs and promoted by the INSTEON Alliance (INSTEON 2013). One of the distinctive features of INSTEON is the fact that it defines a mesh topology composed of RF and power line links. Devices can be RF-only or power-line-only, or can support both types of communication. INSTEON RF signals use frequency shift keying (FSK) modulation at the 904 MHz center frequency, with a raw data rate of 38.4 Kbps.

INSTEON networking has several features (Fig. 1.20):

- INSTEON devices are peers, which means that any of them can play the role of sender, receiver, or relayer.
- Communication between devices that are not within the same range is achieved by means of a multihop approach that differs in many aspects from traditional techniques. All devices retransmit the messages they receive, unless they are the destination of the messages. The maximum number of hops for each message is

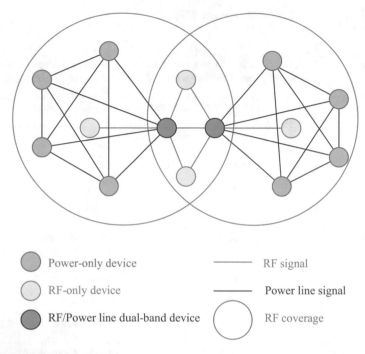

Power-only device   ——— RF signal

RF-only device   ——— Power line signal

RF/Power line dual-band device   RF coverage

Fig. 1.20 INSTEON networking

limited to four (as in Z-Wave). The multihop transmission is performed using a
time slot synchronization scheme, by which transmissions are permitted in certain
time slots, and devices within the same range do not transmit different messages
at the same time. These time slots are defined by a number of power line zero
crossings.

- RF devices that are not attached to the power line can transmit asynchronously,
  but RF devices attached to the power line will retransmit the related messages
  synchronously.
- In contrast to classical collision avoidance mechanisms, devices within the same
  range are allowed to transmit the same message simultaneously. This approach,
  which is called *simulcast*, relies on the very low probability of multiple simultaneous signals being canceled at the receiver.

1.8.11 Wavenis

Wavenis is a wireless protocol stack developed by Coronis Systems for control and
monitoring applications in several environments, including home and building
automation (Gomez and Paradells 2010). Wavenis is currently being promoted and

managed by the Wavenis Open Standard Alliance (Wavenis-OSA). It defines the
functionality of physical, link, and network layers. Wavenis services can be accessed
from upper layers through an application programming interface (API).

Wavenis operates mainly in the 433 MHz, 868 MHz, and 915 MHz bands, which
are ISM bands in Asia, Europe, and the United States. Some products also operate in
the 2.4 GHz band. The minimum and maximum data rates offered by Wavenis are
4.8 Kbps and 100 Kbps, respectively, with 19.2 Kbps being the typical value. Data
are modulated using Gaussian FSK (GFSK). Fast frequency hopping spread spectrum (FHSS) is used over 50 kHz bandwidth channels. The Wavenis MAC sublayer
offers synchronized and non-synchronized schemes:

- In a synchronized network, nodes are provided with a mixed CSMA/TDMA
  mechanism for transmitting in response to a broadcast or multicast message. In
  such a case, a node allocates a time slot that is pseudo-randomly calculated, based
  on its address. Before transmission in that slot, the node performs carrier sense
  (CS). If the channel is busy, the node computes a new time slot for the
  transmission.
- For non-synchronized networks, in applications where reliability is a critical
  requirement (alarms, security, etc.), CSMA/CA is used. The Wavenis logical
  link control (LLC) sublayer manages flow and error control by offering per-frame
  or per-window ACKs.

Wavenis defines only one type of device. The Wavenis network layer specifies a
four-level virtual hierarchical tree. The root of the tree may play the role of a data
collection sink or a gateway. A device that joins a Wavenis network intends to find
an adequate parent, for this purpose, the new device broadcasts a request for a device
of a certain level and a sufficient quality of service (QoS) value. The QoS value is
obtained by taking into consideration parameters such as received signal strength
indicator (RSSI) measurements, battery energy, and the number of devices that are
already attached to this device. Table 1.8 assembles and compares domestic WSNs
as presented in the previous subsections.

1.8.12 ANT

ANT is a proprietary technology featuring a wireless communication protocol stack
thought for ultra-low-power networking applications (ANT 2013). It is designed to
run using low-cost, low-power micro-controllers and transceivers operating in the
2.4 GHz ISM band. The ANT WSN protocol has been engineered for simplicity and
efficiency, resulting in ultra-low-power consumption, maximized battery life, a
minimal burden on system resources, simpler network designs, and lower implementation costs. ANT also features low latency, ability to trade-off data rate against
power consumption, support for broadcast, burst, and acknowledged transactions up
to a net data rate of 20 Kbps (ANTs over the air data rate is 1 Mbps).

Table 1.8 Domestic WSNs compared

Protocol	ZigBee	Z-Wave	EnOcean	UWB	Wibree, BLE	INSTEON	Wavenis
IEEE standard	802.15.4	–	–	802.15.4a	802.15.1	–	–
Frequency band	868/915 MHz, 2.4 GHz	868/908 MHz	868 MHz, 1.70 Mbps, and 6.81 Mbps, and 27.24 Mbps	3.1–10.6 GHz	2.4 GHz	904 MHz	433/868/ 915 MHz
Data rate	20/40/250 Kbps	9.6 Kbps/ 40 Kbps, 200 Kbps	120 Kbps	850 Mbps	1 Mbps	38.4 Kbps	19.2 Kbps
Modulation	BPSK/BPSK/O-QPSK	FSK/GFSK	ASK	BPSK, QPSK	GFSK	FSK	GFSK/PSK
Spreading	DSSS	No	No	DS-UWB, MB-OFDM	FHSS	No	FHSS
Interference risk	Very high (2.4 GHz)	Medium	Very low	Very low	Low	Medium	Low
Communication range (m)	30 (indoor) 100 (outdoor)	30 (indoor) 100 (outdoor)	30 (indoor) 300 (outdoor)	10	10	50 (outdoor)	200 (indoor) 1000 (outdoor)
Energy requirements	Medium	Low	Very low	High	Very low	Medium	Medium
Security	AES	AES-128	Basic	AES	E0 Stream AES-128	Rolling codes, public key	3 DES 128AES
Error control/ reliability	16-bit CRC, ACK, CSMA/CA	8-bit CRC, ACK, CSMA/CA	–	32-bit CRC, CSMA/CA	16-bit CRC	8-bit checksums	BCH (32,21) FEC
Network size	>65,000	232	2^{32}	8	8	256	N/A
Internet connection	Gateway required	Gateway required	Gateway required	Gateway required	Gateway required	Gateway required	Gateway required
Logistic	Standard	Proprietary	Proprietary	Standard	Standard	Proprietary	Proprietary

Acronyms: *ACK* acknowledgment, *AES* advanced encryption standard, *BPSK/QPSK* binary/quadrature phase SK, *BSS/IBSS/ESS* basic/independent basic/ extended service set, *CSMA/CA* carrier sense multiple access with collision avoidance, *COFDM* coded OFDM, *CRC* cyclic redundancy check, *DES* data encryption standard, *FHSS/DSSS* frequency hopping/direct sequence spread spectrum, *GFSK* Gaussian frequency SK, *MB-OFDM* multiband OFDM, *O-QPSK* offset-QPSK, *OFDM* orthogonal frequency division multiplexing

Different topologies could be established, peer-to-peer, star, tree, and other types of mesh network. ANT nodes are capable of acting as slaves or masters within a network and swapping roles at any time. This means that the nodes can act as transmitters, receivers, or transceivers to route traffic to other nodes. ANT is a good protocol for practical networks because of this inherent ability to support ad hoc interconnection of tens or hundreds of nodes. ANT allows a system to spend most of its time in an ultra-low-power sleep mode, wake up quickly, transmit for the shortest possible time, and quickly return back to an ultra-low-power sleep mode. This implies that ANT is one of the energy-efficient available technologies. While Bluetooth is designed for rapid file transfer between devices in a PAN, its average power consumption is 10 times greater with respect to ANT, and the hardware costs are 90% higher. With respect to IEEE 802.15.4, ANT presents a larger data rate of 1 Mbpsec and is relatively less complex. However, being a proprietary technology, ANT lacks interoperability.

ANT+ is a relatively recent addition to ANT. This software function provides interoperability in a managed network; it facilitates the collection, automatic transfer, and tracking of sensor data for monitoring all involved nodes and devices. But what is a managed network? It is a type of communication network that is built, operated, secured, and managed by a third-party service provider; it is an outsourced network that provides some or all the network solutions required by an organization. SensRcore, an extra ANT feature, is a development system that helps developers create low-power sensor networks. ANT transceiver chips are available from Nordic Semiconductor and Texas Instruments.

A number of similarities exist between ANT and BLE (Sect. 1.8.7), but their differences are stark. Both are good choices for very low-power applications (Table 1.9), but:

- ANT has the simplest protocol with minimum overhead, and it supports more different types of network topologies.
- BLE is a star-only format, while ANT supports all types including mesh.

More vendors offer Bluetooth chips and modules versus ANT, though.

Table 1.9 ANT and BLE compared

	ANT	Wibree BLE
Frequency	2.4–2.483 GHz	2.4–2.483 GHz
Network topology	Peer to peer, tree, mesh	Peer to peer, star
Modulation	GFSK	GFSK
Channel width	1 MHz	1 MHz
Protocol	Simplest	More complex
Data rate	1 Mbps	1 Mbps
Range	50 m	50 m
Security	64-bit encryption	128-bit AES

Acronyms: *AES* advanced encryption standard and *GFSK* Gaussian frequency shift keying

1.8.13   MyriaNed

MyriaNed is a wireless sensor network (WSN) platform developed by DevLab
(Wateren and Van 2008). It uses an epidemic communication style based on
standard radio broadcasting. This approach reflects the way humans interact,
gossiping. Messages are sent periodically and received by adjoining neighbors.
Each message is repeated and duplicated toward all nodes that span the network, it
spreads like a virus, hence the term epidemic communication. This is a very efficient
and robust protocol for two reasons:

*   The nodes do not need to know who is in their neighborhood at the time of
    sending a message, there is no notion of an a priori planned routing, data is just
    shared instantaneously.
*   The network is implicitly reliable since messages may follow different commu-
    nication routes in parallel. The loss of a message between two nodes does not
    mean that the data is lost.

Nodes can be added, removed or may be physically moving without the need to
reconfigure the network. The GOSSIP protocol is a self-configuring network solution. The network may even be heterogeneous, where several types of nodes
communicate different pieces of information with each other at the same time.
This is possible due to the fact that no interpretation of the message content is
required in order to be able to forward it to other nodes. Message communication is
fully transparent, providing a seamless communication platform, where new functionality can be added later, without the need to change the installed base. Furthermore, MyriaNed is enabled to update the wireless sensor nodes software by means of
"over the air" programming of a deployed network.

Radios used by MyriaNed nodes are operating at an ISM frequency of 2.4 GHz.
Since the nodes of the network are mostly battery-powered, low energy consumption
is necessary. By orchestrating the exchange of information between the nodes
periodically, the nodes can go into standby modes to save energy, during the radio
silence. The drawback of the low energy consumption is that the nodes have a low
send rate; this makes it important for nodes to minimize data exchange. To accomplish this, the network has a built-in feature, which ensures that messages that were
already seen/send by a particular node will be discarded to save energy. The same
mechanism also avoids flooding of the network. In Table 1.10, the characteristics of
MyriaNed are summarized.

1.8.14   EnOcean

EnOcean has commercially pioneered the concept of energy scavenging in the field
of building automation (Reinisch et al. 2007; EnOcean 2013). Entirely solar-
powered modules are available as well as pushbutton sensors driven by piezoelectric

Table 1.10 MyriaNed characteristics

Large-scale networks	No limitation on the number of concurrent nodes in the network
Ad hoc networks	The network is suitable for unreliable and dynamic environments, because of gossiping
No hierarchy	The MyriaNed WSN has no hierarchy; this removes the single point of failure
Low energy consumption	The network is designed for low energy consumption. Most of the time the nodes are in standby mode
Reliability	Redundancy and the multipath communication ensure that the network is reliable even in dynamic environments

Table 1.11 EnOcean energy harvesting wireless standard

High reliability	Use of regulated frequency ranges with highest air time availability (approved for pulsed signals only)—868 MHZ according to R&TTE regulation en 300,220 (GOV.UK 2012), and 315 MHz according to FCC regulation CFR-47 Part 15 (U.S. Government Printing Office 2013) Multiple telegram transmission with checksum Short telegrams (approx. 1 ms) for little probability of collision Long range: Up to 30 m (indoor), and 300 m (outdoor) Repeater available for range extension One-way and bidirectional communication
Low energy need	High data transmission rate for sensor information of 120 Kbps Small data overhead ASK modulation
Interoperability	Wireless protocol defined and integrated in the modules Sensor profiles specified and implemented by users Unique transmission ID (32 bits)
Coexistence with other wireless systems	No interference with DECT, WLAN, PMR systems, etc System design verified in industrial environment

Acronyms: *ASK* amplitude shift keying, *DECT* digital enhanced cordless telecommunications, *FCC* Federal communications commission, *PMR* private/professional mobile radio, *R&TTE* radio and telecommunications terminal equipment, *WLAN* wireless LAN.

elements. EnOcean operates at 868.3 MHz, using amplitude shift keying (ASK) modulation. A high data rate of 120 Kbps together with a maximum payload of 6 Bytes ensures a short frame transmission time (below 1 ms); this minimizes power consumption, but also results in a low statistical probability for collisions. Also, EnOcean transceivers use a novel RF oscillator that can be switched on and off in less than 1 µs; thus, it can be switched off at every "zero" Bit transmission, further reducing energy consumption. The low collision probability is also presented as a key argument that the protocol will scale toward networks with a large number of nodes. The available radio modules do not appear to support security mechanisms. Table 1.11 summarizes EnOcean harvesting wireless standard.

1.9   Conclusion for a Beginning

Sensing is life, WSNs are acquiring snowballing interests in research and industry, and they are infiltrated in day-to-day use. Owing to their requirement of low device complexity as well as slight energy consumption, proper standards are devised to ensure impeccable communication and meaningful sensing. This chapter takes care of enlightening the special features of WSNs and differentiates WSNs from MANETs and mesh networks. Care is also accorded to the different WSN standards that adapt to home and industry applications.

Concerns that WSNs have been unreliable and difficult to use are lessening. But to put a WSN together, a potential user or developer has to be adept in multiple disciplines, hardware, embedded software programming, RF, and enterprise integration, which creates a gap between application concept and deployed network. What is constantly needed is a way to abstract the complexity of setting up and commissioning a WSN from the ongoing management and data mining of the sensor data itself. As much as WSNs are made easily accessible over conventional IP-based networks, their potential user base will become far vaster and more diverse.

A key attribute of WSNs, and the reason they represent the future of intelligent embedded devices, is their ability to be deployed in diverse and varied physical world environments. With no computer-based map of sensor locations, users may be left alone to remember (or guess) where their sensors had been deployed. Sensor network applications, that bind the physical to the logical positioning, allow users to upload an existing floor plan, map or image into the WSN user interface, and then survey an individual sensor node positioned on that map. Once the nodes begin to monitor and collect data on a particular space, thing, or interaction, the map provides context, meaning and the ability to easily manage the WSN deployment.

The critical requirement of any WSN deployment strategy is to gather and export the collected into an enterprise application, or a spreadsheet. Embedded WSN-to-Internet integration is implemented via some kind of gateway device seated between the IEEE 802.15.4 network and the IP network. The gateway server's role is to translate the sensor network traffic and provide it in a consumable form for another network, either IP or an industrial network. Also, the 6LoWPAN working group of the Internet Engineering Task Force (IETF) submitted the implementation of IP for low-power, low-bandwidth networks. 6LoWPAN defines IP communication over low-power wireless IEEE 802.15.4 personal area networks. The proposed standard, approved by the IETF in March 2007, incorporates IPv6 version of the IP protocol. Because of IP pervasiveness as a global communication standard across industries, vendors can create sensor nodes that can communicate directly with other IP devices, whether those devices are wired or wireless, local, or across the Internet, on Ethernet, WiFi, 6LoWPAN, or other networks. Network managers are thus able to gain direct real-time access to sensor nodes and are able to apply a broad range of Internet management and security tools. More important, the WSN can be viewed and managed as just another IP device, making it accessible and familiar to many more people and applications.

Once the network is formed and the sensor nodes start collecting data. Collected data need to be accessible, either in a database or directly to an application for display or analysis. This is where the WSNs have taken experience from the enterprise IT world. Modern enterprise applications communicate and share information using the Web services model, which provides a convenient and scalable way for WSNs to pass collected data to an end-user application or remote database. Sensor data can be accessed from a corporate IT network using Web services that build Web pages and API calls to collect data from the WSNs and return them in a well-formed XML to the requestor.

The ability to access the data in a number of different ways through Web services APIs or by running SQL queries against a database allows data to be used for trending analysis or fed directly into an existing Enterprise Resource Planning (ERP). Information from the physical world can accordingly be used to drive decisions and actions by the now offered increased visibility. However, because WSNs are distributed and largely unattended, the network that supports them must be robust and the data integration schema well thought out and sufficiently generalized to accommodate the diverse sources of information being generated, such as temperature, humidity, light, motion, pressure, etc.

There is no shortage of current and potential applications for sensor networks, and as a result, a wide array of sensor and actuator devices have come on the market. Accommodating that variety of devices, however, is not a trivial challenge. Users should look for an embedded operating system that supports a wide range of leading hardware platforms, while preserving the full capabilities of each. The leading open-source embedded operating system designed for wireless sensor networks, TinyOS, is such an operating system. The OS should include a simple driver framework to support the incorporation of new sensors across multiple platforms. External expansion ports and drivers should also be available to add new kinds of sensors after installation. Accommodation for analog sensors of different types (resistive or inductive) as well as digital sensors (contact switches) is crucial. This makes sensor nodes and WSNs ideal for proof-of-concept and pilot networks where functionality and Return on Investment (ROI) must be proved before finalizing industrial design and appropriate enclosure in the deployment environment.

WSNs entail numerous basics and details, one would be bulky for them all, the dose of this chapter is enough, following chapters will carry on, one after one, but not as fast as the 36.84 sec record of Jamaica 4*100 m relay team in the 2012 London Olympic Games.

1.10   Exercises

1. What are the components of a wireless sensor node?
2. Detail the specs of a sensor node.
3. How are sensor nodes deployed in a terrain? What are the deployment phases?
4. Define MANETs and explain symmetric and asymmetric links.

5. Describe the architecture of WSNs.
6. Determine the differences between MANETs and WSNs.
7. Detail the characteristics of WMNs.
8. Compare between WSNs and WMNs.
9. What are the types of WSNs.
10. Illustrate the characteristics of UASNs.
11. How is WSNs mobility useful?
12. Identify and compare WSN standards used for PANs.
13. Identify and compare WSN standards used for home applications.
14. Look for WSNs models used for personal applications, determine their functions, specs, and manufacturers.
15. Look for WSNs models used for home applications, determine their functions, specs, and manufacturers.
16. Identify and compare the most energy-efficient WSN standards.

References

Ahna, Joon, Affan Syedb, Bhaskar Krishnamacharia, and John Heidemannb. 2011. Design and Analysis of a Propagation Delay Tolerant ALOHA Protocol for Underwater Networks. *Ad Hoc Networks Journal* 9 (5): 752–766.

Akyildiz, I.F., W. Su, Y. Sankarasubramaniam, and E. Cayirci. 2002a. A Survey on Sensor Networks. *Communications Magazine* 40 (8): 102–114.

———. 2002b. Wireless Sensor Networks: A Survey. *Computer Networks* 38 (4): 393–422.

Akyildiz, I.F., T. Melodia, and K.R. Chowdury. 2007. Wireless Multimedia Sensor Networks: A Survey. *IEEE Wireless Communication* 14 (6): 32–39.

Anastasi, G., M. Conti, Di Francesco, and M. 2009. Reliable and Energy-Efficient Data Collection in Sparse Sensor Networks with Mobile Elements. *Performance Evaluation* 66 (12): 791–810.

ANT. 2013. *ANT Message Protocol and Usage.* ANT. www.thisisant.com/resources/ant-message-protocol-and-usage. Accessed 5 Sept 2013.

Birds & Blooms. 2013. *Can Birds Smell or Taste?* http://www.birdsandblooms.com/Birds/Summer/Can-Birds-Smell-or-Taste/. Accessed 22 Aug 2013.

Bouckaert, Stefan, Eli De Poorter, Benôit Latré, Jeroen Hoebeke, Ingrid Moerman, and Piet Demeester. 2010. Strategies and Challenges for Interconnecting Wireless Mesh and Wireless Sensor Networks. *Wireless Personal Communications* 53 (3): 443–463.

Callaway, E., et al. 2002. Home Networking with IEEE 802.15.4: A Developing Standard for Low-Rate Wireless Personal Area Networks. *Communications Magazine* 40 (8): 70–77.

Castillo-Effen, M., D.H. Quintela, R. Jordan, W. Westhoff, and W. Moreno. 2004. Wireless Sensor Networks for Flash-Food Alerting. In *The 5th International Caracas Conference on Devices, Circuits, and Systems, Dominican Republic*, 142–146. Caracas: IEEE.

Chandrasekhar, V., W.K. Seah, Y.S. Choo, and H.V. Ee. 2006. Localization in Underwater Sensor Networks: Survey and Challenges. In *First ACM International Workshop on Underwater Networks (WuWNet)*, 33–40. Los Angeles.

Cordeiro, C.M., and D.P. Agrawal. 2002. Mobile Ad Hoc Networking. *20th Brazilian Symposium on Computer Networks*. 125–186.

Costa, Márcio S., and Jorge L. M. Amaral. 2012. *Analysis of Wireless Industrial Automation Standards: ISA-100.11a and WirelessHART.* http://www.isa.org/InTechTemplate.cfm?template=/ContentManagement/ContentDisplay.cfm&ContentID=93257. Accessed 28 Aug 2013.

Di Francesco, M., S.K. Das, and G. Anastasi. 2011. Data Collection in Wireless Sensor Networks with Mobile Elements: A Survey. *ACM Transactions on Sensor Networks (TOSN)* 8 (1): 1–31.

EnOcean. 2013. *Radio Technology*. http://www.enocean.com/en/radio-technology/. Accessed 5 Sept 2013.

Erol-Kantarci, Melike, Hussein T. Mouftah, and Sema Oktug. 2011. A Survey of Architectures and Localization Techniques for Underwater Acoustic Sensor Networks. *Communications Surveys & Tutorials* 13 (3): 487–502.

Frenzel, Lou. 2012, November 29. *What's The Difference Between Bluetooth Low Energy and ANT?*. http://electronicdesign.com/mobile/what-s-difference-between-bluetooth-low-energy-and-ant. Accessed 31 Aug 2013.

Gao, T., D. Greenspan, M. Welsh, R.R. Juang, and A. Alm. 2005. Vital Signs Monitoring and Patient Tracking over a Wireless Network. In *27th IEEE EMBS Annual International Conference*, 102–105. Shanghai: IEEE.

Gomez, Carles, and Josep Paradells. 2010. Wireless Home Automation Networks: A Survey of Architectures and Technologies. *Communications Magazine* 48: 92–101.

GOV.UK. 2012, September 18. *Radio and Telecommunications Terminal Equipment*. www.gov.uk. Accessed 27 Sept 2013.

Gutierrez, J.A., M. Naeve, E. Callaway, M. Bourgeois, V. Mitter, and B. Heile. 2001. IEEE 802.15.4: A Developing Standard for Low-Power, Low-Cost Wireless Personal Area Networks. *Network* 15 (5): 12–19.

Heidemann, J., W. Ye, J. Wills, A. Syed, and Y. Li. 2006. Research Challenges and Applications for Underwater Sensor Networking. In *Wireless Communications and Networking Conference (WCNC)*, 228–235. Las Vegas: IEEE.

Howitt, Ivan, and Jose A. Gutierrez. 2003. IEEE 802.15.4 Low Rate – Wireless Personal Area Network Coexistence Issues. In *Wireless Communications and Networking (WCNC)*, 1481–1486. IEEE.

IEEE. 2013a. *IEEE 802.15 Working Group for WPAN*. http://grouper.ieee.org/groups/802/15/. Accessed 15 Sept 2013.

—. 2013b. *IEEE 802.15 WPAN Task Group 3 (TG3)*. www.ieee802.org/15/pub/TG3.html. Accessed 15 Sept 2013.

IEEE. 2019. *IEEE 802.15 Working Group for WPAN*. 2019, January 1. http://www.ieee802.org/15/. Accessed 10 June 2019.

INSTEON. 2013. The Details – Version 2.0. White Paper, INSTEON.

ITU. 1947. ITU. http://www.itu.int/dms_pub/itu-s/oth/02/01/S020100002B4813PDFE.pdf. Accessed 7 Sept 2013.

Jornet, M., M. Stojanovic, and M. Zorzi. 2008. Focused Beam Routing Protocol for Underwater Acoustic Networks. In *Third ACM International Workshop on Underwater Networks (WUWNET)*, 75–82. San Francisco: ACM.

Khan, J.M., R.H. Katz, and K.S.J. Pister. 1999. Next Century Challenges: Mobile Networking for Smart Dust. In *5th Annual ACM/IEEE International Conference on Mobile Computing and Networking (MobiCom)*, 271–278. ACM.

Kim, A.N., F. Hekland, S. Petersen, and P. Doyle. 2008. When HART Goes Wireless: Understanding and Implementing the WirelessHART Standard. In *IEEE International Conference on Emerging Technologies and Factory Automation (ETFA)*, 899–907. Hamburg: IEEE.

Lee, Jin-Shyan, Yu-Wei Su, and Chung-Chou Shen. 2007. A Comparative Study of Wireless Protocols: Bluetooth, UWB, ZigBee, and Wi-Fi. In *The 33rd Annual Conference of the IEEE Industrial Electronics Society (IECON)*, 46–51. Taipei: IEEE.

Lee, U., P. Wang, Y. Noh, L. Vieira, M. Gerla, and J.-H. Cui. 2010. Pressure Routing for Underwater Sensor Networks. In *29th Conference on Information Communications (INFOCOM)*, 1676–1684. San Diego: IEEE.

Li, M., and Y. Liu. 2007. Underground Structure Monitoring with Wireless Sensor Networks. In *The 6th International Conference on Information Processing in Sensor Networks (IPSN)*, 69–78. Cambridge: ACM/IEEE.

Li, Mo, and Yunhao Liu. 2009. Underground Coal Mine Monitoring with Wireless Sensor Networks. *Transactions on Sensor Networks* V (2): 1–29.

Li, L., M.C. Vuran, and I.F. Akyildiz. 2007. Characteristics of Underground Channel for Wireless Underground Sensor Networks. In *The 6th Annual Mediterranean Ad Hoc Networking Work-Shop*, 32–39. Corfu: Ionian University.

Montenegro, G., N. Kushalnagar, J. Hui, and D. Culler. 2007, September. *Transmission of IPv6 Packets Over IEEE 802.15.4 Networks-RFC4944*. http://www.hjp.at/doc/rfc/rfc4944.html. Accessed 29 Aug 2013.

Mulligan, G. 2007. The 6LoWPAN Architecture. In *4th Workshop on Embedded Networked Sensors (EmNets)*, 78–82. ACM.

National Museum Scotland. 2013. *Animal Senses*. 2013, January 1. http://www.nms.ac.uk/our_museums/national_museum/explore_the_galleries/natural_world/animal_senses.aspx. Accessed 22 Aug 2013.

Negron, V. 2020, July 4. *How Fish Sense and 'Feel'*. PETMED. https://www.petmd.com/fish/care/evr_fi_fish_senses. Accessed 15 Nov 2010.

O. Orkin Insect Zoo. 2013. *Basic Facts: Insect Senses*. Mississippi State University. 2013, January 1. http://insectzoo.msstate.edu/Students/basic.senses.html. Accessed 22 Aug 2013.

Pakzad, S.N., G.L. Fenves, S. Kim, and D.E. Culler. 2008. Design and Implementation of Scalable Wireless Sensor Network for Structural Monitoring. *Journal of Infrastructure Systems* 14: 89–101.

Pei, Zhongmin, Zhidong Deng, Bo Yang, and Xiaoliang Cheng. 2008. Application-Oriented Wireless Sensor Network Communication Protocols and Hardware Platforms: A survey. In *IEEE International Conference on Industrial Technology (ICIT)*, 1–6. Chengdu: IEEE.

Petersen, S., and S. Carlsen. 2011. WirelessHART Versus ISA100.11a: The Format War Hits the Factory Floor. *IEEE Industrial Electronics Magazine* 5 (4): 23–34.

Raffaele, Bruno, Marco Conti, and Enrico Gregori. 2005. Mesh Networks: Commodity Multihop Ad Hoc Networks. *Communications Magazine* 43: 123–131.

Reinisch, C., W. Kastner, G. Neugschwandtner, and W. Granzer. 2007. Wireless Technologies in Home and Building Automation. In *The 5th IEEE International Conference on Industrial Infromatics*, 93–98. Vienna: IEEE.

Safaric, Stanislav, and Kresimir Malaric. 2006. ZigBee Wireless Standard. In *48th International Symposium ELMAR*, 259–262. Zadar: IEEE.

Shelby, Zach, and Carsten Bormann. 2011. *6LoWPAN: The Wireless Embedded Internet*. Wiley.

Song, J., et al. 2008. WirelessHART: Applying Wireless Technology in Real-Time Industrial Process Control. In *Real-Time and Embedded Technology and Applications Symposium (RTAS)*, 377–386. St. Louis: IEEE.

Tilak, S., N. Abu-Ghazaleh, and W. Heinzelman. 2002. A Taxonomy of Wireless Micro-Sensor Network Models. *ACM SIGMOBILE Mobile Computing and Communications Review* VI (2): 28–36.

Tseng, Yi-Hsien, E.H.-K. Wu, and Gen-Huey Chen. 2003. Maximum Traffic Scheduling and Capacity Analysis for IEEE 802.15.3 High Data Rate MAC Protocol. In *IEEE 58th Vehicular Technology Conference*, 1678–1682. IEEE.

U.S. Government Printing Office. 2013, September. *TITLE 47 – Telecommunication*. www.ecfr.gov. Accessed 27 Sept 2013.

UCSB ScienceLine. 2020. *How Do Plants Sense a Change in the Environment?* UCSB ScienceLine. 2020, July 4. http://scienceline.ucsb.edu/getkey.php?key=1014. Accessed 25 Oct 2005.

Vuran, M.C., and I.F. Akyildiz. 2008. Cross-Layer Packet Size Optimization for Wireless Terrestrial, Underwater, and Underground Sensor Networks. In *The 27th Conference on Computer Communications (INFOCOM)*. Phoenix: IEEE.

Warneke, B., and K.S.J. Pister. 2002. MEMS for Distributed Wireless Sensor Networks. In *9th International Conference oon Electronics, Circuits and Systems*, 291–294. Dubrovnik: IEEE.

Der Wateren, and Frits Van. 2008. *The Art of Developing WSN Applications with Myrianed.* Technical Report, Chess Company, The Netherlands.

Wener-Allen, G., et al. 2006. Deploying a Wireless Sensor Network on an Active Volcano. *Internet Computing* 10: 18–25.

Xiao, D., M. Wei, and Y. Zhou. 2006. Secure-SPIN: Secure Sensor Protocol for Information via Negotiation for Wireless Sensor Networks. In *First IEEE Conference on Industrial Electronics and Applications*, 1–4. Singapore: IEEE.

Yick, J., B. Mukherjee, and D. Ghosal. 2005. Analysis of a Prediction-based Mobility Adaptive Tracking Algorithm. In *The 2nd International Conference on Broadband Networks (BROADNETS)*, 753–760. Boston: IEEE.

Yick, Jennifer, Biswanath Mukherjee, and Dipak Ghosal. 2008. Wireless Sensor Network Survey. *Computer Networks* (52):2292–2330.

Zhang, J., P.V. Orlik, A.F. Molisch, and P. Kinney. 2009. UWB Systems for Wireless Sensor Networks. *Proceedings of the IEEE* 97 (2): 313–331.

Zhou, Zhong, Zheng Peng, Jun-Hong Cui, and Zhijie Shi. 2011. Scalable Localization with Mobility Prediction for Underwater Sensor Networks. *IEEE Transactions on Mobile Computing* 10 (3): 335–348.

ZigBee Alliance. 2013. *Specifications.* ZigBee Alliance. 2013, January 1. http://zigbee.org/ Specifications.aspx. Accessed 28 Aug 2013.

Z-Wave Alliance. 2012. *Products.* http://www.Z-Wavealliance.org/. Accessed 30 Aug 2013.

Chapter 2
Protocol Stack of WSNs

Etiquette is protocol, rules of behavior. How a gentleman opens the door for a lady, how he smiles and handshakes.

2.1 Introduction

A *protocol*, etiquette, code of conduct, is a set of rules that govern a certain behavior, in social or diplomatic activities, at work, when driving, etc. Socially, there is a dress for night parties; there is a way to put off a coat, to sit, eat, and speak. Diplomatic activities are framed in strict protocol rules that determine who comes first, who is next, who will be to the right, who speaks; deviating from such rules is a serious breach of job duties. A protocol is also the draft of a treaty or agreement. At work, there are limitations to what can be said in public, to what can be worn, to where to eat or smoke. When driving, there are rules to follow a lane or to change lanes, to surpass, to honk, to speed limits. Protocol rules may be imposed by administrative regulations, or by social habits, either way they are followed, and monitored, a person is appreciated with regard to how far he clings to protocol guidelines. In communication networks, protocols govern, determine the functioning specifications and guidelines, and guarantee how networks fulfill their intended use.

A wireless sensor network is an ad hoc arrangement of multifunctional sensor nodes in a sensor field, disseminated to gather information regarding some phenomenon. Sensor nodes can be densely distributed over a large and may be remote area and collaborate their efforts to the benefit of the network to the extent that even if a number of nodes malfunction, the network will continue to function. There are two main layouts for wireless sensor networks. The first is a star layout where the nodes communicate, in a single hop, directly to the sink whenever possible and peer-to-peer communication is minimal. In the second, information is routed back to the sink via data passing between nodes. This multihop communication is expected to consume less power than single-hop communication because nodes in the sensor field are densely distributed and are relatively close to each other.

As previously stated, wireless sensor networks differ from traditional ad hoc networks in a few very significant ways:

- Power awareness. Because nodes are placed in remote, hard to reach places, it is not feasible to replace dead batteries. All protocols must be designed to minimize energy consumption and preserve the life of the network.
- Sensor nodes lack global identifications (IDs), so that the networks lack the usual infrastructure. Attribute-based naming and clustering are used instead. Querying WSNs is done by asking for information regarding a specific attribute of the phenomenon, or asking for statistics about a specific area of the sensor field. This requires protocols that can handle requests for a specific type of information, as well as data-centric routing and data aggregation.
- Position of the nodes may not be engineered or predetermined and, therefore, must provide data routes that are self-organizing.

A protocol stack for WSNs must support their typical features and singularities. According to (Akyildiz et al. 2002), the sensor network protocol stack is much like the traditional protocol stack, with the following layers: application, transport, network, data link, and physical. The physical layer is responsible for frequency selection, carrier frequency generation, signal detection, modulation, and data encryption. The data link layer is responsible for the multiplexing of data streams, data frame detection, medium access, and error control. It ensures reliable point-to-point and point-to-multipoint connections in a communication network. The network layer takes care of routing the data supplied by the transport layer. The network layer design in WSNs must consider the power efficiency, data-centric communication, data aggregation, etc. The transport layer helps to maintain the data flow and may be important if WSNs are planned to be accessed through the Internet or other external networks. Depending on the sensing tasks, different types of application software can be set up and used on the application layer.

WSNs must also be aware of several management planes in order to function efficiently, specifically, mobility, power, task, quality of service (QoS), and security management planes. Among them, the functions of task, mobility, and power management planes have been elaborated in (Akyildiz et al. 2002; Wang and Balasingham 2010). The protocol stack and the associated planes used by the sink, cluster head, and sensor nodes are shown in Fig. 2.1. The power management plane is responsible for minimizing power consumption and may turn OFF functionality in order to preserve energy. The mobility management plane detects and registers movement of nodes so that a data route to the sink is always maintained. The task management plane balances and schedules the sensing tasks assigned to the sensing field and thus only the necessary nodes are assigned with sensing tasks and the remainder are able to focus on routing and data aggregation. QoS management in WSNs (Howitt et al. 2006) can be very important if there is a real-time requirement with regard to the data services. QoS management also deals with fault tolerance, error control, and performance optimization in terms of certain QoS metrics. Security management is the process of managing, monitoring, and controlling the security-related behavior of a network. The primary function of security management is in controlling access points to critical or sensitive data. Security management also includes the seamless integration of different security function modules, including encryption, authentication, and intrusion detection.

Fig. 2.1 Protocol stack
of WSNs (Wang and
Balasingham 2010)

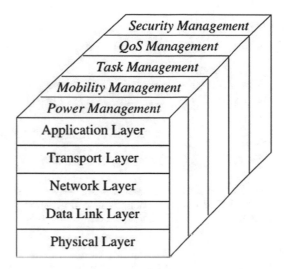

2.2   Physical Layer

In many wireless sensor networks, the number and location of nodes make recharging or replacing the batteries infeasible. For this reason, energy consumption is a universal design issue for wireless sensor networks. Much work has been done to minimize energy dissipation at all levels of system design, from the hardware to the protocols to the algorithms. Hence to the network, it is important to appropriately set parameters of the protocols in the network stack. At the physical layer, the parameters open to the network designer include, modulation scheme, transmit power, and hop distance. The optimal values of these parameters will depend on the channel model. When a wireless transmission is received, it can be decoded with a certain probability of error, based on the ratio of the signal power to the noise power of the channel (i.e., the SNR). As the energy used in transmission increases, the probability of error goes down, and thus, the number of retransmissions goes down. Thus, there exists an optimal tradeoff between the expected number of retransmissions and the transmit power to minimize the total energy dissipated to receive the data (Holland et al. 2011).

At the physical layer, there are two main components that contribute to energy loss in a wireless transmission, the loss due to the channel and the fixed energy cost to run the transmission and reception circuitry (Heinzelman et al. 2002). The loss in the channel increases as a power of the hop distance, while the fixed circuitry energy cost increases linearly with the number of hops. This implies that there is an optimal hop distance where the minimum amount of energy is expended to send a packet across a multihop network. Similarly, there is a tradeoff between the transmit power and the probability of error. In this tradeoff, there are two parameters that a network

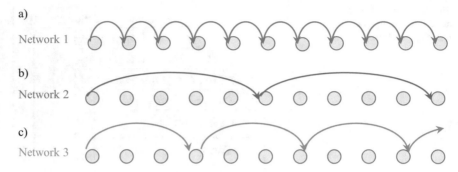

Fig. 2.2 Instances of a linear wireless network. (**a**) Network 1 has a short hop distance. (**b**) Network 2 has a long hop distance. (**c**) Network 3 has the optimal hop distance (Holland et al. 2011)

designer can change to optimize the energy consumed, transmit power and hop distance. The third option for physical layer parameter selection is much broader than the other two. The coding/modulation of the system determines the probability of transmission success, changes in the probability of a successful transmission lead to changes in the optimal values for the other physical layer parameters (Wang et al. 2001).

To illustrate these physical layer tradeoffs, consider the linear network shown in Fig. 2.2 (Holland et al. 2011). In this network, a node must send data back to the basestation. The first physical layer consideration is hop distance. In the first case (Network 1), the hop distance is very small, which translates to low per-hop energy dissipation. Because the transmit energy must be proportional to d^n where $n \geq 2$ and d is the distance between the transmitter and receiver, the total transmit energy to get the data to the basestation will be much less using the multihop approach than a direct transmission (Heinzelman et al. 2002). However, in this network, the main factor in the energy dissipation of the transmission is the large number of hops. The fixed energy cost to route through each intermediate hop will cause the total energy dissipation to be high.

In the second case (Network 2), the hop distance is very large. With so few hops, there is little drain of energy on the network due to the fixed energy cost. However, there is a large energy drain on the nodes due to the high energy cost to transmit data over the long individual hop distances. With a large path loss factor, the total energy in this case will far exceed the total energy in the case of short hops. Thus, it is clear that a balance must be struck, as shown in Network 3, so that the total energy consumed in the network is at a minimum.

Several standards that enhance low power communication, as required for WSNs, are laid out in Chap. 1.

2.3   Data Link Layer

The responsibilities of the data link layer are the multiplexing of data streams, data frame detection, medium access (MAC), and error control. A wireless sensor network must have a specialized MAC protocol to address the issues of power conservation and data-centric routing. The MAC protocol must meet two goals. The first is to create a network infrastructure, which includes establishing communication links between may be thousands of nodes, and providing the network self-organizing capabilities. The second goal is to fairly and efficiently share communication resources between all the nodes. Existing MAC protocols fail to meet these two goals because power conservation is only a secondary concern in their development. Also, wireless sensor networks have no central controlling agent and a much larger number of nodes than traditional ad hoc networks. Any MAC protocol for wireless sensor networks must also take into account the ever-changing topology of the sensor network due to node failure and redistribution.

Since sensor nodes are usually operated by batteries and left unattended after deployment, power saving is a critical issue in WSNs. Many research efforts in the recent years have focused on developing power-saving schemes for wireless sensor networks. These schemes include power-saving hardware design, power-saving topology design (Salhieh et al. 2001; Chakrabarti et al. 2003), power-efficient MAC layer protocols (Ye et al. 2002; Zheng et al. 2005; Rajendran et al. 2006; Pang et al. 2012), and network layer routing protocols (Sohrabi et al. 2000; Akkaya and Younis 2005). Designing power-efficient MAC protocols is one of the techniques that prolong the lifetime of the network. In addition to energy efficiency, latency and throughput are also important features for consideration in MAC protocol design for WSNs. Commercial standards like IEEE 802.11 have a power management scheme for ad hoc networks, wherein the nodes remain in idle listening state at low traffic to conserve power, significant power is wasted even in the idle listening mode. Hence, IEEE 802.11 is not suitable for sensor networks. A properly designed MAC protocol allows the nodes to access the channel in a way that saves energy and support QoS.

2.4   Network Layer

The network layer in a WSN must be designed with typical considerations in mind, ever-existing power efficiency, WSNs are data-centric networks, and WSNs have attribute-based addressing and location awareness. The link layer handles how two nodes talk to each other, while the network layer is responsible for deciding which node to talk to.

The simplest design is flooding. When using flooding, each node receiving data repeats it by broadcasting the data to every neighbor unless the max hop lifetime of the data has been reached or the receiving node is the destination. The major support for flooding is the simplicity. It requires no costly topology maintenance or complex route discovery. The shortcomings, however, are substantial:

- Implosion, it occurs when two nodes (A and B) share multiple (n) neighbors. Node A will broadcast data to all n of these neighbors. Node B will then receive a copy of the data from each of them.
- Overlap, when two nodes share the same sensing region. If a stimulus occurs within this overlap, both nodes will report it.
- The last and most crucial problem is resource blindness. Flooding does not take into account available energy resources.

Gossiping is an enhancement to flooding. In gossiping, when a node receives data, it randomly chooses a neighbor and sends the data to it. Gossiping avoids the problem of implosion, but does not address the other two concerns and contributes to the latency of the network.

A step up from flooding and gossiping is ideal dissemination. In this algorithm, data are sent along a shortest path route from the originating node. Such approach guarantees that every node will receive every piece of information exactly once. No energy is wasted in sending or receiving redundant data. However, the overhead involved in keeping track of the shortest paths is substantial. Also, ideal dissemination does not take into account that some node may not need a particular piece of information, nor does it allow for resource awareness.

A little more sophisticated family of protocols is sensor protocols for information via negotiation (SPIN). The SPIN family addresses the deficiencies of classic flooding by negotiation and resource adaptation. With more sophisticated and energy aware techniques for data dissemination, it reduces the amount of energy expended, solves the problems of implosion, overlap, and resource blindness, and ensures that only interested nodes will expend energy to receive data (Kulik et al. 2002; Rehena et al. 2011). Negotiation helps to overcome the problems of implosion and overlap and ensures only useful and desired information is disseminated. In order for negotiation to work, nodes must describe the data to be sent using meta-data. In order for SPIN to be efficient, the meta-data must be significantly shorter than the data being described. Also, meta-data describing two distinguishable pieces of data must be different. Likewise, if two pieces of data are indistinguishable, they will share the same meta-data. The format of the meta-data is not specified by SPIN, but rather application specific.

SPIN-2 is an implementation of SPIN that employs a low-energy threshold. When energy is abundant, the node functions as normal. However, when the resource manager detects that a node power supply is reaching the low-energy threshold, the node will not participate in later stages of the protocol. This prolongs the life of the node and allows it to perform only high priority functions.

SPIN is a more sophisticated and energy aware schema for data dissemination. It reduces the amount of energy expended, solves the problems of implosion, overlap, and resource blindness, and ensures that only interested nodes will expend energy to receive data.

Chapter 4 focuses on the energy and lifetime aware routing protocols designed to maintain sustained WSN functionality.

2.5   Transport Layer

Transport control protocol for WSNs should account for several concerns (Wang et al. 2005):

- Congestion control and reliability. The more data streams flow from sensor nodes to sinks in WSNs, the more congestion might occur around sinks. Also, there are some high-bandwidth data streams produced by multimedia sensors. Therefore, it is necessary to design effective congestion detection, congestion avoidance, and congestion control mechanisms for WSNs. Although MAC protocol can recover packets loss from bit-error, it has no way to handle packets loss from buffer overflow. Then, the transport protocol for WSNs should have mechanism for packets loss recovery such as ACK and selective ACK as used in TCP protocol so as to guarantee reliability.

    Reliability under WSNs may have different meaning from traditional networks that generally guarantee correct transmission of every packet. For some application, WSNs only need to correctly receive packets from a certain area, not from every sensor nodes in this area, or may be contempt with some ratio of successful transmission from a sensor node. These modified reliability concept motivates the design of different transport control protocols. It would be better to use hop-by-hop mechanism for congestion control and loss recovery since it can reduce packet dropping and conserve energy. The hop by hop mechanism can simultaneously lower buffer requirement at intermediate nodes, which suits the limited memory sensor nodes.

- Simplifying initial connecting process or use connectionless protocol so as to speed up start and guarantee throughput and lower transmission delay. Most of applications in WSNs are reactive, that is passively monitor and wait for event occurring before reporting to sink. These applications may have only few packets for each reporting, and the simple and short initial setup process is more effective and efficient.

- Avoiding packets dropping as possible to lessen energy wastage. In order to avoid packet dropping, the transport protocol can use active congestion control at the cost of a lower link utility. The active congestion control (ACC) can trigger congestion avoidance before congestion occurs. An example of ACC is to make sender (or intermediate nodes) reduce sending (or forwarding) rate when the buffer size of their downstream neighbors overruns a threshold.

- Guaranteeing fairness for different sensor nodes so that each sensor node can achieve a fair throughput. Otherwise, the loaded sensor nodes cannot properly report events in their area, which leads to erroneous monitoring, tracking, and control.

- Enabling cross-layer interaction. If a routing algorithm can notify route failure to the transport protocol, the transport protocol will know that packet loss is not from congestion but from route failure, and consequently, the sender will regulate its current sending rate to guarantee high throughput and low delay.

Chapter 5 of this book exhaustively considers transport control protocols for WSNs.

2.6   Application Layer

To address application layer protocols, it is primordial to address some functions that are to be implemented, specifically, data fusion and management, clock synchronization, and positioning. A WSN is intended for deployments in environments where sensors can be exposed to circumstances that might interfere with provided measurements. Such circumstances include strong variations of pressure, temperature, radiation, and electromagnetic noise. Thus, measurements may be imprecise in such scenarios. *Data fusion* is used to overcome sensor failures, technological limitations, and spatial and temporal coverage problems. Data fusion is generally defined as the use of techniques that combine data from multiple sources and gather this information in order to achieve inferences, which will be more efficient and potentially more accurate than if they were achieved by means of a single source. The term efficient, in this case, can mean more reliable delivery of accurate information, more complete, and more dependable. The data fusion can be implemented in both centralized and distributed systems. In a centralized system, all raw sensor data would be sent to one node, and the data fusion would all occur at the same location. In a distributed system, the different fusion modules would be implemented on distributed components (Abdelgawad and Bayoumi 2012).

Communications in wireless sensor networks are data-centric, with the objective of delivering collected data in a timely fashion. Also, such networks are resource-constrained, in terms of sensor nodes processing power, communication bandwidth, storage space, and energy. This gives rise to new face-offs in information processing and data management in wireless sensor networks. In-network data processing techniques, from simple reporting to more complicated collective communications, such as data aggregation, broadcast, multicast, and gossip, are challenging. On the other hand, data collected by sensors can intrinsically be viewed as signals. By exploiting signal processing techniques, collective communications can be done in more energy-efficient ways. Several work deal with data management (Xu et al. 2009) investigate in-network query processing strategies for K nearest neighbor (KNN) queries in location aware wireless sensor networks. Also, Brayne et al. (2008) propose an adaptive query processing mechanism to dynamically adjust query processing in wireless sensor networks. Moreover, Akcan and Brönnimann (2007) develop a distributed, weighted sampling algorithm to sample sensing data to reduce energy consumption. By exploring the adaptive model selection algorithms, Le Borgne et al. (2007) derive an adaptive, lightweight, and online algorithm for prediction sensing data.

Sensed data are of limited usage if it is not accompanied by the coordinates of the sensor position and a timestamp; this is a primary motive for *clock synchronization* in WSNs. Data fusion is a prime function that depends also on clock synchronization. For instance, a vehicle going through acoustic sensors can be detected, throughout its path, by different sensor nodes at different moments. A fusion node receiving the raw information from the sensor nodes can refine it by estimating the speed and the direction of the sensed vehicle. For this application, among others, synchronized timestamps together with position information are essential. Also, WSNs are expected to have very small form factors and be cheap such that they can be deployed in very large numbers. Once deployed, WSNs are usually unattended, so battery replacement is impractical, but since they are typically expected to work for extended periods of time, there is no better way to conserve energy but to put the nodes to sleep and to wake up at the same time to be able to exchange information. Clock synchronization in WSNs is the subject of extensive work (Elson and Römer 2003; Sundararaman et al. 2005; Sun et al. 2006; Sommer and Wattenhofer 2009; Wu et al. 2011).

Positioning, knowledge of the position of the sensing nodes in a WSN is an essential part of many sensor network operations and applications. Sensors reporting monitored data need to also report the location where the information is sensed, and hence, sensors need to be aware of their position. In addition, many network protocols such as routing require location information in order to provide the specific protocol service. WSNs may be deployed in hostile environments where malicious adversaries attempt to spoof the locations of the sensors by attacking the localization process. For example, an attacker may alter the distance estimations of a sensor to several reference points, or replay beacons from one part of the network to some distant part of the network, thus providing false localization information. Hence, there is a need to ensure that the location estimation is performed in a robust way, even in the presence of attacks. Furthermore, adversaries can compromise the untethered sensor devices and force them to report a false location to the data collection points. Therefore, a secure positioning system must have a mechanism to verify the location claim of any sensor. Positioning in WSN is a topic of extensive research, leading to numerous positioning systems that provide an estimation of the sensor location (Lazos et al. 2005; Akkaya et al. 2007; Kim et al. 2007; Tennina et al. 2008; Younis and Akkaya 2008; Tennina et al. 2009).

System administrators interact with WSNs using sensor management protocol (SMP). Unlike many other networks, WSNs consist of nodes that do not have global IDs, and they are usually infra-structureless. Therefore, SMP needs to access the nodes by using attribute-based naming and location-based addressing. SMP is a management protocol that provides the software operation needed to perform several administrative tasks (Akyildiz et al. 2002):

- Introducing to the sensor nodes the rules related to data aggregation, attribute-based naming, and clustering.
- Exchanging data related to the location-finding algorithms.
- Time synchronization of the sensor nodes.
- Moving sensor nodes.
- Turning sensor nodes on and off.
- Querying the sensor network configuration and the status of nodes, and re-configuring the sensor network.
- Authentication, key distribution, and security in data communications.

2.7   Cross-Layer Protocols for WSNs

The severe energy constraints of battery-powered sensor nodes necessitate energy-efficient communication protocols in order to fulfill the application objectives of wireless sensor networks (WSNs). It is much more resource-efficient, according to some research, to have a unified scheme which melts common protocol layer functionalities into a cross-layer module for resource-constrained sensor nodes. A unified cross-layer communication protocol, for efficient and reliable event communication, considers the effects on WSNs of replacing transport, routing, medium access functionalities, and physical layers (wireless channel).

A unified cross-layering is such that both the information and the functionalities of traditional communication layers are melted in a single protocol. The objective of the proposed cross-layer protocol is highly reliable communication with minimal energy consumption, adaptive communication decisions, and local congestion avoidance. Protocol operation is governed by the concept of initiative determination. Based on this concept, the cross-layer protocol performs received based contention, local congestion control, and distributed duty cycle operation in order to realize efficient and reliable communication in WSN. Performance evaluation reveals that the proposed cross-layer protocol significantly improves the communication efficiency and outperforms the traditional layered protocol architectures (Akyildiz et al. 2006).

Chapter 6 presents the different cross-layering approaches for WSNs.

2.8   Conclusion for Continuation

Several considerations must be taken when developing protocols for wireless sensor networks. Traditional thinking where the focus is on quality of service is somehow revised. In WSNs, QoS is compromised to conserve energy and preserve the life of the network. WSNs are a kind of "totalitarian" system, everyone is for the good of all, no individualism, the whole network must survive even at the expense of falling sensors. Concern must be accorded at every level of the protocol stack to conserve energy and to allow individual nodes to reconfigure the network and modify their set of tasks according to the resources available.

The protocol stack for WSNs consists of five standard protocol layers trimmed to satisfy typical sensors features, namely, application layer, transport layer, network layer, data link layer, and physical layer. These layers address network dynamics and energy efficiency. Functions such as localization, coverage, storage, synchronization, security, and data aggregation and compression are network services that enable proper sensor functioning. Implementation of WSN protocols at different layers in the protocol stack aims at minimizing energy consumption, and end-to-end delay, and maintaining system efficiency. Traditional networking protocols are not designed to meet these WSN requirements; hence, new energy-efficient protocols have been proposed for all layers of the protocol stack. These protocols employ cross-layer optimization by supporting interactions across the protocol layers. Specifically, protocol state information at a particular layer is shared across all the layers to meet the specific requirements of the WSN.

As sensor nodes operate on limited battery power, energy usage is a very important concern in a WSN, and there has been significant research focus that revolves around harvesting and energy conservation by minimizing energy consumption. When a sensor node is depleted of energy, it will fade out and disengage from the network, which may significantly impact the performance of the application. Sensor network lifetime depends on the number of active nodes and network connectivity, so energy must be used efficiently in order to maximize the network lifetime.

Energy harvesting involves nodes replenishing their energy from an energy source (Gilbert and Balouchi 2008; Galperti and Alippi 2008; Seah et al. 2009; Vullers et al. 2010). Potential energy sources include solar cells (Hande et al. 2007), vibration (Lei and Yuan 2008), fuel cells, acoustic noise, and a mobile supplier such as a robot to replenish energy. The robots charge themselves with energy and then deliver energy to the nodes.

Energy conservation in a WSN maximizes network lifetime and is addressed through efficient reliable wireless communication, smart sensor placement to achieve adequate coverage, security and efficient storage management, and data aggregation and data compression. Such approaches satisfy both the energy constraint and provide QoS. For reliable communication, services such as congestion control, active buffer monitoring, acknowledgments, and packet-loss recovery are necessary to guarantee packet delivery. Communication strength depends on the placement of sensor nodes. Sparse sensor placement may result in long-range transmission and higher energy usage, while dense sensor placement may result in short-range transmission and less energy consumption. Coverage is interrelated to sensor placement. The total number of sensors in the network and their placement determine the degree of network coverage. Depending on the application, a higher degree of coverage may be required to increase the accuracy of the sensed data.

One for all, and all for all, that is the main objective of all layers in the WSN protocol stack.

2.9 Exercises

1. Define protocol.
2. What are the considerations and concerns of the WSNs protocol stack?
3. Elaborate on the physical layer for WSN typical features.
4. How is the data link layer for WSNs different?
5. Explain how is the network layer in WSNs different.
6. What is positioning and clock synchronization?
7. How is data fusion crucial in WSNs?

8. What is the importance of data aggregation for WSNs?
9. Determine the functions of the transport layer in WSNs.
10. How does the typical usage of WSNs affect the application layer?

References

Abdelgawad, A., and M. Bayoumi. 2012. Data Fusion in WSN. In *Resource-Aware Data Fusion Algorithms for Wireless Sensor Networks*, ed. R.-A. D. Networks, vol. 118, 17–35. New York: Springer.

Akcan, H., and H. Brönnimann. 2007. A New Deterministic Data Aggregation Method for Wireless Sensor Networks. *Signal Processing* 87 (12): 2965–2977.

Akkaya, K., and M. Younis. 2005. A Survey on Routing Protocols for Wireless Sensor Networks. *Ad Hoc Networks* 3 (3): 325–349.

Akkaya, K., M. Younis, and W. Youssef. 2007. Positioning of Base Stations in Wireless Sensor Networks. *Communications Magazine* 45 (4): 96–102.

Akyildiz, I., W. Su, Y. Sankarasubramaniam, and E. Cayirci. 2002. A Survey on Sensor Networks. *Communications Magazine* 40 (8): 102–114.

Akyildiz, I., M.C. Vuran, and O. Akan. 2006. A Cross-Layer Protocol for Wireless Sensor Networks. In *40th Annual Conference on Information Sciences and Systems*, 1102–1107. Princeton: IEEE.

Brayne, A., A. Lopes, D. Meira, R. Vasconcelos, and R. Menezes. 2008. An Adaptive in-Network Aggregation Operator for Query Processing in Wireless Sensor Networks. *Journal of Systems and Software* 81 (3): 328–342.

Chakrabarti, A., A. Sabharwal, and B. Aazhang. 2003. Using Predictable Observer Mobility for Power Efficient Design of Sensor Networks. In *Information Processing in Sensor Networks*, ed. F. Zhao and L. Guibas, 129–145. Berlin, Heidelberg: Springer.

Elson, J., and K. Römer. 2003. Wireless Sensor Networks: A New Regime for Time Synchronization. *ACM SIGCOMM Computer Communication Review* 33 (1): 149–154.

Galperti, C., and C. Alippi. 2008. An Adaptive System for Optimal Solar Energy Harvesting in Wireless Sensor Network Nodes. *IEEE Transactions on Circuits and Systems I* 55 (6): 1742–1750.

Gilbert, J.M., and F. Balouchi. 2008. Comparison of Energy Harvesting Systems for Wireless Sensor Networks. *International Journal of Automation and Computing* 5 (4): 334–347.

Hande, A., T.W. Polk, and D. Bhatia. 2007. Indoor Solar Energy Harvesting for Sensor Network Router Nodes. *Microprocessors and Microsystems* 31 (6): 420–432.

Heinzelman, W., A. Chandrakasan, and H. Balakrishnan. 2002. An Application Specific Protocol Architecture for Wireless Microsensor Networks. *IEEE Transactions on Wireless Communication* 1 (4): 660–670.

Holland, M., T. Wang, B. Tavli, A. Seyedi, and W. Heinzelman. 2011. Optimizing Physical Layer Parameters for Wireless Sensor Networks. *ACM Transactions on Sensor Networks (TOSN)* 7 (4): 28:1–28:20.

Howitt, I., W.W. Manges, P. Kuruganti, G. Allgood, J.A. Gutierrez, and J.M. Conrad. 2006. Wireless industrial sensor networks: Framework for QoS assessment and QoS management. *ISA Transactions* 45 (3): 347–359.

Kim, S., J. Ko, J. Yoon, and H. Lee. 2007. Multiple-Objective Metric for Placing Multiple Base Stations in Wireless Sensor Networks. In *The 2nd International Symposium on Wireless Pervasive Computing (ISWPC)*. San Juan, Puerto Rico: IEEE.

Kulik, J., W. Heinzelman, and H. Balakrishnan. 2002. Negotiation–Based Protocols for Disseminating Information in Wireless Sensor Networks. *Wireless Networks* 8 (2/3): 169–185.

Lazos, L., R. Poovendran, and S. Čapkun. 2005. ROPE: Robust Position Estimation in Wireless Sensor Networks. In *The 4th International Symposium on Information Processing in Sensor Networks (IPSN)*. Los Angeles: ACM.

Le Borgne, Y., S. Santini, and G. Bontempi. 2007. Adaptive Model Selection for Time Series Prediction in Wireless Sensor Networks. *Signal Processing* 87 (12): 3010–3020.

Lei, W., and F.G. Yuan. 2008. Vibration energy harvesting by magnetostrictive material. *Smart Materials and Structures* 17 (4): 045009.

Pang, B.M., H.S. Shi, and Y.X. Li. 2012. An Energy-Efficient MAC Protocol for Wireless Sensor Network. In *Future Wireless Networks and Information Systems*, ed. Y. Zhang, vol. 143, 163–170. Springer.

Rajendran, V., K. Obraczka, and J.J. Garcia-Luna-Aceves. 2006. Energy-Efficient, Collision-Free Medium Access Control for Wireless Sensor Networks. *Wireless Networks* 12 (1): 63.

Rehena, Z., S. Roy, and N. Mukherjee. 2011. A Modified SPIN for Wireless Sensor Networks. In *The 3rd International Conference on Communication Systems and Networks (COMSNETS)*, 1–4. Bangalore: IEEE.

Salhieh, A., J. Weinmann, M. Kochhal, and L. Schwiebert. 2001. Power Efficient Topologies for Wireless Sensor Networks. In *International Conference on Parallel Processing*, 156–163. Valencia: IEEE.

Seah, W., Z. Eu, and H. Tan. 2009. Wireless Sensor Networks Powered by Ambient Energy Harvesting (WSN-HEAP) - Survey and Challenges. In *The 1st International Conference on Wireless Communication, Vehicular Technology, Information Theory and Aerospace & Electronic Systems Technology (Wireless VITAE)*, 1–5. Aalborg: IEEE.

Sohrabi, K., J. Gao, V. Ailawadhi, and G. Pottie. 2000. Protocols for Self-Organization of a Wireless Sensor Network. *Personal Communications* 7 (5): 16–27.

Sommer, P., and R. Wattenhofer. 2009. Gradient Clock Synchronization in Wireless Sensor Networks. In *The 8th International Conference on Information Processing in Sensor Networks (IPSN)*, 37–48. San Francisco: ACM/IEEE.

Sun, K., P. Ning, and C. Wang. 2006. Secure and Resilient Clock Synchronization in Wireless Sensor Networks. *IEEE Journal on Selected Areas in Communications* 24 (2): 395–408.

Sundararaman, B., U. Buy, and A.D. Kshemkalyani. 2005. Clock Synchronization for Wireless Sensor Networks: A Survey. *Ad Hoc Networks* 3 (3): 281–323.

Tennina, S., M. Di Renzo, F. Graziosi, and F. Santucci. 2008, September 19. *Locating Zigbee® Nodes using the Ti®S cc2431 Location Engine: A Testbed Platform and New Solutions for Positioning Estimation of WSNs in Dynamic Indoor Environments*. The First ACM International Workshop on Mobile Entity Localization and Tracking in GPS-Less Environments (MELT), 37–42.

———. 2009. ESD: A Novel Optimisation Algorithm for Positioning Estimation of WSNs in GPS-Denied Environments – From Simulation to Experimentation. *International Journal of Sensor Networks* 6 (3/4): 131–156.

Vullers, R., R. Schaijk, H. Visser, J. Penders, and C. Hoof. 2010. Energy Harvesting for Autonomous Wireless Sensor Networks. *Solid-State Circuits Magazine* 22 (2): 29–38.

Wang, Q., and I. Balasingham. 2010. Wireless Sensor Networks - An Introduction. In *Wireless Sensor Networks: Application-Centric Design*, ed. Y.K. Tan, 1–13. InTech.

Wang, A., S. Cho, C. Sodini, and A. Chandrakasan. 2001. Energy Efficient Modulation and MAC for Asymmetric RF Microsensor Systems. In *The International Symposium on Low Power Electronics and Design (ISLPED)*, 106–111. Huntington Beach: ACM.

Wang, C., K. Sohraby, Y. Hu, B. Li, and W. Tang. 2005. Issues of Transport Control Protocols for Wireless Sensor Networks. In *International Conference on Communications, Circuits and Systems. 1*, 422–426. Honk Kong: IEEE.

Wu, Y.C., Q. Chaudhari, and E. Serpedin. 2011. Clock Synchronization of Wireless Sensor Networks. *Signal Processing Magazine* 28 (1): 124–138.

Xu, B., F. Vafaee, and O. Wolfson. 2009. In-Network Query Processing in Mobile P2P Databases. In *The 17th ACM SIGSPATIAL International Conference on Advances in Geographic Information Systems (GIS)*, 207–216. Seattle: ACM.

Ye, W., J. Heidemann, and D. Estrin. 2002. An Energy-Efficient MAC Protocol for Wireless Sensor Networks. In *The 21st Annual Joint Conference of the IEEE Computer and Communications Societies (INFOCOM)*, vol. 3, 1567–1576. New York: IEEE.

Younis, M., and K. Akkaya. 2008. Strategies and Techniques for Node Placement in Wireless Sensor Networks: A Survey. *Ad Hoc Networks* 6 (4): 621–655.

Zheng, T., S. Radhakrishnan, and V. Sarangan. 2005. PMAC: An Adaptive Energy-Efficient MAC Protocol for Wireless Sensor Networks. In *The 19th IEEE International Parallel and Distributed Processing Symposium*, 65–72. Denver: IEEE.

Chapter 3
WSNs Applications

Many can do ... Few innovate.

3.1 Applications Categories, Challenges, and Design Objectives

Research in many scientific areas, like physics, microelectronics, control, material science, and the focused collaboration of scientists which used, traditionally, to work toward totally different directions, has led to the creation of the micro-electro-mechanical systems, commonly referred to as MEMS (Gardner and Varadan 2001). MEMS have succeeded in augmenting the limits of what was considered to be a system-on-a-chip (SoC). Indeed, MEMS have enabled chips, which were formerly assumed to carry only logic functions, to sense the real word and even to react. Measuring of physical parameters and actuating is now possible via integration of sensors and actuators to silicon. MEMS are not the only part of the silicon industry that has made astonishing strides. RF technology and digital circuits have also progressed spectacularly. Lower power and higher frequency transceivers are implemented on chips, while digital circuits tend to shrink and be fabricated more and more densely.

The collaboration and synergy of sensing, processing, communication, and actuation is the must-follow step to exploit the inheritance of this technology. The possibilities and challenges offered by this field both in theory and in practice are widely recognized, and many research teams and companies are active in the design and implementation of units that encompass these four attributes. Devices of this kind, which are created either as prototypes or as commercial products, are generally referred to as "motes." A mote is an autonomous, compact device, a sensor unit that also has the capability of processing and communicating wirelessly (Arampatzis et al. 2005). Despite the autonomy they present, the big strength of motes is that they can form networks and cooperate according to various models and architectures. These networks, known as wireless sensor networks (WSNs), have been the focus of considerable research efforts in the areas of communications (protocols, routing, coding, error correction, etc.), electronics (energy efficiency, miniaturization), and control (networked control system, theory, and applications).

The unique characteristics and applications of WSNs pose several challenges that are to be tackled by researchers at both academia and industry. Technologies, schemes, and protocols must be developed to make WSNs as efficient as possible. Military, industrial, and environmental applications impose disseminating sensor motes in harsh surroundings, which necessitate adequate challenges face off. The coming subsections layout such challenges and the design objectives that may technically get over them.

3.1.1   Functional Challenges of Forming WSNs

The major technical challenges for the buildup of industrial WSNs will be laid out in what follows (Gungor and Hancke 2009). These challenges can be extended to the military and environmental WSN applications due to the similarities in the surroundings and functional requirements:

- Resource constraints. The design and implementation of WSNs are constrained by three limited resources, energy, memory, and processing.
- Dynamic topologies and harsh environmental conditions. In military, industrial, and environmental applications, the topology and connectivity of the network may vary due to link and sensor node failures. Furthermore, sensors may be subject to RF interference, highly caustic or corrosive environments, high humidity levels, vibrations, dirt and dust, or other conditions that challenge performance. These harsh environmental conditions and dynamic network topologies may cause a portion of industrial sensor nodes to malfunction (Gungora et al. 2007).
- Quality-of-service (QoS) requirements. The wide variety of applications envisaged on WSNs will have different QoS requirements and specifications. The QoS provided by WSNs refers to the discrepancy between the data reported to the sink node (the control center) and what is actually occurring in the industrial environment. In addition, since sensor data are typically time-sensitive, e.g., alarm notifications for the industrial facilities, it is important to receive the data at the sink in a timely manner. Data with long latency due to processing or communication may be outdated and lead to wrong decisions in the monitoring system.
- Data redundancy. Due to the high density in the network topology, sensor observations are highly associated. In addition, the nature of the physical phenomenon constitutes a temporal association between each consecutive observation of the sensor node.
- Packet errors and variable-link capacity. Compared to wired networks, in WSNs, the attainable capacity of each wireless link depends on the interference level perceived at the receiver, as well high bit error rates (BER $= 10^{-2} - 10^{-6}$) are observed in communication. In addition, wireless links exhibit widely varying

characteristics over time and space due to obstructions and noisy environment. Thus, capacity and delay attainable at each link are location-dependent and vary continuously, making QoS provisioning a challenging task.

- Security. Security should be an essential feature in the design of WSNs to make the communication safe against external denial-of-service (DoS) attacks and intrusion. WSNs have special characteristics that enable new ways of security attacks. Passive attacks are carried out by eavesdropping on transmissions, including traffic analysis or disclosure of message contents. Active attacks consist of modification, fabrication, and interruption, which in WSN cases may include node capturing, routing attacks, or flooding.
- Large-scale deployment and ad hoc architecture. Most WSNs contain a large number of sensor nodes (hundreds to thousands or even more), which might be spread randomly over the deployment field. Moreover, the lack of predetermined network infrastructure necessitates the WSNs to establish connections and maintain network connectivity autonomously.
- Integration with Internet and other networks. It is of prime importance for the commercial development of WSNs to provide services that allow the querying of the network to retrieve useful information from anywhere and at any time. For this reason, WSNs should be remotely accessible from the Internet and, hence, need to be integrated with the Internet Protocol (IP) architecture, either through gateways or thorough IP connectivity (Montenegro et al. 2007; Akyildiz et al. 2007).

3.1.2 Design Objectives of WSNs

The existing and potential applications of WSNs span a very wide range of applications. To deal with the technical challenges and meet the diverse WSN application requirements, specially industrial, and by resemblances military and environmental applications, several design goals need to be adopted (Gungor and Hancke 2009):

- Low-cost and small sensor nodes. Compact and low-cost sensor devices are essential to accomplish large-scale deployments of WSNs. The system owner should consider the cost of ownership (packaging requirements, modifications, maintainability, etc.), implementation costs, replacement and logistics costs, and training and servicing costs as well as the per unit costs (Howitt et al. 2006).
- Scalable architectures and efficient protocols. WSNs support heterogeneous applications with different requirements, especially in industry and environment. It is necessary to develop flexible and scalable architectures that can accommodate the requirements of all these applications in the same infrastructure. Modular and hierarchical systems can enhance the system flexibility, robustness, and reliability. In addition, interoperability with existing legacy solutions, such as Fieldbus (Thomesse 2005) and Ethernet-based systems, is required.

- Data fusion and localized processing. Instead of sending the raw data to the sink node directly, sensor nodes can locally filter the sensed database and transmit only the processed data, which is in-network processing. Thus, only necessary information is transported to the end user and communication overhead can be significantly reduced.
- Resource-efficient design. In WSNs, energy efficiency is important to maximize the network lifetime while providing the QoS required by the application. Energy saving can be accomplished in every component of the network by integrating network functionalities with energy-efficient protocols, e.g., energy-aware routing on network layer, and energy-saving mode on MAC layer.
- Self-configuration and self-organization. In WSNs, the dynamic topologies caused by node failure/mobility/ temporary power-down and large-scale node deployments necessitate self-organizing architectures and protocols. Note that, with the use of self-configurable WSNs, new sensor nodes can be added to replace failed sensor nodes in the deployment field, and existing nodes can also be removed from the system without affecting the general objective of the application.
- Adaptive network operation. The adaptability of WSNs is extremely crucial, since it enables end users to cope with dynamic/varying wireless channel conditions in military, industrial, environmental applications, and with new connectivity requirements driven by new processes. To balance the trade-offs among resources, accuracy, latency, and time synchronization requirements, adaptive signal processing algorithms and communication protocols are needed.
- Time synchronization. In WSNs, large numbers of sensor nodes need to collaborate to perform the sensing task, and the collected data are usually delay-sensitive (Howitt et al. 2006; Akyildiz et al. 2007). Thus, time synchronization is one of the key design goals for communication protocol design to meet the deadlines of the application. However, due to resource and size limitations and lack of a fixed infrastructure, as well as the dynamic topologies in WSNs, existing time synchronization strategies designed for other traditional wired and wireless networks may not be appropriate for WSNs. Adaptive and scalable time synchronization protocols are required for WSNs.
- Fault tolerance and reliability. In WSNs, based on the application requirements, the sensed data should be reliably transferred to the sink node. Similarly, the programming/retasking data for sensor operation, command, and queries should be reliably delivered to the target sensor nodes to ensure the proper functioning of WSNs. However, for many WSN applications, the sensed data are exchanged over time-varying and error-prone wireless medium. Thus, data verification and correction on each communication layer and self-recovery procedures are extremely critical to provide accurate results to the end user.
- Application-specific design. In WSNs, there exists no one-size-fits-all solution; instead, the alternative designs and techniques should be developed based on the application-specific QoS requirements and constraints.
- Secure design. When designing the security mechanisms for WSNs, both low-level (key establishment and trust control, secrecy and authentication,

privacy, robustness to communication denial of service (DoS), secure routing, resilience to node capture), and high-level (secure group management, intrusion detection, secure data aggregation) security primitives should be addressed (Perrig et al. 2004). In addition, because of resource limitations in WSNs, the overhead associated with security protocols should be balanced against other QoS performance requirements. It is very challenging to meet all aforementioned design goals simultaneously. Fortunately, most WSN designs have different requirements and priorities on design objectives. Therefore, the network designers and application developers should balance the trade-offs among the different parameters when designing protocols and architectures for WSNs.

In the sections to come, several applications of WSNs are presented. Applications are categorized into military, industrial, environmental, health, daily life, and multimedia. Moreover, the collaboration between robotics and WSNs is presented as promising and necessary applications booster. Some applications are detailed to focus on concepts and to guide to other applications that are available in each WSNs category of use. Military applications will be the start, human instinct of self-defense is the motive behind innovation, how to combat and overwhelm enemies mean no less than existence, survival of societies, and civilizations. Wars write history, war-related industries create new ideas, some are released to civilian life when overpassed by newer releases; non-military, peaceful domestic applications, might have originated in the loud, fiery military industry. When placed in a field, sensors either monitor or track, they help decisions making based on clear facts and potentiate preventing events that may turn out to be catastrophic.

3.2 Military Applications

Several areas of research are encompassed in the use of WSNs in military applications. Acoustic detection and recognition have been under research since the early 50's. An analysis of the complex near-field pressure waves that occur within a foot of the muzzle blast is presented in (Fansler 1998); it gives a good idea of the ideal muzzle blast pressure wave without contamination from echoes or propagation effects. Experiments with greater distances from the muzzle were conducted in (Stoughton 1997).

Another area of research is the signal processing of gunfire acoustics. The focus is on the robust detection and length estimation of small caliber acoustic shockwaves and muzzle blasts. The edges of the shockwave are typically well defined, and the shockwave length is directly related to the bullet characteristics. The work in (Sadler et al. 1998) compares two shockwave edge detection methods, a simple gradient-based detector and a multiscale wavelet detector. It also demonstrates how the length of the shockwave, as determined by the edge detectors, can be used to estimate the caliber of a projectile.

A related topic is the research and development of experimental and prototype shooter location systems. Researchers at BBN[1] (Raytheon 2014) have developed the Bullet Ears system, which has the capability to be installed in a fixed position or worn by soldiers (Duckworth et al. 2001). The problem with this approach and similar centralized systems is the need of the one or handful of microphone arrays to be in line of sight of the shooter. A sensor networked-based solution has the advantage of widely distributed sensing for better coverage, multipath effect compensation, and multiple simultaneous shot resolution (Lédeczi et al. 2005).

Each of the coming sections, as based and named on a leading paper, will elaborate further on such topics. Chapter 10 of this book includes the datasheets of the hardware used in the proposed approaches and projects.

3.2.1 Countersniper System for Urban Warfare

In (Lédeczi et al. 2005), an ad hoc WSN-based system is presented; it detects and accurately locates shooters even in urban environments. The presented sensor network-based solution surpasses the traditional approach because it can mitigate acoustic multipath effects prevalent in urban areas and it can also resolve multiple simultaneous shots. These unique characteristics of the system are made possible by employing novel sensor fusion techniques that utilize the spatial and temporal diversity of multiple detections.

Countersniper systems can use several different physical phenomena related to the shot or the weapon itself, such as acoustic, visual, or electromagnetic signals. A detectable visual event is the muzzle flash, as in case of the Viper system (Moroz et al. 1999), or the reflection from the sniper's scope (Vick et al. 2002). The electromagnetic field or heat generated by the projectile can also be used for detection (Vick et al. 2002). In spite of the wide range of possibilities, so far acoustic signals, such as the muzzle blast and the ballistic shockwave, provide the easiest and most accurate way to detect shots, and hence, the majority of existing countersniper systems use them as the primary information source (Duckworth et al. 2001).

The most obvious acoustic event generated by the firing of any conventional weapon is the muzzle blast. The blast is a loud, characteristic noise originating from the end of the muzzle, and propagating spherically away at the speed of sound, making it ideal for localization purposes (Fig. 3.1). A less favorable property of the blast is that it can be suppressed by silencers or rendered ambiguous by acoustic propagation effects.

---

[1]Raytheon BBN Technologies delivers innovative solutions in quantum sensing, quantum communications, quantum computing, multisensor processing systems, speech recognition, software systems. Their solutions are widely used in the U.S Navy, the UK Royal Air Force, and the Canadian Navy.

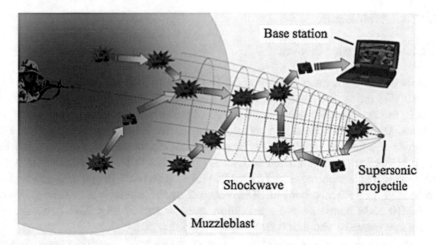

Fig. 3.1 The sensor network based shooter localization system using shockwave and muzzle blast time of arrival data. (Ledeczi et al. 2005)

Typical rifles fire projectiles at supersonic velocities to increase both the range and accuracy, producing acoustic shocks along their trajectory. The shockwave is the result of the air being greatly compressed at the tip and expanded at the end of the bullet, as it slices through the air. Under ideal circumstances, the pressure signal detected by a microphone has a characteristic and distinctive waveform, called N-wave referring to its shape. Because of its very fast rise time (<1 μs), it cannot be produced by any other natural phenomenon. The ideal shockwave front is a cone (the Mach cone) moving along the trajectory of the projectile. The angle of the cone depends on the speed of the bullet. Note that this angle is continuously increasing as the bullet decelerates producing a distorted conical shape, as shown in Fig. 3.1. Since N-waves can be accurately detected, shockwaves provide excellent means to determine projectile trajectories.

The proposed sensor networking approach allows the use of possibly several orders of magnitude higher number of inexpensive sensor units, but requires quite different processing approach because of the very limited communication bandwidth. Some of the processing must be allocated to the sensor units, while the sensor fusion needs to be carried out on a more powerful computer. The concept is illustrated in Fig. 3.1. The sensors accurately detect shockwave and/or muzzle blast events and measure their time of arrival (TOA). These timestamps of detected events are sent to a central basestation, where the fusion algorithm calculates the shot trajectory and/or the shooter location, based on the TOA measurements and the known sensor locations. The communication in the network is provided by ad hoc routing protocols, incorporating the time synchronization service as well.

3.2.1.1 Architecture

3.2.1.1.1 Hardware Platform

The hardware platform is built upon the UC Berkeley MICA2 mote device running the TinyOS embedded operating system (Crossbow 2002a; Hill and Culler 2002), a widely used component-based architecture targeting wireless sensor network applications. Open interfaces at the software and hardware levels made it possible to integrate specialized smart sensor elements and supporting middleware services. Each MICA2 mote is furnished with an ATmega 128 L 8-bit microcontroller with 128 Bytes instruction memory (Atmel 2011a), 4 KBytes data memory and typical embedded peripherals built-in. The onboard radio transceiver operates in the 433 MHz ISM band and has a maximum transfer rate of 38.4 Kbits/sec with the maximum range of about 300 feet (Moog Crossbow 2013).

Real-time detection, classification, and correlation of acoustic events require processing power and buffer sizes not present in standard microcontroller-based embedded devices. To overcome these limitations, application-specific sensorboards have been designed and built at Vanderbilt University. The different architectures reflect the current dilemma faced by many signal processing engineers:

- The first version of the sensorboard (Fig. 3.2) utilizes a Xilinx XC2S100 FPGA chip (Xilinx 2008) with three independent analog channels exploiting the inherent parallelism of the hardware. The algorithms implemented in VHDL are focusing on precise time domain analysis of acoustic signals captured at high sample rates (one million Sample/sec). Hardware and software interfaces (I^2C bus, interrupts, led display, and serial A/D) are implemented as custom intellectual property core (IP cores) in the same gate array. While this approach offers

Fig. 3.2 The FPGA-based acoustic sensorboard

Fig. 3.3 The DSP-based acoustic sensorboard

very appealing features, that is, high accuracy (note that onboard angle of arrival estimate is possible), high speed (though not fully utilized for audio purposes), and efficient resource utilization, the size of the field-programmable gate array (FPGA) component severely constrains the complexity of the applicable algorithms. Suboptimal power consumption of the processing unit and the lack of effective power management modes are among the handicaps of the sensor network domain.

- To overcome these limitations, another sensorboard has been developed, where customized analog signal paths and an energy-efficient, powerful digital signal processor (DSP) make the unit uniquely suitable for power-constrained applications. At the heart of the second platform (Fig. 3.3) is a low-power fixed-point ADSP-218x (Analog Devices 1998) digital signal processor running at 50 MHz. Its internal program (48 KByte) and data (56 KByte) memory buffers with advanced addressing modes and direct memory access (DMA) controllers enable sophisticated signal processing and advanced power management methods.

Two independent analog input channels with low-cost electret microphones (Pui Audio 2008) pick up the incoming acoustic signals utilizing 2-stage amplification with software programmable gain (0–54 dB). The analog-to-digital (A/D) converters sample at up to 100 KSample/sec at 12-bit resolution. Analog comparators with software adjustable thresholds can be used to wake up the signal processor from low-power sleep mode, enabling continuous deployment for weeks on two AA batteries.

The FPGA and the DSP boards running the detection algorithms continuously draw 30 and 31 mA, respectively. In the power saving mode on the DSP board, this number drops to 1–5 mA, depending on the sleep mode. For comparison, the MICA2 mote draws 15 mA on average running the countersniper application.

3.2.1.1.2 Software Structure

As the system evolved, different versions of the system architecture were described in detail in (Simon et al. 2004) and (Ledeczi et al. 2005). Figure 3.4 presents a summary of the latest software architecture. The Muzzle Blast and Shockwave Detectors are implemented in VHDL on the FPGA of the first-generation sensorboard and in C on the DSP board. The TOA data from either board are sent through the inter-integrated circuit (I^2C) interface to the mote. A separate software component translates the time from the clock of the sensorboard to that of the mote. The Acoustic Event Encoder assembles a packet containing the TOA data and passes it to the Message Routing service.

In addition to transporting the packets to the basestation through multiple hops, the Message Routing service also performs implicit time synchronization. Additional software components running on the mote include a remote control service

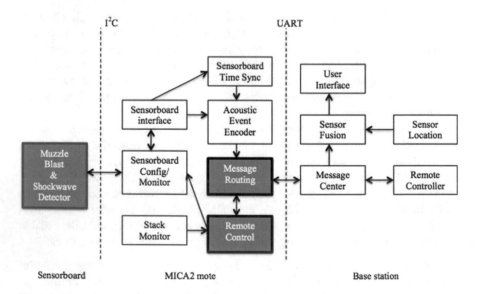

Fig. 3.4 The software architecture of the system. (Ledeczi et al. 2005)

enabling the configuration/polling of a single node, a group of or all of the nodes from the basestation. A Stack Monitor makes sure that the limited memory of the mote is not exhausted.

The basestation runs the sensor fusion algorithm utilizing the known sensor positions and displays the results on the user interface. The accuracy and/or range of existing sensor self-localization methods are not satisfactory for the shooter localization application (Sallai et al. 2004). Hence, all tests of the system were performed utilizing hand-placed motes on surveyed points. Later systems use localization based on the accurate radio interferometric geolocation technique (Maróti et al. 2005).

3.2.1.2 Detection

A block diagram of the signal processing algorithm is shown in Fig. 3.5. The incoming raw acoustic signal is compressed using zero-crossing (ZC) coding. The coded signals are used to detect possible occurrences of shockwave and muzzle blast patterns by the shockwave detection (SWD) and muzzle blast detection (MBD) blocks, respectively. Although the operation of the two detection blocks is mainly independent, the SWD block can provide information (time of arrival of the detected shockwave) for the MBD block to facilitate the detection of a muzzle blast after a shockwave. Both blocks measure the TOA of the detected acoustic event using the onboard clock and then notify the mote. The MICA2 mote reads the measurement data (TOA and optionally signal characteristics) and performs time synchronization between its own clock and that of the acoustic board. The measurement data are then propagated back to the basestation using middleware services of the sensor network.

The signal detection algorithm proved to be quite robust. It recognized 100% of the training events and more than 90% of the other recorded shot events. It is to be noted that, in reality, a shot may be detected by some sensors and may not be

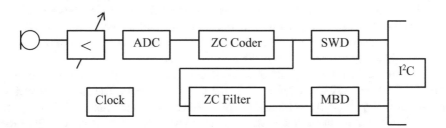

Fig. 3.5 Block diagram of the signal processing algorithm. The raw sampled signal is compressed and coded in the zero-crossing (ZC) coder. The shock wave detector (SWD) utilizes the ZC-coded signal. The muzzle blast detector (MBD) uses a filtered version. The detectors can communicate with the mote through an I²C interface. (Ledeczi et al. 2005)

recognized by others, depending on the location of the sensor. False positives could be produced only by physical contact with the microphone itself (Simon et al. 2004).

3.2.1.3   Routing Integrated Time Synchronization

An integrated time synchronization and routing algorithm is proposed; that is, no additional radio messages, no support for power management, and the low imposed overhead on message size make the algorithm suitable for many power-aware data collecting applications (Kusy et al. 2006b).

To find the position of the shooter(s), a consistency function on the four-dimensional space-time space is defined. A fast multiresolution search algorithm recursively finds the maxima of this function, which corresponds to the location and time of possible shots. Then, these maxima are further analyzed to eliminate false positives caused by echoes. The consistency function is defined in such a way that it automatically classifies and eliminates erroneous measurements and multipath effects.

3.2.1.4   Sensor Fusion

The TOA measurements originating from either the muzzle blast or the shockwave can be used in the estimation process. Muzzle blasts are extremely useful in near-field position estimation (Simon et al. 2004), while shockwaves provide effective means to determine the direction of a distant shooter (Balogh et al. 2005).

Muzzle blast and shockwave detections carry information about the shooter location, and the projectile trajectory, respectively. Either type of events, or both, combined can be used for localization purposes. The muzzle blast fusion algorithm works very well when the shooter is located within the sensor field and there are enough (at least 8–10) line-of-sight measurements. Once the shooter is shooting outside of the sensor field, the accuracy starts to decrease. One reason is that the angle of the sensor field from the shooter (field of view) is getting smaller, and hence, individual measurement errors have larger effects on the result. The other reason is that, as the distance to the shooter increases, fewer and fewer sensors are able to detect the muzzle blast at all. Once the shooter is beyond 50 m or so, muzzle blast alone is typically not enough to make accurate localization.

Shockwave fusion alone cannot determine the exact location of the shooter, but provide the trajectory of the bullet, even for long-range shots. The sensor network is presumably deployed in and around the protected area, and as long as the bullet goes through this region, the sensors can detect shockwave events, independently of the distance from the shooter. Naturally, shockwave trajectory estimation and muzzle blast ranging can be combined to provide accurate localization, if at least a few muzzle blast detections are available.

3.2.1.4.1 Range Estimation

Using the estimated trajectory based upon shockwave measurements and at least a few muzzle blast detections, the range of the source can also be estimated. Again, special care must be taken of potential multipath measurements and multiple shots.

Once the trajectory is estimated, the projectile location (X,Y,Z) at time instant t_0 is available, along with the elevation, azimuth, and the speed of the projectile. Thus, the complete timeline of the bullet can be computed, provided the speed of the projectile is constant, by associating a time instant t with each bullet location (x,y,z) on the trajectory. As the unknown location of the shooter is also on the trajectory, a straightforward solution is to correlate the timeline of the bullet and the muzzleblast TOA data, using a simplified consistency function approach. It is to be noted that the search in this case is reduced to the one-dimensional space along the estimated trajectory, as to be clarified. A sliding window is moved backward on the trajectory to find the shooter position. The simplified consistency function at the trajectory position (x,y,z) and its corresponding time t is defined as:

$$C_\delta(x, y, z, t) = \text{count}_{i=1,\ldots,N}\left(d^i_{\min} < (t_i - t)v_{\text{sound}} < d^i_{\max}\right) \tag{3.1}$$

where $d^i_{\min}$ and $d^i_{\max}$ are the minimum and maximum distances, respectively, between sensor position (x_i, y_i, z_i) (with TOA measurement t_i) and the sliding window of width δ, centered at (x,y,z) along the trajectory. The width of the window is determined by the estimated detection errors, a typical number being 1 m. The window position (x,y,z) with the highest consistency value $C_\delta(x,y,z,t)$ gives the estimated origin of the shot and, hence, the range. Similarly to the muzzle blast case, this solution automatically eliminates the erroneous measurements and also the measurements corresponding to other shots.

3.2.1.5 Experimentation

During the development period, several field tests were conducted in two U.S. Army facilities to evaluate the performance and accuracy of the shooter localization system. The data collected in the field tests were used to determine the accuracy of the system and its sensitivity to various sources of errors. For fusion technologies based on the muzzle blast, the localization error of the system is studied in 3D and 2D. In 3D, error is the total localization error; in 2D error, the elevation information is omitted. The system accuracy is remarkably good in 2D, the average error was 0.6 m, 83% of shots had less than 1 m of error, and 98% had less than 2 m of error. In 3D, the average error was 1.3 m, 46% of the shots had less than 1 m, and 84% of shots had less than 2 m of localization error.

The shockwave fusion algorithm was tested in the U.S. Army Aberdeen Test Center, in December 2004. Various targets were placed inside and outside the sensor field opposite from the shooter positions. In one particular experiment, 12 shots were fired over the middle of the network shooting approximately 100 m from the edge of the sensor field, so there were sensors on each side of the trajectory. The average azimuth error was 0.66 degrees, the average elevation error was 0.61 degrees, and the average range error was 2.56 m. Another 11 shots were fired from the same distance near the edge of the network, so there were no or only a few sensors on one side of the trajectory. The average error increased to 1.41 degrees in azimuth, to 1.11 degrees in elevation and to 6.04 m in range.

For fusion technologies based on the shockwave, multiple simultaneous shots have also been tested with mixed results. The typical test involved two shots only. About half the time, the system correctly localized both trajectories. There were cases, however, when three trajectories were found. Two of these were typically very close to each other and to one of the true trajectories. This error can happen when there is more than double the number of detections needed for localizing a single trajectory. In such a case, the error value corresponding to a subset of the detections may be smaller than the one involving all the detections for one trajectory. The error function and the genetic algorithm need to be adjusted to avoid this situation.

The latency of the shockwave-based fusion algorithm is somewhat greater than that of the muzzle-blast-based technique; the calculation of a single-shot trajectory takes about 3–4 seconds on a 3 GHz PC.

3.2.2   Shooter Localization and Weapon Classification with Soldier-Wearable Networked Sensors

This work as presented in (Volgyesi et al. 2007) moves from a static sensor network-based solution to a highly mobile one, which presents significant challenges. Specifically, the sensor positions and orientation need to be constantly monitored. Also, as soldiers may work in groups of as little as four people, the number of sensors measuring the acoustic phenomena may be an order of magnitude smaller than before. Moreover, the system should be useful to even a single soldier. Finally, additional requirements may be provided for caliber estimation and weapon classification in addition to source localization.

The firing of a typical military rifle, such as the AK47 or M16, produces two distinct acoustic phenomena. The muzzle blast, that is generated at the muzzle of the gun, and travels at the speed of sound. Also, the supersonic projectile generates an acoustic shockwave, a kind of sonic boom. The wavefront has a conical shape, the angle of which depends on the Mach number, the speed of the bullet relative to the speed of sound. The shockwave has a characteristic shape resembling a capital N. The rise time at both the start and end of the signal is very fast, under 1 μsec. The length is determined by the caliber and the miss distance, the distance between the trajectory and the sensor. It is typically a few hundred μsec. Once a trajectory estimate is available, the shockwave length can be used for caliber estimation.

The proposed system is based on four microphones connected to a sensorboard. The board detects shockwaves and muzzle blasts and measures their time of arrival (TOA). If at least three acoustic channels detect the same event, its angle of arrival (AOA) is also computed. If both the shockwave and muzzle blast AOA are available, a simple analytical solution gives the shooter location. As the microphones are close to each other, typically 5–10 cm, very high precision is not expected. Also, this method does not estimate a trajectory. However, the sensorboards are also connected to COTS MICAz motes (Crossbow 2006a; Koh 2006) and they share their AOA and TOA measurements, as well as their own location and orientation, with each other using a multihop routing service (Maróti 2004). A hybrid sensor fusion algorithm then estimates the trajectory, the range, the caliber, and the weapon type based on all available observations.

The sensorboard is also Bluetooth capable for communication with the soldier's PDA or laptop computer. A wired USB connection is also available. The sensor fusion algorithm and the user interface get their data through one of these channels.

The orientation of the microphone array at the time of detection is provided by a 3-axis digital compass. Currently, the system assumes that the soldier's PDA is GPS-capable and it does not provide self-localization service itself. However, the accuracy of GPS is a few meters degrading the overall accuracy of the system. The latest generation sensorboard features a Texas Instruments CC-1000 radio (Texas Instruments 2007a) enabling the high-precision radio interferometric self-localization approach (Kusy et al. 2006a).

3.2.2.1   Hardware

The board utilizes a powerful Xilinx XC3S1000 FPGA chip (Xilinx 2013) with various standard peripheral IP cores, multiple soft processor cores, and custom logic for the acoustic detectors (Fig. 3.6). The onboard Flash (4 MByte) and the pseudostatic random access memory (PSRAM) (8 MByte) modules allow storing raw samples of several acoustic events, which can be used to build libraries of various acoustic signatures and for refining the detection cores off-line. Also, the external memory blocks can store program code and data used by the soft processor cores on the FPGA.

The sensorboard supports four independent analog channels sampled at up to one million Sample/sec. These channels, featuring an electret microphone Panasonic WM- 64PNT (Panasonic Corporation 2013), amplifiers with controllable gain (30–60 dB), and a 12-bit serial ADC AD7476 (Analog Devices 2013), reside on separate tiny boards which are connected to the main sensorboard with ribbon cables. This partitioning enables the use of truly different audio channels (e.g., slower sampling frequency, different gain or dynamic range) and also results in less noisy measurements by avoiding long analog signal paths.

The sensor platform offers a rich set of interfaces and can be integrated with existing systems in diverse ways. An RS232 port and a Bluetooth Bluegiga WT12 (Bluegiga 2013; Glyn Store 2013) wireless link with virtual universal asynchronous

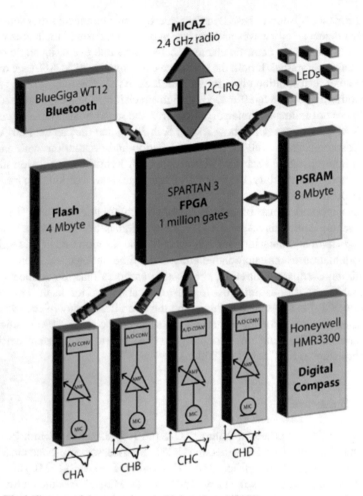

Fig. 3.6 Block diagram of the sensorboard. (Volgyesi et al. 2007)

receiver/transmitter (UART) emulation are directly available on the board and provide simple means to connect the sensor to personal computers (PCs) and personal digital assistants (PDAs). The mote interface consists of an I²C bus along with an interrupt and general purpose input/output (GPIO) line (the latter one is used for precise time synchronization between the board and the mote). The motes are equipped with IEEE 802.15.4 compliant radio transceivers and support ad hoc wireless networking among the nodes and to/from the basestation. The sensorboard also supports full-speed universal serial bus (USB) transfers (with custom USB dongles) for uploading recorded audio samples to the PC. The onboard JTAG chain, directly accessible through a dedicated connector, contains the FPGA part and configuration memory and provides in-system programming and debugging facilities.

The integrated Honeywell HMR3300 digital compass module (Honeywell 2012) provides heading, pitch, and roll information with 1° accuracy, which is essential for calculating and combining directional estimates of the detected events.

The first prototype of the proposed system employed ten sensor nodes. Some of these nodes were mounted on military kevlar helmets (Dupont 2013) with the microphones directly attached to the surface at about 20 cm separation. The rest of the nodes were mounted in plastic enclosures with the microphones placed near the corners of the boxes to form approximately 5 cm × 10 cm rectangles.

3.2.2.2 Software Architecture

The sensor application relies on three subsystems exploiting three different computing paradigms. Although each of these execution models suits their domain-specific tasks, this diversity presents a challenge for software development and system integration. The sensor fusion and user interface subsystem are running on PDAs and were implemented in Java. The sensing and signal processing tasks are executed by an FPGA, which also acts as a bridge between various wired and wireless communication channels. The ad hoc internode communication, time synchronization, and data sharing are the responsibilities of a microcontroller-based radio module.

3.2.2.3 Detection Algorithm

There are several characteristics of acoustic shockwaves and muzzle blasts, which distinguish their detection and signal processing algorithms from regular audio applications. Both events are transient by their nature and present very intense stimuli to the microphones. The detection algorithms have to be robust enough to handle severe nonlinear distortion and transitory oscillations. Since the muzzle blast signature closely follows the shockwave signal and because of potential automatic weapon bursts, it is extremely important to settle the audio channels and the detection logic as soon as possible after an event. Also, precise angle of arrival estimation necessitates high sampling frequency (in the MHz range) and accurate event detection. Moreover, the detection logic needs to process multiple channels in parallel (four channels on the proposed hardware).

The most conspicuous characteristics of an acoustic shockwave are the steep rising edges at the beginning and end of the signal (Fig. 3.7). Also, the length of the N-wave is fairly predictable and is relatively short (200–300 µs). The shockwave detection core is continuously looking for two rising edges within a given interval. The only feature calculated by the core is the length of the observed shockwave signal.

In contrast to shockwaves, the muzzle blast signatures are characterized by a long initial period (1–5 msec) where the first half period is significantly shorter than the

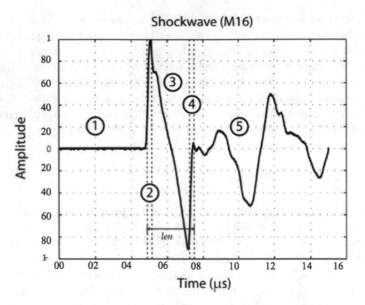

Fig. 3.7 Shockwave signal generated by 5.56 × 45 mm NATO projectile. (Volgyesi et al. 2007)

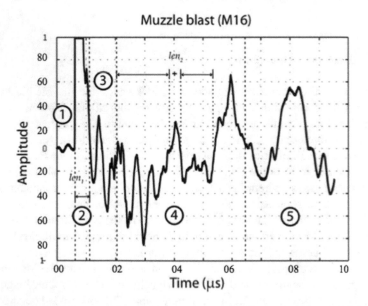

Fig. 3.8 Muzzle blast signature produced by an M16 assault rifle. (Volgyesi et al. 2007)

second half (Fansler 1998). Due to the physical limitations of the analog circuitry, irregular oscillations and glitches might show up within this longer time window as they can be clearly seen in Fig. 3.8. Therefore, the real challenge for the matching detection core is to identify the first and second half periods properly.

The detection cores were originally implemented in Java and evaluated on prerecorded signals because of much faster test runs and more convenient debugging facilities. Later on, they were ported to Verilog hardware description language (VHDL) and synthesized using the Xilinx ISE tool suite.

3.2.2.4   Sensor Fusion

The sensor fusion algorithm receives detection messages from the sensor network and estimates the bullet trajectory, the shooter position, the caliber of the projectile, and the type of the weapon. The algorithm consists of the distinct computational tasks outlined below:

1. *Compute* muzzle blast and shockwave directions of arrivals for each individual sensor.
2. *Compute* range estimates. This algorithm can analytically fuse a pair of shockwave and muzzle blast AoA estimates.
3. *Compute* a single trajectory from all shockwave measurements.
4. *If* trajectory available, *then* compute range, *else* compute shooter position first and then trajectory based on it.
5. If trajectory available, *then* compute caliber.
6. *If* caliber available, *then* compute weapon type.

3.2.2.5   Results

An independent evaluation of the system was carried out by a team from the National Institute of Standards and Technology (NIST) at the US army Aberdeen test center in April 2006 (Weiss et al. 2006). The experiment was set up on a shooting range with mock-up wooden buildings and walls for supporting elevated shooter positions and generating multipath effects. Ten sensor nodes, statistically placed, were deployed on surveyed points in an approximately 30×30 m^2 area. There were five fixed targets behind the sensor network. Several firing positions were located at each of the firing lines at 50, 100, 200, and 300 m. These positions were known to the evaluators, but not to the operators of the system. Six different weapons were utilized: AK47 and M240 firing 7.62 mm projectiles, M16, M4, and M249 with 5.56 mm ammunition and the 0.50 caliber M107.

The tests outcome may be outlined in what follows:

- During a 1-day test, there were 196 shots fired. The system detected all shots successfully. Since a ballistic shockwave is a unique acoustic phenomenon, it makes the detection very robust. There were no false positives for shockwaves, but there were a handful of false muzzle blast detections due to parallel tests of artillery at a nearby range.
- The caliber and weapon estimation accuracy rates are based on the 189 shots that were successfully localized. For four of the weapons (AK14, M16, M240, and

M107), the classification rate is almost 100%. There were only two shots out of approximately 140 that were missed. The M4 and M249 proved to be too similar and they were mistaken for each other most of the time.

- There are several disadvantages of the single sensor case compared to the networked system. Namely, there is no redundancy to compensate for other errors and to perform outlier rejection, the localization rate is markedly lower, and a single sensor alone is not able to estimate the caliber or classify the weapon.
- Errors in the time synchronization, node localization, and node orientation degrade the overall accuracy of the system.
- Successful localization goes down from almost 100% to 50% when tests go from ten sensors to two even without additional errors. This is primarily caused by geometry, for a successful localization, the bullet needs to pass over the sensor network, that is, at least one sensor should be on the side of the trajectory other than the rest of the nodes.
- There is hardly any difference in the data for six, eight, and ten sensors. This means that there is little advantage of adding more nodes beyond six sensors as far as the accuracy is concerned.
- The speed of sound depends on the ambient temperature. It would be straight-forward to employ a temperature sensor to update the value of the speed of sound periodically during operation. Also, wind may adversely affect the accuracy of the system. The sensor fusion, however, could incorporate wind speed into its calculations.
- Silencers reduce the muzzle blast energy, and hence, the effective range the system can detect it at. However, silencers do not effect the shockwave and the system would still detect the trajectory and caliber accurately. The range and weapon type could not be estimated without muzzle blast detections. Subsonic weapons do not produce a shockwave. However, this is not of great significance, since they have shorter range, lower accuracy, and much less lethality. Hence, their use is not widespread and they pose less danger in any case.
- Irregular armies may use substandard, even hand-manufactured bullets, this affects the muzzle velocity of the weapon. For weapon classification to work accurately, the system would need to be calibrated with the typical ammunition used by the given adversary.

3.2.3   Shooter Localization Using Soldier-Worn Gunfire Detection Systems

This paper (George and Kaplan 2011) presents the development of a sensor fusion module that would take full advantage of the team aspect of a small combat unit (SCU) to provide a fused solution that would be highly accurate and suitable for a command and control geographic information system (C2 GIS) map display compared to the individual soldier's solution. The objective is to improve accuracy across an entire SCU so even soldiers in non-ideal settings (out of range, bad angle, etc.) can exploit the good solutions from their neighbors to come up with improved solutions, both geo-rectified and relative. The individual soldier-wearable gunfire detection systems (SW-GDSs) considered in this work is composed of a passive microphones array that is able to localize a gunfire event by measuring the direction of arrival for both the acoustic wave generated by the muzzle blast and the shockwave generated by the supersonic bullet (Duckworth et al. 2001; Bedard and Pare 2003). After detecting a gunfire, the individual sensors report their solution along with their global positioning system (GPS) positions to a central node. At the central node, the individual solutions are fused along with the GPS positions to yield a highly accurate, geo-rectified solution, which is then relayed back to individual soldiers for added situational awareness.

It is recognized that there is an eminent need for highly precise small-arms gunfire detection systems on individual soldiers for added battlefield situational awareness and threat assessment. Today, several acoustic shooter localization systems are commercially available (Duckworth et al. 2001; Bedard and Pare 2003). Currently, operational SW-GDSs can provide an appropriate level of localization accuracy as long as the soldier is at an ideal location (range, attitude, etc.) when incoming fire is received (Maroti et al. 2004; Kuckertz et al. 2007; Ash et al. 2010). The localization system suffers severe performance degradation when the soldier is at a non-ideal location. Moreover, when a relative solution, i.e., the shooter location relative to the sensor, is transformed into a geo rectified solution using a magnetometer and GPS, the solution often becomes unusable due to localization errors. Geo-rectified solutions are necessary when displaying hostile fire icons on a C2 GIS map display.

SW-GDSs use acoustic phenomena analysis of small-arms fire to localize the source of incoming fire, usually with a bearing and range relative to the user (Kaplan et al. 2008). These individual SW-GDSs operate separately and are not designed to exploit the sensor network layout of all the soldiers within a SCU to help increase accuracy. Researchers are exploring some novel solutions that utilize the team aspect of these SCUs by exploiting all SW-GDSs in a squad/platoon to increase detection rates and accuracy (Ledeczi et al. 2005; Volgyesi et al. 2007; Lindgren et al. 2009).

3.2.3.1 Mathematical Formulation

Consider a SCU consisting of n individual soldiers equipped with the SW-GDS. In order to set up the problem and develop a sensor model, a first scenario is considered where there is only one shooter and the SW-GDS receives both the muzzle blast and the shockwave. The shooter or the target location and the soldier or the ith sensor location are defined as T and S_i, respectively. For simplicity, the problem is formulated in R^2, i.e., $T \in R \equiv \begin{bmatrix} T_x \\ T_y \end{bmatrix}$ and $S_i \in R^2 \equiv \begin{bmatrix} S_{ix} \\ S_{iy} \end{bmatrix}$. Now define the individual range, r_i, and bearing, φ_i, between the ith sensor node and the target as:

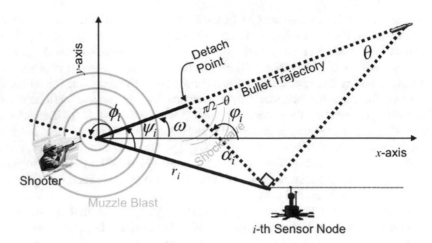

Fig. 3.9 Geometry of the bullet trajectory and propagation of the muzzle blast and shockwave to the sensor node. (Volgyesi et al. 2007)

$$r_i = \sqrt{\left(T_x - S_{i_x}\right)^2 + \left(T_y - S_{i_y}\right)^2} \qquad (3.2)$$

$$\emptyset_i = \arctan 2\left(T_y - S_{i_y}, T_x - S_{i_x}\right) \qquad (3.3)$$

When a gun fires, the blast from the muzzle produces a spherical acoustic wave that can be heard in any direction. The bullet travels at supersonic speeds and produces an acoustic shockwave that emanates as a cone from the trajectory of the bullet. Because the bullet is traveling faster than the speed of sound, the shockwave arrives at the sensor node before the wave from the muzzle blast, which is simply referred to as the muzzle blast. Figure 3.9 illustrates the geometry of the shockwave and the muzzle blast for the ith sensor node when the orientation of the bullet trajectory is ω with respect to the horizontal axis. As the bullet pushes air, it creates an impulse wave.

The wavefront is a cone whose angle θ with respect to the trajectory is:

$$\theta = \arcsin\left(\frac{1}{m}\right) \qquad (3.4)$$

where m is the Mach number. The Mach number is assumed to be known since the typical value for a Mach number is $m = 2$ (Kaplan et al. 2008). Since the Mach number directly influences the range estimates, uncertainty in bullet speed may be treated as range estimation error. As indicated in Fig. 3.9, the angle φ_i indicates the direction of arrival (DOA) of the muzzle blast, and φ_i indicates the DOA of the shockwave. The muzzle blast DOA is measured counter-clockwise such that $0 \le \varphi_i \le 2\pi$. Details are available in (Kaplan et al. 2008). Figure 3.10 indicates the field of view (FOV) for both the muzzle blast and the shockwave. Note that the FOV of the muzzle blast is 2π, i.e., omnidirectional, and the FOV for the shockwave

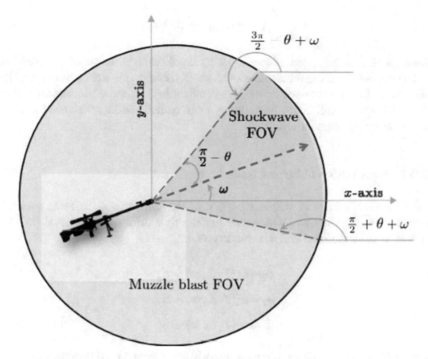

Fig. 3.10 Muzzle blast and shockwave field of view. (Volgyesi et al. 2007)

is $\pi - 2\theta$. SW-GDS receives the shockwave only if the muzzle blast DOA is within the bounds:

$$\pi/2 + \theta + \omega < \phi_i < 3\pi/2 - \theta + \omega \qquad (3.5)$$

Now the DOA angle for the shockwave can be written as:

$$\varphi_i = \begin{cases} -\pi/2 - \theta + \omega, & \text{if } \pi + \omega < \phi_i < 3\pi/2 - \theta + \omega; \\ \pi/2 + \theta + \omega, & \text{if } \pi/2 + \theta + \omega < \phi_i < \pi + \omega \end{cases} \qquad (3.6)$$

The first case $\pi + \omega < \phi_i < 3\pi/2 - \theta + \omega$ corresponds to the scenario where the sensor is located above the bullet trajectory and the case $\pi/2 + \theta + \omega < \phi_i < \pi + \omega$ corresponds to the scenario where the sensor is located below the bullet trajectory (as shown in Fig. 3.9). The case where $\phi_i = \pi + \omega$ corresponds to the scenario when the sensor is located on the bullet trajectory and here such a scenario is not considered.

If ϕ_i is outside the bound given in Eq. 3.5, the sensor node only receives the muzzle blast and it is outside the FOV of the shockwave. Under the assumptions that the bullet maintains a constant velocity over its trajectory, the time difference between the shockwave and the muzzle blast can be written as in (Bedard and Pare 2003):

$$\tau_i = \frac{r_i}{c}[1 - \cos|\phi_i - \varphi_i|], \ \forall \phi_i \neq \varphi_i \tag{3.7}$$

where c indicates the speed of sound. Using Eq. 3.6, the bullet trajectory angle, ω, can be obtained from the shockwave DOA angle. Though this work assumes that the bullet speed is constant over its trajectory, others have proposed localization algorithms (Lindgren et al. 2009) that employ more realistic bullet speed models at the expense of computational efficiency.

3.2.3.2   Data Fusion at Sensor Node Level

When the sensor node is within the FOV of the shockwave, the three available measurements are the two DOA angles and the time difference of arrival (TDOA) between the muzzle blast and the shockwave, i.e.,

$$\widehat{\phi}_i = h_1(T, S_i, \omega) + \eta_\phi \tag{3.8}$$

$$\widehat{\varphi}_i = h_2(T, S_i, \omega) + \eta_\varphi \tag{3.9}$$

$$\widehat{\tau}_i = h_3(T, S_i, \omega) + \eta_\tau \tag{3.10}$$

where $h_1(\cdot)$ is given in Eq. 3.3, $h_2(\cdot)$ is given in Eq. 3.6, and $h_3(\cdot)$ is given in Eq. 3.7. The measurement noise is assumed to be zero mean Gaussian white noise, i.e., $\eta_\phi \sim \mathbb{N}\left(0, \sigma_\phi^2\right)$, $\eta_\varphi \sim \mathbb{N}\left(0, \sigma_\varphi^2\right)$, and $\eta_\tau \sim \mathbb{N}(0, \sigma_\tau^2)$. Let $\widehat{T}_i = \begin{bmatrix} \widehat{\phi}_i & \widehat{r}_i & \widehat{\omega}_i \end{bmatrix}$ denote the individual sensor level estimates on the target bearing, range, and the bullet trajectory. Data fusion at the sensor node involves calculating these individual estimates based on the three sensor measurements.

Using Eq. 3.6, the bullet trajectory angle, ω, can be obtained from the shockwave DOA measurements. Thus, the observations on the trajectory angle can be written as:

$$\widehat{\omega}_i = \omega_i + \eta_\varphi \tag{3.11}$$

For a sensor located in the FOV of the shockwave, the target location can be estimated as:

$$\widehat{T}_{x_i} = \widehat{S}_{i_x} + \widehat{r}_i \cos \widehat{\phi}_i \tag{3.12}$$

$$\widehat{T}_{y_i} = \widehat{S}_{i_y} + \widehat{r}_i \sin \widehat{\phi}_i \tag{3.13}$$

When the sensor is located outside the shockwave FOV, the only estimate would be the bearing angle. After individual estimates are obtained at the sensor node level, the measured information is transmitted to a central node where it is fused to obtain a more accurate estimate of shooter location.

3.2.3.3 Data Fusion at the Central Node

While sensors in the FOV of the muzzle blast and the shockwave yield a range, bearing, and trajectory angle estimates, the gunfire detection systems outside the FOV of the shockwave yield a muzzle blast DOA. Also, GPS measurements are available on each sensor locations. At the central node, this information from the individual sensor nodes is fused to obtain an accurate estimate of the shooter location, bullet trajectory angle, and the sensor location.

3.2.3.4 Results

Simulation results reveal that:

- The fused estimate is superior to the individual sensor estimates, and the uncertainty associated with the fused estimates is much less than the uncertainty associated with the individual sensor estimates.
- The fusion algorithm was able to improve the sensor location accuracy by reducing the GPS uncertainties.

3.3 Industrial Applications

The application of WSN technology to the design of field-area networks for industrial communication and control systems has the potential to provide major benefits in terms of flexible installation and maintenance of field devices, support for monitoring the operations of mobile robots, and reduction in costs and problems due to wire cabling. This section targets such applications with emphasis on their features, constraints, and requirements. It is well known that industrial applications are more than plentiful, and the goal of this section is to zoom in on the used WSN techniques and types of sensors used, at the research and practice levels. Additionally, getting acquainted with the big players who manufacture and design WSN building blocks is a major asset to be released after this section.

3.3.1 On the Application of WSNs in Condition Monitoring and Energy Usage Evaluation for Electric Machines

The work in (Lu et al. 2005) proposes a scheme for applying WSNs in energy usage evaluation and condition monitoring for electric machines. The importance of this scheme lies in its non-intrusive, intelligent, and low-cost nature as will be elucidated in the coming sections.

3.3.1.1 Energy Evaluation and Condition Monitoring

Energy usage evaluation and motor condition monitoring are two basic functions of an energy management system for an industrial plant. They have their unique motivations and requirements, but also share many common needs, such as data collection.

3.3.1.1.1 Energy Usage Evaluation

It is estimated by the Department of Energy (DOE) that motor-driven systems use over two-third of the total electric energy consumed by industry in the United States. In industry, motors below 200 hp. make up 98% of the motors in service and consume 85% of the energy used. On average, these motors operate at no more than 60% of their rated load because of oversized installations or underloaded conditions and thus at reduced efficiency which results in wasted energy. As the global energy shortage and the greenhouse effect worsen, the improvement of energy usage in industry is drawing more attention. Obviously, to improve energy efficiency, an evaluation of the energy usage condition of the industrial plant is required.

Among all the energy usage evaluation functions, motor efficiency estimation is the most important. In the literature, many motor efficiency estimation methods have been proposed (Electric Machines Committee 1997; Wallace et al. 2001). A common problem of these methods is either expensive speed and/or torque transducers are needed for rotor speed and shaft torque measurements, or a highly accurate motor equivalent circuit needs to be developed from the motor parameters. Generally, these methods are too intrusive and are often not feasible for in-service motor testing. To overcome these problems, (Lu et al. 2006) present a complete survey on motor efficiency estimation methods, specifically considering the advances in sensorless speed estimation and in-service stator resistance estimation techniques. Three candidate methods for non-intrusive efficiency estimation are modified for in-service motor testing. The non-intrusive characteristic of these methods enables efficiency evaluation with a WSN.

3.3.1.1.2 Condition Monitoring

Motor condition monitoring gives the health condition of running electric motors and avoids economical losses resulting from unexpected motor failures. Sharing many common requirements with energy usage evaluation in terms of data collection, motor condition monitoring could be naturally added into an energy management system considering that the necessary data are readily available. For example, the motor stator currents need to be measured in the energy management system since they are required by almost all efficiency estimation methods. On the other hand, many condition monitoring algorithms, such as the detection of the stator winding turn faults, broken rotor bars, worn bearings, and air-gap eccentricities, use the motor current spectral analysis (MCSA) technique, which also requires the stator current waveforms to be sampled and collected (Habetler et al. 2002). Therefore, it would be natural to incorporate these condition monitoring functions in the energy management system without additional cost for data collection.

3.3.1.1.3 Additional Requirements

Generally, the measurements needed for each efficiency estimation and condition monitoring method are different, but essentially all require the input line voltages and the line currents. Some methods require the nameplate data (rated voltage, current, horsepower, speed, etc.), stator resistance, RS, or rotor speed, ω_r. Among these, the measurements or estimates of stator resistance and speed have for years been regarded as stumbling blocks. However, research has made great progress in the area of stator resistance and speed estimation (Lu et al. 2006). Most of these estimators utilize the terminal voltages and currents, which are available in the energy evaluation and condition monitoring system.

WSNs target primarily the very low-cost and ultra-low-power consumption applications, with data throughput and reliability as secondary considerations. Fueled by the need to enable inexpensive WSNs for monitoring and control of non-critical functions in the residential, commercial, and industrial applications, the concept of a standardized low-rate wireless personal area networks (LR-WPANs) has emerged (Gutiérrez et al. 2004). In October 2003, the LR-WPANs standard finally became the IEEE 802.15.4 standard (Callaway et al. 2002). The unique characteristics of the IEEE Std. 802.15.4/LR-WPANs such as the flexibility, inherent intelligence, fault tolerance, high sensing fidelity, low-cost, and rapid deployment make WSN the ideal structure for low-cost energy evaluation and planning system incorporating energy usage evaluation and condition monitoring functions together, and furthermore constructing a high-level intelligent power management system in industrial plants.

3.3.1.2 Energy Evaluation and Condition Monitoring Using WSNs

Due to the unique characteristics of WSN, it could be applied as the backbone structure of the low-cost electric machine energy management system incorporating energy usage evaluation and condition monitoring functions. The deployment of WSN results in a sensor-rich environment, which allows for a high-level intelligent power management system for industrial plants.

3.3.1.2.1   System Description

In an industrial plant, motor control centers (MCCs) provide power for motors of all different sizes. The motor terminal data are collected and processed in the central supervisory station (CSS). Based on the reports from CSS, the user can assess the plant operational cost and make decisions. It is necessary to point out that a typical plant usually has more than one MCC and CSS. Traditionally, communication cables need to be installed to collect data from the MCCs or motors and send them to the CSSs. These communication cables could be eliminated by deployment of WSNs.

Due to the challenges of WSN technology, such as the relatively long latency, and limited reliability and security, the objective of engaging WSN in an industrial plant is not to replace the existing wired communication and control systems completely. Rather, the objective is to form a wireless and wired coexisting system; wherein the non-critical tasks such as efficiency estimation, operating cost evaluation, and diagnosis are carried out by the wireless part to reduce the overall cost, while the critical tasks (in terms of time requirement and cost) such as real-time motor controls and overload protection are still performed by the wired system for reliability reasons.

The WSN sensor node has both sensing and communication capabilities and can work as a transmitter node, a receiver node, or a relay node. Figure 3.11 illustrates a WSN transmitter node. It first measures the motor terminal quantities (i.e., line voltages, line currents, and temperature, if available) and scales them into analog signals in the range of 0–5 volts; then, these scaled signals are passed through an analog-to-digital conversion (ADC) unit; finally, the digitized signals are passed via the serial peripheral interface bus (SPI) to the radio unit and the data packets are transmitted through the WSN.

Figure 3.12 illustrates a WSN receiver node. It first receives the data packets from WSN, then the raw packets are reconstructed into the original digitized signals in the interface unit; finally, these digital signals are sent to the CSS through an RS232 link. When a sensor works as a relay node, it does nothing but receives data packets and sends the same packets out.

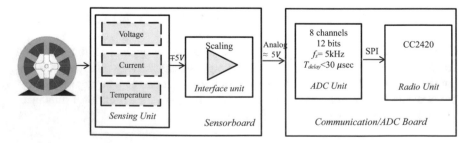

Fig. 3.11 WSN transmitter node. (Lu et al. 2005)

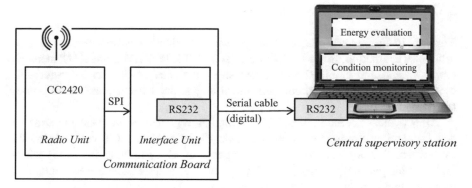

Fig. 3.12 WSN receiver node. (Lu et al. 2005)

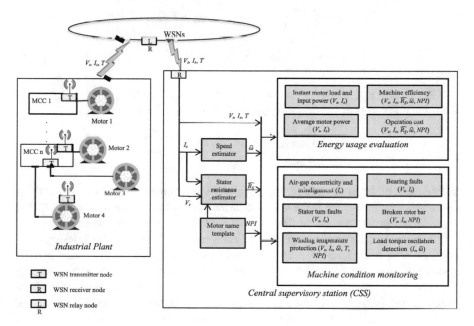

Fig. 3.13 Energy usage evaluation and condition monitoring system for electric machines using WSNs. (Lu et al. 2005)

Figure 3.13 shows the proposed energy usage evaluation and condition monitoring system for electric machines with a WSN architecture. The terminal quantities of each motor are measured at the MCCs and transmitted to the CSSs through the WSN. Using these data, non-intrusive methods are used to estimate the energy usage and health condition of each motor in the plant. These results from the CSSs are finally reported to the plant manager to evaluate the operational cost of the whole plant and make planning decision such as replacing oversized or malfunctioning motors.

3.3.1.2.2   Energy Usage Evaluation

The key to the electric machine energy usage evaluation is non-intrusive motor efficiency estimation. In (Lu et al. 2006), the ORMEL96, OHME, and AGT methods are suggested and modified as non-intrusive candidates for in-service motor efficiency estimation that only relies on line voltages, line currents, and motor nameplate information.

Among these methods, the AGT method is regarded as the best in terms of accuracy and ease of implementation. For simplicity, only the AGT method is briefly introduced. The original AGT method proposed in (Hsu and Scoggins 1995) calculates the air-gap torque using Eq. 3.14 from the motor instantaneous input line voltages, line currents, and stator resistance:

$$T_{\text{air}-\text{gap}} = \frac{\text{poles}}{2\sqrt{3}} \left\{ (i_A - i_B). \int \left[v_{CA} - R_s(i_C - i_A) \right] dt \right.$$

$$\left. - (i_C - i_A). \int [v_{AB} - R_s(i_A - i_B)] dt \right\} \tag{3.14}$$

The friction and windage loss, W_{fw}, and rotor stray-load loss, W_{LLr}, are obtained from the no load test. Finally, the motor efficiency is calculated using Eq. 3.15:

$$\eta = \frac{T_{\text{shaft}}.\omega_r}{P_{\text{input}}} = \frac{T_{\text{air}-\text{gap}}.\omega_r - W_{fw} - W_{LLr}}{P_{\text{input}}} \tag{3.15}$$

where poles is the number of poles, i_A, i_B, and i_C are three line currents, v_{CA} and v_{AB} are two line voltages, RS is the stator resistance per phase, and ω_r is the rotor speed.

A significant advantage of this method is that it considers the losses associated with the unbalances in the voltages and currents, which reflects the working environment of a real motor. It is reported in (Hsu et al. 1998) that the AGT method shows high accuracy ($\pm 0.5\%$ error) and ease of implementation, however, it requires the no load test and direct measurements of stator resistance and rotor speed, which makes it highly intrusive. In the proposed system, the following modifications and assumptions are added to improve the non-intrusiveness of the original AGT method:

- The friction and windage loss are assumed to be a constant percentage of the rated output power, e.g., 1.2% for 4-pole motors below 200 hp., similar to the ORMEL96 method in (Hsu et al. 1998).
- The stray-load loss at rated load is assumed to be a constant percentage of the rated output power depending on the motor sizes.
- The rotor speed is estimated from motor current spectrum, using methods summarized in (Lu et al. 2006).
- The stator resistance is estimated from induction motor model-based or signal injection-based stator resistance estimation methods, as summarized in (Lu et al. 2006).

3.3.1.2.3   Motor Condition Monitoring

Motor condition monitoring includes the detection of air-gap eccentricities and misalignment, worn bearings, stator winding turn faults, broken rotor bars, winding overheating, and load torque oscillations. As pointed out in Sect. 3.3.1.1, motor condition monitoring could conveniently be added to the energy evaluation system using the data from the WSN. Similar with the requirements of efficiency estimation, non-intrusive methods are required for motor condition monitoring using only motor terminal quantities. Various current-based condition monitoring techniques have already been developed and summarized in (Habetler et al. 2002).

3.3.1.2.4   Applicability Analysis

The risk of success of the proposed scheme is minimized by the fact that several major concerns of WSN, energy evaluation, and condition monitoring are no longer problems of this specific integrated application:

*   Power Consumption. Power consumption or battery life is the dominating factor that affects the design of WSN. In this application, power consumption constraints can be simply neglected because in industrial plants the WSN sensor nodes are installed in the MCCs and the power can be obtained easily from very inexpensive ac/dc converters. This also eliminates the implementation of complicated protocols and routing algorithms of WSN, which are primarily intended to resolve the power constraints.
*   Accuracy. The motor energy usage evaluation and condition monitoring results are mainly provided for the industrial plant managers to make their planning decisions. To do this, even rough estimates of motor efficiency and health conditions are of great value. This greatly relieves the requirements for the accuracy of the algorithms in the proposed system.

3.3.1.3   Experimentation Results

The proposed system has been implemented by combining WSN data transmission, energy usage evaluation, and condition monitoring functions. For the time being, the stator resistance is assumed to be a constant for simplicity.

3.3.1.3.1   Energy Usage Evaluation—Motor Efficiency Estimation

As a key function of the energy evaluation system, the motor efficiency estimation algorithm is evaluated by both computer simulations and real experiments. In the experimental setup, a 3-phase induction motor is line connected to a 230 V mains supply. The motor has the following nameplate information: 4-pole, NEMA-A,

7.5 hp., 230 V, 18.2 A, and 89.5% nominal efficiency. A dc generator connected to resistor boxes serves as the dynamometer. An inline rotary torque transducer measures the shaft torque, while an optical encoder measures the speed.

The voltages and currents are slightly unbalanced and reflect the actually motor working condition. The estimated motor efficiency and shaft torque are calculated using the AGT method, which uses only line voltages and currents. The air-gap flux is obtained through the integral of the stator voltages subtracting the stator IR drop (voltage drop) with zero initial conditions. Then, the DC offset in the air-gap flux is removed by a 3 cycle (50 msec) moving average window. The actual efficiency is directly calculated from measured speed and shaft torque.

It has been shown that the estimated motor efficiency has a very good agreement ($\pm 2\%$ error) with the measured efficiency, especially in the normal motor load range (40–90% of rated load). Besides, the proposed system also gives relatively accurate efficiency estimates at underloaded and overloaded conditions, which are useful for industrial energy management. If the estimated temperature-varying stator resistance is used, the error in the stator copper loss estimation will be reduced. As a result, the accuracy of the estimated efficiency and shaft torque will be improved.

3.3.1.3.2   Condition Monitoring—Detection of Air-Gap Eccentricities

A substantial portion of induction motor faults is air-gap eccentricity related (Huang et al. 2007). Basically, there are two types of eccentricities: static eccentricity and dynamic eccentricity. In practice, they tend to coexist due to an inherent level of either static or dynamic eccentricity even in a new motor. In general, online condition monitoring of air-gap eccentricity primarily depends on the detection of fundamental side band harmonics located at $f_e \pm f_{rm}$, where f_e is the fundamental excitation frequency and f_{rm} is the rotor rotational frequency.

As an example motor condition monitoring technique, an air-gap eccentricity detection algorithm is investigated using only a single-phase motor line current. The same experimental setup adopted in the previous section is used. The static eccentricity is created by first machining the bearing housings of the end bell eccentrically and then placing a 0.01-inch shim in the end bell to offset the rotor. The dynamic eccentricity is created by first machining the motor shaft under the bearings eccentrically and then inserting a 0.01-inch offset sleeve under the bearings. A FFT is applied to the measured single-phase stator current to obtain its spectrum. Experimentation was carried on when the motor was running at 1752 rpm ($f_e \approx 60$ Hz and $f_{rm} = 29.2$ Hz), the air-gap eccentricity fault was detected.

3.3.2 Breath: An Adaptive Protocol for Industrial Control Applications Using WSNs

This work (Park et al. 2011) proposes the Breath protocol, a self-adapting efficient solution for reliable and timely data transmission. The protocol adapts to the network variations by enlarging or shrinking next-hop distance, sleep time of the nodes, and transmit radio power, just like a breathing organism. The system model considers nodes that have to send packets to the sink via multihop routing under tunable reliability and delay requirements. The protocol is based on randomized routing, CSMA/CA MAC and randomized sleep discipline that are jointly optimized for energy consumption. The protocol contribution entails what follows:

- It provides explicit analytical relations of the reliability, delay, and total energy consumption as a function of MAC, routing, physical layer, duty cycle, and radio power. The approach is based on simple yet good approximations whose accuracy is systematically verified.
- The analytical relations allow developing and solving a mixed integer-real optimization problem where the energy minimization is achieved under tunable reliability and delay requirements.
- Based on this optimization, a novel algorithm is developed to allow for rapid deployment and self-adaptation of the network to traffic variations and channel conditions, and to guarantee the application requirements without heavy computation or communication overhead.
- The protocol is implemented on a testbed using Tmote sensors (Moteiv 2006). Analysis and experimental evaluation show the benefits of the proposed solution.

3.3.2.1 System Setup

The system scenario is quite general, because it applies to any interconnection of a plant by a multihop WSN to a controller that tolerates a certain degree of data loss and delay (Zhang et al. 2001; Schenato et al. 2007; Witrant et al. 2007). Figure 3.14

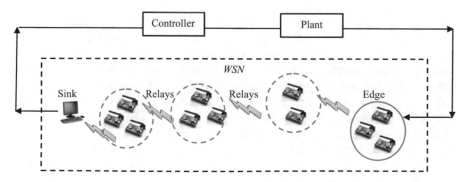

Fig. 3.14 WSN control loop

depicts a scenario where a plant is remotely controlled over a WSN (Hespanha et al. 2007; Schenato et al. 2007). The following assumptions and considerations are adopted in Breath:

- Outputs of the plant are sampled at periodic intervals by the sensors with total packet generation rate of λ packets/sec.
- It is assumed that packets associated to the state of the plant are transmitted to a sink, connected to the controller, over a multihop network of uniformly and randomly distributed relaying nodes. No direct communication is possible between the plant and the sink. Relay nodes forward incoming packets.
- The measurements received by a controller are used in a control algorithm to compensate the control output. The control law induces constraints on the communication delay and the packet loss probability. Packets must reach the sink within some minimum reliability and maximum delay. These boundaries are application requirements that are chosen by the control algorithm designers, they can be changed from one control algorithm to another, or a control algorithm can modify the application requests from time to time.
- Nodes of the network cannot be recharged, so the operations must conserve energy.

Sensor nodes uniformly distributed over the walls or the ceiling form a WSN infrastructure that supports control of the states of the robots in a manufacturing cell. A cell is a stage of an automation line; its physical dimensions range from 10 to 20 m on each side. Several robots cooperate in the cell to manipulate and transform the same production piece. The state of a robot is monitored by observing its vibration pattern; if the values of the vibrations are above a given threshold, a controller sends a control message to the robot. Hence, each node senses vibrations and reports the data to the controller within a delay. The decision-making algorithm runs on the controller, which is usually a processor placed outside the cell. Multihop communication is needed to overcome the deep attenuations of the wireless channel due to moving metal objects and to limit energy consumption.

3.3.2.2   The Breath Protocol

The Breath protocol groups all N nodes between the cluster of nodes attached to the plant and the sink with h-1 relay clusters. Data packets can be transmitted only from a cluster to the next cluster closer to the sink. Clustered network topology is supported in networks that require energy efficiency, since transmitting data through relays consumes less energy than routing directly to the sink (Heinzelman et al. 2002). In (Ma and Aylor 2004), a dynamic clustering method adapts the network parameters. In (Heinzelman et al. 2002) and (Younis and Fahmy 2004), a cluster header is selected based on the residual energy levels for clustered environments. However, the periodic selection of clustering may not be energy-efficient and does not ensure the flexibility of the network to a time-varying wireless channel environment. A simpler geographic clustering is instead used in Breath. Nodes in the forwarding region send short beacon messages when they are available to receive data packets. Beacon messages carry information related to the control parameters of the protocol. When a node receives a beacon message with the updated number of clusters h-1, then the node adapts to its cluster based on a rough knowledge of its location.

In the two coming sections, the protocol stack and state machine of Breath are described.

3.3.2.3 The Breath Protocol Stack

Breath uses a randomized routing, a CSMA/CA mechanism at the MAC, radio power control at the physical layer, and sleeping disciplines. In many industrial environments, the wireless conditions vary heavily because of moving metal obstacles and radio disturbances. In such situations, routing schemes that use fixed routing tables cannot provide the flexibility necessary for mobile equipment, physical design limitations, and reconfiguration characteristic of an industrial control application. Fixed routing is inefficient in WSNs due to the cost of building and maintaining routing tables. To overcome this limitation, routing through a random sequence of hops has been introduced in (Zorzi and Rao 2003). The Breath protocol is built on an optimized random routing, where next-hop route is efficiently selected at random, nodes route data packets to next-hop nodes randomly selected in a forwarding region. Randomized routing allows reducing overhead because no node coordination or routing state needs to be maintained by the network; it also considerably increases robustness to node failures.

The MAC of Breath is based on a CSMA/CA mechanism similar to the IEEE 802.15.4. Both data packets and beacon packets are transmitted using the same MAC. Specifically, the CSMA/CA checks the channel activity by performing clear channel assessment (CCA) before the transmission can commence. Each node maintains a variable NB for each transmission attempt, which is initialized to 0 and counts the number of additional backoffs the algorithm does while attempting the current transmission of a packet. Each backoff unit has duration T_{ca} msec. Before performing CCAs, a node takes a backoff of random $(0, W$-$1)$ backoff units, i.e., a random number of backoffs uniformly distributed over $0,1,\ldots, W$-1. If the CCA fails, i.e., the channel is busy, NB is increased by one and the transmission is randomly delayed $(0, W$-$1)$ backoff periods. This operation is repeated at most M_{ca} times, after which a packet is discarded.

Each node, whether transmitter or receiver, does not stay in an active state all time, but goes to sleep for a random amount of time, which depends on the traffic and channel conditions. Since traffic, wireless channel, and network topology may be time varying, the Breath protocol uses a randomized duty-cycling algorithm. Sleep disciplines turn off a node whenever its presence is not required for the correct operation of the network. GAF (Xu et al. 2001), Span (Chen et al. 2002), and S-MAC (Ye et al. 2004) focus on controlling the effective network topology by selecting a connected set of nodes to be active and turning off the rest of the nodes.

These approaches require extra communication, since nodes maintain partial knowledge of the state of their individual neighbors. In Breath, each node goes to sleep for an amount of time that is a random variable dependent on traffic and network conditions. Let μ_c be the cumulative wakeup rate of each cluster, i.e., the sum of the wakeup rates that a node sees from all nodes of the next cluster. The cumulative wakeup rate of each cluster must be the same for each cluster to avoid congestions and bottlenecks.

The Breath protocol assumes that each node has a rough knowledge of its location. This information, which is commonly required by the applications (Willig 2008), can be obtained by running a coarse positioning algorithm, or by using the received signal strength indicator (RSSI), typically provided by off-the-shelf sensor nodes (Texas Instruments 2005b). Some radio chips already provide a location engine based on RSSI (Texas Instruments 2005c). Location information is needed for tuning the transmit radio power and to change the number of hops. The energy spent in radio transmission has a tangible role in the energy budget and for the interference in the network. Breath, therefore, includes an effective radio power-control algorithm.

3.3.2.4   State Machine Description

Breath distinguishes between three node classes: edge nodes, relays, and the sink:

- The edge nodes wake up as soon as they sense packets generated by the plant to be controlled. Before sending packets, the edge node waits for a beacon message from the cluster of nodes closer to the edge. Upon the reception of a beacon, the node sends the packet.
- The detailed behavior of a relay node k is illustrated by the state machine of Fig. 3.15:
    - Calculate Sleep state. The node calculates the parameter μ_k for the next sleeping time and generates an exponentially distributed random variable having average $1/\mu_k$. Then, the node goes back to the Sleep state. μ_k is computed such that the cumulative wakeup rate of the cluster μ_c is ensured.
    - Sleep state. The node turns off its radio and starts a timer whose duration is an exponentially distributed random variable with average $1/\mu_k$. When the timer expires, the node goes to the Wakeup state.
    - Wakeup state. The node turns its beacon channel on, and broadcasts a beacon indicating its location. Then, it switches to listen to the data channel, and goes to the Idle Listen state.
    - Idle Listen state. The node starts a timer for a fixed duration that must be long enough to receive a packet. If a data packet is received, the timer is discarded, the node goes to the Active-TX state, and its radio is switched from the data channel to the beacon channel. If the timer expires before any data packet is received, the node goes to the Calculate Sleep state.

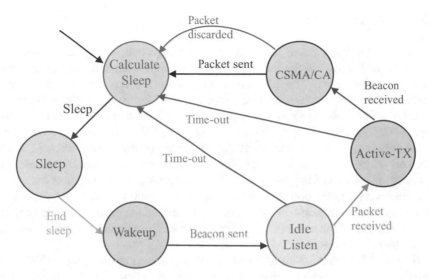

Fig. 3.15 State machine of a relay node executing Breath. (Park et al. 2011)

- Active-TX state. The node starts a waiting timer for a fixed duration. If the node receives the first beacon coming from a node in the forwarding region within the waiting time, it retrieves the node ID and goes to the CSMA/CA state. Otherwise, if the waiting timer expires before receiving a beacon, the node goes to the Calculate Sleep state.
- CSMA/CA state. The node switches its radio to hear the data channel, and it tries to send a data packet to a node in the next cluster by the CSMA/CA MAC. If the channel is not free within the maximum number of tries, the node discards the data packet and goes to the Calculate Sleep state. If the channel is free within the maximum number of attempts, the node transmits the data packet using an appropriate level of radio power and goes to the Calculate Sleep state.

- The sink node sends periodically beacon messages to the last cluster of the network to receive data packets. Such a node regularly estimates the traffic rate and the wireless channel conditions. By using this information, the sink runs an algorithm to optimize the protocol parameters. Once the results of the optimization are achieved, they are communicated to the nodes by beacons.

According to the Breath protocol, the packet delivery depends on the traffic rate, the channel conditions, the number of forwarding regions, and the cumulative wakeup time.

3.3.2.5 Results and Experimentation

Breath is a protocol based on a system-level approach to explicitly guarantee reliability and delay requirements in wireless sensor networks for control and actuation applications. The protocol considers duty cycle, routing, MAC, and physical layers all together to maximize the network lifetime, by taking into account the trade-off between energy consumption and application requirements for control applications. Analytical expressions are developed for the total energy consumption of the network, as well as for reliability and delay of the packet delivery. These relations allow to pose a mixed real-integer constrained optimization problem to optimize the number of hops in the multihop routing, the wakeup rates of the nodes, and the transmit radio power as a function of the routing, MAC, physical layer, traffic, and hardware platform. An algorithm is devised for the dynamic and continuous adaptation of the network operations to the traffic and channel conditions, and application requirements.

A testbed implementation of the protocol is provided by building a WSN with TinyOS and Tmote sensors. Experimentation was conducted to test the validity of Breath in an indoor environment with both AWGN and Rayleigh fading channels (Goldsmith and Varaiya 1997). Experimental results illustrate that Breath achieves the reliability and delay requirements, while minimizing the energy consumption. It outperformed a standard IEEE 802.15.4 implementation in terms of both energy efficiency and reliability. In addition, Breath reveals good load balancing performance and is scalable with the number of nodes. Given its satisfactory performance, Breath is a convenient candidate for many control and industrial applications, since these applications require both reliability and delay requirements in the packet delivery. A practical application of the protocol is disclosed in (Witrant et al. 2007).

3.3.3   Requirements, Drivers, and Analysis of WSN Solutions for the Oil and Gas Industry

Oil and gas industry extends in harsh landless and weatherless environments, working is hard, and research is stiff; life is non-friendly for humans, backbreaking even for bulky sturdy instruments and is definitely unlivable for miniature tiny feeble sensors. This explains the scarcity of published work on how WSNs may be involved in such industry. WSN in the oil and gas industry has been studied in (Carlsen et al. 2008; Petersen et al. 2008). The detailed study in (Petersen et al. 2007) will be the focus of this section.

The IEEE 802.15.4 specification has enabled low-power, low-cost WSNs capable of robust and reliable communication. The main objective of the work in (Petersen et al. 2007) was to investigate whether or not currently available IEEE 802.15.4-based WSN solutions fulfill the technical requirements for WSNs within the boundaries of the oil and gas industry. From an oil and gas standpoint, switching from wired to wireless sensors will enable a cost-efficient means to provide additional measurement points through the elimination of cables; furthermore, it will extend the reach of data collection into areas that are too remote or hostile for wires. Also, for temporary installations, the use of wireless technology will reduce costs related to installation, personnel, and equipment. However, there are concerns related to the use of WSNs, of which reliability, power consumption, and standardization are most important. Experiments have been performed to determine whether or not currently available technologies fulfill these requirements. The conclusion was that an open, and energy-efficient, standard is needed before WSNs can be fully utilized in the oil and gas industry.

This work was experimented in the laboratory facilities at Statoil's Research Centre in Trondheim, Norway. Statoil ASA (Statoil 2014) is a Norwegian-based integrated oil and gas company with international activities in more than 33 countries.

The physical environment of offshore and onshore oil and gas production facilities provides some interesting challenges for RF transmissions. Typical installations have several layers of process decks, constructed of reinforced steel and concrete. The steel cage structure of such decks is expected to demonstrate some degree of Faraday cage effects. Since deck space is a limiting factor, as much equipment as possible is placed in each process deck.

In addition to coping with the offshore steel environment, WSNs will also have to coexist with other typical offshore systems such as large power generators, UHF/VHF radios, radars, and safety and automation systems (SAS). However, early results from offshore spectrum analysis do not indicate any significant background noise in the 2.4 GHz frequency band. On the other hand, future deployments of WLAN (IEEE 802.11) and WiMax (IEEE 802.16) systems may change this picture.

3.3.3.1 Technical Requirements

A number of technical requirements have been identified by the oil and gas industry for competent WSN deployment.

3.3.3.1.1 Long Battery Lifetime

The oil and gas industry is pushing for battery lifetimes in excess of 5 years (at a 1-minute update rate) for wireless sensors. A key driver for this requirement is the maintenance effort needed to replace either the batteries or, the device, if the battery is encapsulated or otherwise non-rechargeable. An enormous effort, if affordable, is needed to maintain thousands of deployed wireless sensors.

3.3.3.1.2   Quantifiable Network Performance

The performance of WSNs is more susceptible to environmental changes than their wired counterparts. Thus, it is central to be able to reasonably quantify the expected and operational reliability and availability of wireless communication links and networks. Moving equipment or personnel and even fluctuations in temperature and humidity can influence the quality of a wireless link. Several techniques can be employed to improve the performance of a WSN, specifically, the use of redundant paths, self-healing formations, and link quality-aware nodes, which can deliver reliability and availability close to that expected from a wired system.
    Network performance quality also includes:

- Predictable behavior when the system is scaled up or down.
- Easy commissioning and engineering.
- Fail-safe operation in the event of intentional (e.g., jamming) and unintentional (e.g., interference or propagation problems) loss of wireless links.

3.3.3.1.3   Friendly Coexistence with WLAN

It is certain that WLAN is a vital technology within the oil and gas sector. The promotion of integrated operations models and the continuous requirements to connect offshore facilities with onshore experts is leading to necessities for mobility in the field. These services are delivered using WLANs; nevertheless, for WSN solutions to set foot in oil and gas industry, it is crucial that they coexist neatly with WLANs. This means no noticeable degradation in WLAN or WSN performance when operating within the same area.

3.3.3.1.4   Security

As wireless data are transmitted over the air, it is more susceptible to eavesdropping and security breaches than wired transmissions. The two most common threats are toward privacy and access. Encryption techniques are employed to protect the privacy of the data being communicated, while access threats are counteracted by using tools for transmitter authentication and data consistency validation.

3.3.3.1.5   Open Standardized Systems

The use of standardized, open communication protocols over proprietary protocols provides the industry with the freedom to choose between suppliers with guaranteed interoperability. Standardized solutions usually have a much longer lifespan than proprietary solutions. Furthermore, a standardized solution allows a single wireless infrastructure to deliver a communications medium to many devices, and potentially many applications.

On the other hand, as the creation of an international standard is a slow and time-consuming process, standardized solutions normally enter the market later than their proprietary counterparts. In addition, the security mechanism employed in standardized solutions is habitually published and obtainable, making standardized protocols somewhat more vulnerable to attacks than closed, proprietary systems.

As such, the need for long-term and technologically stable solutions makes open standardized systems highly preferable for the oil and gas industry, provided that the disadvantages are duly handled.

3.3.3.2   Proprietary Solutions Based on IEEE 802.15.4

Two proprietary solutions are available, SmartMesh (Linear Technology 2014) and SensiNet (Wireless Sensors, LLC 2011). Dust Networks have created time synchronized mesh protocol (TSMP), a time-driven (i.e., scheduled) solution based on guaranteed time slots (GTS) ensuring low-power and low-bandwidth reliable networking (Pister and Doherty 2008). A SmartMesh network consists of two types of devices: one network coordinator and up to 250 sensor nodes. Each sensor node is also a router, which enables a full-mesh network topology.

Sensicast Systems have a slightly different approach, using dedicated mesh routers and sensor nodes in a star-mesh network topology. Therefore, a SensiNet network consists of one network coordinator, and multiple mesh routers and sensor nodes.

Both solutions implement frequency hopping schemes; SmartMesh with a pseudo-random predetermined hop sequence and SensiNet with an adaptive algorithm with blacklisting of channels with low link quality. Due to the plenty of its details, SmartMesh will be presented in the coming section.

3.3.3.3   SmartMesh Experimentation and Interpretations

To investigate whether or not proprietary available WSN solutions are able to fulfill the technical requirements of the oil and gas industry, a SmartMesh network and a SensiNet network were deployed in the laboratory facilities at Statoil Research Centre. The laboratory contains real-size replicas of equipment used in Statoil installations, providing a test environment that is almost identical to an on-site installation.

Of the five technical requirements identified in Sect. 3.3.3.1, battery lifetime, quantifiable network performance, and coexistence with IEEE 802.11b/g were supposed suitable for laboratory investigation.

The SmartMesh software enables the user to extract summarized average network statistics (latency, stability and reliability) for 15-minute intervals. Latency is the average time it takes for a data packet from a sensor node to reach the network coordinator. Stability is the percentage of successfully transmitted data packets in the network (the inverse of packet loss). Reliability is the percentage of expected data packets that are received by the network coordinator. Due to retransmissions of lost data packets and the inherent redundancy in a mesh network topology, the reliability of a network can be high even when the stability is low. The SmartMesh network used for the network performance and coexistence tests consisted of one network coordinator and ten sensor nodes.

3.3.3.3.1 Network Performance

As shown in Fig. 3.16, the self-healing mechanism of the SmartMesh network can be observed by the slow increase in stability and latency occurring during the first few hours of the test. This is due to changes in network routing and formations related to path optimization and the creation of new communication paths involving more hops than the old ones. If the increased latency were due to packet loss, one would also observe a decrease in stability. For the latter part of the test, a stable network is achieved with 93–94% stability and 340–360 msec latency. The reliability of the network remained at 100% for the duration of the test.

From the accomplished tests, several suggestions are made to realize reliability and availability for a WSN in an oil and gas environment:

- Mesh networking allows for the creation of redundant paths, which combined with link-to-link acknowledge-based retransmissions, will ensure that data is not lost even if one or more nodes temporary lose their connection to the network.
- Self-healing algorithms are also an important feature, as the network will dynamically adjust routing paths to combat weak or missing links due to noise and interference. Using frequency hopping can also reduce the effects of frequency and time variant noise and interference.

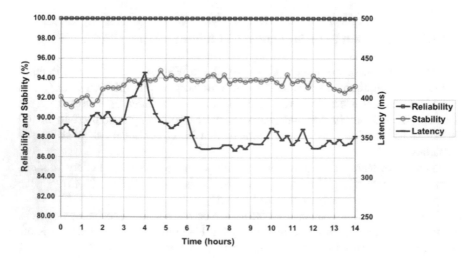

Fig. 3.16 SmartMesh network performance. (Petersen et al. 2007)

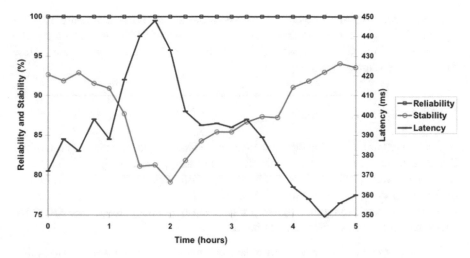

Fig. 3.17 SmartMesh coexistence statistics. (Petersen et al. 2007)

3.3.3.3.2 Coexistence with IEEE 802.11b

From Fig. 3.17, the IEEE 802.11b networks were activated after 1 hour and turned off shortly before 4 hours had passed. The outcomes of the experimentation can be outlined in what follows:

- When the IEEE 802.11b networks are enabled, the introduced interference from the 802.11b data traffic causes packet loss in the communication paths in the SmartMesh network. As a result, there are an increased number of retransmissions, which justifies the decrease in stability and increase in latency occurring as the 802.11b networks are activated.
- The TSMP protocol uses a self-healing algorithm, which at all times attempts to optimize the performance of the network. If a path suffers from high packet loss, alternate paths will be considered in order to increase to overall stability and decrease latency. The continuous self-healing capabilities of the SmartMesh network can be observed after approximately 2 hours into the test, lasting until the IEEE 802.11b networks are turned off. In this period, the stability increases gradually from around 79% to 87%, and the latency decreases from 450 msec to 375 msec.
- When the IEEE 802.11b networks are disabled, there is an immediate increase in stability and decrease in latency due to the sudden absence of the interference from the IEEE 802.11b network traffic.
- It is worth noting that the reliability of the SmartMesh network remained at 100% throughout the entire coexistence test.

Thus, friendly coexistence with IEEE 802.11b network(s) can be achieved by using adaptive frequency hopping to limit communication in the channels occupied by the IEEE 802.11b network(s). However, this is not a necessary feature, as the laboratory experiments showed that even with the non-adaptive frequency hopping scheme utilized by the SmartMesh network, the impact of IEEE 802.11b networks on a WSN is limited to a slight reduction in stability and increase in latency, while the reliability remained at 100%.

3.3.3.3.3 Power Consumption

In order to examine how the network topology affects the power consumption of the SmartMesh sensor nodes, two topologies are set: tree (hierarchical) and linear. Two test series were performed on each network topology, each series featuring a different reporting rate (6 sec and 60 sec). The calculations for the estimated battery lifetime are based on a typical high-capacity ANSI AA battery delivering 2250 mAh. Measurements revealed interesting findings:

- The sensor end nodes have deterministic behavior with equal power consumption characteristics.
- The power consumption of a sensor node depends on both the reporting rate and its number of children and grandchildren. A high reporting rate results in a high radio activity as sensor data are transmitted more often, and with higher radio activity, the power consumption is increased. As a sensor node is also a router, it has to relay sensor data from each of its children and grandchildren. The power consumption will thus increase for each of the node children and grandchildren.
- The topology of the children and grandchildren of a sensor node does not influence its power consumption. However, the power consumptions of each of the children and grandchildren are influenced by their relative topology.

As an inference, to ensure long battery lifetime of the sensor nodes, the following is proposed:

- Nodes must be able to enter a low-power sleep mode. The utilization of the sleep-mode feature requires the network to be completely time synchronized so that the sensor nodes know when, and for how long, they can sleep before they have to transmit or receive data.
- Battery life can be prolonged by employing efficient routing protocols that optimize network traffic, so that the number of hops and retransmissions are kept to a minimum.

The power consumption tests showed that current WSN technologies are not capable of fulfilling the oil and gas industry demand for battery lifetimes in excess of 5 years.

3.3.3.3.4 Security

The communication protocol of a WSN in an oil and gas facility must be secure, and as resistant as possible to eavesdropping and denial-of-service attacks. These security issues have been addressed in the 2006 edition of the IEEE 802.15.4 standard.

3.3.3.3.5 Open Standardized Systems

A WSN in oil and gas industry should be based on a standardized open solution in order to provide long-term and technologically stable systems with guaranteed interoperability between different vendors. Adequate open standards are found in the WirelessHART (Kim et al. 2008a; Song et al. 2008), the ZigBee Pro (ZigBee Alliance 2013), and the ISA SP100 (Costa and Jorge 2012).

3.4 Environmental Applications

Environment is defined as (Oxford Dictionaries 2014):

- The surroundings or conditions in which a person, animal, or plant lives or operates.
- The natural world, as a whole or in a particular geographical area, especially as affected by human activity.

According to this definition, environment-related applications comprise agriculture, farming, mining, seismology, climatology, volcanology, wildlife surveillance, and many others.

Environmental sensor networks (ESNs) facilitate the study of fundamental processes and the development of hazard response systems (Fig. 3.18). They have evolved from passive logging systems that require manual downloading, into intelligent sensor networks that comprise a network of automatic sensor nodes and communications systems, which actively communicate their data to a sensor network server (SNS) where these data can be integrated with other environmental datasets. The sensor nodes can be fixed or mobile and may range in scale and function appropriate to the environment being sensed. Following is the scale and function classification (Hart and Martinez 2006):

- Large-scale single-function networks. They tend to use large single purpose nodes to cover a wide geographical area.
- Localized multifunction sensor networks. They typically monitor a small area in more detail, often with wireless ad hoc systems.
- Biosensor networks. They use emerging biotechnologies to monitor environmental processes as well as developing proxies for immediate use.

It is envisaged that ESNs will become a standard research tool for Earth System and Environmental Science. Not only do they provide a virtual connection with the environment, they allow new field and conceptual approaches to the study of environmental processes to be developed.

Environmental monitoring applications can be broadly categorized into indoor and outdoor monitoring (Arampatzis et al. 2005):

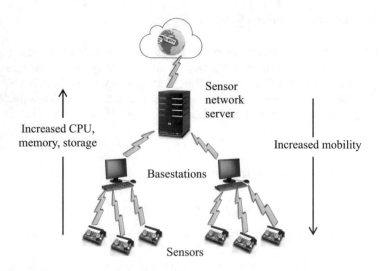

Fig. 3.18 ESN hierarchy

- Indoor monitoring applications typically include buildings and offices monitoring. These applications involve sensing temperature, light, humidity, and air quality. Other important indoor applications may include detection of fire and civil structures.
- Outdoor monitoring applications include chemical hazardous detection, habitat monitoring, traffic monitoring, earthquake detection, volcano eruption, flooding detection, and weather forecasting. Sensor nodes have also found their applicability in agriculture; soil moisture and temperature monitoring is one of the most important application of WSNs in agriculture.

When monitoring the environment, it is not sufficient to have only technological knowledge about WSNs and their protocols, and knowledge about the ecosystem is also necessary. Several projects, with real implementations, have focused on ESNs; the coming two sections will care for introducing some of these applications, conceptually overviewing for some and in illustrative details for an agriculture-related ESN application.

3.4.1   Assorted Applications

3.4.1.1   Large-Scale Habitat Monitoring

Berkeley's habitat modeling at Great Duck Island (GDI) analyzed bird-nesting habits, using camouflaged motes in the birds burrows (Szewczyk et al. 2004a). Two different motes were used: burrow motes for detecting occupancy using non-contact infrared thermopiles and temperature/humidity sensors, and weather motes for monitoring surface microclimates. Microclimate is a climatic condition in a relatively small area, within a few feet above and below the Earth's surface and within canopies of vegetation. Microclimates are affected by such factors as temperature, humidity, wind and turbulence, dew, frost, heat balance, evaporation, the nature of the soil and vegetation, the local topography, latitude, elevation, and season. Weather and climate are sometimes influenced by microclimatic conditions, especially by variations in surface characteristics (Merriam Webster 2014). GDI application comprised 147 Berkeley MICA2DOT sensor nodes equipped with TinyOS (Crossbow 2002b). Readings from sensor nodes are periodically sampled and relayed from the local sink node to a basestation on the island. The basestation sends the data using a satellite link to a server connected to the Internet.

3.4.1.2   Environmental Monitoring

In SECOAS project (Britton and Sacks 2004), a sensor network was deployed at Scroby Sands Wind Farm, off the coast of Great Yarmouth, with the purpose to monitor the impact of a newly developed wind farm on coastal processes. New sensor hardware based on MCU PIC 18F452 (Microchip 2000) was developed in this project, and a new operating system, kOS was adopted (Sacks et al. 2003). The sensor nodes register pressure, turbidity, temperature, and salinity. Sensor nodes, basestations on sea and land stations, form the hierarchical and single hop network. Nodes transmit their data to the sea basestations, which will then transmit the data to the land station. Basestations are sensor nodes equipped with additional functionalities, more power supplies and larger communication range. Data are Internet accessed from the land station.

3.4.1.3   Precision Agriculture

Measurement of the microclimate in potato crops is the main goal of Lofar-Agro project (Baggio 2005; LOFAR 2014). Based on the circumstances within each individual field, the collected information helps improving the decision on how to combat phytophthora within a crop. Phytophthora is a fungal disease in potatoes; their development and attack of the crop strongly depend on the weather conditions within the field. A total of 150 sensorboards, namely TNOdes, very similar to the Mica2 motes from Crossbow are installed in a parcel for monitoring the crop. The nodes are manually localized so that a map of the parcel can be created. The TNOdes are equipped with sensors for registering the temperature and relative humidity.

Earlier deployments have shown that the radio range is dramatically reduced when the potato crop is flowering. To maintain sufficient network connectivity, 30 sensorless TNOdes act as communication relays. To further improve communication, the nodes are installed at a height of 75 cm while the sensors are installed at a height of 20, 40, or 60 cm. In addition to the TNOdes, the field is equipped with a weather station registering the luminosity, air pressure, precipitation, wind strength, and direction. Since humidity of the soil is a major factor in the development of the microclimate, a number of sensors that measure soil humidity are deployed in the field. Finally, an extra sensor measures the height of the groundwater table. A TNOde records the temperature and relative humidity every minute.

For energy efficiency considerations, the nodes are reporting data only once per 10 minutes. To further save energy, the data sent over the wireless links are minimized by using delta encoding. The T-MAC protocol (Van Dam and Langendoen 2003) cares for energy efficiency as well and imposes on the radio a duty cycle of 7%. The TNOdes use TinyOS as operating system. Data are thus sent using the multihop routing protocol MintRoute available within TinyOS. In addition, the nodes are reprogrammable over the air using Deluge (Deluge 2008). The data collected by the TNOdes are gathered at the edge of the field by a so-called field gateway and further transferred via Wi-Fi to a simple PC for data logging, the Lofar gateway. The Lofar gateway is connected via wire to the Internet and data are uploaded to a Lofar server and further distributed to a couple of other servers under XML format.

3.4.1.4   Macroscope in the Redwoods

The WSN, macroscope, enables dense temporal and spatial monitoring of large physical volumes (Tolle et al. 2005). Some refer to sensor networks as "macroscopes" because the dense temporal and spatial monitoring of large volumes that they provide offers a way to perceive complex interactions. This work conducted in Sonoma California presents a case study of a WSN that recorded 44 days in the life of a 70-meter-tall redwood tree, at a density of every 5 minutes in time and every 2 m in space. Each node measured air temperature, relative humidity, and photosynthetically active solar radiation. The network of MICA2DOT TinyOS nodes (Crossbow 2002b) captured a detailed picture of the complex spatial variation and temporal dynamics of the microclimate surrounding a coastal redwood tree.

3.4.1.5   Active Volcano Monitoring

A team of computer scientists at Harvard University collaborated with volcanologists at the University of North Carolina, the University of New Hampshire, and the Instituto Geofísico in Ecuador (Werner-Allen et al. 2006). Studying active volcanoes typically involves sensor arrays built to collect seismic and infrasonic (low-frequency acoustic) signals. In August 2005, they deployed a network on Volcán Reventador in Northern Ecuador. The array consisted of 16 nodes equipped with seismo-acoustic sensors deployed over 3 km. The system routed the collected data through a multihop network and over a long-distance radio link to an observatory.

The WSN 16 stations are equipped with seismic and acoustic sensors. Each station consisted of a Moteiv Tmote Sky node (Moteiv 2006), an 8 dBi 2.4 GHz external omnidirectional antenna, a seismometer, a microphone, and a custom hardware interface board. Every one of 14 nodes was fitted with a Geospace Industrial GS-11 geophone (Geospace Technologies 2014a) a single-axis seismometer with a corner frequency of 4.5 Hz, vertically oriented. Each of the two remaining nodes was equipped with triaxial Geospace Industries GS-1 seismometers (Geospace Technologies 2014b) with corner frequencies of 1 Hz, yielding separate signals in each of the three axes.

The Tmote Sky, designed to run TinyOS (TinyOS 2012), is a descendant of the University of California, Berkeley Mica mote sensor node. The Tmote Sky is chosen because its MSP430 microprocessor (Texas Instruments 2011b) provides several configurable ports that easily support external devices, and the large amount of flash memory is useful for buffering collected data. Over three weeks, the network captured 230 volcanic events, producing useful data that permit to evaluate the performance of large-scale sensor networks for collecting high-resolution volcanic data.

3.4.1.6 Sensor and Actuator Networks on the Farm

Agriculture has two cares, plants and cattle. Managing farms, particularly large-scale extensive farming systems, is hindered by lack of data and increasing shortage of labor. To address these issues, (Sikka et al. 2006) propose the deployment of a large heterogeneous sensor network on a working farm to explore sensor network applications. The network is solar powered and has been running for over 6 months. The implemented deployment consisted of over 40 moisture sensors that provide soil moisture profiles at varying depths, weight sensors to compute the amount of food and water consumed by animals, electronic tag readers, up to 40 sensors that can be used to track animal movement (consisting of GPS, compass, and accelerometers), and 20 sensor/actuators that can be used to apply different stimuli (audio, vibration, and mild electric shock) to the animal. The static part of the network is designed for 24/7 operations and is linked to the Internet via a dedicated solar-powered high-gain radio link.

The initial goals of the deployment are to provide a testbed for sensor network research in programmability and data handling while also being a vital tool for scientists to study animal behavior. The longer-term aim is to create a management system that completely transforms the way farms are managed. Also, in (Huircán et al. 2010), a localization scheme in WSNs for cattle monitoring applications in grazing fields is designed. No additional hardware was required for distance estimation since operation is based on the link quality indication (LQI), which is a standard feature of the ZigBee protocol. Experimental results have shown acceptable localization performance at low-cost and little power consumption. In the sections to come, A^2S planting system is detailed.

3.4.1.7   Cultural Property Protection

In (Sung et al. 2008), WSNs are deployed in Bul-guk-sa Temple, which is one of the most important UNESCO cultural property sites in Korea. Such WSN has two objectives, periodic environmental information collection to monitor any changes in wooden structures, and detecting fire in the surrounding forest outside the wooden temple. Collected data from the surrounding environment include temperature, humidity, light, pressure, flame and carbon monoxide (CO). A software system is developed; it utilizes sensor data on top of the ANTS-EOS (evolvable OS) sensor network operating system (Kim et al. 2005a) and ANTS series of sensor node hardware. The sensor node hardware is classified into two, precinct sensor node and perimeter sensor node. The precinct sensor nodes are used to gather and provide essential information such as temperature, humidity, and pressure in order to check conditions of wooden buildings. These nodes are located inside the temple, and send sensory data over a RF communication.

The processor on the single board is an Atmel ATmega128L 8-bit microcontroller (Atmel 2011a), which is used in tiny sensor nodes. It has 128 KBytes of Flash memory, a 4 KBytes SRAM, and a 4 KBytes EEPROM inside the processor. The RF transceiver is a Chipcon CC2420 (Texas Instruments 2005b) that operates in the 2.4 GHz frequency band. Three types of sensors are used, temperature/humidity sensor, pressure sensor, and light sensor. Since such sensors require low-power consumption, a battery without external power supply is used. The outside sensor nodes consist of separate processor boards and a sensorboard with a 41-pin connector. The processor is an Atmel ATmega128L 8-bit microcontroller, which is the same as the precinct sensor node. The RF transceiver is a Chipcon CC1100 (Texas Instruments 2005a), a low-power single-chip UHF transceiver. It can be easily programmed and configured for operation at frequencies in the 300–348 MHz, 400–464 MHz, and 800–928 MHz bands. To alleviate the power requirements of the sensor and processor boards, a 4.5 V 140 mA solar cell is installed.

Since the temple is located at the center of a deep forest and the mission of the network is to detect and alert of fires that may spread from outside the temple, the suitable topology is to encircle the temple with connected nodes. To achieve reliability and fast delivery of the event data, the network was divided into two separated string-shape networks, each of which is assigned to specific communication channels. Each network was consisting of 16 sensor nodes and had a dedicated relay station that is connected to a basestation using CDMA connections. Three WSNs consisting of 30 sensor nodes were deployed around a real cultural property, and operated for 6 months.

3.4.1.8   Underground Structure Monitoring

Environment monitoring in coalmines is an important application of WSNs. In (Li and Liu 2007), the design of a structure aware self-adaptive WSN system (SASA) is proposed. By regulating the mesh sensor network deployment and formulating a collaborative mechanism based on a regular beacon strategy, SASA is able to rapidly detect structure variations caused by underground collapses. A prototype is deployed with 27 TinyOS Mica2 motes (Crossbow 2002a).

3.4.1.9 Foxhouse Project

Foxhouse (Hakala et al. 2010) project, jointly implemented by The Kokkola University Consortium Chydenius (later Chydenius) and MTT Agrifood Research Finland, gets real-time information about the habitat of foxes in a foxhouse. The WSN for environmental monitoring was implemented in the Fur Farming Research Station at Kannus. The amount of light received is presumed to be the key factor in stimulating reproduction of foxes, so measuring light intensity in different parts of the foxhouse was the focal point of interest. Measurement data for luminosity as well as temperature and humidity were gathered outdoors over a period of 1 year. Also observed are the functionality and usability of the network; some tools for network maintenance were developed during the project.

The WSN produces measurements data, which is delivered to the gateway (sink node). The sink node is connected to a PC server via an RS232 cable. The PC server runs a Java program for reading packets from the serial port. All received information is stored into an SQL database in the PC. Data in the database can be browsed by a web application. Cluster topology appeared to be the most suitable topology for this project; the network nodes included two kind of devices, reduced function device for sensing (RFD) and full function device for routing (FFD).

The WSN in the Foxhouse case consisted of 14 nodes in two clusters, the front cluster and the rear cluster. The nodes used in the Foxhouse network were CiNet nodes. CiNet is a research and development platform for WSN implemented by Chydenius. The hardware is the IEEE 802.15.4 compatible CiNet node specially designed for WSNs and consisting of inexpensive standard off-the-shelf components; it has an ATmega 128 MCU (Atmel 2011a) and a transceiver on board as well as one temperature sensor for testing purposes. More sensors are needed in real monitoring. These sensors can be placed on a special sensorboard, which can be connected to the main board. The sensing nodes are equipped with temperature, humidity, and light sensors.

3.4.1.10 SensorScope for Environmental Monitoring

SensorScope (Ingelrest et al. 2010) is such an environmental monitoring system based on a time-driven WSN, developed in collaboration between two laboratories at EPFL (École Polytechnique Fédérale de Lausanne), LCAV (signal processing and networking), and EFLUM (hydrology and environmental fluid mechanics). SensorScope is a turnkey solution for environmental monitoring systems, based on a WSN and resulting from a collaboration between environmental and network researchers; it aims at providing high-resolution spatio-temporal data for long periods of time.

The mote is a Shockfish TinyNode (Dubois-Ferrière et al. 2006) composed of a Texas Instruments MSP430 16-bit microcontroller running at 8 MHz (Texas Instruments 2011b), and a Semtech XE1205 radio transceiver, operating in the 868 MHz band, with a transmission rate of 76 Kbps (Semtech 2008). The mote has 48 KByte ROM, 10 KByte RAM, and 512 KByte Flash memory. This platform is mainly chosen for the good ratio it offers between communication range and power consumption (up to 1200 m outdoors for 60 mA). The power source is composed of three modules. The first is a 162 × 140 mm MSX-01F polycrystalline module solar panel that provides a nominal power output of 1 W in direct sunlight, with an expected lifetime of 20 years (BP Solar 2014); such low-power solar energy system achieves power autonomy during deployments. The second is a primary 150 mAh NiMH rechargeable battery, NiMH battery is chosen over a supercapacitor due to its higher capacity and its lower price, and it allows for up to 5 days of solar blackout (considering a networking duty cycle of 10%). The third is a secondary Li-Ion battery with a capacity of 2200 mAh. The sensing station is composed of a two-meter-high flagstaff, to which the solar panel and up to seven sensors are fixed.

Seven outdoor deployments were conducted, ranging in size from 6 to 97 stations and from EPFL campus to high-up in the Alps. The measured quantities were air humidity and temperature, precipitation, soil moisture, solar radiation, surface temperature, water content, wind direction, and speed.

3.4.1.11   A Biobotic Distributed Sensor Network for Under-Rubble Search and Rescue

The Cyborg Insect Networks for Exploration and Mapping (CINEMa) project at North Carolina State University targets establishing the fundamental physical and algorithmic building blocks of a sensor network for under-rubble environments (Bozkurt et al. 2016). After a disaster, the survival of victims trapped under-rubble mostly depends on how quickly they are found, rescued, and treated. First responders use search-and-rescue techniques such as canines, listening devices, and radars, which search the rubble's near-surface regions. However, complementary techniques are needed to penetrate the smaller gaps and cavities deep in the rubble. Distributed systems are indisputably superior for tasks such as exploration, mapping, and large-area sensing. Recent achievements in swarm robotics, a new multirobot coordination-system approach, have been inspired thru observations of emergent collective behaviors among insects. In swarm robotics, multiple smaller-scale robots interacting with one another and the environment could coordinate their actions to help manage the under-rubble environment's uncertain and dynamic conditions.

Imagining that a major earthquake hits an urban environment and that several survivors are trapped under the collapsed ruins of a 20-story building. Time is short and limited tools are available for removing debris, a sequence of actions is threaded till a rescue is made:

- A special first responder team arrives with a set of insect-sized robots, which they drop at the edge of the rubble pile. These robotic agents carry tiny radios on their backs that, together with other sensors, they measure the distance between the agents to localize them with respect to a few reference points.
- As the robotic agents crawl through the rubble, their task is to keep moving while staying within a certain distance of one another to maintain the radio network. The agents' tiny environmental sensors, including microphones and infrared or gas sensors, monitor the rubble for dangerous gas leaks or victims' cries for help.
- The swarm moves collectively from one end of the rubble to the other. During this sweeping action to find survivors, each robotic agent's location is used to map the under-rubble environment.
- At last, one of the agents locates a dubious sound signal 30 m deep, but it is almost impossible to transmit this information outside the pile of concrete and steel to the first responders. So, the autonomous radio network determines a multihop route, which allows the information to be sent from one agent to another, all the way to the mission control center outside.
- Mission control establishes a radio link with the survivor, assesses his health situation, and informs him that help is on the way. The first responders determine the shortest route to the victim and concentrate their digging efforts in that direction.

3.4.1.11.1   Mobile Sensor Nodes and Biobotic Agents

In under-rubble reconnaissance, the success of distributed robotic systems depends highly on the robotic agents' capability to cope with the uncertain and dynamic environmental conditions that make navigating through rubble notoriously difficult. Although a number of centimeter-scale insect-like robots have been exhibited successfully (Wilson et al. 2015), current technology falls short in offering mobile robotic agents that function effectively under these adverse conditions. Until synthetic robots become practical for use in disaster situations, an alternative solution is to instrument real insects for this purpose (Bozkurt et al. 2009). Cockroaches, for example, exhibit an unmatched ability to navigate narrow passages under-rubble, climb rough surfaces in any direction, and maintain control and stability during perturbations.

The CINEMa project is built on the latest neural engineering and neuromuscular stimulation techniques to exploit the locomotory advantages of insects as biobotic search-and-rescue agents (Fig. 3.19). From the 12 cockroach species reared in North Carolina State University's Entomology Department, the Madagascar hissing cockroach (Gromphadorhina portentosa) is selected as the model insect (GBIF 2016). Its relatively large size and slower speed allow for larger payload capacities and easier

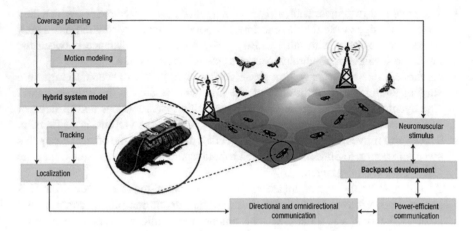

Fig. 3.19 System-level overview of the CINEMa project. (Bozkurt et al. 2016)

biobotic manipulation of its locomotion[2] (Dictionary.com 2017). This kind of species is also relatively easy to rear and maintain and is commercially available.

Several locations within the cockroaches' peripheral nervous system can be stimulated to bias their natural locomotory behavior. The antennae are chosen as the target location for neuro-stimulation. Because the optical cues detected with their eyes might not be processed fast enough, cockroaches naturally formulate their escape responses to avoid obstacles by using their antennae's tactile guidance (Camhi and Johnson 1999).

3.4.1.11.2 Biobotic Control Demonstrations

Fine wire electrodes are surgically implanted into the antennae to apply stimulation pulses wirelessly through a backpack with an RF link. By sending right- and left-turn commands via a manual remote controller, a human operator can guide the insect biobot along an S-shaped line pattern drawn on the floor (Latif and Bozkurt 2012). To automatically and objectively evaluate the insect biobots, a test platform is constructed using a Microsoft Kinect sensor[3] (TechTarget 2017; Whitmire et al. 2013). A PC

---

[2]Locomotion is the movement of an organism from one place to another, often by the action of appendages such as flagella, limbs, or wings. In some animals, such as fish, locomotion results from a wavelike series of muscle contractions.

[3]Kinect is Microsoft's motion sensor add-on for the Xbox 360 gaming console. The device provides a natural user interface (NUI) that allows users to interact intuitively and without any intermediary device, such as a controller. The Kinect system identifies individual players through face recognition and voice recognition. A depth camera, which "sees" in 3D, creates a skeleton image of a player and a motion sensor detects their movements. Speech recognition software allows the system to understand spoken commands and gesture recognition enables the tracking of player movements. Although Kinect was developed for playing games, the technology has been applied to real-world applications as diverse as digital signage, virtual shopping, education, telehealth service delivery, and other areas of health IT.

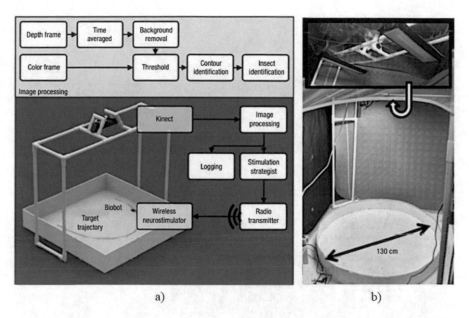

Fig. 3.20 Biobot evaluation platforms. (**a**) Kinect-based automated biobot evaluation platform. (**b**) Anechoic chamber in which the biobot evaluation experiments are performed. (Bozkurt et al. 2016)

connected to Kinect locates the insects using the video feed and automatically steers the insects along a predetermined test path by sending neuro-stimulation pulses wirelessly (Figs. 3.20 and 3.21). The Kinect program, by providing infrared-based depth images, also allows conducting experiments in the dark, when the insects are more active.

Using this setup, the insects' biobotic capabilities were tested in a maze environment; their task was to follow a defined path between the start and finish points (Fig. 3.21). The walls and corners of the maze served as an extra challenge to distract the insect biobot from completing the task, because cockroaches are naturally attracted to cool, dark areas, and tend to stay close to wall corners (Latif et al. 2014). This demonstrates that the biobots could complete the assigned tasks despite external environmental factors and their corresponding natural behavioral response.

Because insects are astonishing at navigating the small gaps and cavities under-rubble, a useful biobotic search strategy is to keep the insects moving naturally and in random directions within a defined region. Holding the biobot's position within a particular region of interest and gradually moving the location of this region enable the biobotic swarm to scan the entire pile of rubble. Virtual fences are defined in the test platform, much like the invisible fences designed to keep pets in a yard (Latif et al. 2014). The Kinect sensor-based computer program detects the insect's position with respect to the virtual perimeter. Using a PC, the insect is turned around through a set of automated stimulation pulses (Figs. 3.20 and 3.21). In real-life scenarios, localization technologies (Sect. 3.4.1.11.5) will replace the Kinect sensor to guide

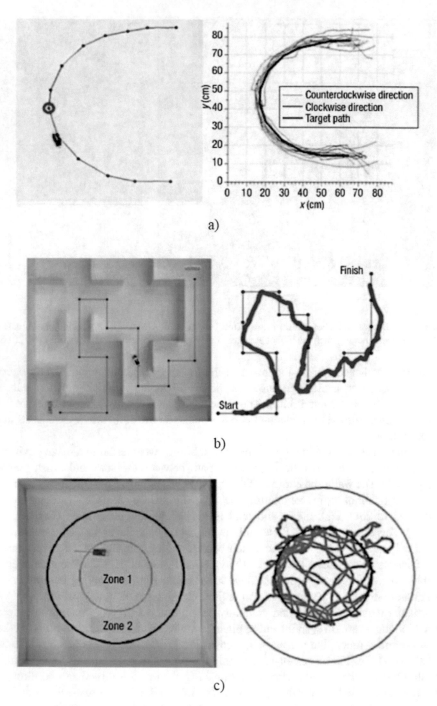

Fig. 3.21 Sample results of three automated Kinect-based evaluation experiments. For the invisible fence, the Kinect-connected computer commands the insect biobot to make a U-turn when it is outside the boundary of Zone 1 to ensure that it will always stay in Zone 2. (**a**) U-line in an empty arena. (**b**) Maze arena with erected walls. (**c**) Invisible fence in an empty arena. (Bozkurt et al. 2016)

insect biobots to regions of interest. This capability is also critical to ensure that the biobots stay within the reception range of one another's radios to maintain the communication network.

3.4.1.11.3 Backpack Technologies for Biobots

For neuro-stimulation-based locomotion control, environmental sensing, localization of biobotic agents, and communication with first responders, a miniaturized wireless electronic backpack was developed (Figs. 3.19 and 3.22). To minimize the cost of a multiagent swarm, a Texas Instruments CC2530 is adopted; it is a suitable commercial off-the-shelf system-on-chip (SoC) solution that combines analog, digital, and mixed signals with RF functions on a single substrate level (Texas Instruments 2011a). As described in Chap. 10, CC2530 is tailored for applications such as IEEE 802.15.4 ZigBee and ZigBee RF4CE[4] (Zigbee Alliance 2017); it combines an 8051 microcontroller with a high-performance RF transceiver while providing 8 KByte RAM and up to 256 KByte Flash memory. This solution is adequate for the CINEMa particular application due to the availability of 21 general purpose I/O pins, and eight channel 12-bit analog-to-digital converters for connecting to external devices.

Experiments indicate that Gromphadorhina portentosa biobots have a payload-carrying capacity up to 15 g (Latif et al. 2016). The assembled backpack weighs 300–800 mg without batteries, depending on the number of sensors used. Based on the experiment duration, the pickup was a lithium-polymer battery with a 90 mAh capacity and 2 g weight, or a 20 mAh capacity and 400 mg weight. The backpack's power consumption ranges from 5 to 75 mW, based on the duty cycling of the radio transmission and the number of sensors deployed. Also implemented as shown in Fig. 3.22, backpacks with solar-charging capabilities; using the aforementioned invisible-fence strategy, the biobots can be guided to a bright light source, which recharges the batteries in 10–30 minutes (Latif et al. 2014).

3.4.1.11.4 Sensors for Distributed Sensing and Localization

A miniature custom microphone array was mounted on the biobot backpack to allow detection and localization of sound sources to autonomously localize victims under-rubble (Fig. 3.22). Three-directional microphones were connected to the backpack and the signal was sampled at a rate nearing 2 kHz. As Fig. 3.23 illustrates, the sound's relative intensity is used to determine sound direction; the biobot is then

---

[4]The ZigBee RF4CE specification offers an immediate, low-cost, easy-to-implement networking solution for control products based on ZigBee remote control and ZigBee input device. The ZigBee RF4CE specification is designed to provide low-power, low-latency control for a wide range of products including home entertainment devices, garage door openers, keyless entry systems, and many more.

Fig. 3.22 CINEMa experimental setups. (**a**) Unidirectional backpacks for sound localization. (**b**) Omnidirectional backpacks for sound localization. (**c**) Miniature solar cells used to charge the backpack batteries. (**d**) Interrupting the experiments. (Bozkurt et al. 2016)

autonomously steered toward the sound source. Furthermore, voice-transmission capabilities are demonstrated by replacing this three-microphone directional array with a single omnidirectional micro-electromechanical systems microphone using a 6.25 kHz sampling rate (Latif et al. 2016).

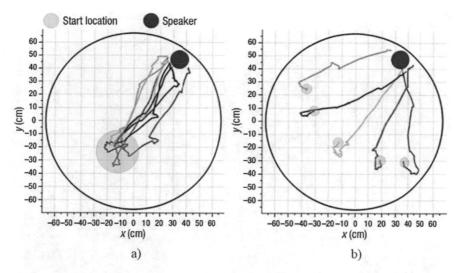

Fig. 3.23 Sample trajectories resulting from automated steering. A miniature custom microphone array mounted on the biobots backpacks allows detection and localization of sound sources. (**a**) Varying orientations from the same start location. (**b**) Varying start locations and an orientation away from the speaker. (Bozkurt et al. 2016)

For agent localization, extra sensors are added to the backpack to supplement the received signal strength indicator (RSSI) inherent within the ZigBee-enabled SoC. These sensors include a six-axis inertial measurement unit, a three-axis magnetometer, and a microphone-buzzer couple for acoustic time of flight measurements (ToF)[5] (TeraRanger 2017).

3.4.1.11.5 Localization Technologies and Algorithms

Localizing an under-rubble survivor in absolute coordinates involves two subproblems, localizing the survivor relative to the biobot, and localizing the biobot in either an absolute or relative frame of reference, as below illustrated:

- To localize a survivor relative to a biobot, an array of microphones is employed for finding the direction of a signal's arrival. By comparing the sound's signal strength at the three microphones, it was possible to reliably establish the direction of a sound source.
- The second part of the problem is localizing the biobots in the pile of rubble. GPS is unlikely to work because of signal attenuation through the rubble, as well as the

---

[5]Time-of-flight principle is a method for measuring the distance between a sensor and an object, based on the time difference between the emission of a signal and its return to the sensor, after being reflected by an object. Various types of signals, also called carriers, can be used with ToF, the most common being sound and light.

Fig. 3.24 Localization and communication system that relies on anchor nodes on the surface and on range measurements among biobots above and below rubble. (Bozkurt et al. 2016)

heavy weight and high-power budgets. The proposed distributed solution relies on range measurements among neighboring biobots, and among biobots and fixed reference points (anchors) placed around the perimeter of the rubble pile (Fig. 3.24). It is assumed that the anchors can localize themselves in an absolute frame of reference using GPS or other methods; their main purpose in the system is to help localize the biobots in the rubble and to provide communication to those biobots. For range measurements, two technologies are considered, RSSI and ultrasonic ranging:

– RSSI is available in virtually every transceiver on the market today; availability and communication range make it a very attractive option. RSSI's main disadvantage is inaccuracy; for Zigbee transceivers, indoor RSSI measurements might vary by as much as 20 dB while maintaining a fixed distance between the sender and receiver, even when having a line of sight (Xiong et al. 2016). Multipath fading is likely the main source of this variation. In rubble, RSSI variability is expected to be higher.
– Ultrasonic ranging involves measuring the propagation time of an ultrasonic signal from a transmitter to a receiver. To eliminate the need for network-wide synchronization, a packet can be transmitted to indicate to the receiver when the ultrasonic pulse was sent, thus allowing measurement of propagation time, which can be accurately translated to distance because of sound's relatively slow speed in air. The acoustic measurements rely on the biobots' buzzers and microphones. The main advantage of ultrasonic range measurements is their significant precision as submillimeter accuracy is possible. The main

disadvantages are their range, especially under-rubble, and their need for additional hardware, which entails additional weight and power consumption. Fortunately, the speakers and microphones can be reused for detecting and communicating with survivors.

Devising a localization method that can provide the biobots' location given the range measurements is a tricky challenge. Although several research have been published on localization in WSNs, their approaches are generally not appropriate for disaster conditions. To begin with, any approach relying on fingerprinting is excluded, because prior calibration is impossible in most disaster scenarios. Also, centralized approaches are unsuitable, because the network might be loosely connected or even disconnected at times, while the biobots' mobility implies that the localization process must be repeated periodically. Furthermore, the extremely limited resources on the biobots reveal that the approaches relying on significant amounts of communication or computation are also unsuitable, thus arises the need for fully distributed, low-computation, localization approaches that are resilient to range measurement errors. Errors in range measurements will inevitably result in inaccurate localization; luckily, tens of centimeters of error appear tolerable for this project.

3.4.1.11.6   Mapping and Exploration Strategies

One of the biobotic swarm's essential tasks is to create a map of the environment indicating which areas have been explored and determining the location of individuals requiring assistance. This task becomes extremely challenging in disaster scenarios because of hardware shortcomings, such as energy, locomotion, and communication bandwidth limitations, and the unstructured nature of the environment. Power and computational resource constraints prohibit using continuous control schemes for the agents' locomotion and exclude exploiting onboard imaging techniques for their localization. Furthermore, because the location could be indoors or even underground in cluttered environments, signal propagation-based localization techniques, such as GPS, signal strength computation, or time of flight, might be unreliable; also, odometry[6] (FOLDOC 2006)-based information might include a high amount of uncertainty due to irregular terrain conditions. Therefore, traditional mapping and exploration techniques for simultaneous localization and mapping will not perform well under these adverse conditions.

Since obtaining an accurate metric map of the environment might not be possible in disaster scenarios, there is a need to construct topological maps of unknown

---

[6]Odometry is the use of data from motion sensors to estimate change in position over time. It is used in robotics by some legged or wheeled robots to estimate their position relative to a starting location. This method is sensitive to errors due to the integration of velocity measurements over time to give position estimates. Rapid and accurate data collection, instrument calibration, and processing are required in most cases for odometry to be used effectively.

environments using biobotic mobile sensor networks under the constraint of limited sensing information. The proposed approach entails several considerations:

- Extracting a sketch of the environment rather than a fully detailed map. This sketch includes robust topological information obtained from a minimal amount of sensing (Dirafzoon and Lobaton 2013).
- Instead of providing continuous control feedback to the agents, there is an exploration of how the insects natural behavior as modeled using stochastic motion, and on how the weak encounter information in the form of identification of nearby agents, can be exploited for mapping.
- Making up for the platforms' hardware limitations, tools are employed from algebraic topology. It is thus possible, without need for localization data, to extract spatial information about the environment based on neighbor-to-neighbor interactions among the agents. This information is used to build a map of the environment's persistent and robust topological features.

The adopted topological mapping approach assumes no information about the agent location. Instead, only encounters between nearby agents are recorded, together with the involved agents unique IDs and the time at which the event occurred. The encounter information is used to construct a graphical structure that captures approximate metric information about the environment, and to extract robust topological features from the environment, such as the number of connected components and holes in the space (Dirafzoon et al. 2014). Robustness is quantified using persistence diagrams that capture the birth and death of topological features. This approach is validated using simulation as well as real robotic platforms. Such simple characterization of the space can become a building block for mapping larger environments.

The CINEMa project has the potential to provide the infrastructure for a robust, flexible, mass-producible, and self-sustaining mobile sensor network for reconnaissance and survivor localization. Moreover, it provides a testbed for understanding the interfaces formed with insects by modeling individual and collective responses to various stimuli for further cyber-physical biobotic applications. The technologies and algorithms developed for search-and-rescue biobots, as well as the new findings on insect locomotory biology, can be applied to future centimeter-scale synthetic robotic swarms that physically and behaviorally mimic insects.

3.4.1.12 Efficient Data Collection and Tracking with Flying Drones

Data collection is an important mechanism for WSNs to be viable in applications such as environmental monitoring or surveillance. Within the context of the Internet of mobile things, the focus is on observing a set of mobile targets (sensors), using a fleet of unmanned aerial vehicles (UAVs), the flying drones. Specific applications can gain significant benefits from this scenario, such as wildlife monitoring, vehicle observation and tracking in smart cities, and monitoring sporting events. The flying drones form a wireless mobile backbone, covering the targets on the ground and

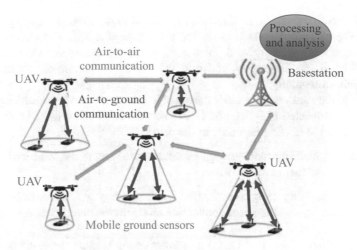

Fig. 3.25 Multitier network architecture for the ADCP. (Caillouet et al. 2019)

collecting their information. Gathered data are continuously sent to a basestation for further processing via multihop communication through the wireless backbone. The problem, named the aerial data collection problem (ADCP), can be described as an optimization of spacial and temporal coverage with mobile flying drones; it is a complex time and space coverage, and connectivity problem (Caillouet et al. 2019).

Several objectives motivate the aerial data collection problem (ADCP) from a set of mobile wireless sensors located on the ground, using a fleet of flying drones:

- Deploying a set of unmanned aerial vehicles (UAVs) in a 3D space, to cover and collect data from all the mobile wireless sensors, at each time step through a ground-to-air communication.
- Sending these data to a central basestation using multihop wireless air-to-air communication through the network of UAVs.
- Minimizing the total cost (communication and deployment) over time.

The use of mobile flying devices to cover mobile ground targets has become an important topic in the past few years. It allows the monitoring system to optimally adapt to the evolving space and time placement of the ground targets, as well to the changing requirements of the application. Therefore, deployment efficiency and cost can be jointly optimized for a wide adoption of wireless mobile monitoring systems. Figure 3.25 shows the multitier network architecture composed of mobile sensors, flying drones, and a fixed basestation. The functioning follows several principles:

- The sensor nodes are located on the ground and produce data. Their mobility pattern is usually unknown.
- Data are then gathered by a fleet of UAVs whose mobility is fully controlled in order to track the sensor nodes. Each ground sensor must be covered at any time by at least one UAV for ground-to-air communications.

- Each UAV moves in a 3D-space and its altitude must be managed for the coverage and data collection. When a UAV is at a high altitude, the covered area is wider on the ground. However, altitude increases the distance between UAVs and ground sensors and thus reduces the ground-to-air communication transmission quality.
- Finally, data gathered by UAVs are sent to a basestation via a multihop connection. A connected backbone of UAVs and the basestation must be formed and maintained at anytime to transmit the data.

Regarding mobility management of sensor nodes, there are three main mobility categories (Caillouet et al. 2019):

- Random mobility (not controllable). In this category, results cannot provide guarantees on data gathering/collection and solutions are often subject to long delays.
- Partially controllable mobility. Trajectories and/or stop positions are constrained. This category is often environment specific, for example, bus trajectories in a city; noticeably, data gathering/collection might suffer high delays.
- Fully controllable mobility. It widens the application usage of WSNs. Focusing on fully controllable mobility is beneficial, since it makes possible the computation of the sensor nodes trajectory that optimizes the collection process.

3.4.2 A^2S: Automated Agriculture System Based on WSN

A^2S (Yoo et al. 2007) consists of WSNs, gateways, and a management subsystem. Twenty-five sensor nodes, one actuator node, and three sink nodes are deployed in greenhouses and operated during harsh cold weather; one of the sensor nodes was plunged in a field near one of greenhouses to endure heavy snows at -15 °C coldness. Three industrial PC-based gateways are installed to transform RS-232 data from sink nodes to TCP/IP server data. WLAN access points (APs) with directional antennas provide the long-range wireless link between WSNs and the management subsystem located at 0.5 km from the greenhouses. The management subsystem manages the WSNs, and provides easy interface to farmers equipped with hand-help devices such as personal data assistants (PDAs).

3.4.2.1 System Architecture

A^2S comprises several functional components and performs a multiplicity of tasks (Fig. 3.26):

- A-node. It is an Agriculture sensor node intended to be deployed in a greenhouse to sense its environment. The A-node embeds in a printed circuit board (PCB) board, an 8-bit microcontroller (MCU), an IEEE 802.15.4 compatible transceiver

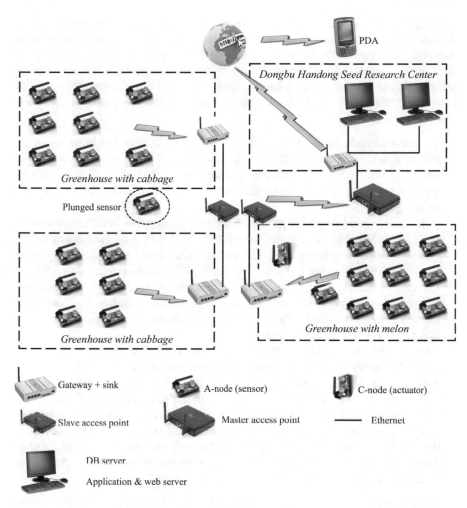

Fig. 3.26 Automated agriculture system architecture

(2.4 GHz band), a complex programmable logic device (CPLD) for a sleep timer to wake up the MCU from power-down mode (Xilinx 2014), and sensors (ambient light, temperature, and humidity). A sensor node is equipped with a lithium-ion rechargeable cells battery whose voltage level is monitored for maintenance purpose. The A-node software is based on the initial version of ANTS-EOS (Kim et al. 2005b). EOS is a lightweight C-based multithreaded operating system developed to support multiple WSNs platforms.

After setting up the network topology, an A-node runs its application software. The application software begins its active period by turning on its sensors, it reads sensed temperature, humidity, and luminance of a greenhouse and reports via A-node parents to the sink. If during its active time the A-node receives packets from its children, it relays them to its parent. After transmitting its sensed data, an A-node waits for another work schedule, sent from the sink, in a sleep order message. The application software then turns off its attached sensors and puts the transceiver to power-down mode; finally, it sets up the internal sleep timer to the sleep period. After timer expiration, the application software restarts its next active periods by turning on the transceiver and the sensors to pursue sensing the greenhouse environment.

- C-node (aCtuator node). Designed to control the illumination intensity of the growing melon greenhouse, it has an additional relay board to A-node to control the light switches in the greenhouse. The application software of C-node waits for a command from the sink, once received, it controls the relay to turn lights on or off.
- Sink node. It is developed to gather the sensing information from A-nodes and to transmit commands to A-nodes and C-node. As the core component of the gateway, the sink node has an additional interface board to A-nodes to provide RS-232 serial link to the gateway. The main MCU module of the sink node has the same hardware specification of A-nodes, except that the sink node is not equipped with sensors. The sink node gets the sensing schedule from the application server and schedules the operation of the internal A-nodes by sending sleep order message every sensing period.
- The gateway. It transforms RS-232 sink data to TCP/IP server data and vice versa. The gateway is implemented on a Pentium-M 1.6 GHz industrial PC. The gateway is connected to AP via a WLAN link or a wired Ethernet link. The AP is connected to the management subsystem 0.5 km distant via WLAN link.
- The management subsystem. It consists of a DB server, an application server, and a web server. The application server receives data from WSN and stores them in the DB server, and configures the WSNs. The sensing schedule is configured by the application server. The whole sensing data are stored in the DB server and can be accessed by users (PC or PDA) via the web server.

A lightweight CSMA-based MAC protocol and a robust multihop ad hoc routing protocol are implemented. When an A-node has a packet to send, it checks the channel. If the channel is idle the packet is transmitted. If there is no acknowledgment from the recipient of the packet, the MAC layer retransmits the packet up to three times. To simplify the route discovery, a tree level which is the same as the hop count from the sink node is preprogrammed in an electrically erasable programmable read-only memory (EEPROM) chip with a unique 16-bit network address.

To conserve power, the order-based sleep scheme is used. When the sensing schedule (sensing period) is set (or ordered) by the application server, the sink node keeps the schedule and spreads the sleep order message over its network every sensing period. Whenever an A-node receives the sleep order message, it sets the expiration time of its sleep timer to the value of the duration field included in the message. When the timer expires, the A-node senses its ambient environment and the voltage level of its battery and sends the data to the sink, and waits for the next sleep order message. A-nodes are placed at a priori planned positions decided by agriculturists in accordance with the cultivation requirements.

3.4.2.2 Experimentation Results

A^2S was operated for 1 month in severe cold winter. From this real deployment, an interesting experience was acquired:

- Line-of-sight (LOS) communication range of A-nodes was up to 70 m. However, the range was reduced to 30 m in greenhouses because of interfering sources such as iron wires and foliage.
- Originally, an A-node was planned to be awaken from the sleep mode by an external timer in onboard CPLD to reduce drawing current. However, the power consumption of the CPLD was found relatively high. Also, some of the sensed data was lost due to battery exhaustions of some sensor nodes. It is recommended to resort to a real-time clock using an internal timer and an external low-frequency crystal oscillator in order to minimize the power consumption.
- Although the order-based sleep scheme does not require complex time synchronization schemes, the parent node should be awaken earlier than the child node to relay the sensing data to the child node. Noteworthy, because the sleep order message is spread out from sink to nodes level-by-level, the parent node goes to sleep mode earlier than the child node.
- Despite the use of accurate sensors, they did not show the same output levels in the same place. It is found that there were interferences from some components such as power-control ICs in the PCB and some components such as batteries in the same enclosure. The recommendation is thus to isolate the sensors from the other interfering components in the PCB and the enclosure.
- The comparison between Korean Meteorological Administration (KMA) standard and WSN measurements revealed a difference up to 4.5 °C and an average of 2.7 °C. Calibrating the sensors leads to more accurate results.

3.4.3 Living IoT: A Flying Wireless Platform on Live Insects

Living IoT is a novel general purpose wireless sensing platform that is low-power, lightweight, and can support computing and communication operations on flying insects like bees (Iyer et al. 2019). In order to meet the bees stringent size and weight requirements, a backscatter-based communication system can be achieved with commercially available microcontrollers. The platform is compatible with insects by its small and lightweight form factor. The design considers the capability to compute as well as sample onboard sensors. Furthermore, backscatter offloads power expensive components to an access point (AP) setup near the hive, enabling low-power operations.

3.4.3.1 Why Live Insects?

Mobility in WSNs has the potential to modernize agriculture by enabling smart farming applications including precision irrigation (Sect. 3.4.2) and environmental sensing (Lee et al. 2010). Sensor mobility significantly reduces the overhead of manual sensor deployment and upkeep, which poses a major barrier to adoption of smart farms. Drones have thus far been the platform of choice for enabling mobility; yet, they are severely power constrained and last for only 5–30 minutes on a single charge due to the energy density limits of existing battery technologies (Thackeray et al. 2012). This is mainly because the motors drones use for mechanical propulsion and control are power-consuming (Escobar-Alvarez et al. 2018).

Thus, exploring the idea of creating mobile WSNs by placing them on live insects offers a worthy alternative. Using live insects such as bees is attractive for two key reasons:

- Flight Time. Unlike drones, flying insects use chemical energy stored in fats and carbohydrates, which have a much higher energy mass density than batteries. This allows for much longer flight times. Flies have been shown to fly for hours without food (Dethier 1976), while worker bees spend most daylight hours foraging for nectar and pollen (Dukas and Visscher 1994) and can fly while carrying payloads of over 100 mg (Hagen et al. 2011). Further, flies evolved to have aerodynamic and musculoskeletal systems that minimize power usage (Dudley 2002).
- Ubiquity. Insects are nearly ubiquitous across the planet and adapted to live in diverse ecosystems, making it easy to find a species well suited for a particular environment or application. Moreover, while some are regarded as pests, others are essential to human activities. For example, bees are needed to pollinate many commercial crops, and are in many cases intentionally introduced for that purpose.

Piggybacking on these insects enables mobility for WSNs applications including smart farms. Using live insects like bees however introduces two key challenges (Iyer et al. 2019):

- They are physically small and can only carry small payloads, which severely limits possibilities for power, computation and communication.
- The flight of small insects like bees cannot be controlled.

The smart farming techniques utilize a variety of sensors to measure plant health. For example, moisture and humidity sensors measure the availability of water, light sensors measure the availability and intensity of sunlight, and temperature sensors can help determine whether conditions are optimal for particular crops. Live insects like bees present an attractive choice for use as mobile sensing platforms for agriculture. Bees are among nature best pollinators and are regularly purchased for use on farms; many fruit crops depend on bees for pollination. These insects fly for hours foraging for food and also fly up to individual plants, which is difficult to do with drones. Further, tracking bees could give important insights about pollination, which is not obtainable from commercial sensors; this includes pollination patterns that can help maintain genetic diversity (Brooks and Flynn 1989).

3.4.3.2 Self-Localization of Insects

To address the lack of flight control, insects are to localize themselves in the 2D space using RF signals transmitted by APs. Such self-localization architecture, similar in spirit to GPS, is attractive because it does not require the AP to estimate the locations and send them back to each bee. Consequently, it can scale well with the number of insects and work at high speeds without requiring the bees to transmit signals. More importantly, this enables location-based mobile sensors where the insect can associate location information with its sensing data as well as perform sensing operation only when they pass over the desired set of locations. The sensor data can then be uploaded to the AP when the bee returns within range of the hive.

Achieving self-localization on the proposed living IoT platform is however challenging for three key reasons:

- It should be low power in nature as it is not possible to run power-hungry radios on the tiny batteries that small insects, such as bees, can carry.
- Accurate wireless tracking relies on phase information requiring a radio receiver at the bee (Vasisht et al. 2016), a challenging task because radios are power-consuming. Further, there is a lack of knowledge about small low-power off-the-shelf radios that provide I/Q samples[7] (NI 2019) of the raw RF signals or CSI data[8] (ScienceDirect 2020).
- Existing localization algorithms are designed for Wi-Fi chips (Vasisht et al. 2016) or software radios that are not constrained by their computational capabilities (Shangguan and Jamieson 2016). In contrast, self-localization at the bee requires algorithms that can operate with a tiny antenna and a low-power microcontroller.

3.4.3.3 Living IoT Project Design

The proposed design eliminates power-hungry radio receiver components, e.g., high-frequency oscillators, by using passive operations to perform envelope

---

[7] I/Q data show the changes in magnitude (or amplitude) and phase of a sine wave. If amplitude and phase changes occur in an orderly, predetermined fashion, these amplitude and phase changes can be used to encode information upon a sine wave, a process known as modulation.

[8] In wireless communications, channel state information (CSI) refers to known channel properties of a communication link. This information describes how a signal propagates from the transmitter to the receiver and represents the combined effect of, for example, scattering, fading, and power decay with distance. The CSI makes it possible to adapt transmissions to current channel conditions, which is crucial for achieving reliable communication with high data rates in multi-antenna systems.

a)   b)

Fig. 3.27 Living IoT platform on a bumblebee. (**a**) Bee carrying platform. (**b**) Bee flying with platform. (Iyer et al. 2019)

detection and extract just the signal amplitude. While this design enables the insect to receive at a low power, it discards the phase information, which is essential for RF localization.

The novel technique extracts the angle to the AP from the amplitude information output by the envelope detector. The AP broadcasts signals to the insects from two of its antennas; by changing the relative phase between the two transmit antennas at the AP, amplitude changes can be created at the insect envelope detector over time. These amplitude changes effectively give multiple equations that allow solving for the angle to each AP. Combining the angle information from two APs allows the platform to localize itself on a farm using a passive envelope detector. A low-complexity algorithm is designed to work in the presence of multipath as well as at speeds up to 9.1 m/s.

Figure 3.27 shows the hardware including the antenna, backscatter communication, wireless receiver for self-localization and the circuit board that connects the microcontroller and sensors in a lightweight form factor using microfabrication techniques. The 102 mg platform, including a 70-mg battery, is attached to three common species of bumblebees, namely, Bombus impatiens, Bombus vosnesenskii, and Bombus sitkensis (BugGuide 2020).

The contributions of this project are conceptual, technical, and algorithmic:

- Exploring the idea that insects can be used in lieu of drones to explore WSN mobility.
- Presenting a novel general purpose platform that is lightweight and can support computing, communication, and sensing operations on flying insects.
- Introducing the first self-localization technique for small insects, like bees, using a novel algorithm that computes 2D location using only the output of a passive envelope detector.

Two deployment scenarios are envisaged:

- The sensors periodically wake up from a low-power sleep mode to sample sensor data and location, store the data in memory, and then return to sleep mode. As the insect goes about its business foraging for food, samples of data are obtained along its trajectory. This is ideal for low-power sensors such as temperature and humidity that simply require storing a single value.
- There are applications where it is required to acquire more detailed information at a particular location that is programmed when the bees are on the hive. In this case, the system could periodically check its location, and opportunistically sample the sensor if it is close to the destination.

The idea of using flying insects as a mobile sensing platform raises a number of wireless networking and sensing challenges. At a high level, this requires an ultra-miniature sensing and computation package, memory to store data, power source, and wireless connectivity for data transfer. Additionally, it requires a wireless localization method capable of connecting the sensor measurements to specific locations.

The proposed solution consists of two components; explicitly, a lightweight battery-powered electronics package that can be mounted on a flying bee. The package includes a sensor, RF switch used for backscatter communication and microcontroller, and multiantenna access points with dedicated power sources that broadcast the RF signals needed for backscatter and localization. Since large bumblebee hives cover foraging areas of roughly 8000 m^2 (Boyden 1982), an operating range of 50–100 m is targeted for localization. The range requirements for communication however are much smaller as sensor data can be stored in onboard memory and uploaded when the bee returns to the hive. Figure 3.28 illustrates how the insects communicate via the APs.

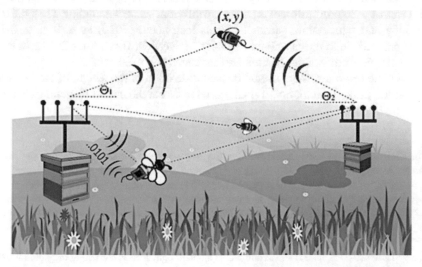

Fig. 3.28 Insert-borne sensor packages. Insect-borne sensor packages can self-locate and collect location-dependent data using onboard sensors. Data are uploaded to AP when the bee is back at the hive. (Iyer et al. 2019)

To fulfill such scenario, the bee-mounted electronics package necessitates the following two components:

- Low-power self-localization. The first requirement is a localization algorithm that runs on the insect-mounted platform and computes its location based on wireless transmissions from APs at known locations. By adopting this broadcast architecture similar to GPS, any number of insects can concurrently compute their location. A novel algorithm is designed; it uses a passive envelope detector to receive the RF signal and compute the location on a microcontroller.
- Communication. A form-factor compatible wireless communication link to download sensor data. The backscatter approach has the advantage of requiring only an antenna connected to RF switches and a microcontroller, all of which can be achieved using off-the-shelf components.

Understanding the form-factor requirements is crucial in this project. Bumblebees are chosen as they can fly while carrying payloads of their own body weight or more. The platform is 102 mg, including 70 mg for the 3 V 1 mAh rechargeable lithium-ion battery. This allows only 32 mg for communication, computing, sensing, and self-localization.

The platform consists of four different elements in ultra-miniature packages, namely, a microcontroller, RF switches, an envelope detector, and sensor. The core of the design is the Kinetis KL03z ARM Cortex M0+ microcontroller[9] (NXP 2020), which is available in a 2×1.61 mm package and weighs only 4.1 mg. This microcontroller is used to sample the output of the envelope detector and the sensor and to toggle the RF switches for backscatter communication. Also, there are two Skyworks 13314-374LF single pole dual throw switches[10] (Skyworks 2020) weighing 3.3 mg each; the first to select between the envelope detector and backscatter, and the second to toggle between the two backscatter impedances. The envelope detector is constructed out of small diodes and capacitors consuming a total area of 7.26 mm^2. Moreover, there are different sensors including TI HDC2010 humidity and temperature sensor[11] (Texas Instruments 2020) as well as an ALS PT19 photodiode to measure light intensity[12] (Everlight Electronics 2013). In total, the whole platform weighs 102 mg and measures 6.1×6.4 mm^2.

Not to be forgotten, there should be a consideration of the weight of the substrate such as the printed circuit board (PCB) used to create the circuit that connects these

---

[9] The Kinetis® KL0x-48 MHz MCU family is the entry point into the Kinetis L series based on the Arm® Cortex®-M0+ core.

[10] The SKY13314-374LF is a pHEMT GaAs I/C SPDT antenna switch operating in the 0.1–6 GHz frequency range. Switching between the antenna and ports is accomplished with 2 control voltages. The low loss, high isolation, high linearity, small size, and low cost make this switch ideal for all dual-band WLAN systems operating in the 2.4–2.5 GHz and 4.9–5.9 GHz bands.

[11] ±2% ultra-low-power, digital humidity sensor with temperature sensor in WCSP.

[12] The ALS-PT19-315C/L177/TR8 is a low-cost ambient light sensor, consisting of phototransistor in miniature SMD.

Fig. 3.29 Electronic package. (Iyer et al. 2019)

components. A lightweight 100-µm thick PCB is constructed by laser micromachining, as illustrated in Fig. 3.29.

3.4.3.4   Realized Outcomes

The acquired results disclose the following:

- Communication. The microcontroller-based backscatter design, placed on the bee, can transmit bits using ON/OFF keying at 1 Kbps, when the bee returns to the hive.
- Self-localization. Across deployments in a soccer field and farm, the envelope detector design on the bees achieves median angular resolutions of 4.6° at ranges up to 80 m. Ground truth benchmarking with drones show that similar angular resolution can be achieved even at speeds of 9.1 m/s.
- Power. With the onboard 70 mg rechargeable battery, the designed system could last up to 7 hours while sampling its location once every 4 seconds. There is also the feasibility of fully recharging the battery at the hive within 6 hours, using RF power.
- Sensing. Prototypes including humidity, temperature, and light intensity sensors are built. These sensors fit within the devised 102 mg prototype, enabling mobile sensing using bees.

This project is not the first to place electronics on insects; prior work in biology uses electronics on bees to understand their foraging behavior (Kissling et al. 2014). However, it is the first time that insects such as bees are used to carry general purpose sensors, localize themselves, and perform location-based sensing operations. Furthermore, it explores the use of insects in lieu of drones. Making this vision pervasive, however, requires addressing additional challenges:

- Weather dependency. Bees hibernate during winter in cooler climates. This however correlates with plant growth and increased activity on farms in warmer seasons.
- E-waste. Piggybacking on insects could lead to electronic waste being scattered around the farms when the insects eventually consume their lifespan. There are three approaches to address this problem:
  - Ensuring that the electronics are removed before the expected lifetime of the bees.
  - Localizing the electronics for a while even after the bees die, which can be used for cleanup purposes.
  - Using biodegradable electronics in the design of living IoT (Hwang et al. 2012).
- Fabrication. The current prototype requires manual fabrication and attachment to the insects. Using commercially available parts allows for easy scaling of the electronics fabrication. Also, gluing the electronics to the insects is similar to the process of attaching tracking markers to bees in commercial hives. Additionally, researchers have shown that insects can survive common microfabrication processes such as deposition of conductive material in a vacuum chamber (Shum et al. 2007) and performing surgeries at different stages of an insect lifecycle (Reissman and Garcia 2008). This suggests potential approaches for mass attachment of electronics or fabricating devices on insects themselves.
- Camera sensing. A future research direction is to integrate cameras (Naderiparizi et al. 2018) with the Living IoT platform. This can be useful for smart farm applications like canopy monitoring. Centeye image sensors (Centeye 2020) and cameras such as the Himax HM01B0[13] (Himax 2015) offer a potential for achieving such a camera-based sensing system within the weight/power budget.
- IACUC requirements. While working with insects and other invertebrates is not governed by the Institutional Animal Care and Use Committee (IACUC) policies, the best has been done to follow the three Rs of animal research, namely, the replacement, reduction, and refinement (Russell and Burch 1959). The use of insects is minimized thru benchmarking each aspect of the proposed system, using drones to simulate flight and robots to simulate flapping wings. For the experiments, no surgical modification was performed; simply, weight is attached to the exoskeleton. No significant changes are observed in behavior after the procedure; additionally, only a small number of insects were used for experiments. After experimentation, electronics package was removed, and wild-caught insects were released back in the area where they were captured.

---

[13] The HM01B0 is an ultra-low power CMOS Image Sensor that enables the integration of an "Always ON" camera for computer vision applications such as gestures, intelligent ambient light and proximity sensing, tracking and object identification. The unique architecture of the sensor allows the sensor to consume very low power of <2 mW at QVGA 30 fps.

3.4.4   Learning from Researching and Trialing

Despite careful design and concerns about possible deployment issues, many problems arise. Learning from one's own experience and from those of others shortens distances and avoids pitfalls and stumbling. This section outlines hands-on experience outcomes, by describing problems and how they may be solved and avoided, and it is worthy to have them shared with the WSN community (Ingelrest et al. 2010).

3.4.4.1   Hardware and Software Development

3.4.4.1.1   Consider Local Conditions

It is not always obvious how, possibly drastic, variations in temperature and humidity will affect hardware devices, so that a lack of testing under real conditions may lead to serious issues. For instance, Li-Ion battery should not be charged when the temperature is below freezing, as it could explode. It is therefore crucial to simulate the anticipated conditions as accurately as possible. Studying the impact of weather conditions may be done by using a climate chamber, in which arbitrary temperature/humidity conditions are created. In most cases, basic tests inside a household freezer will expose potential points of failure.

3.4.4.1.2   Sensor Packaging

Outdoor packaging is difficult, as it must protect electronics from humidity and dust while being unobtrusive (Szewczyk et al. 2004b). International Ingress Protection (IP) codes are used to specify the degree of protection for electrical enclosures (The Engineering ToolBox 2014). The required level for outdoors is IP67, which provides full protection against dust and water, up to an immersion depth of 1 m. Any lesser degree of protection exposes electronics to humidity and atmospheric contaminants, potentially leading to irreparable damages. Corrosion may cause the malfunction of a sensor connector, consequently corrupting the data from that sensor. Even more disastrous, humidity may cause a short circuit in the connector, resulting in permanent damage and/or continuous rebooting of the affected station.

3.4.4.1.3   Keep It Small and Simple

To avoid as much as possible unexpected interactions between software components, Protocols must be well-fitted to the application (Buonadonna et al. 2005). Sometimes, complexity cannot be avoided, but whenever the benefits are questionable, simple solutions should be preferred. For instance, when stations are equipped with a solar panel, an overall positive energy balance is sufficient to achieve long-term autonomy, which helps avoiding complex, ultra-low-power MAC layers, generally requiring high-precision synchronization, which may be difficult to achieve in realistic conditions. Furthermore, because of channel degradation, packet losses with such complex protocols are more likely to occur in harsh conditions (e.g., heavy rain).

3.4.4.1.4 Think Embedded

On a computer, code is easy to debug, using debugging statements or tools. It is more difficult with sensor motes, as the simplest way for them to communicate with the outside world is by blinking their LEDs or using their serial port. These interfaces are not only limited, but also mostly untraceable once a network is deployed. Moreover, embedded programs are more often subject to hardware failures, so that their behavior can be incorrect, even if the code itself is actually fine (Szewczyk et al. 2004b). It is thus important to be practically able to determine what happens inside the network.

3.4.4.1.5 Get All Data You Can

The primary goal must be to develop a working system and to succeed in collecting environmental data. Issues related to publishing obtained measurements to the networking community must go along successful deployments. It is important to gather data related to network conditions, not just environmental conditions. Once a network is deployed, it becomes too late to think about such issues. By planning early on what data are useful for networking issues, like performance analysis, the code to gather needed data can be incorporated into the development process.

3.4.4.1.6 Data that Is Useful

A successful deployment consists not only of gathering data, but also of exploiting it. A WSN exists to transport data from one point (i.e., the targeted site) to another one (e.g., a database), but there is no purpose in gathering data just for the sake of it. The final objective of a WSN deployment is to gather data for an end user. This end user must be present in all the stages of the deployment preparation, from sensor selection, placement, and calibration, to data analysis (Tolle et al. 2005).

3.4.4.2 Testing and Deployment Preparation

3.4.4.2.1 Check for Interferences

When setting up a deployment, the first priority should be to inspect the radio spectrum to detect possible interferences. The optimal way is to use a spectrum analyzer, but due to its size, weight, and power consumption, it can be difficult to use at the deployment site. A simpler way is to run a test program to determine losses over time. There are actually a lot of radio devices that can create interference, which compromises the results and leads to thinking that the code is incorrect.

3.4.4.2.2 Data You Can Trust

Although sensors should be precalibrated, some manipulations (e.g., packaging) can affect their measurements (Buonadonna et al. 2005). Sensors, once packaged, should be tested before deployments, first indoors, then outdoors. Readings are to be compared to high-precision reference stations over several days, and bad sensors must be discarded. Calibration may also be required at the time of deployment; an example of this is the wind direction sensor, which must be North-oriented to provide accurate data. Once a deployment is over, and sensors are back at the laboratory, it is important to repeat the calibration process.

3.4.4.2.3 Be Consistent

At some point, one may be tempted to change some parameters or to switch to new drivers, just before a deployment, to improve a given aspect. With new versions, however, always come new bugs, and it is by far easier to detect them on a testbed rather than during a deployment. The exact same configuration should thus be used during both tests and real deployments. Another possible issue is the "last minute commit," which can kill a complete deployment (Langendoen et al. 2006).

3.4.4.3 Deployments

3.4.4.3.1 Consider Local Conditions—Once Again

Some bugs can be hard to spot before the real deployment, because they do not occur under normal testing conditions. For instance, cellular connectivity may be good on campus but rather poor or non-existing outside, potentially leading to malfunction. Also, the crystal of the sink's mote and the crystal of the general packet radio service (GPRS) chip react differently to temperature variations. Thus, knowing that communications between the mote and the GPRS occur over a serial bus, the drift caused by temperature changes should result in a loss of synchronization between the mote and the GPRS chip, and consequently in packet loss. Moreover, manipulating electronics outdoors must be kept to a minimum to avoid bad weather and dust.

3.4.4.3.2   Get a Watchdog

To promptly detect problems and malfunctions, all data must be scrutinized as soon as they reach the server. However, while some incorrect measurements may be easy to detect, other problems may be subtler such as those due to bugs in a sensor driver.

3.4.4.3.3   Keep all Data

On the back-end server, there will be various programs processing data; nevertheless, all the raw data must be securely archived for future reference. For instance, there may be some needed statistics that were not envisioned at first. Also, one may discover that the equation used to transform raw data into the international system of units (SI) was poorly implemented; if the original data are no longer available, the obtained data may be worthless.

3.4.4.3.4   Data You Can Interpret

Gathering data is a step forward, but interpreting it is a lot of work and reasoning. Sensors provide only a partial view of the real world, which may be insufficient to correctly rationalize their readings. For instance, whether readings are collected while raining or snowing may pose question marks. To better understand gathered data, equipping one or more stations with a camera is likely to provide meaningful visual feedback.

3.4.4.3.5   Traceability

As the software on both server and motes will evolve over time, traceability becomes more and more important. Traceability of individual measurements is required, for instance, when a bad sensor is detected, it is common practice to exchange it. Without any further provision, it is impossible to determine which values from previous deployments should be double-checked. Tagging motes and sensors with radiofrequency identification readers (RFIDs) is a solution. With the corresponding reader, it is possible to scan stations during deployments to associate sensors and stations. Storing this information in a database allows retracing the exact history of all devices and measurements.

3.5   Healthcare Applications

The medical applications of WSNs aim to improve the existing healthcare and monitoring services especially for the elderly, children, and chronically ill. Numerous benefits are achieved with these systems (Alemdar and Ersoy 2010):

- Remote monitoring capability. It is the main benefit of pervasive healthcare systems. With remote monitoring, the identification of emergency conditions for at-risk patients will become easy and the people with different degrees of cognitive and physical disabilities will be assisted to have a more independent and easy life. The little children and babies will also be cared for in a more secure way while their parents are away. Consequently, the special dependability on caregivers will be lessened.
- Real-time identification and action taking. In healthcare applications, a real-time system is actually a soft real-time system, in which some latency is allowed (Shin and Ramanathan 1994). Identifying emergency situations like heart attacks or sudden falls in a few seconds or even minutes will suffice for saving lives considering that, without real-time systems these conditions will not be identified at all. Hence, providing real-time identification and action taking in pervasive healthcare systems are among the main benefits.

The technology advancements in consumer electronics have reduced the production costs and have made it possible to afford inexpensive sensor devices for ordinary users as well. Together with the mature and also inexpensive RFID technology, the costs for pervasive healthcare systems are within the affordable range for many people. In Caregiver's Assistant (Philipose et al. 2004), inexpensive RFID tags are placed on household object and the systems precision can be increased, at very low costs, by tagging more objects with these RFID tags.

- Being able to identify the context is another benefit achieved with pervasive healthcare systems. Context awareness enables understanding the conditions of the people to be monitored constantly and the environments in which they are. This context information is achieved mostly by sensing systems that incorporate more than one type of sensing capabilities. By fusing the information gathered by several sensors, a more clear understanding of the context may be obtained. The context information helps better in identifying the unusual patterns and making more precise inferences about the situation. For instance, during night-time, being in the sleeping room in a lying position may not indicate something serious, whereas lying down in the sleeping room in the middle of the day may indicate an alarm situation. Context awareness provides this useful information.

An array of prototypes and commercial products for WSN-based healthcare is available. When several applications are investigated, it is observed that they have common properties. Precisely, most of the existing solutions include one or more types of sensors carried by the patient, forming a body area network (BAN), and one or more types of sensors deployed in the environment materializing a personal area network (PAN). BAN and PAN are connected to a backbone network via a gateway node. At the application level, the healthcare professionals or other caregivers can monitor the vital health information of the patient in real time via a graphical user interface (GUI). The emergency situations produce alerts by the planned application; these alerts and other health status information can be reached via mobile devices like laptop computers, personal digital assistants (PDAs), and smartphones.

Generally speaking, two types of devices can be distinguished, sensors and actuators. The sensors are used to measure certain parameters of the human body, either externally or internally. Cases include measuring the heartbeat, body temperature, or recording a prolonged electrocardiogram (ECG). The actuators (or actors) on the other hand take some specific actions according to the data they receive from the sensors or through interaction with the user. For instance, an actuator equipped with a built-in reservoir and pump manages the correct insulin dose to be given to diabetics based on the glucose level measurements. Interaction with the user or other persons is usually handled by a personal device, e.g., a PDA or a smartphone, which acts as a sink for data of the wireless devices.

The overview of a WSN-based healthcare application setup is depicted in Fig. 3.30. Based on this observation, in a typical scenario, there are four different categories of actors other than the power users of the systems such as administrators and developers:

- Children. This group consists of young people who are not capable of taking care of themselves like babies, infants, toddlers, or those who are more grown-up but still need to be constantly monitored.
- Elderly and chronically ill. This category includes the chronically ill people who have cognitive difficulties or other medical disorders related with heart,

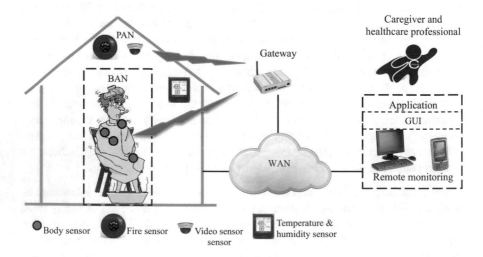

Fig. 3.30 Healthcare application actors and subsystems

respiration, etc., and the elderly people who also may have these symptoms, besides, who are more susceptible to sudden falls.

- Caregivers. They are the parents and the babysitters of the children group, as well the caregivers and other care network of the elderly and the chronically ill.
- Healthcare professionals. These are the professional caregivers like physicians and other medical staff who are responsible for the constant health status monitoring of the elderly and the chronically ill people and are capable of giving the immediate response in case of an emergency situation.

These groups of actors constantly interact with the WSN healthcare systems throughout different subsystems. Characteristically, five subsystems are available in such a scenario:

- BAN subsystem.
- PAN subsystem.
- Gateway to the wide area network (WAN).
- WANs.
- End-user healthcare monitoring application.

The following sections provide a detailed emphasis on the issues to be considered in the design and setup of each subsystem (Alemdar and Ersoy 2010).

3.5.1 Body Area Network Subsystem

The BAN subsystem is the ad hoc sensor network and tags that the children and the elderly carry on their body. The RFID tags, electrocardiogram (ECG or EKG)[14] (WebMD 2005a) sensors, and accelerometers worn by the patient are example components of the BAN. An example of a medical BAN used for patient monitoring is shown in Fig. 3.31. Several sensors are placed in clothes, directly on the body or under the skin of a person and measure the temperature, blood pressure, heart rate, ECG, electro-encephalogram (EEG)[15] (WebMD 2005b), respiration rate, SPO2 levels (oxygen saturation), etc. Next to sensing devices, the patient has actuators, which act as drug delivery systems. The medicine can be delivered on predetermined moments, triggered by an external source (i.e., a doctor who analyzes the data) or immediately when a sensor notices a problem. One example is the monitoring of the glucose level in the blood of diabetics. If the sensor monitors a sudden drop of glucose, a signal can be sent to the actuator in order to start the injection of insulin. Consequently, the patient will experience fewer nuisances from his disease. A BAN can also be used to offer assistance to the disabled. For example, a paraplegic can be equipped with sensors determining the position of the legs or with sensors attached

---

[14]Test in which electrode patches are attached to the skin to monitor the electrical activity of the heart.

[15]Test that measures and records the electrical activity of the brain.

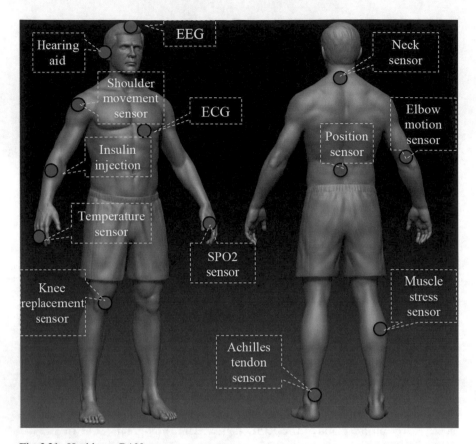

Fig. 3.31 Healthcare BAN sensors

to the nerves. In addition, actuators positioned on the legs can stimulate the muscles. Interaction between the data from the sensors and the actuators makes it possible to restore the ability to move (Latré et al. 2011).

Several issues are pertinent to healthcare BAN devices and distinguish them from issues specific to other applications, as considered in this chapter. The coming subsections will emphasize them all.

3.5.1.1 Power Consumption

One of the main issues about the BAN is the power consumption, since changing the batteries is a burdensome task. Because of this, development of energy-efficient MAC protocols and energy-efficient sensor devices is critical (Omeni et al. 2008). Technologies like Bluetooth and Wi-Fi fail to provide support for energy-efficient systems since they can only offer one- or two-week runtime on a coin like small battery (Bhatia et al. 2007). The proper use of the wireless communication channels during routing ensures energy efficiency (Fariborzi and Moghavvemi 2007; Kumar and Rao 2008). Moreover, by using very low-power pyroelectric infrared (PIR) sensors[16] (Murata Manufacturing Co. 2012; Adafruit Learning Technologies 2014), as in (Lee et al. 2007), a simple yet efficient remote monitoring application can be developed.

3.5.1.2 Output Transmission Power of the Sensor Nodes

The output power must be kept minimal for health issues, which may lead to coverage and communication problems. The study presented in (Ren and Meng 2006) investigates the bioeffects caused by radiofrequency transmission of sensor nodes, including thermal and athermal[17] effects, and proposes a control algorithm to reduce the bioeffects.

3.5.1.3 Unobtrusiveness

Unobtrusiveness must be a design consideration that cannot be overlooked. A design of an adhesive bandage type ECG sensor that is wirelessly powered by a health monitoring chest band is proposed in (Yoo et al. 2010). The sensors do not include batteries and are wirelessly charged via the chest band instead. The network controller on the chest band is able to find the locations of sensors automatically and provide power only to selected sensors. The sensor IC is very small (4.8 mm^2 area) and the network controller is also relatively small (15 mm^2 area). They consume only 12 µW and 5.2 mW average powers, respectively, which makes them power-efficient devices. A watch-shaped activity recorder, which provides six degrees-of-freedom, inertial data is presented in (Barth et al. 2009). The device contains a three-axes accelerometer and two gyroscopes, and data transmission is accomplished via Bluetooth. The system draws 185 mW in full operation mode where all sensors and the radio module are functioning.

3.5.1.4 Mobility and Portability

Mobility and portability are considerations for the physical design of such sensors, since the patients have to wear the BAN devices all the time; thus, mobility limitation is not acceptable. The sensor devices must be designed with the aim of providing the highest degree of mobility for the patients (Seo et al. 2007), which necessitates the integration of several network technologies like RFID and near-field

---

[16]PIR sensors allow to sense motion; they detect whether a human has moved in or out of their range (Adafruit Learning Technologies 2014).

[17]Adjective describing any process that does not involve either heat or a change in temperature.

communication (NFC) as proposed in (Lahtela et al. 2008). In (Lee and Chung 2009), a smart shirt which measures ECG and acceleration signals continuously and transmits the physiological ECG data and physical activity data to IEEE 802.15.4 ad hoc network. The shirt consists of sensors for continuous monitoring of the health data and conductive fabrics to get the body signal as electrodes. The study in (Lee et al. 2009) also makes use of ZigBee and mobile phone for ECG and blood glucose measurement. In a similar study (Morris et al. 2009), a sensor that can be integrated into clothing is designed to measure biochemical changes in sweat, which may indicate some health-related problems. The study has shown that the sensor has the potential to record real-time variations in sweat during exercise.

3.5.1.5 Real-Time Availability and Reliable Communications

The real-time availability and reliable communications of the system is a further major issue since the data gathered by such system may be critical. In (Nyan et al. 2008) there is a design of a wearable fall detection system that can detect the falls with an average lead time of 700 msec before the incident occurs. A framework proposal for high levels of reliability and low delays is available in (Varshney 2008a). The performance results show that reliable message delivery and low monitoring delays can be achieved by using multicast or broadcast-based routing schemes in an ad hoc network.

3.5.1.6 Multihop Design

Additionally, designing multihop systems are of great importance. As in (Lai et al. 2009), continuous monitoring of the patients ECG, without range limit due to the extended communication coverage by multihop wireless connectivity, is one of the basic needs for such critical applications.

3.5.1.7 Security

Finally, security issues also need to be considered since the physiological data of the individuals is highly confidential. In (Bao et al. 2005), the authors design a physiological signal-based entity authentication scheme. The information extracted from physiological signals is used to generate identity information for mutual authentication. The proposed scheme allows the secure identity verification during wireless link setup process. In (Dağtas et al. 2008), a secure key establishment and authentication algorithm is used for transmitting medical data from body sensors like ECG sensors to a handheld device of the mobile patient.

3.5.2   Personal Area Network Subsystem

The PAN subsystem is composed of environmental sensors deployed around, and mobile or nomadic devices that belong to the patient. The issues to be considered for a proper PAN design and installation are tackled in what ensues.

3.5.2.1   Contextual Information Acquisition and Location Tracking

The environmental sensors like RFID readers, video cameras, or sound, pressure, temperature, luminosity, and humidity sensors help providing rich contextual information about the people to be monitored. Location tracking can also be achieved by this subsystem. The smart appliances that are capable of communicating with other devices and taking actions can be included in the system.

3.5.2.2   Modular and Scalable Design

Keeping the design modular, scalable, and allowing the easy integration of new functionalities are key issues. In (García-Sáez et al. 2009), a wireless personal assistant device is proposed; it supports communications with three different medical devices, namely, a glucometer, an insulin pump, and a continuous glucose sensor. With the help of the patient PDA device, medical devices can be controlled, and context awareness with a user-friendly interface is provided to the patient.

3.5.2.3   Efficient Locating Algorithms

Efficient algorithms for locating RFID-tagged objects through appropriate data collection and preventing data interference are needed. RFID-tagged objects are also significant for this subsystem. They can convey detailed contextual information by mature and widely used RFID technology that can also support battery-free operation with the help of inexpensive tags. However, although a tag may be present, it may not necessarily be visible to the tag reader due to blocking by an impenetrable object (Tu et al. 2009).

3.5.2.4   Energy Efficiency of the MAC Layer

The energy efficiency of the MAC layer is a major issue. In (Lamprinos et al. 2006), a MAC protocol is proposed; it aims to improve energy efficiency by giving emphasis on the prevention of main energy consumption sources, such as collision, idle listening, and power outspending. Also in (Chen et al. 2008b) a variable control system is suggested to optimize the measurement resolution, thus saving power. The higher resolutions of sensor devices will consume more energy. By using this system, the users can flexibly set the resolutions in any situation. In their experiments, the signal-to-noise ratio (SNR) of ECG can be promoted from 25 to 73 dB when extraordinary ECG signal occurs.

3.5.2.5   Self-Organization between Nodes

The self-organization between the nodes is essential. In miTag system (Gao et al. 2008), the data collected from a self-organizing wireless network of sensors that operate in mesh mode are relayed to the Internet. By making use of mesh mode operation and distributed processing, the system can be used during serious disaster conditions by deploying repeaters quickly and providing easily extendible coverage. In (Osmani et al. 2008), a self-organizing distributed scheme is proposed for recognition of people activities.

3.5.3   Gateway to the Wide Area Networks

The gateway subsystem is responsible for connecting the BAN and PAN subsystems to the WANs. The gateway can be a mobile device carried by the user like a PDA or a smartphone, or a sensor node deployed in the environment as well as a laptop or a server computer. The issues to be considered for the gateway setup are twofold:

3.5.3.1   Local Processing Capability at the BAN and PAN Subsystems

The main function of the gateway subsystem is to provide the connection between the ad hoc sensor networks to the infrastructure-based WANs. Because of this property, the gateway subsystem can easily become the weakest link of the overall scenario. Therefore, local processing capability at the BAN and PAN subsystems has a great impact on the gateway subsystem, by preventing network congestion through reducing the amount of the transferred data. The study in (Chung et al. 2007) focuses on this issue by collecting the ECG data coming from the sensor and processing the data on the cell phone locally before sending. In (Pawar et al. 2008), context awareness provides the capability of selecting the suitable network interface for the data transfer.

3.5.3.2   Security

Security demands to be handled by the gateway subsystem. The security necessities for this subsystem include verifying the correct identity of the source and not modifying the patient data, except for aggregation or other defined transformations

Table 3.1 WSNs security requirements and solutions (Ng et al. 2006)

Security threats	Security requirement	Possible security solutions
Unauthenticated or unauthorized access	Key establishment and trust setup	Random key distribution Public key cryptography
Message disclosure	Confidentiality and privacy	Link/network-layer encryption Access control
Message modification	Integrity and authenticity	Keyed secure hash function Digital signature
Denial of service (DoS)	Availability	Intrusion detection Redundancy
Node capture and compromised node	Resilience to node compromise	Inconsistency detection and node revocation Tamper-proofing
Routing attacks	Secure routing	Secure routing protocols
Intrusion and high-level security attacks	Secure group management, intrusion detection, secure data aggregation	Secure group communication Intrusion detection

(Leister et al. 2008). The security scheme proposed in (Hu et al. 2008) employs a session key buffer to defeat gateway attacks. The delay between receiving the new session key and using it helps identifying the gateway compromise. The scheme also brings solutions to the man-in-the-middle (MITM) attacks, and fake data injection. An MITM attack is a type of cyberattack where a malicious actor inserts himself into a conversation between two parties, impersonates both parties and gains access to information that the two parties were trying to send to each other. An MITM attack allows a malicious actor to intercept, send, and receive data meant for someone else, or not meant to be sent at all, without either outside party knowing until it is too late (Veracode 2006). In Table 3.1, the summary of security requirements is given and possible solutions are provided.

3.5.4 WANs for Healthcare Applications

For a remote monitoring and tracking scenario, a network infrastructure is inevitable. The gateway can relay information to one or more network systems depending on the application. The models of network systems can vary from cellular networks to ordinary telephone network, or from satellite networks to the Internet. These WANs have their own issues and properties independent of the healthcare application. Whenever the data rate and reliable communication protocols for WANs advance, the ubiquitous healthcare applications will also benefit. The new and existing broadband networking technologies are needed to be integrated into the pervasive healthcare solutions in order to provide coverage up to global scale. In Table 3.2,

Table 3.2 Wireless technologies for healthcare systems (Kuran and Tugcu 2007)

Technology	Candidate subsystem	Data rate	Cell radius	Frequency band
IEEE 802.11g/Wi-Fi	BAN/PAN	54 Mbps	50–60 m	2.4 GHz
IEEE 802.11n/Wi-Fi	BAN/PAN	540 Mbps	50–60 m	2.4 GHz
ETSI HiperLAN/2	BAN/PAN	54 Mbps	50–60 m	5 GHz
IEEE 802.16/WiMAX	WAN	36–135 Mbps for LOS	Up to 70–80 km	2–66 GHz
		75 Mbps for NLOS		
IEEE 802.16e/WiMAX	WAN	30 Mbps	Up to 7080 km	2–6 GHz
ETSI HiperACCESS	WAN	25–100 Mbps	1.8–2.5 km	11–43.5 GHz
ETSI HiperMAN	WAN	25 Mbps	2–4 km	<11 GHz
WiBro	WAN	18 Mbps	1 km	2.3–2.4 GHz
High- Altitude Platforms (HAP)	WAN	Varies	Varies	28–31 GHz and 42–43 GHz
IEEE 802.20	WAN	16 Mbps	>15 km	3.5 GHz
IEEE 802.22	WAN	18 Mbps	40 km	54–862 MHz
Satellite geostationary earth orbit (GEO)	WAN	Up to a few Gbps	Four satellites give global coverage	4–8 GHz (C Band)
				10–18 GHz (Ku band)
				18–31 GHz (Ka band)
				37–50 GHz (Q/V band)
Satellite medium earth orbit (MEO)	WAN	Up to a few Mbps	11 satellites give global coverage	Same as GEO
Satellite low earth orbit (LEO)	WAN	Up to a few Mbps	Varies	Same as GEO

the characteristics of the candidate wireless connections and mobile networking technologies are provided.

In order to extend the healthcare to a global scale, satellite communication systems may be needed as well; satellite-based telemedicine networks, telemedicine applications, and services offer wide opportunities (Wootton et al. 2006; Martinez et al. 2008). These tasks mainly focus on improvement of the healthcare in remote locations, like marine vessels, or healthcare-deficient parts of the world that have no technological infrastructure. With the increased transmission capacity of the network, teleconsultation with remote experts will be possible. Likewise, satellites or high-altitude platforms (HAPs) can provide healthcare services for disaster areas with quick and easy deployment.

3.5.5 *End-User Healthcare Monitoring Application*

The application is at the heart of the system at which the collected data are interpreted and required actions are triggered. The application has a processing part and a graphical user interface (GUI) part. The processing part performs the reasoning with some signal processing algorithms to understand a distorted cardiac signal for instance, and with machine-learning algorithms to identify an unexpected situation from an image or video. The GUI is used for real-time monitoring of the vital sign information together with an alerting mechanism in case of an emergency. The application should also provide an interface for the definition and configuration of the system overall behavior. What kind of alarms will be generated, and via which network the messages will be delivered, and who are the intended users, are cases of the application configuration. For such applications, the issues related to the design and setup are as provided in the coming subsections.

3.5.5.1 Security

Security must be ensured throughout the healthcare application scenario. Therefore, end-to-end security mechanisms are needed. Based on the security requirements provided in Table 3.1, the systems must be resistant to security attacks; the patients sensitive health information must be viewed only by the authorized parties. The mainly studied security issues are related to encryption. There are several proposals for key distribution protocols for encryption mechanisms (Garcia-Morchon et al. 2009; Garcia Morchon and Baldus 2009). Yet, before establishing encryption, the security policies must be addressed in the first place. Moreover, for multimodal systems such as RFID enhanced pervasive healthcare systems, the security issues for every modality should be studied separately and there must be strong mechanisms against all kinds of attacks (Xiao et al. 2006).

3.5.5.2 Privacy

Privacy is of utmost importance, although in a study conducted with randomly selected seniors from several elderly community groups in Australia, it was suggested that privacy of health information is not perceived as a highly significant concern and does not have a significant effect on an elderly person perception or acceptance of WSN systems (Steele et al. 2009). On the other hand, the same study reveals that the seniors strongly reject being monitored by cameras. The users should have autonomy; they must be permitted to have control over their information so that they can decide which information is transferred and during which intervals as in the CareNet display project (Consolvo et al. 2004). As such, the application must guarantee a well-defined degree of privacy with precisely formulated and verified rules. In Smart Home Care Network (Tabar et al. 2006), the use of image sensors is proposed only under emergency conditions and only for verification purposes. This scheme changes a probable privacy flaw into a privacy-respecting mechanism.

3.5.5.3 Reliability

The application should be reliable. Reliability issues are classified into three main categories: reliable data measurement, reliable data communication, and reliable data analysis (Lee et al. 2008). Although the reliable data measurement and communication issues belong to the BAN and PAN subsystems, they are essential for the reliable data analysis at the application layer. A typical architecture handles data cleaning, data fusion, and context and knowledge generation for reliable data analysis. Data analysis is critical for pervasive healthcare systems since the inferences are obtained from the data. The application system should prove to be doing the job it is designed for, and faulty system components and exceptions must not result in system misbehavior.

3.5.5.4 Middleware Design

The combination of sensors from different modalities and standards makes it necessary to develop hardware independent software solutions for efficient application development (Triantafyllidis et al. 2008). In this context, middleware design is a major concern; it helps to manage the inherent complexity and heterogeneity of medical sensor networks. The concept is to isolate common behavior that can be reused by several applications and to encapsulate it as system services. In this way, multiple sensors and applications can be easily supported; thus, resource management and plug-and-play functions become easy (Chatzigiannakis et al. 2007).

3.5.5.5 Context Awareness

Context awareness is defined as providing relevant information and/or services to the user, where relevancy that depends on the user task (Abowd et al. 1999), is the core issue for smart home applications as well as remote health monitoring applications. Context awareness can be provided through the use of different sensing modalities together, while considering previously mentioned issues, such as proper data aggregation and analysis. There are several ways for improving context information. Context models and context management frameworks are proposed in (Paganelli and Giuli 2007; Paganelli et al. 2008). With ontology-based models, analyzing not only the rule-based alarm conditions but also more complex patterns become easier. This semantic representation of WSNs data enables structured information to be understood. Ontology-based semantic collaboration mechanisms provide the cooperation among different persons anytime, anywhere, and with anybody, thus reducing the complexity of making correct decisions and taking correct actions (Kim et al. 2008b).

3.5.5.6   Seamless Healthcare Tracking and Monitoring System

In a probable futuristic design of a multimodal, seamless healthcare tracking, and monitoring system, the chronically ill person lying in his bedroom carries on his body a group of wireless sensors that construct a BAN. These sensors constantly measure the vital signs of the patient and relay this data to the basestation node connected to specially configured home-server computer acting as a gateway between the BAN and the WANs like the Internet and the cellular or fixed telephony networks. The video and audio sensor nodes, RFID tags and other sensors for humidity, temperature, motion, etc., are disseminated in the living place, and they are used to provide detailed context information about not only the patient but also the living place conditions. These sensors form the PAN. In such a scenario, small inexpensive RFID tags can be used for location tracking, i.e., in which room the patients or the other residents are. In this way, when an emergency situation occurs, the location information helps to activate the closest video sensor node to obtain the best scene about the event. These video images can be delivered to the caregivers or healthcare professionals for further exploration via the Internet or cellular network through the gateway server.

By using smart home appliances, the living place can be controlled or the interaction with the residents can be achieved. For instance, if the patient is observed to be sitting on the coach in front of the television at medication time, then the smart set-top box can provide a reminder for the patient to take his or her medications. For the infants and babies, when the temperature of the room where the baby is sleeping goes below or above the optimal value, then the air conditioning device in the room can be automatically activated. Also, when the baby is observed to be crying in his or her room and the caregiver is in another room for some time, the speakers in the room can be activated to alert him. These scenarios are not very far from becoming ordinary for our lives with the development of seamless, pervasive healthcare systems.

3.5.6   Categorization and Design Features of WSN Healthcare Applications

3.5.6.1   Applications Prototypes

There are several prototype and commercial applications for pervasive healthcare monitoring for the elderly, children, and chronically ill people. When these applications are explored, it is observed that the main focus categories include (Alemdar and Ersoy 2010):

- Activities of daily living monitoring. These applications identify and differentiate everyday activities of the patients and the elderly such as watching television, sleeping, ironing, and work on detecting abnormal conditions.

- Fall and movement detection. Such applications are focused on the physiological conditions such as posture and fall detection for persons who need special care, like the elderly who are susceptible to sudden falls, which may lead to death, or the infants and patients recovering from an operation.
- Location tracking. Applications that help cognitively impaired people to survive independently by having their steps watched.
- Medication intake monitoring. Applications to help cognitively impaired people to survive independently by observing how they take their medication.
- Medical status monitoring. Such applications make use of medical and environmental sensors in order to obtain comprehensive health status information of the patients, including ECG, heart rate, blood pressure, skin temperature, and oxygen saturation.

3.5.6.2 Wearable and Implantable Systems

Wearable and implantable systems are such systems that encompass one or more of the aforementioned prototypes. Wearable and in vivo implantable health systems can be used both indoors and outdoors to monitor people 24 hours a day, and 7 days a week (Chan et al. 2008). Such systems are not just for monitoring, they can also affect the vital body functions and deliver therapy. These devices have the potential to greatly enhance comfort, health, and the efficiency of disease prevention. If vital functions are maintained at a normal level, complications and hospitalization can be avoided. Such biomedical sensors are usually worked into textiles, equipped with data storage and a wireless transceiver system. The data are sent to a central processing unit, a medical center, and able to diagnose the situation and organize assistance if needed.

Wearable and in vivo systems must be easy to operate, small in size, and unobtrusive. In addition, they must be waterproof and possess a long battery life. They also must automatically collect their measurements, without the intervention of a third party, and provide total confidentiality and reliable data. There are several forms for these systems, a textile garment (Marques et al. 2004), a wrist-worn device (Ho et al. 2005), a ring (Lopez-Nores et al. 2008), a sensor attached to the belt (Pang et al. 2009), an over-the-shoulder pouch, a small box worn on the patient head (Konstantas and Herzog 2003), a chest belt for stress monitoring (Shnayder et al. 2005), and a glucose sensor with a needle (Wood et al. 2008), etc.

3.5.6.3 Design Features of WSN Healthcare Applications

Research projects and applications in healthcare are distinguishable by several design features:

- They use different sensor types ranging from tiny biosensors to battery-free RFID tags. Some of the proposals deploy only a single type of sensors such as passive

RFID tags or accelerometers while others deploy a combination of sensor types allowing multimodal sensing. Most of the proposed works also present their special sensor hardware designs to provide specific measurements and interfaces to other devices.

- Likewise, nearly all of the projects have special software development with appropriate APIs, middleware and GUI.
- As far as the routing of critical information indicating health status of the patients is considered, there are very few applications that use multihop routing. Multihop routing is important in enlarging the coverage area of the application thus providing flexibility at the cost of complexity.
- Moreover, the level of being unobtrusive differs among applications. Some of them use very tiny sensors integrated into clothing or video cameras that are not carried by the patient. These applications obtrusiveness level is low. As the size and amount of the devices carried by the patients increase, this level goes up.
- Similarly, the context awareness is at different levels among the applications. Some of them provide rich contextual information such as the time, location, and status of different sensors, while others give only limited information about some specific event condition. In that sense, location-tracking utility is important in providing detailed contextual information as well. When deducing meaningful information from sensor data, machine-learning techniques are also important besides signal processing.
- Finally, several applications utilize inexpensive RFID tags to make location tracking and activity classification easier.

A classification of some healthcare-based WSN applications is offered in (Alemdar and Ersoy 2010). In the following section, a representative application proposal is described.

3.5.7   Using Heterogeneous WSNs in a Telemonitoring System for Healthcare

The proposal in (Corchado et al. 2010) presents a distributed telemonitoring system that improves healthcare and assistance to dependent people at their homes. It implements a service-oriented architecture (SOA)-based platform, which allows heterogeneous WSNs to communicate in a distributed way independent of time and location restrictions. This approach provides the system with a higher ability to recover from errors and a better flexibility to modify behavior at execution time.

The proposed system makes use of the Services laYers over Light PHysical devices (SYLPH) platform. SYLPH is based on an SOA model for integrating heterogeneous WSNs into ambient intelligence (AmI) systems; it focuses on distributing the systems functionalities into independent functionalities (i.e., services). This model provides a flexible distribution of resources and facilitates the inclusion of new functionalities in highly dynamic environments. WSNs provide an infrastructure capable of supporting the distributed communication AmI-based telemonitoring system.

AmI-based developments require the use of several sensors and actuators strategically distributed in the environment. This provides the systems with context-aware capabilities in order to automatically change its behavior. The ZigBee standard operates in the frequency range belonging to the radio band known as industrial, scientific, and medical (ISM), especially in the 868 MHz band in Europe, the 915 MHz in the U.S.A., and the 2.4 GHz in almost all over the world (ZigBee Alliance 2013). The underlying IEEE 802.15.4 standard is designed to work with low-power and limited computational resources nodes (IEEE 2013a, b). ZigBee incorporates additional network, application, and security layers over the 802.15.4 standard; it allows up to 65,534 nodes connected in a star, tree, or mesh topology network (Baronti et al. 2007). Bluetooth is another standard to deploy WSNs; it allows multiple WPAN and WBAN applications for interconnecting mobile phones, earphones, personal computers, printers, etc. Bluetooth also operates in the ISM 2.4 GHz band; it creates star topology networks of up to eight devices in which one of them acts as master and the rest as slaves. Several Bluetooth networks can be interconnected by means of Bluetooth devices that simultaneously belong to two or more networks, creating thus more extensive networks.

Although there are plenty of options for creating WSNs, the main problem is the difficulty for integrating devices from different technologies in a single network (Lei et al. 2006). In addition, the lack of a common architecture may lead to additional costs due to the necessity of deploying nontransparent interconnection elements between networks. Moreover, the developed elements (e.g., devices) are too dependent on the application to which they belong, hence complicating their reutilization. Some approaches attempt to integrate devices by implementing middleware layers as reduced versions of virtual machines (e.g., Squawk Java virtual machine); they require devices with high computational power and large memory microcontrollers (Simon and Cifuentes 2005). As such, there is a need for more expensive devices with larger size or more costly miniaturization. These drawbacks are very important regarding WSNs, as it is essential to deploy applications with reduced resources and low infrastructural cost, especially in homecare scenarios. The SYLPH platform integrates an SOA approach for facilitating the distribution and management of resources (i.e., services) into heterogeneous WSNs. There are several attempts to integrate WSNs and an SOA approach (Meshkova et al. 2008; Moeller and Sleman 2008; Song and Lee 2008). In SYLPH, unlike these approaches, services are directly embedded on the WSN nodes and can be invoked from other nodes in the same network or from other connected networks.

Efficient solutions are required to allow building AmI environments for providing dependent people healthcare at their homes. One of the key aspects for the construction of these environments is obtaining context information through sensor networks. There are several healthcare approaches for telemonitoring based on WSNs (Jurik and Weaver 2008; Varshney 2008b). However, they do not take into account their integration with other systems and are difficult to adapt to changing scenarios. The use of SYLPH is proposed in order to face some of these issues while integrating heterogeneous WSNs.

3.5.7.1 SYLPH Platform

The SYLPH platform is a distributed architecture, which integrates an SOA approach over WSNs for building systems based on the AmI paradigm. As previously stated, the SOA approach has been chosen because such architectures are asynchronous and non-dependent on context (i.e., previous states of the system, which must not be confused with context-aware environments). Thus, devices do not continuously spend processing time and are free to do other tasks, which saves energy.

SYLPH can be executed over multiple wireless devices independently of their microcontroller or the programming language they use; being distributed, the application code does not have to reside on a central node. Applications run independent of the lower layers related to the WSNs formation (i.e., network layer), and the radio communication between nodes (i.e., data link and physical layers). SYLPH allows interconnecting several networks from different wireless technologies, such as ZigBee or Bluetooth; thus, a node designed through a specific technology can be connected to a node from a different technology. In this case, both WSNs are interconnected by means of a set of intermediate gateways connected to several wireless interfaces simultaneously.

SYLPH organization is based on a stack of layers. Figure 3.32 shows the different layers added over the application layer of each WSN stack. The SYLPH message layer (SML) offers to the upper layers the possibility of sending asynchronous messages between two wireless devices through the SYLPH services protocol (SSP). SSP is the internetworking protocol of the SYLPH platform, and it has functionalities similar to those of the Internet protocol (IP); that is, it allows sending packets of data from one node to another node regardless of the WSN to which each

Fig. 3.32 SYLPH architecture. (Corchado et al. 2010)

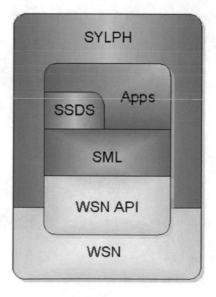

one belongs. The messages specify the origin and target nodes, and the service invocation in a SYLPH services definition language (SSDL) format. The SSDL describes the service itself and its parameters to be invoked. Applications can directly communicate between devices, using the SML layer or by means of the SYLPH services directory sublayer (SSDS) which in turn uses the SML layer. The SSDS offers functionalities related to discovering the services offered by the network nodes. A-node that stores and maintains services tables is called SYLPH directory node (SDN).

The functioning of SYLPH is further described in the coming subsections.

3.5.7.2 SYLPH Services

The behavior of SYLPH is in essence similar to that of any SOA. However, SYLPH has several characteristics and functionalities that make it different from other models. Figure 3.33 shows the basic operation of SYLPH.

As a start, a service registers itself on the SDN and broadcasts its location, the parameters it requires and the type of returned value after its running. The interface definition language (IDL) used by SYLPH is created to work with limited resources nodes. Distributed architectures use an IDL in order to enable communication between software components, regardless of their programming language or hardware implementation. Unlike other IDLs as web service definition language (WSDL), based on extensible markup language (XML) and used on web services (Corchado et al. 2008), SSDL uses a few intermediate separating tags and its service

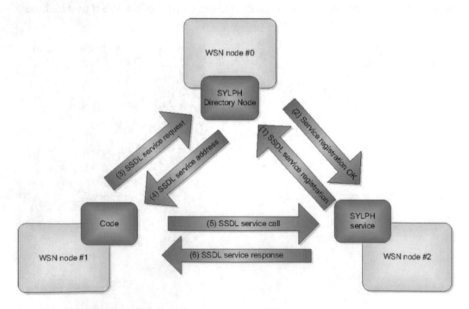

Fig. 3.33 SYLPH basic operation. (Corchado et al. 2010)

descriptions are short binary data sequences. These constraints reduce processing in the device microcontrollers. A simple IDL thus allows utilizing nodes with fewer resources, less power consumption, and lower cost. It is enough with a few floating-point data to inform the status of a sensor; hence, most service definitions require only a few Bytes. SSDL considers the basic types of data (e.g., integer, float, or Boolean), allowing also more complex data structures as variable length arrays or character strings.

Once the service has been registered in the SDN, it can be invoked by any application by means of SYLPH. Both the SDN and the services can be stored in any node of the WSN or in other subsystem connected to the WSN. This system can be, for instance, a simple personal computer connected through a universal serial bus (USB) port to a wireless interface. Thus, developers may decide which nodes or subsystems will implement each part of the distributed application. Using SSDL, any node in the network can ask the SDN for the location of a determined service and its specification.

3.5.7.3   SYLPH Directory Nodes

For the sake of a distributed architecture, it is allowed to have more than one SDN in the same network. The SDN can be stored in a node of the network, in a memory external to the microcontroller, if necessary, or in a computationally higher machine connected to the WSN, as a data server or a personal computer with wireless connection.

For a node (Node 0) to discover available services, it registers itself in the WSN by means of SYLPH. Then, it sends a broadcast message, after connecting to the WSN, searching for existing SDNs. The active node (Node 1) responds by sending a message informing of its situation (i.e., SSP address) and its setup parameters. An example of a setup parameter is whether the SDN will inform periodically of its presence or if the nodes will have to poll it. Accordingly, the requesting node becomes able to communicate with the replying node to obtain information about the services existing in the network. Later, Node 3 registers itself in the WSN, as having SDN functionalities, it informs the rest of the nodes by means of a broadcast message. Node 1 stores this information on its SSDS entries list and informs Node 3 about its role as SDN. Any node in the network cannot only offer or invoke SYLPH services, but also includes SDN functionalities in order to provide services descriptions to other network nodes.

3.5.7.4   Telemonitoring System Implementation

The system makes use of several WSNs in order to gather context information in an automatic and ubiquitous way. Several functionalities are directly embedded on the WSN nodes and can be invoked from other nodes in the same network or in other connected networks by means of the SYLPH platform. SYLPH gateways are used in order to interconnect different heterogeneous WSNs; it is accordingly possible to connect WSNs based on different radio and link technologies (e.g., ZigBee, Bluetooth, and Wi- Fi). In addition, SYLPH focuses specially on devices with small resources in order to save microcontrollers computing time, memory size, and energy.

Two types of sensors are available in the system, biomedical sensors and automation sensors. Biomedical sensors (e.g., electrocardiogram, blood pressure, and body temperature) obtain continuous information about vital signs, whose samples are important and should not be lost. Automation sensors (e.g., building temperature, light, and humidity) collect information at a relatively lower frequency compared to biomedical sensors (Sarangapani 2007). Biomedical sensors should be smaller and easier to wear. It is necessary to interconnect several WSNs from different radio technologies in a telemonitoring scenario to obtain a compatible distributed platform for healthcare applications (Jurik and Weaver 2008).

Figure 3.30 reiterates the basic communication and infrastructure schema of the telemonitoring system. A network of ZigBee devices has been designed to cover the home of each patient to be monitored. A ZigBee remote control carried by the monitored patient has a button that can be pressed for remote assistance or urgent help. Moreover, there is a set of ZigBee sensors that obtain information about the home environment (e.g., light, smoke, temperature, and door states) and that physically respond to the variations (e.g., light dimmers, fire alarms, or door locks). Each of these ZigBee nodes includes a C8051F121 microcontroller (Silicon Laboratories 2004) and a CC2420 IEEE 802.15.4 radiofrequency transceiver (Texas Instruments 2005b). There are also several Bluetooth biomedical sensors on the monitored patient body. Biomedical sensors allow the system to continuously acquire data about the vital signs of the patient. Each patient carries three different biomedical sensors, an ECG monitor, a respiration monitor (implemented by means of an air pressure sensor), and a micro-electro-mechanical systems (MEMS) triaxial accelerometer for detecting possible patient falls. These Bluetooth nodes are Bluetooth 2.0 standard compatible; they use a BlueCore4-Ext chip with a reduced instruction set computer (RISC) microcontroller having a 48 KByte RAM and a 1024 KByte external Flash memory. ZigBee and Bluetooth devices work as SYLPH nodes and can both offer and invoke functionalities (i.e., services) throughout the entire WSN.

A computer is connected to the remote healthcare telemonitoring center via the Internet. Alerts can be forwarded from the patient's homes to the caregivers in the remote center, allowing communication with patients to check for possible incidences and proper reaction. These alerts may detect a patient fall or a high smoke level in his home. The computer acts as a ZigBee master node through a physical wireless interface, it is also the master node of a Bluetooth network formed by the biomedical sensors working as slave nodes. At the SYLPH level, it acts as a gateway that connects WSNs.

Although this system is mainly focused on monitoring tasks, it also provides additional useful facilities to the patients and caregivers. For example, the remote center can consult really simple syndication (RSS) sources from external and internal web servers in order to obtain weather reports or entertainment options for patients and notify them about their scheduled medical staff visits. Such information is shown on a graphical user interface (GUI) on a display connected to the computer at home. The display is a touch-sensitive screen, for an easy and intuitive patient interaction. Moreover, the application includes home automation capabilities, such that a light sensor can switch or dim a lamp.

3.5.7.5   Experimentation Results

Several test cases were satisfactorily conducted over 4 weeks to evaluate the overall system performance, especially the management of emergency situations. The tests involved 13 patients and six caregivers. The evaluation considered the four main objectives, defined for WSNs use in healthcare development, specifically, minimizing error rates, conducting diagnosis with real-time patient data, improving efficiency, and reducing cost.

The use of multimedia WSNs in healthcare applications is presented in Sect. 3.7.

3.6   Daily Life Applications

Daily life applications are plentiful; they are intended to make life easier, more friendly, at home, at work, while shopping, driving, and many more. This section exhibits some of the available applications to enlighten their technologies and how to use, and to stress on the fact that applications overlap on sensors and standards. Two applications are chosen for illustration, an intelligent car park management system, and a WSN of everyday objects in a smart home environment.

3.6.1   An Intelligent Car Park Management System Based on WSNs

This section as built upon (Tang et al. 2006) describes a WSN-based intelligent car parking system. In this system, low-cost wireless sensors are deployed into a car park field, where each parking lot is equipped with one sensor node, which detects and monitors the occupation of the parking lot. The status of the parking field detected by sensor nodes is reported periodically to a database via the deployed wireless sensor network and its gateway.

In preceding work, Irisnet (Campbell et al. 2005) offered a wide area architecture, for pervasive sensing networks, which enables driving users to retrieve the information about available car parking space via distributed accessing methods. In this system, the video cameras (webcams), microphones, and motion detectors are employed to detect and recognize automobiles. The sensory data, for example parking field images captured by webcams, will be processed in a networking environment. The processed data will be published on the web; then, the user can acquire the interesting information by using the web access technologies. However, the video cameras generate a large amount of data. The transmission and processing of these data will consume a great deal of the limited WSNs resources, including communication bandwidth, processing cycles, and energy.

MIT Intelligent Transportation System (Massachusetts Institute of Technology 2014) proposes transportation applications based on WSNs. Automobile sensors are deployed on both sides of a road and into a roadbed to detect the relevant information about automobiles. Although the systems can be effective for traffic and road condition monitoring, they are not designed for car parking management.

An important problem in designing a car parking and transportation system is how to accurately detect the mobility of automobiles, especially when the vehicles move at high speed. There are studies that use magnetic sensors (Cheung et al. 2005); however, these sensors are energy consuming (Anastasi et al. 2004). The widespread deployment of such sensors is still a challenging problem in the energy constraint WSNs.

3.6.1.1 Car Parks Requirements

From a business standpoint, the common goal for all car parks is to attract more drivers to use their facilities that are required to fulfill several conventional necessities:

- The location of the car park should be easy to find in the street network.
- The entrance of the car park should be easy to discover.
- There should be many parking lots, and a parking lot should be spacious.
- A parking lot should be easy to exit and to re-enter on foot.

Also, an intelligent car parking system should provide more convenience and automation to both the parking lot business and the customers:

- The system should provide plenty of informative instructions or guidelines to help drivers find an available parking lot.
- The system should provide effective security measures to prevent cars from crashing, or being stolen, etc.
- The system should provide suitable auto toll methods.
- The system should provide powerful functions to facilitate administrators and managers to manage a car park.

In accordance with the above requirements, an automatic and smart car park management system should minimize human intervention, so as to reduce the cost of manpower and prevent human mistakes, which enhances security and efficiency.

3.6.1.2 System Overview

3.6.1.2.1 Hardware Components

The wireless sensor nodes and gateway used as the underlying hardware platform are from Crossbow Technology Inc., which is one of the suppliers of WSNs. Crossbow series of WSNs are based on Berkeley motes. The products used in the proposed system are as follows:

- Motes. The devices consisting of a processor and a radio chip are commonly referred as motes processor radio boards (MPR). Each of these battery-powered devices is preloaded with the open-source TinyOS (TinyOS 2012) operating system, which provides low-level event and task management services, and the Crossbow's XMesh networking stack (Crossbow 2006c). The MPR2400 mote is selected (Crossbow 2004a). The motes are compatible with IEEE802.15.4 and can be extended to connect with different sensorboards.
- Sensorboards. Sensor and data acquisition boards (MTS and MDA) mate directly to the mote processor radio board (MPR). The sensorboard MTS310 (Crossbow 2007) is equipped with sensors of light, temperature, and acoustic and a sounder.
- Gateways. The mote interface board (MIB), MIB510 (Crossbow 2004a), provides a gateway for the motes and allows the acquisition of sensory data on a PC, as well as, on other standard computer platforms via a RS232 serial interface. Alongside data transferring, the MIB board allows the motes to accept control command from the upper layer application systems.
  Datasheets of the system-building blocks are detailed in Chap. 10.

3.6.1.2.2 Structure of the WSN-s based Application System

The application system based on WSNs adopts a 3-layers framework:

- The first layer is the mote layer, which is a wireless sensor mesh network. The motes are programmed as TinyOS firmware to perform some tasks, such as environment monitoring.
- The second layer is server layer, which provides data logging and database services for sensory data transferred to the basestation and stored on the server.
- Finally, the software at the client layer provides visualizing, monitoring, and analyzing tools to display and interpret sensory data. MOTE-VIEW is a free software tool developed by Crossbow and can be used to perform the above manipulations of sensory data.

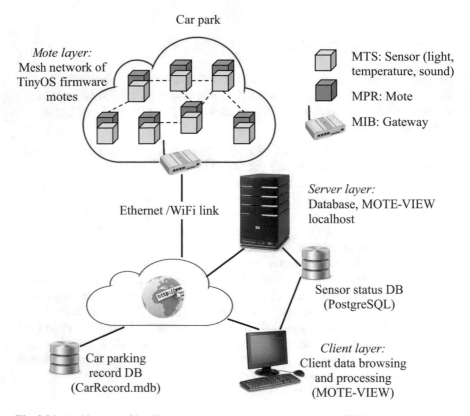

Fig. 3.34 Architecture of intelligent car park management system using WSNs

3.6.1.2.3   Intelligent Car Park Management System

The architecture of the proposed system, as shown in Fig. 3.34, illustrates the relationship between the sensor network, MOTE-VIEW (Crossbow 2006b), PostgreSQL database (PostgreSQL 2013), TinyOS, CarRecord database, and the car park application. The sensor nodes are deployed in a car parking field to collect the real-time information on occupation and vehicles. The collected information can be transmitted to a gateway via wireless communication among the sensor nodes. The gateway is connected to a database server via the Internet. The car park management application operates on top of the database. This architecture effectively decouples the upper layer application from the underlying WSNs. Accordingly, modifications of the underlying WSNs will not lead to changes on the upper layer application system.

3.6.1.3 System Implementation

This section gives a brief introduction of the main functional parts of the intelligent car parking management system and then describes the event-driven processing and interactions of these modules.

3.6.1.3.1 Functional Components of the System

The software system of the proposed application can be divided into three parts, the bottom part includes the motes and the network, the middle part contains the database system, and the top part embodies the application system. The interaction between the bottom part and top part is via the middle part that contains the database system. The application layer focuses on the business logic of the car park administration and the processing of the collected information stored in the database system.

The three parts of the software system are described as follows:

- The bottom part of the software system supports the operations of the WSNs composed of motes. A Mote is loaded with TinyOS, a lightweight operating system for WSNs. The adopted XMesh network protocol is specifically developed for mote networks.
- The middle part is implemented using the PostgreSQL (PostgreSQL 2013) database system. The data stored in the database is updated by the underlying WSNs. The sensory report generated by the mote is transmitted to the database system and used by the upper layer applications.
- The top part, the application system, is divided into four main modules:
  - Parking lot management module that monitors and detects the occupation of parking lots.
  - Auto toll module, which manages the payment of parking fee.
  - Security management module, which alerts the illegal departure of cars, previously parked in a parking lot.
  - Statistic and reporting module that generates various reports to help managers or administrators to understand the running status of the car parking field.

3.6.1.3.2 Event-Driven Processing

The prototype system is implemented using the object-oriented programming approach and is event-driven for processing. In the system operations, there are five major types of events:

- Timer event. The system timer generates this event to refresh the sensor status stored in the PostgreSQL database.
- Car-in event. Indicates that a car has just checked in.

- Driving status. Designates the moving path of the car and its parking status as sensed by wireless sensors.
- Car-out event. Specifies that a car has just checked out from the system.
- Field management event. This event detects a manager that performs the management task of the car parking field.

The events, as such, are intended to trigger the interactions and the operations to be performed by the various function modules described above.

3.6.1.4   System Evaluation

The intelligent car parking system is built for real applications that must be reliable and accurate. Some testing experiments were carried out using a prototype system built upon remote-controlled toy cars; several test scenarios help evaluate its functionalities.

3.6.2   Wireless Sensor Networking of Everyday Objects in a Smart Home Environment

Within a smart home environment the information processing is supposed to be thoroughly integrated into everyday objects that provide functionalities beyond their primary purpose, thereby enhancing their characteristics, properties, and abilities. By correlating the sensor output of such everyday objects, the WSN as a whole can potentially provide functionality that an individual everyday object cannot. Using a middleware, such functionalities include situation and activity awareness of the inhabitants. The backbone of this section is the work presented in (Surie et al. 2008).

3.6.2.1   Requirements for WSNs in Smart Home Environments

From the conducted assessment, the requirements for deploying a WSN in a home environment are found to be:

- Usability, availability, installation (non-functional). The spotted preferences are:
  - The systems must be readily available, off-the-shelf at an affordable price, and easy to install.
  - It is required to carry few devices as part of the wearable outfit (one device is enough).
  - Changing the way one would interact with everyday objects is not welcome.

- Performance (functional). Several likings are recorded:
  - The system to use must be primarily reliable with adequate performance. Hence, the sensing precision and recall values for the system as a whole are important evaluation aspects.
  - The transmission/reception range is an important parameter to consider as well. This issue may be tackled via a mobile receiver that is part of the user's wearable outfit and thus is more often within the range of the sensor nodes activated based on the user's interaction with the concerned object.
  - Battery lives of the sensor nodes were considered more important, since it is difficult to frequently charge all the nodes. On the other hand, the receiver node is expected to have a good battery life, but with a slightly lesser priority.

3.6.2.2 System Overview

The system described in (Surie et al. 2008) consists of a set of everyday objects present in a smart home environment connected to a wearable personal server (Want et al. 2002) worn by the user and running an activity-centered computing middleware (Surie and Pederson 2007).

3.6.2.2.1 Wireless Personal Area Network

The everyday objects are embedded with stick-on nodes that sense the internal states and state changes of the objects, and transmit this information wirelessly using ZigBee (ZigBee Alliance 2013) communication protocol to the user's personal server. ZigBee was preferred over Bluetooth for WPAN due to its usage of low-power digital radios intended for low data rate, long battery life, and secure networking applications. ZigBee supports up to 65,000 nodes, which enhances the possibility to include additional everyday objects.

Generic communication boards are designed with easily replaceable sensor connectors. Maxstream XBee 802.15.4 transceiver (MaxStream 2007) and Atmel ATmega88-20PU microcontroller (Atmel 2011b) are used in such boards. The XBee transceiver operates at ISM 2.4 GHz frequency, 1 mW (0 dBm) power output and allows for data rates of up to 250 Kbps. The average data rate of all the sensor nodes was 20.4 Hz (a maximum of 100 Hz for some nodes and a minimum of 10 Hz for most of them). The microcontroller runs at 8 MHz. The communication boards embedded onto everyday objects include a 2.4 GHz omnidirectional antenna with $1/2 \lambda$ wavelength and a gain of 2.90 dBI. The boards require 3 Volts and are powered by three 1.2 V 2600 mA NiMH batteries in series.

The receiver node connected to the user's personal server includes a Maxstream XBee 802.15.4 transceiver and a circuitry board for USB connection to a Sony Vaio VGN-UX70 notebook with 1 GHz processor and 512 MB RAM. The majority of the sensor nodes (70%) operate at a low sampling rate of 10 Hz, regarding the application where the nodes transmit only when there is a change in the sensor reading range defined by threshold values in the microcontroller. The internal states and state changes of the everyday objects are calibrated in the personal server based on their unique identities. Such a double-step calibration allows for introducing additional internal states for everyday objects within the personal server, based on the requirements from other components within the middleware, and also the applications running above the middleware.

The sensor nodes transmit the sensed data three times (default ZigBee protocol value that could be increased, but was found sufficient for the proposed application) to the receiver node before a time-out. The probability of correctly receiving packet at the receiver node increases with the number of retransmissions. This is important considering channel noise and collisions. However, too many retransmissions can block up the network bandwidth, thereby requiring a threshold for the number of retransmissions. Within the sensor node, the microcontroller sends the data four times (experimentally found out to be the ideal number) through a universal synchronous/asynchronous receiver/transmitter (USART) to the XBee transceiver and awaits an ACK message in return.

The data format used for communication include Object Identity (3 Bytes), Sensor Data (S_1, S_2,...,S_n), and End-of-Frame (1 bit). The length of the sensor data depends on the everyday object. (S_1, S_2,...,S_n) values are set with a minimum of 4 Bytes and a maximum of 17 Bytes. Sensor type information is not included in the data format to reduce the size of the data frame. The Object Identity information is used within the personal server to query a database containing information about everyday objects present in the user's environment.

The star topology is adopted for the WSN, chosen for its simplicity (no need for complex routing or message passing protocols), better performance (no need to pass data packets through unnecessary nodes), power efficiency, and isolation of everyday objects that are not changing their internal state. A star network topology demands that all the sensor nodes be within the vicinity of the receiver node. In the proposed application, the receiver node is worn by the user, creating a mobile context within a home environment. The state changes are created by the user interaction with environmental everyday objects. The range within which the receiver node can receive sensor data with acceptable noise ($<5\%$), was evaluated to be 33 m with a single wall obstruction and 19 m with multiple (greater than two) wall obstruction.

In the proposed smart home application, 81 sensors were distributed (chosen from the eight sensor types described in Table 3.5) onto 42 everyday objects in a live-in laboratory home environment as the one shown in Fig. 3.35. The receiver node receives sensor data from the sensor nodes that are based on a combination of these sensor types. Making minor firmware modifications can easily include additional sensor types with RS232, I2C, or SPI output. Analog sensors can also be encompassed within the sensor node, but with an external circuit that condition the signal to the ADC input voltage range of 3 V.

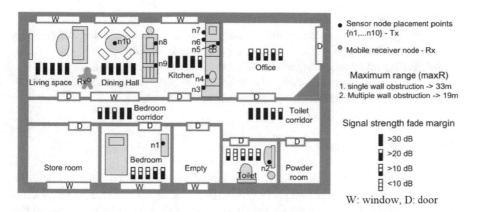

Fig 3.35 A home environment with Tx/Rx node placement point and signal strengths. (Surie et al. 2008)

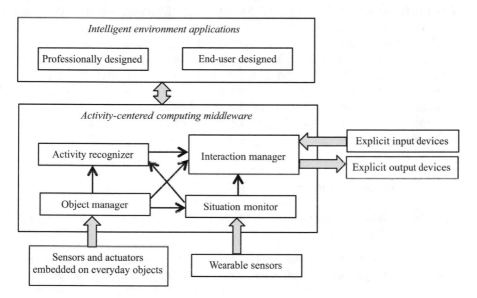

Fig. 3.36 An activity-centered wearable computing infrastructure for intelligent environment applications. (Surie and Pederson 2007)

3.6.2.2.2 Personal Server Running an Activity-Centered Computing Middleware

The sensor data from the everyday objects are received and processed within the object manager, a software component of the activity-centered computing middleware described in (Surie and Pederson 2007) (Fig. 3.36). The middleware object manager is implemented in C#, and is loaded with several tasks:

- Discovering the set of everyday objects present in the user's environment.
- Querying a database during the initialization phase for additional information about the everyday objects based on their unique identities.
- Initializing and managing communication with the everyday objects.
- Transferring information, about the everyday objects, to other middleware components and to the intelligent environment applications.

3.6.2.2.3 Experimental Setup

Four volunteers for a week's duration performed the experiments individually. In addition to the system components (wearable personal server and receiver node), the volunteers were given a wearable camera connected to a mobile digital video recorder (DVR) to obtain the ground truth (state changes of everyday objects). Experience sampling method (ESM) (Intille et al. 2003) is a commonly used method to obtain the ground truth, such that the subject manually enters to a PDA the events he produces while performing activities.

The volunteer subjects were asked to collect ground truth for the set of 10 activities presented in Table 3.4. Time-based synchronization was used to map the sensor firings with the ground truth. The subjects were not restricted on how to perform the activities, however, were briefed on how to use the system. For a week, the activities were performed from a minimum of 7 times to a maximum of 20 times. The subjects were interviewed after the experimentation period for qualitative evaluations.

3.6.2.2.4 System Evaluation

3.6.2.2.4.1 Wireless Communication: Transmission Reception Range and Signal Strength measures

In an outdoor environment, the transmission reception range is usually evaluated within line of sight and free of obstructions. However, in a typical indoor home environment, the transmission reception range is 33 m with a single wall obstruction, and 19 m with multiple wall obstructions. Signal strength is measured with a certain amount of background noise, created when the everyday objects like the fridge, microwave oven, regular oven, vacuum cleaner, etc., are turned on. In Fig. 3.35, one of the subjects marked by Rx is stationed near the dining room. The signal strength from the various sensor nodes was evaluated by a push button (ON/OFF event) and was recorded five times at eight different locations in the home environment. It is revealed that 97.5% of the times; the signal strength values at the eight different locations are acceptable (>10 dB). Table 3.3 lists the signal strength at the eight locations. In both the toilet-toilet corridor case, and the bedroom-bedroom corridor case, the state of the door (closed or open) was a factor to consider. Similarly, the line-of-sight cases (living space, dining hall, and kitchen) have performed better than cases having wall obstructions.

Table 3.3 Signal strengths at eight locations with the subject located near the dining hall (Surie et al. 2008)

Signal strength	Living space (%)	Dining hall (%)	Kitchen (door open) (%)	Bedroom corridor (%)	Bedroom (door closed) (%)	Toilet (door closed) (%)	Toilet corridor (%)	Office (door open) (%)
Best (>30 dB)	100	100	80	60	0	0	60	0
Good (>20 dB)	0	0	20	40	40	0	40	60
Medium (>10 dB)	0	0	0	0	60	80	0	40
Low (<10 dB)	0	0	0	0	0	20	0	0

3.6.2.2.4.2 Sensing Precision and Recall Values

The accuracy in sensing the object state changes based on the user's interaction with those objects is an important factor to evaluate. The sensing system was tested within scenarios where the subjects performed a set of everyday activities. Several measures were adopted while evaluating:

- The sensing system should be evaluated as a whole instead of being a sum of individually isolated parts.
- The sensing system should be evaluated in a realistic setup where the subjects are performing everyday activities by interacting with everyday objects.
- The data collected for activity recognition should be used with the additional information known about the accuracy of the sensing system.

The precision and recall metrics adopted to measure system performance are defined as follows:

$$\text{Precision} = \frac{\text{True Positives}}{\text{True Positives} + \text{False Positives}} \quad (3.16)$$

$$\text{Recall} = \frac{\text{True Positives}}{\text{True Positives} + \text{False Negatives}} \quad (3.17)$$

Table 3.4 shows that the sensing system has an overall precision value of 91.2% and an overall recall value of 98.8%. The results are promising considering the amount of background noise present in a wireless environment and the fact that sensing some of the internal state changes of objects was tricky. For instance, the ambient light present in the user's environment affected the decision of determining if the dustbin, that uses a light sensor, was full or empty. Hence, there was a need for performing ambient light noise cancelation for cases involving the light sensor. The location, number, and type of sensors embedded onto everyday objects are significant factors to address for obtaining good performance measures.

Table 3.4 Percentage precision and recall values after sensing object state changes for a set of 10 everyday activities (Surie et al. 2008)

Activity	Sensing precision	Sensing recall
1. Drinking coffee	84.1	98.8
2. Baking cake, bread, etc.	100.0	98.8
3. Doing the dishes	90.0	100.0
4. Repairing the coffee machine	74.7	88.0
5. Changing clothes	91.0	99.0
6. Heating up the frozen food	93.6	100.0
7. Toilet routine	99.0	99.0
8. Preparing dinner	87.6	99.0
9. Setting up the table	100.0	95.7
10. Having dinner	100.0	100.0
Global	**91.2**	**98.8**

It is worth noting that the installation of the WSN was performed by two individuals separately. One of them needed 65 minutes, while the other one took 45 minutes to install 81 sensors onto 42 objects. The mobile receiver connected to the personal server weighted 0.632 Kg with dimensions of $15 \times 10 \times 4$ cm^3. All other sensor nodes are instrumented in the environment instead of including them in the user's wearable outfit. The wearable camera and DVR were used only for experimentation purpose but are not part of the actual system.

3.6.3 What Else?

What is applicable at home is valid for office, and for activities confined to buildings, with care accorded to scale adaptation when necessary. Noteworthy, smart installations are built upon harmonizing interactions between several technologies, such as sensor nodes, radio communication, and smart outlets. Smart energy saving is a typical application that leans toward less sensor dependence for more of radio communication and smart outlets as to be elucidated.

A scheme for human detection is applicable via the automatic control of home appliances' power consumption; it uses a wireless smart outlets network, and changes of received signal strength indicator (RSSI) between stationary communication nodes (2.4 GHz smart outlets). The main idea is to monitor the changes of RSSI which violate the established radio communication field between nodes inside a room, due to a human's presence (Mrazovac et al. 2012). A person entering into the established radio communication field induces the change of RSSI, which is periodically read during the message exchange between wireless nodes. Based on the detected changes with regard to the initial thresholds, the system detects human presence and responds with the automatic control of power consumption of all appliances connected to the power network. Such an approach saves installation costs, and increases users' awareness by contributing to the energy savings. Instead of integrating various sensors which require complex installation and processing algorithms, use of existing smart outlets and light switches has two roles: (1) they are in charge of controlling the plugged device as well as of giving an overview on energy consumption; (2) they detect the human presence inside a room by using an existing wireless network established for communication between nodes (smart outlets or light switches) and making use of the RSSI change (Table 3.5).

3.7 Multimedia Applications

Scalar sensor networks measure physical phenomena, such as temperature, pressure, humidity, or location of objects that can be conveyed through low-bandwidth and delay-tolerant data streams. More and more, the focus is shifting toward research and practice aimed at revisiting the sensor network paradigm to enable delivery of

Table 3.5 Used sensor types and their performance parameters (Surie et al. 2008)

Sensor Type	Measures	Sensor(s)	Range	Resolution	Power consumption
Temperature	Temperature	Dallas DS18S20 Thermometer[a]	−55 °C to+125 °C	±0.5 °C	Active mode: 1 to 1.5 mA Sleep mode: 750 to 1000 nA
Ambient light	Ambient light intensity	Omnidirectional light sensor (TAOS TSL250R light sensor)[a]	0 to 255	137 mV/ (W/cm^2) at 635 nm	Active mode: 1.1 to 1.7 mA
Pressure pad	Spot(s) where pressure is applied within an area	Network of multiple pressure sensors (Omron microswitch SS-5)[a] spread across a surface	Depends on the size of the pressure pad. A 50 cm × 10 cm pressure pad was used	4 cm^2	Active mode: 1 mA
Touch /Press button	Button touched /Pressed	Push button	Binary: 0 or 1	—	Active mode: 1 mA to 2 mA
Touch /Press pad	Button (s) touched / pressed	Network of multiple push buttons spread across a surface	Depends on the size (no. of push buttons included) of the touch / press pad. 35 cm × 15 cm and 20 cm × 10 cm pads were used	Resolution of the touch /press button	Depends on the power consumption of an individual push button and the number of push buttons included in the network
Appliance feedback light	Feedback light intensity	Light sensor (TAOS TSL250R)[a] made unidirectional	0 to 255	137 mV/ (W/cm^2) at 635 nm	Active mode: 1.1 mA to 1.7 mA
Open-Close	Light intensity	Network of light sensors (TAOS TSL250R)[a] spread across surfaces	Depends on the size (no. of light sensors included) of the object. 15.75 decimeter^3 to 270 decimeter^3 volumes were used	Depends on the light sensor and the ambient light	Depends on the power consumption of an individual light sensor and the number of light sensors included in the network
Containment limiter	Light intensity	Network of light sensors (TAOS TSL250R)[a] spread across surfaces	Depends on the size (no. of light sensors included) of the object. 6.6 decimeter^3 to 44 decimeter^3 volumes were used	Depends on the light sensor and the ambient light	Depends on the power consumption of an individual light sensor and the number of light sensors included in the network

[a]Datasheet is available in Chap. 10

multimedia content, such as audio and video streams and still images, as well as scalar data. This trend led to distributed, networked systems, referred to as wireless multimedia sensor networks (WMSNs).

The integration of low-power wireless networking technologies with inexpensive hardware such as complementary metal-oxide semiconductor (CMOS) cameras and microphones is enabling WMSNs, that is, networks of wireless, interconnected smart devices that enable retrieving video and audio streams, still images, and scalar sensor data. As an example, the Cyclops image capturing and inference module (Rahimi et al. 2005), designed for extremely lightweight imaging, can be interfaced with a host mote such as Crossbow's MICA2 or MICAz, thus realizing an imaging device with processing and transmission capabilities. WMSNs enable the retrieval of multimedia streams and will store, process in real-time, correlate, and fuse multimedia content captured by heterogeneous sources.

The characteristics of a WMSN diverge consistently from traditional network paradigms, such as the Internet and even from scalar sensor networks. Most potential applications of a WMSN require the sensor network paradigm to be rethought to provide mechanisms to deliver multimedia content with a predetermined level of quality of service (QoS). Whereas minimizing energy consumption has been the main objective in sensor network research, mechanisms to efficiently deliver application-level QoS and to map these requirements to network-layer metrics, such as latency and jitter, have not been primary concerns.

3.7.1 Network Architecture

The architecture of WMSNs where users connect through the Internet and issue queries to a deployed sensor network is shown in Fig. 3.27. The functionalities of the various network components are summarized in a bottom-up manner as listed beneath (Akyildiz et al. 2007):

- Scalar sensors. These sensors sense scalar data and physical attributes, such as temperature, pressure, and humidity and report measured values to their cluster-head. They are typically resource-constrained devices in terms of energy, storage capacity, and processing capability.
- Multimedia processing hubs. These devices have comparatively large computational resources and are suitable for aggregating multimedia streams from the individual sensor nodes. They are integral to reducing both the dimensionality and the volume of data conveyed to the sink and storage devices.
- High-end video sensors. Located at the upper tier of the WMSN, they are directly connected to the sink, to send their own data or the aggregated data received from the multimedia processing hubs. They are of medium or high resolution.
- Storage hubs. Depending upon the application, the multimedia stream is desired in real-time or after further processing. These storage hubs allow data-mining and

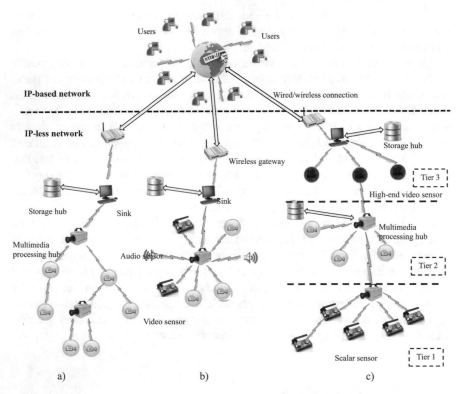

Fig. 3.37 Typical architecture of a WMSN. (**a**) Single-tier flat, homogeneous sensors, distributed processing, centralized storage. (**b**) Single-tier clustered, heterogeneous sensors, centralized processing, centralized storage. (**c**) Multitier, heterogeneous sensors, distributed processing, distributed storage. (Based on (Akyildiz et al. 2007))

feature extraction algorithms to identify the important characteristics of the event, even before the data are sent to the end user.

- Sink. The sink is responsible for packaging high-level user queries to network-specific directives and returning filtered portions of the multimedia stream back to the user. Multiple sinks may be required in a large or heterogeneous network.
- Gateway. It serves as the last mile connectivity by bridging the sink to the Internet, and it is also the only IP-addressable component of the WMSN. A gateway is meant to maintain a geographical estimate of the area covered under its sensing framework to allocate tasks to the appropriate sinks that forward sensed data through it.
- Users. Positioned at the highest ends of the hierarchy, they issue-monitoring tasks to the WMSN based on geographical regions of interest. Users are typically identified through their IP addresses, and run application-level software that assigns queries and displays results obtained from the WMSN (Fig. 3.37).

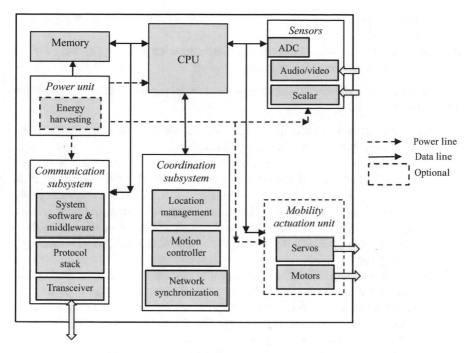

Fig. 3.38 Internal organization of a multimedia sensor. (Akyildiz et al. 2007)

3.7.2 Design Issues of WMSNs

A multimedia sensor device may be composed of several basic components, as shown in Fig. 3.38, a sensing unit, a processing unit (CPU), a communication subsystem, a coordination subsystem, a storage unit (memory), and an optional mobility/actuation unit. Sensing units are usually composed of two subunits, sensors (cameras, microphones, and/or scalar sensors) and analog-to-digital converters (ADCs).

Overall components' functioning is emphasized to be (Akyildiz et al. 2007):

- The analog signals produced by the sensors, based on the observed phenomenon, are converted to digital signals by the ADC and then fed into the processing unit.
- The processing unit executes the system software in charge of coordinating sensing and communication tasks and is interfaced with a storage unit.
- The communication subsystem interfaces the multimedia sensor device to the network and is composed of a transceiver unit and of communication software. The communication software includes a communication protocol stack and system software, such as middleware, operating systems, and virtual machines.
- The coordination subsystem is in charge of coordinating the operation of different network devices, by performing operations such as network synchronization and

location management. An optional mobility/actuation unit can enable movement or manipulation of objects.

- Finally, a power unit that may be supported by an energy-scavenging unit, such as solar cells, powers the whole system.

The design of a WMSN is subject to several factors as itemized below (Akyildiz et al. 2008):

- Resource constraints. Sensor devices are constrained in terms of battery, memory, processing capability, and achievable data rate.
- Variable channel capacity. In multihop wireless networks, the capacity of each wireless link depends on the interference level perceived at the receiver. This, in turn, depends on the interaction of several functions that are distributively handled by all network devices such as power control, routing, and rate policies. Hence, the capacity and the delay attainable on each link are location-dependent, vary continuously, and may be bursty in nature, thus making QoS provisioning a challenging task.
- Cross-layer coupling of functionality. Because of the shared nature of the wireless communication channel, in multihop wireless networks, there is a strict interdependence among functions handled at all layers of the communication stack. This interdependence must be explicitly considered when designing communication protocols aimed at QoS provisioning.
- Application-specific QoS requirements. In addition to data delivery modes that are typical of scalar sensor networks, multimedia data include snapshot and streaming multimedia content. Snapshot-type multimedia data contain event-triggered observations obtained in a short time period (e.g., a still image). Streaming multimedia content is generated over longer time periods and requires sustained information delivery.
- High-bandwidth demand. Multimedia contents, especially video streams, require transmission bandwidth that is orders of magnitude higher than that supported by current off-the-shelf sensors. Hence, high data rate and low-power, consumption-transmission techniques must be leveraged. In this respect, the ultra-wide-band (UWB) transmission technology seems particularly promising for WMSNs.
- Multimedia source coding techniques. State-of-the-art video encoders rely on intra-frame compression techniques to reduce redundancy within one frame, and on inter-frame compression (also predictive encoding or motion estimation) to exploit redundancy among subsequent frames in order to reduce the amount of data to be transmitted and stored. Because predictive encoding requires complex encoders, powerful processing algorithms, and also entails high-energy consumption, it may not be suited for low-cost multimedia sensors. However, it was shown in (Girod et al. 2005) that the traditional balance of complex encoder and simple decoder can be reversed within the framework of so-called distributed source coding. These techniques exploit the source statistics at the decoder and by shifting the complexity at this end, and enable the design of simple encoders. Clearly, such algorithms are very promising for WMSNs, where it may not be

feasible to use existing video encoders at the source node due to processing and energy constraints.

- Multimedia in-network processing. Processing of multimedia content has been approached mainly as a problem isolated from the network design problem, with a few exceptions, such as joint source-channel coding (Johnson et al. 2006) and channel-adaptive streaming (Kurkowski et al. 2005). Similarly, research that addressed the content delivery aspects has typically not considered the characteristics of the source content and has primarily studied cross-layer interactions among lower layers of the protocol stack. However, processing and delivery of multimedia content are not independent, and their interaction has a major impact on the achievable QoS. The QoS required by the application will be provided by means of a combination of cross-layer optimization of the communication process and in-network processing of raw data streams that describe the phenomenon of interest from multiple views, with different media, and on multiple resolutions. Hence, it is necessary to develop application-independent and self-organizing architectures to flexibly perform in-network processing of multimedia contents.

In the ensuing sections, WMSNs applications and hardware platforms are to be detailed.

3.7.3 WMSNs Applications

WMSNs enable several applications, which are broadly classified into five categories (Akyildiz et al. 2008):

- Surveillance. Video and audio sensors enhance and complement existing surveillance systems against crime and terrorist attacks. Large-scale networks of video sensors can extend the ability of law-enforcement agencies to monitor areas, public events, private properties, and borders. Multimedia sensors could infer and record potentially relevant activities (thefts, car accidents, traffic violations) and make video/audio streams or reports available for future query. Multimedia content such as video streams and still images, along with advanced signal processing techniques, may be used to locate missing persons or to identify criminals or terrorists.
- Traffic monitoring and enforcement. Which makes it possible to monitor car traffic in big cities or highways and deploy services that offer traffic routing advice to avoid congestion. Multimedia sensors may also monitor the flow of vehicular traffic on highways and retrieve aggregate information such as average speed and number of cars. Sensors could also detect violations and transmit video streams to law-enforcement agencies to identify the violator, or buffer images and streams in case of accidents for subsequent accident scene analysis. Moreover, smart parking advice systems based on WMSNs (Campbell et al. 2005) allow monitoring available parking spaces and provide drivers with automated parking advice, thus improving mobility in urban areas.

- Personal and healthcare. Multimedia sensor networks can be used to complement scalar sensors described in Sect. 3.5 to monitor and study the behavior of elderly people as a means that identifies the causes of illnesses affecting them such as dementia (Reeves 2005). Networks of wearable or video and audio sensors can infer emergency situations and immediately connect elderly patients with remote assistance services or with relatives. Telemedicine sensor networks are the type of networks that provide ubiquitous healthcare services. Patients will carry medical sensors to monitor parameters such as body temperature, blood pressure, pulse oximetry, electrocardiogram, and breathing activity. Furthermore, remote medical centers will perform advanced remote monitoring of their patients via video and audio sensors, location sensors, and motion or activity sensors, which can also be embedded in wrist devices (Sect. 3.5).
- Gaming. Networked gaming is a popular recreational activity. WMSNs will find applications in future prototypes that enhance the effect of the game environment on the game player. As an example, virtual reality games that assimilate touch and sight inputs of the user as part of the player response (Capra et al. 2005) need to return multimedia data under strict time constraints. In addition, WMSN application in gaming systems will be closely associated with sensor placement and the easiness of carrying the sensors by the players. The growing popularity of such games will undoubtedly propel WMSN research in the design and deployment of pervasive systems involving a rich interaction between the game players and the environment.
- Environmental and industrial. Several projects on habitat monitoring that use acoustic and video feeds are being envisaged, in which information has to be conveyed in a time-critical fashion. For example, arrays of video sensors are already used by oceanographers to determine the evolution of sandbars via image processing techniques (Holman et al. 2003). Multimedia content, such as imaging, temperature, or pressure, among others, may be used for time-critical industrial process control. For example, in quality control of manufacturing processes, final products are automatically inspected to find defects. In addition, machine vision systems can detect the position and orientation of parts of the product to be picked up by a robotic arm. The integration of machine vision systems with WMSNs can simplify and add flexibility to systems for visual inspections and automated actions that require high speed, high magnification, and continuous operation.

3.7.4   Hardware Platforms of WMSNs

In order to have the capability of handling multimedia applications in WMSN, the ability to support their requirements and challenges, and to examine and test the proposed protocols and algorithms developed for WMSNs, the underlying supporting technology and platforms are required to be more efficient and must overcome the drawbacks of the existing hardware designed for WSN that detect

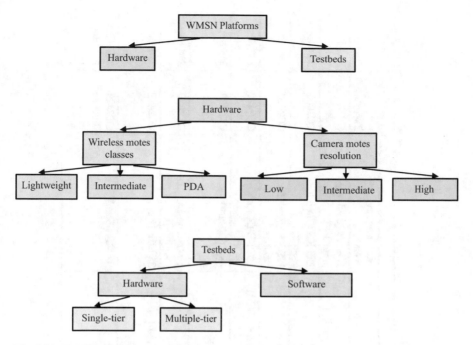

Fig. 3.39 WMSN platforms classification. (Almalkawi et al. 2010)

scalar events. Many efforts have been presented in the literature to modify the existing hardware platform or present new hardware implementation and testbeds. These proposed platform and testbeds are more powerful and have more potential to process and efficiently handle multimedia traffic in terms of processing power, memory, data rate, power consumption, and communication capabilities (Fig. 3.39). In the coming sections, currently off-the-shelf hardware is to be presented. Testbeds for WMSNs are detailed in Chap. 7.

3.7.4.1 Classification of Wireless Motes

There are several commercially available wireless motes that can be used as WMSN motes. Depending on their processing power and storage capacity, these wireless motes can be classified into three groups (Almalkawi et al. 2010):

- Lightweight-class platforms. Devices in this category are initially designed for detecting scalar data, such as temperature, light, and humidity. Their main concern is to consume less amount of energy as possible. Therefore, they have low processing power capability and little storage; most of them are equipped with a basic communication chipset, such as IEEE 802.15.4 on CC2420 radio (Texas Instruments 2005b). The CC2420 chipset just consumes 17.4 and 19.7 mA for sending and receiving, respectively, and has a maximum transmit power of 0 dBm with data rate of 250 Kbps. Table 3.6 illustrates comparative

Table 3.6 Wireless motes classified and compared based on (Almalkawi et al. 2010)

Wireless mote		Microcontroller	Memory			Radio	Data rate
			EEPROM/RAM	Flash memory /ROM			
Lightweight-class	MICA2	ATmega128L (8 bit) 7.37 MHz	4 KB (EEPROM)	128 KB (Flash for programs)		CC1000	38.4 Kbps
	MICA2DOT	ATmega128L (8 bit) 7.37 MHz	4 KB (EEPROM)	128 KB (Flash for programs)		CC1000	38.4 Kbps
	MICAZ	ATmega128L (8 bit) 7.37 MHz	4 KB (EEPROM)	128 KB (Flash for programs)		CC2420	250 Kbps
	FireFly	ATmega1281 (8 bit) 7.37 MHz	2 KB (EEPROM) + 8 KB (RAM)	128 KB (ROM)		CC2420	250 Kbps
Intermediate-class	Tmote Sky	MSP430 F1611 (16 bit) 8 MHz	10 KB (RAM)	48 KB (Flash)		CC2420	250 Kbps
	TelosB	TI MSP430 (16 bit) 8 MHz	10 KB (RAM)	1 MB (External Flash)		CC2420	250 Kbps
PDA-class	Imote2	PXA271 XScale (32 bit) 13–416 MHz	256 KB (SRAM) + 32 MB (SDRAM)	32 MB (Flash)		CC2420	250 Kbps
	Stargate	PXA255 XScale (32 bit) 400 MHz	64 MB (SDRAM)	32 MB (Flash)		CC2420 Bluetooth IEEE 802.11	250 Kbps 1–3 Mbps 1–11 Mbps

exemplars of lightweight-class wireless motes, typically, Mica-family motes (Chap. 10) and FireFly (Mangharam et al. 2007).

- Intermediate-class platforms. The devices in this group have better computational and processing capabilities, and larger storage memory compared to the lightweight-class counterparts. However, they are also equipped with low bandwidth and data rate communication modules (e.g., CC2420 IEEE 802.15.4 compatible chipset). Tmote Sky (Moteiv 2006) is such intermediate-class mote; it was used to implement a camera mote among CITRIC (Chen et al. 2008a) and CMUCam3 (Rowe et al. 2007b).

- PDA-class Platforms. The devices in this category are more powerful in terms of computational and processing power; they are designed for fast and efficient processing of multimedia content. They can run different operating systems (e.g., Linux, TinyOs) and may run Java applications and .NET micro frameworks, and also support several radio standards with different data rates (IEEE 802.15.4, IEEE 802.11, and Bluetooth). However, they relatively consume more power. Stargate and Imote2 are instances of PDA-class platforms (Chap. 10):

  - Stargate board (Crossbow 2004b), designed by Intel and manufactured by Crossbow, uses the 400 MHz 32-bit Intel PXA255 XScale RISC processor with 32 MByte Flash memory and 64 MByte SDRAM and runs Linux operating system. It can be interfaced with Crossbow's MICA2 or MICAz motes for IEEE 802.15.4 wireless communication as well as PCMCIA IEEE 802.11 wireless cards or compact Flash Bluetooth. Consequently, Stargate board can be used as a sensor network gateway, robotics controller card, or distributed computing platform. It forms a camera mote when it is connected with a camera device (e.g., webcam) as shown in the Meerkats testbed (Boice et al. 2005) to be described in Chap. 7, and in the hardware platforms laid out in (Kulkarni et al. 2005; Feng et al. 2005) as presented in Sect. 3.7.4.3.

  - Imote2 (Crossbow 2005), also designed by Intel and manufactured by Crossbow, is a wireless sensor node platform built over the low-power 32-bit PXA271 XScale processor and integrates a CC2420 IEEE802.15.4 radio with a built-in 2.4 GHz antenna. It operates in the range 13–416 MHz with dynamic voltage scaling and includes 256 KByte SRAM, 32 MByte Flash memory, 32 MByte SDRAM, and several I/O options. It can run different operating systems such as TinyOs and Linux with Java applications and it is also available with .NET microframework. Imote2 integrates many I/O options making it extremely flexible for supporting different sensors including cameras, A/Ds, radios, etc. The PXA271 processor includes a wireless MMX coprocessor to accelerate multimedia operations and add media processor instructions to support alignment and video operations. Imote2 uses in address-event imagers (Teixeira et al. 2006) as will be illustrated in Sect. 3.7.4.3; an Imote2 implementation of a dual-camera sensor is presented in (Xie et al. 2008).

A comparison between wireless motes classes is provided in Table 3.6.

3.7.4.2 Camera Motes Features

To reduce the amount of resources required for transmitting multimedia traffic (images, videos) over WMSN, the multimedia content should be intelligently manipulated and processed using appropriate compression and coding algorithms along with other application-specific multimedia processing such as background subtraction and feature extraction. However, most of these algorithms are complex and require high computational and processing power as well as larger memory for buffering frames. Sometimes, these requirements cannot be satisfied with the limited resources offered by the wireless motes as previously mentioned, especially if they require floating-point operations for efficient multimedia processing. Therefore, camera sensors are to be coupled with additional processor (microcontrollers, DSPs, FPGAs, etc.) and memory resources before relaying the processed data to the wireless mote for wireless communication. Yet, the additional processor and memory resources require more energy consumption and cost, which prompts a trade-off between energy consumption and cost against computational power and traffic amount. It has been shown that the time needed to perform relatively complex operations on a 4 MHz 8-bit processor, such as the ATmega128 (Atmel 2011a), is 16 times higher than the time needed with a 48 MHz 32-bit ARM7 device, while the power consumption of the 32-bit processor is only six times higher (Downes et al. 2006). This designates that powerful processors, such as 32-bit ARM7 architecture, are more power-efficient in multimedia applications.

Section 3.7.4.3 surveys the existing WMSN platforms and compares their specifications and use. It is perceived that camera motes have different capabilities; consequently, they have different functionalities that permit them to have different roles. For instance, low-resolution cameras can be used at the lower tier of a multitier network for a simple object detection task that exploits their low-power consumption feature, which allows them to be turned on most of the time (or in duty cycle manner). Cyclops (Rahimi et al. 2005), CMUCam3 (Carnegie Mellon University 2007), and eCam (Park and Chou 2006a) are models of low-resolution cameras. Intermediate and high-resolution cameras can be used at higher tiers of the network for more complex and power-consuming tasks, such as object recognition and tracking. These types of cameras consume more power and hence are awakened on-demand by lower-tier devices, when detecting an object of interest. Webcams, attached with Stargate board or Imote2, can be considered as intermediate-resolution cameras, while PTZ cameras used in (Kulkarni et al. 2005) are typical high-resolution cameras.

3.7.4.3 Available Camera Mote Platforms

3.7.4.3.1 Cyclops

Cyclops (Rahimi et al. 2005) is a small camera device developed for WMSN. It is compatible with the computationally constrained WSNs (motes) and exploits the characteristics of CMOS camera sensors as they are low power, low cost, and of small size. Cyclops platform isolates the camera module requirement for high-speed data transfer from the low-speed capability of the embedded controller and provides still images at low rates. It is designed to interface with the common motes used in WSNs such as MICA2 (Crossbow 2002a) and MICAz (Crossbow 2006a) detailed in Chap. 10. Cyclops hardware architecture consists of (Fig. 3.40):

- An imager, Agilent compact common intermediate format (CIF) CMOS ADCM-1700 (Agilent Technologies 2003b). The clock is set to 4 MHz to give the CPLD enough time to grab an image pixel and copy it into memory. The CIF resolution of the image sensor is 352 × 288. The output of the image array is digitized by a set of ADCs.

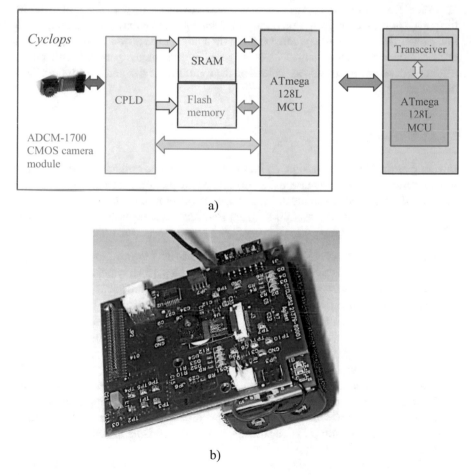

Fig. 3.40 Cyclops platform. (**a**) Cyclops platform-building blocks. (**b**) Cyclops with an attached MICA2 mote. (Rahimi et al. 2005)

- An 8-bit RISC Atmel ATmega128L MCU (Atmel 2011a). It controls Cyclops for images capturing and for interfacing the camera with a lightweight wireless host.
- A Xilinx XC2C256 CoolRunner (Xilinx 2007) complex programmable logic device (CPLD). With its 16 MHz high-speed clock, compared to the 4 MHz imager, it provides synchronization and memory control required for image capturing. CPLD acts as a lightweight frame grabber to provide on-demand access to high-speed clocking at capture time and to perform limited amount of image processing such as background subtraction or frame differentiation.
- An external 64 KByte extended RAM (SRAM) (Toshiba 2002), and an external 512 KByte CMOS Flash programmable and erasable ROM (Atmel 2003).

Cyclops firmware is written in nesC language (UC Berkeley WEBS Project 2004) and runs under TinyOS operating system (TinyOS 2012). In addition to the libraries provided by TinyOS, Cyclops provides primitive structural libraries, such as matrix operation libraries or histogram libraries, and advanced or high-level algorithms libraries, such as coordinate conversion and background subtraction. Cyclops is a low-power device; its energy consumption depends on the power consumption and time duration at different states, like image capturing, memory access, microcontroller processing, sleep, etc., as well as on the input image size and the ambient light intensity.

3.7.4.3.2   Panoptes

In (Feng et al. 2003) the design, implementation, and performance of video-based sensor networking architecture using visual sensor platform, called Panoptes, are introduced. The Panoptes platform was originally developed on the Bitsy board from Applied Data Systems (Liotta 2000). The video sensor as developed is based on the following components:

- Intel StrongARM 206 MHz embedded platform (Intel 2000). The SA-1110 is a general purpose, 32-bit RISC microprocessor with a 16 KByte instruction cache (Icache), an 8 KByte write-back data cache (Dcache), a minicache, a write buffer, a read buffer, a memory-management unit (MMU), a liquid crystal display (LCD) controller, and serial I/O combined in a single component. The SA-1110 provides portable applications with high-end computing performance without requiring users to sacrifice available battery time.
- 64 MBytes of memory.
- Linux 2.4.19 operating system kernel. Linux was adopted because it provides the flexibility necessary to modify parts of the system to specific applications. The functionality of the video sensing itself is split into a number of components including capture, compression, filtering, buffering, adaptation, and streaming.
- A Logitech 3000 USB-based video camera.
- An 802.11-based networking card. While 802.11 is being used, it is possible to replace it with a lower-powered, lower frequency RF radio device.

The complete device including the compression and transmission over 802.11 consumes approximately 5.5 Watts of power while capturing and delivering video of 320 × 240 resolution at 18–20 fps. It is approximately 17.8 cm long (with the 802.11 card inserted) and approximately 10.2 cm wide.

Panoptes was also implemented on the Crossbow Stargate platform (Feng et al. 2005). The Stargate platform (Crossbow 2004b) features the 400 MHz 32-bit Intel PXA255 XScale RISC processor with 32 MB Flash memory and 64 MB SDRAM and runs Linux operating system. It is also equipped with an 802.11b PCMCIA wireless card (11 Mb/s) and a USB camera.

Panoptes accomplished an array of meaningful contributions:

- A low-power, high-quality video-capturing platform is developed, to serve as the basis of video-based sensor networks as well as other application areas such as virtual reality or robotics.
- A prioritizing buffer management algorithm is designed to effectively deal with intermittent network connectivity or disconnected operation to save power.
- A bit-mapping algorithm is designed for the efficient querying and retrieval of video data.

3.7.4.3.3   Address-Event Imagers

An implementation is available in (Teixeira et al. 2006) for a camera mote aimed at behavior recognition in WMSNs based on biologically inspired address-event imagers and sensory grammars. In address-event representation (AER), the camera networks operate on symbolic information rather than on images by filtering out all redundant information at the sensor level and outputting only selected handful of features in address-event representation. This leads to minimizing power consumption and bandwidth; only a few μW of power in active camera state is consumed. A different computation model, that is faster and more lightweight than conventional image processing techniques, is used. The output of the AE imagers can be connected into the sensing grammar that converts low-level sensor measurements to higher level behavior interpretation based on probabilistic context free grammars (PCFGs).

Three different platforms have been developed to experiment the above techniques where each platform is built on top of the XYZ sensor node (Lymberopoulos and Savvides 2005). XYZ uses an Oki ML67Q5002 processor (Oki Semiconductor 2004) based on ARM7TDMI core running at 58 MHz. The processor has 32 KByte of internal RAM and 256 KByte of Flash, and there is an additional 2 Mbit memory available onboard. XYZ platform operates on SOS, which is a lightweight operating system that follows an event-driven design similar to TinyOS. Unlike TinyOS, SOS supports the use of dynamically loadable modules (Han et al. 2005):

a)   b)

Fig. 3.41 XYZ sensor node interfaced to COTS camera modules. (**a**) XYZ-OV7649. (**b**) XYZ-ALOHA. (Teixeira et al. 2006)

- The first platform is XYZ-OV7649 (OmniVision Technologies 2003), an XYZ sensor node with a camera sensor from OmniVision (OmniVision Technologies 2011) that can capture color images at resolution of 640 × 480 VGA and 320 × 240 quarter-VGA (QVGA), and also supports a windowing function that allows the user to acquire images at different resolutions by defining a window on the image plane (Fig. 3.41a).
- The second platform is XYZ-ALOHA, an XYZ sensor node with ALOHA image sensor (Teixeira et al. 2005) that is composed of four quadrants of 32 × 32 pixels, and is able of generating 10,000 events in 1.3 sec with a power consumption of 6 µW per quadrant. The ALOHA image sensor uses the simple ALOHA medium access technique to transmit individual events to a receiver (Fig. 3.41b).
- The third platform consists of a software emulator of AE imagers. It allows quick simulation of AER imager prototypes, as well as the development of algorithms for these prototypes before they are even fabricated. The software is written in Visual C++ and runs under Windows. It takes an 8-bit grayscale input stream from a commercial off-the-shelves (COTS) USB camera and outputs a queue of events to a text file. Additionally, an image may be displayed by constructing it from the output events.

Afterward, Imote2 was used with OmniVision OV7649 camera (Teixeira and Savvides 2007). A lightweight, online people-counter utilizing a novel, AE-friendly motion-histogram is developed. The histogram is robust to pixel intensity fluctuations, gradual lighting changes, and furniture repositioning. Abrupt alterations in lighting may, at times, cause false positives, but they vanish within a few frames.

3.7.4.3.4   eCAM

eCAM (Park and Chou 2006a) is an ultra-compact, high data rate wireless sensor node with a miniature camera. It is constructed by interfacing a video graphics array (VGA) quality digital video camera with the Eco node (Park and Chou 2006b). The Eco sensor node (Fig. 3.42) consists of four subsystems:

Fig. 3.42 Eco sensor node
on the index finger. (Park
and Chou 2006b)

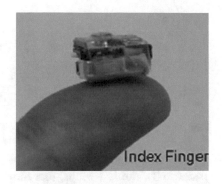

- MCU/Radio. nRF24E1 is a 2.4 GHz RF transceiver with embedded 8051 compatible microcontroller, and 9 input 10 bit 100 KSample/sec ADC (Nordic Semiconductor 2004). The MCU has a 512 Byte ROM for a bootstrap loader, a 4 KByte RAM for the user program, SPI (3-wire), RS-232, and a 9-channel ADC. The ADC is software-configurable for 6–12 bits of resolution. A 32 KByte SPI EEPROM stores the application program. The nRF24E1's 2.4 GHz transceiver uses a Gaussian frequency-shift keying (GFSK) modulation scheme with 125 frequency channels that are 1 MHz apart. The transmission output power is also software-configurable for four different levels: -20 dBm, -10 dBm, -5 dBm, and 0 dBm. The AN9520 RainSun chip antenna (RainSun 2009) has a maximum gain of 1.5 dBi.
- Sensors. Eco has a 3-axial acceleration sensor, Hitachi-Metal H34C (3.4 mm × 3.7 mm × 0.92 mm). It measures acceleration from -3 g to $+3$ g and temperature from 0 to 75° C, while consuming 0.36 mA at 3 V in active mode. Eco has also an S1087 light sensor.
- Power. Eco's power subsystem includes a 3.3 V regulator (LTC3204-3.3 V), battery protection circuitry, and a custom 30 mAh rechargeable Li-Polymer battery.
- Expansion Port. Eco's expansion port has 16 pins, including 4 digital I/Os, one analog input, SPI, RS232, 3.3 V output, and voltage input for a regulator and battery charging. This port enables Eco to interface with other sensing devices such as image sensor, gyroscope, pressure sensor, or compass.

In addition to the Eco node, eCAM contains (Fig. 3.43):

- A C328-7640 camera module, which can operate as either a video camera or a JPEG compressed still camera. It consists of a lens, an OmniVision's OV7640 image sensor (OmniVision Technologies 2003), and an OmniVision's OV528 compression/serial-bridge chip (OmniVision Technologies 2002). The OV7640 is a low-voltage CMOS image sensor that supports various image resolutions (VGA/CIF/SIF/QCIF/160 × 128/80 × 64) as well as various color formats (4 gray/16 gray/256 gray/12-bit RGB/16-bit RGB). It can capture up to 30 frames per second (fps) and provide complete user control over image quality, formatting, and output data transfer.
- 170 mAh PL-052025x1Li-Polymer battery.

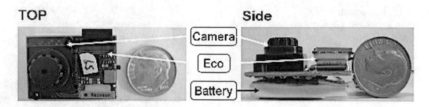

Fig. 3.43 eCAM with US dime coin for scale. (Park and Chou 2006a)

eCAM is capable of much higher data rate than most platforms; its theoretical peak bandwidth is 1 Mbps, four times Zigbee's 250 Kbps, and can reliably deliver the live video feed without further compromising video quality. eCAM achieves higher efficiency by:

- In-camera hardware compression, which is much more power efficient than software implementations.
- High-speed, low-power wireless communication interface with a simple MAC, instead of a complex MAC with much higher power.
- Overall streamlined system-level design from the camera, node, and RF to the basestation and uplink.
- Highly optimized board-level system design for very compact form factor.

3.7.4.3.5 WiSN

A mote architecture with minimal component count was introduced at Stanford's Wireless Sensor Networks Lab (Downes et al. 2006). It deploys several components at its core (Fig. 3.44):

- A 32-bit ARM7 microcontroller operating at clock frequencies up to 48 MHz, and accessing up to 64 KByte of on-chip RAM, and up to 256 KByte on-chip Flash (Atmel 2011c).
- Up to 2 MBytes Flash memory.
- Wireless communication is provided by the Chipcon CC2420 radio (Texas Instruments 2005b), which operates in the 2.4 GHz ISM band and is compliant with the IEEE 802.15.4 standard for low power, low data rate (250 Kbps) communication.
- An integrated USB and serial debug interface allows simple programming and debugging of applications.
- Agilent ADCM-1670, 352 × 288 CIF resolution, CMOS camera module (Agilent Technologies 2003a), and Agilent ADNS-3060 high-performance optical mouse Sensor (Agilent Technologies 2004)

This mote connects to multiple vision sensors as it can host up to four low-resolution Agilent ADNS-3060 sensors, and two Agilent ADCM-1670 camera modules. Both types of vision sensors feature a serial interface thus eliminating the

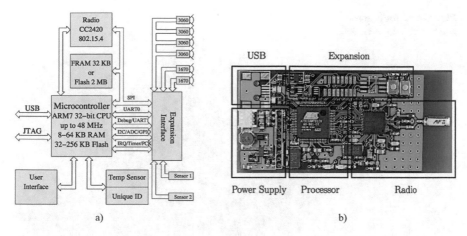

Fig. 3.44 WiSN mote. (**a**) Block diagram, (**b**) Implementation. (Downes et al. 2006)

need for additional FPGA or CPLD devices. In addition to interfacing to cameras, the mote is able to connect to other sensors (passive infrared, temperature, pressure, humidity, etc.).

3.7.4.3.6 FireFly Mosaic

FireFly Mosaic, a vision-enabled wireless sensor platform and image processing framework (Rowe et al. 2007a), uses camera motes consisting of FireFly wireless node coupled with a CMUcam3 camera sensor (Rowe et al. 2007b):

- FireFly sensor node has a low-power Atmel ATmega 128 L 8-bit processor with 8 KByte RAM and 128 KByte Flash memory (Atmel 2011a), connected with Chipcon CC2420 802.15.4 (Texas Instruments 2005b) radio capable of transmitting at 250 Kbps for up to 100 m. The platform has several sensors, namely, light, temperature, sound, passive infrared motion detection, and dual-axis acceleration Fig. 3.45. The FireFly nodes run the Nano-RK real-time operating system (Eswaran et al. 2005) and communicate wirelessly using the RT-link collision-free TDMA-based protocol. FireFly Mosaic is designed to be low-cost, energy-efficient, and scalable compared to the centralized wireless webcam-based solution. The used RT-link provides tight global time synchronization to prevent collisions and to save energy, while Nano-RK operating system provides hooks for globally synchronized task processing and camera frame capturing. While network communication relies on TDMA-based link layer, the internal communication between the camera and the wireless node is based on the serial line IP (SLIP).

Fig. 3.45 FireFly sensor
node. (Eswaran et al. 2005)

Fig. 3.46 CMUcam3
mated with the CMOS
camera board. An MMC
memory card for mass
storage is seen on the right
side of the board. The board
is 5.5 cm × 5.5 cm and about
3 cm deep depending on the
camera module. (Rowe et al.
2007b)

- The CMUcam3 camera of FireFly Mosaic consists of (Fig. 3.46):

  - CMOS OmniVision OV6620 camera chip (OmniVision Technologies 1999)
    capable of capturing 50 352 × 288 color images per second.
  - An AL440B 4 MBits FIFO Field Memory (AverLogic Technologies 2002).

- An ARM7TDMI-S core-based microcontroller that features a 16/32 bit LPC2106 with 128 KBytes Program Flash, a 64 KBytes RAM, a real-time clock (RTC), and up to 60 MHz operation (Philips 2004).
- Four on-chip servo controller outputs, which can be used to actuate a pan-tilt device.

CMUcam3 is an open-source camera that comes with several libraries (named CC3) and example applications such as JPEG compression, frame differencing, color tracking, convolutions, edge detection, connected components analysis, and a face detector. This multiplicity of image processing algorithms can be run at the source; the results are sent over the multihop wireless channel to the FireFly gateway. CMUcam3 can be also interfaced with other type of sensor nodes such as (Moteiv 2004), Telos (Polastre et al. 2005), and Tmote Sky (Moteiv 2006) motes running different operating systems. Table 10.6 compares Berkeley motes, and datasheets are made available in Chap. 10.

3.7.4.3.7 MeshEye

An energy-efficient smart camera mote, MeshEye (Hengstler et al. 2007), is proposed for distributed intelligent surveillance application in WMSN. A "smart" camera is a camera that can do onboard processing itself instead of transmitting all video data to a central controller (Fig. 3.47). The smart camera is motivated by the much lower power consumption of processing compared to transmitting raw data through a wireless link.

MeshEye mote architecture is designed to support in-node image processing with sufficient processing power capabilities, for distributed intelligent algorithms in two tiers WSN, while minimizing component count and power consumption:

- In the first tier, a low-resolution stereovision system is used to determine position, range, and size of moving objects in its field of view.
- The second tier includes high-resolution color cameras that are triggered in case of detecting objects by the first tier.

The MeshEye architecture is hosting several components (Fig. 3.48):

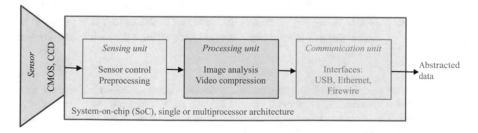

Fig. 3.47 Typical architecture of a smart camera. (Rinner and Wolf 2008)

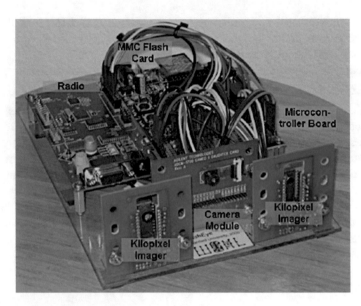

Fig. 3.48 MeshEye platform. (Hengstler et al. 2007)

- An Atmel AT91SAM7S family microcontroller (Atmel 2008a) at the core of the architecture. It features a USB 2.0 full-speed port and a serial interface for wired connection.
- A multimedia card/secure digital (MMC[18]/SD[19]) flash memory card that provides sufficient and scalable non-volatile memory for temporary frame buffering or even image archival.
- The mote can host up to eight kilopixel imagers[20] (The Free Dictionary 2014) and one VGA camera module, equipped with an Agilent Technologies' ADNS-3060 high-performance, 30×30 pixel, 6-bit grayscale, optical mouse sensor (Agilent Technologies 2004), and an Agilent Technologies' ADCM-2700 landscape VGA resolution CMOS camera module (640×480 pixel programmable, grayscale or 24-bit color). A datasheet for the ADCM-2700 is not available, close by features are available in the ADCM-2650 (Agilent Technologies 2003c).
- Wireless connection to other motes in the network can be established through Texas Instruments CC2420, a 2.4 GHz IEEE 802.15.4/ZigBee-ready 250 Kbit/s, 1 mW transmit power, RF transceiver (Texas Instruments 2005b). Although the supported data rate is not high enough for multimedia streaming, it is possible to perform in-node intermediate-level visual processing for efficient image

---

[18] A tiny memory card that uses flash memory to make storage portable among various devices, such as car navigation systems, cellular phones, eBooks, PDAs, ... (TechTarget 2014b)

[19] A tiny memory card used to make storage portable among various devices, such as car navigation systems, cellular phones, eBooks, PDAs, ... (TechTarget 2014a)

[20] An electronic device that records images.

compression and/or descriptive representations, such as axis projection, color histogram, or object shaping.

- The mote can either be powered by a stationary power supply, if available, or it may be battery-operated for mobile applications or ease of deployment.

3.7.4.3.8 MicrelEye

MicrelEye (Kerhet et al. 2007) is a wireless video sensor node, intended for video processing and image classification in WMSNs. The MicrelEye node is used for people detection, where a smart camera positioned at a critical position (e.g., main entrance of a building) discriminates between the objects within its FOV (i.e., whether the object is a human being or not). The main motivation is developing an intelligent system capable of understanding certain aspects of the incoming data by performing image classification algorithms.

MicrelEye hardware platform harbors several components:

- The processor is an Atmel field-programmable system-level integrated circuit (FPSLIC) SoC, composed of an AT40K MCU, which is a field-programmable gate array (FPGA) comprising 40 K gates, and 36 KByte of onboard SRAM (16 KByte can be used for data and 20 KByte is reserved for program storage). The external memory for frame storage is a 1 MB SRAM for multimedia processing and parallelized computation between hardware and software (Atmel 2002).
- An OV7640 (OmniVision Technologies 2003), CMOS VGA image sensor.
- An LMX9820A Bluetooth transceiver for wireless communication (Texas Instruments 2007b). Its range is 100 m, it is few cubic centimeters and is featured by the low-power consumption, the ease to interface MicrelEye with other devices, and its high data rates (up to 704 Kbps). LMX9820A is now obsolete.

MicrelEye has no operating system. The Bluetooth serial port profile is used in MicrelEye to allow the establishment of a virtual serial port between the transceiver and a remote device. The algorithm implemented in MicrelEye is split between the FPGA and the MCU to achieve parallelism, where image processing tasks involving high-speed logic and high computational power (e.g., background subtraction, sub-window transfer) are performed at the FPGA and the higher level operations (e.g., feature extraction, support vector machine operations) are performed at the MCU. The portion of the algorithm run on the MCU is written in C.

The image stream coming from the CMOS imager includes both luminance and chrominance components. After a complete frame is transferred to the FPGA, background subtraction is performed by pixel differencing with the reference frame (which can be updated as needed). The region of interest (ROI) within the frame (128 × 64) is extracted and transferred to the internal memory and the remaining higher level operations are performed by the MCU. The feature vector is extracted from the ROI, which is normalized to [0,1] interval by using a highly efficient algorithm. The feature vector consists of 192 elements (the averages of the rows and the columns). The feature vector is fed into a state vector machine-like (SVM-like) structure (called ERSVM), which is used to recover unknown dependencies. SVM is a "learning from examples" technique that requires a set of training data to be able classify the incoming feature vectors, which is provided before the classification operation starts. The end result is a binary classification of whether the feature vector describes a human being or not. The device targets a power budget of 500 mW and supports people detection at 15 fps at 320×240 quarter-VGA (QVGA) image resolution.

In MicrelEye, having both an MCU and an FPGA block on the same chip eliminates the energy dissipation on the capacitive loading introduced by the inter-chip PCB connections. The main function of the FPGA is to accelerate computationally demanding vision tasks, which cannot be efficiently handled by the MCU (e.g., image capture high-speed logic, SRAM memory access management, most of the image processing tasks for detection, interface between the FPSLIC and the transceiver, the finite state machine governing the overall system operation).

3.7.4.3.9   WiCa

WiCa (Kleihorst et al. 2007b) is another camera mote designed for WMSNs. It is a wireless smart camera based on a single-instruction multiple data (SIMD) processor, a technique employed to achieve data level parallelism. WiCa main components are (Fig. 3.49):

- SIMD processor for low-level image processing and suitable for parallel processing. It is the IC3D (Kleihorst et al. 2007a), a member of the

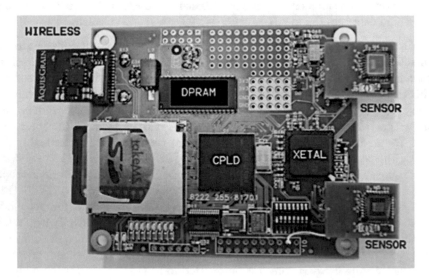

Fig. 3.49 Architecture of WiCa. (Kleihorst et al. 2007a)

non-commercial Xetal family of processors. A key feature of this processor is the use of single-instruction multiple data (SIMD) that allows one instruction to operate in parallel on several data items instead of looping through them individually. This is especially useful in audio and image processing as it considerably shortens the processing time. The IC3D has a linear array of 320 RISC processors, with the function of instruction decoding shared between them. With a 10 Mbit memory, up to 4 VGA-sized video frames can be stored on-chip allowing energy-efficient inter-frame and intra-frame computations. In addition, one of the components, called global control processor (GCP), is equipped to carry out several signal processing functionalities on the entire data. The lower power application consumption (below 100 mW) and the ease of programmability through extended C (XTC) language makes this processor useful for WMSN applications. WiCa is an efficient VSN platform with its unique design and its use of an SIMD processor rather than an FPGA chip for low-level image processing operations.

- Atmel 8051 (Atmel 2008b), a general purpose processor for higher level operations. It includes 256 Bytes of on-chip RAM, 2048 Bytes of on-chip ERAM, 64 KBytes Flash, and 2 KBytes EEPROM to store the parameters and instruction code for the IC3D processor.
- Both processors have access to a 128 KByte dual port RAM that enables them to share a common workspace, which enables both processors to collectively use the data and even pipeline the processing of data in a flexible manner.
- One or two VGA color image sensors.
- Communications module. A Phillips Aquis Grain ZigBee module developed around the CC2420 transceiver (Texas Instruments 2005b).

The multimedia processing in this camera sensor mote is divided into three levels Fig. 3.50:

- Low-level image processing (pixel level) is manipulated by the SIMD processor and is associated with typical kernel operations such as convolutions, data-dependent operations using neighboring pixels, and initial pixel classification.
- The intermediate and high-level image processing (object level) are done by the general purpose processor because it has the flexibility to implement complex software tasks, to run an operating system, and to do networking application.

3.7.4.3.10 CITRIC

A wireless camera network system called CITRIC is developed in (Chen et al. 2008a) for WMSNs to enable in-network processing of images in order to reduce communication overheads. The CITRIC platform consists of a camera daughter board connected to a Tmote Sky board (Fig. 3.51). The Tmote Sky (Moteiv 2006) is a variant of the popular TelosB mote (Moteiv 2004), dedicated for wireless sensor network research, it uses a Texas Instruments MSP430 microcontroller and Chipcon

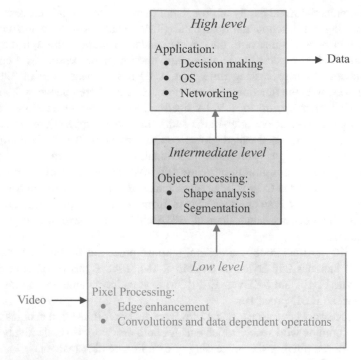

Fig. 3.50 Levels of image processing algorithms. (Based on (Kleihorst et al. 2007a))

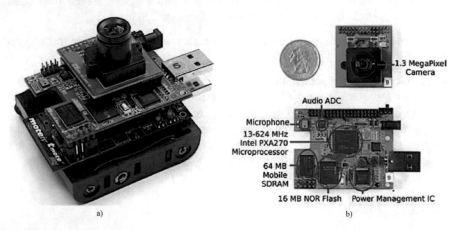

Fig. 3.51 CITRIC mote. (**a**) Assembled camera daughter board with TelosB. (**b**) Camera daughter board with major functional units outlined. (Chen et al. 2013)

CC2420 IEEE 802.15.4-compliant radio, both selected for low-power operation. The camera daughter board is comprised of a 4.6 cm × 5.8 cm processor board and a detachable image sensor board. The design of the camera board uses a small number of functional blocks to minimize size, power consumption, and manufacturing costs.

To choose a proper onboard processor, two options were available, to use a field-programmable gate arrays (FPGAs) or general purpose processors running embedded Linux. Although FPGAs have advantages in terms of speed and low-power consumption, the user would need to program in a hardware description language, making algorithm implementation and debugging a time-consuming process. On the other hand, many image processing and computer vision algorithms have been efficiently coded in C/C++, such as the OpenCV library (OpenCV 2014). Therefore, it was chosen to use a general purpose processor running embedded Linux, as opposed to TinyOS (TinyOS 2012), for the camera board to achieve rapid prototyping and ease of programming and maintenance. The components are as revealed:

- CMOS image sensor. The camera for the CITRIC platform is the OmniVision OV9655 (OmniVision Technologies 2006), a low-voltage super XGA (SXGA) 1.3 Megapixel CMOS image sensor that offers the full functionality of a camera and image processor on a single chip. It supports image sizes SXGA (1280 × 1024), VGA, CIF, and any size scaling down from CIF to 40 × 30 and provides 8-bit/10-bit images. The image array is capable of operating at up to 30 frames per second (fps) in VGA, CIF, and lower resolutions, and 15 fps in SXGA. The OV9655 is designed to perform well in low-light conditions. The typical active power consumption is 90 mW (15 fps @SXGA) and the standby current is less than 20 μA.
- Processor. The PXA270 (Intel 2005a) is a fixed-point processor with a maximum speed of 624 MHz, 256 KByte of internal SRAM, and a wireless MMX coprocessor to accelerate multimedia operations. The processor is voltage and frequency scalable for low-power operation, with a minimum voltage and frequency of 0.85 V and 13 MHz, respectively. Furthermore, the PXA270 features the Intel Quick Capture Interface, which eliminates the need for external preprocessors to connect the processor to the camera sensor. Moreover, the PXA270 is chosen because of its maturity and the popularity of its software and development tools. The current CITRIC platform supports CPU speeds of 208, 312, 416, and 520 MHz.
- External Memory. The PXA270 (Intel 2005a) is connected to 64 MByte of 1.8 V Qimonda Mobile SDRAM (Qimonda AG 2006) and 16 MByte of 1.8 V Intel NOR Flash (Intel 2005b). The SDRAM is for storing image frames during processing, and the Flash is for storing code. 64 MByte of SDRAM is more than sufficient for storing two frames at 1.3 Megapixel resolution (3 Bytes/pixel × 1.3 Megapixel × 2 frames = 8 MByte), the minimal requirement for background subtraction. 64 MByte is also the largest size of the single data rate (SDR) mobile SDRAM components natively supported by the PXA270 currently available on the market. As for the Flash, the code size for most computer vision algorithms falls well under 16 MB. The selection criteria for the types of non-volatile and volatile memory are access speed/bandwidth, capacity, power consumption, cost, physical size, and availability.

The choices for non-volatile memory were NAND and NOR Flash. NAND has lower cost-per-bit and higher density but slower random access, and NOR has the capability to execute code directly out of the non-volatile memory on boot up (eXecution-in-Place, XIP). NOR Flash was chosen not only because it supported XIP, but also because NAND Flash is not natively supported by the PXA270 processor.

The choices for volatile memory were Mobile SDRAM and Pseudo SRAM, both of which consume very little power. Low-power consumption is an important factor when choosing memory because it has been demonstrated that the memory in handsets demands up to 20 percent of the total power budget, equal to the power demands of the application processor. Mobile SDRAM was chosen because of its significantly higher density and speed.

- Microphone. In order to run high-bandwidth, multimodal sensing algorithms fusing audio and video sensor outputs, it was important to include a microphone on the camera daughter board rather than using a microphone attached to the Tmote Sky wireless mote. This simplified the operation of the entire system by dedicating the communication between the Tmote Sky and the camera daughter board to data that needed to be transmitted over the wireless network. The microphone on the board is connected to the Wolfson WM8950 mono audio ADC (Wolfson Microelectronics 2011), which was designed for portable applications. The WM8950 features high-quality audio (at sample rates from 8 to 48 KSample/sec) with low-power consumption (10 mA all-on 48 KSample/sec mode) and integrates a microphone preamplifier to reduce the number of external components.
- Power Management. The camera daughter board uses the NXP PCF50606 (Philips 2002), a power management IC for the XScale application processors, to manage the power supply and put the system into sleep mode. When compared to an equivalent solution with multiple discrete components, the PCF50606 significantly reduces the system cost and size. The entire camera mote, including the Tmote Sky, is designed to be powered by four AA batteries, or a USB cable, or a 5 V DC power adapter cable.
- USB-to-UART bridge. The camera daughter board uses the Silicon Laboratories CP2102 USB-to-UART bridge controller (Silicon Laboratories 2013) to connect the UART port of the PXA270 with a USB port on a personal computer for programming and data retrieval.

A back-end client/server architecture is proposed; it provides a user interface to the system and supports further centralized processing for higher level applications. CITRIC mote enables a wider variety of distributed pattern recognition applications than traditional platforms because it produces more computing power and tighter integration of physical components while still consuming relatively little power. Furthermore, the mote easily integrates with existing low-bandwidth sensor networks because it can communicate over the IEEE 802.15.4 protocol with other sensor network platforms. CITRIC was tested on three applications, image compression, target tracking, and camera localization.

3.7.4.3.11   ACME Fox Board Camera Platform

A little known example of medium-resolution camera mote for WMSN applications
is the embedded camera mote platform (Capo-Chichi and Friedt 2008) based on
ACME (ACME Systems srl 2014) Fox Board. The designed platform has a multiplicity of components (Fig. 3.52):

- The Fox Board LX416 has 100 MHz CPU, 4 MByte Flash, and 16 MByte of
  RAM. It runs GNU/Linux as operating system. Because of such capabilities it
  may be used for a high-level device in a multitier model.
- Webcam QuickCam Zoom (Logitech 2004).
- Several sensors including GPS positioning receiver.
- Current consumption sensor. It is used as an energy analyzer to study energy
  consumption of nodes during image transmission.
- This sensor node relies exclusively on a Bluetooth radio. This radio choice is an
  interesting attempt to strike the balance between a high-power 802.11 (Wi-Fi)
  radio and a limited data rate 802.15.4 (Zigbee-ready) radio with very low energy
  consumption.

The platform can be connected via USB ports with webcam QuickCam Zoom or
Labtec[21] Webcam (Logitech 2001), and Bluetooth dongle. The designed platform
uses Bluetooth IEEE 802.15 for data transmission, rather than the 802.11, as
compared to other high-level platforms like Panoptes and SensEye. It is experimentally shown that image grabbing and transmission need more power than image
routing.

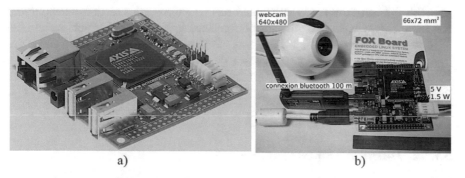

Fig. 3.52 ACME Fox Board camera platform. (**a**) ACME Fox Board LX416. (**b**) The Fox Board
with QuickCam Zoom webcam and Bluetooth dongle. (Capo-Chichi and Friedt 2008)

---

[21]Logitech bought Labtec in February 7, 2001.

3.7.4.3.12   Vision Mesh

Also lightly known, Vision Mesh is a scalable video sensor network (VSN) platform for water conservancy engineering (Zhang and Cai 2010). Vision Mesh is composed of a number of image or video sensor nodes, vision motes, to obtain multiview image or video information of field of view (FOV). A vision mote is built upon several components:

- AT91SAM ARM-based Embedded MPU (Atmel 2011c). It runs 210 MIPS at 190 MHz, and operates under the Linux operating system; it is also responsible for video processing and power management.
- K9F1G08 chip used as 128 MByte NAND Flash (Samsung Electronics 2006).
- K4S561632 chips used as 64 MByte SDRAM (Samsung Electronics 2004).
- Zigbee transceiver module Chipcon CC2430 that operates at 2.4 GHz, the theoretical transmission speed of Chipcon CC2430 can reach 250 Kbps (Texas Instruments 2006).
- CMOS camera used as image or video source. Compared to charge-coupled device (CCD) cameras, CMOS cameras are smaller, lighter, and consume less power. Hence, they constitute a suitable technology to interface sensors with vision motes.
- DS18B20 temperature sensor embedded in Chipcon CC2430 (Maxim Integrated 2008).

OpenCV machine vision lib (OpenCV 2014) is migrated to Vision Mesh platform so as to improve video processing ability. The obtained maximal processing time for 320×240 pixel JPEG coded images is 10 msec, while it is 16 msec for 640×480 pixel images. The wireless transmission rate can approach 35 Kbps in practice.

Table 3.7 compares the WMSN platforms presented in this section, exemplifying the basic components features and performance indicators. On which tier are the WMSN motes installed (Fig. 3.37) and what are their distinctive applications are important closing points not to be left over, Table 3.8 takes charge of this finale. Several interesting efforts compare WMSN motes regarding the used image processing techniques are found in (Seema and Reisslein 2011) and (Tavli et al. 2012).

3.7.4.4   Distributed Smart Cameras

The term "distributed camera" refers in computer vision to a system of physically distributed cameras that may or may not have overlapping fields of view. The images from these cameras are analyzed jointly. Distributed cameras allow viewing a subject of interest from several different angles. This, in turn, helps solving some tedious problems that arise in single-camera systems. Distributed smart cameras helps facing several key issues (Rinner and Wolf 2008).

Table 3.7 Wireless motes platforms compared

Platform	Processor	Memory		OS/Software[a]	Camera & Resolution	Radio	Power consumption[b]
		RAM	Flash				
Cyclops (Sect. 3.7.4.3.1)	7.37 MHz 8-bit Atmel ATmega128L MCU	4 KB SRAM	128 KB	TinyOS / nesC	Agilent compact CIF CMOS ADCM-1700 352 × 288@10 fps	Interfaced with MICA2 that uses TR1000 radio (40 Kbps)	64.8 mW (max) 0.76 mW (sleep)
	CPLD	64 KB 8-bit SRAM (External)	512 KB 8-bit CMOS (External)				
Panoptes (Sect. 3.7.4.3.2)	206 MHz 32-bit Intel StrongARM[b] SA-1110	16 KB of instruction cache and 8 KB of data cache		Linux / C,C++ (OpenCV library), Python	Logitech 3000 USB Camera 160 × 120 @ 30 fps 320 × 240 @ 30 fps 640 × 480@15 fps	Inserted IEEE 802.11 wireless card	5.5 W (max) 58 mW (sleep)
	Bitsy board	64 MB	32 MB				
AER (ALOHA-OV) (Sect. 3.7.4.3.3)	58 MHz 32-bit OKI ML67Q5002 based on ARM7TDMI (XYZ)	32 KB	256 KB	SCS	OmniVision OV7649 VGA (640 × 480) QVGA (320 × 240) @4.1 fps	Integrated CC2420 2.4 GHz IEEE 802.15.4 /ZigBee (XYZ)	145 mA (active) 190 µA (standby) @ 3.3 V
		2 MB (External)					
eCAM (Sect. 3.7.4.3.4)	8051 compatible MCU (Eco node)	4 KB	—	No OS	OmniVision OV7640 VGA (640 × 480) @30 fps	Interfaced with Eco node nRF24EI 2.4 GHz RF radio transceiver	70 mA @ 3.3 V
		32 KB serial (SPI) EEPROM (External)					
WiSN (Sect. 3.7.4.3.5)	48 MHz 32-bit	64 KB	256 KB	No OS	Agilent ADCM-1670	Integrated CC2420 2.4 GHz	30–110 mA @ 3.3 V

(continued)

Table 3.7 (continued)

Platform	Processor	Memory		OS/Software[a]	Camera & Resolution	Radio	Power consumption[b]
		RAM	Flash				
	ARM7TDMI based on Atmel AT91SAM7S				352 × 288 @15 fps Agilent ADNS-3060 30 × 30	IEEE 802.15.4 /ZigBee	
FireFly Mosaic (Sect. 3.7.4.3.6)	7.37 MHz 8-bit Atmel ATmega 128 L	8 KB	128 KB	Nano-RK / CC3 image libraries	CMUCam3 352 × 288 @30 fps	Integrated CC2420 2.4 GHz IEEE 802.15.4 /ZigBee	572.3 mW (active) 132.52 mW (idle) 0.29 mW (sleep)
	60 MHz 32-bit LPC2106 ARM7TDMI	64 KB	128 KB				
Imote2 + Cam (Sect. 3.7.4.3.3)	13–416 MHz 32-bit PXA271 XScale processor (Imote2)	256 KB (Imote2)	32 MB (Imote2)	TinyOS and Linux/ Java and C++	OmniVision OV7649 VGA (640 × 480) @30 fps	CC2420 2.4 GHz IEEE 802.15.4 /ZigBee	322 mW (max) 1.8 mW (sleep)
MeshEye (Sect. 3.7.4.3.7)	55 MHz 32-bit Atmel AT91SAM7S based on ARM7TDMI ARM Thumb processor	64 KB	256 KB	No OS	Agilent ADNS-3060 30 × 30 Agilent ADCM-2700 640 × 480 @10 fps	Integrated CC2420 2.4 GHz IEEE 802.15.4 /ZigBee	88.1 mA (max) 0.594 mA (sleep) using two AA batteries
MicrelEye (Sect. 3.7.4.3.8)	8-bit Atmel FPSLIC SoC composed of AT40K MCU	36 KB onboard SRAM / 1 MB SRAM (External)	—	No OS / C	OmniVision OV7640 320 × 240 @15 fps	LMX9820A Bluetooth 230.4 Kbps	430 mW @ 5 fps (serial implementation) 500 mW @ 9 fps (parallel implementation) 500 mW @ 15 fps (optimized implementation)

Platform	Processor	Memory		OS / Software	Camera	Communication	Power
WiCa (Sect. 3.7.4.3.9)	84 MHz 16-bit SIMD Xetal II / 8051Atmel MCU	10 Mbit on-chip SRAM / 256 Bytes on-chip RAM 2048 Bytes on-chip ERAM 2 KBytes EEPROM / 128 KE dual port RAM (External)	64 KB	No OS / Extended C (XTC)	One or two VGA color camera CIF (320 × 240) VGA (640 × 480) HDTV (1280 × 720)	Aquis Grain ZigBee IEEE 802.15.4	600 mW max
CITRIC (Sect. 3.7.4.3.10)	624 MHz 32-bit Intel XScale PXA270 CPU	64 MB	16 MB	Linux / C,C++ (OpenCV library)	OmniVision OV9655 1280 × 1024 @15 fps 640 × 480 @30 fps	Interfaced with Tmote Sky mote IEEE 802.15.4	970 mW (max) 428 mW (Idle)
ACME Fox Board + Cam (Sect. 3.7.4.3.11)	100 MHz LX416 Fox Board	16 MB	4 MB	(GNU/Linux) / C, C++(OpenCV library)	QuickCam Zoom Labtec Webcam 640 × 480	USB Bluetooth IEEE 802.15 100 m	1.5 W @ 5 V
Vision Mesh (Sect. 3.7.4.3.12)	190 MHz 32-bit AT91SAM ARM-based MCU	64 MB SDRAM	128 MB NAND Flash	Linux / C,C++ (OpenCV library)	CMOS camera 640 × 480 320 × 240	CC2430 (250 Kbps) IEEE 802.15.4/ ZigBee	52.7 mA (max) @ 188 MHz

[a]Depends on the reported data. [b]Values and units are as reported in the source paper

Table 3.8 WMSN tiers and applications

Platform	Tier	Applications
Cyclops (Sect. 3.7.4.3.1)	Tier 1	Limited computation, low-resolution images, extended lifetime applications
Panoptes (Sect. 3.7.4.3.2)	Tier 3	Low-power high-quality video monitoring
AER (ALOHA-OV) (Sect. 3.7.4.3.3)	Tier 1	Ultra-low-power, information-selective, and privacy-preserving sensing suitable for assisted living type applications
eCAM (Sect. 3.7.4.3.4)	Tier 1	Ultra-compact, high data rate and wireless surveillance system
WiSN (Sect. 3.7.4.3.5)	Tier 1, Tier 2	Frequent low amount of data transmissions Limited latency applications Node tracking
FireFly Mosaic (Sect. 3.7.4.3.6)	Tier 2	Distributed vision-sensing tasks, such as assisted living applications
Imote2 + Cam (Sect. 3.7.4.3.3)	Tier 1	Ultra-low-power, information-selective, and privacy-preserving sensing suitable for assisted living type applications
MeshEye (Sect. 3.7.4.3.7)	Tier 1, Tier 2	Energy-efficient distributed intelligent surveillance
MicrelEye (Sect. 3.7.4.3.8)	Tier 1, Tier 2	Cooperative low-power distributed video processing applications that involve image classification
WiCa (Sect. 3.7.4.3.9)	Tier 1, Tier 2	Real-time robust detection of objects and their orientations in uncontrolled lighting conditions Real-time frame-based video processing
CITRIC (Sect. 3.7.4.3.10)	Tier 2	Image compression Single target tracking Camera localization using multitarget tracking
ACME Fox Board + Cam (Sect. 3.7.4.3.11)	Tier 1, Tier 2	Acquisition of scalar measurements, and high-quality acquisition and image transmission over Bluetooth
Vision Mesh (Sect. 3.7.4.3.12)	Tier 1, Tier 2	Video sensor network platform for water conservancy engineering

3.7.4.4.1   Occlusion

Occlusion is a major problem in single-camera systems. A subject may be occluded by another object; if the subject is nonconvex, part of the subject may be occluded by another part. With multiple views of a subject, it is much more likely to be able to see the parts of an object occluded in one view by switching to another camera's view. Occlusion may be static or dynamic. A fixed object, such as a wall or a table, causes occlusion problems that are easier to predict. When one moving object occludes another, such as when two people pass by each other, occlusion events are harder to predict.

3.7.4.4.2   Pixels on Target

The ability to analyze a subject is limited by the amount of information, measured in pixels, that is available about that subject. Not only do distributed cameras give several views, but one camera is more likely to be closer to the subject. A traditional camera setup would use a single camera to cover a large area; subjects at the opposite end of the space would be covered by very few pixels. Distributed camera systems help covering the space more evenly.

3.7.4.4.3   Field of View

The number of cameras needed to cover a space depends on both the field of view (FOV) of the camera and the required number of pixels on target. For a typical camera with a rectangular image sensor, the FOV is a pyramid extending from the lens, as shown in Fig. 3.53. The angular FOV of the lens determines the size of this pyramid. A normal lens provides the same angular FOV as does the human eye, between 25° and 50°. A wider lens covers more area in the scene, spreading a given number of pixels over a larger area in the scene. A longer lens covers less area in the scene, putting more pixels on the target.

The pixels-on-target criterion and the size of the smallest target of interest tell how far this pyramid extends from the camera. For example, a common intermediate format (CIF) image is 362 × 240 pixels. When this array of pixels is placed across the FOV pyramid, as shown in Fig. 3.53, it is feasible to easily calculate the number of pixels that cover a target of a given size at various distances from the camera.

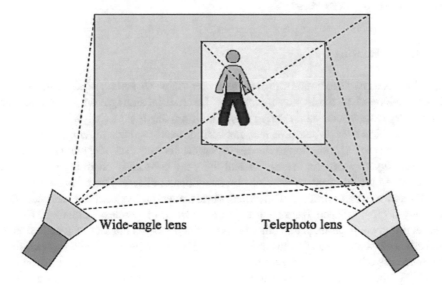

Fig. 3.53   Field of view and pixels on target of a camera. (Rinner and Wolf 2008)

Fig. 3.54 Occlusion and fields of view. (Rinner and Wolf 2008)

Given the FOV volumes dictated by the cameras and application requirements, and given the number of different cameras that should cover any given point in a space, it is possible to determine the number of cameras required to provide that coverage. A simple case is a rectangular room as shown in Fig. 3.54. Once a simple occluding object is added, such as a box, covering the space becomes harder. If the occluding object occupies less volume than the FOV that it blocks, then cameras must be added to maintain the same coverage. A thin occluding object, such as a wall or table, is a worst-case occlusion since it occupies little spatial volume but can block a large FOV volume.

As occlusion may be static or dynamic, when Person 1 walks behind the box, it causes a temporary, dynamic occlusion, while a static object behind the box is statically and permanently occluded. Subjects can also occlude each other, as when Person 3 blocks Person 2.

3.7.4.4.4 Tracking

Tracking is one of the major topics in computer vision. A variety of algorithms have been developed to track moving objects. Local data aggregation is an effective means to save sensor node energy and prolong the lifespan of wireless sensor networks. However, when a sensor network is used to track moving objects, the task of local data aggregation in the network presents a new set of challenges, such as the necessity to estimate, usually in real-time, the constantly changing state of the target based on information acquired by the nodes at different time instants.

To address these issues, distributed object tracking systems are needed; they employ a cluster-based Kalman filter in a network of wireless cameras. When a target is detected, cameras that can observe the same target interact with one another to form a cluster and elect a cluster-head. Local measurements of the target acquired by members of the cluster are sent to the cluster-head, which then estimates the target position via Kalman filtering and periodically transmits this information to a basestation. The underlying clustering protocol allows the current state and uncertainty of the target position to be easily handed off among clusters as the object is being tracked. This allows Kalman filter-based object tracking to be carried out in a distributed manner. An extended Kalman filter is necessary since measurements acquired by the cameras are related to the actual position of the target by nonlinear transformations. In addition, in order to take into consideration the time uncertainty in the measurements acquired by the different cameras, it is compulsory to introduce nonlinearity in the system dynamics. Such object tracking protocol requires the transmission of significantly fewer messages than a centralized tracker that naively transmits all of the local measurements to the basestation (Medeiros et al. 2008).

3.8   Robotic WSNs (RWSNs)

WSNs consist of a set of wireless sensor nodes that sense their surrounding environment and a data center, also known as sink or basestation, which collects the sensory data and uploads it to users for further analysis. In traditional WSNs, both sensor nodes and sinks are static, and the sensory data at a node are transmitted to the sinks via intermediate nodes. Such feature results in the hot-spot/energy-hole issue (Li and Mohapatra 2007), i.e., the nodes near the sinks are usually overloaded because they not only transmit their own sensory data, but also relay data for other nodes far from the sinks. Thus, these nodes usually run out battery quickly, which isolates the rest of the nodes from the sink. Much effort has been made to improve the energy efficiency of network, such as designing energy-efficient communication protocols (Pantazis et al. 2013). However, as long as the sinks and sensor nodes are static, the hot-spot/energy-hole problem cannot be fully solved. Consequently, mobility is introduced to be the solution.

3.8.1   Mobility in WSNs

Mobile nodes have been embraced into WSN as an effective approach to tackle the hot-spot/energy-hole issue and hence improve the network lifetime; they can move closer to some static sensor nodes and help them to relay sensory data. Attached to robots or vehicles, mobile nodes can visit sensor nodes and collect data at short range to reduce the energy consumption at sensor nodes. Although the network lifetime is lengthened, it is still finite, due to the limited energy supply at sensor nodes. Researchers try hard to overcome such a shortcoming and design almost perpetual networks. One idea is to let sensor nodes extract energy from the environment, i.e., energy harvesting (Fahmy 2020); however, this technique depends heavily on the environment. Using wireless power transfer technology, a new approach has been enabled (Kurs et al. 2007). The idea is to install a sufficient battery on a robot, then the robot works as a mobile charger to recharge nodes wirelessly. Considering that the energy consumption of a robot is likely to be the dominant component in the total

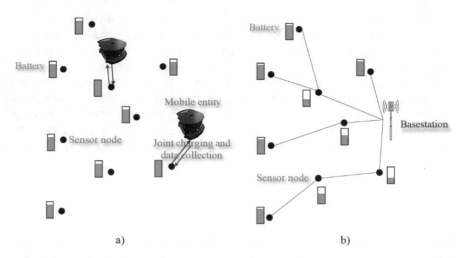

Fig. 3.55 Using MEs to collect data and charge sensor nodes versus the traditional routing approach. (**a**) Using MEs. (**b**) Traditional approach. (Huang et al. 2019)

energy consumption, employing a robot for both data collection and energy charging may consume less energy than calling for two robots, one for data collection and the other for charging. The joint data collection and energy charging by mobile robots is a worthy challenge, as illustrated in Fig. 3.55.

Generally, three main tasks of MEs in WSNs have been widely studied; these are the collection, delivery, and the combination of both, as defined in the following bullets (Huang et al. 2019):

- Collection. MEs collect from sensor nodes.
- Delivery. MEs deliver to sensor nodes.
- Combination. MEs collect from sensor nodes and deliver to them as well.

Collection is the most common task of MEs in WSNs and it refers to sensory data collection mainly (Fig. 3.56).

Under the taxonomy of the tasks of MEs, several approaches describe the mobility pattern of MEs (Khan et al. 2014) and (Gu et al. 2016):

- Random. MEs randomly move in the sensing field or have their own designated routes or destinations but cannot be controlled by the system.
- Partially controllable. MEs have movement constraints in either trajectories or stop locations.
- Fully controllable. MEs are fully controlled by the system. The considered parameters include trajectory, sojourn time at certain places, and speed.

To design an effective WSN with MEs, designers need to consider the application scenarios, and have to fulfill objectives that ensure proper performance, while abiding by the constraints that the applications impose and the limitations inherent in the WSNs and MEs. In this respect, sensor nodes and MEs have various features

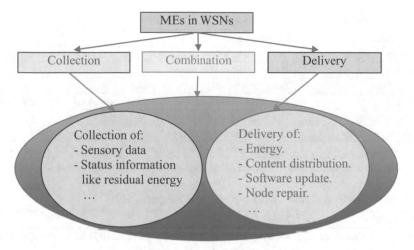

Fig. 3.56 Tasks of MEs in WSNs. (Huang et al. 2019)

Table 3.9 Features of sensor nodes and MEs in different scenarios (Huang et al. 2019)

Feature		Scenario
Sensing rate	Uniform	All sensor nodes have the same task
	Diverse	Sensors near the places where events happen frequently have larger sensing rates
Energy consumption	Uniform	When sensing rate is uniform and uploading sensory data is only after ME arrives
	Diverse	Sensor nodes near basestation have larger energy consumption rates
Node sends data to ME	Single hop	Sensor nodes transmit data to MEs only when they arrive in proximity
	Multihop	Sensor nodes transmit data to MEs current locations
ME uploads data to basestation	In proximity	MEs upload the collected data to basestation only when they are near it
	In real time	MEs can send the collected data through long-distance communication

that adapt to application scenarios; typical such features are the sensing rate, energy consumption, how nodes send data to MEs, and how MEs upload data to the basestation. Table 3.9 assembles the features of sensor nodes/MEs in different applications scenarios. Moreover, while satisfying the application scenarios, several performance metrics are to be fulfilled by WSNs with MEs; such metrics, as detailed in Table 3.10, ensure proper performance. Constraints are the barriers that any successful system must survive; Table 3.11 provides them as imposed by the application scenario or inherent in the WSNs with MEs.

Table 3.10 Performance metrics of using MEs in WSNs (Huang et al. 2019)

Performance metrics	Description
Network lifetime	The duration between the deployment and the first node expiry[a]
Energy efficiency	The efficient energy usage
Throughput	The total amount of collected data
Delay	The duration between data is sensed and uploaded to basestation
Network utility	The summation of utility of all sensor nodes
Charging quantity	The total number of charged nodes
MEs number	The minimum number of MEs to achieve any of the above metrics

[a]The network lifetime can also be defined based on the collective behavior of nodes

Table 3.11 Constraints considered in applications (Huang et al. 2019)

Constraint		Description
ME	Trajectory	Whether MEs have to follow predefined trajectories or trajectories can be predicted
	Stop position	MEs cannot stop anywhere due to terrain restrictions
	Energy capacity	Limited number of nodes to be charged or reduced travel distance
	Travel distance	Reflect the data collection latency
	Number of MEs	Investment of system
	Velocity	Including line speed and angular speed
Sensor node	Storage	The size of memory to buffer data
	Link capacity	The ability of transmission rate
	Energy status	Reflects the node lifetime

3.8.2 Robotics and WSNs

Robotics has been a very important and active field of research over last couple of decades with the main focus on seamless integration of robots in human lives to assist and help in difficult, cumbersome jobs such as search and rescue in disastrous environments and exploration of unknown environments (Murphy 2004; Penders et al. 2011) and Sect. 3.4.1.11. The rapid technological advancements in terms of cheap and scalable hardware with necessary software stacks have provided a significant momentum to this field of research and applications. As part of the increasing stream of investigations into robotics, research have been motivated to look into the collaborative aspects where a group of robots can work in synergy to perform a set of diverse tasks (Şahin and Winfield 2008) and (Gazi and Passino 2011). Nonetheless, most of the research works on collaborative robotics, such as swarming, have remained mostly either theoretical or incomplete practical systems lacking some important pieces of the puzzle such as realistic communication channel modeling and efficient network protocols for interaction among the robots. The term "realistic communication channel model" refers to a wireless channel model that accounts for most of the well-known dynamics of a standard wireless channel such as path loss, fading, and shadowing (Rappaport 2002).

On the other hand, the field of wireless networks, more specifically, WSNs and wireless ad hoc networks, has been explored extensively by communication and network researchers, where the nodes are considered static or mobile. With the availability of cheap easily programmable robots, researches have started to explore the advantages and opportunities granted by the controlled mobility in the context of wireless networks. Nonetheless, the mobility models used by the network researchers remained simple, impractical, and not very pertinent to robotic motion control until last decade (Ghosh et al. 2019).

Over the last decade, a handful of researchers noticed the significant disconnection between the robotics and the wireless network research communities and its bottleneck effects in the full-fledged development of a network of collaborative robots. Consequently, incorporating wireless network technologies in robotics and vice versa opened up a whole new field of research at the intersection of robotics and wireless networks. This new research domain is called by many different names such as "wireless robotics networks," "wireless automated networks," "networked robots," and "robotic wireless sensor networks (RWSNs)"(Ghosh et al. 2019).

To get acquainted with the field of RWSN, some core questions are raised and answered. What is a RWSN? What kind of research works are classified as RWSN related? What are the system components and algorithms required for RWSNs?

3.8.2.1   What Is a RWSN?

RWSN is defined as a wireless network that includes a set of robotic nodes with controlled mobility and a set of nodes equipped with sensors, whereas all nodes have wireless communication capabilities (Ghosh et al. 2019). Ideally, as clarified in Fig. 3.57, each node in a robotic sensor network should have controlled mobility, a set of sensors, and wireless communication capabilities. Such nodes are referred to as "robotic wireless sensors (RWS)." Nonetheless, a RWSN can also have some nodes with just sensing and wireless communication capabilities but without controlled mobility. Following traditional terminology, such nodes are referred to as "wireless sensors." Moreover, a RWSN typically fulfills or guarantees certain communication performance requirements imposed by the application contexts such as minimum achievable bit error rate (BER) in every link of the network.

3.8.2.2   What Kind of Research Works Are RWSN Related?

The existing research works related to RWSN can be subdivided into two broader genres (Ghosh et al. 2019):

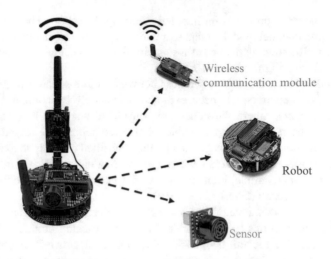

Fig. 3.57 Robotic wireless sensors. (Ghosh et al. 2019)

- The first genre focuses on generic multirobot sensing systems with realistic communication channels, i.e., they include the effects of fading, shadowing, etc., between the robots. These are mostly works in the robotics literature on multirobot systems with practical wireless communication and networking models. One application context of such a RWSN is in robot-assisted firefighting where the robots are tasked to sense the unknown environments inside rubbles to help and guide the firefighters. If the robots fail to maintain a good connectivity among them or to a mission control station, the whole mission is voided. Figure 3.58 illustrates such contexts where a group of robots are sensing an unknown environment to guide the human movements. In the figure, the network consisting of five robots and two firefighters has to be connected all the time and also needs to have properties such as reliability and lower packet delays. Thus, a class of multiobjective motion control is needed to optimize the sensing and exploration task performance and to also ensure the connectivity and performance of the network. Some of the main identifiable challenges in this genre of works are connectivity maintenance, efficient routing to reduce end-to-end delay of packets, and multiobjective motion control and optimization.
- The second genre of works focuses on the application of RWS to create and support a temporary communication backbone between a set of communicating entities. In these contexts, the terms "robotic router" and "robotic wireless sensor" are used synonymously, to put emphasis on the communication and routing goals. The main theme of these works is to exploit the controlled mobility of the robotic routers to perform sensing and communication tasks. There exists a vast literature on multiagent systems in robotics and control community that apply simple disk models for communication modeling and, subsequently, apply graph theory to solve different known problems such as connectivity and relay/repeater node placements. In order to be directly included in the RWSN literature, these existing

Fig. 3.58 Robotic sensors guide firefighters. (Ghosh et al. 2019)

works need to include the effects of fading and shadowing in the communication models which is likely to significantly increase the complexity of the problems as well as the solutions. An instance is in the application of RWS in setting up a temporary communication backbone. While sensing is still involved for the robotic router placement optimization and adaptation, the main purpose of the system is to support communication, not sensing. In Fig. 3.59, a set of two robotic routers form a communication relay path between two humans, e.g., two fire-fighters, who are unable to communicate directly. Some of the main challenges in this genre of RWSN research are link performance guarantee, in terms of signal to interference plus noise ratio (SINR) or bit error rate (BER), optimized robotic router placements and movements in a dynamic network, nonlinear control dynamics due to inclusion of network performance metrics into control loop, and localization. A special case of this would be robot-assisted static relay deployments, where the robots act as carriers of static relay nodes and smartly place/deploy them to form a communication path/backbone.

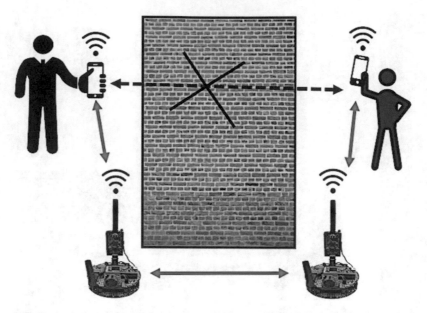

Fig. 3.59 Robotic routers for communication in the presence of obstacles. (Ghosh et al. 2019)

3.8.2.3   What Are the System Components and Algorithms Required for RWSNs?

Several research problems are needed for RWSNs; these topics are individual research problems by themselves and require separate attentions (Ghosh et al. 2019):

- RSSI models, measurements, and RF mapping. In a RWSN, it is important to estimate and monitor the quality of the communication links between the nodes, in terms of BER, SINR, etc., in order to satisfy the communication-related requirements. Noticeably, RF-based communication is the standard mode of communication in RWSN. For practicality, these estimations must be either partly or fully based on online RF sensing such as temporal RSSI measurements in a deployment. Moreover, in some application contexts of RWSN, the sole goal of a robotic sensor network can be to sense and formulate an RF map of an environment to be processed or exploited later.
- Routing protocols. Similar to any WSN, routing and data collection is crucial in a RWSN. The concept of RWSN has opened up the door to a new class of routing protocols that incorporates the controlled mobility of the nodes in the routing decisions. Moreover, end-to-end delay reduction and reliability improvement have become of prime interests.
- Connectivity maintenance. In any collaborative network of robots, it is important to maintain a steady communication path, direct or multihop, between any pair of nodes in the network, for an effective operation. The robotics research

community commonly studies this problem, traditionally referred to as connectivity maintenance problem.

- Communication aware robot positioning and movement control. As mention earlier, one of the application contexts of RWSN is in supporting temporary communication backbones. The most important research question in such contexts is to devise a control system that adapts the positions of the robotic routers throughout the period of deployment to optimize the network performance while optimizing the movements as well. Therefore, the main goal is the continuous joint optimization of the robotic movements and the wireless network performance. Besides, the router placement controller should also be able to support network dynamics such as node failures and change in the set of communication endpoints. Another important application context of RWSN systems is distributed coordinated sensing using multiple robots. In such sensing contexts, the robotic sensing agents should be able to optimally sense the region of interest and route the sensed data to other nodes or a command center.
- Localization. Localization is well known in the field of WSN as well as robotics; thus, it is quite intuitive to include it in the field of RWSN. Moreover, the field of RWSN sometimes requires techniques that permit robots, to follow or maintain proximity, to each other. Consequently, high accuracy relative localization is more important than absolute localization.

3.9 Conclusion for Further

Marathon chapter, that goes by lands, waters, hills, rocks, sands; some run, others jog, but the finish line is reachable. WSNs are infiltrating the environment in its wide sense, indoors and outdoors, in the human body, in unapproachable emplacements; they have found their way into a wide variety of applications and systems with vastly varying requirements and characteristics. Guardian angels? Watchdogs? Whatever, they are intended to work properly, faultlessly, no matter when and where. As a consequence, it is becoming increasingly difficult to forge unique requirements regarding hardware issues and software support. This is particularly important in a multidisciplinary research and practice area such as WSNs, where close collaboration between users, application domain experts, hardware designers, and software developers is needed to implement efficient systems.

In this chapter, who is who in WSNs are identified, motes, building blocks, producers, techniques, applications. A categorization of WSNs applications according to their intended use is presented considering deployment, mobility, resources, cost, energy, heterogeneity, modality, infrastructure, topology, coverage, connectivity, size, lifetime, and QoS. The considered application categories, though non-exclusive, are branded as military, industrial, environmental, healthcare, daily life, and multimedia. Typical applications tasks are:

- Performance monitoring.
- Surveillance.
- Environmental monitoring.
- Process control.
- Tracking of personnel and goods.
- Emergency management.

When compared with conventional Mobile Ad hoc Networks (MANETs), WSNs have different characteristics and present different engineering challenges and considerations:

- WSN protocols and solutions tend to be very application-specific.
- WSNs have great long-term economic potential and are expected to manage daily life in numerous areas.
- WSNs pose many new system-building challenges, which leads to a must explore multiplicity of convenient conceptual and optimization problems, such as localization, deployment, and tracking, where many applications depend on. Coverage in general, tickles the concerns about quality of service that can be provided by a particular sensor network.
- The integration of multiple types of heterogeneous sensors such as seismic, acoustic, optical, etc. in one network platform and the fulfillment of an overall coverage also impose several imperative challenges.
- An integrated and efficient framework for sensor placement must incorporate power management and fault tolerance.
- Routing and data dissemination protocols must be specifically designed for WSNs with energy awareness as a prime issue.
- WSNs have become a dependable tool for military applications involving intrusion detection, perimeter monitoring, information gathering and smart logistics support in an unfriendly deployment area.

The exercises in the next section are a get together with WSNs widespread and pervasive use; they motivate research topics, and new implementations of existing applications, or to incent novel non-tackled applications.

See, hear, smell, taste, touch, and may be foresee are five plus one senses. Any living being can partially survive with some of the senses, but he will never exist without them all. Wireless sensors are the get together with the surroundings in a whole dynamic era becoming more and more machine dependent, automated, but with intelligence.

3.10   Exercises

Note: A technical report follows the template of a peer-reviewed journal paper.

1. What are the applications categories of WSNs?
2. What are the functional challenges of forming WSNs?

3. Elaborate on the design objectives of WSNs.
4. Write an article on Fieldbus.
5. How would DoS affect the performance of WSNs?
6. Define muzzleblast and shockwave.
7. Emphasize the research topics involved in the military applications of WSNs.
8. Compare the military application approaches in WSNs.
9. Describe the Breath protocol.
10. Explain the condition monitoring for electric machines.
11. Describe the energy usage evaluation for electric machines.
12. What are the technical requirements for the deployment of WSNs in the oil and gas industry?
13. Describe the SmartMesh network.
14. Investigate the literature for WSNs use in the oil and gas industry.
15. Describe the cattle monitoring WSNs applications.
16. What are the measures that must be taken when deploying WSNs?
17. What are the considerations of using WSNs in applications?
18. Elaborate on the subsystems of a sensor-based healthcare system.
19. What are the issues pertinent to BANs?
20. What are the issues pertinent to PANs?
21. How can security be compared in the subsystems of a sensor-based healthcare system?
22. Explain unobtrusiveness in a sensor-based healthcare system.
23. Describe the MITM attacks.
24. Illustrate the application prototypes of a sensor-based healthcare system.
25. Compare security, privacy and reliability of a sensor-based healthcare application.
26. Compare home automation sensors and healthcare sensors.
27. How WSNs may be used in car parking systems?
28. What are the requirements for deploying WSNs in smart home applications?
29. Design and implement a WSN-based home application.
30. Design and implement a WSN-based office application.
31. Study the CINEMa project. Search the literature for similar projects.
32. What are the design considerations in WMSNs?
33. Describe the components of a WMSN node.
34. Elaborate on the applications of WMSNs.
35. Why are testbeds used in WMSNs?
36. Compare the WMSN motes based on the camera resolution.
37. Compare the WMSN platforms regarding the used MCU.
38. Compare the different WMSN platforms considering the number of cameras.
39. Compare the WMSN platforms noting the power consumption.
40. Describe the CMUcam3.
41. Explain the issues related to distributed smart cameras.
42. Write a technical report on RWSNs and their applications.

References

Abowd, G.D., A.K. Dey, P.J. Brown, N. Davies, M. Smith, and P. Steggles. 1999. Towards a Better Understanding of Context and Context-Awareness. In *Lecture Notes in Computer Science-Handheld and Ubiquitous Computing*, ed. H.-W. Gellersen, vol. 1707, 304–307. Berlin: Springer.

ACME Systems srl. 2014. *About Us*. ACME Systems srl. January 1, 2014. http://foxlx. acmesystems.it/about_us. Accessed 4 Apr 2014.

Adafruit Learning Technologies. 2014. *PIR Motion Sensor*. Adafruit Learning Technologies. January 1, 2014. http://learn.adafruit.com/pir-passive-infrared-proximity-motion-sensor/. Accessed 6 Mar 2014.

Agilent Technologies. 2003a. *Agilent ADCM-1670 CIF Resolution CMOS Camera Module, UART Output*. Agilent Technologies. May 13, 2003. http://www.digchip.com/datasheets/parts/ datasheet/021/ADCM-1670.php. Accessed 10 Apr 2014.

———. 2003b. *Agilent ADCM-1700-0000: Landscape CIF Resolution CMOS Camera Module*. Agilent Technologies. November 7, 2003.. http://www.zhopper.narod.ru/mobile/adcm1700. pdf. Accessed 23 Mar 2014.

———. 2003c. *Agilent ADCM-2650 Portrait VGA Resolution CMOS Camera Module*. Agilent Technologies. February 19, 2003. http://centerforartificialvision.com/pdf/c4avdcm/ADCM2650 PB.pdf. Accessed 30 Mar 2014.

———. 2004. *Agilent ADNS-3060 High-performance Optical Mouse Sensor*. Agilent Technologies. October 20, 2004. http://datasheet.eeworld.com.cn/pdf/HP/48542_ADNS-3060.pdf. Accessed 30 Mar 2014.

Akyildiz, I.F., T. Melodia, and K.R. Chowdury. 2007. Wireless Multimedia Sensor Networks: A Survey. *IEEE Wireless Communication* 14 (6): 32–39.

Akyildiz, I.F., T. Melodia, and K.R. Chowdhury. 2008. Wireless Multimedia Sensor Networks: Applications and Testbeds. *Proceedings of the IEEE* 96 (10): 1588–1605.

Alemdar, H., and C. Ersoy. 2010. Wireless Sensor Networks for Healthcare: A Survey. *Computer Networks* 54 (15): 2688–2710.

Almalkawi, I.T., M.G. Zapata, J.N. Al-Karaki, and J. Morillo-Pozo. 2010. Wireless Multimedia Sensor Networks: Current Trends and Future Directions. *Europe PubMed Central* 10 (7): 6662–6717.

Analog Devices. 1998. *DSP Microcomputer: ADSP-2181. Analog Devices*. January 1, 1998. http:// www.analog.com/static/imported-files/data_sheets/ADSP-2181.pdf. Accessed 1 Mar 2014.

———. 2013. *AD7476: 1MSPS, 12-BIT ADC IN 6 LEAD SOT-23*. December 19, 2013. http:// www.analog.com/en/analog-to-digital-converters/ad-converters/ad7476/products/product.html. Accessed 19 Dec 2013.

Anastasi, G., A. Falchi, A. Passarella, M. Conti, and E. Gregori. 2004. Performance measurements of motes sensor networks. In *The 7th ACM International Symposium on Modeling, Analysis and Simulation of Wireless and Mobile Systems (MSWiM)*, 174–181. Venice: ACM.

Arampatzis, T., J. Lygeros, and S. Manesis. 2005. A Survey of Applications of Wireless Sensors and Wireless Sensor Networks. In *The 2005 IEEE International Symposium on Intelligent Control, Mediterrean Conference on Control and Automation*, 719–724. Limassol: IEEE.

Ash, J.N., G.T. Whipps, and R.J. Kozick. 2010. Performance of Shockwave-based Shooter Localization under Model Misspecification. In *IEEE International Conference on Acoustics Speech and Signal Processing (ICASSP)*, 2694–2697. Dallas: IEEE.

Atmel. 2002. *Atmel FPSLIC 5K – 40K Gates of AT40K FPGA with 8-bit AVR*. Atmel. January 1, 2002. http://media.digikey.com/pdf/Data%20Sheets/Atmel%20PDFs/AT94K05_10_40AL% 20Complete.pdf. Accessed 31 Mar 2014.

———. 2003. *4-megabit (512K x 8) Single 2.7-Volt Battery-Voltage Flash Memory*. Atmel. 2003. http://pdf.datasheetcatalog.com/datasheet/atmel/doc0383.pdf. Accessed 14 Apr 2014.

————. 2008a. *AT91 ARM Thumb-based Microcontrollers*. Atmel. December 7, 2008. http://www.tme.eu/en/Document/84227da6b1aeb2914051d2f6e285168f/at91sam7sxxx.pdf. Aaccessed March 30, 2014.

————. 2008b. *Atmel AT89C51AC3 Enhanced 8-bit Microcontroller with 64KB Flash Memory*. Atmel. February 1, 2008. http://www.atmel.com/Images/doc4383.pdf. Accessed 1 Apr 2014.

————. 2011a. *8-bit Atmel Microcontroller with 128KBytes In-System Programmable Flash*. Atmel. January 1, 2011. http://www.atmel.com/Images/doc2467.pdf. Accessed 21 Mar 2014.

————. 2011b. *8-bit Atmel Microcontroller with 4/8/16K Bytes In-System Programmable Flash*. Atmel. January 1, 2011. http://www.atmel.com/Images/2545s.pdf. Accessed 9 Mar 2014.

————. 2011c. *AT91SAM ARM-based Embedded MPU*. Atmel. October 3, 2011. http://www.atmel.com/Images/doc6062.pdf. Accessed 5 Apr 2014.

AverLogic Technologies. 2002. *AL440B: 4M-Bit High Speed FIFO Field Memory*. AverLogic Technologies. November 11, 2002. http://www.averlogic.com/pdf/AL440B_Flyer.pdf. Accessed 29 Mar 2014.

Baggio, A. 2005. Wireless Sensor Networks in Precision Agriculture. In *Workshop on Real-World Wireless Sensor Networks (REALWSN)*. Stockholm: Swedish Institute of Computer Science (SICS).

Balogh, G., A. LéDeczi, M. Maróti, and G. Simon. 2005. Time of Arrival Data Fusion for Source Localization. In *The WICON Workshop on Information Fusion and Dissemination in Wireless Sensor Networks (SensorFusion)*. Budapest: IEEE.

Bao, S.D., Y.-T. Zhang, and L.-F. Shen. 2005. Physiological Signal Based Entity Authentication for Body Area Sensor Networks and Mobile Healthcare Systems. In *27th Annual International Conference of the Engineering in Medicine and Biology Society (IEEE-EMBS)*, 2455–2458. Shanghai: IEEE.

Baronti, P., P. Pillai, V.W.C. Chook, S. Chessa, A. Gotta, and Y.F. Hu. 2007. Wireless Sensor Networks: A survey on the State of the Art and the 802.15.4 and ZigBee Standards. *Computer Communications* 30 (7): 1655–1695.

Barth, A.T., M.A. Hanson, H.C. Powell, and J. Lach. 2009. TEMPO 3.1: A Body Area Sensor Network Platform for Continuous Movement Assessment. In *Sixth International Workshop on Wearable and Implantable Body Sensor Networks (BSN)*, 71–76. Berkeley: IEEE.

Bedard, J., and S. Pare. 2003. Ferret: A Small Arms Fire Detection System: Localization Concepts. In *Sensors, and Command, Control, Communications, and Intelligence (C3I) Technologies for Homeland Defense and Law Enforcement II*, 497–509. Orlando: SPIE.

Bhatia, D., I. Estevez, and S. Rao. 2007. Energy Efficient Contextual Sensing for Elderly Care. In *The 29th Annual International Conference of the IEEE Engineering in Medicine and Biology Society*. Lyon: IEEE.

Bluegiga. 2013. *WT12 Bluetooth Class 2 Module*. Bluegiga. January 1, 2013. http://www.bluegiga.com/en-US/products/bluetooth-classic-modules/wt12-bluetooth%2D%2Dclass-2-module/. Accessed 21 Jan 2014.

Boice, J., et al. 2005. *Meerkats: A Power–Aware, Self–Managing Wireless Camera Network for Wide Area Monitoring*. Santa Cruz: Technical, Department of Computer Engineering, University of California.

Boyden, T.C. 1982. The pollination Biology of Calypso Bulbosa var. Americana (Orchidaceae): Initial Deception of Bumblebee Visitors. *Oecologia* 55: 178–184.

Bozkurt, A., R.F. Gilmour, and A. Lal. 2009. Balloon-Assisted Flight of Radio-Controlled Insect Biobots. *Transactions on Biomedical Engineering* 56 (9): 2304–2307.

Bozkurt, A., E. Lobaton, and M. Sichitiu. 2016. A Biobotic Distributed Sensor Network for Under-Rubble Search and Rescue. *Computer* 49 (5): 38–46.

BP Solar. 2014. *MSX-01F*. January 1, 2014. http://au.element14.com/bp-solar/msx-01f/solar-panel-1-2w/dp/654012. Accessed 2 Feb 2014.

Britton, M., and L. Sacks. 2004. The SECOAS Project—Development of a Self-Organising, Wireless Sensor Network for Environmental Monitoring. In *Second International Workshop on Sensor and Actor Network Protocols and Applications (SANPA)*, 1–7. Boston.

Brooks, R.A., and A.M. Flynn. 1989. *Fast, Cheap and Out of Control. Technical*. Cambridge: Massachusetts Institute of Technology.

BugGuide. 2020. *Bombus*. Iowa State University. January 1, 2020. https://bugguide.net/index.php? q=search&keys=Bombus&search=Search. Accessed 20 June 2020.

Buonadonna, P., D. Gay, J.M. Hellerstein, W. Hong, and S. Madden. 2005. TASK: Sensor Network in a Box. In *Proceeedings of the Second European Workshop on Wireless Sensor Networks*, 133–144. Istanbul: IEEE.

Caillouet, C., F. Giroire, and T. Razafindralambo. 2019. Efficient Data Collection and Tracking with Flying Drones. *Ad Hoc Networks* 89: 35–46.

Callaway, E., et al. 2002. Home Networking with IEEE 802.15.4: A Developing Standard for Low-Rate Wireless Personal Area Networks. *Communications Magazine* 40 (8): 70–77.

Camhi, J.M., and E.N. Johnson. 1999. High-Frequency Steering Maneuvers Mediated by Tactile Cues: Antennal Wall-Following in the Cockroach. *Journal of Experimental Biology* 202: 631–643.

Campbell, J., P.B. Gibbons, S. Nath, P. Pillai, S. Seshan, and R. Sukthanka. 2005. IrisNet: An Internet-scale Architecture for Multimedia Sensors. In *The 13th annual ACM international conference on Multimedia (MULTIMEDIA)*, 81–88. Singapore: ACM.

Capo-Chichi, E.P., and J.-M. Friedt. 2008. Design of Embedded Sensor Platform for Multimedia Application. In *First International Conference on Distributed Framework and Applications (DFmA)*, 146–150. Penang: IEEE.

Capra, M., M. Radenkovic, S. Benford, L. Oppermann, A. Drozd, and M. Flintham. 2005. The Multimedia Challenges Raised by Pervasive Games. In *The 13th Annual ACM International Conference on Multimedia (MULTIMEDIA)*, 89–95. Singapore: ACM.

Carlsen, S., A. Skavhaug, S. Petersen, and P. Doyle. 2008. Using Wireless Sensor Networks to Enable Increased Oil Recovery. In *IEEE International Conference on Emerging Technologies and Factory Automation (ETFA)*, 1039–1048. Hamburg: IEEE.

Carnegie Mellon University. 2007. *CMUcam3 Datasheet*. Carnegie Mellon University. September 22, 2007. http://www.superrobotica.com/download/cmucam3/CMUcam3_datasheet.pdf. Accessed 20 Mar 2014.

Centeye. 2020. *Easily to Use Image Sensors for Microcontrollers and DSPs*. Centeye. January 1, 2020. http://www.centeye.com/products/current-centeye-vision-chips/. Accessed 21 June 2020.

Chan, M., D. Estève, C. Escriba, and E. Campo. 2008. A Review of Smart Homes-Present State and Future Challenges. *Computer Methods and Programs in Biomedicine* 91 (1): 55–81.

Chatzigiannakis, I., G. Mylonas, and S. Nikoletseas. 2007. 50 Ways to Build your Application: A Survey of Middleware and Systems for Wireless Sensor Networks. In *IEEE Conference on Emerging Technologies and Factory Automation (ETFA)*, 466–473. Patras: IEEE.

Chen, B., K. Jamieson, H. Balakrishnan, and R. Morris. 2002. Span: An Energy-Efficient Coordination Algorithm for Topology Maintenance in Ad Hoc Wireless Networks. *Wireless Networks* 8 (5): 481–494.

Chen, P., et al. 2008a. CITRIC: A Low-bandwidth Wireless Camera Network Platform. In *Second ACM/IEEE International Conference on Distributed Smart Cameras (ICDSC)*, 1–10. Stanford: ACM/IEEE.

Chen, S.-L., H.Y. Lee, Y.-W. Chu, C.-A. Chen, C.-C. Lin, and C.-H. Luo. 2008b. A Variable Control System for Wireless Body Sensor Network. In *IEEE International Symposium on Circuits and Systems (ISCAS)*, 2034–2037. Seattle: IEEE.

Chen, P., et al. 2013. A Low-Bandwidth Camera Sensor Platform with Applications in Smart Camera Networks. *ACM Transactions on Sensor Networks*: 21:1–21:23.

Cheung, S.Y., S.C. Ergen, and P. Varaiya. 2005. Traffic Surveillance with Wireless Magnetic Sensors. In *12th ITS World Congress*. San Francisco: Intelligent Transport Systems.

Chung, W.-Y., C.-L. Yau, K.-S. Shin, and R. Myllyla. 2007. A Cell Phone Based Health Monitoring System with Self Analysis Processor using Wireless Sensor Network Technology. In *29th*

Annual International Conference of the IEEE Engineering in Medicine and Biology Society (IEEE-EMBS), 3705–3708. Lyon: IEEE.

Consolvo, S., P. Roessler, and B.E. Shelton. 2004. The CareNet Display: Lessons Learned from an In Home Evaluation of an Ambient Display. In *Lecture Notes in Computer Science-UbiComp 2004: Ubiquitous Computing*, ed. N. Davies, E.D. Mynatt, and I. Siio, vol. 3205, 1–17. Berlin: Springer.

Corchado, J.M., J. Bajo, Y. de Paz, and D.I. Tapia. 2008. Intelligent Environment for Monitoring Alzheimer Patients, Agent Technology for Health Care. *Decision Support Systems* 44 (2): 382–396.

Corchado, J.M., J. Bajo, D.I. Tapia, and A. Abraham. 2010. Using Heterogeneous Wireless Sensor Networks in a Telemonitoring System for Healthcare. *IEEE Transactions on Information Technology in Biomedicine* 14 (2): 234–240.

Costa, M.S., and L.M. Jorge. 2012. Amaral. *Analysis of Wireless Industrial Automation Standards: ISA-100.11a and WirelessHART*. http://www.isa.org/InTechTemplate.cfm?template=/ContentManagement/ContentDisplay.cfm&ContentID=93257. Accessed 28 Aug 2013.

Crossbow. 2002a. *MICA2*. January 1, 2002. http://www.eol.ucar.edu/isf/facilities/isa/internal/CrossBow/DataSheets/mica2.pdf. Accessed 3 Feb 2014.

———. 2002b. *MICA2DOT*. January 1, 2002. http://www.eol.ucar.edu/isf/facilities/isa/internal/CrossBow/DataSheets/mica2dot.pdf. Accessed 5 Feb 2014.

———. 2004a. *MPR/MIB User's Manual*. August 1, 2004. http://www-db.ics.uci.edu/pages/research/quasar/MPR-MIB%20Series%20User%20Manual%207430-0021-06_A.pdf. Accessed 7 Jan 2014.

———. 2004b. *Stargate: X-Scale, Processor Platform*. Crossbow. January 1, 2004. http://www.eol.ucar.edu/isf/facilities/isa/internal/CrossBow/DataSheets/stargate.pdf. Accessed 19 Mar 2014.

———. 2005. *Imote2: High performance Wireless Sensor Network Node*. Crossbow. January 1, 2005. http://web.univ-pau.fr/~cpham/ENSEIGNEMENT/PAU-UPPA/RESA-M2/DOC/Imote2_Datasheet.pdf. Accessed 19 Mar 2014.

———. 2006a. *MICAz*. January 1, 2006. http://www.openautomation.net/uploadsproductos/micaz_datasheet.pdf. Accessed 5 Feb 2014.

———. 2006b. *MOTE-VIEW 1.2 User's Manual*. January 1, 2006. http://www.willow.co.uk/MOTE-VIEW_User_Manual_.pdf. Accessed 20 Jan 2014.

———. 2006c. *XMesh User's Manual*. March 1, 2006. http://eps2009.dj-inod.com/docs/09-03-09/XMesh_Users_Manual_march_2006.pdf. Accessed 8 Jan 2014.

———. 2007. *MTS/MDA Sensor Board Users Manual*. August 1, 2007. http://www.investigacion.frc.utn.edu.ar/sensores/Equipamiento/Wireless/MTS-MDA_Series_Users_Manual.pdf. Accessed 7 Jan 2014.

Dağtas, S., G. Pekhteryev, Z. Sahinoğlu, H. Çam, and N. Challa. 2008. Real-time and Secure Wireless Health Monitoring. *International Journal of Telemedicine and Applications-Pervasive Health Care Services and Technologies (ACM)*.

Deluge. 2008. *Deluge 1.3.6*. January 1, 2008. http://deluge-torrent.org. Accessed 31 Jan 2014.

Dethier, V.G. 1976. *The Hungry Fly: A Physiological Study of the Behavior Associated with Feeding*. Harvard University Press (American Psychological Association).

Dictionary.com. 2017. *Locomotion*. Dictionary.com, LLC. January 1, 2017. http://www.dictionary.com/browse/locomotion. Accessed 28 Jan 2017.

Dirafzoon, A., and E. Lobaton. 2013. Topological Mapping of Unknown Environments Using an Unlocalized Robotic Swarm. In *IEEE/RSJ International Conference on Intelligent Robots and Systems (IROS)*, 5545–5551. Tokyo: IEEE/RSJ.

Dirafzoon, A., J. Betthauser, J. Schornick, D. Benavides, and E. Lobaton. 2014. Mapping of Unknown Environments using Minimal Sensing from a Stochastic Swarm (IROS). In *IEEE/RSJ International Conference on Intelligent Robots and Systems*, 3842–3849. Chicago: IEEE/RSJ.

Downes, I., L.B. Rad, and H. Aghajan. 2006. Development of a Mote for Wireless Image Sensor Networks. In *Cognitive Systems and Interactive Sensors (COGIS)*. Paris: France.

Dubois-Ferrière, H., L. Fabre, R. Meier, and P. Metrailler. 2006. TinyNode: A Comprehensive Platform for Wireless Sensor Network Applications. In *The 5th International Conference on Information Processing in Sensor Networks (IPSN)*, 358–365. Nashville: ACM/IEEE.

Duckworth, G.L., J.E. Barger, and D.C. Gilbert. 2001. *Acoustic Counter-Sniper System*. US Patent 6178141 B1. January 23, 2001.

Dudley, R. 2002. *The Biomechanics of Insect Flight: Form, Function, Evolution*. Princeton: Princeton University Press.

Dukas, R., and P.K. Visscher. 1994. Lifetime Learning by Foraging Honey Bees. *Animal Behaviour* 48 (5): 1007–1012.

Dupont. 2013. *Better, Stronger and Safer with Kevlar® Fiber*. December 19, 2013. http://www. dupont.com/products-and-services/fabrics-fibers-nonwovens/fibers/brands/kevlar.html. Accessed 19 Dec 2013.

Electric Machines Committee. 1997. *112-1996 – IEEE Standard Test Procedure for Polyphase Induction Motors and Generators*. Standards, Electric Machines Committee, IEEE Power Engineering Society, IEEE.

Escobar-Alvarez, H.D., et al. 2018. R-ADVANCE: Rapid Adaptive Prediction for Vision-based Autonomous Navigation, Control, and Evasion. *Journal of Field Robotcs*: 91–100.

Eswaran, A., A. Rowe, and R. Rajkumar. 2005. Nano-RK: An Energy-aware Resource-centric RTOS for Sensor Networks. In *26th IEEE International Real-Time Systems Symposium (RTSS)*. Miami: IEEE.

Everlight Electronics. 2013. *Ambient Light Sensor Surface-Mount ALS-PT19-315C/L177/TR8*. Everlight Electronics. December 24, 2013. https://www.everlight.com/file/ProductFile/20140 7061531031645.pdf. Accessed 21 June 2020.

Fahmy, Hossam M.A. 2020. *Wireless Sensor Networks: Energy Harvesting and Management for Research and Industry*. Cham: Springer.

Fansler, Kevin S. 1998. Description of Muzzle Blast by Modified Ideal Scaling Models. *Shock and Vibration* 5 (1): 1–12.

Fariborzi, H., and M. Moghavvemi. 2007. Architecture of a Wireless Sensor Network for Vital Signs Transmission in Hospital Setting. In *International Conference on Convergence Information Technology*, 745–749. Gyeongju: IEEE.

Feng, W.-C., B. Code, E. Kaiser, M. Shea, W.-C. Feng, and L. Bavoil. 2003. Panoptes: Scalable Low-Power Video Sensor Networking Technologies. In *ACM Multimedia Conference*, 652–571. Berkeley: ACM.

Feng, W.-C., E. Kaiser, W.C. Feng, and M. Le Baillif. 2005. Panoptes: Scalable Low-power Video Sensor Networking Technologies. *ACM Transactions on Multimedia Computing, Communications, and Applications (TOMCCAP)* 1 (2): 151–167.

FOLDOC. 2006. *Odometry*. The Free On-line Dictionary of Computing, ©Denis Howe 2010. September 11, 2006. http://foldoc.org/odometry. Accessed 28 Jan 2017.

Gao, T., et al. 2008. Wireless Medical Sensor Networks in Emergency Response: Implementation and Pilot Results. In *IEEE Conference on Technologies for Homeland Security*, 187–192. Waltham: IEEE.

Garcia-Morchon, O., and H. Baldus. 2009. The ANGEL WSN Security Architecture. In *Third International Conference on Sensor Technologies and Applications (SENSORCOMM)*, 430–435. Glyfada: IEEE.

Garcia-Morchon, O., T. Falck, T. Heer, and K. Wehrle. 2009. Security for Pervasive Medical Sensor Networks. In *6th Annual International Mobile and Ubiquitous Systems: Networking & Services (MobiQuitous)*, 1–10. Toronto: IEEE.

García-Sáez, G., et al. 2009. Architecture of a Wireless Personal Assistant for Telemedical Diabetes Care. *International Journal of Medical Informatics* 78 (6): 391–403.

Gardner, J.W., and V.K. Varadan. 2001. *Microsensors, Mems and Smart Devices*. New York: Wiley.

Gazi, V., and K.M. Passino. 2011. *Swarm Stability and Optimization*. Springer Science.

GBIF. 2016. *Gromphadorhina Portentosa*. Global Biodiversity Information Facility. January 1, 2016. http://www.gbif.org/species/113191345. Accessed 30 Jan 2017.

George, J., and L.M. Kaplan. 2011. Shooter Localization using Soldier-Worn Gunfire Detection Systems. In *The 14th International Conference on Information Fusion*, 398–405. Chicago: Office of Naval Research and U.S. Army Research Laboratory.

Geospace Technologies. 2014a. *Geophones GS-11D*. January 1, 2014. http://www.geospace.com/geophones-gs-11d/. Accessed 1 Feb 2014.

———. 2014b. *GS-1 Low Frequency Seismometer*. January 1, 2014. http://www.geospace.com/tag/gs-1-low-frequency-seismometer/. Accessed 1 Feb 2014.

Ghosh, P., A. Gasparri, J. Jin, and B. Krishnamachari. 2019. Robotic Wireless Sensor Networks. In *Mission-Oriented Sensor Networks and Systems: Art and Science. Studies in Systems, Decision and Control*, ed. H. Ammari, vol. 164, 345–595. Springer.

Girod, B., A.M. Aaron, S. Rane, and D. Rebollo-Monedero. 2005. Distributed Video Coding. *Proceedings of the IEEE* 93 (1): 71–83.

Glyn Store. 2013. *Bluegiga WT12 – Class 2 Bluetooth 2.1 + EDR Module*. December 19, 2013. http://www.glynstore.com/bluegiga wt12 class 2 bluetooth-2-1-edr-module/. Accessed 19 Dec 2013.

Goldsmith, A.J., and P.P. Varaiya. 1997. Capacity of Fading Channels with Channel Side Information. *IEEE Transactions on Information Theory* 43 (6): 1986–1992.

Gu, Y., F. Ren, Y. Ji, and J. Li. 2016. The Evolution of Sink Mobility Management in Wireless Sensor Networks: A Survey. *Communications Surveys & Tutorials* 18 (1): 507–524.

Gungor, V.C., and G.P. Hancke. 2009. Industrial Wireless Sensor Networks: Challenges, Design Principles, and Technical Approaches. *IEEE Transactions on Industrial Electronics* 56 (10): 4258–4265.

Gungor, V.C., M.C. Vurana, and O.B. Akan. 2007. On the Cross-Layer Interactions between Congestion and Contention in Wireless Sensor and Actor Networks. *Ad Hoc Networks* 5 (6): 897–909.

Gutiérrez, J.A., E.H. Callaway, and R.L. Barrett. 2004. *Low-Rate Wireless Personal Area Networks: Enabling Wireless Sensors with IEEE 802.15.4*. New York: IEEE.

Habetler, T.G., R.G. Harley, R.M. Tallam, S. Lee, R. Obaid, and J. Stack. 2002. Complete Current-based Induction Motor Condition Monitoring: Stator, Rotor, Bearings, and Load. In *8th IEEE International Power Electronics Congress (CIEP)*, 3–8. Guadalajara: IEEE.

Hagen, M., M. Wikelski, and W.D. Kissling. 2011. Space Use of Bumblebees (Bombus spp.) Revealed by Radio-Tracking. *PLoS ONE* 6 (5).

Hakala, I., J. Ihalainen, I. Kivelä, and M. Tikkakoski. 2010. Evaluation of Environmental Wireless Sensor Network – Case Foxhouse. *International Journal on Advances in Networks and Services* 3 (1,2): 29–39.

Han, C.-C., R. Kumar, R. Shea, E. Kohler, and M. Srivastava. 2005. A Dynamic Operating System for Sensor Nodes. In *The 3rd International Conference on Mobile Systems, Applications, and Services (MobiSys)*, 163–176. Seattle: ACM.

Hart, J.K., and K. Martinez. 2006. Environmental Sensor Networks: A revolution in the earth system science? *Earth-Science Reviews* 78 (3/4): 177–191.

Heinzelman, W., A.P. Chandrakasan, and H. Balakrishnan. 2002. An Application Specific Protocol Architecture for Wireless Microsensor Networks. *IEEE Transactions on Wireless Communication* 1 (4): 660–670.

Hengstler, S., D. Prashanth, S. Fong, and H. Aghajan. 2007. MeshEye: A Hybrid-resolution Smart Camera Mote for Applications in Distributed Intelligent Surveillance. In *The 6th International Conference on Information Processing in Sensor Networks (IPSN)*, 360–369. Cambridge: ACM/IEEE.

Hespanha, J.P., P. Naghshtabrizi, and Y. Xu. 2007. A Survey of Recent Results in Networked Control Systems. *Proceedings of the IEEE* 95 (1): 138–162.

Hill, J.L., and D.E. Culler. 2002. Mica: A Wireless Platform for Deeply Embedded Networks. *IEEE Micro* 22 (6): 12–24.

Himax. 2015. *HM01B0 Ultra Low Power CIS*. Himax Technologies, Inc. January 1, 2015. https://www.himax.com.tw/products/cmos-image-sensor/image-sensors/hm01b0/.   Accessed 21 June 2020.

Ho, L., M. Moh, Z. Walker, T. Hamada, and C.-F. Su. 2005. A prototype on RFID and Sensor Networks for Elder Healthcare: Progress Report. In *The ACM SIGCOMM Workshop on Experimental Approaches to Wireless Network Design and Analysis (E-WIND)*, 70–75. Philadelphia: ACM.

Holman, R., J. Stanley, and T. Ozkan-Haller. 2003. Applying Video Sensor Networks to Nearshore Environment Monitoring. *IEEE Pervasive Computing* 2 (4): 14–21.

Honeywell. 2012. *Digital Compass Solutions HMR3300*. 2012. http://www51.honeywell.com/aero/common/documents/myaerospacecatalog-documents/MissilesMunitions-documents/HMR3300_Datasheet.pdf. Accessed 19 Dec 2013.

Howitt, I., W.W. Manges, P.T. Kuruganti, G. Allgood, J.A. Gutierrez, and J.M. Conrad. 2006. Wireless industrial sensor networks: Framework for QoS assessment and QoS management. *ISA Transactions* 45 (3): 347–359.

Hsu, J.S., and B.P. Scoggins. 1995. Field Test of Motor Efficiency and Load Changes Through Air-Gap Torque. *IEEE Transactions on Energy Conversion* 10 (3): 477–483.

Hsu, J.S., J.D. Kueck, M. Olszewski, D.A. Casada, P.J. Otaduy, and L.M. Tolbert. 1998. Comparison of Induction Motor Field Efficiency Evaluation Methods. *IEEE Transactions on Industry Applications* 34 (1): 117–125.

Hu, F., M. Jiang, L. Celentano, and Y. Xiao. 2008. Robust medical Ad hoc Sensor Networks (MASN) with Wavelet-based ECG Data Mining. *Ad Hoc Networks* 6 (7): 986–1012.

Huang, X., T.G. Habetler, and R.G. Harley. 2007. Detection of Rotor Eccentricity Faults in a Closed-Loop Drive-Connected Induction Motor Using an Artificial Neural Network. *IEEE Transactions on Power Electronics* 22 (4): 1552–1559.

Huang, H., A.V. Savkin, M. Ding, and C. Huang. 2019. Mobile Robots in Wireless Sensor Networks: A Survey on Tasks. *Computer Networks* 148: 1–19.

Huircán, J.I., et al. 2010. Zigbee-based Wireless Sensor Network Localization for Cattle Monitoring in Grazing Fields. *Computers and Electronics in Agriculture* 74 (2): 258–264.

Hwang, S.-W., et al. 2012. A Physically Transient Form of Silicon Electronics. *Science* 337 (6102): 1640–1644.

IEEE. 2013a. *IEEE 802.15 Working Group for WPAN*. 2013. http://grouper.ieee.org/groups/802/15/. Accessed 15 Sept 2013.

———. 2013b. *IEEE 802.15 WPAN Task Group 3 (TG3)*. 2013. www.ieee802.org/15/pub/TG3.html. Accessed 15 Sept 2013.

Ingelrest, F., G. Barrenetxea, G. Schaefer, M. Vetterli, O. Couach, and M. Parlange. 2010. SensorScope: Application-Specific Sensor Network for Environmental Monitoring. *ACM Transactions on Sensor Networks (TOSN)* 6 (2): 17.1–17.32.

Intel. 2000. *Intel StrongARM* SA-1110 Microprocessor*. Intel. April 1, 2000. http://access.ee.ntu.edu.tw/course/SoC_Lab_961/reference/StrongARM%20Datasheet.pdf. Accessed 23 Mar 2014.

———. 2005a. *Intel PXA270 Processor*. Intel. January 1, 2005. http://www.armkits.com/download/PXA270datasheet.pdf. Accessed 2 Apr 2014.

———. 2005b. *Intel StrataFlash Embedded Memory (P30)*. Intel. November 1, 2005. http://www.xilinx.com/products/boards/ml505/datasheets/30666604.pdf. Accessed 3 Apr 2014.

Intille, S.S., et al. 2003. Tools for Studying Behavior and Technology in Natural Settings. In *UbiComp 2003: Ubiquitous Computing*, ed. A.K. Dey, A. Schmidt, and J.F. McCarthy, vol. 2864, 157–174. Berlin: Springer.

Iyer, V., R. Nandakumar, A. Wang, S. Fuller, and S. Gollakota. 2019. Living IoT: A Flying Wireless Platform on Live Insects. In *The 25th International Conference on Mobile Computing and Networking (Mobicom)*. Los Cabos: ACM.

Johnson, D., T. Stack, D.M. Flickinger, L. Stoller, R. Ricci, and J. Lepreau. 2006. Mobile Emulab: A Robotic Wireless and Sensor Network Testbed. In *The 25th Conference on Computer Communications (INFOCOM)*. Barcelona: IEEE.

Jurik, A.D., and A.C. Weaver. 2008. Remote Medical Monitoring. *Computer* 41 (4): 96–99.

Kaplan, L.M., T. Damarla, and T. Pham. 2008. QoI for Passive Acoustic Gunfire Localization. In *The 5th IEEE International Conference on Mobile Ad Hoc and Sensor Systems (MASS)*, 754–759. Atlanta: IEEE.

Kerhet, A., M. Magno, F. Leonardi, A. Boni, and L. Benini. 2007. A Low-power Wireless Video Sensor Node for Distributed Object Detection. *Journal of Real-Time Image Processing* 2 (4): 331–342.

Khan, A.W., A.H. Abdullah, M.H. Anisi, and J.I. Bangash. 2014. A Comprehensive Study of Data Collection Schemes Using Mobile Sinks in Wireless Sensor Networks. *Sensors* 14 (2): 2510–2548.

Kim, D., T. Sanchez Lopez, S. Yoo, and J. Sung. 2005a. ANTS platform for Evolvable Wireless Sensor Networks. In *The 2005 IFIP International Conference on Embedded and Ubiquitous Computing (EUC)*. Nagasaki: Springer.

Kim, D., T.S. López, S.Y. Sung, J. Kim, Y. Kim, and Y. Doh. 2005b. ANTS: An Evolvable Network of Tiny Sensors. In *Lecture Notes in Computer Science-Embedded and Ubiquitous Computing (EUC)*, ed. L.T. Yang, M. Amamiya, Z. Liu, M. Guo, and F.J. Rammig, vol. 3824, 142–151. Berlin: Springer.

Kim, A.N., F. Hekland, S. Petersen, and P. Doyle. 2008a. When HART Goes Wireless: Understanding and Implementing the WirelessHART Standard. In *IEEE International Conference on Emerging Technologies and Factory Automation (ETFA)*, 899–907. Hamburg: IEEE.

Kim, J.-H., H. Kwon, D.-H. Kim, H.-Y. Kwak, and S.-J. Lee. 2008b. Building a Service-Oriented Ontology for Wireless Sensor Networks. In *Seventh IEEE/ACIS International Conference on Computer and Information Science (ICIS)*, 649–654. Portland: IEEE.

Kissling, W.D., D.E. Pattemore, and M. Hagen. 2014. Challenges and Prospects in the Telemetry of Insects. *Biological Reviews* 89: 511–530.

Kleihorst, R., A. Abbo, B. Schueler, and A. Danilin. 2007a. Camera Mote with a High-Performance Parallel Processor for Real-Time Frame-Based Video Processing. In *First ACM/IEEE International Conference on Distributed Smart Cameras (ICDSC)*, 109–116. Vienna: ACM/IEEE.

Kleihorst, R., B. Schueler, and A. Danilin. 2007b. Architecture and Applications of Wireless Smart Cameras. In *IEEE International Conference on Acoustics, Speech, and Signal Processing (ICASSP)*, 1373–1376. Singapore: IEEE.

Koh, S.J. 2006. *RF Characteristics of Mica-Z Wireless Sensor Network Motes*. Thesis, Naval Postgraduate School.

Konstantas, D., and R. Herzog. 2003. Continuous Monitoring of Vital Constants for Mobile Users: The MobiHealth Approach. In *The 25th Annual International Conference of the IEEE Engineering in Medicine and Biology Society, 2003 (IEEE-EMBS)*, 3728–3731. Kyoto: IEEE.

Kuckertz, P., J. Ansari, J. Riihijarvi, and P. Mahonen. 2007. Sniper Fire Localization using Wireless Sensor Networks and Genetic Algorithm based Data Fusion. In *Military Communications Conference (MILCOM)*, 1–8. Orlando: IEEE.

Kulkarni, P., D. Ganesan, P. Shenoy, and Q. Lu. 2005. SensEye: A Multi-tier Camera Sensor Network. In *The 13th Annual ACM International Conference on Multimedia (MULTIMEDIA)*, 229–238. Singapore: ACM.

Kumar, P., and T.V. Rao. 2008. Communication in Personal Healthcare. In *The 3rd International Conference on Communication Systems Software and Middleware and Workshops (COMSWARE)*, 66–73. Bangalore: IEEE.

Kuran, M.S., and T. Tugcu. 2007. A Survey on Emerging Broadband Wireless Access Technologies. *Computer Networks* 51 (11): 3013–3046.

Kurkowski, S., T. Camp, and M. Colagrosso. 2005. MANET Simulation Studies: The Incredibles. *ACM SIGMOBILE Mobile Computing and Communications Review – Special Issue on Medium*

Access and Call Admission Control Algorithms for Next Generation Wireless Networks 9 (4): 50–61.

Kurs, A., A. Karalis, R. Moffatt, J.D. Joannopoulos, P. Fisher, and M. Soljačić. 2007. Wireless Power Transfer via Strongly Coupled Magnetic Resonances. *Science* 317 (5834): 83–86.

Kusy, B., A. Ledeczi, M. Maroti, and L. Meertens. 2006a. Node Density Independent Localization. In *The 5th International Conference on Information Processing in Sensor Networks (IPSN)*, 441–448. Nashville: ACM/IEEE.

Kusy, B., P. Dutta, P. Levis, M. Maroti, A. Ledeczi, and D. Culler. 2006b. Elapsed Time on Arrival: A Simple and Versatile Primitive for Canonical Time Synchronisation Services. *International Journal of Ad Hoc and Ubiquitous Computing* 1 (4): 239–251.

Lahtela, A., M. Hassinen, and V. Jylha. 2008. RFID and NFC in Healthcare: Safety of Hospitals Medication Care. In *Second International Conference on Pervasive Computing Technologies for Healthcare*, 241–244. Tampere: IEEE.

Lai, C.-C., R.-G. Lee, C.-C. Hsiao, H.-S. Liu, and C.-C. Chen. 2009. A H-Qos-Demand Personalized Home Physiological Monitoring System over a Wireless Multi-Hop Relay Network for Mobile Home Healthcare Applications. *Journal of Network and Computer Applications* 32 (6): 1229–1241.

Lamprinos, I.E., A. Prentza, E. Sakka, and D. Koutsouris. 2006. Energy-efficient MAC Protocol for Patient Personal Area Networks. In *27th Annual International Conference of the Engineering in Medicine and Biology Society (IEEE-EMBS)*, 3799–3802. Shanghai: IEEE.

Langendoen, K., A. Baggio, and O. Visser. 2006. Murphy Loves Potatoes: Experiences from a Pilot Sensor Network Deployment in Precision Agriculture. In *The 20th International Parallel and Distributed Processing Symposium (IPDPS)*. Rhodes Island.

Latif, T., and A. Bozkurt. 2012. Line Following Terrestrial Insect Biobots. In *34th Annual International Conference of the IEEE Engineering in Medicine and Biology Society (EMBC)*, 972–975. San Diego: IEEE.

Latif, T., E. Whitmire, T. Novak, and A. Bozkurt. 2014. Towards Fenceless Boundaries for Solar Powered Insect Biobots. In *The 36th Annual International Conference of the IEEE Engineering in Medicine and Biology Society (EMBC)*, 1670–1673. Chicago: IEEE.

———. 2016. Sound Localization Sensors for Search and Rescue Biobots. *Sensors Journal* 16 (10): 3444–3453.

Latré, B., B. Braem, I. Moerman, C. Blondia, and P. Demeester. 2011. A Survey on Wireless Body Area Networks. *Wireless Networks* 17 (1): 1–18.

Lédeczi, A., et al. 2005. Countersniper System for Urban Warfare. *ACM Transactions on Sensor Networks (TOSN)* 1 (2): 153–177.

Ledeczi, A., et al. 2005. Multiple Simultaneous Acoustic Source Localization in Urban Terrain. In *Fourth International Symposium on Information Processing in Sensor Networks (IPSN)*, 491–496. Los Angeles: ACM/IEEE.

Lee, Y.-D., and W.-Y. Chung. 2009. Wireless Sensor Network based Wearable Smart Shirt for Ubiquitous Health and Activity Monitoring. *Sensors and Actuators B: Chemical* 140 (2): 390–395.

Lee, S.W., Y.J. Kim, G.S. Lee, B.O. Cho, and N.H. Lee. 2007. A Remote Behavioral Monitoring System for Elders Living Alone. In *2007 International Conference on Control, Automation and Systems*, 2725–2730. Warsaw.

Lee, H., K. Park, B. Lee, J. Choi, and R. Elmasri. 2008. Issues in Data Fusion for Healthcare Monitoring. In *The 1st International Conference on Pervasive Technologies Related to Assistive Environments (PETRA)*. Arlington: ACM.

Lee, H.J., et al. 2009. Ubiquitous Healthcare Service using Zigbee and Mobile Phone for Elderly Patients. *International Journal of Medical Informatics* 78 (3): 193–198.

Lee, W.S., V. Alchanatis, C. Yang, M. Hirafuji, D. Moshou, and C. Li. 2010. Sensing Technologies for Precision Specialty Crop Production. *Computers and Electronics in Agriculture* 74 (1): 2–33.

Lei, S., H. Xu, W. Xiaoling, Z. Lin, J. Cho, and S. Lee. 2006. VIP Bridge: Integrating Several Sensor Networks into One Virtual Sensor Network. In *International Conference on Internet Surveillance and Protection (ICISP)*. Cote d'Azur: IEEE.

Leister, W., Abie, A.-K. Groven, T. Fretland, and I. Balasingham. 2008. Threat Assessment of Wireless Patient Monitoring Systems. In *3rd International Conference on Information and Communication Technologies: From Theory to Applications (ICTTA)*, 1–6. Damascus: IEEE.

Li, M., and Y. Liu. 2007. Underground Structure Monitoring with Wireless Sensor Networks. In *The 6th International Conference on Information Processing in Sensor Networks (IPSN)*, 69–78. Cambridge: ACM/IEEE.

Li, J., and P. Mohapatra. 2007. Analytical Modeling and Mitigation Techniques for the Energy Hole Problem in Sensor Networks. *Pervasive and Mobile Computing* 3 (3): 233–254.

Lindgren, D., O. Wilsson, F. Gustafsson, and H. Habberstad. 2009. Shooter Localization in Wireless Sensor Networks. In *The 12th International Conference on Information Fusion (FUSION)*, 404–411. Seattle: IEEE.

Linear Technology. 2014. *Wireless Sensor Networks-Dust Networks*. Linear Technology. January 1, 2014. http://www.linear.com/products/wireless_sensor_networks_-_dust_networks. Accessed 5 Mar 2014.

Liotta, B. 2000. *Bitsy' 3 x 4-inch Board Totes StrongARM*. EE Times. UBM Tech. September 27, 2000. http://www.eetimes.com/document.asp?doc_id=1289616. Accessed 14 Apr 2014.

LOFAR. 2014. *Agriculture*. January 1, 2014. http://www.lofar.org/agriculture/fighting-phytophtora-using-micro-climate/fighting-phytophtora-using-micro-climate. Accessed 31 Jan 2014.

Logitech. 2001. *Logitech to Acquire Labtec Inc*. Logitech. February 7, 2001. http://ir.logitech.com/releasedetail.cfm?ReleaseID=174502. Accessed 7 Apr 2014.

———. 2004. *QuickCam Zoom*. Logitech. October 27, 2004. http://www.logitech.com/en-us/support/3377. Accessed 7 Apr 2014.

Lopez-Nores, M., J. Pazos-arias, J. Garcia-Duque, and Y. Blanco-Fernandez. 2008. Monitoring Medicine Intake in the Networked Home: The iCabiNET Solution. In *Second International Conference on Pervasive Computing Technologies for Healthcare*, 116–117. Tampere: IEEE.

Lu, B., L. Wu, T.G. Habetler, R.G. Harley, and J.A. Gutierrez. 2005. On the Application of Wireless Sensor Networks in Condition Monitoring and Energy Usage Evaluation for Electric Machines. In *31st Annual Conference of IEEE Industrial Electronics Society (IECON)*. Raleigh: IEEE.

Lu, B., T.G. Habetler, and R.G. Harley. 2006. A Survey of Efficiency-Estimation Methods for In-Service Induction Motors. *IEEE Transactions on Industry Applications* 42 (4): 924–933.

Lymberopoulos, D., and A. Savvides. 2005. XYZ: A Motion-enabled, Power Aware Sensor Node Platform for Distributed Sensor Network Applications. In *Fourth International Symposium on Information Processing in Sensor Networks (IPSN)*, 449–454. Los Angeles: ACM/IEEE.

Ma, Y., and J.H. Aylor. 2004. System Lifetime Optimization for Heterogeneous Sensor Networks with a Hub-spoke Technology. *IEEE Transactions on Mobile Computing* 3 (3): 286–294.

Mangharam, R., A. Rowe, and R. Rajkumar. 2007. FireFly: A Cross-layer Platform for Real-time Embedded Wireless Networks. *Real-Time Systems* 37 (3): 183–231.

Maróti, M. 2004. Directed Flood-Routing Framework for Wireless Sensor Networks. In *The 5th ACM/IFIP/USENIX International Conference on Middleware*, 99–114. Toronto: ACM/IFIP/USENIX.

Maroti, M., G. Simon, A. Ledeczi, and J. Sztipanovits. 2004. Shooter Localization in Urban Terrain. *Computer* 37 (8): 60–61.

Maróti, M., et al. 2005. Radio Interferometric Geolocation. In *The 3rd International Conference on Embedded Networked Sensor Systems (SenSys)*, 1–12. San Diego: ACM.

Marques, O., P. Chilamakuri, S. Bowser, and J. Woodworth. 2004. Wireless Multimedia Technologies for Assisted Living. In *Second LACCEI International Latin American and Caribbean Conference for Engineering and Technology (LACCEI)*. Miami.

Martinez, A.W., S.T. Phillips, E. Carrilho, S.W. Thomas, H. Sindi, and G.M. Whitesides. 2008. Simple Telemedicine for Developing Regions: Camera Phones and Paper-Based Microfluidic Devices for Real-Time, Off-Site Diagnosis. *Analytical Chemistry* 80 (10): 3699–3707.

Massachusetts Institute of Technology. 2014. *Transportation Research Center (ITRC)*. January 7, 2014. http://www-mtl.mit.edu/researchgroups/itrc/. Accessed 7 Jan 2014.

Maxim Integrated. 2008. *DS18B20 Programmable Resolution 1-Wire Digital Thermometer*. Maxim Integrated. Janaury 1, 2008. http://datasheets.maximintegrated.com/en/ds/DS18B20. pdf. Accessed 6 Apr 2014.

MaxStream. 2007. *XBeeTM/XBee-PROTM OEM RF Modules*. MaxStream. January 1, 2007. https://www.sparkfun.com/datasheets/Wireless/Zigbee/XBee-Manual.pdf.   Accessed 9 Mar 2014.

Medeiros, H., J. Park, and A.C. Kak. 2008. Distributed Object Tracking Using a Cluster-Based Kalman Filter in Wireless Camera Networks. *IEEE Journal of Selected Topics in Signal Processing* 2 (4): 448–463.

Merriam Webster. 2014. *microclimate*. January 1, 2014. http://www.merriam-webster.com/ dictionary/microclimate. Aaccessed 31 Jan 2014.

Meshkova, E., J. Riihijarvi, F. Oldewurtel, C. Jardak, and P. Mahonen. 2008. Service-Oriented Design Methodology for Wireless Sensor Networks: A View through Case Studies. In *IEEE International Conference on Sensor Networks, Ubiquitous and Trustworthy Computing (SUTC)*, 146–153. Taichung: IEEE.

Microchip. 2000. *PIC18 Microcontroller Family*. January 1, 2000. http://ww1.microchip.com/ downloads/en/DeviceDoc/30327b.pdf. Accessed 26 Jan 2014.

Moeller, R., and A. Sleman. 2008. Wireless Networking Services for Implementation of Ambient Intelligence at Home. In *7th International Caribbean Conference on Devices, Circuits and Systems (ICCDCS)*. Cancun: IEEE.

Montenegro, G., N. Kushalnagar, J. Hui, and Culler. 2007. Transmission of IPv6 Packets over IEEE 802.15.4 Networks. *Rfc 4944, Internet Engineering Task Force*.

Moog Crossbow. 2013. *Aerospace and Defense Systems*. Moog Crossbow. http://xbow.com/ products/inertial-products/. Accessed 7 Dec 2013.

Moroz, S.A., R.B. Pierson, M.C. Ertem, and D.A. Burchick Sr. 1999. *Airborne Deployment of and Recent Improvements to the Viper Counter Sniper System*. Online, Naval Research Lab,. Washington: Online Information for the Defense Community.

Morris, D., S. Coyle, Y. Wu, K.T. Lau, G. Wallace, and D. Diamond. 2009. Bio-sensing Textile based Patch with Integrated Optical Detection System for Sweat Monitoring. *Sensors and Actuators B: Chemical* 138 (1): 231–236.

Moteiv. 2004. *Telos: Ultra Low Power IEEE 802.15.4 Compliant Wireless Sensor Module*. Moteiv. May 12, 2004. http://www2.ece.ohio-state.edu/~bibyk/ee582/telosMote.pdf. Accessed 27 Mar 2014.

———. 2006. *Tmote Sky Data Sheet: Ultra Low Power IEEE 802.15.4 Compliant Wireless Sensor Module*. June 2, 2006. http://www.eecs.harvard.edu/~konrad/projects/shimmer/references/ tmote-sky-datasheet.pdf. Accessed 3 Jan 2014.

Mrazovac, B., M.Z. Bjelica, D. Kukolj, B.M. Todorovic, and D. Samardzija. 2012. A Human Detection Method for Residential Smart Energy Systems based on Zigbee RSSI Changes. *IEEE Transactions on Consumer Electronics* 58 (3): 819–824.

Murata Manufacturing Co. 2012. *Pyroelectric Infrared Sensors*. Murata Manufacturing Co. October 1, 2012. http://www.murata.com/products/catalog/pdf/s21e.pdf. Accessed 6 Mar 2014.

Murphy, R.R. 2004. Trial by Ffire [Rescue Robots]. *Robotics & Automation Magazine* 11 (3): 50–61.

Naderiparizi, N., M. Hessar, V. Talla, S. Gollakota, and J.R. Smith. 2018. Towards Battery-Free HD Video Streaming. In *The 15th Symposium on Networked Systems Design and Implementation (NSDI)*, 233–247. Renton: USENIX.

Ng, H.S., M.L. Sim, and C.M. Tan. 2006. Security Issues of Wireless Sensor Networks in Healthcare Applications. *BT Technology Journal* 24 (2): 138–144.

NI. 2019. *What is I/Q Data?* National Instruments, Inc. October 16, 2019. http://www.ni.com/ tutorial/4805/en/. Accessed 21 June 2020.

Nordic Semiconductor. 2004. *nRF24E1: 2.4GHz RF transceiver with embedded 8051 compatible micro-controller and 9 input, 10 bit ADC.* Nordic Semiconductor. June 1, 2004. http://www. datasheetarchive.com/dlmain/Datasheets-23/DSA-442985.pdf. Accessed 21 Mar 2014.

NXP. 2020. *KL0x: Kinetis® KL0x-48 MHz, Entry-Level Ultra-Low Power Microcontrollers (MCUs) based on Arm® Cortex®-M0+ Core.* January 1, 2020. https://www.nxp.com/ products/processors-and-microcontrollers/arm-microcontrollers/general-purpose-mcus/kl-series-cortex-m0-plus/kinetis-kl0x-48-mhz-entry-level-ultra-low-power-microcontrollers-mcus-based-on-arm-cortex-m0-plus-core:KL0x. Accessed 21 June 2020.

Nyan, M.N., F.E.H. Tay, and E. Murugasu. 2008. A Wearable System for Pre-impact Fall Detection. *Journal of Biomechanics* 41 (16): 3475–3481.

Oki Semiconductor. 2004. *ML675K Series.* Oki Semiconductor. February 1, 2004. http://www.keil. com/dd/docs/datashts/oki/ml675xxx_ds.pdf. Accessed 21 Mar 2014.

Omeni, O., A. Wong, A.J. Burdett, and C. Toumazou. 2008. Energy Efficient Medium Access Protocol for Wireless Medical Body Area Sensor Networks. *IEEE Transactions on Biomedical Circuits and Systems* 2 (4): 251–259.

OmniVision Technologies. 1999. *OV6620/OV6120.* OmniVision Technologies. June 1, 1999. http://coecsl.ece.illinois.edu/ge423/datasheets/DS-OV6620-1.2.pdf. Accessed 29 Mar 2014.

———. 2002. *OVT OV528: Single Chip Camera-to-Serial Bridge.* OmniVision Technologies. October 10, 2002. http://www.datasheet-pdf.com/datasheet-html/O/V/5/OV528-OmniVision. pdf.html. Accessed 29 Mar 2014.

———. 2003. *OV7640 Color CMOS VGA (640 x 480) CAMERACHIP, OV7140 B&W CMOS VGA (640 x 480) CAMERACHIP.* OmniVision Technologies. January 15, 2003. http://www. datasheet4u.com/datasheet/O/V/7/OV7640_OmniVision.pdf.html. Accessed 27 Mar 2014.

———. 2006. *OV9655/OV9155 CMOS SXGA (1.3 MegaPixel) CameraChip Sensor with OmniPixel Technology.* OmniVision Technologies. November 21, 2006. http://www.surveyor. com/blackfin/OV9655-datasheet.pdf. Accessed 2 Apr 2014.

———. 2011. *About Us.* OmniVision Technologies. January 1, 2011. http://www.ovt.com/ aboutus/. Accessed 27 Mar 2014.

OpenCV. 2014. *OpenCV (Open Source Computer Vision).* OpenCV. January 1, 2014. http:// opencv.org. Aaccessed April 2, 2014.

Osmani, V., S. Balasubramaniam, and D. Botvich. 2008. Human Activity Recognition in Pervasive Health-care: Supporting Efficient Remote Collaboration. *Journal of Network and Computer Applications* 31 (4): 628–655.

Oxford Dictionaries. 2014. *environment.* January 1, 2014. http://www.oxforddictionaries.com/ definition/english/environment?q=environment. Accessed 22 Jan 2014.

Paganelli, F., and D. Giuli. 2007. An Ontology-based Context Model for Home Health Monitoring and Alerting in Chronic Patient Care Networks. In *21st International Conference on Advanced Information Networking and Applications Workshops (AINAW)*, 838–845. Niagara Falls: IEEE.

Paganelli, F., E. Spinicci, and D. Giuli. 2008. ERMHAN: A Context-Aware Service Platform to Support Continuous Care Networks for Home-Based Assistance. *International Journal of Telemedicine and Applications-Pervasive Health Care Services and Technologies* 2008: 1–13.

Panasonic Corporation. 2013. *Omnidirectional Back Electret Condenser Microphone Cartridge.* Panasonic Corporation. December 19, 2013. http://industrial.panasonic.com/www-data/pdf/ ABA5000/ABA5000CE11.pdf. Accessed 19 Dec 2013.

Pang, Z., Q. Chen, and L. Zheng. 2009. A Pervasive and Preventive Healthcare Solution for Medication Noncompliance and Daily Monitoring. In *2nd International Symposium on Applied Sciences in Biomedical and Communication Technologies (ISABEL)*, 1–6. Bratislava: IEEE.

Pantazis, N.A., S.A. Nikolidakis, and D.D. Vergados. 2013. Energy-Efficient Routing Protocols in Wireless Sensor Networks: A Survey. *Communications Surveys & Tutorials* 15 (2): 551–591.

Park, C., and P.H. Chou. 2006a. eCAM: Ultra Compact, High Data-Rate Wireless Sensor Node with a Miniature Camera. In *The 4th International Conference on Embedded Networked Sensor Systems (SenSys)*, 359–360. Boulder: ACM.

———. 2006b. Eco: Ultra-Wearable and Expandable Wireless Sensor Platform. In *International Workshop on Wearable and Implantable Body Sensor Networks (BSN)*, 162–165. Cambridge: IEEE.

Park, P., C. Fischione, A. Bonivento, K.H. Johansson, and A. Sangiovanni-Vincent. 2011. Breath: An Adaptive Protocol for Industrial Control Applications Using Wireless Sensor Networks. *IEEE Transactions on Mobile Computing* 10 (6): 821–838.

Pawar, P., B.-J. Van Beijnum, M. Van Sindere, A. Aggarwal, P. Maret, and F. De Clercq. 2008. Performance Evaluation of the Context-Aware Handover Mechanism for the Nomadic Mobile Services in Remote Patient Monitoring. *Computer Communications – Performance Evaluation of Communication Networks (SPECTS 2007)* 31 (16): 3831–3842.

Penders, J., et al. 2011. A Robot Swarm Assisting a Human Fire-Fighter. *Advanced Robotics* 25 (1/2): 93–117.

Perrig, A., J. Stankovic, and D. Wagner. June 2004. Security in Wireless Sensor Networks. *Communications of the ACM – Wireless Sensor Networks* 47 (6): 53–57.

Petersen, S., P. Doyle, S. Vatland, C.S. Aasland, T.M. Andersen, and D. Sjong. 2007. Requirements, Drivers and Analysis of Wireless Sensor Network Solutions for The Oil & Gas Industry. In *IEEE Conference on Emerging Technologies and Factory Automation (ETFA)*, 219–226. Patras: IEEE.

Petersen, S., et al. 2008. A Survey of Wireless Technology for the Oil and Gas Industry. *SPE Projects, Facilities & Construction* 3 (4): 1–8.

Philipose, M., et al. 2004. Fast, Detailed Inference of Diverse Daily Human Activities. In *The Seventeenth Annual ACM Symposium on User Interface Software and Technology*. Santa Fe: ACM.

Philips. 2002. *PCF50606/605: Single-chip Power Management Unit+*. Philips. May 1, 2002. http://www.datasheetarchive.com/dl/Datasheet-03/DSA0051772.pdf. Accessed 3 Apr 2014.

———. 2004. *Philips LPC2106 (ARM) Board, RS232, 16x2 LCD, Relay & Buzzer*. Philips. June 18, 2004. http://microcontrollershop.com/product_info.php?products_id=644. Accessed 29 Mar 2014.

Pister, K.S.J., and L. Doherty. 2008. TSMP: Time Synchronized Mesh Protocol. In *Parallel and Distributed Computing and Systems (PDCS)*. Orlando: Acta.

Polastre, J., R. Szewczyk, and D. Culler. 2005. Telos: Enabling Ultra-low Power Wireless Research. In *Fourth International Symposium on Information Processing in Sensor Networks (IPSN)*, 364–369. Los Angeles: ACM/IEEE.

PostgreSQL. 2013. *PostgreSQL 2013-12-05 Update Release*. December 5, 2013. http://www.postgresql.org. Accessed 7 Jan 2014.

Pui Audio. 2008. *Electret Condenser Microphone Basics*. Pui Audio. January 1, 2008. http://www.digikey.com/Web%20Export/Supplier%20Content/PUI_668/PDF/PUI_ElectretCondenserMicrophone%20Basics.pdf?redirected=1. Accessed 5 Mar 2014.

Qimonda AG. 2006. *HYB18L512160BF-7.5 DRAMs for Mobile Applications 512-Mbit Mobile-RAM*. Qimonda AG. December 1, 2006. http://www.datasheet4u.com/datasheet/H/Y/B/HYB1 8L512160BF-75_QimondaAG.pdf.html. Accessed 2 Apr 2014.

Rahimi, M., et al. 2005. Cyclops: In Situ Image Sensing and Interpretation in Wireless Sensor Networks. In *The 3rd International Conference on Embedded Networked Sensor Systems (SenSys)*, 192–204. San Diego: ACM.

RainSun. 2009. *AN9520: Multilayer Chip Antenna for 2.4GHz Wireless Communication*. RainSun. August 1, 2009. http://www.compotek.com/fileadmin/Datenblaetter/Antennen/9_SMD/RainSun_AN9520_ds_2.3.pdf. Accessed 21 Mar 2014.

Rappaport, T.S. 2002. *Wireless Communications: Principles and Practice*. New Jersey: Prentic Hall.

Raytheon. 2014. *Raytheon BBN Technologies*. Raytheon. January 1, 2014. http://www.raytheon. com/ourcompany/bbn/index.html. Accessed 27 Dec 2014.

Reeves, A.A. 2005. Remote Monitoring of Patients Suffering from Early Symptoms of Dementia. In *International Workshop on Wearable and Implantable Body Sensor Networks*. London.

Reissman, T., and E. Garcia. 2008. Surgically Implanted Energy Harvesting Devices for Renewable Power Sources in Insect Cyborgs. In *The International Mechanical Engineering Congress and Exposition*, 645–653. Boston: The American Society of Mechanical Engineers (ASME).

Ren, H., and M.Q.-H. Meng. 2006. Bioeffects Control in Wireless Biomedical Sensor Networks. In *The 3rd Annual IEEE Communications Society on Sensor and Ad Hoc Communications and Networks (SECON)*, 896–904. Reston: IEEE.

Rinner, B., and W. Wolf. 2008. An Introduction to Distributed Smart Cameras. *Proceedings of the IEEE* 96 (10): 1565–1575.

Rowe, A., D. Goel, and R. Rajkumar. 2007a. FireFly Mosaic: A Vision-Enabled Wireless Sensor Networking System. In *28th IEEE International Real-Time Systems Symposium (RTSS)*, 459–468. Tucson: IEEE.

Rowe, A.G., A. Goode, D. Goel, and I. Nourbakhsh. 2007b. *CMUcam3: An Open Programmable Embedded Vision Sensor*. Pittsburgh: Technical, Robotics Institute, Carnegie Mellon University.

Russell, W.M.S., and R.L. Burch. 1959. *The Principles of Humane Experimental Technique*. Baltimore: Johns Hopkins University.

Sacks, L., et al. 2003. "*The Development of a Robust, Autonomous Sensor Network Platform for Environmental Monitoring*." *IOP Sensors and their applications XII*. Limerick: IOP Institute of Physics.

Sadler, B.M., T. Pham, and L.C. Sadler. 1998. Optimal and Wavelet-Based Shock Wave Detection and Estimation. *The Journal of the Acoustical Society of America* 104: 955–963.

Şahin, E., and A. Winfield. 2008. Special Issue on Swarm Robotics. *Swarm Intelligence* 2: 69–72.

Sallai, J., G. Balogh, M. Maróti, and Á. Lédeczi. 2004. *Acoustic Ranging in Resource Constrained Sensor Networks*. Nashville: Technical, Institute for Software Integrated Systems, Vanderbilt University.

Samsung Electronics. 2004. *SDRAM 256Mb E-die (x4, x8, x16)*. Samsung Electronics. May 1, 2004. http://pdf1.alldatasheet.com/datasheet-pdf/view/37074/SAMSUNG/K4S561632E. html. Accessed 5 Apr 2014.

———. 2006. *K9F1G08R0B – Flash Memory*. Samsung Electronics. December 28, 2006. http:// www.ic-on-linc.cn/view_download.php?id=1664076&file−0333\k9f1g08r0b_2077601.pdf. Aaccessed April 5, 2014.

Sarangapani, J. 2007. *Wireless Ad hoc and Sensor Networks: Protocols, Performance, and Control*. Boca Raton: CRC Press.

Schenato, L., B. Sinopoli, M. Franceschetti, K. Poolla, and S.S. Sastry. 2007. Foundations of Control and Estimation Over Lossy Networks. *Proceedings of the IEEE* 95 (1): 163–187.

ScienceDirect. 2020. *Channel State Information*. Elsevier B.V. 2020. https://www.sciencedirect. com/topics/engineering/channel-state-information. Accessed 21 June 2020.

Seema, A., and M. Reisslein. 2011. Towards Efficient Wireless Video Sensor Networks: A Survey of Existing Node Architectures and Proposal for A Flexi-WVSNP Design. *IEEE Communications Surveys & Tutorials* 13 (3): 462–486.

Semtech. 2008. *XE1205*. December 10, 2008. http://www.semtech.com/images/datasheet/xe1205. pdf. Accessed 2 Feb 2014.

Seo, J., J. Heo, S. Lim, J. Ahn, and W. Kim. 2007. A Study on the Implementation of a Portable u-Healthcare System using SNMP and AODV. In *29th Annual International Conference of the IEEE on Engineering in Medicine and Biology Society (EMBS)*. Lyon: IEEE.

Shangguan, L., and K. Jamieson. 2016. The Design and Implementation of a Mobile RFID Tag Sorting Robot. In *The 14th Annual International Conference on Mobile Systems, Applications, and Services (MobiSys)*, 31–42. Singapore: ACM.

Shin, K.G., and P. Ramanathan. 1994. Real-time Computing: A New Discipline of Computer Science and Engineering. *Proceedings of the IEEE* 82 (1): 6–24.

Shnayder, V., B.-R. Chen, K. Lorincz, T.R.F. Fulford-Jones, and M. Welsh. 2005. *Sensor Networks for Medical Care*. Cambridge: Technical Report TR-08-05, Division of Engineering and Applied Sciences, Harvard University.

Shum, A.J., J. Crest, G. Schubiger, and B.A. Parviz. 2007. Drosophila as a Live Substrate for Solid-State Microfabrication. *Advanced Materials* 19: 3608–3612.

Sikka, P., P. Corke, P. Valencia, C. Crossman, D. Swain, and G. Bishop-Hurley. 2006. Wireless Adhoc Sensor and Actuator Networks on the Farm. In *The 5th International Conference on Information Processing in Sensor Networks (IPSN)*, 492–499. Nashville: ACM/IEEE.

Silicon Laboratories. 2004. *C8051F121: 100 MIPS, 128 kB Flash, 12-Bit ADC, 64-Pin Mixed-Signal MCU*. Silicon Laboratories. June 15, 2004. https://www.silabs.com/Support%20 Documents/TechnicalDocs/C8051F121-Short.pdf. Accessed 28 Feb 2014.

———. 2013. *CP2102/9: Single-Chip USB to UART Data*. Silicon Laboratories. December 1, 2013. http://www.silabs.com/Support%20Documents/TechnicalDocs/CP2102-9.pdf. Accessed 3 Apr 2014.

Simon, D., and C. Cifuentes. 2005. The Squawk Virtual Machine: Java™ on the Bare Metal. In *OOPSLA '05 Companion to the 20th annual ACM SIGPLAN conference on Object-oriented Programming, Systems, Languages, and Applications*, 150–151. San Diego: ACM.

Simon, G., et al. 2004. Sensor Network-Based Countersniper System. In *The 2nd International Conference on Embedded Networked Sensor Systems (SenSys)*, 1–12. Baltimore: ACM.

Skyworks. 2020. *SKY13314-374LF: 0.1-6.0 GHz GaAs SPDT Switch*. Skyworks. January 1, 2020. https://www.skyworksinc.com/en/Products/switches/SKY13314-374LF.   Accessed 21 June 2020.

Song, E.Y., and K.B. Lee. 2008. STWS: A Unified Web Service for IEEE 1451 Smart Transducers. *IEEE Transactions on Instrumentation and Measurement* 57 (8): 1749–1756.

Song, Jianping, et al. 2008. WirelessHART: Applying Wireless Technology in Real-Time Industrial Process Control. In *Real-Time and Embedded Technology and Applications Symposium (RTAS)*, 377–386. St. Louis: IEEE.

Statoil. 2014. *About Statoil*. Statoil. January 1, 2014. http://www.statoil.com/en/Pages/default.aspx. Accessed 5 Mar 2014.

Steele, R., A. Lo, C. Secombe, and Y.K. Wong. 2009. Elderly Persons' Perception and Acceptance of using Wireless Sensor Networks to Assist Healthcare. *International Journal of Medical Informatics-Mining of Clinical and Biomedical Text and Data (Special Issue)* 78 (12): 788–801.

Stoughton, R. 1997. Measurements of Small-Caliber Ballistic Shock Waves in Air. *The Journal of the Acoustical Society of America* 102: 781–787.

Sung, J., et al. 2008. Wireless Sensor Networks for Cultural Property Protection. In *22nd International Conference on Advanced Information Networking and Applications – Workshops (AINAW)*, 615–620. Okinawa: IEEE.

Surie, D., and T. Pederson. 2007. An Activity-Centered Wearable Computing Infrastructure for Intelligent Environment Applications. In *Lecture Notes in Computer Science-Embedded and Ubiquitous Computing*, ed. T.-W. Kuo, E. Sha, M. Guo, L.T. Yang, and Z. Shao, vol. 4808, 456–465. Berlin: Springer.

Surie, D., O. Laguionie, and T. Pederson. 2008. Wireless Sensor Networking of Everyday Objects in a Smart Home Environment. In *International Conference on Intelligent Sensors, Sensor Networks and Information Processing (ISSNIP)*, 189–194. Sydney: IEEE.

Szewczyk, R., A. Mainwaring, J. Polastre, J. Anderson, and D. Culler. 2004a. An Analysis of a Large Scale Habitat Monitoring Application. In *The 2nd International Conference on Embedded Networked Sensor Systems (SenSys)*, 214–226. Baltimore: ACM.

Szewczyk, R., J. Polastre, A. Mainwaring, and D. Culler. 2004b. Lessons from a Sensor Network Expedition. In *First European Workshop on Wireless Sensor Networks and Applications (EWSN)*, 307–322. Berlin: Springer.

Tabar, A.M., A. Keshavarz, and H. Aghajan. 2006. Smart Home Care Network using Sensor Fusion and Distributed Vision-Based Reasoning. In *The 4th ACM International Workshop on Video surveillance and Sensor Networks (VSSN)*, 145–154. Santa Barbara: ACM.

Tang, V.W.S., Y. Zheng, and J. Cao. 2006. An Intelligent Car Park Management System based on Wireless Sensor Networks. In *1st International Symposium on Pervasive Computing and Applications*, 65–70. Urumqi: IEEE.

Tavli, B., K. Bicakci, R. Zilan, and J.-M. Barcelo-Ordinas. 2012. A Survey of Visual Sensor Network Platforms. *Multimedia Tools and Applications* 60 (3): 689–726.

TechTarget. 2014a. *MultiMediaCard (MMC)*. TechTarget. January 1, 2014. http://whatis. techtarget.com/definition/MultiMediaCard-MMC. Accessed 31 Mar 2014.

———. 2014b. *Secure Digital card (SD card)*. TechTarget. January 1, 2014. http://searchstorage. techtarget.com/definition/Secure-Digital-card. Accessed 31 Mar 2014.

———. 2017. *Kinect*. TechTarget. January 1, 2017. http://searchhealthit.techtarget.com/definition/ Kinect. Accessed 28 Jan 2017.

Teixeira, T., and A. Savvides. 2007. Lightweight People Counting and Localizing in Indoor Spaces Using Camera Sensor Nodes. In *First ACM/IEEE International Conference on Distributed Smart Cameras (ICDSC)*, 36–43. Vienna: ACM/IEEE.

Teixeira, T., A.G. Andreou, and E. Culurciello. 2005. Event-Based Imaging with Active Illumination in Sensor Networks. In *IEEE International Symposium on Circuits and Systems (ISCAS)*, 644–647. Kobe: IEEE.

Teixeira, T., E. Culurciello, J.H. Park, D. Lymberopoulos, A. Barton-Sweeney, and A. Savvides. 2006. Address-event Imagers for Sensor networks: Evaluation and Modeling. In *The 5th International Conference on Information Processing in Sensor Networks (IPSN)*, 458–466. Nashville: ACM/IEEE.

TeraRanger. 2017. *Time-of-Flight Principle*. TeraRanger. January 1, 2017. http://www.teraranger. com/technology/time-of-flight-principle/. Accessed 28 Jan 2017.

Texas Instruments. 2005a. *CC1100: Low-Cost Low-Power Sub-1 GHz RF Transceiver*. Texas Instruments. January 1, 2005. http://datasheet.octopart.com/CC1100RTKR-Texas-Instru ments-datasheet-10422951.pdf. Accessed 3 Feb 2014.

———. 2005b. *CC2420: First Single-chip 2.4 GHZ IEEE 802.15.4 Compliant And Zigbeetm Ready RF Transceiver*. January 1, 2005. http://www.ti.com/lit/ds/symlink/cc2420.pdf. Accessed 3 Jan 2014.

———. 2005c. *CC2431: System-on-Chip for 2.4 GHz ZigBee®/ IEEE 802.15.4 with Location Engine*. January 1, 2005. http://www.ti.com/lit/ds/symlink/cc2431.pdf. Accessed 3 Jan 2014.

———. 2006. *CC2430: A True System-on-Chip solution for 2.4 GHz IEEE 802.15.4/ZigBee*. Texas Instruments. January 1, 2006. http://www.ti.com.cn/cn/lit/ds/symlink/cc2430.pdf. Accessed 6 Apr 2014).

———. 2007a. *CC1000: Single Chip Very Low Power RF Transceiver*. Texas Instruments. 2007. http://www.ti.com/lit/ds/symlink/cc1000.pdf. Accessed 23 Dec 2013.

———. 2007b. *LMX9820A Bluetooth Serial Port Module*. Texas Instruments. February 1, 2007. http://www.ti.com/lit/ds/snosaf5j/snosaf5j.pdf. Accessed 31 Mar 2014.

———. 2011a. *CC2530F32, CC2530F64, CC2530F128, CC2530F256: A True System-on-Chip Solution for 2.4-GHz IEEE802.15.4 and ZigBee Applications*. Texas Instruments Inc. February 1, 2011. http://www.ti.com/lit/ds/symlink/cc2530.pdf. Accessed 30 Jan 2017.

———. 2011b. *MSP430C11x1, MSP430F11x1A Mixed Signal Microcontroller*. Texas Instruments. January 1, 2011. http://www.ti.com/lit/ds/symlink/msp430f1611.pdf. Accessed 1 Feb 2014.

———. 2020. *HDC2010: ±2% ultra-low-power, digital humidity sensor with temperature sensor in WCSP*. Texas Instruments, Inc. January 1, 2020. https://www.ti.com/product/HDC2010. Accessed June 2020.

Thackeray, M.M., C. Wolverton, and E.D. Isaacs. 2012. Electrical Energy Storage for Transportation—Approaching the Limits of, and Going Beyond, Lithium-Ion Batteries. *Energy & Environmental Science* (7): 7854–7863.

The Engineering ToolBox. 2014. *IP – Ingress Protection Ratings*. January 1, 2014. http://www.engineeringtoolbox.com/ip-ingress-protection-d_452.html. Accessed 6 Feb 2014.

The Free Dictionary. 2014. *Imager*. The Free Dictionary. January 1, 2014. http://www.thefreedictionary.com/imager. Accessed 30 Mar 2014.

Thomesse, J.P. 2005. Fieldbus Technology in Industrial Automation. *Proceedings of the IEEE* 93 (6): 1073–1101.

TinyOS. 2012. *TinyOS*. August 20, 2012. http://www.tinyos.net. Accessed 9 Oct 2013.

Tolle, G., et al. 2005. A Macroscope in the Redwoods. In *The 3rd International Conference on Embedded Networked Sensor Systems (SenSys)*, 51–63. San Diego: ACM.

Toshiba. 2002. *TC55VCM208ASTN40,55: 524,288-Word by 8-bit Full CMOS Static RAM*. Toshiba. July 4, 2002. http://pdf.datasheetcatalog.com/datasheet2/1/03ace2kfgo489cs3i32xflsi1xcy.pdf. Accessed 14 Apr 2014.

Triantafyllidis, A., V. Koutkias, I. Chouvarda, and N. Maglaveras. 2008. An Open and Reconfigurable Wireless Sensor Network for Pervasive Health Monitoring. In *Second International Conference on Pervasive Computing Technologies for Healthcare*, 112–115. Tampere: IEEE.

Tu, Y.-J., W. Zhou, and S. Piramuthu. 2009. Identifying RFID-embedded Objects in Pervasive Healthcare Applications. *Decision Support Systems* 46 (2): 586–593.

UC Berkeley WEBS Project. 2004. *nesC: A Programming Language for Deeply Networked Systems*. UC Berkeley. December 14, 2004. http://nescc.sourceforge.net. Accessed 23 Mar 2014.

Van Dam, T., and K. Langendoen. 2003. An Adaptive Energy-Efficient MAC Protocol for Wireless Sensor Networks. In *The 1st International Conference on Embedded Networked Sensor Systems (SenSys)*, 171–180. Los Angeles: ACM.

Varshney, U. 2008a. A Framework for Supporting Emergency Messages in Wireless Patient Monitoring. *Decision Support Systems* 45 (4): 981–996.

———. 2008b. Improving Wireless Health Monitoring Using Incentive-based Router Cooperation. *Computer* 41 (5): 56–62.

Vasisht, D., S. Katabi, and D. Kumar. 2016. Decimeter-Level Localization with a Single WiFi Access Point. In *The 13th Symposium on Networked Systems Design and Implementation (NSDI)*, 165–178. Santa Clara: USENIX.

Veracode. 2006. *Man in the Middle Attack*. January 1, 2006. https://www.veracode.com/security/man-in-the-middle-attack. Accessed 14 Feb 2014.

Vick, A., et al. 2002. *Aerospace Operations in Urban Environments: Exploring New Concepts*. RAND Corporation.

Volgyesi, P., G. Balogh, A. Nadas, C.B. Nash, and A. Ledeczi. 2007. Shooter Localization and Weapon Classification with Soldier-Wearable Networked Sensors. In *The 5th International Conference on Mobile Systems, Applications and Services (MobiSys)*, 113–126. San Juan: ACM.

Wallace, A., A. Von Jouanne, E. Wiedenbrug, E. Matheson, and J. Douglass. 2001. A Laboratory Assessment of In-Service and Nonintrusive Motor Efficiency Testing Methods. *Electric Power Components and Systems* 29 (6): 517–529.

Want, R., T. Pering, G. Danneels, M. Kumar, M. Sundar, and J. Light. 2002. The Personal Server: Changing the Way We Think about Ubiquitous Computing. In *Lecture Notes in Computer Science-UbiComp 2002: Ubiquitous Computing*, ed. G. Borriello and L.E. Holmquist, vol. 2498, 194–209. Berlin: Springer.

WebMD. 2005a. *Epilepsy Health Center*. January 1, 2005. http://www.webmd.com/epilepsy/electroencephalogram-eeg-21508. Accessed 12 Feb 2014.

———. 2005b. *Heart Disease Health Center*. January 1, 2005. http://www.webmd.com/heart-disease/electrocardiogram-ecg-ekg-directory. Accessed 12 Feb 2014.

Weiss, B.A., C. Schlenoff, M. Shneier, and A. Virts. 2006. Technology Evaluations and Performance Metrics for Soldier-Worn Sensors for Assist. In *Performance Metrics for Intelligent Systems Worshop (PerMIS)*. Gaithersburg: NIST and DARPA.

Werner-Allen, G., et al. 2006. Deploying a Wireless Sensor Network on an Active Volcano. *IEEE Internet Computing* 10 (2): 18–25.

Whitmire, E., T. Latif, and A. Bozkurt. 2013. Kinect-Based System for Automated Control of Terrestrial Insect Biobots. In *35th Annual International Conference of the IEEE Engineering in Medicine and Biology Society (EMBC)*, 1470–1473. Osaka: IEEE.

Willig, A. 2008. Recent and Emerging Topics in Wireless Industrial Communications: A Selection. *IEEE Transactions on Industrial Informatics* 4 (2): 102–124.

Wilson, S.P., P. Verschure, A. Mura, and T.J. Prescott. 2015. *Biomimetic and Biohybrid Systems: Living Machines*. Springer.

Wireless Sensors, LLC. 2011. *The SensiNet Handbook*. Wireless Sensors, LLC. January 1, 2011. http://www.wirelesssensors.com/images/pdfs/SensiNet_Handbook_rev2_2011.pdf. Aaccessed March 5, 2014.

Witrant, E., P.G. Park, M. Johansson, C. Fischione, and K.H. Johansson. 2007. Predictive Control over Wireless Multi-hop Networks. In *IEEE International Conference on Control Applications (CCA)*, 1037–1042. Singapore: IEEE.

Wolfson Microelectronics. 2011. *WM8950: ADC with Microphone Input and Programmable Digital Filters*. Wolfson Microelectronics. November 1, 2011. http://www.wolfsonmicro.com/documents/uploads/data_sheets/en/WM8950_1.pdf. Aaccessed April 3, 2014.

Wood, A., et al. 2008. Context-Aware Wireless Sensor Networks for Assisted Living and Residential Monitoring. *Network* 22 (4): 26–33.

Wootton, R., J. Craig, and V. Patterson. 2006. *Introduction to Telemedicine. 2nd. Edited by Richard Wootton, John Craig and Victor Patterson*. London: Royal Society of Medicine Press.

Xiao, Y., X. Shen, B. Sun, and L. Cai. 2006. Security and Privacy in RFID and Applications in Telemedicine. *IEEE Communications Magazine* 44 (4): 64–72.

Xie, D., T. Yan, D. Ganesan, and A. Hanson. 2008. Design and Implementation of a Dual-Camera Wireless Sensor Network for Object Retrieval. In *The 7th International Conference on Information Processing in Sensor Networks (IPSN)*, 469–480. St. Louis: ACM/IEEE.

Xilinx. 2007. *XC2C256 CoolRunner-II CPLD*. Xilinx. March 8, 2007. http://www.xilinx.com/support/documentation/data_sheets/ds094.pdf. Accessed 23 Mar 2014.

———. 2008. *Spartan-II FPGA Family Data Sheet*. Xilinx. June 13, 2008. http://www.xilinx.com/support/documentation/data_sheets/ds001.pdf. Accessed 1 Mar 2014.

———. 2013. *Spartan-3 FPGA Family Data Sheet*. Xilinx. June 27, 2013. http://www.xilinx.com/support/documentation/data_sheets/ds099.pdf. Aaccessed December 19, 2013.

———. 2014. *CPLD*. Xilinx. January 1, 2014. http://www.xilinx.com/cpld/. Accessed 22 Jan 2014.

Xiong, H., T. Latif, E. Lobaton, A. Bozkurt, and M.L. Sichitiu. 2016. Characterization of RSS Variability for Biobot Localization Using 802.15.4 Radios. In *IEEE Topical Conference on Wireless Sensors and Sensor Networks (WiSNet)*, 1–3. Austin: IEEE.

Xu, Y., J. Heidemann, and D. Estrin. 2001. Geography-informed Energy Conservation for Ad Hoc Routing. In *The 7th Annual International Conference on Mobile Computing and Networking (MobiCom)*, 70–84. Rome: ACM.

Ye, W., J. Heidemann, and D. Estrin. 2004. Medium Access Control with Coordinated Adaptive Sleeping for Wireless Sensor Networks. *IEEE/ACM Transactions on Networking* 12 (3): 493–506.

Yoo, S.E., J.E. Kim, T. Kim, S. Ahn, J. Sung, and D. Kim. 2007. A2S: Automated Agriculture System based on WSN. In *IEEE International Symposium on Consumer Electronics (ISCE)*, 1–5. Irving: IEEE.

Yoo, J., L. Yan, S. Lee, Y. Kim, and H.-J. Yoo. 2010. A 5.2 mW Self-Configured Wearable Body Sensor Network Controller and a 12 W Wirelessly Powered Sensor for a Continuous Health Monitoring System. *IEEE Journal of Solid-State Circuits* 45 (1): 178–188.

Younis, O., and S. Fahmy. 2004. HEED: A Hybrid, Energy-efficient, Distributed Clustering Approach for Ad hoc Sensor Networks. *IEEE Transactions on Mobile Computing* 3 (4).

Zhang, M., and W. Cai. 2010. Vision Mesh: A Novel Video Sensor Networks Platform for Water Conservancy Engineering. In *3rd IEEE International Conference Computer Science and Information Technology (ICCSIT)*, 106–109. Chengdu: IEEE.

Zhang, W., M.S. Branicky, and S.M. Phillips. 2001. Stability of Networked Control Systems. *Control Systems* 21 (1): 84–99.

ZigBee Alliance. 2013. *Specifications*. ZigBee Alliance. January 1, 2013. http://zigbee.org/ Specifications.aspx. Accessed 28 Aug 2013.

———. 2017. *ZigBee RF4CE*. Zigbee Alliance. January 1, 2017. http://www.zigbee.org/zigbee-for-developers/network-specifications/zigbeerf4ce/. Accessed 30 Jan 2017.

Zorzi, M., and R.R. Rao. 2003. Energy and Latency Performance of Geographic Random Forwarding for Ad Hoc and Sensor Networks. In *IEEE Wireless Communications and Networking (WCNC)*, 1930–1935. New Orleans: IEEE.

Part II
Network and Transport Layers, Cross-Layering

Chapter 4
Energy and Lifetime Aware Routing Protocols for WSNs

The shortest path is not necessarily to be followed.

4.1 WSNs Energy-Driven Considerations

As Chap. 3 precisely disclosed, WSNs are key constituents of modern life with the multiplicity of applications they span. In these applications, the sensor nodes are envisaged for deployment in sizeable numbers in unattended areas, which makes it difficult if not infeasible to replace these nodes after being deployed and put to operation. Therefore, as sensor nodes are predominantly battery-powered devices, the energy consumption of the nodes must be properly managed in order to prolong the network lifetime and functionality to a satisfactory level that allows fulfilling the intended operation.

Wireless sensors are diverse; there are light sensors, barometric pressure sensors, accelerometers, humidity sensors, temperature sensors, GPS modules, acoustic sensors, magnetic RPM sensors, magnetometers, wind speed sensors, moisture sensors, solar radiation sensors, temperature sensors, seismic sensors, rainfall meters, and more and more. Sensor nodes perform three basic energy-consuming tasks (Akyildiz et al. 2002):

- Sampling a physical quantity from the specified surrounding environment.
- Processing and storing, if required, of the sensed data.
- Transferring the sensed data to a data collection point, basestation or sink node, all the way through wireless communication.

Sensor nodes communicate with the basestation (gateway unit) that is proficient in communicating with other computers via other networks, such as LAN, WLAN, WPAN, and the Internet. The basestations are connected to other nodes forming the infrastructure of WSNs. At the basestation, a snapshot of the monitored region is obtained by assembling the received sampling values from the corresponding locations. The sensor nodes communicate with each other and with the basestation through radios to exchange the data that accomplish the application functions.

© The Author(s), under exclusive license to Springer Nature Switzerland AG 2023
H. M. A. Fahmy, *Concepts, Applications, Experimentation and Analysis of Wireless Sensor Networks*, Signals and Communication Technology,
https://doi.org/10.1007/978-3-031-20709-9_4

Table 4.1 Design factors for WSNs

Design factor	Description
Network connectivity	Sensor nodes are connected with each other and basestation via wireless links to share information
Area coverage	Region where the sensor nodes are deployed for monitoring and data collection. It needs maximum coverage by sensor node/basestation transmission range
Node deployment	Strategy to deterministically place the sensor nodes in order to meet the desired performance goals such as coverage of monitored region
Fault tolerance	Ability of sensor nodes to provide functionality even when few sensor nodes may fail and stop the data transmission due to power shortage, physical damage, or environment interferences that restrain the purpose of the network
WSN lifetime	The time until the energy in any sensor node in the network is depleted, or the time until the energy in all network nodes is depleted The time until the energy in a defined percentage of the sensor nodes in the network is depleted It is based on the network application and formulation. So, the network lifetime of WSNs is the maximum time that the network is capable of measuring a physical value or event
Data aggregation	Guiding principle to combine the sensed data from numerous sensor nodes with elimination of redundancy and provide collective information to the basestation
Clustering	Course of action to group sensor nodes having similar role with an assigned cluster head in a densely deployed large-scale WSN
Routing	The process of determining a network path for a packet from source node to its destination according to set criteria
Network dynamics	Changes in the topology of a WSN due to sensor node failure or its mobility. It makes periodic monitoring difficult as path stability, bandwidth, and energy estimation are required

Based on Yick et al. (2008)

The major concern in a WSN is to extend the network lifetime in addition to robustness as compared to traditional wired networks where the focus is to maximize channel throughput with minimum node exploitation. WSNs are to be deployed with a primary objective of replacing wired networks in areas where either deployment of wired network is complicated due to geographic locations or costly in terms of its utilization and random requirements. As described in Table 4.1, several essential factors should be taken into consideration while designing a WSN and its protocols.

While the basestation in WSNs can have continuous power supply, a sensor node has limited resources in terms of processing capability and memory, restricted battery backup along with unpredictable harvesting capabilities that determine its longevity and whole network lifetime. A sensor node lifetime is constrained by backup available at a time, amount of computation, as well as communication it carries out depending on the distance with communicating nodes. Once power backup depletes below a threshold, the node gets disconnected. Therefore, energy management and conservation techniques present a major research challenge for WSNs (Fahmy 2020).

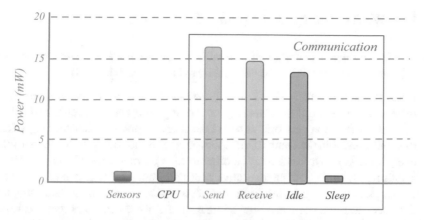

Fig. 4.1 Energy consumption distribution of a sensor node (Estrin 2002)

As exhibited in Fig. 4.1, the communication required from a typical sensor node is the major power consumer; its wireless communication module has four states: typically, send, receive, idle, and sleep. Transmitting signals, i.e., sending and receiving, take about two thirds of its total energy consumption, while the number of transmitted data packets of a node depends to a great extent on the routing strategy. An efficient routing protocol can balance the energy consumption levels among WSN nodes; given the same hardware conditions, it can help prolong the lifetime of a WSN, as well as improve the quality of data transmission. However, traditional routing protocols tend to focus on how to make the data packets reach fastest their destination along the shortest transmission path. This may not be the best from the viewpoint of WSN lifetime owing to their energy-constrained sensor nodes. Furthermore, due to the minimal energy consumption at the sleep and idle states, consideration is mostly accorded to the energy consumption of the sending and receiving states only.

A model describing the energy consumption of wireless sensor nodes is developed in (Zhou et al. 2011).

4.2 WSNs Energy and Lifetime Terminology, Models, and Metrics

This section is concerned with introducing the WSN energy and lifetime terminology and the energy efficiency metrics that are mandatory for the design of energy and lifetime aware routing protocols for WSNs. Respectively, Sects. 4.2.1 and 4.2.2 will assume this task.

4.2.1 WSNs Energy and Lifetime Terminology

Several terms related to energy efficiency in WSNs are used as metrics to evaluate the performance of routing protocols (Alazzawi and Elkateeb 2008):

- Network lifetime. In a WSN, it is important to maximize its lifetime, which increases the network survivability or prolongs the battery life of nodes. The common practice in networks is to use the shortest routes to transfer the packets; this could result in the death of the nodes along the shortest path. Since in a WSN every node has to act as a relay in order to forward messages, if some nodes die sooner due to the lack of energy, other nodes may not be able to communicate any more. Hence, the network will get disconnected, the energy consumption becomes non-balanced and the lifetime of the whole network gets seriously affected. Therefore, a combination of the shortest path and the extension of the network lifetime is most suitable for use in WSNs. Moreover, the lifetime of a node is effectively determined by its battery life; the main battery drainage is due to transmitting and receiving data among nodes.

    Thus, the main purpose for the design of energy-efficient and energy-balanced routing protocols for WSNs is to extend the network lifetime, which in turn preserves the network functionality. The network lifetime is a main evaluation metric that is mostly adopted to measure the energy efficiency of a WSN. There are several definitions for the lifetime of a WSN, though the main theme remains the same (Ogundile and Alfa 2017):

- The time until the energy in any sensor node in the network is depleted, or the time until the energy in all network nodes is depleted.
- The time until the energy in a defined percentage of the sensor nodes in the network is depleted.
- It is based on the network application and formulation. So, the lifetime of WSNs is the maximum time that the network is capable of measuring a physical value or event.
- Sensor nodes residual energy (RE). The lifetime of a sensor node is usually measured by its energy level after a given WSN data transmission round. The energy level in a node after each data transmission phase is referred to as its residual energy. Given that the energy of a node at the initial deployment is defined as E_{in}, and the energy consumed by the node after a particular data transmission round is E_r, the residual energy of the node can be defined as (Ogundile and Alfa 2017):

$$E_{re} = E_{in} - \sum_{r=0}^{R} E_r \qquad (4.1)$$

where $E_{re} = E_{in}$ for $r = 0$, r is the current round, and R is the maximum number of rounds.

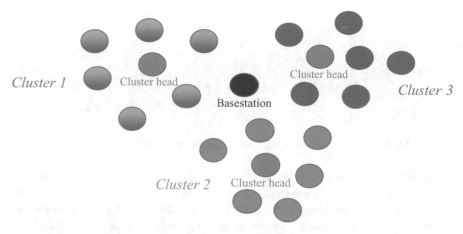

Fig. 4.2 A clustered WSN. (Based on Ogundile and Alfa (2017))

From Eq. (4.1) when $E_{re} = 0$ J, the node completely depletes its energy and cannot participate in further network activities. Thus, the energy level of sensor nodes is a prime factor to be considered in order to maintain the functionality of the network for a reasonable time. E_{re} is a key decision metric used in the design of most of the energy-efficient and energy-balanced routing protocols. Knowing the energy level of the node becomes more important for the design of clustering-based routing protocols (Tripathy and Chinara 2012). Typically, if the energy level of the cluster-head node in Cluster 3 is zero ($E_{re} = 0J$), as shown in Fig. 4.2, all nodes in that cluster cannot participate in the next network activity. Accordingly, in order to avoid untimely network partitioning, most clustering-based routing protocols consider E_{re} as an essential decision metric in the routing algorithms.

- Average network energy (AE). WSN developers use AE to optimize the energy consumption of sensor nodes. The average network energy at any given data transmission round can be expressed as (Ogundile and Alfa 2017):

$$E_{av} = \frac{\sum\limits_{1}^{n} E_{re_n}}{n} \tag{4.2}$$

where n is the total number of sensor nodes in the WSN, and E_{re_n} is the residual energy at node n. E_{av} is widely used as a decision metric in designing energy-efficient and energy-balanced routing protocols for WSNs.

In clustering-based routing protocols, E_{av} is defined as the energy threshold for selecting the cluster-heads for a particular data transmission phase in order to avoid network segregation. Other energy-efficient and energy-balanced routing protocols use E_{av} as the defined energy threshold in choosing the relay nodes toward the basestation during a transmission.

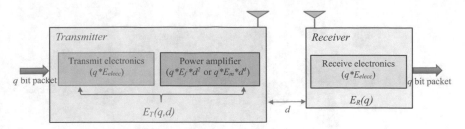

Fig. 4.3 Energy dissipation model. (Based on Heinzelman et al. 2002)

- Energy dissipation model. In the energy model, both free-space and multipath fading channel models are used, depending on the distance between the transmitter and receiver (Heinzelman et al. 2002). As shown in Fig. 4.3, the transmitting sensor node consumes energy to drive its radio subsystem, which includes the radio electronics, and power amplifier. The receiving node also dissipates energy to drive its radio electronics. The distance d between the transmitting node and receiving node is computed in the form of a distance metric that specifies the channel model used by the power amplifier. If the distance d is greater than a set threshold d_0, the multipath model is assumed. Otherwise, the free-space channel model is used for $d < d_0$. Therefore, the required energy for a sensor node to transmit a q bit message is defined as:

$$E_T(q, d) = \begin{cases} q * E_{elec} + q * E_f * d^2 & d < d_0 \\ q * E_{elec} + q * E_m * d^4 & d \geq d_0 \end{cases} \tag{4.3}$$

where.

E_f and E_m are the free space and multipath power loss at the amplifier, respectively; they depend on the acceptable bit rate and distance to the receiver,

E_{elec} is the energy dissipated to drive the radio electronics; it depends on factors such as the digital coding, modulation, filtering, and spreading of the signal,

d_0 is the transmission distance threshold expressed as:

$$d_0 = \sqrt{\frac{E_f}{E_m}} \tag{4.4}$$

On the other hand, the energy consumed by the receiving node to receive a q bit packet is given by:

$$E_R(q) = q * E_{elec} \tag{4.5}$$

4.2.2 Energy Efficiency Metrics

In addition to the energy and lifetime-related terminology detailed in Sect. 4.2.1, more metrics linked to the energy efficiency on WSNs are used to evaluate the performance of their routing protocols (Alazzawi and Elkateeb 2008):

- Energy per packet. The amount of energy that is spent while sending a packet from a source to a destination.
- Energy and reliability. It refers to how to achieve a trade-off between different application requirements. In some applications, emergency events may justify an increased energy cost to speed up the reporting of such events or to increase the redundancy of the transmission by using several paths.
- Average energy dissipated. This metric is related to the WSN lifetime and expresses the average dissipation of energy per sensor node over time, as it performs various functions such as transmitting, receiving, sensing, and aggregation of data.
- Low-energy consumption. A low-energy protocol has to consume less energy than traditional protocols. This means that a low-energy protocol takes into consideration the remaining energy level of the nodes and selects routes that maximize the network lifetime.
- Total number of alive nodes. This metric is also linked to the WSN lifetime as it observes the area coverage of the network over time.
- Time until the first node dies. It is an indication of the duration where all WSN nodes are alive. In some protocols, there may be a first node that runs out of energy earlier than in other protocols, but it manages to keep the WSN operational much longer.
- Half of the nodes alive. It denotes an estimated value for the half-life period of a WSN.
- Last node dies. It gives an estimated value for the overall lifetime of a WSN.
- Total number of data signals received at the basestation. It is equivalent to the energy the protocol saves by not continuously transmitting needless data packets, for instance HELLO messages.
- Average packet delay. The average one-way latency that is spotted between the transmission and reception of a data packet at the sink. It measures the temporal accuracy of a packet.
- Packet delivery ratio. It is the ratio of the number of distinct packets received at sinks to the number originally sent from source sensor nodes. It indicates the reliability of data delivery.
- Energy spent per round. The total energy spent in routing messages in a round. It is a short-term measure designed to reflect the energy efficiency of a proposed protocol in a particular round.
- Idle listening. A sensor node that is in idle listening mode does not send or receive data, though it can still consume a substantial amount of energy. Therefore, this node should not stay in idle listening mode, but should be powered OFF.

- Packet size. The size of a packet determines the time that its transmission will last; hence, it is effective in determining the energy consumption. The packet size has to be reduced or compressed.
- Distance. The distance between the transmitter and receiver affects the power that is required to send and receive packets. The routing protocols can select the shortest paths between nodes and reduce energy consumption.
- Hop-count. It is widely used as a decision metric in the design of energy-efficient and energy-balanced routing protocols for WSNs (Kim et al. 2010; Capone et al. 2010). The hop-count is defined as the number of relay nodes traversed by a packet from the source to the destination node. It is desirable to minimize the hop-count during the data transmission phase in order to extend as possible the lifetime of WSNs. There should be a compromise between reducing the hop-count and maximizing the lifetime of WSNs. By reducing the hop-count from the source node to the destination, some relay nodes are compelled to convey a large amount of data, hence depleting their energy. Consequently, the hop-count is a decision metric when designing an energy-balanced routing protocol for WSNs.

These metrics are used with more details when presenting the energy and lifetime aware routing protocols for WSNs in Sects. 4.7, 4.8, and 4.9.

4.3 Traffic Patterns, Data Collection and Aggregation, and Clustering in WSNs

From the energy standpoint, data collection and aggregation are the compilation, from multiple sensor nodes, of sensed data that are to be transmitted to the basestation for further processing. Since the wireless sensor nodes are distance and energy-constrained; it is inefficient for all nodes to transmit the data directly to the basestation. Moreover, if multiple nodes are sensing the same event, the data will be redundant and massive. Hence, techniques are needed to combine redundant data into valuable information at the nodes, to reduce the total number of packets transmitted to the basestation, which reduces energy consumption and bandwidth.

Clustering algorithms play a crucial role in achieving the targeted design goals for data collection and aggregation with maximum area coverage. In a densely deployed large-scale WSN, grouping sensor nodes, with an assigned cluster-head, form a cluster. The energy limitation on sensor nodes results in a limited WSN lifetime; consequently, appropriate clustering can reduce the overall energy utilization in a WSN and improve its lifetime.

Moreover, when a sensor node drains out energy, it becomes unable to receive or transmit packets and thus gets disconnected from the WSN and loses its coverage area. When data are aggregated, relaying of data from the sensor nodes to the basestation should be done by means of energy-efficient route detection so that the

Table 4.2 Design requirements for energy-efficient WSN protocols (Yadav and Yadav 2016)

Requirement	Design constraint
Data collection and aggregation	Restricted redundant data Low computational overhead Minimized memory storage
Clustering	Maximum supported area coverage Distributed sensor nodes Connection density on cluster-heads Adding new nodes
Routing	Shortest path Maximum network connectivity Maximum supported WSN size Low communication overhead

WSN lifetime is maximized. This implies that for increased lifetime, WSNs must have the following design consideration:

- Maximum area coverage.
- Maximum connectivity.
- Minimum energy consumption.

More on traffic patterns, data collection and aggregation, and clustering in WSNs are presented in Sects. 4.3.1, 4.3.2, and 4.3.3, respectively.

Table 4.2 lists the various design considerations concerning the three major activities performed by WSNs; specifically, data collection and aggregation, clustering, and routing.

4.3.1 Traffic Patterns in WSNs

Different from traditional networks, WSNs exhibit unique asymmetric traffic patterns; this is mainly due to their function, which is to collect data. Sensor nodes persistently send their data to the basestation, while the basestation occasionally sends control messages to the sensor nodes. Moreover, the varied applications can cause a wide range of traffic patterns. The traffic of WSNs can be either single-hop or multihop (Sects. 4.5 and 4.8). The multihop traffic patterns can be further divided, as illustrated in Fig. 4.4, depending on the number of send and receive nodes, or on whether the network supports in-network processing:

- Local communication. It is used to broadcast the status of a node to its neighbors, also to transmit data between two nodes directly.
- Point-to-point. It is adopted to send a data packet from an arbitrary node to another arbitrary node. It is usual in a wireless LAN environment.
- Convergence. The data packets of multiple nodes are routed to a single basestation. It is commonly used for data collection in WSNs.

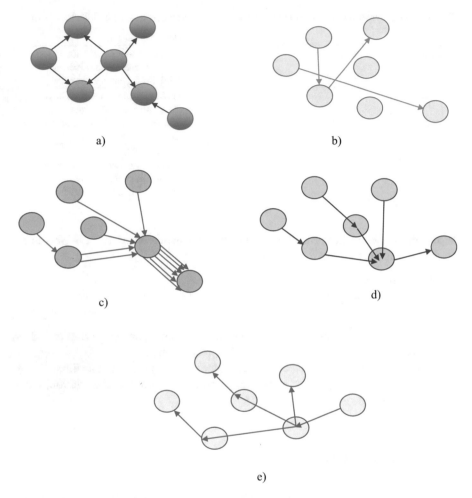

Fig. 4.4 Traffic patterns in WSNs. (**a**) Local communication. (**b**) Point-to-point. (**c**) Convergence. (**d**) Aggregation. (**e**) Divergence (Pantazis et al. 2013)

- Aggregation. The data packets can be processed in the relaying nodes and the aggregate value is routed to the basestation rather than the raw data.
- Divergence. It is used to send a command from the basestation to other sensor nodes.

It is interesting to investigate the traffic patterns in WSNs along with the mobility of the nodes, as node mobility has been developed in several WSN applications such as healthcare monitoring (Chap. 3).

4.3.2 Data Collection and Aggregation in WSNs

WSNs are mainly utilized for data collection and aggregation. Data collection is defined as the organized integration of sensed data, from multiple sensors, transmitted ultimately to the basestation for processing. However, data generated from neighboring sensor nodes is persistently redundant and highly interrelated. In such scenarios, sensor nodes can transmit data to a local collector or assigned head, which combines data from all the sensor nodes that reside in its connectivity, and transmits the concise packet to the basestation, consequentially reducing the total number of packets transmission, thus conserving bandwidth and energy. This can be accomplished by data aggregation. Data aggregation can be defined as the guiding principle to combine the sensed data from numerous sensor nodes with elimination of redundancy and to provide collective information to the basestation (Rajagopalan and Varshney 2006).

Data aggregation protocols could be categorized into two groups: structure-free and structured, as shown in Fig. 4.5 (Fan et al. 2007):

- Structure-free data aggregation protocols. In structure-free WSNs, sensors node deployed in a specific region have similar responsibility and holds out almost same kind of battery. In such networks, data aggregation is done by data-centric routing, where the basestation regularly broadcasts a query message to the sensor nodes. Such structure-free networks are generally deployed for crisis management, as they are meant for short duration varying from few hours to few days. Structure-free approaches do not pay out energy on establishing any structure and provide solutions with benefits of reduced average delay, reduced maintenance overhead, and improved robustness when experiencing sensor node failures. Such structure-free protocols have a major role in providing fault tolerance to energy-constrained time-critical applications. Consequently, they tend to manage aggregation such as the incident information during a crisis of the like of fire evolution, safe zones discovery, gas diffusion, natural disaster, all along with tracking and orientation of rescue persons and intrusion robots. In an unstructured WSN, aggregation node has excessive communication and computation load; hence, its battery exhausts faster; the depletion of such a crucial sensor node breaks down the functionality of the network. Hence, in vision of scalability and energy efficiency, several structured data aggregation approaches have been proposed.

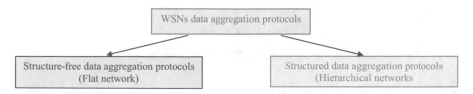

Fig. 4.5 Taxonomy of energy-based data aggregation protocols for WSNs. (Based on Yadav and Yadav (2016))

- Structured data aggregation protocols. For efficient data aggregation, structured approaches employ either tree-based or cluster-based structure at the WSN initialization phase. Such structured protocols are well suited to a stable environment where sensor nodes function continually. Nevertheless, in a more practical environment where sensor nodes may move or fail unexpectedly, there are too much necessary construction and maintenance overhead associated with structured mechanism, which cannot be reimbursed with the advantage from structured data aggregation. On the contrary, structure-free approaches do not spend energy for constructing any structure.

4.3.3 Clustering in WSNs

WSNs face many challenges, as specified in Sect. 4.6, that are to be confronted while considering their mutual conflicts. To overcome the energy restrictions of WSNs, their main challenge, many methods have been proposed to reduce the energy consumption of sensor nodes and increase the network lifetime, such as data gathering, data correlation, energy harvesting, beam forming, resource allocation using cross-layer design, opportunistic transmission schemes (sleep-wake scheduling), mobile relays and sinks, optimal deployment, clustering, and multihop routing (Yetgin et al. 2017). Clustering is an example of these methods and is considered as an efficient and scalable energy method for WSNs (Sect. 4.8.1.1). For clustering in WSNs, sensors are divided into groups or clusters, each of which has a cluster-head. The sensors in each cluster transmit the relevant information to cluster-heads periodically or after an event. Then, cluster-heads transmit the information to a basestation directly or via a multihop path. Clustering has significant benefits (Frye and Cheng 2009):

- Maintaining communication bandwidth and preventing redundancy of exchange messages.
- Stabilizing the WSN topology at the sensor nodes level and reducing communication overhead due to nodes interactions only with cluster-heads.
- Implementing optimized management strategies in the WSN.

In addition to the clustering mechanism, routing for sending data plays a major role in reducing energy consumption and, consequently, increasing the lifetime of the WSN. More on clustering is detailed in Sect. 4.8.1.1.

4.4 Homogeneous and Heterogeneous WSNs

WSNs can be classified based on their infrastructure as homogeneous or heterogeneous. Different energy-efficient and energy-balanced routing protocols have been proposed for both WSN structures:

- Homogeneous WSNs where all sensor nodes have similar hardware components, such as sensing components, processing unit, radio module, and power supply (Yu et al. 2012; Yilmaz et al. 2012; Wang et al. 2012; Zaatouri et al. 2017; Lin and Wang 2017). Traditional routing protocols assume static and homogeneous WSNs.
- Heterogeneous WSNs wherever the hardware components of two or more sensor nodes are different. A major reason for the design of heterogeneous WSNs is to equip some nodes with more powerful processors and higher battery power to cover larger areas (Kumar et al. 2009; Khalil and Attea 2011; Kuila and Jana 2014b; Zaatouri et al. 2017).

Although deploying homogeneous WSNs can be easier, heterogeneous WSNs are more adaptable to practical sensing scenarios.

4.5   Single-Hop and Multihop Transmission

Figure 4.6 depicts a WSN where all the nodes are heterogeneous and static. The coordinates of the nodes and basestation are known, as well as their residual energy. Node A attempts to transmit a q bit packet to the basestation at its allocated time-

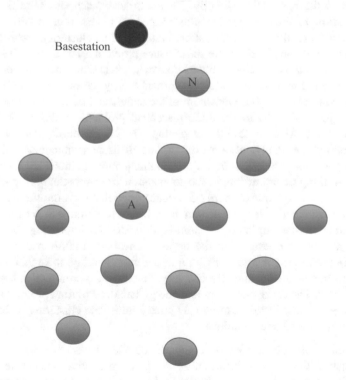

Fig. 4.6 WSN transmission scenario

division multiple-access (TDMA) slot. Note that the TDMA schedule ensures that there are no data collisions during transmissions. Hence, the TDMA schedule conserves energy by allowing the sensor nodes to sleep except during the node transmission time. Node A can convey the q bit packet to the basestation either by single-hop or multihop communication.

Exploring scenarios in single-hop and multihop transmissions:

- In single-hop communication, it transmits the message to the basestation directly without using a relay node. If the distance d between node A and the basestation is large, then from Eq. (4.3), considerable energy will be consumed to transfer the packet; a clear disadvantage of single-hop communication. However, this can be advantageous in scenarios where the node is very close to the basestation, for instance node N. A main characteristic of WSNs is that the participating nodes are randomly placed in the sensor field; hence, those that are at a large distance from the basestation will certainly deplete their battery energy through single-hop transmission, which undermines the network functionality.

  To reduce the energy consumed by the nodes that are distant from the basestation, some energy-balanced routing protocols such as low-energy adaptive clustering hierarchy (LEACH) clustered the WSN and rotated the duty cycle of the cluster-heads (Heinzelman et al. 2002). This method is somewhat useful because the distant non-cluster-head nodes need not to send their sensed information to the basestation directly. Then, cluster-head nodes send their sensed information to their respective cluster-head with less transmission power as compared to sending to the basestation directly. This clustering technique, however, does not completely solve the distance problem associated with single-hop communication as the cluster-heads convey packets from their cluster members directly to the basestation, dissipating their energy as much as they are distant from the basestation. The placement of the nodes, including the cluster-heads, is challenging; solutions to the node placement problem in WSNs are proposed (Younis and Akkaya 2008). Depleting the cluster-heads energy leads to partitioning the WSN; as a result of this distance problem associated with single-hop communication, multihop communication methods are favorable.

- With multihop communication, the source node transmits the q bit packet to the basestation through one or more relay nodes. In such a way, the energy dissipated by the sensor nodes largely distant from the basestation will be minimized in comparison to the single-hop communication. Besides improving the connectivity of a WSN and extending the network coverage, multihop communication helps in prolonging the network lifetime and functionality; in such a way, higher transmission data rates and effective use of the wireless communication channels are realized. The energy-efficient and energy-balanced routing protocols based on multihop communication can be categorized under the clustering technique and the load-balanced tree technique:

  - In the multihop clustering technique, detailed in Sect. 4.8.1.1, the nodes transmit the sensed information to their respective cluster-heads using multihop communication. Similarly, the cluster-head forward the aggregated

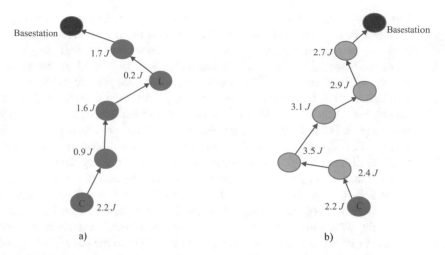

Fig. 4.7 Multihop transmission. Residual energy is indicated in Joules besides each node. (**a**) Shortest path from node C to basestation. (**b**) Energy-efficient path from node C to basestation. (Based on Ogundile and Alfa (2017))

data to the basestation via relay nodes (the relay node can either be a cluster-head or non-cluster-head) or directly (single-hop communication) depending on the cluster-head distance to the basestation.

- The load-balanced tree technique, presented in Sect. 4.8.1.2, follows from the source node to the basestation the route that balances the energy consumption in the WSN. This technique can be via a single-path or a multipath.

Nonetheless, multihop communication methods do not fully solve the energy consumption concerns in WSNs for several reasons:

- The trade-off between finding the most distance-efficient path to the basestation and the energy-efficient and/or energy-balanced path to the basestation. For instance, as illustrated in Fig. 4.7a, following the assumed shortest path, in terms of hop-count, from node C to the basestation, node L has consumed more of its energy in comparison with the other nodes in the path. Hence, it will deplete if used as one of the relay nodes when transmitting the packet from node C to the basestation. In such a scenario, this distance-efficient route from node C is not the most energy-balanced. From Eqs. (4.3)– (4.5), the distance between sensor nodes and the basestation is key to mini-mizing their energy consumption; yet, the distance-efficient route does not ensure a balance in the energy dissipated by the involved nodes. Finding the most distance-efficient route to the basestation that optimizes the energy consumption can be an NP-hard problem.

- The energy-efficient route to the basestation may not be the most energy-balanced. Some energy-efficient routing protocols use the current energy level (residual energy) of the sensor nodes to find the most energy-efficient route (s) to the basestation (Sect. 4.2.1). The nodes with high residual energy are picked up to form the relay nodes chain from the source node to the basestation. Figure 4.7b shows the energy-efficient route from node C to the basestation. Clearly, the relay nodes have high residual energy and can amply convey packets for many data transmission rounds before fully depleting their energy. This depicted energy-efficient route solves the concerns associated with the distance-efficient route of Fig. 4.7a. However, using the nodes with high residual energy does not necessarily balance the energy consumption in the WSN. As Fig. 4.7b displays, if the sensed information from node C is transmitted over a distance longer than that in Fig. 4.7a, then from Eqs. (4.3)–(4.5), the total energy consumed using the energy-efficient route will be higher. In such a scenario, this energy-efficient route to the basestation is not the most energy-balanced route despite transmitting through relay nodes with high residual energy. Therefore, there must be a route from node C to the basestation that balances the energy dissipated by the nodes and the total energy dissipated in the WSN. This route can be unforeseeable and might change every data transmission round depending on the energy level on the nodes and the distance traveled in conveying the data.
- There must be a route with an optimal hop-count to the basestation that balances the energy consumed by the nodes and the total energy dissipated in the WSN. In Fig. 4.7b, the energy-efficient route uses more relay nodes (hop-count = 6) in comparison with the shortest path route (hop-count = 5) of Fig. 4.7a. From Eq. (4.5), this implies that the energy-efficient route will dissipate more energy. The hop-count is a significant decision metric in WSNs routing protocols to optimize the node energy consumption. The smaller the hop-count, the less energy is consumed in receiving and relaying from the source node. As such, it is desirable to keep the hop-count as small as possible, but this does not necessarily guarantee a balance in the energy consumption. For instance, assuming that the route in Fig. 4.7a offers the smallest hop-count from node C to the basestation. Node L is among the relay nodes chain; however, its energy level is remarkably low compared with other nodes in the WSN; so, to maintain full network functionality, it should not act as a relay node. In fact, node L should sleep all the time except when it is transmitting its own sensed information. Consequently, despite the fact that the route in Fig. 4.7a offers the smallest hop-count from node C to the basestation, it is not the most energy-balanced. Thus, there must be a route with an optimal hop-count from node C to the basestation that balances the energy consumed by the nodes and the total energy dissipated in the WSN.
- The hot-spot/energy-hole problem is a considerable holdup. In multihop communication, sensor nodes closer to the basestation consume more energy in receiving and forwarding messages toward it; as most traffic load passes through them, they tend to dissipate more energy in comparison with farther

nodes (Wu et al. 2006). For illustration, in Fig. 4.6, most of the traffic flow toward the basestation will pass through node N implying that it will deplete its battery faster. An imbalance in the nodes energy consumption will cause the untimely depletion of some nodes. An energy-balanced multihop route from the source node to the basestation must, at all times, solve the hot-spot/ energy-hole problem. An example energy-efficient protocol that mitigates this issue using unequal clustering is presented in (Singh et al. 2018).

A taxonomy of reliable energy-efficient and energy-balanced routing protocol for WSNs, built on multihop and single-hop transmissions, is explored in Sect. 4.8.

4.6 Design Issues of Energy and Lifetime Aware Routing Protocols for WSNs

The design issues of energy and lifetime aware routing protocols for WSNs must be understood before developing any such protocols. In Sect. 4.6.1, it is exposed why routing protocols for WSNs are different, while Sect. 4.6.2 lays out the factors that influence their design, and Sect. 4.6.3 reveals the goals that must be kept.

4.6.1 Why Routing Protocols for WSNs Are Different?

For several reasons, routing in WSNs may be more demanding than in other wireless networks, like mobile ad hoc networks or cellular networks (Pantazis et al. 2013):

- Sensor nodes demand careful resource management because of their severe constraints in energy, processing, and storage capacities.
- Almost all applications of WSNs require the flow of sensed data from multiple sources to a particular basestation.
- Design requirements of a WSN depend on the application, because WSNs are application-specific.
- The nodes in WSNs are mostly stationary after their deployment, which results in predictable and non-frequent topological changes.
- Data collection is, under normal conditions, based on the location; therefore, position awareness of sensor nodes is important. The position of the sensor nodes is detected by using methods based on triangulation, e.g., radio strength from a few known points. For the time being, it is possible to use global positioning system (GPS) hardware for this purpose. Moreover, it is favorable to have solutions independent of GPS for the location problem in WSNs (Savvides and Srivastava 2004).
- In WSNs, there is a high probability that collected data may present some undesirable redundancy, which is necessary to be exploited by the routing protocols to improve energy and bandwidth utilization.

4.6.2   Factors That Influence the Design of Energy and Lifetime Aware Routing Protocols for WSNs

One of the most significant design goals of WSNs is to go into data communication via aggressive energy management techniques, while maintaining the WSN lifetime and precluding connectivity degradation. The design of energy-efficient routing protocols in WSNs is influenced by many factors that must be considered meticulously (Junhai et al. 2008):

- Node deployment. It is an application-dependent operation affecting the routing protocol performance and can be either deterministic or randomized.
- Node/link heterogeneity. The existence of a heterogeneous set of sensors gives rise to many technical problems related to data routing.
- Data reporting model. Data sensing, measurement and reporting, in WSNs depend on the application and the time criticality of the data reporting. Data reporting can be time-driven (continuous), event-driven, query-driven, or hybrid.
- Energy consumption without compromising accuracy. Consequently, energy-conserving mechanisms of data communication and processing are remarkably urgent.
- Scalability. WSN routing protocols should be scalable enough to respond to events, such as the huge increase of sensor nodes in the environment.
- WSN dynamics. Mobility of sensor nodes is necessary in many applications; despite the fact that most of the WSN architectures assume that sensor nodes are stationary.
- Fault tolerance. The overall task of the WSN should not be affected by the failure of some sensor nodes.
- Connectivity. Sensor nodes connectivity depends seriously on the random distribution of nodes.
- Transmission media. In a multihop WSN, communicating nodes are linked by a wireless medium. One approach of MAC design for WSNs is to use TDMA-based protocols that conserve more energy compared to contention-based protocols like CSMA.
- Coverage. In WSNs, a given sensor view of the environment is limited both in range and in accuracy; it can only cover a limited physical area.
- Quality of service. Data should be delivered within a certain period of time. However, in a number of applications, conservation of energy, which is directly related to network lifetime, is considered relatively more important than the quality of data sent. Hence, energy-aware routing protocols are required to capture this requirement.
- Data aggregation. Data aggregation, as clarified in Sect. 4.3.2, is the combination of data from different sources according to a certain aggregation function, e.g., duplicate suppression.

Worth observing, routing protocols for WSNs differ in their scalability and performance characteristics. Several routing protocols are designed for small WSNs. Also, there are routing protocols that work best in a static environment and are hard converging to a new topology.

4.6.3   Goals of Energy and Lifetime Aware Routing Protocols for WSNs

Basically, all routing protocols have the same general goal, that is to share and achieve WSN reachability information among routers; thus, they may send a complete routing table to other routers or send specific information on the status of directly connected links. Alternative routing protocols may send periodic HELLO packets to maintain their status with peer routers or may include advanced information such as a subnet mask or prefix length with route information. Most routing protocols share dynamic (learned) information, but in some cases, static configuration information is more appropriate. However, the major goals for developing routing protocols for WSNs are (Pantazis et al. 2013):

- Improvement of WSN lifetime, availability (up-time), and service.
- Increasing the WSN nodes battery lifetime.
- Providing fault tolerance and guaranteeing connectivity under various applications scenarios.
- Ensuring efficient energy consumption.
- Minimization of the transfer delay of the mission critical information.
- Reduction of routing protocols complexity for easier and affordable functioning regarding the sensor nodes limitations.
- Enhancement of WSN performance regarding throughput, reliability and latency.

Sections 4.7, 4.8, and 4.9 classify in detail some of the energy and lifetime aware routing protocols for WSNs and clarify how their design issues are embraced.

4.7   Energy-Efficient Routing Protocols

Because of the specialties of WSNs specified in Sects. 4.3 and 4.6, several routing mechanisms have been developed and proposed. These routing mechanisms have taken into account the inherent features of WSNs along with the applications and architecture requirements. A high-efficient routing scheme offers significant power cost reductions and prolongs WSN lifetime. Finding and maintaining routes in WSNs is a major issue since energy constraints and unexpected changes in node status, such as inefficiency or failure, give rise to frequent and unforeseen topological alterations. In (Pantazis et al. 2013), the classification initially proposed by

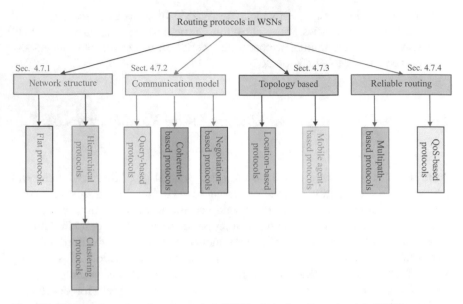

Fig. 4.8 Classification of routing protocols in WSNs. (Based on Pantazis et al. (2013))

(Al-Karaki and Kamal 2004) is expanded with detailed performance comparison. Thus, as detailed in Sects. 4.7.1, 4.7.2, 4.7.3, and 4.7.4, the classified routing protocols are categorized into four main approaches, respectively, network structure, communication model, topology-based, and reliable routing (Fig. 4.8).

4.7.1 Network Structure-Based Approach

The structure of a WSN can be classified according to node uniformity. The nodes in some WSNs are considered to be deployed uniformly and to be homogeneous; while in other WSNs the nodes are heterogeneous (Sect. 4.4). More specifically, the main attribute of the routing protocols belonging to the network structure approach is how nodes are connected and how they route information. This addresses two types of node deployments, nodes with the same level of connection and nodes with different hierarchies. Therefore, the schemes on this category can be further subclassified as:

- Flat networks routing protocols. All nodes in the WSN play the same role. Flat network architecture offers the main advantage of including minimal overhead to maintain the infrastructure between communicating nodes.
- Hierarchical networks routing protocols. The routing protocols in this scheme impose a structure on the WSN to achieve energy efficiency, stability, and scalability. In this subclass of network structure protocols, the nodes are

organized into clusters where a node with, possibly, higher residual energy assumes the role of a cluster-head (Sects. 4.3.3 and 4.8.1.1). The cluster-head is responsible for coordinating activities within the cluster and forwarding information between clusters. Clustering has the potential to reduce energy consumption and extend the WSN lifetime; a high delivery ratio is secured, as well as scalability and balanced energy consumption. The nodes around the basestation or cluster-head might deplete their energy sources faster than other nodes. WSN disconnectivity is a common problem where certain sections of the network might be unreachable. If only one node, connecting a section of the WSN to the rest, fails, then this section would cut off.

The main attribute of the routing protocols belonging to the network structure approach is how nodes are connected and how information is routed. Typically, in the hierarchical structure approach, the lower level nodes transmit information to the upper level nodes, resulting in a balanced energy structure of the WSN.

Sections 4.7.1.1 and 4.7.1.2 present, correspondingly, the flat networks routing protocols and the hierarchical networks routing protocols.

4.7.1.1 Flat Networks Routing Protocols

In general, flat networks routing protocols for WSNs can be classified, according to the routing strategy; they perform differently, although they have been designed for the same underlying WSN. Their sub-classification is bifold, either proactive (or table-driven routing protocols) or reactive (or source-initiated on-demand routing protocols), respectively, presented in Sects. 4.7.1.1.1 and 4.7.1.1.2.

4.7.1.1.1 Proactive or Table-Driven Routing Protocols

Proactive (or table-driven routing protocols) function in a way similar to that in wired networks. Based on the periodic exchange of routing information between the different nodes, each node builds its own routing table that can be used to find a path to a destination. Each node is required to maintain one or maybe more tables for storing routing information. Nodes also respond to any changes in WSN topology by sending updates throughout the wireless network and thus maintaining a consistent network view. Therefore, when a path to some destination is needed at a node, or a packet needs to be forwarded, the route is already known and there is no extra delay due to route discovery. However, keeping the information up-to-date may require more bandwidth and extra battery power, which are limited in WSNs; thus, information may still be outdated. Some of the existing table-driven routing protocols are the topology dissemination based on reverse-path forwarding (TBRPF) protocol (Bellur and Ogier 1999; Ogier et al. 2004).

4.7.1.1.2   Reactive or Source-Initiated on-Demand Routing Protocols

Unlike proactive (table-driven) routing protocols, reactive protocols (on-demand protocols) only start a route discovery procedure when needed (Pucha et al. 2007). When a route from a source to a destination is needed, a kind of global search procedure is started. This task does not request the constant updates to be sent through the WSN, as in proactive protocols, but this process does cause delays, since the requested routes are not available and have to be found. In some cases, the desired routes are still in the route cache maintained by the sensor nodes; when this is the case, there is no additional delay since routes do not have to be discovered. The whole process is completed as soon as a route is found or all possible route combinations have been examined. Typical such on-demand routing protocols are the temporarily ordered routing algorithm (TORA) (Park and Corson 1997) and the energy-aware temporarily ordered routing algorithm (E-TORA) that is an alteration of TORA with the main focus of minimizing the energy consumption of the nodes (Yu et al. 2007).

Flooding, gossiping, and rumor routing are three more protocols under this sub-classification:

- Flooding. Flooding is an old and very simple technique, which is also used for routing in WSNs; it is a reactive technique that does not require costly topology maintenance and complex route discovery algorithms (Lim and Kim 2001). In flooding, copies of incoming packets are sent by every link except the one from which the packets arrived; this procedure generates an enormous amount of superfluous traffic. Flooding is an extremely robust technique as long as there is a route from source to destination that guarantees packet delivery. However, it has several drawbacks:

  - Implosion. It is the situation where duplicated messages are broadcasted to the same node.
  - Overlay. If two nodes share the same monitored region, both of them may sense the same stimuli at the same time. As a result, neighbor nodes receive duplicated messages.
  - Resource blindness. The flooding protocol does not take into consideration all the available energy resources. An energy resource aware protocol must take into account the amount of energy available at all nodes all the time.

  Flooding has two interesting characteristics, which arise from the fact that all possible routes are tried:

  - As long as there is a route from source to destination, the delivery of the packet is guaranteed. Flooding is particularly suitable for a battlefield situation due to its robustness.
  - One copy of the packet will arrive by the quickest possible route, which might be useful for route learning.

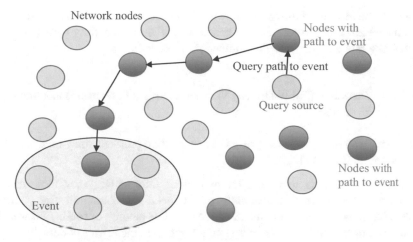

Fig. 4.9 Rumor routing protocol. Braginsky and Estrin (2002)

There have been some protocols that use flooding as part of their routing mechanism (Ma et al. 2006). However, flooding consumes much energy, since for each data packet, all nodes that are in the broadcast domain will receive packets that they will forward to their neighbors. Thus, a large amount of power is required causing a prohibitively short WSN lifetime.

- Gossiping and broadcasting are two situations of information dissemination related to a group of individuals connected by means of a communication network (Hedetniemi et al. 1988). In gossiping, every person in the network knows a unique item of information and needs to communicate it to everyone else while, in broadcasting, one individual has an item of information, which needs to be communicated to everyone else. Actually, gossiping is a derivative of flooding where nodes do not broadcast but send the incoming packets to a randomly selected neighbor. Although this approach avoids the implosion problem by just having one copy of a message at any node, it takes long to propagate the message to all nodes in the WSN.

- Rumor routing (RR). It is a compromise between flooding queries and flooding event notifications (Braginsky and Estrin 2002). The main idea of this protocol is to create paths that lead to each event (Fig. 4.9), unlike event flooding which creates a network-wide gradient field. Thus, in case that a query is generated, it can be sent on a random walk until it finds the event path, instead of flooding it throughout the network. As soon as the event path is discovered, it can be further routed directly to the event. On the other hand, if the path cannot be found, the application can try resubmitting the query or flooding it. The rumor routing can be suitable for delivering queries to events in large networks for a wide range of conditions, due to its lower energy requirements than the alternatives. It is designed to be tunable to different application requirements, while being adjustable to support different query to event ratios, successful delivery rates, and route repair. Furthermore, it may handle node failure gracefully by degrading its delivery rate linearly with the number of failed nodes.

4.7.1.1.3 Proactive (Table-Driven) Versus Reactive (On-Demand) Routing Protocols

The main differences between proactive and reactive protocols may be recapped in several points (Lee et al. 2004):

- Proactive protocols require a lot of routing information; they also maintain routing information independently of the need for communication, whereas reactive protocols are on-demand and require less amount of routing information for each node and thus less energy consumption at the sensor nodes.
- In proactive protocols, there is no latency in route discovery, so they are suitable for real-time traffic. In reactive protocols, there is a delay due to route discovery, called route acquisition delay, which may not be appropriate for real-time communication.
- Proactive protocols waste bandwidth and energy to periodic updates; in comparison with reactive protocols that do not require periodic updating, so they save energy and bandwidth during inactivity.
- Proactive protocols obtain and maintain the routing information for all the nodes in a WSN. They require a powerful processing capacity to keep WSN information current. In reactive protocols, intermediate nodes do not have to make routing decisions; hence, there is no need to have information about nodes.
- Proactive protocols send update messages throughout the WSN periodically or when the topology changes. In case of reactive protocol, there is no need to send the update messages when topology changes.
- Proactive protocols are adequate for heavy loads but are not as good for light loads, while reactive protocols are good for light loads and collapse during large loads.
- Proactive protocols are never bursty but reactive protocols can be bursty as there is congestion during high activity.
- If routing information changes frequently, then the proactive protocols would not exert any impact on the packet delivery. But for reactive protocols, if routing information changes frequently, as in MANETs, and if route discoveries are needed for those changing routes, then a large volume of messaging overhead may result due to on-demand global broadcast.

Hence, the availability of routing information is a key advantage of proactive (table-driven) routing protocols, since faster routing decisions and consequently less delay in route setup process can be made (Layuan et al. 2007). On the other hand, this significant advantage requires periodic routing updates of the routing tables, which costs higher signaling traffic than in the reactive (on-demand) routing protocols. Consequently, sensor nodes spend more energy for their periodic update

Table 4.3 Proactive and reactive routing protocols compared (Pantazis et al. 2013)

	Proactive protocols	Reactive protocols
On-demand protocols		✓
Update routes continuously	✓	
Route acquisition delay		✓
Periodic updating	✓	
Maintain the routing information for all nodes in the network	✓	
Send update messages when the topology changes		✓
Proper for heavy loads	✓	
Bursty		✓
Result in a large volume of messaging	✓	

messages. However, for other functions, like path reconfiguration after link failures, there are variations between the protocols of each class. For example, TORA is a reactive (on-demand) routing protocol, but at the same time, it uses local route maintenance schemes, which reduces signaling overhead (Park and Corson 1997).

In Table 4.3, a comparison between proactive and reactive protocols is presented.

4.7.1.2 Hierarchical Networks Routing Protocols

Unlike flat protocols, where each node has its unique global address and all the nodes are peers, in hierarchical protocols nodes are grouped into clusters (Sects. 4.3.3 and 4.8.1.1). Every cluster has a cluster-head whose election is based on different election algorithms. The cluster-heads are used for higher-level communication, thus reducing the traffic overhead. Clustering may be extended to more than just two levels, keeping the same concepts of communication in every level. The use of routing hierarchy has the advantages of reducing the size of the routing tables, hence providing better scalability. Typical such protocols are low-energy adaptive clustering hierarchy (LEACH) (Heinzelman et al. 2002; Handy et al. 2002), low-energy adaptive clustering hierarchy centralized (LEACH-C) (Heinzelman et al. 2002), power-efficient gathering in sensor information systems (PEGASIS) (Lindsey and Raghavendra 2002), adaptive threshold sensitive energy-efficient sensor network (APTEEN) (Manjeshwar and Agrawal 2002; Wang et al. 2020), and distributed hierarchical agglomerative clustering (DHAC) (Lung and Zhou 2010). In (Lu et al. 2014), two secure and efficient data transmission (SET) protocols for clustered WSNs are proposed. They are called SET-IBS and SET-IBOOS, for using the identity-based digital signature (IBS) scheme and the identity-based online/offline digital signature (IBOOS) scheme, respectively.

4.7.2   Communication Model-Based Approach

The communication model adapted in a routing protocol determines how packets are to be routed in the WSN. The protocols of this category can deliver more data for a given amount of energy. Also, in terms of dissemination rate and energy usage, such protocols can perform close to the theoretical optimum in point-to-point and broadcast networks. The problem with communication model protocols is that they do not offer high delivery ratio of the data sent to a destination; thus, they do not guarantee the delivery of data. The protocols on this scheme can be subclassified as:

- Query-based routing protocols. The destination nodes propagate a query for data (sensing task) through the WSN nodes, the node having the data that match the query sends them back to the initiating node.
- Coherent and non-coherent-based routing protocols. In coherent routing, the data are forwarded to aggregators after a minimum processing. In non-coherent data processing routing, nodes locally process the raw data before they are sent to other nodes for further processing.
- Negotiation-based routing protocols. They use meta-data negotiations to reduce redundant transmissions in the WSN.

Sections 4.7.2.1, 4.7.2.2, and 4.7.2.3 describe, respectively, the query-based routing protocols, the coherent and non-coherent-based routing protocols, and the negotiation-based routing protocols.

4.7.2.1   Query-Based Routing Protocols

In query-based routing protocols, the destination nodes propagate through the WSN a query for data (sensing task). The node having the data that matches the query sends them back to the requesting node (Sadagopan et al. 2005). These queries are usually described in natural language, or in high-level query languages. For example, client C1 may submit a query to node N1 asking: Are there moving vehicles in battle space region 1? All nodes have tables consisting of the sensing tasks queries they receive; upon receiving queries they send the data matching these sensing tasks. Directed diffusion (DD) is an example of this type of routing (Intanagonwiwat et al. 2003). Directed diffusion is data-centric in that all communication is for named data. All nodes in a directed diffusion-based network are application aware. This enables diffusion to achieve energy savings by selecting empirically good paths and by caching and processing data in-network (data aggregation). Data generated by sensor nodes are named by attribute-value pairs. A node (basestation) requests data by sending interests for named data. Data matching the interest is then "drawn" down toward that node. Intermediate nodes can cache, or transform data and may direct interests based on previously cached data. To lower energy consumption, data aggregation, such as duplicate suppression, is performed en route. Other such protocols are Cougar (Yao and Gehrke 2002) and active query forwarding in sensor networks (ACQUIRE) (Sadagopan et al. 2005).

4.7.2.2 Coherent and Non-Coherent-Based Routing Protocols

In WSNs, the processing of data is required at the node level. In this approach, the sensor nodes make a collaborative effort to process the data within the WSN. The routing mechanism, which initiates the data processing module, is proposed in (Sohrabi et al. 2000). This mechanism is divided into two categories:

- Coherent data processing-based routing. This category is an energy-efficient mechanism where the sensor node does only the minimum processing. Time stamping and duplicate suppression are the tasks accomplished in minimum processing. After the minimum processing, the data are forwarded to the aggregators.
- Non-coherent data processing-based routing (Jolly and Latifi 2006). In this category, the sensor nodes locally process the actual data and then send them to the other nodes for further processing. The nodes that perform further processing are called the aggregators. There are three phases of data processing in non-coherent routing:

  - Target detection, data collection, and preprocessing. In this stage, an event is detected and its information is collected and preprocessed.
  - Membership declaration. The sensor node chooses to participate in a cooperative function and declares this intention to all neighbors.
    Central node election. A central node is chosen to perform more refined information processing.

4.7.2.3 Negotiation-Based Routing Protocols

Negotiation-based routing protocols or sensor protocols for information via negotiation (SPIN) are among the early works to pursue a data-centric routing mechanism. The SPIN class of protocols rests upon two basic ideas (Kulik et al. 2002):

- To operate efficiently and to conserve energy, sensor applications communicate with each other about the data they already have and the data they still need to obtain.
- Nodes in a WSN must monitor and adapt to changes in their own energy resources to extend the operation lifetime of the system.

SPIN is based on data-centric routing. The main idea of SPIN is to name the data using high-level descriptors or meta-data. Meta-data negotiations are used to reduce redundant transmissions in the WSN. SPIN is a 3-stage protocol, ADV, REQ, and DATA:

- Advertise (ADV). If a node has some data, it will advertise by sending an advertise packet that it has sensed an event or received data from another node.
- Request (REQ). If some other node has received the advertised packet and is interested in that data, it will send a request packet.
- DATA. Upon receiving the request packet, the advertising node will send the actual data in a packet.

SPIN provides scalability in a sense that each node needs to know only its single-hop neighbors; so, any changes in the topology would be local. The problem with SPIN is that it does not guarantee delivery of data, like the situation when an interested node is too far from the advertising node; then, it will not get any data, if the nodes in-between are not interested in that data.

Several protocols belong to the SPIN class of protocols, typically SPIN with energy conservation (SPIN-EC), SPIN for broadcast networks (SPIN-BC), SPIN for point-to-point communication (SPIN-PP), and SPIN with reliability (SPIN-RL) (Cordeiro and Agrawal 2011).

4.7.3 Topology-Based Approach

Topology-based routing protocols use the principle that every node in a WSN maintains topology information and that the main protocol operation is based on the topology. The protocols on this scheme can be further subclassified as:

- Location-based routing protocols. They take advantage of the position information in order to relay the received data to certain regions and not to the whole WSN. The protocols of this subclass can find a path from a source to a destination and minimize the energy consumption of the sensor nodes. Yet, they have limited scalability if the nodes are mobile. Also, a node must know or learn about the locations of other nodes.
- Mobile agent-based routing protocols. Mobile agent protocols are used in WSNs to route data from the sensed area to the destination. The main component is a mobile agent, which migrates among the WSN nodes to perform a task autonomously and intelligently, based on the environment conditions. Mobile agent protocols might provide extra flexibility to the WSN, as well as more capabilities in contrast to the conventional WSN operations that are based on the client–server computing model.

Location-based routing protocols and mobile agent-based routing protocols are described in Sects. 4.7.3.1 and 4.7.3.2, correspondingly.

4.7.3.1 Location-Based Routing Protocols

The location-aided or position-based routing acknowledges the influence of physical distances and distribution of nodes on areas, as significant to WSN performance. Location-based routing protocols are based on two principal assumptions:

- Every node knows its own network neighbor positions.
- The source of a message is informed about the position of the destination.

This technique for localized broadcasting of queries in geo-aware WSNs makes use of the existing query routing tree and does not create additional communication channels. The algorithms require the nodes to periodically transmit HELLO messages to allow the neighbors know their positions. The location-based routing technique is interesting because it operates without routing tables. Furthermore, once the position of the destination is known, all operations become strictly local, that is, every node is required to keep track only of its direct neighbor.

The main disadvantages of such algorithms are:

- Efficiency depends on balancing the geographic distribution versus occurrence of traffic.
- Performance degrades if the traffic load distribution disregards distance.

Multiple routing algorithms belong to this subclass of location-based protocols. Among many, there are distance routing effect algorithm for mobility (DREAM) (Basagni et al. 1998), geographic and energy-aware routing (GEAR) (Yu et al. 2001), graph embedding for routing (GEM) (Newsome and Song 2003), cross-link detection protocol (CLDP) (Kim et al. 2005), greedy distributed spanning tree routing (GDSTR) (Leong et al. 2006), on-demand geographic forwarding (OGF) (Chen and Varshney 2007), and energy-efficient beaconless geographic routing (EBGR) (Blum et al. 2010).

4.7.3.2 Mobile Agent-Based Routing Protocols

In most cases, the application-specific nature of WSNs requires the sensor nodes to have multiple capabilities. Yet, it is impractical for sensors to run every possible application and to store all the needed programs in their tight local memory. One of the main focus areas, related to WSNs, is the design, development, and deployment of mobile agent systems (Chen et al. 2007). Mobile agent systems have as main component a mobile agent, software that migrates among the nodes of a WSN, to perform a task autonomously and intelligently, based on the environment conditions. Mobile agent systems employ migrating codes in order to facilitate flexible application retasking, local processing, and collaborative signal and information processing. This may provide the WSN with extra flexibility, as well as new capabilities in contrast to the conventional WSN operations that are based on the client–server computing model.

Thus, it is beneficial to design mobile agents and develop protocols in WSNs to route data from the sensed area to the destination. The design issues of mobile agents in WSNs are numerous:

- Architecture. The architecture is based on the topology of the WSN and is further subdivided into flat or hierarchical.
- Itinerary planning. The itinerary is the route followed during mobile agent migration. The itinerary planning is related to the selection of the set of source nodes, to be visited by the mobile agent, and the determination of a source-visiting sequence in an energy-efficient manner. The itinerary planning may be static, dynamic, or hybrid.

- Middleware system design. Mobile agents are often implemented as middleware. Middleware is used to bridge the gap between the operating system and high-level components and to facilitate the development and deployment of applications.
- Agent cooperation. Mobile agents can work either as single processing units or as a distributed collection of components. The requirement to provide the means for agent cooperation is an important consideration in multimedia WSN design to reduce energy consumption.

In most cases, applying mobile agent systems may lead to reducing bandwidth consumption and high flexibility on the WSN. Moving the data processing elements to the location of the sensed data may reduce the energy expenditures of the nodes. However, finding the optimal itinerary is NP-hard.

Typical mobile agent-based protocols are multiagent-based itinerary planning (MIP) (Chen et al. 2009), itinerary energy minimum for first-source-selection (IEMF), and itinerary energy minimum algorithm (IEMA) (Chen et al. 2011).

4.7.4   Reliable Routing Approach

The protocols on this scheme are more resilient to route failures either by achieving load-balancing routes or by satisfying certain QoS metrics, as delay, energy, and bandwidth. WSN nodes may suffer from the overhead of maintaining routing tables and keeping the QoS metrics at each node. These protocols are subclassified as:

- Multipath-based routing protocols. They achieve load balancing and are more resilient to route failures.
- QoS-based routing protocols. The WSN has to balance between energy consumption and data quality. Whenever a sink requests data from the sensed nodes, the transmission has to meet specific quality metrics.

In Sects. 4.7.4.1 and 4.7.4.2, respectively, multipath-based routing protocols and QoS-based routing protocols are presented.

4.7.4.1   Multipath-Based Routing Protocols

Multipath routing is an interesting routing approach for WSNs for its advantage of achieving load balancing and its resilience to route failures (Tarique et al. 2009). Several multipath routing protocols are available in the literature; they show lower routing overhead, lower end-to-end delay, and less congestion in comparison with single-path routing protocols. Some of these protocols are routing on-demand acyclic multipath (ROAM) (Raju and Garcia-Luna-Aceves 1999), gradient broadcast (GRAB) (Ye et al. 2005), hierarchy-based multipath routing protocol (HMRP) (Wang et al. 2006), and cluster-based multipath routing (CBMPR) (Zhang et al. 2008a).

4.7.4.2 QoS-Based Routing Protocols

In QoS-based routing protocols, the WSN has to balance between energy consumption and data quality (Akkaya and Younis 2005; Shafiullah et al. 2008). In particular, certain QoS metrics are to be satisfied; such as delay, energy, bandwidth, when delivering data to the basestation. In the best-effort routing, the main concerns are the throughput and average response time. In a connection-oriented communication, QoS routing is usually performed through resources reservation that meets the QoS requirements for each individual connection. While many mechanisms have been proposed for QoS routing in real-time multimedia wired networks, they cannot be typically applied to WSNs due to the limited sensor nodes resources, such as processing, storage, bandwidth, and energy. Among these QoS-based routing protocols for WSNs, there are sequential assignment routing (SAR) (Sohrabi et al. 2000), SPEED (He et al. 2003), multipath and multi-SPEED (MMSPEED) (Felemban et al. 2006), energy-efficient, and QoS aware multipath routing (EQSR) protocol (Ben-Othman and Yahya 2010).

Energy-efficient routing techniques for multimedia WSNs with QoS assurances are surveyed in (Ehsan and Hamdaoui 2012) together with the highlights of the performance issues of each strategy. In (Han et al. 2016), routing protocols that can balance out the trade-off between network lifetime and QoS requirements are called "green routing protocols"; a detailed survey and comparison of such routing protocols for multimedia WSNs is presented along with a classification into two categories based on network structures. Also, QoS assurances for both best-effort data and real-time multipath routing protocols for multimedia WSNs are surveyed in (Hasan et al. 2017).

4.7.5 Notable Outlines

The protocols subclassified and described throughout Sect. 4.7 have their own characteristics and applications, but joining features is a practice to realize better protocols performance:

- The main attribute of the protocols belonging to the network structure-based approach is the way the nodes are connected and exert influence on the routing of information (Sect. 4.7.1). For example, in the hierarchical network structure the lower level nodes transit information to the upper lever nodes, resulting in a balanced energy structure of the WSN.
- In the communication model-based approach, the main characteristic of the protocols is how to make a routing decision, irrespective of the WSN structure (Sect. 4.7.2). Thus, as a well-defined technique, the negotiation performed between nodes before transmitting data, to route information from the source to the destination.

- Some protocols, besides the communication model they adopt, take into consideration the topology of the WSN for data transmission (Sects. 4.7.2 and 4.7.3). They operate without routing tables, by periodically transmitting HELLO messages to allow neighbors know their positions.
- Some protocols use mobile agents to move the data processing elements to the location of the sensed data so as to reduce the energy expenditures of the nodes (Sect. 4.7.3.2).
- Several protocols, in addition to the energy efficiency they provide, tend to secure reliable routing of data (Sect. 4.7.4). This is achieved either by providing multipath from source to destination or by applying QoS on their main routing activity.

In (Bhushan and Sahoo 2019), a categorization of various routing protocols in WSNs into three major subclasses is presented, explicitly the flat networks routing protocols, the hierarchical networks routing protocols, and the QoS aware routing protocols.

4.8 Energy-Efficient and Energy-Balanced Routing Protocols

The energy-efficient routing protocols strive to increase the network lifetime through minimizing the energy consumption in each sensor node. On the other hand, the energy-balanced routing protocols protract the network lifetime by uniformly balancing the energy consumption among the nodes in the network. This section is based on the elaborate taxonomy of the energy-efficient and energy-balanced routing protocols for WSNs presented in (Ogundile and Alfa 2017). As pictured in Fig. 4.10, these protocols are classified, based on their mode of communication toward the

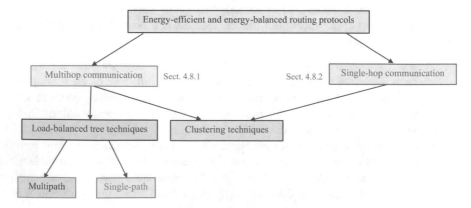

Fig. 4.10 Classification of energy-efficient and energy-balanced routing protocols for WSNs (Ogundile and Alfa 2017)

basestation, into the multihop approach (Sect. 4.8.1) and the single-hop approach (Sect. 4.8.2).

Basically, the design of a reliable energy-efficient and energy-balanced routing protocol for WSNs involves several decision metrics (Sects. 4.2.1 and 4.2.2). There should be a trade-off in setting these decision metrics. Energy-balanced routing protocols for WSNs must extend the network lifetime to a duration that maintains full network functionality as long as possible. These protocols also consider other design objectives, specifically, scalability, reliability, and accuracy (Iqbal et al. 2015).

4.8.1   Multihop Communication

Different routing protocols have been developed using the multihop communication approach in order to optimize several single and multiobjective problems associated with WSNs. With attention to prolonging the network lifetime and functionality, the concept of multihop communication is adopted to develop different energy-efficient and energy-balanced routing protocols for WSNs. The main goal is to reduce the distance when transmitting the sensed information from the source node to the destination. The transmission distance plays a vital role in either reducing or increasing the energy dissipated by the WSN nodes. Nonetheless, solving the distance issue through using the concept of multihop communication does not necessarily guarantee an optimal use of the limited battery energy in the WSN, as clarified in Sect. 4.5. To resolve the distance concern associated with WSNs, the multihop communication technique raises other design matters such as selecting the route with optimal hop-count to the basestation and/or the route to the basestation with optimal energy consumption, both in terms of the energy consumed by the participating nodes and the total energy dissipated following that route.

Therefore, several multihop techniques are proposed to tackle these glitches, with the ultimate goal of conserving energy in WSNs. These multihop techniques are classified into two main categories: namely, multihop clustering techniques and load-balanced tree techniques as detailed in Sects. 4.8.1.1 and 4.8.1.2, respectively.

4.8.1.1   Multihop Clustering Techniques

Over couple of decades, clustering-based protocols continue to be the best for heterogeneous WSNs because they work on the principle of divide and conquer. Clustering techniques are widely used in designing routing protocols for WSNs due to their numerous advantages such as scalability, efficient data aggregation, fault tolerance, latency reduction, robustness, and reduced energy consumption (Afsar and Tayarani-N 2014; Rostami et al. 2018; Fanian and Rafsanjani 2019). As exhibited in Fig. 4.11, a clustered WSN contains two sets of nodes, specifically, the coordinating nodes usually referred to as the cluster-heads, and the member

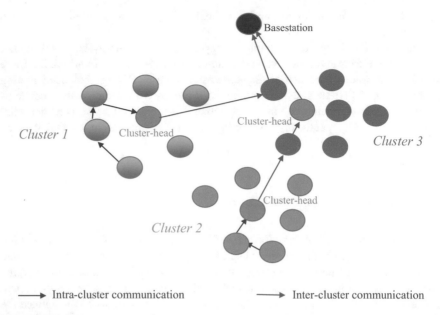

Fig. 4.11 Multihop clustering in WSNs. (Based on Ogundile and Alfa (2017))

nodes known to be the non-cluster-heads. The cluster-heads in the WSN are used to process the information before sending to the basestation, while the non-cluster-heads forward the sensed information to their respective cluster-heads. The clustering routing solution can be seen as a two-layer hierarchy where the cluster-heads operate in the upper layer and the non-cluster-heads operate in the lower layer. As such, clustering routing protocols are usually referred to in the literature as hierarchical routing protocols (Fig. 4.8). In some cases, because the cluster-heads perform more functions than the non-cluster-heads, the cluster-heads are equipped with superior sensing, processing, and radio subsystems, as well as a power supply unit, in comparison with the non-cluster-heads. If the components of the cluster-heads are different from those in the non-cluster-heads, the clustering WSN is denoted as a heterogeneous clustering WSN; otherwise, it is a homogeneous clustering WSN (Sect. 4.4).

Clustering algorithms have many objectives as portrayed in Fig. 4.12.

In the multihop clustering technique, the number of clusters or the size of the cluster is specified in the routing algorithm. As Fig. 4.11 clarifies, the non-cluster-heads send the sensed information to their respective cluster-head directly or via one or more relay nodes within their cluster. This clustering mode of communication is regarded as intra[1]-cluster communication (Merriam-Webster 2022). Via inter-cluster

---

[1] The prefix "intra" means "within" (as in happening within a single thing), while the prefix "inter" means "between" (as in happening between two things.)

Clustering objectives	
Reducing the number of control messages	Scalability
Maintaining network coverage	Fault-tolerance
Removing hot-spots/energy-holes problem	Data aggregation/fusion
Utilizing sleeping schemes	Load balancing
Avoiding collision	Stability of network topology
Decreasing delay	Maximizing network lifetime
Increasing connectivity	Reducing energy consumption

Fig. 4.12 Clustering objectives (Fanian and Rafsanjani 2019)

communication, the cluster-heads aggregate all the sensed information from their respective cluster members and forward such aggregated information to the basestation, either directly or via one or more relay nodes. For the cluster-heads, the relay nodes can either be a non-cluster-head node or a cluster-head node. This means that any of the nodes in the WSN can serve as a relay node to convey the aggregated message from the cluster-heads to the basestation as specified in the routing algorithm. The communication between a cluster-head and other clusters, in order to convey the message from its cluster members to the basestation, is referred to as inter[1]-cluster communication (Merriam-Webster 2022). Moreover, most clustering algorithms split the network operation into rounds and periodically recluster the network in order to alternate the cluster-head duty.

The cluster-heads consume more energy because they perform further functions in comparison with the non-cluster-heads. Hence, in order to avoid untimely network partitioning, it is always desirable to rotate the cluster-head duty and select nodes, with higher residual energy, to operate as the cluster-head in each round. By so doing, the load in the network is uniformly distributed among the nodes. Furthermore, because the source node does not have to convey the q bit packet over a long distance to the destination, the multihop clustering technique technically solves the distance problem associated with WSNs, thereby conserving the nodes energy and the total energy in the WSN.

The multihop clustering technique involves three phases: typically, cluster-head selection (Sect. 4.8.1.1.1), forming the cluster (Sect. 4.8.1.1.2), and WSN partitioning into optimal clusters (Sect. 4.8.1.1.3).

4.8.1.1   Cluster-Head Selection

Despite the numerous advantages offered by the multihop clustering routing protocols (Younis et al. 2006; Abbasi and Younis 2007; Zaatouri et al. 2017), designing an energy-efficient and energy-balanced multihop clustering routing protocol for WSNs can be challenging. A reliable energy-balanced clustering routing protocol carries some important attributes. Besides the hitches in designing multihop routing protocols as introduced in Sect. 4.5, the design of a reliable energy-balanced multihop clustering routing protocol has its own peculiar energy dissipation issues. For instance, assuming a multihop clustering routing protocol for the WSN of Fig. 4.6. The first major load-balancing issue is to select the set of nodes that will perform the cluster-head duty. As simple as it may sound, the cluster-head selection phase strongly indicates the network lifetime performance of the proposed clustering routing protocol. As depicted in Fig. 4.2, if the cluster-head for a particular cluster is inactive or it has completely depleted its battery energy, all the nodes in that cluster or subnetwork get segregated from the WSN. The nodes in that cluster will not be able to access the basestation for that particular data transmission round despite having sufficient battery energy left. Thus, it is desirable that a node, with sufficient battery energy, be always selected as cluster-head in order to avoid WSN partitioning and to maintain the full WSN functionality. Hence, the cluster-heads must be selected based on some design metrics:

- Residual energy (RE). The cluster-heads are selected based on the residual energy of the sensor nodes (Wei et al. 2011; Li et al. 2013a; Hoang et al. 2014). The nodes with the highest residual energy are chosen as cluster-heads, mainly because the cluster-head consumes more energy than the non-cluster-heads. This approach sounds reasonable but might not be as efficient. If the nodes selected to act as cluster-heads are distant from the basestation and the clustering routing algorithm assumes that the cluster-head communicates directly with the basestation, then the cluster-head node will use more energy in forwarding the aggregated information from its cluster members to the basestation.
- Distance to the basestation (DBS). The cluster-heads are chosen based on their proximity to the basestation, which is computed in the form of a distance metric (Mandala et al. 2008; Xie and Jia 2014). It is assumed that since the transmission distance d plays an important role in the energy dissipation model, the nodes with the shortest distances to the basestation will consume less energy when forwarding packets to the basestation (Sect. 4.2.1). These selected nodes serving as cluster-heads might be low on battery energy, and if used continuously, they will fully drain their energy, thereby resulting in untimely WSN partitioning.
- Hop-count to the basestation (HCBS). The cluster-heads are picked up based on their hop-counts to the basestation (Amis and Prakash 2000). As detailed in Sect. 4.5, a small hop-count from the source node to the basestation helps minimizing the energy consumption in the WSN. However, this approach might suffer the same fate as using the shortest distance to the basestation approach.
- Threshold-based (TB). The cluster-heads are randomly chosen based on some set threshold (Muruganathan et al. 2005; Pal et al. 2015). A threshold is set such that any sensor node that does not meet this condition cannot be a cluster-head. The majority of these threshold-based cluster-head selection methods define this threshold based on the estimated average energy or residual energy in the WSN.
- The WSN is firstly divided into a virtual grid. Subsequently, the gateways or cluster-heads are chosen based on their location in the grid; they are those closer to the boundary of the other grids in the WSN (Sabbineni and Chakrabarty 2005). This approach tends to prolong the WSN lifetime because more than one node

can be used as gateways; consequently, the gateway function is shared. A setback with this approach is that the nodes acting as gateways are not rotated and might deplete their energy fully at some point during data transmission. As such, the grid(s) will be segregated from the whole WSN.

The majority of these clustering routing protocols alternate the duty of the cluster-head in order to balance the load in the WSN. Rotating the duty of the cluster-head has shown to improve the lifetime of clustering-based WSNs, as well as functionality because the functions of the cluster-head are shared among all the nodes. Additionally, this tactic avoids untimely network partitioning that might occur as in Fig. 4.2. Cluster-head rotation is more common in homogeneous clustering routing protocols since all nodes in the WSN assume the same battery energy at initial deployment. Rotating the cluster-head duty in such a scenario will ensure that all nodes share the hard task of the cluster-head, thereby evenly distributing the traffic load in the WSN.

On the other hand, some of the heterogeneous clustering routing protocols assume larger battery energy for the predefined cluster-heads, thus preventing them from alternating the cluster-head task. This approach tends to extend the network lifetime and functionality. However, the disadvantage is that, at some point, the cluster-heads will entirely deplete their battery energy due to performing more energy demanding functions. To overcome this concern, some routing protocols use a multisink technique where two or more sinks perform the cluster-head duty (Wu et al. 2008; Cheng and Chang 2012). The sinks forward the aggregated information from their cluster members to the basestation. These sinks are powered in the same way as the main basestation; thus, they do not have energy consumption hitches. Although this approach solves the energy consumption issues with cluster-heads, it may not be practical due to the cost implicated if deploying more than one basestation to monitor an environment, specifically when monitoring a small geographical area.

4.8.1.1.2 Forming the Cluster

How the non-cluster-heads join a cluster is equally important to choosing the cluster-heads. The assumptions, concepts, and techniques used to distribute the nodes into clusters play a significant role in effectively balancing the traffic load in WSNs. Several approaches are embraced:

- Distance to the cluster-head (DCH). The non-cluster-heads join the cluster where the distance to the cluster-head node is shortest so as to reduce the energy consumed in transmission. However, since the nodes are randomly placed in the sensor field, the obvious drawback of this approach is that some clusters might have excessive members. Therefore, their cluster-heads might carry more traffic load, hence depleting their battery energy causing an imbalance in the load distribution throughout the WSN. In lieu of this, the nodes are distributed to cover the whole sensor field by assuming that they are aware of their location (Zhang et al. 2008b). Despite enabling better sensing of the physical values as the nodes are evenly distributed over the sensed environment, this approach might not be viable because it is bound by the availability of location services (Wang et al. 2012). Several works are proposed to solve the node placement problems in WSNs (Younis and Akkaya 2008).

- Hop-count to the cluster-head (HCCH). The hop-count can be used as a decision metric to specify the cluster that a non-cluster-head node joins (Kuila and Jana 2014b; Hammoudeh and Newman 2015). The non-cluster-head node compares the hop-count to convey information to every cluster-head in the network and joins the cluster with the smallest hop-count. The hop-count approach reduces the energy consumed in receiving and relaying the information from the source node to the cluster-head, but it suffers the same drawback as the shortest distance selection in the DCH approach.

- Transmission power to the cluster-head (TPCH) is a decision metric where the non-cluster-head node estimates the transmission power required to convey a packet to all cluster-heads in the WSN and joins the cluster whose transmission power is smallest (Li et al. 2013a; Anisi et al. 2013). Recalling, from Eq. (4.3), that the transmission power is a function of the distance between two nodes. This technically means that a non-cluster-head node will join the cluster whose cluster-head node is closest. Likewise, this approach suffers the setback of following the shortest distance in the DCH approach.

- Some mathematical formulations might be derived using a combination of one or more decision metrics to effectively distribute both the inter-cluster and intra-cluster traffic load so as to identify the cluster that a non-cluster-head joins (Liu et al. 2016).

4.8.1.1.3 WSN Partitioning into Optimal Clusters

Having selected the cluster-heads and specified how the non-cluster-heads should join a cluster, the next load-balanced requirements for the clustering routing protocols are:

- How to partition the network into an optimal number of subnetworks or clusters. Some literature divides the network randomly into equal sizes while other works partition the network into different shapes. In order to balance the burdens among clusters in a WSN, some clustering algorithms, proposed for energy saving, design clusters to be of equal sizes. However, in homogeneous WSNs, the directional and uneven data traffic toward the data sink places a greater burden on the clusters near the sink, which makes them drain energy more rapidly than other clusters. The areas covered by these clusters are not monitored once they have depleted prematurely (Wei and Chan 2008).

For heterogeneous WSNs, two clustering-based protocols are proposed, namely, a single-hop energy-efficient clustering protocol (S-EECP) and a multihop energy-efficient clustering protocol (M-EECP). The nodes are assumed to have different battery energies and are uniformly dispersed within a square field (Kumar 2014).

Besides, some work divides the WSN by assuming a virtual grid-based WSN structure. The ultimate goal is to find out solutions for how to efficiently divide the WSN into clusters, while prolonging its lifetime. An information dissemination protocol termed location-aided flooding (LAF) uses location information to reduce redundant transmissions, thereby saving energy (Sabbineni and Chakrabarty 2005). The WSN is divided into virtual grids and each sensor node associates itself with a virtual grid based on its location. Sensor nodes within a virtual grid are classified as either gateway nodes or internal nodes. While gateway nodes are responsible for forwarding the data across virtual grids, internal nodes forward the data within a virtual grid. The proposed approach achieves energy savings by reducing the redundant transmissions of the same packet by a node. In (Liu et al. 2016), there is a suggestion for a grid-based load-balanced routing method (GLRM) that aims at using a controlled sink to achieve load-balance in a non-uniform distributed WSN. Cluster-head election of each cell is based on three parameters: specifically, the number of data packets that nodes need to relay, the Euclidean distance to the mid-point of cells, and the residual energy of each node. The GLRM also considers other factors that waste battery power, such as packet collision. Simulation results confirm that the proposed routing method has better performance.

- Finding the optimal cluster size. The focus is on determining the number of clusters and the optimal cluster size (Ghiasi et al. 2002; Chang and Kuo 2006; Amini et al. 2012; Li et al. 2013a). A general assumption is that an equal cluster size helps balancing the traffic load in the WSN, particularly in the intra-cluster.[1] Contrarily, it is argued that an equal cluster size results in an imbalance in the energy consumption of the nodes (Wei and Chan 2008; Kuila and Jana 2014b; Malathi et al. 2015; Arjunan and Sujatha 2019). Basically, it is perceived that equally distributing the nodes to clusters is a suitable approach to balance the intra-cluster traffic load; however, it suffers during inter-cluster communications. Specifically, the clusters closer to the basestation consume more energy in receiving and forwarding the traffic load from other distant clusters during inter-cluster communications, consequently causing an imbalance in the energy consumption during inter-cluster communications, which prompts a possible WSN partitioning; this phenomenon known as the hot-spot/energy-hole problem is discussed in Sect. 4.5.

Several approaches are suggested to mitigate the hot-spot/energy-hole problem when developing a clustering-based routing protocol:

- Dividing the network into unequal clusters. The size of each cluster depends on how close the cluster is to the basestation. That is, the cluster size increases as the distance of the cluster-heads from the basestation increases. Multiple protocols are proposed in this regard, specifically, (Kuila et al. 2013; Kuila and Jana 2014a; Vijayan and Raaza 2016; Singh et al. 2018).

- Providing mobile basestation or sink. The basestation moves across the sensor field in order to change the direction of the traffic load (Wang et al. 2013; Tunca et al. 2015; Liu et al. 2016). In so doing, it is assumed that all nodes in the WSN randomly participate in forwarding the traffic to the basestation; thus, the energy consumption is distributed equally among all nodes. With this approach, the basestation moves back and forth around the sensor field so as to collect the aggregated information conveyed from each cluster-head. While solving the hot-spot/energy-hole problem, this approach helps prolonging the network lifetime and functionality. However, a major deficiency is that the basestation has to continuously broadcast its current position to the nodes in the WSN, resulting in high network delay.

4.8.1.2   Load-Balanced Tree Techniques

The load-balanced tree technique finds energy-balanced and energy-efficient path (s) from the source node to the basestation either through the multipath technique (Sect. 4.8.1.2.1) or the single-path technique (Sect. 4.8.1.2.2).

4.8.1.2.1   Multipath Approach

Multipath routing is a load-balanced tree technique broadly used to develop routing protocols that improve WSN performance. The multipath routing approach has been used extensively in the past decade for different optimization problems in WSNs, such as load balancing, fault tolerance, bandwidth aggregation, security, data transmission reliability, and congestion control. Multipath routing chooses multiple paths to convey data from the source node to the destination in the form of a tree as shown in Fig. 4.13. As for multihop clustering technique, multipath routing protocols fall under the multihop communication method (Fig. 4.10). The source node uses different relay nodes to convey its sensed information to the sink. However, due to the dense placement of the nodes, several paths can be constructed from the source node to the destination. Noteworthy, multipath routing techniques are established to propose alternatives to the low data transmission rate associated with the constrained single-path routing. With care to load balancing, multipath routing achieves load balancing by distributing the traffic load from the source node over more nodes toward the destination. From Eqs. (4.3)–(4.5), the q bits traffic load is directly proportional to the energy consumed; that is, the more the traffic q, the higher the energy consumed by the nodes in transmitting or receiving. Thus, with multipath routing, the traffic load is shared among the nodes from the source toward the destination; in such a way, the WSN lifetime and functionality are prolonged for a reasonable period.

In spite of the numerous advantages offered by multipath routing protocols, designing such protocols is challenging. A good multipath routing protocol adopts several considerations to select the multiple paths and evenly distribute the traffic

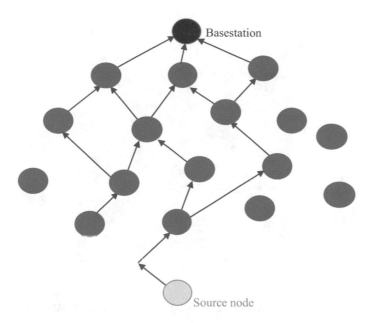

Fig. 4.13 Multipath load-balanced tree routing. (Based on Ogundile and Alfa (2017))

load over the selected paths. The main concerns to be addressed in designing a multipath routing protocol include path discovery, path selection/load distribution, and path maintenance (Radi et al. 2012):

- Path discovery. Since multipath routing is designed based on the multihop communication principle, the task of the path selection stage is to find the set of relay nodes that will form the multiple paths from the source node toward the destination.
- Path selection/load distribution. After selecting the set of intermediate nodes, the number of paths that distribute the traffic load is a main concern.
- Path maintenance. Furthermore, as a result of resources constraints and the low battery power of sensor nodes, the selected paths are usually susceptible to errors. As a plus of multipath routing protocols, path maintenance is an essential focus to help ensuring reliable data transmission from the source node to the destination.

4.8.1.2.2 Single-Path Approach

The single-path technique is a class of the load-balanced tree routing protocol, where the source node finds a single energy-efficient and energy-balanced route to the basestation, as depicted in Fig. 4.14. This technique is easy to implement and exhibits low complexity in comparison with multihop communication methods. Moreover, it avoids delay in conveying the sensed information to the basestation,

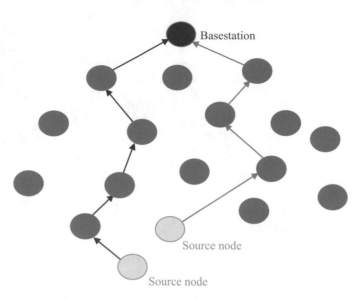

Fig. 4.14 Single-path load-balanced tree routing. (Based on Ogundile and Alfa (2017))

which is suitable in real-time communication. Other advantages of the single-path load-balanced tree technique include easier congestion and interference control, as well as simpler resource utilization.

Although the design of a single-path load-balanced tree routing protocol can be quite simple in comparison to multihop communication methods, selecting the design decision metrics used in the path creation can be challenging. Besides choosing an energy-efficient and energy-balanced path from the source node to the basestation, the chosen path must guarantee reliable data transmission. As such, the design decision metrics must be carefully selected to ensure non-faulty data transmission and to extend the network lifetime for a reasonable period. Hence, most single-path routing protocols select the path to the basestation using different mathematical formulations that involve various decision metrics that guarantee an energy-efficient and energy-balanced path. Typical such metrics, as clarified in Sects. 4.2.1 and 4.2.2, are the total energy (Eghbali et al. 2009), the average network energy (Zytoune et al. 2010), the hop-count (Hsiao et al. 2001; Ren et al. 2011), the residual energy (Zhang et al. 2014), and the distance to the sink (Kacimi et al. 2013; Yao et al. 2015).

These metrics have pros and cons; hence, trade-offs are watched in their selection. A trade-off is considered regarding the residual energy and the distance between nodes when selecting the path from the source node to the basestation (Shah and Rabaey 2002; Zhang et al. 2014). The sensed information is forwarded to the nearest node with high residual energy; hence, the transmission distance is reduced and the forwarding node has sufficient energy to convey the sensed data, which also guarantees a reliable data transmission. Using the residual energy and shortest distance between nodes as criteria do not necessarily balance the load in the network, since it might involve many hops or relay node chains to reach the basestation. Thus, more energy will be consumed in transmitting (Eq. 4.3) and receiving (Eq. 4.5) the sensed information to/from the basestation, respectively. This implies that the total energy consumed in the network might increase. So, it is desirable to keep the hop-count between the source node and the basestation as small as possible. With this in mind, some literature combines the residual energy, distance to the sink, and hop-count to select the energy-efficient and energy-balanced path from the source node to the basestation (Puccinelli and Haenggi 2008; Eghbali et al. 2009; Ren et al. 2011). A metric, that explores the diversity in link qualities, takes into account the transmission and reception costs for a specific radio to choose an energy-efficient radio. The metric also uses the remaining energy of nodes in order to regulate the traffic so that critical nodes are avoided (Moad et al. 2011).

4.8.2 Single-Hop Communication

In the single-hop communication method, the sensed information is conveyed to the basestation from the source node directly; this method was largely used in the early discovery phase of WSNs. Figure 4.15 depicts a typical example of a single-hop clustering WSN, where the non-cluster-heads send their sensed information to the cluster-head directly without the help of any intermediate node. Similarly, the cluster-heads forward the aggregated data from their cluster members directly to the basestation.

The single-hop communication method offers several advantages:

- Avoiding delay in data transmission, a useful attribute in real-time data transmission.
- Preventing the hot-spot/energy-hole phenomenon, an important asset in designing routing protocols.

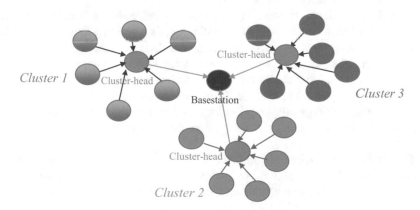

Fig. 4.15 Small-scale single-hop clustering WSN. (Based on Ogundile and Alfa (2017))

- Ensuring reliable data transmission while all the sensor nodes are still alive. Since packets are sent directly to the basestation, reliability helps avoiding the loss of parts of the sensed information. In the multihop communication method, some of the sensed information could be missed while conveying the packet from one node to another toward the basestation.

Since both the intra-cluster and inter-cluster[1] communications in single-hop clustering-based routing protocols require no intermediate nodes, as in multihop clustering routing protocols, the cluster formation stages should be carefully designed in order to evenly distribute the traffic load, so that the WSN lifetime and functionality can be extended for an acceptable period (Younis and Akkaya 2008; Khalil and Attea 2011; Yuea et al. 2012; Wang et al. 2012).

The single-hop clustering approach does not necessarily solve the distance problem associated with the single-hop communication method. For example, in a large-scale WSN, some non-cluster-heads might be at a large hop-distance from their respective cluster-heads. Moreover, some of the cluster-heads that are at a large distance from the basestation will continuously send the aggregated data, which might rapidly deplete their battery energy, consequently resulting in early WSN partitioning as clarified in Fig. 4.2.

Still, the single-hop communication method is not suitable in a large-scale WSN, where the distant nodes will always send their sensed information to the basestation over a long distance. So, they will deplete their battery energy quickly in comparison with the nearby nodes, hence hampering the network lifetime and functionality. To extend the network lifetime and functionality for a reasonable period, different single-hop clustering routing protocols are proposed to optimize several single and multiobjective problems associated with WSNs.

In an algorithm proposed to improve LEACH performance, the metrics measured are WSN lifetime, number of cluster-heads, energy, and number of packets (Wang et al. 2012):

- For WSN lifetime, three metrics are used; typically, first node dies, half of the nodes alive, and last node dies (Sect. 4.2.2).
- When there are fewer clusters, the non-cluster-head nodes often have to transmit data notably far to reach the cluster-head node, thus draining their energy. When there are more clusters, there is not as much local data aggregation being performed. The proposed algorithm can keep stable the expected number of clusters per round that attains a better balance of energy consumption. Uniform energy consumption is crucial for WSN load balancing.
- More uniform energy consumption implies less possibility of nodes premature depletion; the less energy consumption per round the better is the WSN performance.
- The number of packets determines how many packets are received at the basestation. The more amount of packets received discloses less depletion rate of nodes and efficient energy expenditure. Moreover, the proportion of residual energy can be evaluated from the number of packets.

Also, from a different algorithm, the throughput against energy consumed reflects the energy efficiency on average; in general, the bigger it is, the longer the lifetime of the WSN (Lin and Wang 2017).

4.9   Energy-Efficient Routing Protocols for Homogeneous and Heterogeneous WSNs

The network layer aims to realize the communication among sensors, between sensors and observers, data routing and cooperative sensing. WSN routing protocols must be designed to meet the desired performance requirements in energy efficiency, scalability, robustness, and convergence (Sect. 4.6). Their main goal is to establish reliable and energy-efficient paths for WSN nodes and achieve the longest lifetime for the entire WSN. Energy consumption in routing is caused by neighborhood discovery, communication, and computation. Based on (Yan et al. 2016), this section presents the energy-efficient routing protocols according to whether they are intended for homogenous or heterogeneous WSNs. Also, distinctive from other classifications, care is aimed at static and mobile nodes. Figure 4.16 illustrates the meant classification.

Routing protocols for homogeneous WSNs deal with identical nodes; their energy-efficient performances are proposed from several points of view. They can be subdivided into static and mobile routing protocols. Static routing protocols include opportunistic, cross-layer, cooperative, and biologically inspired optimal routing protocols, depending on their design principles. Mobile routing protocols deal with energy issues in mobile scenarios where mobile nodes can be sources and sinks.

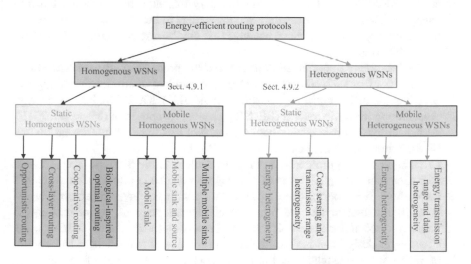

Fig. 4.16 Classification of energy-efficient routing protocols for WSNs. (Based on Yan et al. (2016))

Routing protocols for heterogeneous WSNs tackle heterogeneity and energy issues as well. According to (Yarvis et al. 2005), heterogeneity can triple the average delivery rate and provide a fivefold increase in network lifetime when properly deployed. The heterogeneity is reflected through energy, computation, WSN protocol, and/or links. Note that cluster-based routing is not classified as a class in Fig. 4.16, because it is widely used in both homogeneous and heterogeneous WSNs. The characteristics of cluster-based routing are profoundly described in Sects. 4.3.3 and 4.8.1.1.

Energy-efficient routing protocols for homogeneous WSNs are presented in Sect. 4.9.1 and energy-efficient routing protocols for heterogeneous WSNs are presented in Sect. 4.9.2.

4.9.1   Routing Protocols for Homogeneous WSNs

In this class, routing protocols are subcategorized as those for static homogeneous WSNs and those for mobile homogeneous WSNs, respectively, detailed in Sects. 4.9.1.1 and 4.9.1.2.

4.9.1.1   Static Homogeneous WSNs

Routing protocols for static homogeneous WSNs are subclassified into opportunistic routing, cross-layer routing, cooperative routing, and biological-inspired routing:

- Opportunistic routing protocols. Opportunistic routing is proposed to solve unreliable link problems and reduce unnecessary retransmissions (Biswas and Morris 2005). Thus, it can improve not only transmission reliability, but also energy efficiency; it involves multiple forwarders to increase WSN communication throughput by taking advantage of the broadcast nature of wireless communication.

  The extremely opportunistic routing (ExOR) protocol in (Biswas and Morris 2005) is the first such scheme; the MAC-independent opportunistic routing and encoding (MORE) protocol (Chachulski et al. 2007) is its extension. An energy-efficient opportunistic routing (EEOR) protocol is proposed in (Mao et al. 2011). Also, a simple but robust multipath routing protocol, the energy-efficient routing (E^2R), is introduced in (Zhu and Towsley 2011).

- Cross-layer routing protocols. Because controllers at the network layers interact with each other, the parameters of each layer should be jointly decided to achieve the optimal WSN performance. Apart from a stringent layered protocol, cross-layer design that breaks the principle of layered protocol, by permitting the interaction of non-adjacent layers, has received much attention (Chap. 6). Cross-layer design can realize flexible and intelligent management and control of WSNs so as to extend their lifetime and achieve high-energy efficiency.

Among these protocols there are the joint routing, power control and random access (JRPRA) algorithm (He et al. 2012), the distributed algorithm lifetime maximization cooperative routing with truncated automatic repeat request (LMCRTA) (Zhai et al. 2012), and cross-layer optimal design (CLOD) (Li et al. 2013b).

- Cooperative routing protocols. Cooperative communication routing can mitigate channel fading, achieve high spectral efficiency and improve transmission capacity; it is mainly a part of cross-layer routing (Nosratinia et al. 2004). It is developed from the traditional multiple-input and multiple-output (MIMO) techniques that can reduce the transmission power and extend the transmission coverage. The basic idea is to allow multiple nodes to form a virtual MIMO system to share their antennas and resources, thereby gaining the advantage of space diversity in a multinode scenario instead of equipping each node with multiple antennas.

  As an energy-efficient cooperative routing scheme with space diversity, there is relay selection-based cooperative routing (RBCR) (Ben Nacef et al. 2012), an energy-balanced cooperative routing (EBCR) (Chen et al. 2013), and the multiparametric mixed-integer linear program (mp-MILP) (Habibi et al. 2013).

- Biologically inspired optimal routing protocols. Biologically inspired principles have led to various technological innovations in different fields of research (Archana and Saravanan 2014). Their applications to routing sensor networks guided to several protocols. A biologically inspired self-organized secure autonomous routing protocol (BIOSARP) (Saleem et al. 2014), BeeSensor inspired by the honey-bee colony (Saleem et al. 2012), and Bee-Sensor-C that enhances BeeSensor where C stands for "cluster" (Cai et al. 2015).

As stated in Sect. 4.4, traditional routing protocols assume static and homogeneous WSNs. Opportunistic, cross-layer, cooperative, and biologically inspired optimal routing protocols are used to achieve high energy efficiency and improve the performance of static and homogeneous WSNs. Based on the presentation made to typical such protocols, the following specifics are worth remembering:

- Opportunistic routing is proposed to deal with unreliable link problems and reduce unnecessary retransmissions.
- The idea of cross-layer routing is straightforward. But its main drawback is its high computation complexity, which may be substantial for most sensor devices.
- The main advantage of cooperative routing is that it is robust and can achieve high WSN throughput. Clearly, reducing the size of a forward list of opportunistic routing or the scope of cooperative relay can decrease communication overhead and energy consumption, which achieves the purpose of prolonging the WSN lifetime.
- Biologically inspired optimal routing is more suitable to large-scale WSNs. However, it faces the difficulty of achieving global optimal results. Note that, protocols in this class do not assume special energy models, except JRPRA (He et al. 2012) and Bee-Sensor-C (Cai et al. 2015). JRPRA describes the energy

model in sending and receiving states, but does not give more details to explain how to compute the energy consumption, while Bee-Sensor-C only describes the free-space energy model in sending or receiving one bit message.

4.9.1.2 Mobile Homogeneous WSNs

Routing protocols for mobile homogeneous WSNs have several subclasses. Typically, mobile sink routing, mobile sink and source routing, and multiple mobile sinks routing:

- Mobile sink routing protocols. Termite-hill is a protocol that balances the energy consumption among sensor nodes of WSNs to avoid the emergence of hot-spots/energy-holes, by adopting one mobile sink that can move without constraints (Zungeru et al. 2012). It can avoid hot-spots/energy-holes caused by the excessive energy consumption at the nodes near the sinks in a static WSN. Termite-hill is an intelligent algorithm, inspired by the behaviors of termites[2] (Britannica 2022).
- Mobile sink and source routing protocols. Trace-announcing routing scheme (TARS) focus on some applications that need the support of both mobile sinks and targets (Chi and Chang 2012). Since sinks and targets can move freely in the WSN, TARS is a virtual grid-based routing scheme; it extends the tracking-assisted routing scheme for WSNs (Chi and Chang 2009).
- Multiple mobile sinks routing protocols. As an energy-efficient distributed clustering protocol with path predicable mobile sinks, MobiCluster is introduced in (Konstantopoulos et al. 2012). It is proposed to deal with the isolated "sensor islands" where mobile nodes cannot move through. Its cluster-heads need only communicate with so-called "rendez-vous nodes" which are located near a mobile sink trajectory, and they take turns in such communication. Moreover, an energy-efficient distance aware routing algorithm with multiple mobile sinks is proposed in (Wang et al. 2014).

Obvious features characterize routing techniques in this subclass:

- They can deal with not only the hot-spot/energy-hole problem but also with sparse and disconnected WSNs. Mobility makes the signal transmission distance shorter, which saves energy, while easily balancing the energy consumption among nodes.
- The major advantage of these protocols is flexibility and scalability.

Some tricky issues do occur, though, due to mobility:

---

[2]Termite, (order Isoptera), any of a group of cellulose-eating insects, the social system of which shows remarkable parallels with those of ants and bees, although it has evolved independently. Even though termites are not closely related to ants, they are sometimes referred to as white ants.

- A higher packet loss rate due to topology changes and increased data latency.
- Since the cost of mobile sinks is much higher as compared to static nodes, an all-mobile sensors option is unlikely. The performance of a WSN using multiple mobile sinks is superior to that using a single sink, if the cost is not a major issue. Otherwise, choosing an optimal number of mobile sinks becomes an important question to be considered.
- The trajectory of mobile nodes has a great influence on a WSN topology, and consequently on the routing performance. Both protocols in (Konstantopoulos et al. 2012; Wang et al. 2014) adopt a fixed trajectory, which is simple and convenient. But, energy consumption of nodes near the sink is relatively large, which cannot fundamentally solve the problem of uneven energy consumption among nodes. On the other hand, since the sinks move without constraints in Termite-hill (Zungeru et al. 2012) and TARS (Chi and Chang 2012), a path can be selected in real time according to the WSN conditions. This flexibility is to be weighted against more complicated implementation and perhaps more uncertainty.
- In a large WSN with a limited moving speed of mobile nodes, the contradiction between the speed of mobile nodes and the requirements for data collection is critical. Efforts are required to design reliable and real-time routing protocols that can be effective in energy conservation while providing delay-guaranteed services.
- The protocols in this class do not assume special energy models, except (Wang et al. 2014) that uses the energy model in (Heinzelman et al. 2002).

4.9.2   Routing Protocols for Heterogeneous WSNs

Routing protocols in this class are subcategorized as those for static heterogeneous WSNs and those for mobile heterogeneous WSNs, respectively, presented in Sects. 4.9.2.1 and 4.9.2.2.

4.9.2.1   Static Heterogeneous WSNs

This subclass of heterogeneous WSNs is subdivided into energy heterogeneity-focused protocols and cost, sensing, and transmission range heterogeneity caring protocols, as clarified in the next points:

- Energy heterogeneity routing protocols. An energy and coverage aware distributed clustering (ECDC) protocol (Gu et al. 2014) is for area coverage and point coverage in heterogeneous WSNs; it aims at prolonging the lifetime of WSNs. ECDC divides sensor nodes, in terms of their energy, into three types: explicitly, cluster-head, cluster member, and plain node. Its obvious advantage is that it elects a cluster-head based on residual energy and coverage; thus, its cluster sizes are even. In addition, lifetime is defined from the initial time to the time when more than 30% of nodes are not alive.

Also, an energy-efficient multilevel heterogeneous routing (EEMHR) protocol (Tanwar et al. 2014) aims at saving energy by partitioning all nodes into k level normal nodes and k level advanced nodes, where k represents the level of energy. The bigger k, the higher energy level. Since the level k advanced nodes have the highest energy, they become cluster-heads and thus may cause the "hot-spot/energy-hole." EEMHR uses weighted election probabilities to elect cluster-heads to avoid such holes. A lifetime extended multilevels heterogeneous routing (LE-MHR) protocol (Tyagi et al. 2015) is an enhancement of EEMHR.

- Cost, sensing, and transmission range heterogeneity routing protocols. The protocol coverage, sink location, and routing problem (CSLRP) (Güney et al. 2012) addresses these three basic issues in the design of heterogeneous WSNs. All sensors are divided into L types, where L denotes the set of sensor types with different deployment costs. Each type also has a different sensing and transmission range. Two mixed-integer linear programs are designed. One is to consider the total routing energy consumption on the arcs. The other aims to minimize the total routing energy that consists of the sensor-to-sink assignment. CSLRP is only applicable to a small-size WSN with total node count not exceeding 49. Coverage threshold is also considered as additional QoS metric.

Well-designed routing proposed for static heterogeneous WSNs could effectively prolong the WSNs lifetime, improve their reliability, and meet diverse application requirements. Numerous features of these protocols are noticed:

- Most existing protocols are based on a cluster topology (Sects. 4.3 and 4.8.1.1) while differing in their cluster-head selection.
- Besides energy efficiency, WSN heterogeneity improves computational capability, and effects protocols design. Future works have to deal with more aspects of heterogeneity.
- Protocols in this subclass do not assume special energy models, except EEMHR (Tanwar et al. 2014) and LE-MHR (Tyagi et al. 2015) that are using the energy model in (Heinzelman et al. 2002).
- Cluster-based routing protocols, like those in (Gu et al. 2014; Tanwar et al. 2014; Tyagi et al. 2015), adopt single-hop intra-cluster routing methods and multihop inter-cluster routing to achieve energy efficiency.

4.9.2.2   Mobile Heterogeneous WSNs

This subclass of heterogeneous WSNs is subdivided into energy heterogeneity-motivated protocols and energy, transmission range, and data rate heterogeneity-focused protocols, as revealed in the coming points:

- Energy heterogeneity routing protocols. A hierarchical adaptive and reliable routing protocol (HARP) (Atero et al. 2011) partitions the nodes into two types, normal nodes and cluster nodes, according to their residual energy capacities. Next, cluster-head selection is performed based on the residual energy of nodes. Its main idea is to build a hierarchical tree in two layers intra-cluster and inter-cluster. Moreover, HARP introduces a local recovery mechanism and mobility management to rebuild trees when link failures occur.

A routing algorithm for heterogeneous mobile network (RAHMoN) (Vilela and Araujo 2012) divides all sensors into static and mobile; the energy of a static sensor is less than that in a mobile sensor. Mobile sensors can be cluster-heads or sink nodes, with different mobility models. It is also assumed that all sensors can be elected as a cluster-head. The selection of a cluster-head depends on mobility level, energy, and distance to the sink. RAHMoN is efficient with respect to overhead messages and transmitted data packets.

- Energy, transmission range, and data rate heterogeneity routing protocols. A clustered heterogeneous sensor network (HSN) (Sudarmani and Kumar 2013) is proposed with a mobile sink. The nodes in the WSN are divided into three categories according to their energy: typically, H-nodes (high-energy level), L-nodes (low-energy level), and the sink with unlimited energy. H-nodes provide a longer transmission range and higher data rate than L-nodes. Compared with HARP and RAHMoN, its cluster-head is fixed and provides a single-hop data transmission. Particle swarm optimization is adopted to optimize the sink-moving trajectory among cluster-heads; the protocol is thus applicable to large-scale WSNs. It is more energy-efficient than WSNs with a static sink. Noticeably, there is a loss of data when the speed of the mobile sink increases.

Routing protocols for mobile heterogeneous WSNs have their characteristics:

- Similar to routing protocols for homogeneous WSNs, introducing mobile nodes to heterogeneous WSN can avoid hot-spots/energy-holes, achieve high-energy efficiency, and balance energy consumption among nodes.
- Protocols in this subclass do not assume special energy models, except HSN (Sudarmani and Kumar 2013) that uses the energy model in (Heinzelman et al. 2002).

4.9.3   Recapitulation

To recap Sect. 4.9, it is important to have several notes in mind (Yan et al. 2016):

- Routing protocols for homogeneous WSNs are more widely investigated than those for heterogeneous WSNs. More routing protocols for heterogeneous WSNs are foreseen to meet evolving application requirements.
- Compared with static WSNs, routing protocols for mobile WSNs promise to bring more benefits to real-time delivery and to guarantee high coverage, energy efficiency, and energy balance but at the expense of more complex implementation and higher deployment cost.

- Creating and using a reliable routing metric is principal to routing design. It should measure routing overhead and routing capability from different aspects due to the diversity of WSNs. New routing metrics such as spatial reusability should be considered to increase WSN throughput with affordable energy overhead (Meng et al. 2016). For heterogeneous WSNs, besides energy heterogeneity, link heterogeneity is also essential and requires further study.
- Many existing QoS routing protocols are constrained to some particular applications and only take one or two QoS metrics, hence tending to lose the balance between QoS guarantee and energy efficiency. In this regard, energy-efficient routing with QoS guarantee in different applications or diverse WSNs can be viewed as an interesting area for investigation (Ernst et al. 2014).
- In WSNs, each node acts in a twofold role, a sensing role and a router, which makes it vulnerable to attacks. Secure routing, thus, needs attention (Zhang et al. 2007; Sakharkar et al. 2014). Clearly a security mechanism incurs additional energy cost; consequently, there should be a proper trade-off between security levels and energy consumption for different applications.
- Since there is wide spectrum of WSN applications, the process of routing implementation varies significantly from a WSN to another. Thus, application-specific routing protocols are needed for such situations as IoT, vehicles, underwater, space, volcanoes, explorations, epidemic, human body, water and oil pipelines, microgrid, system monitoring and diagnosis, and robots (Fortino et al. 2014; Eshghi et al. 2015; Cheng et al. 2015; Zhou et al. 2016) (Table 4.4).

4.10   Conclusion for Good Paths

This chapter has seen and lived events that can never be dropped from history remembrance. In sports, the 126th French Open, one of four Grand Slam tennis tournaments, played on outdoor clay courts, was held at the Stade Roland Garros in Paris, France, from May 22 to June 5, 2022, comprising singles, doubles, and mixed doubles play. Spaniard Rafael Nadal won the men's single and Polish Iga Świątek won the women's single (WIKIPEDIA 2022a).

The Wimbledon Championships, the oldest tennis tournament has been held at the All England Club in Wimbledon, London, since 1877 and is played on outdoor grass courts, with retractable roofs over the two main courts since 2019. The 135th edition was held from June 27th till July 10th, 2022. The Serbian Novak Djokovic won the men's title and Elena Andreyevna Rybakina the Russian-born Kazakhstani won the women title (WIKIPEDIA 2022b).

The 2022 World Athletics Championships (WAC) in its 18th edition was held at Hayward Field in Eugene, Oregon, United States, from July 15–24, 2022. Hayward Field, a track and field stadium at the University of Oregon, hosted the championships. Some major records were set. Noah Lyles broke a 26-year-old American record as he won the men's 200 m final on Thursday, July 21st, 2022, when achieving a 19.31 sec, leading a United States sweep as Kenneth Bednarek

Table 4.4 Classification of energy and lifetime aware routing protocols for WSNs

	Taxonomy	
Energy-efficient routing protocols (Sect. 4.7)	Network structure-based approach (Sect. 4.7.1): Flat networks routing protocols (Sect. 4.7.1.1): Proactive or table-driven routing protocols Reactive or source-initiated on-demand routing protocols Hierarchical networks routing protocols (Sect. 4.7.1.2) Topology-based approach (Sect. 4.7.3): Location-based routing protocols (Sect. 4.7.3.1) Mobile agent-based routing protocols (Sect. 4.7.3.2)	Communication model-based approach (Sect. 4.7.2): Query-based routing protocols (Sect. 4.7.2.1) Coherent and non-coherent-based routing protocols (Sect. 4.7.2.2) Negotiation-based routing protocols (Sect. 4.7.2.3) Reliable routing approach (Sect. 4.7.4): Multipath-based routing protocols (Sect. 4.7.4.1) QoS-based routing protocols (Sect. 4.7.4.2)
Energy-efficient and energy-balanced routing protocols (Sect. 4.8)	Multihop communication (Sect. 4.8.1): Multihop clustering techniques (Sect. 4.8.1.1) Load-balanced tree techniques (Sect. 4.8.1.2): Multipath approach Single-path approach	Single-hop communication (Sect. 4.8.2): Single-hop clustering techniques
Energy-efficient routing protocols for homogeneous and heterogeneous WSNs (Sect. 4.9)	Routing protocols for homogeneous WSNs (Sect. 4.9.1): Static homogeneous WSNs (Sect. 4.9.1.1): Opportunistic routing protocols Cross-layer routing protocols Cooperative routing protocols Biologically inspired optimal routing protocols Mobile homogeneous WSNs (Sect. 4.9.1.2): Mobile sink routing protocols Mobile sink and source routing protocols Multiple mobile sinks routing protocols	Routing protocols for heterogeneous WSNs (Sect. 4.9.2): Static heterogeneous WSNs (Sect. 4.9.2.1): Energy heterogeneity routing protocols Cost, sensing and transmission range heterogeneity routing protocols Mobile heterogeneous WSNs (Sect. 4.9.2.2): Energy heterogeneity routing protocols Energy, transmission range and data rate heterogeneity routing protocols

(19.77 sec) and Erriyon Knighton (19.80 sec) finished second and third, respectively (The Athletic 2022). Also, home favorite Sydney McLaughlin has destroyed her own 400-m hurdles world record, setting a time of 50.68 sec, Femke Bol of the Netherlands won silver and the USA's Dalilah Muhammad bronze (AP 2022). The first three countries ranked by gold medals are USA (13 medals), Ethiopia (4 medals), and Jamaica (2 medals) (The Sporting News 2022).

At Wembley, on Sunday, July 31st, 2022, England created history by winning their first major women's tournament in a dramatic Euro 2022 final against eight-time champions Germany, under the hearts and eyes of a record crowd of 87,192 in the history of men's or women's Euros (BBC 2022).

But the most shaking is the Russian-Ukrainian crisis that erupted on Thursday, February 24th, 2022, at 0:40 am GMT (8:40 pm EDT), shocking brutally the daily life of millions and millions even thousands miles away from the shelling. Living turned over to a nightmare hardship of finding essentials of food, fuel, and before all currency. How should it end??

On Thursday, September 8th, 2022, at 6:35 pm London Time (12:35 pm EST), Buckingham Palace announced the death of Queen Elizabeth II, the UK's longest-serving monarch, aged 96, at Balmoral, after reigning for 70 years. Her elder son is on the throne as King Charles III. The state funeral was held at Westminster Abbey, at 11:00 am London Time (5:00 am EST) on Monday, September 19th.

Getting to the focal chapter topic may be a relief from being absorbed by disturbing news eruptions. Section 4.6 is making obvious that WSNs have several restrictions that impact the design of the protocols that maintain their functioning, such as limited energy supply, reduced computing power, and small bandwidth of the wireless links connecting sensor nodes. One of the main design goals of WSNs is to carry out data communication while targeting to prolong the lifetime of the network and prevent connectivity degradation by employing aggressive energy management techniques. Depending on the application and the size of the WSN, different architectures and design goals constraints have been considered; clearly, the performance of a routing protocol is closely related to the architectural model.

As made obvious throughout this chapter, multiple critical factors influence the selection or design of a routing protocol:

- WSN dynamics. The main components in a WSN are sensor nodes, sink and monitored events. In most of the WSN architectures, sensor nodes are assumed to be stationary. On the other hand, supporting the mobility of sinks or cluster-heads is sometimes required. Routing messages sent or received from nodes become more challenging since route stability turns to be an important optimization factor, in addition to energy and bandwidth. Furthermore, the sensed event can be dynamic or static, depending on the application; thus, in a target detection application, the event is dynamic, while forest monitoring for early fire prevention is a static event.
- Nodes deployment. This affects the performance of the routing protocol; the deployment may be deterministic or self-organizing. In deterministic situations, the sensors are placed manually and all the data are routed through predefined

paths. In self-organizing systems, the sensor nodes are scattered randomly and create an infrastructure in an ad hoc manner.

- Nodes mobility. Nodes in the WSN are assumed to be static; yet, the interest has grown in applications that support nodes mobility. As an example, the medical care applications where mobile sensors attached to the patients send continuous data to the medical professionals (Chap. 3).

- Energy considerations. The setup of a route is greatly influenced by energy considerations. Since the transmission power of a wireless radio depends on distance squared, or in higher order, due to obstacles, multihop routing consumes less energy than single-hop communication (Sect. 4.2.1). However, multihop routing may add significant overhead for topology management and medium access control (Sects. 4.5 and 4.8.1). In contrast, single-path routing performs well enough if all nodes are very close to the sink (Sects. 4.5 and 4.8.2).

- Data delivery models. Depending on the application of the WSN, the data delivery model to the sink can be continuous, event-driven, query-driven, or hybrid. In the continuous delivery model, each sensor sends data periodically. In event-driven and query-driven models, the transmission of data is triggered when an event occurs, or when the sink generates a query. Moreover, there are some WSNs that apply a hybrid model using a combination of continuous, event-driven, and query-driven data delivery. Routing protocols are based on the data delivery model, especially with regard to the minimization of energy consumption and route stability.

- Node capabilities. In a WSN, different functionalities can be associated with the sensor nodes. In most WSNs, a node can be dedicated to a particular function, which might be relaying, sensing or aggregation, since engaging the three functionalities on a single node might quickly drain its energy.

- Data aggregation/fusion. Sensor nodes might generate similar packets; aggregating similar packet reduces the number of transmissions. Data aggregation is the combination of data from different sources (Sect. 4.3.2). This can be fulfilled through using functions such as suppression, min, max, and average. These functions can be performed either partially or fully in each sensor node. Computation can be less energy-consuming than communication and substantial energy savings can be obtained through data aggregation (Sect. 4.1). This technique can achieve energy efficiency and traffic optimization in a number of routing protocols (Sects. 4.7, 4.8 and 4.9). In many WSN architectures, all aggregation functions are assigned to more powerful and specialized nodes.

After designing energy-efficient routing protocols for WSNs, several questions are to be answered to check the capabilities of the protocols and to which extent did they satisfy the application needs:

- Is the protocol energy-balanced? When developing an energy-efficient routing protocol, the load balancing of the energy that the sensors consume should be one of the main targets. This means that the routing protocols need to minimize the energy consumption of the WSN by selecting not only the shortest routes but also the routes that will lead to the extension of the network lifetime (Sects. 4.5 and 4.8.1.2).

- Does the protocol consider network security? Importantly, apart from energy consumption, the security that protocols can offer is to protect against eavesdropping and malicious behavior (Sect. 4.9.3).
- How is the performance on real environment? Most of the protocols for WSNs have been evaluated through simulations. However, it is important to evaluate their performance in real environments and via testbeds (Chap. 7).
- Are both real-time application and QoS considered? There is an ongoing need to develop real-time application that offer high level of QoS to the end users. Hence, it is important to devote efforts for the development of routing protocols that satisfy this objective (Sect. 4.9.3).
- How is QoS considered and which metrics are adopted? The QoS is indispensable in the delivery of data in critical applications such as military, real time, and healthcare. Thus, the development of routing protocols that consider both energy efficiency and accurate delivery of data is highly required (Sect. 4.9.3).
- Are fixed and mobile WSNs integrated? Most of the applications, for instance in healthcare monitoring, require the data collected from the sensor nodes to be transmitted to a server for medical professionals access and diagnosis, or to be sent as medication to the patients. In this scenario, the routing requirements of each environment are different; further work is necessary for tackling this kind of situations (Sect. 4.9.3).

Done in this chapter with energy and lifetime aware routing protocols for WSNs routing protocols at the network layer, the coming chapter will move to the transport protocols tailored for WSNs.

4.11   Exercises

Note: A technical report follows the template of a peer-reviewed journal paper

1. Based on Sect. 4.2.2, write a technical report on the design metrics for energy-efficient routing protocols for WSNs.
2. Guided by Sect. 4.3.2, write a technical report on data collection and aggregation.
3. Guided by Sect. 4.4 and literature reviews, compare homogeneous and heterogeneous WSNs.
4. Steered by Sect. 4.5, differentiate between single-hop and multihop transmission.
5. Section 4.6.1 describes how routing protocols for WSNs are different. Write a technical report on this topic.
6. Section 4.6.3 presents the design goals of energy and lifetime aware routing protocols for WSNs. Write a technical report on this topic.

7. Considering Sect. 4.6, what are the factors that influence the design of routing protocols for WSNs?
8. Based on Sect. 4.7 write a technical report that compares the subclasses presented in Sects. 4.7.1, 4.7.2, 4.7.3, and 4.7.4.
9. Write a technical report on the network structure approach detailed in Sect. 4.7.1.
10. Write a technical report on the reliable routing approach presented in Sect. 4.7.4.
11. Based on Sects. 4.5 and 4.8, differentiate between energy-efficient path and energy-balanced path.
12. Based on Sect. 4.8.1.1, write a technical report on multihop clustering techniques.
13. Based on Sect. 4.8.1.2, write a technical report on load-balanced tree techniques.
14. Following Sect. 4.9 and literature reviews, compare static and mobile WSNs.
15. Look up for newly introduced energy and lifetime aware routing protocols for WSNs. Care should be accorded to their classification and their features and functionality.

References

Abbasi, A.A., and M. Younis. 2007. A Survey on Clustering Algorithms for Wireless Sensor Networks. *Computer Communications* 30 (14–15): 2826–2841.

Afsar, M.M., and M.H. Tayarani-N. 2014. Clustering in Sensor Networks: A Literature Survey. *Network and Computer Applications* 46: 198–226.

Akkaya, K., and M. Younis. 2005. Energy and QoS Aware Routing in Wireless Sensor Networks. *Cluster Computing* 8: 179–188.

Akyildiz, I.F., W. Su, Y. Sankarasubramaniam, and E. Cayirci. 2002. A Survey on Sensor Networks. *Communications Magazine* 40 (8): 102–114.

Alazzawi, L., and A. Elkateeb. 2008. Performance Evaluation of the WSN Routing Protocols Scalability. *Journal of Computer Systems, Networks, and Communications* 14 (2): 1–9.

Al-Karaki, J.N., and A.E. Kamal. 2004. Routing Techniques in Wireless Sensor Networks: A Survey. *Wireless Communications* 11 (6): 6–28.

Amini, N., A. Vahdatpour, W. Xu, M. Gerla, and M. Sarrafzadeh. 2012. Cluster Size Optimization in Sensor Networks with Decentralized Cluster-based Protocols. *Computer Communications* 35 (2): 207–220.

Amis, A.D., and R. Prakash. 2000. Load-Balancing Clusters in Wireless Ad Hoc Networks. In *The 3rd Symposium on Application-Specific Systems and Software Engineering Technology*, 25–32. Richardson: IEEE.

Anisi, M.H., A.H. Abdullah, and S. Abd Razak. 2013. Energy-Efficient and Reliable Data Delivery in Wireless Sensor Networks. *Wireless Networks* 19: 495–505.

AP. 2022. *Sydney McLaughlin Shatters 400m Hurdles Record with 50.68.* AP. July 23, 2022. https://apnews.com/article/0449f2692457e8cc799fdbeb8cb0ca08. Accessed 23 July 2022.

Archana, S., and N.P. Saravanan. 2014. Biologically Inspired QoS Aware Routing Protocol to Optimize Lifetime in Sensor Networks. In *International Conference on Recent Trends in Information Technology (ICRTIT)*, 1–6. Chennai: IEEE.

Arjunan, S., and P. Sujatha. 2019. A Survey on Unequal Clustering Protocols in Wireless Sensor Networks. *Journal of King Saud University – Computer and Information Sciences* 31 (3): 304–317.

Atero, F.J., J.J. Vinagre, J. Ramiro, and M. Wilby. 2011. A Low Energy and Adaptive Routing Architecture for Efficient Field Monitoring in Heterogeneous Wireless Sensor Networks. In *The 9th International Conference on High Performance Computing & Simulation (HPCS)*, 449–455. Istanbul: IEEE.

Basagni, S., I. Chlamtac, V.R. Syrotiuk, and B.A. Woodward. 1998. A Distance Routing Effect Algorithm for Mobility (DREAM). In *The 4th Annual International Conference on Mobile Computing and Networking (MobiCom)*, 67–84. Dallas: ACM/IEEE.

BBC. 2022. *BBC Sport at Wembley.* BBC. July 31, 2022. https://www.bbc.co.uk/sport/football/62339532. Accessed 31 July 2022.

Bellur, B., and R.G. Ogier. 1999. A Reliable, Efficient Topology Broadcast Protocol for Dynamic Networks. In *The 18th Annual Joint Conference of the IEEE Computer and Communications Societies (INFOCOM)*, 178–186. New York: IEEE.

Ben Nacef, A., S.M. Senouci, Y. Ghamri-Doudane, and A.-L. Beylot. 2012. A Combined Relay-Selection and Routing Protocol for Cooperative Wireless Sensor Networks. In *The 8th International Wireless Communications and Mobile Computing Conference (IWCMC)*, 293–298. Limassol: IEEE.

Ben-Othman, J., and B. Yahya. 2010. Energy Efficient and QoS Based Routing Protocol for Wireless Sensor Networks. *Journal of Parallel and Distributed Computing* 70 (8): 849–857.

Bhushan, B., and G. Sahoo. 2019. Routing Protocols in Wireless Sensor Networks. In *Computational Intelligence in Sensor Networks*, ed. B. Mishra, S. Dehuri, B. Panigrahi, A. Nayak, B. Mishra, and H. Das, vol. 776, 215–248. Berlin: Springer.

Biswas, S., and R. Morris. 2005. ExOR: Opportunistic Multi-Hop Routing for Wireless Networks. In *Conference on Applications, Technologies, Architectures, and Protocols for Computer Communications (SIGCOMM)*, 133–144. Philadelphia: ACM.

Blum, B., T. He, S. Son, and J. Stankovic. 2010, June. Energy-Efficient Beaconless Geographic Routing in Wireless Sensor Networks. *Transactions on Parallel and Distributed Systems (TPDS)* 21 (6): 881–896.

Braginsky, D., and D. Estrin. 2002, September. First ACM International Workshop on Wireless Sensor Networks and Applications 2002 Atlanta Georgia USA. In *The 1st International Workshop on Wireless Sensor Networks and Applications*, 22–31. ACM.

Britannica. 2022. *Termite.* Britannica. January 1, 2022. https://www.britannica.com/animal/termite. Accessed 9 Aug 2022.

Cai, X., Y. Duan, Y. He, J. Yang, and C. Li. 2015. Bee-Sensor-C: An Energy-Efficient and Scalable Multipath Routing Protocol for Wireless Sensor Networks. *International Journal of Distributed Sensor Networks* 11 (3): 976127.

Capone, A., M. Cesana, D. De Donno, and I. Filippini. 2010. Deploying Multiple Interconnected Gateways in Heterogeneous Wireless Sensor Networks: An Optimization Approach. *Computer Communications* 33 (10): 1151–1161.

Chachulski, S., M. Jennings, S. Katti, and D. Katabi. 2007. Trading Structure for Randomness in Wireless Opportunistic Routing. *Computer Communication Review* 37 (4): 169–180.

Chang, R.S., and C.-J. Kuo. 2006. An Energy Efficient Routing Mechanism for Wireless Sensor Networks. In *The 20th International Conference on Advanced Information Networking and Applications (AINA)*, 1–5. Vienna: IEEE.

Chen, D., and P.K. Varshney. 2007. On-demand Geographic Forwarding for Data Delivery in Wireless Sensor Networks. *Computer Communications* 30 (14–15): 2954–2967.

Chen, M., S. Gonzalez, and V.C.M. Leung. 2007. Applications and Design Issues for Mobile Agents in Wireless Sensor Networks. *Wireless Communications* 14 (6): 20–26.

Chen, M., S. Gonzalez, Y. Zhang, and V.C.M. Leung. 2009. Multi-Agent Itinerary Planning for Wireless Sensor Networks. In *Lecture Notes of the Institute for Computer Sciences, Social Informatics and Telecommunications Engineering (LNICST)*, ed. N. Bartolini, S. Nikoletseas, P. Sinha, V. Cardellini, and A. Mahanti, vol. 22, 584–597. Berlin, Heidelberg: Springer.

Chen, M., L.T. Yang, T. Kwon, L. Zhou, and M. Jo. 2011. Itinerary Planning for Energy-Efficient Agent Communications in Wireless Sensor Networks. *Transactions on Vehicular Technology* 60 (7): 3290–3299.

Chen, S., Y. Li, M. Huang, Y. Zhu, and Y. Wang. 2013. Energy-Balanced Cooperative Routing in Multihop Wireless Networks. *Wireless Networks* 19: 1087–1099.

Cheng, S.T., and T.U. Chang. 2012. An Adaptive Learning Scheme for Load Balancing with Zone Partition in Multi-Sink Wireless Sensor Network. *Expert Systems with Applications* 39 (10): 9427–9434.

Cheng, J., J. Cheng, M. Zhou, F. Liu, S. Gao, and C. Liu. 2015. Routing in Internet of Vehicles: A Review. *Transactions on Intelligent Transportation Systems* 16 (5): 2339–2352.

Chi, Y.-P., and H.-P. Chang. 2009. TRENS: A Tracking-Assisted Routing Scheme for Wireless Sensor Networks. In *The 10th International Symposium on Parallel Architectures, Algorithms and Networks (ISPAN)*, 190–195. Kaoshiung: IEEE.

———. 2012. TARS: An Energy-Efficient Routing Scheme for Wireless Sensor Networks with Mobile Sinks and Targets. In *The 26th International Conference on Advanced Information Networking and Applications (AINA)*, 128–135. Fukuoka: IEEE.

Cordeiro, C.M., and D.P. Agrawal. 2011. *Ad Hoc and Sensor Networks: Theory and Applications* Singapore: World Scientific Publishing.

Eghbali, A.N., N.T. Javan, A. Dareshoorzadeh, and M. Dehghan. 2009. An Energy Efficient Load-Balanced Multi-Sink Routing Protocol for Wireless Sensor Networks. In *The 10th International Conference on Telecommunications (ConTEL)*, 229–234. Zagreb: IEEE.

Ehsan, S., and B. Hamdaoui. 2012. A Survey on Energy-Efficient Routing Techniques with QoS Assurances for Wireless Multimedia Sensor Networks. *Communications Surveys & Tutorials* 14 (2): 265–278.

Ernst, J.B., S.C. Kremer, and J.P.C. Rodrigues. 2014. A Survey of QoS/QoE Mechanisms in Heterogeneous Wireless Networks. *Physical Communication* 13 (Part B): 61–72.

Eshghi, S., M.H.R. Khouzani, S. Sarkar, N.B. Shroff, and S.S. Venkatesh. 2015. Optimal Energy-Aware Epidemic Routing in DTNs. *Transactions on Automatic Control* 60 (6): 1554–1569.

Estrin, D. 2002. Wireless Sensor Networks Tutorial Part IV: Sensor Network Protocols. In *The 8th International Conference on Mobile Computing and Networking (MobiCom)*, 23–28. Atlanta: ACM SIGMOBILE.

Fahmy, Hossam M.A. 2020. *Wireless Sensor Networks: Energy Harvesting and Management for Research and Industry*. Cham: Springer.

Fan, K.-W., S. Liu, and P. Sinha. 2007. Structure-Free Data Aggregation in Sensor Networks. *Transactions on Mobile Computing* 6 (8): 929–942.

Fanian, F., and M.K. Rafsanjani. 2019. Cluster-based Routing Protocols in Wireless Sensor Networks: A Survey Based on Methodology. *Journal of Network and Computer Applications* 142 (15): 111–142.

Felemban, E., C.-G. Lee, and E. Ekici. 2006. MMSPEED: Multipath Multi-SPEED Protocol for QoS Guarantee of Reliability and Timeliness in Wireless Sensor Networks. *Transactions on Mobile Computing* 5 (6): 738–754.

Fortino, G., G. Di Fatta, M. Pathan, and A.V. Vasilakos. 2014. Cloud-Assisted Body Area Networks: State-of-the-Art and Future Challenges. *Wireless Networks* 20: 1925–1938.

Frye, L., and L. Cheng. 2009. Topology Management for Wireless Sensor Networks. In *Uide to Wireless Sensor Networks. Computer Communications and Networks*, ed. S. Misra, I. Woungang, and S. Misra, 27–45. London: Springer.

Ghiasi, S., A. Srivastava, X. Yang, and M. Sarrafzadeh. 2002. Optimal Energy Aware Clustering in Sensor Networks. *Sensors* 2 (7): 258–269.

Gu, X., J. Yu, D. Yu, G. Wang, and Y. Lv. 2014. ECDC: An Energy and Coverage-Aware Distributed Clustering Protocol for Wireless Sensor Networks. *Computers & Electrical Engineering* 40 (2): 384–398.

Güney, E., N. Aras, İ.K. Altınel, and C. Ersoy. 2012. Efficient Solution Techniques for the Integrated Coverage, Sink Location and Routing Problem in Wireless Sensor Networks. *Computers & Operations Research* 39 (7): 1530–1539.

Habibi, J., A. Ghrayeb, and A.G. Aghdam. 2013. Energy-Efficient Cooperative Routing in Wireless Sensor Networks: A Mixed-Integer Optimization Framework and Explicit Solution. *Transactions on Communications* 61 (8): 3424–3437.

Hammoudeh, M., and R. Newman. 2015. Adaptive Routing in Wireless Sensor Networks: QoS Optimisation for Enhanced Application Performance. *Information Fusion* 22: 3–15.

Han, G., J. Jiang, M. Guizani, and J.P.C. Rodrigues. 2016. Green Routing Protocols for Wireless Multimedia Sensor Networks. *Wireless Communications* 23 (6): 140–146.

Handy, M.J., M. Haase, and D. Timmermann. 2002. Low Energy Adaptive Clustering Hierarchy with Deterministic Cluster-Head Selection. In *The 4th International Workshop on Mobile and Wireless Communications Network (MWCN)*, 368–372. Stockholm: IEEE.

Hasan, M.Z., H. Al-Rizzo, and F. Al-Turjman. 2017. A Survey on Multipath Routing Protocols for QoS Assurances in Real-Time Wireless Multimedia Sensor Networks. *Communications Surveys & Tutorials* 19 (3): 1424–1456.

He, T., J.A. Stankovic, C. Lu, and T. Abdelzaher. 2003. SPEED: A Stateless Protocol for Real-Time Communication in Sensor Networks. In *The 23rd International Conference on Distributed Computing Systems (ICDCS)*, 46–55. Providence: IEEE.

He, S., J. Chen, D.K.Y. Yau, and Y. Sun. 2012. Cross-Layer Optimization of Correlated Data Gathering in Wireless Sensor Networks. *Transactions on Mobile Computing* 11 (11): 1678–1691.

Hedetniemi, S.M., S.T. Hedetniemi, and A.L. Liestman. 1988. A Survey of Gossiping and Broadcasting in Communication Networks. *Networks* 18 (4): 19–349.

Heinzelman, W., A.P. Chandrakasan, and H. Balakrishnan. 2002. An Application Specific Protocol Architecture for Wireless Microsensor Networks. *IEEE Transactions on Wireless Communication* 1 (4): 660–670.

Hoang, D.C., P. Yadav, R. Kumar, and S.K. Panda. 2014. Real-Time Implementation of a Harmony Search Algorithm-based Clustering Protocol for Energy-Efficient Wireless Sensor Networks. *Transactions on Industrial Informatics* 10 (1): 774–783.

Hsiao, P.-H., A. Hwang, H.T. Kung, and D. Vlah. 2001. Load-Balancing Routing for Wireless Access Networks. In *The 20th Annual Joint Conference: INFOCOM, IEEE Computer and Communications Societies*, 986–995. Anchorage: IEEE.

Intanagonwiwat, C., R. Govindan, D. Estrin, J. Heidemann, and F. Silva. 2003. Directed Diffusion for Wireless Sensor Networking. *Transactions on Networking (TON)* 11 (1): 2–16.

Iqbal, M., M. Naeem, A. Anpalagan, A. Ahmed, and M. Azam. 2015. Wireless Sensor Network Optimization: Multi-Objective Paradigm. *Sensors* 15 (7): 17572–17620.

Jolly, V., and S. Latifi. 2006. Comprehensive Study of Routing Management in Wireless Sensor Networks – Part-2. In *International Conference on Wireless Networks (ICWN)*, 49–62. Las Vegas: CSREA Press.

Junhai, L., X. Liu, and Y. Danxia. 2008. Research on Multicast Routing Protocols for Mobile Ad Hoc Networks. *Computer Networks* 52 (5): 988–997.

Kacimi, R., R. Dhaou, and A.-L. Beylot. 2013. Load Balancing Techniques for Lifetime Maximizing in Wireless Sensor Networks. *Ad Hoc Networks* 11 (8): 2172–2186.

Khalil, E.A., and B.A. Attea. 2011. Energy-Aware Evolutionary Routing Protocol for Dynamic Clustering of Wireless Sensor Networks. *Swarm and Evolutionary Computation* 1 (4): 195–203.

Kim, Y.-J., R. Govindan, B. Karp, and S. Shenker. 2005. Geographic Routing Made Practical. In *The 2nd Symposium on Networked System Design & Implementation (NSDI)*. USENIX Association.

Kim, J., X. Lin, N.B. Shroff, and P. Sinha. 2010, April. Minimizing Delay and Maximizing Lifetime for Wireless Sensor Networks with Anycast. *Transactions on Networking* 18 (2): 515–528.

Konstantopoulos, C., G. Pantziou, D. Gavalas, A. Mpitziopoulos, and B. Mamalis. 2012. A Rendezvous-Based Approach Enabling Energy-Efficient Sensory Data Collection with Mobile Sinks. *Transactions on Parallel and Distributed Systems (TPDS)* 23 (5): 809–817.

Kuila, P., and P.K. Jana. 2014a. Approximation Schemes for Load Balanced Clustering in Wireless Sensor Networks. *The Journal of Supercomputing* 68: 87–105.

———. 2014b. Energy eEfficient Clustering and Routing Algorithms for Wireless Sensor Networks: Particle Swarm Optimization Approach. *Engineering Applications of Artificial Intelligence* 33: 127–140.

Kuila, P., S.K. Gupta, and P.K. Jana. 2013. A Novel Evolutionary Approach for Load Balanced Clustering Problem for Wireless Sensor Networks. *Swarm and Evolutionary Computation* 12: 48–56.

Kulik, J., W.R. Heinzelman, and H. Balakrishnan. 2002. Negotiation–Based Protocols for Disseminating Information in Wireless Sensor Networks. *Wireless Networks* 8 (2/3): 169–185.

Kumar, D. 2014. Performance Analysis of Energy Efficient Clustering Protocols for Maximising Lifetime of Wireless Sensor Networks. *Wireless Sensor Systems* 4: 9–16.

Kumar, D., T.C. Aseri, and R.B. Patel. 2009. EEHC: Energy Efficient Heterogeneous Clustered Scheme for Wireless Sensor Networks. *Computer Communications* 32 (4): 662–667.

Layuan, L., L. Chunlin, and Y. Peiyan. 2007. Performance Evaluation and Simulations of Routing Protocols in Ad Hoc Networks. *Computer Communications* 30 (8): 1890–1898.

Lee, S., Y. Yu, S. Nelakuditi, Z.-L. Zhang, and C.-N. Chuah. 2004. Proactive vs Reactive Approaches to Failure Resilient Routing. In *The 23rd Annual Joint Conference of the IEEE Computer and Communications Societies (INFOCOMM)*, 176–186. Hong Kong: IEEE.

Leong, B., B. Liskov, and R. Morris. 2006. Geographic Routing without Planarization. In *The 3rd Symposium on Networked Systems Design & Implementation (NSDI)*, 25–39. San Jose: USENIX Association.

Li, H., et al. 2013a. COCA: Constructing Optimal Clustering Architecture to Maximize Sensor Network Lifetime. *Computer Communications* 36 (3): 256–268.

Li, M., Y. Jing, and C. Li. 2013b. A Robust and Efficient Cross-Layer Optimal Design in Wireless Sensor Networks. *Wireless Personal Communications* 72: 1889–1902.

Lim, H., and C. Kim. 2001. Flooding in Wireless Ad Hoc Networks. *Computer Communications* 24 (3–4): 353–363.

Lin, D., and Q. Wang. 2017. A Game Theory based Energy Efficient Clustering Routing Protocol for WSNs. *Wireless Networks* 23: 1101–1111.

Lindsey, S., and C.S. Raghavendra. 2002. PEGASIS: Power Efficient Gathering in Sensor Information Systems. In *Aerospace Conference*, 1125–1130. Big Sky: IEEE.

Liu, Q., K. Zhang, J. Shen, Z. Fu, and N. Linge. 2016. GLRM: An Improved Grid-based Load-balanced Routing Method for WSN with Single Controlled Mobile Sink. In *The 18th International Conference on Advanced Communication Technology (ICACT)*, 1–2. PyeongChang: IEEE.

Lu, H., J. Li, and M. Guizani. 2014. Secure and Efficient Data Transmission for Cluster-Based Wireless Sensor Networks. *Transactions on Parallel and Distributed Systems (TPDS)* 25 (3): 750–761.

Lung, C.-H., and C. Zhou. 2010. Using Hierarchical Agglomerative Clustering in Wireless Sensor Networks: An Energy-Efficient and Flexible Approach. *Ad Hoc Networks* 8 (3): 328–344.

Ma, M., Y. Yang, and C. Ma. 2006. Single-Path Flooding Chain Routing in mobile Wireless Network. *InternaInternational Journal of Sensor Networks* 1 (1–2): 11–19.

Malathi, L., R.K. Gnanamurthy, and K. Chandrasekaran. 2015. Energy Efficient Data Collection Through Hybrid Unequal Clustering for Wireless Sensor Networks. *Computers & Electrical Engineering* 48: 358–370.

Mandala, D., X. Du, F. Dai, and C. You. 2008. Load Balance and Energy Efficient Data Gathering in Wireless Sensor Networks. *Wireless Communications and Mobile Computing* 8: 645–659.

Manjeshwar, A., and D.P. Agrawal. 2002. APTEEN: A Hybrid Protocol for Efficient Routing and Comprehensive Information Retrieval in Wireless Sensor Networks. In *The International Parallel and Distributed Processing Symposium (IPDPS)*, 195–202. Fort Lauderdale: IEEE.

Mao, X., S. Tang, X. Xu, X.-Y. Li, and H. Ma. 2011. Energy-Efficient Opportunistic Routing in Wireless Sensor Networks. *Transactions on Parallel and Distributed Systems (TPDS)* 22 (11): 1934–1942.

Meng, T., F. Wu, Z. Yang, G. Chen, and A.V. Vasilakos. 2016. Spatial Reusability-Aware Routing in Multi-Hop Wireless Networks. *Transactions on Computers* 65 (1): 244–255.

Merriam-Webster. 2022. *'Intra-' and 'Inter-': Getting Into It.* Merriam-Webster, Incorporated. January 1, 2022. https://www.merriam-webster.com/words-at-play/intra-and-inter-usage. Accessed 27 May 2022.

Moad, S., M.T. Hansen, R. Jurdak, B. Kusy, and N. Bouabdallah. 2011. Load Balancing Metric with Diversity for Energy Efficient Routing in Wireless Sensor Networks. *Procedia Computer Science* 5: 804–811.

Muruganathan, S.D., D.C.F. Ma, R.I. Bhasin, and A.O. Fapojuwo. 2005. A Centralized Energy-Efficient Routing Protocol for Wireless Sensor Networks. *Communications Magazine* 43 (4): S8–S13.

Newsome, J., and D. Song. 2003. GEM: Graph EMbedding for Routing and Data-entric Storage in Sensor Networks without Geographic Information. In *The 1st International Conference on Embedded Networked Sensor Systems (SenSys)*, 76–88. California: ACM.

Nosratinia, A., T.E. Hunter, and A. Hedayat. 2004. Cooperative Communication in Wireless Networks. *Communications Magazine* 42 (10): 74–80.

Ogier, R., F. Templin, and M. Lewis. 2004. *Topology Dissemination Based on Reverse-Path Forwarding (TBRPF)*, Request for Comments: 3684, The RFC Series. Menlo Park: The Internet Society.

Ogundile, O.O., and A.S. Alfa. 2017. A Survey on an Energy-Efficient and Energy-Balanced Routing Protocol for Wireless Sensor Networks. *Sensors* 17 (5): 1084.

Pal, V., G. Singh, and R.P. Yadav. 2015. Balanced Cluster Size Solution to Extend Lifetime of Wireless Sensor Networks. *Internet of Things Journal* 2 (5): 399–401.

Pantazis, N.A., S.A. Nikolidakis, and D.D. Vergados. 2013. Energy-Efficient Routing Protocols in Wireless Sensor Networks: A Survey. *Communications Surveys & Tutorials* 15 (2): 551–591.

Park, V.D., and M.S. Corson. 1997. A Highly Adaptive Distributed Routing Algorithm for Mobile Wireless Networks. In *The 16th Annual Joint Conference of The IEEE Computer and Communications Societies (INFOCOMM)*, 1405–1413. Kobe: IEEE.

Puccinelli, D., and M. Haenggi. 2008. Arbutus: Network-Layer Load Balancing for Wireless Sensor Networks. In *Wireless Communications and Networking Conference (WCNC)*, 2063–2068. Las Vegas: IEEE.

Pucha, H., S.M. Das, and Y.-C. Hu. 2007. The Performance Impact of Traffic Patterns on Routing Protocols in Mobile Ad Hoc Networks. *Computer Networks* 51 (12): 3595–3616.

Radi, M., B. Dezfouli, K. Abu Bakar, and M. Lee. 2012. Multipath Routing in Wireless Sensor Networks: Survey and Research Challenges. *Sensors* 12 (1): 650–685.

Rajagopalan, R., and P.K. Varshney. 2006. Data Aggregation Techniques in Sensor Networks: A survey. *IEEE Communications Surveys & Tutorials* 8 (4): 48–63.

Raju, J., and J.J. Garcia-Luna-Aceves. 1999. A New Approach to on-Demand Loop-Free Multipath Routing. In *The 8th International Conference on Computer Communications and Networks (ICCCN)*, 522–527. Boston: IEEE.

Ren, F., J. Zhang, T. He, C. Lin, and S.K. Das Ren. 2011. EBRP: Energy-Balanced Routing Protocol for Data Gathering in Wireless Sensor Networks. *Transactions on Parallel and Distributed Systems* 22 (12): 2108–2125.

Rostami, A.S., et al. 2018. Survey on Clustering in Heterogeneous and Homogeneous Wireless Sensor Networks. *The Journal of Supercomputing* 74: 277–323.

Sabbineni, H., and K. Chakrabarty. 2005. Location-Aided Flooding: An Energy-Efficient Data Dissemination Protocol for Wireless Sensor Networks. *Transactions on Computers* 54 (1): 36–46.

Sadagopan, N., B. Krishnamachari, and A. Helmy. 2005. Active Query Forwarding in Sensor Networks. *Ad Hoc Networks* 3 (1): 91–113.

Sakharkar, S.M., R.S. Mangrul, and M. Atique. 2014. A Survey: A Secure Routing Method for Detecting False Reports and Gray-Hole Attacks Along with Elliptic Curve Cryptography in Wireless Sensor Networks. In *Students' Conference on Electrical, Electronics and Computer Science (SCEECS)*, 1–5. Bhopal: IEEE.

Saleem, M., I. Ullah, and M. Farooq. 2012. BeeSensor: An Energy-Efficient and Scalable Routing Protocol for Wireless Sensor Networks. *Information Sciences* 200: 38–56.

Saleem, K., N. Fisal, and J. Al-Muhtadi. 2014. Empirical Studies of Bio-Inspired Self-Organized Secure Autonomous Routing Protocol. *Sensors* 14 (7): 2232–2239.

Savvides, A., and M.B. Srivastava. 2004. Location Discovery. In *Mobile Ad Hoc Networking*, ed. S. Basagni, M. Conti, S. Giordano, and I. Stojmenovic, 231–254. Piscataway: IEEE Press.

Shafiullah, G.M., A. Gyasi-Agyei, and P.J. Wolfs. 2008. A Survey of Energy-Efficient and QoS-Aware Routing Protocols for Wireless Sensor Networks. In *Novel Algorithms and Techniques in Telecommunications, Automation and Industrial Electronics*, ed. T. Sobh, K. Elleithy, A. Mahmood, and M.A. Karim, 352–357. Dordrecht: Springer.

Shah, R.C., and J.M. Rabaey. 2002. Energy Aware Routing for Low energy Ad Hoc Sensor Networks. In *Wireless Communications and Networking Conference (WCNC)*, 350–355. Orlando: IEEE.

Singh, S.K., P. Kumar, and J.P. Singh. 2018. An Energy Efficient Protocol to Mitigate Hot Spot Problem Using Unequal Clustering in WSN. *Wireless Personal Communications* 101: 799–827.

Sohrabi, K., J. Gao, V. Ailawadhi, and G.J. Pottie. 2000. Protocols for Self-Organization of a Wireless Sensor Network. *Personal Communications* 7 (5): 16–27.

Sudarmani, R., and K.R.S. Kumar. 2013. Particle Swarm Optimization-Based Routing Protocol for Clustered Heterogeneous Sensor Networks with Mobile Sink. *American Journal of Applied Sciences* 10 (3): 259–269.

Tanwar, S., N. Kumar, and J.-W. Niu. 2014. EEMHR: Energy-Efficient Multilevel Heterogeneous Routing Protocol for Wireless Sensor Networks. *International Journal of Communication Systems-Special Issue: Complex Communication Networks* 27 (9): 1289–1318.

Tarique, M., K.E. Tepe, S. Adibi, and S. Erfani. 2009. Survey of Multipath Routing Protocols for Mobile Ad Hoc Networks. *Journal of Network and Computer Applications* 32 (6): 1125–1143.

The Athletic. 2022. *2022 World Athletics Championships*. The Athletic Media Company. July 22, 2022. https://theathletic.com/news/noah-lyles-american-record-worlds/MwylxEuWAdA7/. Accessed 22 July 2022.

The Sporting News. 2022. *World Thletics Hampionships 2022 Results*. The Sporting News. July 25, 2022. https://www.sportingnews.com/uk/other-sports/news/world-athletics-championships-2022-results-winners/rrykhnwytjgze9fcyc8dmt11. Accessed 25 July 2022.

Tripathy, A.K., and S. Chinara. 2012. Comparison of Residual Energy-based Clustering Algorithms for Wireless Sensor Network. *ISRN Sensor Networks Volume 2012*: 1–10.

Tunca, C., S. Isik, M.Y. Donmez, and C. Ersoy. 2015. Ring Routing: An Energy-Efficient Routing Protocol for Wireless Sensor Networks with a Mobile Sink. *Transactions on Mobile Computing* 14 (9): 1947–1960.

Tyagi, S., S. Tanwar, S.K. Gupta, N. Kumar, and J.P.C. Rodrigues. 2015. A Lifetime Extended Multi-Levels Heterogeneous Routing Protocol for Wireless Sensor Networks. *Telecommunication Systems Volume* 59: 43–62.

Vijayan, K., and A. Raaza. 2016. A Novel Cluster Arrangement Energy Efficient Routing Protocol for Wireless Sensor Networks. *Indian Journal of Science and Technology* 9 (2): 1–9.

Vilela, M.A., and R.B. Araujo. 2012. RAHMoN: Routing Algorithm for Heterogeneous Mobile Networks. In *The 2nd Brazilian Conference on Critical Embedded Systems (CBSEC)*, 24–29. Sao Paulo: IEEE.

Wang, Y.-H., C.-H. Tsai, and H.-J. Mao. 2006. HMRP: Hierarchy-Based Multipath Routing Protocol for Wireless Sensor Networks. *Journal of Applied Science and Engineering* 9 (3): 255–264.

Wang, A., D. Yang, and D. Sun. 2012. A Clustering Algorithm based on Energy Information and Cluster Heads Expectation for Wireless Sensor Networks. *Computers & Electrical Engineering* 38 (3): 662–671.

Wang, J., Z. Zhang, F. Xia, W. Yuan, and S. Lee. 2013. An Energy Efficient Stable Election-based Routing Algorithm for Wireless Sensor Networks. *Sensors* 13 (11): 14301–14320.

Wang, J., B. Li, F. Xia, C.-S. Kim, and J.-U. Kim. 2014. An Energy Efficient Distance-Aware Routing Algorithm with Multiple Mobile Sinks for Wireless Sensor Networks. *Sensors* 14 (8): 15163–15181.

Wang, M., S. Wang, and B. Zhang. 2020. APTEEN Routing Protocol Optimization in Wireless Sensor Networks Based on Combination of Genetic Algorithms and Fruit Fly Optimization Algorithm. *Ad Hoc Networks* 102: 102–138.

Wei, D., and H.A. Chan. 2008. Equalizing Cluster Lifetime for Sensor Networks with Directional Data Traffic to Improve Energy Efficiency. In *The 5th Consumer Communications and Networking Conference (CCNC)*, 714–718. Las Vegas: IEEE.

Wei, D., Y. Jin, S. Vural, K. Moessner, and R. Tafazolli. 2011. An Energy-Efficient Clustering Solution for Wireless Sensor Networks. *Transactions on Wireless Communications* 10 (11): 3973–3983.

WIKIPEDIA. 2022a. *2022 French Open.* WIKIPEDIA. June 5, 2022. https://en.m.wikipedia.org/wiki/2022_French_Open. Accessed 1 July 2022.

———. 2022b. *2022 Wimbeldon Championships.* WIKIPEDIA. July 10, 2022. https://en.m.wikipedia.org/wiki/2022_Wimbledon_Championships. Accessed 12 July 2022.

Wu, X., G. Chen, and S.K. Das. 2006. On the Energy Hole Problem of Nonuniform Node Distribution in Wireless Sensor Networks. In *International Conference on mobile Ad Hoc and Sensor Systems (MASS)*, 180–187. Vancouver: IEEE.

Wu, C., R. Yuan, and H. Zhou. 2008. A Novel Load Balanced and Lifetime Maximization Routing Protocol in Wireless Sensor Networks. In *Vehicular Technology Conference (VTC)*, 113–117. Marina Bay: IEEE.

Xie, R., and X. Jia. 2014. Transmission-Efficient Clustering Method for Wireless Sensor Networks Using Compressive Sensing. *Transactions on Parallel and Distributed Systems* 25 (3): 806–815.

Yadav, S., and R.S. Yadav. 2016. A Review on Energy Efficient Protocols in Wireless Sensor Networks. *Wireless Networks* 22: 335–350.

Yan, J., M. Zhou, and Z. Ding. 2016. Recent Advances in Energy-Efficient Routing Protocols for Wireless Sensor Networks: A Review. *Access* 4: 5673–5686.

Yao, Y., and J. Gehrke. 2002. The Cougar Approach to In-Network Query Processing in Sensor Networks. *Newsletter* 31 (3): 9–18.

Yao, Y., Q. Cao, and A.V. Vasilakos. 2015. EDAL: An Energy-efficient, Delay-aware, and Lifetime-balancing Data Collection Protocol for Heterogeneous Wireless Sensor Networks. *Transactions on Networking* 23 (3): 810–823.

Yarvis, M., N. Kushalnagar, H. Singh, A. Rangarajan, Y. Liu, and S. Singh. 2005. Exploiting Heterogeneity in Sensor Networks. In *The 24th Annual Joint Conference of the IEEE Computer and Communications Societies (INFOCOM)*, 878–890. Miami: IEEE.

Ye, F., G. Zhong, S. Lu, and L. Zhang. 2005. GRAdient Broadcast: A Robust Data Delivery Protocol for Large Scale Sensor Networks. *Wireless Networks* 11: 285–298.

Yetgin, H., K.T.K. Cheung, M. El-Hajjar, and L.H. Hanzo. 2017. A Survey of Network Lifetime Maximization Techniques in Wireless Sensor Networks. *Communications Surveys & Tutorials* 19 (2): 828–854.

Yick, J., B. Mukherjee, and D. Ghosal. 2008. Wireless Sensor Networks Survey. *Computer Networks* 52 (12): 2292–2330.

Yilmaz, O., S. Demirci, Y. Kaymac, S. Ergun, and A. Yildirim. 2012. Shortest Hop Multipath Algorithm for Wireless Sensor Networks. *Computers & Mathematics with Applications* 63 (1): 48–59.

Younis, M., and K. Akkaya. 2008. Strategies and Techniques for Node Placement in Wireless Sensor Networks: A Survey. *Ad Hoc Networks* 6 (4): 621–655.

Younis, O., M. Krunz, and S. Ramasubramanian. 2006. Node Clustering in Wireless Sensor Networks: Recent Developments and Deployment Challenges. *Network* 20 (3): 20–25.

Yu, Y., R. Govindan, and D. Estrin. 2001. *Geographical and Energy Aware Routing: a Recursive Data Dissemination Protocol for Wireless Sensor Networks.* Technical Report, Computer Science, UCLA, California, LA: UCLA, 1–11.

Yu, F., Y. Li, F. Fang, and Q. Chen. 2007. A New TORA-based Energy Aware Routing Protocol in Mobile Ad Hoc Networks. In *The 3rd IEEE/IFIP International Conference in Central Asia on Internet (ICI)*, 1–4. Tashkent: IEEE.

Yu, J., Y. Qi, G. Wang, and X. Gu. 2012. A Cluster-based Routing Protocol for Wireless Sensor Networks with Nonuniform Node Distribution. *International Journal of Electronics and Communications (AEU)* 66 (1): 54–61.

Yuea, J., W. Zhang, W. Xiao, D. Tang, and G. Tang. 2012. Energy Efficient and Balanced Cluster-based Data Aggregation Algorithm for Wireless Sensor Networks. *Procedia Engineering* 29: 2009–2015.

Zaatouri, I., A.B. Guiloufi, N. Alyaoui, and A. Kachouri. 2017. A Comparative Study of the Energy Efficient Clustering Protocols in Heterogeneous and Homogeneous Wireless Sensor Networks. *Wireless Personal Communications* 97: 6453–6468.

Zhai, C., J. Liu, L. Zheng, H. Xu, and H. Chen. 2012. Maximise Lifetime of Wireless Sensor Networks via a Distributed Cooperative Routing Algorithm. *Transactions on Emerging Telecommunications Technologies* 23 (5): 414–428.

Zhang, C., M. Zhou, and M. Yu. 2007. Ad hoc network routing and security: A review. *International Journal of Communication Systems* 20 (8): 909–90s.

Zhang, J., C.-K. Jeong, G.-Y. Lee, and H.-J. Kim. 2008a. Cluster-Based Multi-Path Routing for Multi-Hop Wireless Networks. *Journal of the Institute of Electronics Engineers of Korea CI* 45 (6): 114–121.

Zhang, Z., M. Ma, and Y. Yang. 2008b. Energy-efficient Multihop Polling in Clusters of Two-layered Heterogeneous Sensor Networks. *Transactions on Computers* 57 (2): 231–245.

Zhang, D., G. Li, K. Zheng, X. Ming, and Z.-H. Pan. 2014. An Energy-Balanced Routing Method based on Forward-Aware Factor for Wireless Sensor Networks. *Transactions on Industrial Informatics* 10 (1): 766–773.

Zhou, H.-Y., D.-Y. Luo, Y. Gao, and D.-C. Zuo. 2011. Modeling of Node Energy Consumption for Wireless Sensor Networks. *Wireless Sensor Network* 3 (1): 18–23.

Zhou, M., G. Fortino, W. Shen, J. Mitsugi, J. Jobin, and R. Bhattacharyya. 2016. Guest Editorial Special Section on Advances and Applications of Internet of Things for Smart Automated Systems. *Transactions on Automation Science and Engineering* 13 (3): 1225–1229.

Zhu, T., and D. Towsley. 2011. E2R: Energy Efficient Routing for Multi-Hop Green Wireless Networks. In *Conference on Computer Communications Workshops (INFOCOM WKSHPS)*, 265–270. Shanghai: IEEE.

Zungeru, A.M., L.-M. Ang, and K.P. Seng. 2012. Termite-hill: Performance Optimized Swarm Intelligence Based Routing Algorithm for Wireless Sensor Networks. *Journal of Network and Computer Applications* 35 (6): 1901–1917.

Zytoune, O., M. El Aroussi, and D. Aboutajdine. 2010. A Uniform Balancing Energy Routing Protocol for Wireless Sensor Networks. *Wireless Personal Communications* 55: 147–161.

Chapter 5
Transport Protocols for WSNs

A good business is reliable. . . . A successful person is dependable.

5.1 Presumptions and Considerations of Transport Protocols in WSNS

Wireless sensor networks (WSNs) generally consist of one or more sinks (or basestations) and from tens to thousands of sensor nodes scattered in a physical space. With the integration of information sensing, computation, and wireless communication, the sensor nodes can sense physical information, process crude information, and report them to the sink. The sink in turn queries the sensor nodes for information. WSNs have several distinctive usually recalled features:

- Unique network topology. Sensor nodes are generally organized in a multihop star-tree topology that is either flat or hierarchical. The sink at the root of the tree is responsible for data collection and relaying to external networks. This topology can be dynamic due to the time-varying link conditions and dynamic node status.
- Diverse applications. WSNs may be used in different environments, supporting diverse applications, such as habitat monitoring, target tracking, security surveillance, industrial control, and home automation. These applications may focus on different sensory data and therefore impose different requirements in terms of quality of service (QoS) and reliability.
- Traffic characteristics. In WSNs, the primary traffic is in the *upstream* direction from the sensor nodes to the sink, although the sink may occasionally generate certain *downstream* traffic for the purposes of query and control. In the upstream, this is a many-to-one type of communication. Depending on specific applications, the delivery of upstream traffic may be event-driven, continuous delivery, query-driven delivery, or hybrid delivery.

- Resource constraints. Sensor nodes have limited resources, specifically, low computational capability, small memory, low wireless communication bandwidth, and a limited, usually non-rechargeable battery.
- Small message size. Messages in sensor networks usually have a small size compared with the existing networks. As a result, there is usually no concept of segmentation in most applications in WSNs.

These unique features pose distinct challenges in the design of WSNs that should meet application requirements and operate for the longest possible period of time. Typically, care should be accorded to issues such as energy conservation, reliability, and QoS.

Transport protocols are used to mitigate congestion and reduce packet loss, to provide fairness in bandwidth allocation, and to guarantee end-to-end reliability. However, the traditional transport protocols that are currently used for the Internet, i.e., UDP and TCP, cannot be directly implemented for WSNs. UDP has several pitfalls:

- It does not provide delivery reliability, which is often needed for many sensor applications.
- It does not offer flow and congestion control, which leads to packet loss and unnecessary energy consumption.

On the other hand, TCP has several other drawbacks:

- The overhead associated with TCP connection establishment might not be justified for data collection in most event-driven applications.
- Flow and congestion control mechanisms in TCP can discriminate against sensor nodes that are far away from the sink, resulting in inequitable bandwidth allocation and data collection.
- TCP has a degraded throughput in wireless systems, especially in situations with a high packet loss rate, because TCP assumes that packet loss is due to congestion and triggers rate reduction whenever packet loss is detected.
- In contrast to *hop-by-hop* control, *end-to-end* congestion control in TCP has a tardy response, meaning it requires a longer time to alleviate congestion and in turn leads to higher packet loss when congestion occurs.
- TCP relies on end-to-end retransmission to provide reliable data transport, which consumes more energy and bandwidth than hop-by-hop retransmission.
- TCP guarantees the successful transmission of packets, which is not always necessary for event-driven applications in sensor networks.

Sensory data, as the main concern in WSNs, may be categorized in many ways (Rahman et al. 2008):

- Based on the direction, they are named *upstream* sensory data traffic (Wan et al. 2003; Stann and Heidemann 2003; Iyer et al. 2005; Wang et al. 2006a; Hull et al. 2004; Wan et al. 2005b) and *downstream* sensory data traffic (Wan et al. 2002; Park et al. 2004; Levis et al. 2004; Tezcan and Wang 2007). When the sensory data flows from the sensing nodes to the basestation, it is called upstream sensory data traffic, and the reverse scenario is referred to as downstream data flow. Some literature refers to upstream data flow as many-to-one, sensor-to-sink, or converge-cast and to downstream data flow as one-to-many, sink-to-sensor, or multicasting.

- Based on the traffic pattern experienced by any sensor node, the net traffic seen by any sensor node is received from two sources (Wang et al. 2006a; Wang et al. 2007). The first source is the sensed data captured by a sensor node and injected within the WSN. The second source are the neighbors of a sensor node whose data are routed upstream or downstream. Downstream traffic is sometimes referred to as route-thru, en route or transit traffic.
- Based on traffic source density, *dense* sources produce a high traffic rate, *sparse* sources generate low rates, and sparse sources deliver high rates (Wan et al. 2003).
- Applications types and network topology also shape the nature of traffic flowing within the network. A traffic pattern may be *bursty*, *continuous*, time interval based, or query based (Hull et al. 2004; Iyer et al. 2005; Akan and Akyildiz 2005). Event-based applications generally produce bursty traffic. Some WSN applications need continuous delivery of captured sensory data. Other applications require timely dissemination of data, while there are applications requesting reactive responsive data from the sensor network based on the query sent.

Section 5.2 presents all aspects that are stringent to transport layer protocols tailored for WSNs.

5.2 Obsessions of Transport Protocols for WSNs

Obsession is a persistent preoccupation, idea, or feeling. It is a sign of life, and protocols are alive and caring for satisfactory network performance. The transport protocol runs over the network layer. It enables end-to-end message transmission, where messages may be fragmented into several segments at the transmitter and reassembled at the receiver. Transport protocol stocks several functions: orderly transmission, flow and congestion control, loss recovery, and possibly QoS guarantees such as timing and fairness. In WSNs, several new factors, such as the convergent nature of upstream traffic and limited wireless bandwidth, can result in congestion. Congestion impacts normal data exchange and may lead to packet loss. In addition, wireless channels introduce packet loss due to a higher bit error rate, which not only affects reliability but also wastes energy. As a result, congestion and packet loss (Park et al. 2004) are two major problems that WSN transport protocols need to cope with in a performance metrics frame as enlightened in what is coming.

5.2.1 Transport Protocols Performance Metrics

Transport protocols for WSNs should provide end-to-end reliability and end-to-end QoS in an energy-efficient manner. Performance of transport protocols for WSNs can be evaluated using metrics such as energy efficiency, reliability, QoS (e.g., packet loss ratio, packet delivery latency), and fairness.

5.2.1.1 Energy Efficiency

Sensor nodes have limited energy. As a result, it is important for the transport protocols to maintain high energy efficiency in order to maximize system lifetime. Packet loss in WSNs can be common due to bit error and/or congestion. For loss-sensitive applications, packet loss leads to retransmission and the inevitable consumption of additional battery power. Therefore, several factors need to be carefully considered, including the number of packet retransmissions, the distance (e.g., hop) for each retransmission, and the overhead associated with control messages.

5.2.1.2 Reliability

Reliability in WSNs can be classified into several categories:

- Packet reliability. Applications are loss-sensitive and require successful transmission of all packets or at a certain success ratio.
- Event reliability (Tezcan and Wang 2007). Applications require only successful event detection, but not the successful transmission of all packets:

$$R(v) = \frac{\sum_{k=1}^{K} \text{Prob (success of } v_k)}{K} \tag{5.1}$$

  where v is a message, K is the total number of events defined by the application, k is the event that needs to be delivered reliably, and v_k is the message containing the event k.
- Query reliability (Tezcan and Wang 2007). The end-to-end query transfer is referred to as all queries being received by essential nodes successfully. If there is a number of K' queries to be sent during a time interval, then query reliability in an update interval is defined as:

$$R(q) = \frac{\sum_{k=1}^{K'} \text{Prob (success of } q_k)}{K'} \tag{5.2}$$

- Destination-related reliability (Park et al. 2004). Messages might need to be delivered to sensor nodes according to one of four patterns: (i) delivery to the entire field, which is the default, (ii) delivery to sensors in a subregion of the field, which is representative of location-based delivery, (iii) delivery to sensors such that the entire sensing field is covered, which is representative of redundancy-aware delivery, and (iv) delivery to a probabilistic subset of sensors, which corresponds to applications that perform resolution scoping.

5.2.1.3 QoS Metrics

QoS metrics include bandwidth, latency or delay, and packet-loss ratio. Depending on the application, these metrics or their variants could be used for WSNs. For example, sensor nodes may be used to transmit continuous images for target tracking. These nodes generate high-speed streams and require higher bandwidth than most event-based applications. For a delay-sensitive application, WSNs may also require timely data delivery.

5.2.1.4 Fairness

Sensor nodes are scattered in a geographical area. Due to the many-to-one convergent nature of upstream traffic, it is difficult for sensor nodes that are far away from the sink to transmit data. Therefore, transport protocols need to allocate bandwidth fairly among all sensor nodes so that the sink can obtain a fair amount of data from all the sensor nodes.

5.2.2 Congestion Control

Congestion as a term is the overcrowding or overfilling that terminates with clogging, harmful in health, in traffic, everywhere. In WSNs, there are two main causes of congestion (Wang et al. 2006b):

- The first is due to packet arrival rate exceeding the packet service rate. This is more likely to occur at sensor nodes close to the sink, as they usually carry more combined upstream traffic.
- The second cause is link-level performance aspects such as contention, interference, and bit-error rate. This type of congestion occurs on the link.

Congestion in WSNs has a direct impact on energy efficiency and QoS. Typically:

- Congestion can cause buffer overflow that may lead to larger queuing delays and higher packet loss. Packet loss not only degrades reliability and application QoS but also can waste the limited node energy.
- Congestion can also degrade link utilization.

- Link level congestion results in transmission collisions if contention-based link protocols such as Carrier Sense Multiple Access (CSMA), are used to share radio resources. Transmission collision in turn increases packet service time and wastes energy.

  Therefore, congestion in WSNs must be efficiently controlled, either to avoid it or to appease it. Typically, there are three mechanisms that can deal with this problem:

- Congestion Detection. In TCP, congestion is observed or inferred at the end nodes based on a timeouts or redundant acknowledgments. In WSNs, proactive methods are preferred. A common mechanism would be to use queue length (Wan et al. 2003; Stann and Heidemann 2003) packet service time (Iyer et al. 2005), or the ratio of packet service time over packet interarrival time at the intermediate nodes (Wang et al. 2006a). For WSNs using CSMA-like medium access control (MAC) protocols, channel loading can be measured and used as an indication of congestion (Stann and Heidemann 2003).

- Congestion Notification. After detecting congestion, transport protocols need to propagate congestion information from the congested node to the upstream sensor nodes or the source nodes that contribute to congestion. The information can be transmitted, using a single binary bit (called congestion notification (CN) bit in (Stann and Heidemann 2003), or more information such as allowable data rate, as in (Iyer et al. 2005), or the congestion degree, as in (Wang et al. 2006a). Disseminating congestion information has a twofold categorization:

  - The explicit congestion notification that uses special control messages, such as suppression messages, to notify the involved sensor nodes of congestion.
  - The implicit congestion notification that piggybacks congestion information in normal data packets. By receiving or overhearing such packets, sensor nodes can access the piggybacked information. For instance, the sensor nodes that detect congestion will set a CN bit in the header of data packets to be forwarded. After receiving packets with the CN bit set, the sink learns the network status, congestion or no congestion.

- Rate Adjustment. Upon receiving a congestion indication, a sensor node can adjust its transmission rate. If a single CN bit is used, additive-increase/multiplicative-decrease (AIMD) schemes or their variants are usually applied (Stann and Heidemann 2003). On the other hand, if additional congestion information is available, accurate rate adjustment can be implemented (Iyer et al. 2005; Wang et al. 2006a).

5.2.3   Loss Recovery

Losing is in nobody's fair plan, in exams, games, economics, politics, and definitely networks are not different. In wireless environments, both congestion and bit error can cause packet loss, which deteriorates end-to-end reliability and QoS, and lowers energy efficiency. Other factors that result in packet loss include node failure, wrong or outdated routing information, and energy depletion. Also, interference as the main reason for packet loss is also to be outlined:

1. If the transmitting sensor is far from the receiving sensor, the signal will attenuate significantly by the time it reaches the receiver. The signal attenuation is tricky to model since the radio signal strength is not uniform at the same distance from a sensor in all directions.
2. If more than one sensor in the sensor network is transmitting simultaneously, interference will occur at the listening sensor, which is within range of the transmitting sensors. In general, sensors can be far away to be considered neighbors, but they are still close enough to interfere with reception. This type of interference is also complex to model due to the previous point.

Causes 1 and 2 of packet loss, due to interference, are dependent on the exact location and environment in which the sensors are deployed, as well as on the radio technology implemented. For instance, sensors placed less than 1 m (3 feet) apart on a wall may not be able to hear each other due to reflections off the wall, which makes it hard to have any control over such losses.

3. Self-interference, that is, a node transmission interferes with itself at the receiver. The key challenge in implementing full-duplex-based wireless systems is the self-interference caused by the coupling of the transceiver's own transmit signal to the receiver while attempting to receive a signal sent by another equipment in a WSN.
4. Loss occurs when a packet is successfully received by a sensor but has to be dropped due to queue overflow. This type of loss is due to congestion within the network. The correct implementation of congestion control will minimize it.

In order to overcome packet loss, one can increase the source sending rate or introduce retransmission-based loss recovery:

- Increasing the source sending rate, which is also used in event-to-sink reliable transport (ESRT), works well for guaranteeing event reliability for event-driven applications that require no packet reliability. However, this method is not energy efficient compared to loss recovery.
- The retransmission-based loss recovery method is more active and energy effi- cient, and it can be implemented at both the link and transport layers. Link layer loss recovery is hop-by-hop, while the transport layer recovery is usually done end-to-end. Loss recovery consists of loss detection and notification and retransmission recovery as will be presented in the following sections.

5.2.3.1 Loss Detection and Notification

Since packet loss can be far more common in WSNs than in wired networks, loss detection mechanisms have to be carefully designed. A common mechanism is to include a sequence number in each packet header; the continuity of sequence numbers can be used to detect packet loss. Loss detection and notification can be either end-to-end or hop-by-hop. In the end-to-end approach, such as in the TCP protocol, the end points (destination or source) are responsible for loss detection and notification. In the hop-by-hop method, intermediate nodes detect and notify packet loss.

For several reasons, the end-to-end approach is not that effective for WSNs:

- The control messages that are used for end-to-end loss detection would utilize a return path consisting of several hops, which is not energy efficient.
- Control messages travel through multiple hops and could be lost with a high probability due to either link error or congestion.
- End-to-end loss detection inevitably leads to end-to-end retransmissions for loss recovery. It is worth noting that end-to-end retransmission consumes more energy than hop-by-hop retransmission.

In hop-by-hop loss detection and notification, a pair of neighboring nodes are responsible for loss detection and can enable local retransmission that is more energy efficient, as compared to the end-to-end approach. Hop-by-hop loss detection can further be categorized as receiver-based or sender-based, depending on where packet loss is detected:

- In sender-based loss detection, the sender detects packet loss using either a timer-based or overhearing mechanism. In timer-based detection, a sender starts a timer each time it transmits a packet. If it does not receive an acknowledgment from the targeted receiver before the timer expires, it infers the packet has been lost. Taking advantage of the broadcast nature of wireless channels, the sender can listen to the targeted receiver (passively and in an indirect manner so as to detect packet loss) in order to determine if the packet has been successfully forwarded.
- In receiver-based loss detection, a receiver infers packet loss when it observes out-of-sequence packet arrivals. There are three ways to notify the sender: ACK (Acknowledgment), NACK (Negative ACK), and IACK (Implicit ACK). Both ACK and NACK rely on special control messages, while IACK piggybacks ACK in the packet header. In IACK, if a packet is heard on the link, this implies that it has been successfully received and thus simultaneously acknowledged. However, the application of IACK depends on whether the sensor nodes have the capability to overhear the physical channel. In the case where the transmission is corrupt or the channel is not bidirectional or the sensor nodes access the physical channel using Time Division Multiple Access (TDMA)-based protocols, IACK may not be feasible.

Loss detection and notification can also indicate the reason for packet loss, which can be further used to improve system performance. Specifically, if packet loss is caused by buffer overflow, source nodes need to reduce the sending rate. However, if channel error is the cause, then it is unnecessary to reduce the sending rate in order to maintain high link utilization and throughput.

5.2.3.2 Retransmission-Based Loss Recovery

Retransmission of lost or damaged packets can also be either end-to-end or hop-by-hop. In the end-to-end approach, the source performs retransmission. In hop-by-hop retransmission, an intermediate node that intercepts loss notification searches its local buffer. If it finds a copy of the lost packet, it retransmits the packet. Otherwise, it relays loss information upstream to other intermediate nodes. If the node with a cached packet is considered a cache point and the node where the lost packets are detected as a loss point, the hop number between them can be referred to as the retransmission distance. The retransmission distance is an indication of retransmission efficiency in terms of energy consumed in the process of retransmission. Hence, it is to be noticed that:

- In end-to-end retransmission (such as in TCP), the cache point is the source node. However, in hop-by-hop retransmission, the cache point could be the predecessor node of the loss point.
- The end-to-end retransmission has a longer retransmission distance, while the hop-by-hop approach is more energy efficient, and it requires intermediate nodes to cache packets.
- The end-to-end approach allows for application-dependent variable reliability levels, like that realized by ESRT. On the other hand, the hop-by-hop recovery approach is preferred if 100 percent packet reliability is required, although some applications in WSNs, such as event-driven applications, may not require 100 percent reliability from sensor nodes. It is to be noted, though, that hop-by-hop loss recovery cannot assure message delivery in the presence of node failure.

Since end-to-end and hop-by-hop retransmissions require the caching of transmitted packets at cache points for possible future retransmission requests, the following question would arise: How long should a cache point buffer? This is especially important if the cache point does not receive an acknowledgment. For end-to-end retransmission, the cache duration should be close to round-trip-time (RTT). In wireless systems that use NACK-based acknowledgments, NACK messages could be lost or corrupted on the reverse channel, and the destination would be required to send NACK more than once. In such a case, source nodes need to buffer a packet for a time duration longer than RTT. For hop-by-hop, the cache duration is only influenced by the total local packet service time and one-hop packet transmission time.

Hop-by-hop retransmission in WSNs entails further considerations:

- When to trigger retransmission? Retransmission can be triggered immediately upon the detection of a packet loss. This results in shorter delay, which is desirable for time-sensitive applications. However, if packet loss is caused by congestion, immediate retransmission could aggravate the congestion situation and cause more packet losses.
- Where to cache the transmitted packets? In the hop-by-hop approach, each packet could be cached at each and every intermediate node. Given the limited memory

in sensor nodes, packets may only need to be cached at selected nodes. The central issue is how to distribute cached packets among a set of nodes. Distributed TCP Cache (DTC) (Dunkels et al. 2004) balances the buffer constraints and retransmission efficiency by using probability-based selection for cache points. In order to optimize retransmission efficiency, another possible approach is to cache packets at the intermediate node that is closer to the potential congested node where packet loss is more likely to arise.

5.3   Transport Protocols for WSNs

5.3.1   COngestion Detection and Avoidance (CODA)

CODA maintains an upstream congestion control mechanism (Wan et al. 2003)· To do so, it introduces three schemes: congestion detection, open-loop hop-by-hop back-pressure, and closed-loop end-to-end multisource regulation. CODA senses congestion by taking a look at each sensor node's buffer occupancy and wireless channel load. If they exceed a predefined threshold value, a sensor node will notify its neighbor source node(s) to decrease the sending rate through an open-loop hop-by-hop back-pressure. Receiving a back-pressure signal, the neighbor nodes simply decrease the packet sending rate and also replay the back-pressure continuously.

CODA regulates the multisource rate by the closed-loop end-to-end approach, which works as follows. Before sending a packet, a sensor node probes the channel at a fixed interval, and if it finds the channel busy more than a predefined number, it enables a control bit, called the congestion bit, in the outgoing packet header to inform the basestation that it is experiencing congestion. When the basestation receives a packet with the congestion bit enabled, it sends back an ACK control message to the source node(s) informing them to decrease their sending rate. When the congestion is cleared, the sink actively sends an ACK control message to the source nodes to inform them to increase their data rate.

Despite its satisfactory performance, CODA shows humble congestion handling as the number of source nodes and data rate increase. It does not have a reliability mechanism either, and the latency time of closed-loop multisource regulation increases under heavy congestion.

5.3.2   Event-to-Sink Reliable Transport (ESRT)

ESRT aims at providing both upstream event reliability and congestion control while maintaining the minimum energy expenditure (Akan and Akyildiz 2005). ESRT can also reliably deliver multiple concurrent events to the basestation. ESRT guarantees only the end-to-end reliable delivery of individual events, not individual packets from each sensor node. The notion of reliability is defined with respect to the number of data packets originated by any event that are reliably received at the basestation. The basestation node runs the ESRT algorithm to decide whether the event is reliably detected at the basestation or not. To do this, the basestation tracks the event reporting frequency (f) of the successfully received packets, originated by a particular event within a time interval, and matches it with the required reliability metric. In ESRT, the WSN can stay in one of five states: "No congestion, low reliability (NC, LR)," "no congestion, high reliability (NC, HR)," "congestion, high reliability (C, HR)," "congestion, low reliability (C, LR)," and "optimal operating region (OOR)." According to each scenario, ESRT reacts:

- If the current calculated reliability at the basestation falls below the required reliability and there is no congestion (NC, LR), ESRT increases f abruptly.
- If there is no congestion and the reliability level is high (NC, HR), ESRT decreases f cautiously.
- When congestion is detected despite high reliability levels (C, HR), ESRT decreases f to get rid of congestion without compromising the reliability.
- In case congestion is detected and reliability falls (C, LR), ESRT exponentially decreases the value of f.
- In OOR, the required reliability is attained with minimum energy expenditure. ESRT tries to operate at the optimum point where any event is reliably reported to the basestation without causing congestion to the network.

ESRT assumes that the basestation has a high-power radio and can reach all the sensor nodes in a single broadcast message. The basestation broadcasts the newly calculated value of f to the whole sensor network. Upon receiving the event reporting frequency, each sensor node calculates its event reporting duration and checks the buffer level at the end of each reporting interval to guess any possible congestion.

However, ESRT has some performance glitches:

- ESRT assumes that the basestation is one hop away from all the sensor nodes, which might not be applicable to many WSN applications.
- ESRT floods the value of f to the whole network to override their event sensing rate, which is unfair because different portions of the network or different individual sensor nodes might face different traffic and therefore contribute different levels of congestion.

5.3.3   Reliable Multi-Segment Transport (RMST)

RMST guarantees upstream packet reliability using in-network processing (Stann and Heidemann 2003). It adopts a cross-layer synergy by working in cooperation with the underlying routing protocol at the network layer and the MAC protocol at the link layer in order to guarantee hop-by-hop reliability. RMST uses the term fragmentation/reassembly, which simply means the packets originating from a source node (called an RMST entity) are fragmented and then reassembled at the basestation. Fragmentation is necessary to adjust the size of the maximum transmission unit (MTU) permissible by the transit nodes. The notion of reliability adopted by RMST is the reliable delivery of fragments originating from any particular RMST entity to the basestation.

RMST introduces two modes of operations, cached and non-cached:

- In caching mode, the nodes between the source and the basestation cache the fragments, and any RMST node can initiate recovery for missing fragments along the path toward the source.
- In non-cached mode, only the source and the basestation maintain the cache, and the basestation monitors the integrity of an RMST entity in terms of the received fragments. RMST uses a selective NACK-based protocol to detect a fragment loss and sends NACK from the detecting RMST node to the source node. Each RMST entity receiving the NACK first looks at its cache to find out the missing segment. In the negative case, it forwards the NACK to the RMST entity down the hierarchy toward the source node. RMST is evaluated via simulation.

RMST has some hitches, though. RMST is only suitable for applications that send large-sized sensory data, such as JPEG images, which take advantage of fragmentation at the source and reassembly at the basestation (Stann and Heidemann 2003). Also, RMST might not be suitable for reliably delivering fragments from multiple RMST entities to the same basestation since it cannot ensure the orderly delivery of fragments to the basestation. More number of fragments will cause more contention for the channel, i.e., more in-network data flow will happen. Moreover, RMST does not provide any real-time reliability guarantee or congestion control.

5.3.4 Pump Slowly Fetch Quickly (PSFQ)

PSFQ is designed to provide downstream reliability where the control message from the basestation is sent to the downstream sensor nodes at a relatively slow pace. It allows any intermediate sensor node suffering packet loss to quickly recover any lost segment from immediate neighbors (Wan et al. 2002; Wan et al. 2005a). This protocol is suitable for the timely dissemination of code segments to a group of specific target sensor nodes for retasking their jobs. PSFQ employs a hop-by-hop error recovery mechanism in which intermediate nodes also cache fragments and share responsibility for loss detection and recovery.

PSFQ introduces three operations to maintain reliability:

- Pump operation. The basestation slowly broadcasts a packet containing control scripts to its neighbors every T unit of time interval.
- Fetch operation. It is triggered as soon as a sequence number gap is found by any downstream node. In this mode, a sensor node halts its regular data routing operation to its downstream nodes and issues a NACK message to its immediate upstream neighbors to recover missing fragments. PSFQ supports the term called

loss aggregation, in which case the fetch operation deals with more than one
packet loss.
- Report operation. Each node along the path toward the source node piggybacks
its status information in the report message and then propagates the aggregated
report.

However, PSFQ has several minuses:

- PSFQ cannot recover the loss of every single packet due to congestion because it
uses only NACK.
- Both pump and fetch operations are performed through broadcast, which might
be expensive in terms of energy consumption.
- The slow nature of pump operation in PSFQ results in large delay.
- PSFQ does not allow any out-of-order delivery of packets, which poses a greater
challenge to cache management by the intermediate nodes.
- It is only intended for retasking the sensor node applications and thus might not
be suitable for upstream data reliability.
- It does not provide a congestion control mechanism.

5.3.5   GARUDA

Garuda, in small letters, is a large mythical bird or birdlike creature that appears in
Hindu and Buddhist mythology. Garuda Indonesia is the flag carrier of Indonesia.
GARUDA of this section is a reliable downstream data delivery transport protocol
for WSNs (Park et al. 2004). It addresses the problem of reliable data transfer from
the sink to the sensors. Reliability is defined in four categories:

- Guaranteed delivery to the entire field
- Guaranteed delivery to a subregion of sensors
- Guaranteed delivery to a minimal set of sensors to cover the sensing region
- Guaranteed delivery to a probabilistic subset of sensors

GARUDA design is a loss-recovery core infrastructure with a two-stage NACK-
based recovery process. The core infrastructure is constructed using the first packet
delivery method. The first packet delivery method guarantees first packet delivery
using a Wait-for-First-Packet (WFP) pulse. A WFP pulse is a small finite series of
short-duration pulses sent periodically by the sink. Sensor nodes within the transmission range of the sink will receive this pulse and wait for the transmission of the
first packet. The first packet delivery determines the hop count from the sink to the
node. Nodes along the path can become candidates for the core. A core candidate
elects itself to be a core node if it has not heard from neighboring core nodes. In this
manner, all core nodes are elected in the network. An elected core node must then
connect itself to at least one upstream core node.

GARUDA uses an out-of-order forwarding strategy to overcome the problem of under-utilization in the event of packet losses. Out-of-order forwarding allows subsequent packets to be forwarded even when a packet is lost. GARUDA uses a two-stage loss recovery process:

- The first stage involves core nodes recovering the packet. When a core node receives an out-of-sequence packet, it sends a request to an upstream core node notifying that there are missing packets. The upstream core node receiving that message will respond with a unicast retransmission of the available requested packet.
- The second stage is the noncore recovery phase, which involves noncore nodes requesting retransmission from the core nodes. A noncore node listens to all retransmissions from its core node and waits for completion before sending its own retransmission request.

However, the approach followed by GARUDA might not be suitable for upstream data reliability. In the case of a very large WSN, the core construction and loss recovery might be overly lengthy. GARUDA only offers reliable transfer of the very first packet without guaranteeing the remaining packets of a particular message. Also, it does not provide a congestion control mechanism, and is only evaluated through simulation, not through testbed experimentation.

5.3.6   Tiny TCP/IP

Tiny TCP/IP tends to modify the TCP/IP protocol suite to make it viable for WSNs, it also provides reliability that is a blend of end-to-end and hop-by-hop reliability (Dunkels et al. 2003, Dunkels et al. 2004). The protocol assumes that each sensor node knows its spatial location a priori and falls into any of the predefined subnets. Each sensor node obtains the first two octets from the subnet and calculates the last two octets based on its spatial location within the subnet.

Tiny TCP/IP proposes four modifications to the existing TCP/IP protocols, spatial IP address assignment, shared context header compression, application overlay routing, and distributed TCP caching (DTC). Sensor nodes on the same IP subnet do not need to transmit a full IP header. Hence, the IP header can be compressed and shared among the sensor nodes of the same subnet. Local IP broadcasting of UDP datagrams is used to form an application layer overlay network on top of the physical sensor network. Finally, DTC provides packet reliability using a distributed approach.

Tiny TCP/IP proposes a novel idea of TCP packet caching within the in-network sensor nodes to minimize the burden of the end-to-end retransmission of fragments in case packet loss occurs. Figure 5.1 shows a simplified example of how DTC works. In this example, a TCP sender transmits three TCP segments. Segment 1 is cached by node 5 before being dropped in the network, and segment 2 is cached by node 7 before being dropped. When receiving segment 3, the TCP receiver sends an

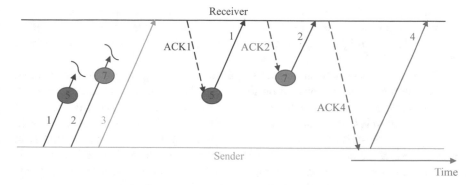

Fig. 5.1 Distributed caching in Tiny TCP/IP (Rahman et al. 2008)

acknowledgment (ACK 1). When receiving ACK 1, node 5, which had previously cached segment 1, performs a local retransmission. Node 5 also refrains from forwarding the acknowledgment so that it does not have to travel all the way to the TCP sender. When receiving the retransmitted segment 1, the TCP receiver acknowledges this segment by transmitting ACK 2. Upon reception of ACK 2, node 7, which previously had cached segment 2, performs a local retransmission. This way, the TCP receiver obtains the two dropped segments by local retransmissions from sensor nodes in the network, without requiring retransmissions from the TCP sender. When the acknowledgment ACK 4 is forwarded to the TCP sender, sensor nodes on the way can clear their cache and are thus ready to cache new TCP segments (Rahman et al. 2008). In the worst case, the receiver fetches the lost packet from the sender if this packet is not cached in any intermediate node. The protocol is evaluated through both simulation and an actual WSN.

Tiny TCP/IP likely experiences some performance issues:

- The assumption of static spatial subnet IP makes it unsuitable for many of the mobility-supported WSN applications.
- The protocol reliability performance depends on the efficiency of caching the last seen packets. Which node will cache which packets thus makes a complex design issue for this protocol.
- It does not explicitly define any congestion control mechanism.
- Finally, it does not explicitly address the design challenges of upstream or downstream reliability.

5.3.7 Sensor TCP (STCP)

STCP is a generic end-to-end upstream transport protocol for a wide variety of WSN applications. STCP provides both congestion detection and avoidance and a variable degree of reliability based on the application requirements (Iyer et al. 2005). STCP uses three types of packets: session initiation, data, and ACK. The session initiation packet is meant to synchronize any sensor node with the basestation. A sensor node is capable of originating multiple types of flow, such as event-driven, continuous, etc.

STCP data packets play an important role in maintaining congestion information. STCP functionalities are focused on the basestations. The basestation uses NACK for applications requiring continuous end-to-end sensory data flow; hence, clock synchronization is to be maintained between the basestation and the source nodes. In the case of event-driven sensory data flow applications, source nodes use ACK to make sure that the basestation has successfully received the packets. Each packet is kept in the source node cache until it gets an ACK from the basestation. Intermediate nodes detect congestion based on queue length and notify the basestation by setting a bit in the data packet headers. STCP was evaluated through simulation.

STCP assumes that all the sensor nodes within the WSN have strict clock synchronization with the basestation, which might cause a performance problem. Sensor nodes waiting for the ACK reply from the basestation suffer from long latency in large-scale multihop WSNs.

5.3.8   SenTCP

SenTCP is an open-loop hop-by-hop congestion control protocol intended for upstream traffic flow (Wang et al. 2005). SenTCP measures the degree of congestion in every intermediate sensor node by taking a look at the average local packet servicing time, local packet inter-arrival time, and buffer occupancy. To combat congestion, SenTCP makes each intermediate sensor issue a feedback signal to its neighbors, carrying the local congestion degree and the buffer occupancy ratio. To adjust the local data sending rate, SenTCP adopts a mechanism to process the received feedback signal. The use of hop-by-hop feedback control regulates congestion quickly and reduces packet dropping, which in turn preserves energy and increases throughput.

However, SenTCP only provides congestion control without loss recovery and does not guarantee reliability. The efficiency of SenTCP is tested via simulation, its suitability needs to be verified on a physical testbed.

5.3.9   Trickle

Trickle facilitates WSN reprogramming by allowing downstream nodes to intelligently infer any new code availability and subsequently pushing the actual code in a hop-by-hop approach (Levis et al. 2004). Trickle uses the concept of polite gossip to propagate metadata regarding any updated code that needs to be pushed downstream. Trickle focuses on metadata propagation rather than on actual code propagation inside the network. When a sensor node detects any older metadata from its neighbors, it updates its neighbors by broadcasting the appropriate code. Conversely, if any sensor node receives newer metadata from its neighbors, it broadcasts its own metadata, which makes the receiving sensor node broadcast its own new code.

Trickle is evaluated through simulation and is experimented upon in a testbed. The empirical results show that Trickle imposes an overhead of 3 packets/hour and can reprogram the entire network in 30 seconds. Although Trickle guarantees the delivery of metadata about the code, it does not guarantee reliable delivery of the code itself. Trickle does not provide a mechanism for knowing the current code version from any one or a set of sensor nodes, which makes the basestation unaware of the current status of the WSN.

5.3.10   Fusion

Fusion provides an upstream congestion control mechanism that *fuses* three techniques, hop-by-hop flow control, rate limiting of source traffic in the transit sensor nodes to provide fairness, and a prioritized MAC protocol (Hull et al. 2004):

- Using hop-by-hop flow control, a sensor node performs congestion detection and congestion mitigation. Congestion is detected through both queue occupancy and channel sampling techniques. A node signals local congestion to its neighbors by setting a congestion bit in the header of every outgoing packet, which thus dampens any sensor node from sending to a neighbor who is overrunning its queue.
- Rate limiting of source traffic in transit sensor nodes tries to maintain the fairness of allocating resources in en route sensor nodes so that a packet crossing a significant number of hops gets a proper treatment.
- Prioritized MAC protocol help a sensor node, under congestion, to drain its output queue by granting prioritized access to the physical channel. Thus, a sensor node experiencing congestion makes its back-off window one-fourth the size of a normal sensor back-off window, so that a sensor node experiencing congestion has a higher probability of winning the contention race.

The efficiency of Fusion was tested with a physical WSN testbed composed of 55 sensor nodes. The congestion handling capacity of Fusion was tested for both event-based and periodic data traffic. Frequent channel probing is a cause of energy depletion. Fusion lacks a packet recovery mechanism and, hence, does not provide for reliability measures.

5.3.11   Asymmetric and Reliable Transport (ART)

ART is an asymmetric and reliable transport mechanism that provides end-to-end reliability in two directions based on energy-aware node classification and a congestion control mechanism (Tezcan and Wang 2007). ART protocol operations include three main functions:

- Reliable query transfer
- Reliable event transfer
- Distributed congestion control

In ART, sensor nodes are classified as essential (E-nodes) and nonessential (N-nodes). End-to-end reliable communications are provided by using asymmetric acknowledgment (ACK) and negative acknowledgment (NACK) signaling between E-nodes and the sink node. A distributed energy-aware congestion control mechanism, which relies on receiving ACK packets from the sink, is devised. When congestion is detected, ART simply regulates data traffic by temporarily squelching the traffic of N-nodes. When there is no congestion, both E-nodes and N-nodes participate in relaying messages to the sink. However, only E-nodes are responsible for providing end-to-end event and query reliability by recovering the lost messages.

5.3.11.1   Reliable Query Transfer

Reliable query (sink-to-sensors) transfer is provided using negative acknowledgments sent from E-nodes to the sink if there is a query loss. Since the queries sent by the sink are in order, sensors can detect the lost message by use of sequence numbers in the query messages. A NACK message is sent if a gap is detected, i.e., an out-of-sequence number, when a sink sends a new query message to the E-nodes. When an E-node detects a gap in the sequence number of the new query, it sends a NACK back to the sink to recover the previous query. However, lost query messages can be detected when E-nodes receive a new query message. This may result in two problems:

- Loss of the last query message cannot be detected. Consider the last message q_k with sequence number k is lost. E-node may not handle the lost message since there is no consecutive query.
- The query transmission frequency might be very low such that lost queries cannot be recovered before timeout.

To differentiate the final query message, an extra Poll/Final (P/F) bit can be set by the sink node. P/F bit is set either when a message is the last query or the next query will not be sent before timeout. The sink retransmits this message until an ACK is received because ACK mechanism is used in reliable event transfer. Therefore, E-nodes, which receive a query with P/F bit set, send an ACK to the sink indicating a successfully received query. An example query transmission scenario is illustrated

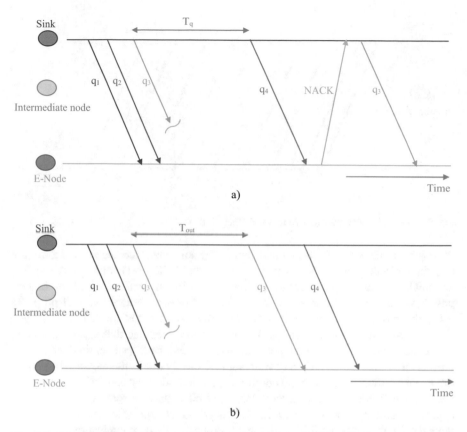

Fig. 5.2 Reliable query transfers. (**a**) Connectionless query reliability. (**b**) Connection-oriented query reliability (Tezcan and Wang 2007)

in Fig. 5.2. In Fig. 5.2a, the P/F bit is not used. The sink sends queries 1, 2, and 3 consecutively where q_3 is lost. After q_3, the sink decreases the query transmission frequency and sends q_4 after a time period T_q. In this case, q_3 is recovered when q_4 is received. If the loss recovery period T_q is very long, even though q_3 can be recovered, the long recovery period may affect the performance. Instead, the same scenario is depicted in Fig. 5.2b, when P/F bit is set, where q_3 is recovered before the next query since an ACK is not received at the sink. This method is helpful when the query traffic pattern is not uniformly distributed, in which case the interarrival times between queries are not constant; hence, the use of P/F bit makes the transport protocol flexible and reliable.

5.3.11.2 Reliable Event Transfer

The NACK mechanism used in query transfer does not work for reliable event transfer because event information is sent by individual sensors and it is usually out

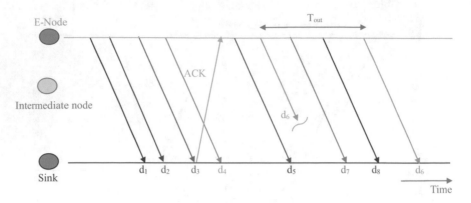

Fig. 5.3 Reliable event transfer (Tezcan and Wang 2007)

of sequence. Hence, NACKs cannot handle the lost event messages by finding the gap in sequence numbers. However, using an ACK mechanism that requires acknowledgment for each message may result in an inefficient use of battery power, which is a very scarce resource in WSNs. For event reliability, ART proposes a lightly loaded ACK mechanism between the E-nodes and the sink node. Each E-node waits for acknowledgment for only the first message that reports an event, i.e., event alarm. When a new sensing value is obtained, an E-node decides if it reports an event or not. If it is an event alarm, it simply marks the message by setting the event notification (EN) bit. Therefore, the sink node sends an ACK for only the messages that are marked as event alarm. EN bit is used to force the sink to send acknowledgment. The event-alarm rate depends on the distribution of events detected in the sensing field. Similar to downstream communications, only the E-nodes are responsible for waiting the acknowledgment and they may retransmit if necessary.

In Fig. 5.3, an event transfer scenario is illustrated where v_3 and v_6 are event-alarm messages and their EN bits are set. In this example, the first event alarm message is received at the sink, and the ACK is transmitted. However, the next alarm message v_6 is lost. Since the sender is responsible for loss detection and recovery, E-node retransmits v_6 after retransmission timeouts. Therefore, loss recovery is triggered only for event-alarm messages by the E-node, which is very effective in energy saving.

5.3.11.3 Distributed Congestion Control

In ART, congestion control is handled by the E-nodes in a distributed manner. It is based on monitoring the ACK packets of event reports. If an ACK is not received during a timeout period by the E-node, the traffic of nonessential sensors is reduced by sending congestion alarm messages, which will temporarily make them stop sending their measurements. When an ACK is received, a congestion-safe message is announced to resume the normal operation of the network.

5.3.12 Congestion Control and Fairness for Many-to-One Routing in Sensor Networks (CCF)

CCF exactly adjusts traffic rate based on packer service time along with fair packet scheduling algorithms, while Fusion performs stop-and-start non-smooth rate adjustment to mitigate congestion (Ee and Bajcsy 2004). CCF was proposed as a distributed and scalable algorithm that eliminates congestion within a sensor network and ensures the fair delivery of packets to a sink node. CCF exists in the transport layer and is designed to work with any MAC protocol in the data-link layer. In the CCF algorithm, each node measures the average rate r at which it can send packets, divides the rate r among the number of children nodes, adjusts the rate if queues are overflowing or about to overflow, and propagates the rate downstream. Figure 5.4 displays a typical congestion control scenario.

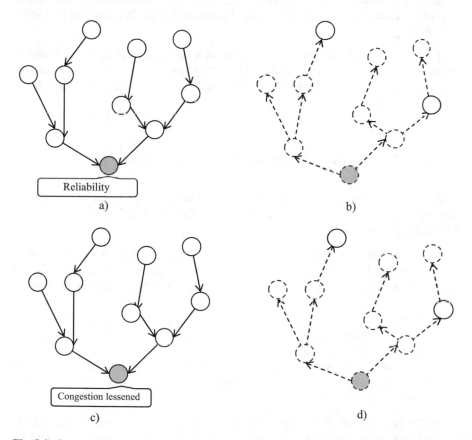

Fig. 5.4 Congestion control scenario. (**a**) The gray-colored node suffers congestion. (**b**) It informs downstream nodes to reduce transmission rates. (**c**) Congestion is reduced. (**d**) It informs downstream nodes to increase transmission rates (Ee and Bajcsy 2004)

CCF uses packet service time to deduce the available service rate; congestion information is thus implicitly reported. Congestion is controlled in a hop-by-hop manner, and each node uses an exact rate adjustment based on its available service rate and child node number. CCF guarantees simple fairness, but it has some shortcomings:

- The rate adjustment in CCF relies only on packet service time, which could lead to low utilization when some sensor nodes do not have enough traffic or if there is a significant packet error rate.
- Furthermore, it cannot effectively allocate the remaining capacity, and as it uses a work conservation scheduling algorithm, it has a low throughput in the case that some nodes do not have packets to send.
- Allocating equal resources to each sensor node to provide an equal opportunity might be inefficient in many scenarios. For instance, some sensor nodes might be capturing events more often than others, a video sensor capturing 10 frames per second needs a higher bandwidth and channel access than that required by a static sensor.

CCF is evaluated through simulation and in a real WSN environment; it implements the congestion control algorithm in the transport layer and is independent of the underlying network and MAC layers. CCF does not provide a reliability mechanism.

5.3.13 Priority-Based Congestion Control Protocol (PCCP)

PCCP is designed to avoid/reduce packet loss while guaranteeing weighted fairness and supporting multipath routing with lower control overhead (Wang et al. 2007). PCCP design motives are:

- In WSNs, sensor nodes might have different priorities based on their function or location. Therefore congestion control protocols need to guarantee weighted fairness so that the sink can get different, but in a weighted fair way, throughput from sensor nodes.
- With the fact that multipath routing is used to improve the system performance of WSNs, congestion control protocols need to be able to support both single-path routing and multipath routing.
- Congestion control protocols need to support QoS in terms of packet delivery latency, throughput, and packet loss ratio, which is required by multimedia applications in WMSNs.

PCCP consists of three components: intelligent congestion detection (ICD), implicit congestion notification (ICN), and priority-based rate adjustment (PRA), as detailed below:

- ICD detects congestion based on packet interarrival time t_a^i and packet service time t_s^i. The joint participation of interarrival and service times reflects the current congestion level and therefore provides helpful and rich congestion information. A congestion degree $d(i)$, over a specified time interval in each sensor node i as calculated:

$$d(i) = \frac{t_s^i}{t_a^i} \tag{5.3}$$

The congestion degree is intended to reflect the current congestion level at each sensor node. When the interarrival time is less than the service time, $d(i)$ is larger than 1, meaning a node experiences congestion, when $d(i)$ is less than 1, congestion abates. Based on $d(i)$, the child nodes adjust their transmission rate.

- In ICN, congestion information is piggybacked in the header of data packets. PCCP uses implicit congestion notification to avoid transmission of additional control messages and therefore helps improve energy efficiency. Taking advantage of the broadcast nature of wireless channels, child nodes can capture such information when packets are forwarded by their parent nodes toward the sink, assuming that there is no power control and the omnidirectional antenna is used.

- PCCP designs a priority-based algorithm, PRA, executed in each sensor node for rate adjustment, in order to guarantee both flexible fairness and throughput. In PRA, each sensor node is given a priority index, which is designed to guarantee that:

    - The node with higher priority index gets more bandwidth, proportional to the priority index.
    - The nodes with the same priority index get equal bandwidth.
    - A node with sufficient traffic gets more bandwidth than a node that generates less traffic. The use of priority index provides PCCP with high flexibility in realizing weighted proportional fairness.

The performance of congestion control protocols mostly depends on whether congestion can be detected in time or even correctly predicted in advance, whether congestion degree can be accurately measured, whether the detected or predicted congestion can be reported quickly to the nodes generating heavy traffic, and whether these nodes can trigger correct rate adjustment. PCCP uses hop-by-hop implicit congestion notification, Eq. (5.3) captures the congestion degree at intermediate nodes much more precisely than the queue-length-based congestion detection in existing work (Wan et al. 2003; Hull et al. 2004; Wan et al. 2005b). However the speed with which PCCP detects congestion is dependent on how quickly packet interarrival and service times can be correctly measured.

PCCP has been evaluated through simulation. With many attractive features, PCCP has some limitations. PCCP does not provide a packet loss recovery mechanism, and it lacks the notion of a reliability guarantee.

5.3.14   Siphon

Siphon is an upstream congestion control protocol that aims at maintaining application fidelity, congestion detection, and congestion avoidance by introducing some virtual sinks (VS) with a longer-range multi-radio within the sensor network (Wan et al. 2005b). Generally, congestion control schemes are effective at mitigating congestion through rate control and packet drop mechanisms, but at the cost of significantly reducing application fidelity measured at the sinks. To address this problem, Siphon exploits the availability of a small number of all-wireless, multi-radio virtual sinks that can be randomly distributed or selectively placed across the sensor field. Virtual sinks are capable of siphoning off (drawing off or taking out) data from regions of the sensor field that are beginning to show signs of high traffic load.

Siphon comprises a set of fully distributed algorithms that support virtual sink discovery and selection, congestion detection, and traffic redirection in sensor networks. Siphon is based on a Stargate implementation of virtual sinks that uses a separate longer-range radio network (based on IEEE 802.11) to siphon events to one or more physical sinks, and a short-range mote radio to interact with the sensor field at siphon points.

VSs can be dynamically distributed so that they can tunnel traffic events from regions of the sensor field that are about to show signs of a high traffic load. At the point of congestion, these VSs divert the extra traffic through them to maintain the required throughput at the basestation. The siphon algorithm mainly aims at addressing VS discovery, operating scope control, congestion detection, traffic redirection, and congestion avoidance.

VS discovery works as follows:

- The physical sink periodically sends out a control packet with an embedded signature byte. The signature byte contains the hop count of the sensor nodes that should use any particular VS.
- Each ordinary sensor node maintains a list of neighbors through which it can reach its parent VS.
- Finally each VS maintains a list of its neighbor VSs. Each VS has a dual radio interface, a long-range radio interface to communicate with other VSs or with a physical sink (if applicable), and a regular low-power radio to communicate with the regular sensor nodes. In the case of congestion, a sensor node enables the redirection bit in its header and forwards the packet to its nearest VS. When the VS finds the redirection bit enabled, it routes the packets using its own long-range communication network toward the physical sink, bypassing the underlying sensor network routing protocols.

Siphon uses a combination of hop-by-hop and end-to-end congestion control depending on the location of congestion. If there is no congestion, it uses a hop-by-hop data delivery model. In case of congestion, it uses a hop-by-hop data delivery model between source nodes and the VS at the point of congestion and an end-to-end approach between the VS handling the congestion and the physical sink.

Several metrics have been devised to analyze the performance of Siphon on sensing applications:

$$\text{Energy tax} = \frac{\text{Total packets dropped in the network}}{\text{Total packets received at the physical sink}} \qquad (5.4)$$

Since packet transmission/reception consumes the main portion of the energy of a node, the average number of wasted packets per received packet directly indicates the energy-saving aspect of Siphon.

Energy tax savings

$$= \frac{((\text{Average energy tax without siphon}) - (\text{Average energy tax with syphon}))}{(\text{Average energy tax without siphon})}$$

$$(5.5)$$

This metric indicates the average energy tax improvement or degradation from using Siphon.

$$\text{Fidelity ratio} = \frac{(\text{Packets received at the physical sink with siphon})}{(\text{Packets received at the physical sink without siphon})} \qquad (5.6)$$

The ratio indicates the average fidelity improvement or degradation from using Siphon.

$$\text{Residual energy} = \frac{\text{Remaining energy}}{\text{Initial energy}} \qquad (5.7)$$

The ns-2 energy model for IEEE 802.11 networks is used to measure the remaining energy of each node at the end of a simulation. The residual energy distribution allows to examine the load balancing feature of Siphon and to estimate the effective network lifetime.

Siphon has been implemented on a real sensor network using TinyOS (TinyOS 2012) on Mica2 motes (Hill et al. 2000). Results from the analysis, ns-2 simulation (The Network Simulator – NS2 2013), and from an experimental testbed show that virtual sinks can scale mote networks by effectively managing growing traffic demands while minimizing the impact on application fidelity. The optimality of Siphon depends on the optimality of the number of VSs. Although Siphon addresses the congestion detection and avoidance mechanism, it does not provide for packet recovery. Also, Siphon does not address reliability.

5.3.15   Reliable Bursty Convergecast (RBC)

RBC addresses the challenges of bursty convergecast in multihop wireless sensor networks (Zhang et al. 2007), where a large burst of packets from different locations needs to be transported reliably and in real-time to a basestation (convergecast). RBC attempts to overcome issues related to hop-by-hop control mechanisms, specifically:

- They do not schedule packet retransmissions appropriately; as a result, retransmitted packets further increase the channel contention and cause more packet loss.
- Due to in-order packet delivery and conservative retransmission timers, packet delivery can be significantly delayed in existing hop-by-hop mechanisms, which leads to packet backlogging and reduction in network throughput.

On the other hand, the new network and application models of bursty convergecast in WSNs offer unique opportunities for reliable and real-time transport control:

- First, the broadcast nature of wireless channels enables a node to determine, by snooping the channel, whether its packets are received and forwarded by its neighbors.
- Second, time synchronization and the fact that data packets are time-stamped relieve the transport layer from the constraint of in-order packet delivery since applications can determine the order of packets by their timestamps.

Addressing the highlighted challenges and taking advantage of the unique WSN models, RBC features some mechanisms:

- For improved channel utilization, RBC uses a window-less block acknowledgment scheme that enables continuous packet forwarding in the presence of packet and ACK loss. The block acknowledgment also reduces the probability of ACK loss, by replicating the acknowledgment for a received packet. Given that the number of packets competing for channel access is less in implicit-ACK-based schemes than in explicit-ACK-based schemes, RBC is based on the paradigm of implicit-ACK (i.e., piggybacking control information in data packets).
- For a better retransmission incurred channel contention, RBC introduces differentiated contention control, which ranks nodes by their queuing conditions as well as the number of times that the queued packets have been transmitted. A node ranked the highest within its neighborhood accesses the channel first.

In addition, RBC embodies techniques that address the challenges of timer-based retransmission control in bursty convergecast:

- To deal with continuously changing ACK-delay, RBC uses an adaptive retransmission timer which adjusts itself as the network state changes.
- To reduce delay in timer-based retransmission and to expedite the retransmission of lost packets, RBC uses block-NACK, retransmission timer reset, and channel utilization protection.

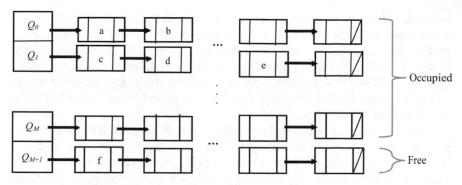

Fig. 5.5 RBC virtual queues at a node (Zhang et al. 2007)

RBC priority queue management for window-less block management merits further elaboration. An arbitrary pair of nodes S and R, where S is the sender and R is the receiver, is considered. The sender S organizes its packet queue as $(M + 2)$ linked lists, as shown in Fig. 5.5, where M is the maximum number of retransmissions at each hop. For convenience, the linked lists are called *virtual queues*, and are denoted as $Q_0, \ldots, Q_{M+1}$. The virtual queues are ranked such that a virtual queue Q_k ranks higher than Q_j if $k < j$.

Virtual queues $Q_0, Q_1, \ldots,$ and Q_M buffer packets waiting to be sent or to be acknowledged, and Q_{M+1} collects the list of free queue buffers. The virtual queues are maintained as follows:

- When a new packet arrives at S to be sent, S detaches the head buffer of Q_{M+1}, if any, stores the packet into the queue buffer, and attaches the queue buffer to the tail of Q_0.
- Packets stored in a virtual queue Q_k $(k > 0)$ will not be sent unless Q_{k-1} is empty; packets in the same virtual queue are sent in FIFO order.
- After a packet in a virtual queue Q_k $(k \geq 0)$ is sent, the corresponding queue buffer is moved to the tail of Q_{k+1}, unless the packet has been retransmitted M times in which case the queue buffer is moved to the tail of Q_{M+1}.
- When a received packet is acknowledged, the buffer holding the packet is released and moved to the tail of Q_{M+1}.

The above rules help identify the relative freshness of packets at a node, which is useful in differentiated contention control; they also help maintain without using sliding windows the order in which unacknowledged packets have been sent, thus providing the basis for window-less block acknowledgment. Moreover, newly arrived packets can be sent immediately without waiting for the previously sent packets to be acknowledged, which enables continuous packet forwarding in the presence of packet and ACK loss.

For block acknowledgment and reduced ACK loss, each queue buffer at S has an ID that is unique at S. When S sends a packet to the receiver R, S attaches the ID of the buffer holding the packet as well as the ID of the buffer holding the packet to be sent next. In Fig. 5.5, for example, when S sends the packet in buffer a, S attaches the values a and b. Given the queue maintenance procedure, if the buffer holding the packet being sent is the tail of Q_0 or the head of a virtual queue other than Q_0, S also attaches the ID of the head buffer of Q_{M+1}, if any, since one or more new packets may arrive before the next queued packet is sent in which case the newly arrived packet(s) will be sent first. For instance, when the packet in buffer c of Fig. 5.5 is sent, S attaches the values c, d, and f.

When the receiver R receives a packet p_0 from S, R learns the ID n' of the buffer holding the next packet to be sent by S. When R receives a packet p_n from S next time, R checks whether p_n is from buffer n' at S: if p_n is from buffer n', R knows that there is no packet loss between receiving p_0 and p_n from S; otherwise, R detects that some packets are lost between p_0 and p_n.

RBC is evaluated by experimenting with an outdoor testbed of 49 MICA2 motes (Drexel University 2013) and with realistic traffic traces from a previous project called Lites (Arora et al. 2004), where a typical event generates up to 100 packets within a few seconds and the packets need to be transported from different network locations to a basestation over multihop routes.

5.3.16 More TCP Protocols for WSNs

The technology wheel never stops. It goes on spinning in all dimensions. New protocols will continue to emerge and catching them should persist. This section will go on introducing more protocols, while Fig. 5.6 and Table 5.1 compare those protocols focused throughout the previous sections.

Segmented data reliable transport (SDRT) is proposed to achieve reliable data transfer in underwater sensor network scenarios (Xie et al. 2010). SDRT is essentially a hybrid approach of automatic repeat request (ARQ) and forward error correction (FEC). It adopts efficient erasure codes and random forward-error

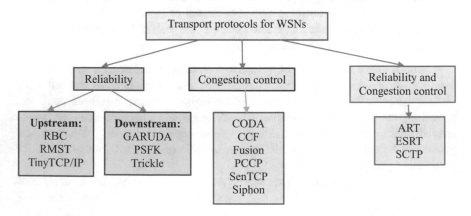

Fig. 5.6 Classification of TCP protocols for WSNs

Table 5.1 Comparison of TCP protocols for WSNs

Protocol	Upstream	Congestion control mechanism			
		Congestion detection	Congestion notification	Rate adjustment	End-to-end/Hop-by-hop
ART	Y	Service time	Implicit	Regulates traffic by decreasing active N-nodes	Y/ -
CCF	Y	Service time	Implicit	Rate adjustment by explicit feedback	- /Y
CODA	Y	Queue length and channel loading	Explicit	AIMD	Y/Y
ESRT	Y	Queue length	Implicit	Rate adjustment by controlling f at the basestation	Y/ -
Fusion	Y	Queue length	Implicit	Explicit feedback and draining congested nodes and MAC layer transmission priority for congested sensors	- /Y
PCCP	Y	Service/ interarrival times	Implicit	Rate adjustment depending on measured congestion degree and priority index	- /Y
SenTCP	Y	Queue length and service/ interarrival times	Explicit	Rate adjustment by explicit feedback	- /Y
Siphon	Y	Queue length and application fidelity		Traffic redirection	Y/Y
STCP	Y	Queue length	Implicit	AIMD	Y/ -
Protocol	Upstream/ downstream	Reliability mechanism			
		Type	Loss recovery	Loss detection and notification	End-to-end/Hop-by-hop
ART	Y/Y	Event/query	Packet retransmission	ACK and NACK and time out	Y/ -
ESRT	Y/ -	Event	Increase reporting frequency	Time out	Y/ -
GARUDA	- /Y	Code/packet	Packet retransmission	NACK and out of sequence	- /Y
PSFQ	- /Y	Packet			- /Y

(continued)

Table 5.1 (continued)

Protocol	Upstream	Congestion control mechanism			
		Congestion detection	Congestion notification	Rate adjustment	End-to-end/Hop-by-hop
			Packet retransmission	NACK and time out and out of sequence	
RBC	Y/ -	Event/packet		IACK and NACK	- /Y
RMST	Y/ -	Packet	Packet retransmission	NACK and time out	- /Y
STCP	Y/ -	Packet/event	Packet retransmission	ACK and NACK and time out	Y/ -
TinyTCP/ IP	Y/ -	Packet	Packet retransmission	ACK and out of sequence	Y/Y
Trickle	- /Y	Metadata	Packet retransmission	Out of sequence	- /Y

correction codes, transferring encoded packets block by block and hop by hop. Compared with traditional reliable data transport protocols, SDRT can reduce the total number of transmitted packets, improve channel utilization, and simplify protocol management.

Tunable reliability with congestion control for information transport in WSNs (TRCCIT) provides for a tunable reliability with congestion control for information transport in WSNs (Shaikh et al. 2010). TRCCIT provides desired application reliability, despite evolving network conditions, by adaptive retransmissions and suppression of unnecessary information. Reliability of information transport is achieved by a hybrid acknowledgment (HACK) mechanism aided by localized retransmission timer management. TRCCIT efficiently monitors the information flow and adapts between single path and multiple paths in order to alleviate congestion such that desired application reliability is maintained.

Energy-efficient and Reliable Transport Protocol for Wireless Sensor Networks (ERTP) is designed for data streaming applications, in which sensor readings are transmitted from one or more sensor sources to a basestation (or sink) (Le et al. 2009). ERTP uses a statistical reliability metric, which ensures that the number of data packets delivered to the sink exceeds a defined threshold. Extensive discrete event simulations and experimental evaluations show that ERTP is significantly energy efficient and can reduce energy consumption. Consequently, sensor nodes are more energy efficient, and the lifespan of the unattended WSN is increased.

Distributed transport for sensor networks (DTSN) is a reliable transport protocol for convergecast and unicast communications in WSNs (Marchi et al. 2007). In DTSN, the source completely controls the loss recovery process in order to minimize the overhead associated with control and data packets. The basic loss recovery algorithm is based on automatic repeat request (ARQ), employing both positive ACK and NACK delivery confirmation. Consequently, DTSN is able to detect when all packets of a session are lost, besides scattered gaps in the data packet sequence. Caching at intermediate nodes is used to avoid the inefficiency of the strictly end-to-end transport reliability TCP-like model, commonly employed in broadband networks. Reliability differentiation is achieved by means of the smart integration of partial buffering at the source, integrated with erasure coding and caching at intermediate nodes. The simulation results attest to the effectiveness of both the total reliability and the reliability differentiation mechanisms in DTSN.

5.4 Conclusion for Enrichment

As shown throughout this chapter, the transport layer's main task is to ensure the reliability and quality of data at the source and the sink. Transport layer protocols in WSNs should support multiple applications, variable reliability, packet-loss recovery, and congestion control. A transport layer protocol should be generic and independent of the application. Transport protocols are quite abundant, with varying design goals to match their intended use.

Depending on their functions, WSN applications can tolerate different levels of packet loss. Packet loss may be due to bad radio communication, congestion, packet collision, full memory capacity, and node failures. Packet loss can result in wasted energy and a degraded quality of service (QoS) in data delivery. Detection of packet loss and correctly recovering missing packets can improve throughput and energy expenditure. There are two approaches to packet recovery: hop-by-hop and end-to-end. Hop-by-hop retransmission requires that an intermediate node cache the packet information in its memory. This method is more energy efficient since the retransmission distance is shorter. For end-to-end retransmission, the source caches all the packet information and performs retransmission when there is a packet loss. End-to-end retransmission allows for variable reliability, whereas hop-by-hop retransmission performs better when reliability requirements are high.

A congestion control mechanism monitors and detects congestion, thereby conserving energy. Before congestion occurs, the source is notified to reduce its sending rate. Congestion control helps reduce retransmission and prevents sensor buffer overrun. As in packet-loss recovery, there are two approaches to congestion control: hop-by-hop and end-to-end. The hop-by-hop mechanism requires every node along the path to monitor buffer overflows and lessens congestion at a faster rate than the end-to-end mechanism. When a sensor node detects congestion, all nodes along the path change their behavior. The end-to-end mechanism relies on the end nodes to detect congestion. Congestion is flagged when timeouts or redundant acknowledgments are received. There are tradeoffs between hop-by-hop and end-to-end approaches for packet-loss recovery and congestion control mechanisms. Depending on the type, reliability, and time sensitivity of the application, one approach may be better than the other. As presented in detail all over this chapter, transport layer protocols in WSNs address, with different interests, the above design issues.

As time goes by, knowledge and experience accumulate, what is worthy persists; protocols are not different, as much they are useful, they are recalled and cited.

5.5   Exercises

1. State the distinctive features of WSNs.
2. Differentiate between traditional transport control and those for WSNs.
3. What are the performance metrics of transport protocols for WSNs?
4. How data may be categorized in WSNs?
5. Discuss the obsessions of transport protocols in WSNs.
6. How could reliability be evaluated in WSNs?
7. Define congestion and explain how it affects energy consumption.
8. What are the mechanisms that may be used to control congestion in WSNs?
9. What are the causes of packet loss in WSNs? How to overcome packet loss?
10. Categorize the transport protocols for WSNs into end-to-end and hop-by-hop.
11. Compare the WSN transport protocols that account for reliability.
12. Compare the WSN protocols that control congestion.
13. Compare the WSN protocols that consider both reliability and congestion.
14. Which of the transport protocols considered in this chapter are experimentally tested?
15. Look up the newly introduced transport protocols for WSNs. Determine their features and functionality.

References

Akan, O., and I. Akyildiz. 2005. Event-to-Sink Reliable Transport in Wireless Sensor Networks. *Transactions on Networking* 13 (5): 1003–1016.

Arora, A., et al. 2004. A Line in the Sand: a Wireless Sensor Network for Target Detection, Classification, and Tracking. *Computer Networks* 46 (5): 605–634.

Drexel University. 2013. Beginner's Guide to Crossbow Motes. *Course & Web Tools*. October 12, 2013. Drexel University. http://www.pages.drexel.edu/~kws23/tutorials/motes/motes.html. Accessed 12 Oct 2013.

Dunkels, A., Voigt, T., and J. Alonso. 2003. Making TCP/IP Viable for Wireless Sensor Networks. Technical Report T2003:23, Swedish Institute of Computer Science. Stockholm: SICS.

Dunkels, A., J. Alonso, T. Voigt, H. Ritter, and J. Schiller. 2004. Connecting Wireless Sensornets with TCP/IP Networks. In *Second International Conference on Wired/Wireless Internet (WWIC)*, 143–152. Frankfurt: Springer.

Ee, C., and R. Bajcsy. 2004. Congestion Control and Fairness for Many-to-One Routing in Sensor Networks (SenSys). In *2nd International Conference on Embedded Networked Sensor Systems*, 148–161. Baltimore: ACM.

Hill, J., R. Szewczyk, A. Woo, S. Hollar, D. Culler, and K. Pister. 2000. System Architecture Directions for Networked Sensors. In *The Ninth International Conference on Architectural Support for Programming Languages and Operating Systems (ASPLOS IX)*, 93–104. Cambridge: ACM.

Hull, B., K. Jamieson, and H. Balakrishnan. 2004. Mitigating Congestion in Wireless Sensor Networks. In *2nd International Conference on Embedded Networked Sensor Systems (SenSys)*, 134–147. Baltimore: ACM.

Iyer, Y., S. Gandham, and S. Venkatesan. 2005. STCP: A Generic Transport Layer Protocol for Wireless Sensor Networks. In *14th International Conference on Computer Communications and Networks (ICCCN)*, 449–454. San Diego: IEEE.

Le, T., W. Hu, P. Corke, and S. Jha. 2009. ERTP: Energy-Efficient and Reliable Transport Protocol for Data Streaming in Wireless Sensor Networks. *Computer Communications* 32 (7–10): 1154–1171.

Levis, P., N. Patel, D. Culler, and S. Shenker. 2004. Trickle: A Self-Regulating Algorithm for Code Propagation and Maintenance in Wireless Sensor Networks. In *First Symposium on Networked Systems Design and Implemention (NSDI)*. San Francisco: USENIX.

Marchi, B., A. Grilo, and M. Nunes. 2007. DTSN: Distributed Transport for Sensor Networks. In *12th IEEE Symposium on Computers and Communications (ISCC)*, 165–172. Aveiro: IEEE.

Park, S.-J., R. Vedantham, R. Sivakumar, and I.F. Akyildiz. 2004. A Scalable Approach for Reliable Downstream Data Delivery in Wireless Sensor Networks. In *Fifth ACM International Symposium on Mobile Ad Hoc Networking and Computing (MobiHoc)*, 78–89. Tokyo: ACM.

Rahman, M., A. El Saddik, and W. Gueaieb. 2008. Wireless Sensor Network Transport Layer: State of the Art. *Lecture Notes in Electrical Engineering* 21: 221–245.

Shaikh, F., A. Helil, A. Ali, and N. Suri. 2010. TRCCIT: Tunable Reliability with Congestion Control for Information Transport in Wireless Sensor Networks. In *The 5th Annual ICST Wireless Internet Conference (WICON)*, 1–9. Singapore: ICST.

Stann, F., and J. Heidemann. 2003. RMST: Reliable Data Transport in Sensor Networks. In *First IEEE International Workshop on Sensor Network Protocols and Applications*, 102–112. Anchorage: IEEE.

Tezcan, N., and W. Wang. 2007. An Asymmetric and Reliable Transport for wireless Sensor Networks. *Journal of Wireless Sensor Neworks* 2 (3/4): 188–200.

The Network Simulator – NS2. 2013. http://www.isi.edu/nsnam/ns/. Accessed 9 Oct 2013.

TinyOS. 2012, August 20. http://www.tinyos.net. Accessed 9 Oct 2013.

Wan, C.-Y., A. Campbell, and L. Krishnamurthy. 2002. PSFQ: A Reliable Transport Protocol for Wireless Sensor Networks. In *First ACM International Workshop on Wireless Sensor Networks and Applications (WSNA)*, 1–11. ACM.

Wan, C.-Y., S. Eisenman, and A. Campbell. 2003. CODA: Congestion Detection and Avoidance in Sensor Networks. In *First International Conference on Embedded Networked Sensor Systems (SenSys'03)*, 266–279. San Diego: ACM.

Wan, C.-Y., A. Campbell, and L. Krishnamurthy. 2005a. Pump-Slowly, Fetch-Quickly (PSFQ): A Reliable Transport Protocol for Sensor Networks. *Selected Areas in Communications* 23 (4): 862–872.

Wan, C.-Y., S. Eisenman, A. Campbell, and J. Crowcroft. 2005b. Siphon: Overload Traffic Management using Multi-Radio Virtual Sinks in Sensor Networks. In *The 3rd International Conference on Embedded Networked Sensor Systems (SenSys)*, 116–129. San Diego: ACM.

Wang, C., K. Sohraby, and B. Li. 2005. SenTCP: A Hop-by-Hop Congestion Control Protocol for Wireless Sensor Networks. In *24th Annual Joint Conference of the IEEE Computer and Communication Societies (INFOCOM)*. Miami: IEEE.

Wang, C., K. Sohraby, V. Lawrence, B. Li, and Y. Hu. 2006a. Priority-based Congestion Control in Wireless Sensor Networks. In *IEEE International Conference on Sensor Networks, Ubiquitous, and Trustworthy Computing*. Taichung: IEEE.

Wang, C., K. Sohraby, B. Li, M. Daneshmand, and Y. Hu. 2006b. A Survey of Transport Protocols for Wireless Sensor Networks. *Network* 20 (3): 34–40.

Wang, C., B. Li, K. Sohraby, M. Daneshmand, and Y. Hu. 2007. Upstream Congestion Control in Wireless Sensor Networks through Cross-Layer Optimization. *IEEE Journal on Selected Areas in Communications* 25 (4): 786–795.

Xie, P., Z. Zhou, Z. Peng, J. Cui, and Z. Shi. 2010. SDRT: A Reliable Data Transport Protocol for Underwater Sensor Networks. *Ad Hoc Networks* 8 (7): 708–722.

Zhang, H., A. Arora, Y. Choi, and M. Gouda. 2007. Reliable Bursty Convergecast in Wireless Sensor Networks. *Computer Communications* 30 (13): 2560–2576.

Chapter 6
Cross-Layer Protocols for WSNs

The truth is deceiving ... it is not what appears to be.

6.1 Why Cross-Layering in WSNs

WSNs achieved a collaborative sensing notion to overcome resource constraints thru adopting the networked deployment of sensor nodes. Moreover, spatiotemporal correlation is a significant characteristic of sensor networks (Vuran and Akyildiz 2006):

- Dense deployment of sensor nodes makes the sensor observations highly correlated in the space domain with noticeable effect of internode proximity.
- Some of the WSN applications such as event tracking require sensor nodes to periodically sample and communicate the sensed event features, yielding temporal correlation between each consecutive observation of a sensor node.

Most of the proposed communication protocols exploiting the collaborative nature of WSNs and its correlation characteristics improve energy efficiency. However, they follow the traditional layered protocol architectures; specifically, the majority of these communication protocols are individually developed for different networking layers, i.e., transport, network, medium access control (MAC), and physical layers. While they may realize high performance in terms of the metrics related to each of these individual layers, they are not jointly optimized to maximize the overall network performance while minimizing the energy expenditure. Considering the scarce energy and processing resources of WSNs, joint optimization and design of networking layers, i.e., cross-layer design, stands as the most promising alternative to inefficient traditional layered protocol architectures.

The basic principle of cross-layer design is to make information available to all levels of the protocol stack. It allows the definition of protocols or mechanisms that do not meet the isolation layers of the OSI model (van der Schaar and Shankar 2005; Srivastava and Motani 2005). In fact, cross-layer integration and design techniques result in significant improvement in terms of energy conservation in WSNs (van

© The Author(s), under exclusive license to Springer Nature Switzerland AG 2023   345
H. M. A. Fahmy, *Concepts, Applications, Experimentation and Analysis of Wireless Sensor Networks*, Signals and Communication Technology,
https://doi.org/10.1007/978-3-031-20709-9_6

Hoesel et al. 2004; Yetgin et al. 2015). Several researches started by focusing on the cross-layer interaction and design to develop new communication protocols (Melodia et al. 2005). Yet these works either provide analytical results without communication protocol design or perform pairwise cross-layer design within limited scope, e.g., only MAC and network layers, which do not consider all of the networking layers involved in WSN communication, such as transport, network, MAC, and physical layers.

Considering the scarce energy and processing resources of WSNs, joint optimization and design of networking layers, i.e., cross-layer design, stands as the most promising alternative to inefficient traditional layered protocol architectures. There are considerable benefits of rethinking the protocol functions of networking layers in a unified way so as to provide a single communication module for efficient communication in WSNs.

Accordingly, an increasing number of recent papers have focused on the cross-layer development of WSN protocols. Researches on WSNs, as detailed in this chapter, reveal that cross-layer integration and design techniques result in significant improvement in terms of energy conservation. Three main reasons stand behind this improvement (Melodia et al. 2005):

- The stringent energy, storage, and processing capabilities of wireless sensor nodes necessitate such an approach. The significant overhead of layered protocols results in high inefficiency.
- Recent empirical studies necessitate that the properties of low-power radio transceivers and the wireless channel conditions be considered in protocol design.
- The event-centric approach of WSNs requires application-aware communication protocols.

On the other hand, a cross-layer solution generally decreases the level of modularity, which may waste the decoupling between design and development process, making it more difficult to add design improvements and innovations. Moreover, it increases the risk of instability caused by functional dependencies, which are not easily foreseen in a non-layered architecture.

In the literature, three main approaches are followed for cross-layering in WSNs (Melodia et al. 2005; Pompili and Akyildiz 2010):

- Layer interactions. The cross-layer interaction is considered, where the traditional layered structure is preserved, while each layer is informed about the conditions of other layers. However, the mechanisms of each layer stay intact. Studies are classified in terms of interactions or modularity among physical (PHY), medium access control (MAC), network, and transport layers. Resource allocation problems are treated thru considering simple interactions between two communication layers. Section 6.2.1 comprehensively presents this approach. Layer interaction does not consider, though, the tight coupling among functionalities handled at all layers of the protocol stack distinctive, for instance, of mutihop underwater networks (Pompili and Akyildiz 2010).

- Single-layer integrated module. These approaches integrate different communication functionalities into a coherent mathematical framework and provide a unified foundation for cross-layer design and control in mutihop wireless networks. Solutions in this category seek optimality based on an application-dependent objective function and provide guidelines and tools to develop mathematically sound distributed solutions. Section 6.2.2 takes care of detailing this approach.

The heuristic approach is the third to be proposed with little literature adopting it (Pompili and Akyildiz 2010). Resource allocation problems following this approach consider interactions between several communication functionalities at different layers, since it is not always possible to model and control the interactions between functionalities. Solutions in this category rely on heuristics, which often leads to suboptimal performance. Cross-layer interactions can have a negative effect on desirable properties of the software architecture such as modularity. If cross-layer interactions are not performed in a controlled fashion, it might not be possible to exchange a module without (major) changes to others. In addition, when modules interact closely, they cannot be developed independently. Therefore, if cross-layer interactions are needed, as it is the case in WSNs, they have to be used thoughtfully (Lachenmann et al. 2005).

Sections 6.2.1 and 6.2.2, respectively, detail the layers interaction approach and the single-layer integrated module approach. Table 6.1 offers the taxonomy of cross-layering approaches offered in this chapter.

6.2 Cross-Layer Design Approaches

6.2.1 Layers Interactions

Targeting the minimization of energy consumption and the enhancement of the WSN lifetime, several works in WSN have revealed meaningful interactions between different layers of the protocol stack. This has led to several propositions for the cross-layer design. There are protocol designs based on interactions between physical and MAC layers (Venkitasubramaniam et al. 2003; Haapola et al. 2005), between physical and network layers (Zamalloa et al. 2008), between physical and transport layers (Chiang 2005), between MAC and application layers (Song and Hatzinakos 2007), between MAC and network layers (Chilamkurti et al. 2009; Hefeida et al. 2013; Petrioli et al. 2014), and between physical, MAC, and network layers (Madan et al. 2006; Bai et al. 2008). More implementations are highlighted all over Sects. 6.2.1.1 and 6.2.1.2.

6.2.1.1   Cross-Layering MAC and Network Layers

To minimize energy consumption in WSNs, several energy-efficient MAC protocols have been proposed to reduce the wasted energy due to the idle listening, by turning *off* the sensor node radio (Ye et al. 2002; van Dam and Langendoen 2003) or by scheduling the transmissions of control packets and data packets to avoid data packet collision (Xie and Cui 2007).

At the network layer, energy-efficient routing protocols were proposed. Initially, routing protocols have focused on consuming low power by finding the minimum energy path, or by finding the path with nodes having the maximum residual energy, or by combining both. The flaw in these protocols is using the optimal path for all node communications; consequently, the energy of nodes along such paths quickly drains out causing a network disconnection. As a way out, there has been a focus on maximizing the network lifetime by delaying as possible the occurrence of disconnection. This is implemented by balancing the traffic throughout several suboptimal paths, so that the nodes consume the energy more equitably. In order to ensure the energy consumption balancing, multipath routing protocols have been proposed (Bouabdallah et al. 2009; Semchedine et al. 2012).

However, the multipath approach is not that sufficient to balance the energy consumption. In fact, a node balances the use of the multiple paths by considering only its data packets without those of other neighboring nodes. Hence, the node has no knowledge about the real amount of the transmitted data by its forwarding nodes. This limitation is treated by devising routing protocols that exploit the interaction between the MAC and the network layers giving rise to a cross-layer approach, as presented in this section. Worth noticing, several works in WSNs have early revealed important interactions between the MAC and network layers, leading to several propositions for the cross-layer design (Melodia et al. 2005; Pompili et al. 2006).

6.2.1.1.1   Cross-Layer Network Formation for Energy-Efficient IEEE 802.15.4/ZigBee WSNs (PANEL)

In a WSN the position of the nodes can be predefined to realize optimal coverage and connectivity constraints. The PAN coordinator election (PANEL) is proposed for optimal node positioning to minimize the required number of nodes, yet guaranteeing perfect sensing coverage of the monitored field (Cuomo et al. 2013). However, in several contexts, the position of the nodes cannot be predefined. When WSNs are used for pervasive data collection in urban areas or to provide real-time monitoring in emergency scenarios, sensors may be randomly scattered and possibly redeployed over time in the considered area.

The development of self-managing, self-configuring, and self-regulating network and communication infrastructures is interesting, both in the research and industrial communities (Cardei and Du 2005; Dobson et al. 2006; Cipollone et al. 2007). There is a need to identify optimal topology formation and efficient PAN election for urban sensing or disaster recovery applications. The IEEE 802.15.4 personal area network (PAN) standard addresses the formation and management of low-energy and low-cost WSNs (LAN/MAN Standards Committee 2006); it defines the physical and MAC layers, while the upper layers of the protocol stack (network and applications) are specified by the ZigBee Alliance guidelines (Zigbee Alliance 2020). An overview is made for the energy efficiency, communication, data management, and security solutions that can be adopted for the IEEE 802.15.4 (Baronti et al. 2007). Other works have been proposed to address networking issues in the IEEE 802.15.4/ ZigBee, such as efficient data broadcasting (Ding et al. 2006), coexistence of multiple colocated networks under different interference conditions (Lo Bello and Toscano 2009), and device localization (Pichler et al. 2009).

The position of the PAN coordinator strongly affects the network energy consumption for both network formation and data routing (Cipollone et al. 2007; Abbagnale et al. 2008). PANEL is resource aware and energy efficient for PAN coordinator election in IEEE 802.15.4/ZigBee WSNs. Adopting a cross-layer approach, PANEL operates at the network layer of a WSN that relies on the IEEE 802.15.4 MAC layer for the network formation, as displayed in Fig. 6.1; it reconfigures the network topology in order to achieve optimal PAN coordinator placement, improve energy savings, and reduce routing delay. The topology of a WSN formed according to the IEEE 802.15.4 MAC layer is a cluster tree where the PAN coordinator is at the root of this tree.

Functionally, as the cluster tree topology is formed, PANEL assigns the role of PAN coordinator to different nodes in the tree. To do so, a distributed approach is adopted to determine for the network a new tree topology where the maximum and

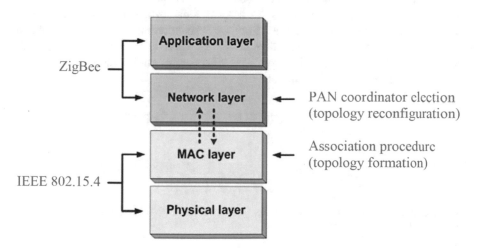

Fig. 6.1 Cross-layer approach for PAN coordinator election in IEEE 802.15.4/ZigBee WSNs (Cuomo et al. 2013)

the average number of hops to the PAN coordinator are minimized. PANEL reduces the energy cost of the network and improves its performance by minimizing the number of hops between the source of the sensor readings and the sink and lessening the packet drop rate due to fewer packet collisions at the MAC layer. Experimental evaluation disclosed that PANEL successfully prolongs WSN lifetime and achieves optimal PAN coordinator election.

PANEL can be easily integrated on top of the IEEE 802.15.4/ZigBee stack; it is fully compliant since it relies on data structures and transmission packets defined in the IEEE 802.15.4 standard, such as the beacon packets. By exploiting the beacons, PANEL is able to:

- Obtain information about the cluster tree topology formed by the IEEE 802.15.4 MAC layer
- Leverage the beacon packet fields to enclose topological information that are needed to optimally reconfigure the network topology

An IEEE 802.15.4 WSN is composed of one PAN coordinator, denoted as p, and a set of nodes (LAN/MAN Standards Committee 2006). The network topology defined is called cluster tree, where nodes associated with p establish parent-child relationships and form a tree rooted at p. Two node typologies can be identified in an IEEE 802.15.4 network:

- Full function devices (FFDs), which are allowed to associate with other nodes in the network. The PAN coordinator and the intermediate nodes, which perform data relay, belong to the FFD class.
- Reduced function devices (RFDs), which are not allowed to associate with other nodes. The nodes that are leaves in the cluster tree are part of the RFD category.

The PAN coordinator is the controller of the network and is responsible for initiating the network setup. All nodes relaying data traffic can be also considered coordinators, named c to distinguish from the PAN coordinator p. Two logical and hierarchical layers can be identified in the network, as illustrated in Fig. 6.2, specifically, the FFD and RFD layers. PANEL operates only at the FFD layer, i.e., only on nodes acting as coordinators.

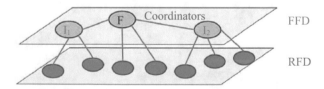

Fig. 6.2 FFD as PANEL operation layer. I_1, I_2 represent initial PAN coordinators. F denotes final PAN coordinators. (Cuomo et al. 2013)

IEEE 802.15.4 defines the steps taken by p and the other nodes to initialize and form the network. Node p starts by selecting a suitable communication channel. This selection is performed by the energy detection scan, which assesses the level of interference on each channel thru measuring the peak energy on each available channel (16 channels in the 2.4-GHz ISM band). Nodes join the network according to the association procedure, where each node performs several tasks:

- Searching for available networks
- Selecting a node coordinator (either a p or c coordinator), if a network is available
- Starting an exchange of signaling packets with the chosen coordinator to complete the association

Discovery of available WSNs is accomplished thru monitoring the beacon frames broadcasted by coordinators (LAN/MAN Standards Committee 2006). After scanning the channels, a node selects the coordinator to join and then sends an association request message to this coordinator. The coordinator communicates its decision to accept or reject the node by responding with an association response command frame. The result of the association procedure, having established parent-child relationships between nodes, is a tree-shaped topology rooted at the PAN coordinator p. Following the network formation, each node in the tree stores a neighbor table built during the association procedure, which includes a description of its children (cardinality and type) and its parent (Zigbee Alliance 2020). The neighbor table allows identifying all one-hop distant nodes.

Level l of a node is the number of hops from the top node. The tree depth L is defined as the maximum value of l within the network, so it identifies the longest path between one of the nodes in the network and p. The mean level of nodes $\bar{L}$ is defined as:

$$\bar{L} = \frac{\sum_{l=1}^{L} l * x_l}{N} \tag{6.1}$$

where

x_l is the number of nodes at level l within the tree
L is the tree depth of the cluster tree
N is the total number of nodes in the cluster tree

L and $\bar{L}$ are affected by the position of the PAN coordinator; they have an impact on the energy consumption of the network during data delivery, as to be clarified in Eq. (6.2).

After the initial setup, the PAN coordinator role could be assigned to different nodes in the network to balance the high-energy drain incurred by a single PAN coordinator (Kim et al. 2003; Sulaiman et al. 2007). Dynamic PAN coordinator election is then needed to achieve several goals:

- Improving the energy efficiency of the network topology
- Boosting the network resiliency by promptly selecting a new PAN coordinator if the current PAN coordinator runs out of battery power, fails, or becomes unsuitable because of movement
- Supporting the interconnection between different colocated WSNs

For instance, in Fig. 6.2, assume that the IEEE 802.15.4 association procedure assigns the role of PAN coordinator to node I_1. Among all the coordinators at the FFD layer, node F would be more suitable than node I_1 to take this role. This is because node F is at a more central position than node I_1, hence minimizing the number of hops with respect to all the other nodes in the network. PANEL identifies node F as the new PAN coordinator and reconfigures the tree where the root is the new PAN coordinator F.

The energy cost associated with the data delivery from a node generating traffic and the sink can be divided into two components:

- The energy necessary to transmit the data packets along a routing path, denoted as E_{routing}
- The energy spent at the MAC layer to access the medium, E_{mac}, which includes the energy for data retransmission in case of packet corruption

IEEE 802.15.4/ZigBee WSNs typically employ hierarchical routing protocols; these are simple to implement, do not require high node computational capability, and do not need additional overhead to establish paths from source to destination within the cluster tree network (Cuomo et al. 2007). According to these protocols, data is relayed by intermediate nodes along the cluster tree paths from the source to the sink identified by the parent-child relationship. Therefore, the energy spent to send data on the routing paths is proportional to the number of hops along the path from source to the PAN coordinator, as destination, since every node on the path spends energy to receive data from its children and to forward the same data to its parent. All nodes are assumed to generate one packet directed to the PAN coordinator without performing data aggregation. Then the E_{routing} component for a single packet can be calculated as:

$$E_{\text{routing}} = (E_{\text{TX}} + E_{\text{RX}}) * \overline{L} * N \qquad (6.2)$$

where E_{TX} and E_{RX} are the energy spent by a node to transmit and to receive a data packet at one-hop distance, respectively.

Equation (6.2) suggests that E_{routing} can be reduced if $\overline{L}$ is minimized. It is to be noticed that reducing L also reduces $\overline{L}$; this can be obtained by properly controlling the PAN coordinator position. Figure 6.3 shows the impact of the PAN coordinator position on L and $\overline{L}$. Figure 6.3a shows the topology formed by the IEEE 802.15.4 association procedure, where the PAN coordinator is node A, $L = 4$ and $\overline{L} = 2.3$. The goal of PANEL is to elect a new PAN coordinator that guarantees smaller values of L and $\overline{L}$ without changing the association relationships already established among the nodes. The resulting topology after running PANEL is shown in Fig. 6.3b, where

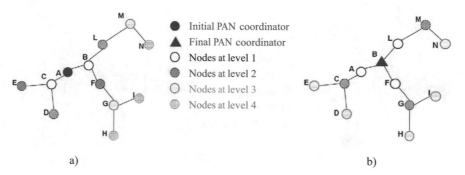

Fig. 6.3 Cluster tree instance without/with PANEL. (**a**) Without PANEL. (**b**) With PANEL
(Cuomo et al. 2013)

the new PAN coordinator role is assigned to node B. In this new scenario $L = 3$ and
$\bar{L} = 2$. By reducing L and $\bar{L}$, the value of E_{routing}, driven by Eq. (6.2), decreases.

The cost of E_{mac} is proportional to the number of transmissions at the MAC layer.
Therefore, E_{mac} depends on the sensor nodes' spatial distribution and transmission
range. Transmission collisions are due to overlapping transmission regions between
nodes, given the broadcast nature of the wireless medium. The goal of PANEL is to
minimize such collisions by positioning the PAN coordinator so as to balance the
utilization of the paths in the network. In Fig. 6.4, after moving the PAN coordinator
from node A to node B, radio overlapping regions are not modified; the nodes are not
physically moved from the configuration of Fig. 6.4a to the configuration of
Fig. 6.4b. The modification instead is on the utilization pattern of some of the
links, e.g., the link between nodes A and B. While three data flows are converging
onto node A according to the configuration of Fig. 6.4a, only two data flows traverse
the A-B link in the configuration of Fig. 6.4b, thus reducing data transmission
collisions on this link. PANEL design revolves around driving down the E_{routing}
cost by minimizing L and $\bar{L}$ while reducing the radio transmission collisions over the
paths toward the PAN coordinator.

Experimentation is carried out through the ns-2 simulator implementing the
physical and MAC layers of the IEEE 802.15.4 standard, including the association
procedure. It is unveiled that PANEL on average decreases the tree depth and mean
level of nodes by determining a new network topology; hence, a new PAN coordinator starting from the initial IEEE802.15.4/ZigBee cluster trees. It is also shown
that the network lifetime increases over the basic IEEE 802.15.4/ZigBee configuration when PANEL is applied to a network with different data traffic patterns. The
experimentation outcomes depict improvement on the tree depth and mean level of
nodes, as well as a network lifetime increase due to energy saving:

- Tree depth and mean level of node improvement. In terms of topological characteristics, i.e., tree depth and mean level of nodes, a comparison is made between
  cluster trees formed by the IEEE 802.15.4 association procedure and the network

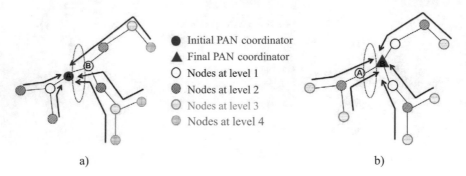

Fig. 6.4 PANEL effect at the MAC layer. (**a**) PAN coordinator is node A. (**b**) PAN coordinator is node B (Cuomo et al. 2013)

topology obtained with PANEL. The network scenario consists of N nodes, with a number, N_F, of FFDs randomly deployed in a square area of side S. The number of FFDs is $N_F = \lceil 0.33 * N \rceil$. The choice of this value of N_F is justified by the fact that in real WSNs only a fraction of nodes is usually assigned the role of coordinator. At the beginning of each experiment, the initial PAN coordinator is randomly selected.

First, the effect of PANEL is illustrated when applied to an IEEE 802.15.4/ ZigBee cluster tree topology. In the simulation model, both physical layer propagation effects and MAC layer radio collisions, including the collisions occurring during the association procedure, are modeled. For the physical layer, ns-2 uses the two-ray ground propagation model. Each packet received at the physical layer should be above the receive threshold value, assumed equal to -97 dBm. As for the collisions, when two packets are received simultaneously, the receiver chooses the strongest among them, based on the capture threshold.

Figure 6.5 illustrates the findings for two scenarios, where $N = 50$ and 100. It is displayed that PANEL improves the topology configuration, as seen from the reduced width and increased height of the node-level probability density function curve. Clearly, with PANEL the probability to find nodes with smaller number of levels is higher than in the IEEE 802.15.4 scenario; shorter routing paths are established between any node and the PAN coordinator. From Eq. (6.2), shorter routing paths imply node energy savings and reduced data delivery delay.

Figure 6.6 shows a comparison of the average tree depth and mean level values for both the IEEE 802.15.4/ZigBee and the PANEL topology configurations.

- Energy considerations and network lifetime increase. The impact of PANEL on the energy consumption of the network is obtained by theoretical considerations taking into account the energy needed to execute PANEL over a basic IEEE 802.15.4 configuration. A thorough energy evaluation requires accounting for the energy needed by each node to execute PANEL, which is directly proportional to the number of iterations each node is required to perform. The energy needed to run PANEL can be divided into two components:

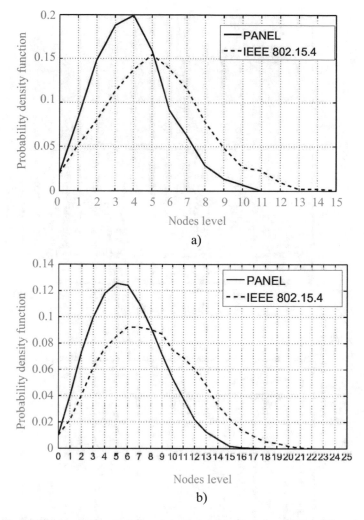

Fig. 6.5 Probability density function of node level. (**a**) For 50 nodes. (**b**) For 100 nodes (Cuomo et al. 2013)

- The energy, E_p, spent by the PAN coordinator at an iteration beginning
- The energy, E_c, spent by the PAN coordinator children at an iteration beginning

- Given these two terms, the energy required by a WSN to run PANEL can be expressed as:

$$E_{\text{PANEL}} = E_p * N_p + E_c * N_c \tag{6.3}$$

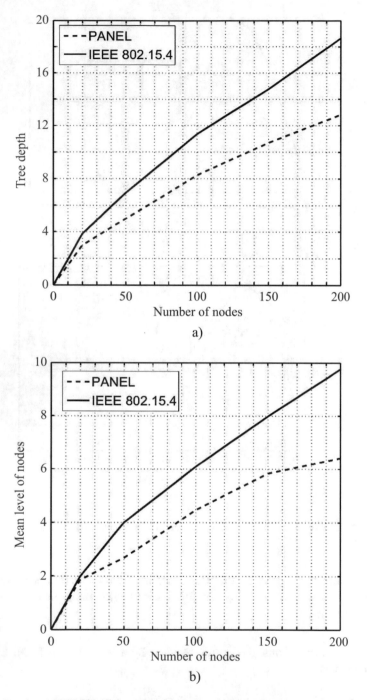

Fig. 6.6 Panel and IEEE 802.15.4 compared in terms of tree depth and mean level of nodes. (**a**) Tree depth comparison. (**b**) Mean level of nodes comparison (Cuomo et al. 2013)

where, given a network of N nodes:

E_p is the energy spent by the PAN coordinator at the beginning of a PANEL iteration
E_c is the energy spent by the child becoming PAN coordinator at the beginning of a PANEL iteration
N_p is the number of nodes that temporarily take the role of PAN coordinator during the execution of PANEL
N_c is the total number of children of all the temporary PAN coordinators during the execution of PANEL

The values of E_p and E_c are related to the number and type of instructions that can be found in the PANEL algorithms. Some of the instructions involve data processing, others comprise data transmission and reception. Data processing instructions require less energy than data transmission and reception. E_p and E_c can be expressed as:

$$E_p = \left(E_{\mathrm{TX}} * L_{\mathrm{payload}}\right) + \left[E_{\mathrm{RX}} * L_{\mathrm{payload}} * (N-1)\right] + \left(E_{\mathrm{instr}} * \mathrm{num}_{\mathrm{instr}}^p\right) \quad (6.4)$$

$$E_c = \left(E_{\mathrm{RX}} * L_{\mathrm{payload}}\right) + \left(E_{\mathrm{instr}} * \mathrm{num}_{\mathrm{instr}}^c\right) + \left(E_{\mathrm{TX}} * L_{\mathrm{payload}}\right) \quad (6.5)$$

where

E_{TX} and E_{RX} are, respectively, the energy spent by a node to transmit and receive one bit at one-hop distance
L_{payload} is the size, expressed in bits, of packets exchanged to transmit data between the PAN coordinator and its children, during each iteration of PANEL
N is the total number of nodes in the WSN
E_{instr} is the energy spent by a node for data processing, e.g., addition, bitwise operations, etc.
$\mathrm{num}_{\mathrm{instr}}^p$ indicates the number of different data processing instructions that the PAN coordinator executes at each iteration
$\mathrm{num}_{\mathrm{instr}}^c$ designates the number of different data processing instructions every PAN coordinator child executes at each iteration

The size of the packet payload used by the PAN coordinator and the children is assumed constant and equal to L_{payload}.

To exhibit the impact on the energy cost when the network runs PANEL, simulations were run with N randomly deployed nodes in a square area of side S. A subset of these N nodes is composed of a number, N_F, of FFDs. The nodes have a transmission range T_R. The basic IEEE 802.15.4 association procedure returns topologies called $\{T_{802.15.4}\}$, while PANEL produces topologies called $\{T_{\mathrm{PANEL}}\}$. For both topologies, a constant bit rate traffic (CBR) pattern is applied to all nodes sending one data packet every superframe. All packets flow toward the PAN coordinator. Under these conditions, the average energy, needed to send the data within a superframe for both $\{T_{802.15.4}\}$ and $\{T_{\mathrm{PANEL}}\}$, is calculated. Moreover, E_{PANEL} is calculated from the execution of PANEL using Eqs. (6.3)–(6.5).

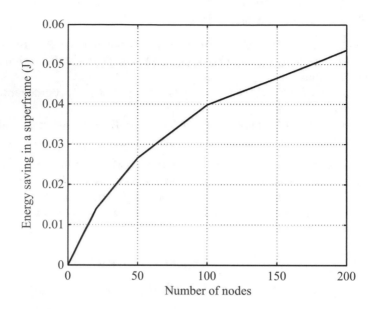

Fig. 6.7 Mean value of the difference between energy usage in $\{T_{802.15.4}\}$ and $\{T_{PANEL}\}$ (Cuomo et al. 2013)

The cost of E_{PANEL} is spread across all the superframes to obtain the total energy consumption for $\{T_{PANEL}\}$. Figure 6.7 shows the mean value of the difference between the energy usage of $\{T_{802.15.4}\}$ and of $\{T_{PANEL}\}$ for a single superframe. As it can be noted, the value of this difference is always greater than zero. Therefore, the energy consumption in a superframe related to topologies reconfigured with PANEL is always smaller than the energy consumed in a superframe by topologies obtained applying the basic IEEE 802.15.4 association procedure. Gains in the range 20–27% are realized, if, in the same scenario, performance gain is computed as a percentage with respect to the average energy needed to send a superframe in case of $\{T_{802.15.4}\}$.

The energy consumption of $\{T_{PANEL}\}$ also includes the consumption from the execution of PANEL in the worst-case scenario, according to Eqs. (6.4) and (6.5). As illustrated in Fig. 6.7, the energy saving is a lower bound on the energy saving achievable with PANEL. It can also be seen that PANEL performance increases as the number N of nodes grows, signifying that PANEL scales well with the size of the network.

Figure 6.8 illustrates the percentage increase of network lifetime with PANEL over IEEE 802.15.4 topologies. This experiment involves two different fractions of FFDs. In both cases, the network lifetime with PANEL increases over 50% when $N = 200$. The performance boost is more evident with $N_F = \lceil 0.5 * N \rceil$ than with $N_F = \lceil 0.33 * N \rceil$, because a higher value of N_F guarantees greater density of FFDs, which implies larger availability of coordinators eligible for the role of PAN

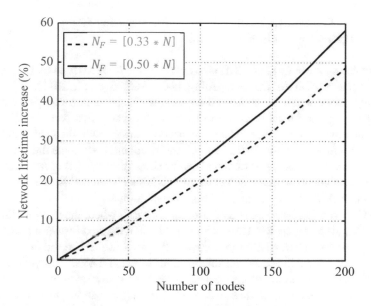

Fig. 6.8 Network lifetime increase for two values of the number of FFDs (Cuomo et al. 2013)

coordinator. This is a significant result considering the need to maintain low-power usage of WSN in order to maximize its operational lifetime.

Lastly, the development of self-managing, self-configuring, and self-regulating protocols for PAN coordinator election for IEEE 802.15.4/ZigBee in WSNs needs further probing. In this direction, PANEL is a cross-layer approach that addresses the election of the PAN coordinator node. By selecting the best PAN coordinator, PANEL allows to reduce the energy consumption in WSNs. The PANEL distributed solution performs the following operations:

- Reconfiguring a network topology previously formed by the IEEE 802.15.4 association procedure
- Introducing little or no overhead in the wireless packet transmission
- Selecting PAN coordinators that are in "balanced" positions at the network, i.e., where the tree depth and the average mean level of nodes are minimized

These operations result in limiting the energy depletion of the network, thus extending its lifetime. PANEL is flexible, in that it transparently cooperates with the IEEE 802.15.4 standard without modifying the underlying 802.15.4 basic functionalities; it outperforms the standard IEEE 802.15.4 association procedure, providing a viable solution for the developments of the low-power WSN technology.

6.2.1.1.2   A Cross-Layer Routing Protocol for Balancing Energy
             Consumption in WSNs (CLB)

The purpose of the suggested CLB is to enhance the WSN lifetime by balancing the
energy consumption in the forwarding task (Yessad et al. 2015). The energy
efficiency concern can be addressed by finding the minimum energy path, while
load balancing can be achieved by using multiple suboptimal paths. CLB exploits
the interaction between the MAC and the network layers, with the goal of enhancing
the WSN lifetime. A mathematical model and simulations evaluate the performance
of the proposed protocol. The obtained results show that the CLB cross-layer routing
protocol uses all forwarding nodes in an equitable manner; this enables avoiding
network partitioning and enhances the network lifetime.

The CLB routing protocol is a bottom-up approach where the network layer uses
information given by the MAC layer for the choice of the next forwarding hop. It just
defines a communication interface between the MAC layer and the network layer;
this is intended to facilitate enhancements in the MAC and network layers separately
and easily consider other design concerns such as delay, quality of service, and
congestion avoidance. CLB has two phases, the initialization phase and the data
transmission phase. In the initialization phase, the sink broadcasts a route request
message to find suboptimal routes from each source node to the sink. Then, in the
data transmission phase, the MAC layer informs the network layer about all the
overheard communications of the neighboring nodes. With this information, a
sending node can know how many times each forwarding node has routed data.
Accordingly, to balance the energy consumption of the forwarding nodes, a sending
node chooses its next hop among the less used. Differently from several multipath
routing protocols, the choice of the next hop is not probabilistic and leads to better
balancing of energy consumption.

Before introducing CLB in details, a differentiation from other approaches is
highlighted in what follows:

- The adaptive load-balanced algorithm (ALBA) was designed to consider load
  balancing and congestion; it is a packet forwarding protocol for ad hoc and sensor
  networks (Casari et al. 2006). ALBA follows the integrated approach that combines geographic routing and medium access control (MAC), thus exploiting the
  knowledge of node positions in order to achieve energy-efficient data forwarding
  (Sect. 6.2.2). The considered scenario is critical for medium and high traffic, as
  contentions for channel access and the resulting collisions lead to performance
  degradation. To counter this effect, leverage on network density favors the choice
  of relay candidates that are not overloaded. With ALBA, nodes strive to chan-
  nelize traffic toward uncongested network regions, rather than just maximizing
  the advancement toward the final destination. All eligible forwarder nodes of a
  source node calculate two indices in their path toward the sink, namely, the
  geographic priority index (GPI) and the priority queue index (QPI). Noticeably,
  the load balancing viewed in ALBA focuses on congestion avoidance, not energy
  consumption balancing as studied in CLB.

- A many-to-one real-time sensor network is considered, where sensing nodes are to deliver their measurements to a basestation under a time constraint, with the overall target of minimizing the energy consumption at the sensing nodes (Puccinelli et al. 2006). The quality of the links and the remaining energy in the nodes are the primary factors that shape the network graph; link quality may be measured directly by most radios, whereas residual energy is related to the node battery voltage, which may be measured and fed into the microcontroller. These quantities may be used to form a cost function for the selection of the most efficient route. Moreover, the presence of a time constraint requires the network to favor routes over a short number of hops in order to minimize delay. Hop number information may be incorporated into the cost function to bias route selection toward minimum-delay routes. Based on the integrated approach, a cross-layer cost function that mixes the physical and network layers is obtained (Sect. 6.2.2); it includes raw hardware information (remaining energy), physical layer data (channel quality), and a network layer metric (number of hops). A route selection scheme based on these principles intrinsically performs node energy control for the extension of the lifetime of the individual nodes and for the achievement of energy balancing in the network; intuitively, the long-hop approach permits the time-sharing of the critical area among more nodes. As a comparison, CLB balances energy consumption based on specific information about the amount of routed data from each node; hence, it is not probability based and does not rely on heuristics.

- CLB aims to enhance the network lifetime by balancing energy consumption; other approaches target energy efficiency by maximizing the sleep time. Achieving minimum energy consumption is the goal of MAC-CROSS (Suh et al. 2006). Based on the layer interaction approach (Sect. 6.2.1), MAC-CROSS exploits the interactions between MAC and network layers to achieve energy efficiency for WSNs (Suh et al. 2006). Routing information at the network layer is used in the MAC layer such that it can maximize the sleep duration of each node. Through implementation on a MICA mote platform (Crossbow 2002) and simulation study using ns-2 simulator (Chap. 8), the performance of MAC-CROSS is evaluated.

  Also, based on the integrated approach (Sect. 6.2.2), the cross-layer energy-efficient protocol (CLEEP) targets prolonging the nodes' sleep time by adopting a strategy that considers the physical, MAC, and network layers (Liu et al. 2008). In the physical layer, CLEEP coordinates the transmission power between two nodes and maintains the node neighbor tables periodically to save the transmission energy. Then, the optimal routing path is constructed by exploiting the transmission power and neighbor tables of the physical layer, which minimizes the total energy consumption. Finally, in order to prolong the node sleep time, the MAC layer makes use of the routing information to determine the node duty cycle.

  Moreover, integrating MAC and network layers (Sect. 6.2.2), an enhanced cross-layer protocol (ECLP) is designed for energy efficiency and latency in WSNs to realize efficient data delivery (Kim, Lee and Kim 2009). To reduce energy wastage due to idle listening and overhearing and to alleviate long delay,

ECLP uses an adaptive duty-cycle scheme with the adaptive time-out and reservation request-to-send (RRTS). Moreover, a tree-based energy-aware routing algorithm is developed in ECLP to minimize overhead cost and prolong the network lifetime.

Basically, to augment the WSN lifetime by balancing energy consumption during the forwarding task, sensor nodes must be used fairly. If two nodes have the same cost for routing source node data, they must be used as relays for the same number of times. Based on this idea, two routing protocols—fair energy-aware routing (FEAR) (Yessad et al. 2011) and balanced energy-efficient routing (BEER) (Yessad et al. 2012)—were previously proposed as improvements of the energy-aware routing (EAR) (Shah and Rabaey 2002). FEAR improved the network lifetime by reducing the probability of using the highly demanded sensor nodes in the network, i.e., those belonging to several routes. BEER reduces the probability of using forwarding nodes belonging to a source node in a unique route. Even though FEAR and BEER provide more equity among sensor nodes when compared to EAR, they suffer from increased overhead. On the other hand, in EAR, FEAR, and BEER, the suboptimal paths are probabilistically chosen; so, load balancing is probabilistic.

In order to achieve an accurate energy consumption balancing without overhead, CLB routing at the network layer exploits information given by the MAC layer. CLB routing operates in two phases:

- First phase. After the network deployment, sensor nodes establish their forwarding tables as in EAR. The sink broadcasts a route request message with a field cost initialized to 0. Each node, receiving the route request message, updates the cost field according to its residual energy and the power required for the communication between that node and the sender of the route request, and then it broadcasts the route request message. If a given node i receives a route request from a node j with the cost field $cost_j$, it calculates $cost_{ij}$ as follows:

$$\text{cost}_{ij} = \text{cost}_j + C_{ij} \qquad (6.6)$$

such that

$$C_{ij} = e_{ij}^{\alpha} * R_i^{\beta} \qquad (6.7)$$

where

e_{ij} is the power required for the communication between nodes i and j
R_i is the residual energy of node i normalized with respect to its initial energy
α and β, the weighting factors, can be chosen to find the minimum energy path, or the
    path with nodes having the maximum residual energy, or the combination of both
After the reception of the route request message from all neighbors, a node can
    establish its forwarding table by adding neighbors with minimal cost. Then, for

forwarding the route request message, it calculates the average cost that it sets in the cost field:

$$\text{cost}_i = \frac{\sum\limits_{k \in \text{FT}_i} \text{cost}_{ik}}{|\text{FT}_i|} \tag{6.8}$$

where $|FT_i|$ is the number of routes recorded in the forwarding table of node i.

This phase ends when the route request message is broadcasted over the whole network and all nodes have set their forwarding tables with routes to the sink.

Second phase. Sensor nodes sense phenomena in the field of interest according to their application and send data to the sink. Initially, nodes send data over the neighbor in their forwarding table having the minimal cost. Progressively, the network layer of each node will have information about the use of its neighbors for routing data. Nodes are supposed to use the CSMA/CA-based MAC protocols, S-MAC (Ye et al. 2002) or T-MAC (van Dam and Langendoen 2003) with RTS/CTS sequence. When a sensor node wants to send data over a neighbor node, it sends an RTS message and then receives a CTS message from the forwarding node; CTS is received by all its neighbors. In CLB, if the node receiving a CTS message is the destination, it sends the data packet to the sender of the CTS message. Otherwise, instead of dropping the CTS as it is the case in layered protocol, it sends the source address of the CTS message to the network layer. Upon receiving the CTS message, the network layer increments the variable N associated to the sender of the message. The variable N is a field in the forwarding table, which counts the number of times that each neighbor node has routed data. Whenever a given node j has data to send, it calculates the value of B associated with each forwarding node i. Then, it sends its data via the neighbor having the greater value of B.

$$B_i = \frac{1/_{\text{cost}_{ji}} * N_i}{\sum\limits_{k \in \text{FT}_j} 1/_{\text{cost}_{jk}} * N_k} \tag{6.9}$$

To evaluate the performance of CLB routing, a mathematical model evaluates and compares its performance with those of EAR, FEAR, and BEER. The WSN model under study has the following properties:

- The sensor network is composed of M sensor nodes scattered in a field of interest in flat manner, i.e., all sensor nodes play the same role in the network.
- There are k source nodes that send the sensed data in the environment to the sink. The network can be divided into levels of k nodes; the first level L_1 is the one composed of source nodes. The ith node of the jth level, L_j, represented as N_{ji}, has i forwarding nodes in the next level L_{j+1} (Fig. 6.9).
- There is one sink that gathers the sensed data.

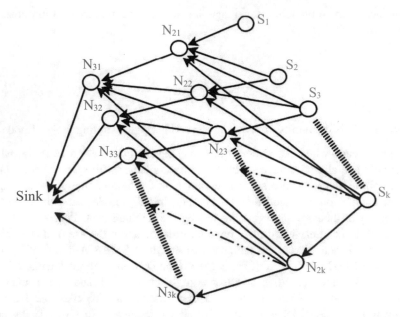

Fig. 6.9 Network model. Except L_4, the number of nodes in L_1, L_2, and L_3 is 4. (Yessad et al. 2015)

- The nodes and the sink are not mobile.
- The sensor nodes are not rechargeable.
- There is no method of getting location information about sensor nodes.
- The network application can be query driven, event driven, time driven, or the hybridization of the three.

The energy E consumed by a node N_{ji}, in the network model, to route S packets of data to the sink is calculated for EAR, FEAR, and BEER. Node N_{ji} can route data sent by nodes $N_{(j-1)(i)}$, $N_{(j-1)(i+1)}$, $N_{(j-1)(i+2)}, \ldots$, and $N_{(j-1)(k)}$, respectively, with the probabilities $P_{N(j-1)(i)\ Nji}$, $P_{N(j-1)(i+1)\ Nji}$, $P_{N(j-1)(i+2)\ Nji}, \ldots$, and $P_{N(j-1)(k)\ Nji}$. Accordingly:

$$E = P_{N_{(j-1)(i)}N_{ji}} * (E_r + E_t) + P_{N_{(j-1)(i+1)}N_{ji}} * (E_r + E_t) + \ldots + P_{N_{(j-1)(k)}N_{ji}}$$
$$* (E_r + E_t) \tag{6.10}$$

where

E_r and E_t are the energy required for the reception and transmission of data by node N_{ji}, respectively.

In EAR, E is calculated as (Shah and Rabaey 2002):

$$E = (E_r + E_t)^* \left[\frac{\frac{1}{c}}{i * \frac{1}{c}} + \frac{\frac{1}{c}}{(i+1) * \frac{1}{c}} + \cdots + \frac{\frac{1}{c}}{(k) * \frac{1}{c}} \right]$$

$$= (E_r + E_t)^* \left[\frac{1}{i} + \frac{1}{i+1} + \cdots + \frac{1}{k} \right]$$

$$= (E_r + E_t)^* \sum_{m=i}^{k} \frac{1}{m}$$

(6.11)

where all nodes in the network are assumed to have the same cost C. This assumption holds in the calculation of E in FEAR, BEER, and CLB.

FEAR evaluates E as follows (Yessad et al. 2011):

$$E = (E_r + E_t) * \frac{1}{k-i+1} * \left[\frac{1}{\sum_{m=k-i+1}^{k} 1/m} + \frac{1}{\sum_{m=k-i}^{k} 1/m} + \ldots + \frac{1}{\sum_{m=1}^{k} 1/m} \right]$$

(6.12)

For BEER, E is computed as (Yessad et al. 2012):

$$E = \begin{cases} (E_r + E_t) * \dfrac{1}{k-i+1} * \left[\dfrac{1}{\sum_{m=k-i+1}^{k} 1/m} + \dfrac{1}{\sum_{m=k-i}^{k} 1/m} + \ldots + \dfrac{1}{\left(\sum_{m=2}^{k} 1/m\right) + k} \right], & \text{for } i < k \\[20pt] (E_r + E_t) * \left[\dfrac{1}{\left(\sum_{m=2}^{k} 1/m*k\right) + 1} \right], & \text{for } i = k \end{cases}$$

(6.13)

In CLB, node N_{ji} can route data sent by nodes $N_{(j-1)(i)}$, $N_{(j-1)(i+1)}$, $N_{(j-1)(i+2)}$, ..., and $N_{(j-1)(k)}$. As in Eqs. (6.11)–(6.13), the cost C of all nodes in the network is the same. For the first packet, the performance is influenced by the position of the node which first sends data. The nodes are supposed to send data in the order of their identifications. This means that if node N_{ji} starts communication, the second will be $N_{j((i+1)\bmod k)}$, and then the third will be $N_{j((i+2)\bmod k)}$, and so forth, until the last node $N_{j((i+k-1)\bmod k)}$ sends its data.

If node $N_{(j-1)(m)}$ starts the communication, node N_{ji} can be solicited by the nodes:

- $N_{(j-1)(i)}$ and $N_{(j-1)(m+i-1)}$, if $i < m$ and $m < k - i + 2$
- $N_{(j-1)(m+i-1)}$, if ($i \geq m$ and $m < k - i + 2$)
- $N_{(j-1)(i)}$, if $i < m$ and $m \geq k - i + 2$
- None, if $i \geq m$ and $m \geq k - i + 2$

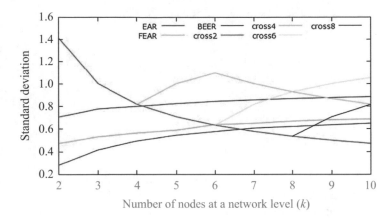

Fig. 6.10 Standard deviation versus number of nodes at a network level. The number of packets sent, $S = 1$. *cross 2*, *cross 4*, *cross 6*, and *cross8* represent the CLB routing graphs for m $= 2, 4, 6$, and 8, respectively. (Yessad et al. 2015)

As there is no probabilistic routing, a node can route data from two, one, or no node. So, the energy consumed by a given node N_{ji} is calculated to be:

$$
E = \begin{cases} (E_r + E_t) * 2, & \text{if } i < m \text{ and } m < k - i + 2 \\ (E_r + E_t), & \text{if } (i \geq m \text{ and } m < k - i + 2) \text{ or } (i < m \text{ and } m \geq k - i + 2) \\ 0, & \text{if } i \geq m \text{ and } m \geq k - i + 2 \end{cases}
$$

$$(6.14)$$

For the energy, as driven in Eqs. (6.11)–(6.14), the standard deviation is drawn versus the number of nodes, k, at a certain level (Fig. 6.10) and the number of packets sent, S (Fig. 6.11). The following is remarked:

- Sending one data packet ($S = 1$) for $k = 1$ to 10, where for each value of k, i takes values in the interval $[1,k]$. For CLB routing, multiple graphs are plotted according to the value of m, the identification of the node starting the communication (Fig. 6.10).
- Calculating the standard deviation for the number of packets sent, $S = 2$–5, where k is set to 10. It is revealed from Fig. 6.11 that in CLB routing, the standard deviation is constant, while for EAR, FEAR, and BEER, it increases with the increase of S. As noticed, the standard deviation of the energy levels of nodes, in CLB throughout the communication, remains the same. This is because after a certain amount of communications in a given zone, all sending nodes in that zone gain information about the amount of data packets forwarded by each node and hence can easily balance the use of their forwarding nodes. As shown in the

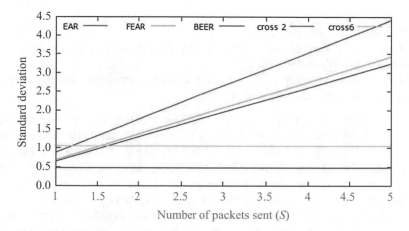

Fig. 6.11 Standard deviation versus number of packets sent (Yessad et al. 2015)

figure, CLB uses sensor nodes equitably, while EAR, FEAR, and BEER use some nodes more than others in the routing task.

Using SenSim (Chap. 8), more comparative simulations of CLB versus EAR, FEAR, and BEER are performed.

6.2.1.2   Cross-Layering Physical and MAC and Network Layers

6.2.1.2.1   Cross-Layer Optimized Routing in WSNs with Duty Cycle and Energy Harvesting (TPGFPlus)

The two-phase geographic greedy forwarding (TPGFPlus) is a cross-layer optimized routing algorithm, where the physical, MAC, and network layers cooperate to choose the optimized transmission route (Han et al. 2015). TPGFPlus is the first cross-layer optimized work to consider two-hop-based geographic routing for duty-cycled and energy renewable WSNs. Specifically, an energy harvesting model is introduced, such that each node has the ability to adjust its transmission power depending on its current energy level. Also, the energy consumed uniformly connected k-neighborhood (EC-CKN) algorithm is applied for sleep scheduling at the MAC layer (Yuan et al. 2011). The physical layer provides the remaining energy and transmission radius information for sleep scheduling at the MAC layer, which dynamically schedules the sleep rate of the network. Then, the network layer chooses the routes that reduce energy consumption, while the physical layer adjusts the transmission power accordingly.

TPGFPlus, which is inherently loop-free, addresses several issues:

- How to find a good approach to facilitate the geographic routing in duty-cycled WSNs with energy harvesting?
- How is the geographic forwarding policy with one-hop neighbor information suitable for duty-cycled WSNs?
- How to design an efficient geographic node-disjoint multipath routing algorithm that allows higher sleep rate in WSNs?

On the basis of extensive simulations, it was found that geographic routing in duty-cycled WSNs should be two-hop based, not one-hop based. The two-hop-based geographic forwarding policies achieve better routing performance than the previously proposed one-hop TPGF (Shu et al. 2010), in terms of both the average number of explored paths and the average path length. What is more, the cross-layer optimized routing allows considerable higher sleep rate in the network.

To describe the interactions of the multiple layers, Fig. 6.12 illustrates how a cross-layer optimized framework underlies the proposed TPGFPlus scheme. The framework takes into account the physical, the MAC, and the network layers:

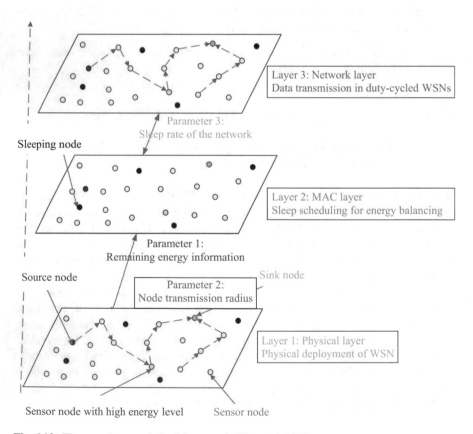

Fig. 6.12   The cross-layer optimized framework (Han et al. 2015)

- The physical layer. The network model is first to be introduced. Let $N(v_i)$ and $N(v_i)'$ be the sets of node v_i one-hop and two-hop neighbor nodes, respectively; that is, v_i two-hop neighbors are the neighbors of v_i one-hop neighbors after removing the duplicated nodes. The considered simple scenario embodies only one source node in the outdoor network.

  The network embodies N sensor nodes randomly deployed. The locations of sensor nodes and the basestation are fixed and can be obtained using the GPS. Each node knows its own location and the position information of its one-hop and two-hop neighbors. The Euclidean distance between any two node v_i and v_j is denoted as $\text{Dist}_{i,j}$. Both the source and the sink nodes are assumed with unlimited power supply. Each sensor node is powered through rechargeable batteries with the capability of harvesting solar power from one additional solar cell. The energy harvesting rate and energy consumption rate in each individual sensor node are different and unpredictable, since many factors can affect the energy level, such as the unstable local weather, the different number of one-hop neighbor nodes, and the unexpected query tasks from users.

  Also, each node uses an adjustable transmission power depending on its battery level. An example of such sensor nodes is Berkeley Motes (Hill et al. 2000). Instead of transmitting at maximum power, nodes take their energy resource into consideration and collaboratively adjust their transmission power accordingly. Any node, if rich in residual energy, will have the ability of enlarging its maximum transmission radius; thus, network connectivity and network sleep rate are directly affected.

- MAC layer. To balance energy consumption and prolong network lifetime, all nodes are assumed to operate with EC-CKN-based sleep/awake duty cycling. The two-hop neighbors are gathered when executing EC-CKN for sleep scheduling in WSNs. Each sensor node switches the radio *on* and *off* in turn based on the two-hop neighbors' remaining energy information. Time is divided into epochs, where each epoch is represented by T. In each epoch, the node will first transmit packets and then run the EC-CKN sleep/awake scheduling algorithm to schedule the state of the next epoch, whether sleep or awake. Sensor nodes take their current energy level information, as the parameter, to decide whether a node is to be active/sleep and to dynamically adjust the network sleep rate. The parameter k, which is considered for controlling the sleep rate of the network, can directly affect the number of awake nodes for geographic routing.

- Network layer. In this layer, a two-hop geographic multipath routing, TPGFPlus, is proposed.

In the proposed routing algorithm, each node locally maintains its one-hop and two-hop neighbors' information such as location, residual energy, energy harvested rate, and energy consumed rate. The algorithm is a two-hop geographic forwarding for a cross-layer optimized multipath routing. The gathering of two-hop neighbors is not an extra overhead for TPGFPlus algorithm, as the two-hop neighborhood information is already gathered when executing EC-CKN. TPGFPlus consists of two phases, namely, two-hop geographic forwarding and path optimization:

• Two-hop geographic forwarding. In this phase, two policies are introduced: explicitly, the greedy forwarding and step back and mark. For the greedy forwarding policy, a forwarding node always chooses its next-hop node that is closest to the basestation, among all its one-hop and two-hop neighbor nodes, and the next hop node to the basestation can be farther than itself. Once the forwarding node is done choosing its next-hop node among its two-hop neighbor nodes that have not been labeled, it finds an intermediate one-hop direct neighbor that has not been labeled according to some selection policy. A digressive number-based label is given to the chosen sensor node along with a path number, which will be kept during the path exploration period. Thus, it is feasible even for resource-constrained sensors. This greedy forwarding principle is different from the greedy forwarding principle in (Karp and Kung 2000; Leong et al. 2005; Holland et al. 2011), where a forwarding node always chooses the one-hop neighbor node that is closer to the basestation than itself. Moreover, it is characterized by the absence of the local minimum problem.

Among candidate nodes with similar progress to the destination, the one with a higher energy harvesting rate and residual energy will be chosen first. Figure 6.13 describes the geographic forwarding process of TPGFPlus. Noticeably, in one-hop routing, forwarding packets are always to the one-hop neighbor that is nearest to the sink. While in two-hop routing, a forwarding node always chooses its next hop node to be the closest to the basestation, among all its one-hop and two-hop neighbors. Once the forwarding node chooses its next-hop node among its two-hop neighbor nodes that have not been labeled, it finds an intermediate one-hop direct neighbor that has not been labeled according to some selection policy.

Although such a method does not have a well-known local maximum problem, there may be block situations (Shu et al. 2010). During a path discovery, if any forwarding node has no one-hop neighbors except its previous hop node, this node is marked as a block node and the situation as a block situation; in such situation, the step back and mark course will start. The block node steps back to its previous-hop node, which will attempt to find another available neighbor as

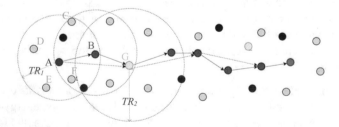

Fig. 6.13 TPGFPlus two-hop geographic forwarding example. Node A chooses a two-hop neighbor node G as its next-hop node that is closest to the sink among all A's one-hop and two-hop neighbors. Once node G is chosen, node A must first select node B as an intermediate one-hop direct neighbor based on a certain selection policy. As shown, if node G has a considerable higher energy level, it can increase its transmission power. TR_1, TR_2 are transmission ranges. (Han et al. 2015)

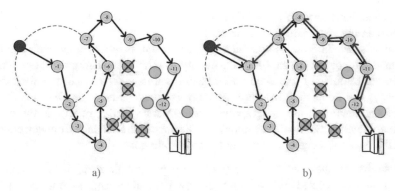

Fig. 6.14 Path optimization. (**a**) The found routing path with circles. (**b**) The optimized routing path after eliminating circles (Han et al. 2015)

next-hop node. This procedure is repeated until a node successfully finds a next-hop node to convert back to the greedy forwarding course.

TPGFPlus does not include the face routing concept, making it different from existing geographic routing algorithms.

- Path optimization. As shown in Fig. 6.14a, a path circle occurs if for any given routing path, two or more nodes in a path are neighbor nodes of another node in this path. To optimize the found routing path, unnecessary circles are to be removed; for this purpose, the label-based optimization is suggested. The principle is that any node in a path only relays the acknowledgement to its one-hop neighbor node that has the same path number and the largest node number. In Fig. 6.14b, after path optimization, a shorter path is obtained. Once the optimized path is found, a release command is sent to all other nodes in the path that are not used for transmission. These released nodes can be reused for exploring additional paths. After receiving the successful acknowledgment, the source node starts to send sensed data to the successful path with the preassigned path number.

  The greedy forwarding mechanism, the main component of geographic routing, usually follows the principle that each node forwards packets to the neighbor node that is closest to the destination with the assumption of highly reliable links. However, this assumption is not realistic. To optimize the forwarding choices, the performance of TPGFPlus is observed under three different forwarding mechanisms:

- Policy 1. Finding an intermediate one-hop direct neighbor node that is closest to the two-hop neighbor node. For this forwarding policy, TPGFPlus finds a neighbor closest to the two-hop neighbor node and makes the maximum progress toward the destination, which is commonly employed. But while assuming the wireless channel reliable, this policy may not work well, as it may choose a neighbor farthest from the current node with a poor link (Seada et al. 2004).

- Policy 2. Finding an intermediate one-hop direct neighbor node that forwards packet from the current node to its two-hop neighbor node with the shortest distance. This forwarding policy attempts to minimize total geographic distance between the source node and the sink.

- Policy 3. Finding an intermediate one-hop direct neighbor node with the most remaining energy, or the best link quality (interference minimized), or even the optimal multifactor weighted cost function value, and so on. In this forwarding policy, the adopted forwarding strategy is Resi _ Energy * Distance for the adopted energy-aware geographic routing, where Resi _ Energy is the current energy level of the candidate node and Distance is the distance from current node to the two-hop neighbor node via the candidate node.

In summary, the goal is to study two-hop geographic node-disjoint multipath routing in duty-cycled WSNs with energy harvesting and to find an optimal forwarding strategy for different application requirements.

Duty-cycle characteristics in the proposed TPGFPlus need to be introduced. The duty cycle is defined typically as the ratio between the active period and the full active/sleep period (Wang et al. 2003; Yan et al. 2003; Liu et al. 2004; Gu et al. 2007). Sensor nodes alternate between sleep and active states. In the sleep state, they go to sleep and thus consume little energy, while in the active state, they actively perform sensing tasks and communications, consuming significantly more energy (Wang and Liu 2009). CKN and EC-CKN are different from other existing duty-cycle algorithms as there is no waiting delay and for the simplicity of their synchronization mechanism:

- Avoidance of waiting delay. In a time-varying connectivity network, a message can either be forwarded over the currently awake nodes by using opportunistic routing algorithms or be temporarily buffered in enroute nodes until a better next-hop node wakes up. In opportunistic routing, the number of hops may increase significantly, which incurs high-energy overheads. Using temporary buffering, the end-to-end latency may increase significantly, e.g., if the next-hop node is not scheduled to wake up for many epochs, and the buffering requirements and waking times also increase (Nath and Gibbons 2007).

  MAC designs for WSNs mostly elect to reduce energy consumption but at the expense of increased latency, since a sender must wait for the receiver to wake up before it can send data; this is the "sleep delay" due to the receiver being in sleep state (Ye et al. 2004). However, in the proposed sleep scheduling, once the source node is within the transmission radius of the destination node, it simply waits until the destination wakes up and then hops to it directly, or the next-hop node is chosen from currently awake neighbors. Thus, waiting delays are avoided.

- Simplification of synchronization. If the number of nodes in a network is small, it may be possible to wake up all nodes for broadcasting through global synchronization with customized active/sleep schedules. However, for larger-scale WSNs, synchronization remains an open problem (Wang and Liu 2009). As shown in Fig. 6.15, to synchronize the sleep schedules of neighboring nodes, each node wakes up at the beginning of each epoch, which reduces latency and control overhead. Therefore, EC-CKN is simple and feasible even for large-scale networks.

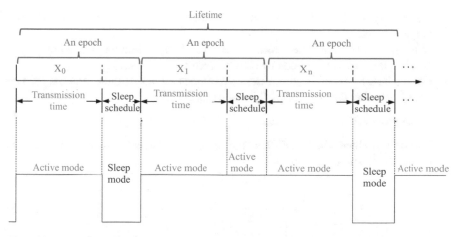

Fig. 6.15 Dynamics of duty cycle in an energy-balanced sleep scheduling scheme (Han et al. 2015)

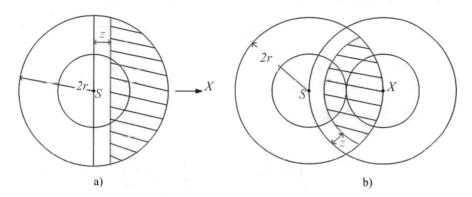

Fig. 6.16 Two limiting scenarios. (**a**) Scenario 1 (**b**) Scenario 2 (Han et al. 2015)

Two theorems describe the relationship between the latency of greedy geographic routing and the number of awake neighbors k, that is, the bounds of expected number of rounds to reach within a specified distance from the destination in TPGFPlus algorithm under an EC-CKN-based network. Figure 6.16 shows the two limiting scenarios, in which the destination X is as far away and as close as possible. Nodes are assumed uniformly located and the disk-r communication model is adopted.

Before introducing the theorems, the following notations are used:

- For given k in the EC-CKN algorithm, $E(v_i)$ and $E(v_i)'$ are the subsets of $N(v_i)$ and $N(v_i)'$ having $\mathrm{Erank}_{v_j} > \mathrm{Erank}_{v_i}, v_j \in N(v_i)$.
- $|N(v_i)|, |N(v_i)'|, |E(v_i)|,$ and $|E(v_i)'|$ are the number of elements in $N(v_i)$, $N(v_i)'$, $E(v_i)$, and $E(v_i)'$, respectively.

- Let $m_{v_i} = |E(v_i)| + |E(v_i)'|$ be the total number of $v_i's$ one-hop and two-hop neighbor nodes.
- D is the Euclidean distance from source node S to its destination X and $D > r$.
- $E_k(D)$ is the expected rounds needed to reach the destination.
- For the TPGFPlus algorithm, the range of possible forwarding progress for two hops is divided into $t \geq 2$ equally spaced segments.

Straightaway, the theorems state:

- Theorem 1. Under the TPGFPlus algorithm, the expected rounds to reach within r from the destination is at most:

$$\frac{D}{r} * \frac{1}{\left(\frac{1}{t} * \sum_{i=1}^{t-1} i * P_i^{(m_{v_i})}\right) - 2^{-m_{v_i}}} \tag{6.15}$$

where $q2hop$ is the probability of the neighbor moving farther from the destination:

$$q2hop = 2^{-m_{v_i}} \tag{6.16}$$

$P_i^{(m_{v_i})}$ is the probability of the neighbor, among the m_{v_i} random one-hop or two-hop neighbors of S, moving at least $\frac{1}{t} * 2 * r$ but at most $\frac{i+1}{t} * 2 * r, (i = 1, \ldots, t-2)$

closer to the destination:

$$P_i^{(m_{v_i})} = \prod_{m_{v_i}} (1 - f_{i+1}) * \left(1 - \prod_{m_{v_i}} \left(\frac{1 - f_i}{1 - f_{i+1}}\right)\right) \tag{6.17}$$

in which:

$$f_i = \frac{2}{\pi} * \left(\cos^{-1}\left(\frac{1}{t}\right) - \frac{i}{t} * \sqrt{1 - \left(\frac{1}{t}\right)^2}\right) \tag{6.18}$$

D is the Euclidean distance to the destination.
Theorem 2. For the setup in the theorem, the expected rounds to reach within r from the destination is at least:

$$\frac{D}{r} * \frac{1}{\frac{1}{t} * \sum_{i=0}^{t-1} (i+1) * P_i^{(m_{v_i})}} \tag{6.19}$$

where

$$q = \left(\frac{1}{3} + \frac{\sqrt{3}}{2*\pi} \right)^{m_{v_i}} \tag{6.20}$$

$$P_i^{(m_{v_i})} = \prod_{m_{v_i}} (1 - f_{i+1}) * \left(1 - \prod_{m_{v_i}} \left(\frac{1 - f_i}{1 - f_{i+1}} \right) \right) \tag{6.21}$$

$$f_i = \frac{1}{\pi} * \left(2 * \cos^{-1} \left(1 - \frac{w_i}{2} \right) + 2 * \cos^{-1} \left(\frac{\sqrt{w_i}}{2} \right) - x_i \right) \tag{6.22}$$

where

$$x_i = \sqrt{w_i} * (4 - w_i) \tag{6.23}$$

Extensive simulations were run on the new sensor network simulator NetTopo (Chap. 8). The studied WSN was 800×600 m^2, and the number of nodes ranged from 100 to 1000. A source node was positioned at the location (50, 50), and the sink node at the location (750, 550). The transmission radius for each node was initially set to 60 m. Each node is initialized with 500 energy units and has power harvesting capability.

To evaluate the overall performance of the cross-layer optimized routing protocol, TPGFPlus, multiple performance metrics were adopted and evaluated under the three aforementioned forwarding policies. Two test setups were formed, namely, fixed transmission power and adjustable transmission power.

For the fixed transmission power setup, the chosen metrics confirm better TPGFPlus performance:

- Average number of found paths. TPGFPlus finds more transmission paths than TPGF.
- Optimized average hops of found paths. TPGFPlus paths are shorter than those in TPGF in terms of the number of hops.
- Network sleep rate. Compared with TPGF, TPGFPlus allows more nodes to sleep while achieving the same average number of paths and average path length.

Considering adjustable transmission power setup, each node transmission power is adjustable depending on its current energy level. A node can adjust its transmission power to a discrete value, e.g., 60 m, 70 m, or 80 m. When the current energy of a node is more than the preset value, the node amplifies its transmission power. Thus, it consumes more energy, balances the whole network energy consumption, and gets more neighbor nodes to choose from. In this way, adjustable transmission power affects network connectivity and network sleep rate. For TPGFPlus on EC-CKN-based WSN with energy harvesting, useful findings were obtained thru comparing the adjustable transmission power setup with the fixed transmission power setup. The comparison metrics are the average number of paths, the optimized average hops of paths, the network sleep rate, and the network average residual energy:

- The average number of paths. The adjustable transmission power setup performs better by finding more transmission paths in a higher density network.
- The optimized average hops of paths. The fixed transmission power setup performs better, as paths have fewer hops, than the adjustable power when the network is sparse. But when the network is dense, the adjustable power setup has better performance, especially under policy 3.
- The network sleep rate. The network sleep rate for adjustable transmission power is much higher than for the fixed transmission power. Thus, a more balanced network can be obtained through increasing the transmission power for nodes that have much more energy. Consequently, nodes with less energy will have more chance to sleep and thus having enough time to recharge.
- The network average residual energy. For the adjustable transmission power, the residual average energy is slightly lower, because enlarging transmission radio incurs more energy consumption. Also, the network average residual energy is more balanced than in the fixed transmission power, with less fluctuation.

Concluding, the proposed TPGFPlus realized several main findings:

- Geographic routing in duty-cycled WSNs should be two-hop based, not one-hop based, because in most existing sleep scheduling algorithms, it is mandatory to gather two-hop neighborhood information. Simulation results further support this argument.
- Cross-layer optimized routing allows more nodes to sleep while achieving the same desired routing performance.
- The performance of the fixed and adjustable transmission power scenarios was evaluated under three forwarding policies. Routing decisions were taken locally by considering the progress to the destination, the shortest distance toward the destination, the residual energy level of nodes, and the environmental energy harvesting.

6.2.2 Single-Layer Integrated Module

Geographic random forwarding (GeRaF, pronounced as "giraffe") is used to enable nodes to be put to sleep and waken up without coordination and to integrate physical, MAC, and network layers into a single layer (Zorzi and Rao 2003). GeRaF is based on the assumption that sensor nodes have a means to determine their location and that the positions of the final destination and of the transmitting node are explicitly included in each message. In this scheme, a node, which hears a message, is able, based on its position toward the final destination, to assess its own priority in acting as a relay for that message. All nodes that received a message may volunteer to act as relays and do so according to their own priority. This mechanism tries to choose the best positioned nodes as relays. In addition, since the selection of the relays is done a posteriori, neither topological knowledge nor routing tables are needed at each node, but the position information is enough.

An implementation of this approach was in the development of a multimedia cross-layer protocol for underwater acoustic sensor networks; a resource allocation framework is built to accurately model every aspect of the layered network architecture (Pompili and Akyildiz 2010). The proposed cross-layer communication solution can adapt to different application requirements and seek optimality in several different situations. The solution relies on a distributed optimization problem to jointly control the routing, MAC, and physical functionalities in order to achieve efficient communications in the underwater environment. In particular, it combines a 3D geographical routing algorithm (network layer functionality), a novel hybrid distributed CDMA/ALOHA-based scheme to access the bandwidth-limited high-delay shared acoustic medium (MAC layer functionality), and an optimized solution for the joint selection of modulation, FEC, and transmit power (physical layer functionalities). The proposed solution is tailored for the characteristics of the underwater acoustic physical channel, e.g., it takes into account the very high propagation delay, which may vary in horizontal and vertical links due to multipath, the different components of the transmission loss, the impairment of the channel, the scarce and range-dependent bandwidth, the high bit error rate, and the limited battery capacity. These characteristics lead to very low utilization efficiencies of the underwater acoustic channel and high energy consumptions when common MAC and routing protocols are adopted in this environment.

The interaction of key underwater communication functionalities was explored and the developed cross-layer communication solution allows for the efficient utilization of the bandwidth-limited high-delay underwater acoustic channel. It was shown that end-to-end network performance improves in terms of both energy and throughput when highly specialized communication functionalities are integrated in a cross-layer module.

The coming section provides a different implementation of the single-layer integrated module.

6.2.2.1 A Cross-Layer Protocol for Efficient Communication in WSNs (XLP)

A cross-layer protocol (XLP) is proposed to achieve congestion control, routing, and medium access control in a cross-layer fashion (Vuran and Akyildiz 2010). XLP integrates functionalities from all layers, starting from the physical layer up to the transport layer, into a single cross-layer protocol. To realize efficient and reliable communication in WSNs, the design principle of XLP is based on a proposed cross-layer concept of "initiative determination," which enables receiver-based contention, initiative-based forwarding, local congestion control, and distributed duty-cycle operation. The "initiative determination" requires simple comparisons against thresholds; thus, it is simple to implement, even on computationally impaired devices. XLP was shown to significantly improve the communication performance and outperforms the traditional layered protocol architectures in terms of both network performance and implementation complexity.

The design principle of XLP is a unified cross-layering such that both the information and the functionalities of three fundamental communication paradigms are considered in a single protocol operation while considering the channel effects; explicitly, medium access, routing, and congestion control are the targeted layers. Multiple concepts form the backbone of XLP, namely, initiative determination, transmission initiation, receiver contention, and angle-based-routing. Each of these concepts will be clarified in the following items:

- Initiative determination. The initiative determination concept coupled with the receiver-based contention mechanism grants each node the freedom to participate in communication. In WSNs, the major goal of a communication suite is to successfully transport event information by constructing, as possible, mutihop paths to the sink. The cross-layer initiative determination concept constitutes the core of the XLP and implicitly incorporates the intrinsic communication functionalities required for successful communication in WSNs. A node i initiates transmission by informing its neighbors that it has a packet to send. This is achieved by broadcasting a request to send (RTS) packet. Upon receiving this packet, each neighbor of node i decides whether to participate in the communication or to abstain. This decision is made through the initiative determination based on the current state of the node. The initiative determination is a binary operation where a node decides to participate in communication if its initiative is 1. Denoting the initiative as $\Im$, it is determined as follows:

$$
\Im = \begin{cases} 1, & if \begin{cases} \xi_{RTS} \geq \xi_{Th} \\ \lambda_{relay} \leq \lambda_{relay}^{Th} \\ \beta \leq \beta^{max} \\ E_{rem} \geq E_{rem}^{min} \end{cases} \\ 0, & \text{otherwise} \end{cases} \tag{6.24}
$$

The initiative is set to 1 if all four conditions in Eq. (6.24) are satisfied, where each condition constitutes certain communication functionality in XLP:

- The first condition, $\xi_{RTS} \geq \xi_{Th}$, ensures reliable links constructed for communication based on the current channel conditions. For this purpose, for a node to participate in communication, it is required that the received signal to noise ratio (SNR) of an RTS packet, ξ_{RTS}, be above some threshold ξ_{Th}. The effect of this threshold on routing and energy consumption performance is analyzed and its most efficient value will be chosen as illustrated in Figs. 6.21 and 6.22.
- The conditions $\lambda_{relay} \leq \lambda_{relay}^{Th}$ and $\beta \leq \beta^{max}$ are used for local congestion control in XLP. As will be elaborated later in Eqs. (6.25)–(6.31), the condition $\lambda_{relay} \leq \lambda_{relay}^{Th}$ prevents congestion by limiting the traffic a node can relay. More specifically, a node participates in the communication if its relay input rate, λ_{relay}, is below some threshold λ_{relay}^{Th}.

- The condition $\beta \le \beta^{\max}$ ensures that the buffer occupancy level, β, of a node does not exceed a specific threshold, $\beta^{\max}$, so that the node does not experience buffer overflow, thus preventing congestion.
- The last condition, $E_{\mathrm{rem}} \ge E_{\mathrm{rem}}^{\min}$, ensures that the remaining energy, E_{rem}, of a node stays above a minimum value, $E_{\mathrm{rem}}^{\min}$. This constraint helps preserving uniform distribution of energy consumption throughout the network.

The cross-layer functionalities of XLP lie in these constraints that define the initiative of a node to participate in communication.

- Transmission initiation involves a transmitting node and receiving nodes:
  - When a node i has a packet to transmit, it first listens to the channel for a specific period of time. If the channel is occupied, the node performs backoff based on its contention window size, CW_{RTS}. When the channel is idle, the node broadcasts an RTS packet, which contains the location information of the sensor node i and the sink. This packet also serves as a link quality indicator that helps the neighbors to perform receiver contention, as will be clarified.
  - When a neighbor of node i receives an RTS packet, it first checks the source and destination locations. The region where the neighbors of a node that are closer to the sink reside is the feasible region, and the remaining neighborhood is in the infeasible region. A node receiving a packet first checks if it is inside the feasible region. To save energy, nodes inside the infeasible region switch to sleep for the duration of the communication.

The nodes inside the feasible region perform initiative determination as detailed earlier. If a node decides to participate in communication, it performs receiver contention as to be clarified in the following bullet.

- Receiver contention involves several functionalities:
  - The receiver contention operation of XLP leverages the initiative determination concept with the receiver-based routing approach (Akyildiz et al. 2006). After an RTS packet is received, if a node has an initiative to participate in the communication, i.e., $\Im = 1$, it performs receiver contention to forward the packet. The receiver contention depends on the routing level of each node, which is determined according to the progress a packet would make if the node forwards the packet. The feasible region is divided into N_p priority regions, i.e., A_i, $i = 1, \ldots, N_p$. Nodes with longer progress have higher priority over other nodes.

    According to the location information, each node determines its priority region and performs contention for medium access. As Fig. 6.17 portrays, each priority region, A_i, corresponds to a backoff window size, CW_i. Depending on its location, a node backs off for $\sum_{j=1}^{i-1} CW_j + cw_i$, where cw_i is randomly chosen such that $cw_i \in [0, CW_{\max}]$, where $CW_{\max} = CW_i - CW_{i-1}$, $\forall\, i$. The backoff scheme helps to differentiate nodes of different progress into different

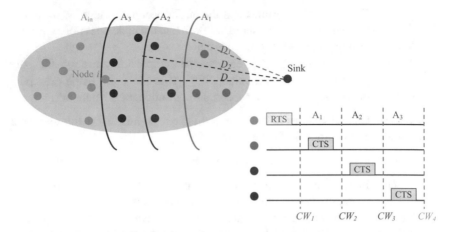

Fig. 6.17 Prioritization mechanism. For $N_p = 3$ priority regions, based on its potential advancement, each feasible node corresponds to one of the three priority regions A_1, A_2, or A_3. The backoff scheme determines the possible times to send a CTS packet. For instance, if a node in A_2 satisfies the initiative function, it first waits for CW_2 in addition to a random cw_2 value. Consequently, the node in A_2 can transmit the CTS packet only if no node in A_1 transmits a CTS packet. D is the distance from node i to the sink. D_1, D_2 are the distances from the farthest end of A_1, A_2, respectively (based on (Vuran and Akyildiz 2010))

     prioritization groups. Only nodes inside the same group contend with each other. The winner of the contention sends a CTS packet to node i indicating that it will forward the packet. On the other hand, if during backoff, a potential receiver k receives a CTS packet, it determines that another potential receiver j with a longer progress has accepted to forward the packet, and node k switches to sleep for the duration of the communication.

– When node i receives a CTS packet from a potential receiver, it determines that the receiver contention has ended and sends a DATA packet with the position of the winner node in the header. The CTS and DATA packets both inform the other contending nodes about the transmitter-receiver pair. Hence, other nodes stop contending and switch to sleep. In the case of two nodes sending CTS packets without hearing each other, the DATA packet sent by node i can resolve the contention. It may happen that multiple CTS packets from the same priority region can collide and a node from a lower-priority region can be selected. XLP does not try to resolve this hitch, as its probability is significantly low since the contention region is already divided into multiple regions and the cost of trying to resolve outweighs the gains.

It is to be noted that node i may not receive a CTS packet due to any of three reasons:
CTS packets collide.
There exist no potential neighbors with $\Im = 1$.
There exist no nodes in the feasible region.

However, node i cannot differentiate these three cases by the lack of a CTS packet. Hence, the neighbors of node i send a keep-alive packet after $\sum_{j=1}^{N_p} CW_j + cw$ if no communication is overheard, where cw is a random number, such that $cw \in [0, CW_{max}]$ and N_p is the number of priority regions. The existence of a keep-alive packet notifies the sender that there are nodes closer to the sink, but the initiative in Eq. (6.24) is not met for any of these nodes. With the reception of this packet, the sender performs retransmission; however, if a keep-alive packet is not received, the node continues retransmission if there is a CTS packet collision. If no response is received after m retries, node i determines that a local minimum is reached and switches to angle-based routing mode as described next.

- Angle-based routing. Since the routing decisions depend, partly, on the locations of the receivers, there may be cases where the packets reach local minima; i.e., a node may not find feasible nodes that are closer to the sink than itself. This situation is known as communications void in geographical routing-based approaches and is generally resolved through face routing techniques (Karp and Kung 2000; Leong et al. 2005; Chen and Varshney 2007; Urrutia 2007). Although localized, face routing requires a node to communicate with its neighbors to establish a planarized graph and construct routes to traverse around the void. This entails information exchange between the neighbors of a node. Since such communication increases the protocol overhead, the angle-based routing technique is a stateless solution proposed to face routing.

  The main principle of the angle-based routing can be seen in Fig. 6.18. When a packet reaches node i, which is a local minimum toward the sink, the packet has to be routed around the void either in clockwise direction (through node j) or in

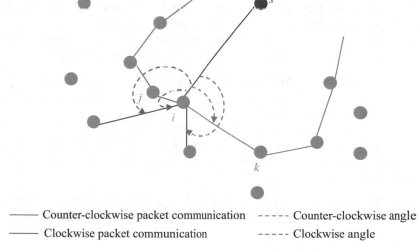

———— Counter-clockwise packet communication   - - - - - Counter-clockwise angle
———— Clockwise packet communication   - - - - - Clockwise angle

Fig. 6.18 Angle-based routing (based on (Vuran and Akyildiz 2010))

counterclockwise direction (through node k). Lines are drawn between node i and the sink, s, as well as between node i and its neighbors. Comparing the angles between the line i,s and the other lines, the angle $\angle sij$ (angle $\angle sik$) has the smallest angle in the counterclockwise (clockwise) routing direction. Using this geometric property, routes can be constructed around the void. Once a direction is set (clockwise or counterclockwise), the packet traverses around the void along the same direction; hence, for angle-based routing, the term traversal direction indicates this direction[1].

When a node switches to angle-based routing mode, it also sets the traversal direction to clockwise and sends an RTS packet, which indicates both the routing mode and the traversal direction. The nodes that receive this packet calculate their angle relative to the source-sink direction. Denoting the angle by θ_{ij}, node j sets its contention window to $c * \theta_{ij} + cw_i$, where cw_i is a random number, and c is a constant that can be selected according to the latency requirements and the density of the network. The node with the smallest angle, hence the smallest contention window, sends a CTS packet; then, data communication takes place. This procedure is repeated until the packet reaches a local minimum; in this case, the traversal direction is set to counterclockwise and the procedure is repeated. Angle-based routing is terminated and the basic XLP is performed when the packet reaches a node that is closer to the sink than the node that initiated the angle-based routing.

After the illustration of the basic concepts that form the XLP backbone, the new hop-by-hop local cross-layer congestion control component is to be introduced. This component is devised based on the buffer occupancy analysis presented in Eqs. (6.25)–(6.33). The objective of this component is to perform hop-by-hop congestion control thru exploiting the local information in the receiver contention and avoiding the need for end-to-end congestion control. It also exploits the local reliability measures taken by the channel access functionality and, consequently, does not necessitate traditional end-to-end reliability mechanisms.

In WSNs, a sensor node has two duties: explicitly, source duty and router duty. Accordingly, there are two sources of traffic as input to each node buffer:

- Generated packets. The first source is the application layer, i.e., the sensing unit of a node, which senses the event and generates the data packets to be transmitted. The rate of the generated packets is denoted by λ_{ii}.
- Relay packets. In addition to generated packets, as a part of its router duty, a node also receives packets from its neighbors to be forwarded to the sink due to the mutihop nature of WSNs. The rate at which node i receives relay packets from node j is denoted as λ_{ji}.

Since the sensor nodes utilize a duty-cycle operation, their buffer occupancy builds up while they sleep because of the generated packets, unless appropriate

---

[1] Note that the clockwise (counterclockwise) traversal direction refers to the traversal direction of the packets rather than the way the angles are measured.

actions are taken. The local cross-layer congestion control component of XLP has two main measures to regulate congestion: in router duty, by providing the sensor nodes with the freedom of deciding whether or not to participate in the forwarding of the relay packets based on the current load on the node and, in source duty, by explicitly controlling the rate of the generated packets.

The upper bound for the total relay packet rate that will prevent congestion is analyzed first. Accordingly, a decision bound is derived for local congestion at each node. More specifically, this bound, denoted by $\lambda_{\text{relay}}^{\text{Th}}$, is used in the XLP initiative determination as presented in Eq. (6.24).

The overall input packet rate at node i, λ_i, can be represented as:

$$\lambda_i = \lambda_{ii} + \lambda_{i,\text{relay}} = \lambda_{ii} + \sum_{j \in \mathbb{N}_i^{in}} \lambda_{ji} \tag{6.25}$$

where

λ_{ii} represents the generated packet rate
λ_{ji} is the relay packet rate from node j to node i
$\mathbb{N}_i^{in}$ depicts the set of nodes from which node i receives relay packets
$\lambda_{i,\text{relay}}$ is the overall relay packet rate of node i

Node i aims to transmit all the packets in its buffer, and hence, the overall output rate of node i is given by:

$$\mu_i = (1 + e_i) * \left(\lambda_{ii} + \lambda_{i,\text{relay}}\right) \tag{6.26}$$

where

e_i is the packet error rate
$1 + e_i$ approximates the retransmission rate since the routes are selected by considering a high SNR value through the initiative determination process

Noticeably, since the node retransmits the packets that are not successfully sent, the output rate is higher than the input rate.

According to Eqs. (6.25) and (6.26), in a long enough interval, T_∞, the average times node i spends in transmitting and receiving are given, respectively, by:

$$T_{rx} = \lambda_{i,\text{relay}} * T_\infty * T_{\text{PKT}} \tag{6.27}$$

$$T_{tx} = (1 + e_i) * \left(\lambda_{ii} + \lambda_{i,\text{relay}}\right) * T_\infty * T_{\text{PKT}} \tag{6.28}$$

where T_{PKT} is the average duration to transmit a packet to another node including the medium access overhead.

To prevent congestion at a node, the generated and received packets should be transmitted during the time the node is active. Because of the duty-cycle operation, on the average, a node is active $\delta * T_\infty$ sec. Therefore:

$$\delta * T_\infty \geq \left[(1 + e_i) * \lambda_{ii} + (2 + e_i) * \lambda_{i,\text{relay}}\right] * T_\infty * T_{\text{PKT}} \tag{6.29}$$

Consequently, the input relay packet rate, $\lambda_{i,\text{ relay}}$ is bounded by:

$$\lambda_{i,\text{relay}} \leq \lambda_{i,\text{relay}}^{\text{Th}} \tag{6.30}$$

where the relay rate threshold, $\lambda_{i,\text{relay}}^{\text{Th}}$, is given by:

$$\lambda_{i,\text{relay}}^{\text{Th}} = \frac{\delta}{(2 + e_i) * T_{\text{PKT}}} - \frac{1 + e_i}{2 + e_i} * \lambda_{ii} \tag{6.31}$$

The above analysis shows that by throttling the input relay rate, congestion at a node can be prevented. This result is incorporated into XLP through a hop-by-hop congestion control mechanism, where nodes participate in routing packets as long as Eq. (6.30) is satisfied. The implementation of Eq. (6.30) necessitates a node to calculate the parameters e_i, T_{PKT}, and λ_{ii}. The generated packet rate, λ_{ii}, is easily extracted from the rate of injected packets from the sensing boards to the communication module. The packet error rate, e_i, is stored as a moving average of the packet loss rate encountered by the node. Similarly, T_{PKT} is determined by using the delay encountered in sending the previous packet by the node. Consequently, each node updates these values after successful or unsuccessful transmission of a packet.

According to Eq. (6.31), the relay rate threshold, $\lambda_{i,\text{relay}}^{\text{Th}}$, is directly proportional to the duty-cycle parameter, δ, suggesting that the capacity of the network decreases as δ is reduced. Moreover, Eq. (6.30) ensures that the input relay rate of source nodes, i.e., nodes with $\lambda_{ii} > 0$, is lower than that of the nodes that are only relays, i.e., $\lambda_{ii} = 0$. This provides homogeneous distribution of traffic load in the network, where source nodes relay less traffic.

The inequality, Eq. (6.30), controls the congestion in the long term. However, in some cases, the buffer of a node can still be full due to short-term changes in the traffic. To prevent buffer overflow in these cases, nodes use the third inequality in Eq. (6.24) to determine their initiative. More specifically, the inequality $\beta \leq \beta^{\text{max}}$ ensures that the buffer level, β, is lower than the threshold, β^{max}, which is the maximum buffer length of a node. Consequently, a node does not participate in communication, if its buffer is full.

In addition to regulating the relay functionality as discussed above, the XLP local congestion control component also takes an active control measure by directly regulating the amount of traffic generated and injected into the network. During the receiver contention mechanism described earlier, node i may not receive any CTS packets but receive keep-alive packets; in this case, it decides that there is congestion in the network. Then, it reduces its transmission rate by decreasing the amount of traffic it generates. In other words, since the traffic injected by any node due to its router duty is controlled based on Eq. (6.30), the active congestion control is performed by controlling the rate of generated packets λ_{ii} at the node i.

In case of congestion, the XLP node reduces the rate of generated packets λ_{ii} multiplicatively:

$$\lambda_{ii} = \lambda_{ii} * \frac{1}{v} \tag{6.32}$$

where v is defined to be the transmission rate throttle factor.

If there is no congestion, then the packet generation rate can be increased conservatively to prevent oscillation in the local traffic load. Therefore, the XLP node increases its generated packet rate linearly for each ACK packet received. Hence:

$$\lambda_{ii} = \lambda_{ii} + \alpha \tag{6.33}$$

The XLP adopts a rather conservative rate control approach, mainly because it has two functionalities to control the congestion for both the source and the router duties of a sensor node. As the node decides to take part in the forwarding based on its buffer occupancy level and relay rate, it already performs congestion control as a part of the XLP forwarding mechanism. Hence, an XLP node does not apply its active congestion control measures, i.e., linear increase and multiplicative decrease, to the overall transmission rate. Instead, only the generated packet rate, λ_{ii}, is updated.

Since the local congestion control is specific to certain regions and may not apply to the entire event area, nodes inside a congested region may reduce their transmission rates and the overall event reliability may still be met at the sink from other node data due to the sheer amount of correlated data flows (Akan and Akyildiz 2005). Thus, instead of an inefficient end-to-end reliability mechanism, the local cross-layer congestion control exploits the local congestion control and reliability to maintain high network utilization and overall reliability in a distributed manner. In fact, this is also clearly observed in the performance evaluation results portrayed in Figs. 6.20, 6.21, and 6.22.

The choice of the duty-cycle value, δ, is main for XLP performance; consequently, the effect of the duty cycle on the network energy consumption is clarified in Figs. 6.19, 6.20, 6.21, and 6.22. In this respect, the energy consumed by the network for a packet sent to the sink as a function of the distance of its source to the sink is investigated. In the network model of XLP operation, each node performs a distributed duty-cycle operation such that the transceiver circuit of the node is *on* for a certain fraction of the time and is switched *off* for the remaining fraction, where the sensors can still sample data. The ON/OFF periods are managed through a duty-cycle parameter, δ, which defines the fraction of the time a node is active. More specifically, each node is implemented with a sleep frame of length T_S sec. A node is active for $\delta * T_S$ sec and is in asleep state for $(1 - \delta) * T_S$ sec.

Note that the start and end times of each node sleep cycle are not synchronized. Consequently, a distributed duty-cycle operation is employed. Furthermore, each node is assumed to be aware of its location through either an onboard GPS or a localization algorithm (Moore et al. 2004). This assumption is motivated by the fact that WSN applications inherently require location information to associate the observed information by each node to a physical location. Thus, it is ordinary to leverage this information for communication. The network model is also geared toward event-based information flow, where nodes send information to a single stationary sink if an event occurs in their vicinity. The area where an event occurs is denoted as the event area and the nodes in this area generate event information.

The total energy consumed from a source node at distance D from the sink can be found to be:

$$E_{\text{flow}}(D) = E_{\text{per\_hop}} * E\left[n_{\text{hops}}(D)\right] \tag{6.34}$$

where

$E_{\text{per\_hop}}$ is the average energy consumed in one hop for transmitting a packet
$E[n_{\text{hops}}(D)]$ is the expected hop count from a source at distance D to the sink.
   An accurate approximation for the expected hop count is given in (Akyildiz et al. 2006):

$$E\left[n_{\text{hops}}(D)\right] \cong \frac{D - R_{\text{inf}}}{E\left[d_{\text{next\_hop}}\right]} + 1 \tag{6.35}$$

where

$E[d_{\text{next\_hop}}]$ is the expected hop distance
R_{inf} is the approximated transmission range

The energy consumed in one hop has three components as given by:

$$E_{\text{per\_hop}} = E_{\text{TX}} + E_{\text{RX}} + E_{\text{neigh}} \tag{6.36}$$

where

E_{TX} is the energy consumed by the node transmitting the packet (Eq. 6.37)
E_{RX} represents the energy consumed by the node receiving the packet (Eq. 6.41)
E_{neigh} is the energy consumed by the neighbors of the transmitter and receiver nodes (Eq. 6.42)

To successfully transmit the packet, a pair of nodes needs to accomplish the four-way handshaking. The distance between the pair of nodes is $d_h = E$ $[d_{\text{next\_hop}}]$. Moreover, the probabilities to successfully receive a data packet and a control packet at this distance are $p_s^D(d_h)$ and $p_s^C(d_h)$, respectively. The lengths of the RTS, CTS, and ACK packets are assumed to be equal. When a transmitter node sends an RTS packet, it is received by the receiver node with probability $p_s^C(d_h)$, which then replies with a CTS packet. If the CTS packet is received, also with probability $p_s^C(d_h)$, the transmitter node sends a DATA packet, and the communication is completed with an ACK packet. In every failure event, the node begins retransmission. Therefore, the expected energy consumed by the transmitting node is:

$$E_{TX} = \frac{K}{\left(p_s^C\right)^3 * p_s^D} \tag{6.37}$$

where

$$K = E_{\text{sense}} + \left(p_s^C\right)^2 * \left[E_{tx}^R + E_{\text{wait}}^C + E_{rx}^C\right] + \left(1 - \left(p_s^C\right)^2\right) * E_{t/o}^C + \left(p_s^C\right)^3$$
$$* p_s^D * \left[E_{tx}^D + E_{rx}^A\right] + \left(p_s^C\right)^2 * \left(1 - p_s^C * p_s^D\right) * E_{t/o}^A \tag{6.38}$$

such that

E_{sense} is the energy consumption spent sensing the region
$E_{tx}^R, E_{rx}^C, E_{tx}^D,$ and E_{rx}^A are the transmission and reception energies spent for RTS, CTS, DATA, and ACK packets, respectively
E_{wait}^{CTS} is the expected energy consumed waiting for a receiver CTS
$E_{t/o}$ is the energy consumed before the transmitter node times out, deciding that a suitable relay node does not exist
E_{wait}^C and $E_{t/o}^C$, in Eq. (6.37), are the only system-dependent terms

According to the previous discussion on the receiver contention, on the average, each node in priority region, A_i, waits for $CW_{\max}/2$ in its priority slot in addition to waiting for the previous priority slots. Denoting the probability that the next hop, $\mathcal{N}_i$, for node i exists in A_k by $P_i = P\{\mathcal{N}_i = j, \text{ such that } j \in A_k\}$, the average waiting time for the next hop is:

$$E_{\text{wait}}^C = e_{rx} * \left\{ \sum_{i=1}^{N_p} \left[\left(\sum_{k=1}^{i-1} CW_k \right) + \frac{CW_{\max}}{2} \right] * P_i \right\} \tag{6.39}$$

where

$$P_i = \left(1 - P_{[A(\gamma_i - 1), \xi_{\text{Th}}]}\right) * P_{[A(\gamma_i), \xi_{\text{Th}}]} \tag{6.40}$$

$$P_{[A(\gamma_i), \xi_{\text{Th}}]} = 1 - p_i$$

p_i is given in (Vuran and Akyildiz 2009)
e_{rx} is the energy consumption for receiving
γ_k is maximum distance from the sink for nodes in A_k

Using the same approach, the energy consumption of the receiver node can be calculated as follows:

$$E_{RX} = \frac{1}{\left(p_s^C\right)^3 * p_s^D} * \left\{E_{rx}^R + E_{wait}^C + E_{tx}^C + E_{rx}^D + E_{rx}^A\right\} \tag{6.41}$$

$$E_{Neigh} = \begin{aligned} &\frac{1}{\left(p_s^C\right)^2 * p_s^D} * \left\{\rho * \delta * \left(\pi * R_{inf}^2 - 2\right) * p_s^C * E_{rx}^R \right. \\ &\left. + \left(\rho * \delta * A(D, R_{inf}, D) - 2\right) * \left(E_{wait}^C + E_{rx}^C + \frac{E_{rx}^D}{2}\right)\right\} \end{aligned} \tag{6.42}$$

where, for the two summed terms in the brackets, the first is the energy consumption for the RTS packet reception by the neighbors of the transmitter node residing in the area $\pi * R_{inf}^2$; the second term models the remaining neighbors of the receiver node residing in the area $A(D, R_{inf}, D)$ while listening only to the CTS message it sends

R_{inf} is the approximated transmission range
D, the distance from a source node to the sink
δ, the duty-cycle parameter
ρ, the density of a 2-D Poisson distribution of the sensor nodes over the network
p_s^C and p_s^D are given in Eqs. (6.43) and (6.44), respectively

The probabilities of receiving a control or a DATA packet are given, respectively, by (Zuniga and Krishnamachari 2004):

$$p_s^C = \left(1 - \frac{1}{2} * e^{-\frac{\xi}{1.28}}\right)^{16l_c} \tag{6.43}$$

$$p_s^D = \left(1 - \frac{1}{2} * e^{-\frac{\xi}{1.28}}\right)^{16l_D} \tag{6.44}$$

where

Mica2 architecture is assumed with Manchester encoding
ξ is the received SNR
l_c and l_D are the control and DATA packet lengths in bits, for p_s^C and p_s^D, respectively

The total energy consumed from a source node at distance D from the sink, $E_{flow}(D)$, as described in Eq. (6.34), is fully computed from Eqs. (6.35)–(6.44). Using numerical integration methods, the effect of the distance, D, on the energy consumption of a flow is shown in Fig. 6.19. Clearly, the energy consumption of a flow is minimal for duty-cycle parameter $\delta \cong 0.002$. However, in relatively small-sized networks of less than 1000 nodes, this operating point may not provide network connectivity. On the other hand, the energy consumption has a local minimum around $\delta \cong 0.2$.

To gain more insight into the protocol operation, the effects of XLP parameters on the overall network performance are to be investigated. XLP is evaluated on a cross-layer simulator (XLS) developed at the laboratory in C++. XLS consists of a realistic channel model based on (Zuniga and Krishnamachari 2004) and ns-2 and an event-

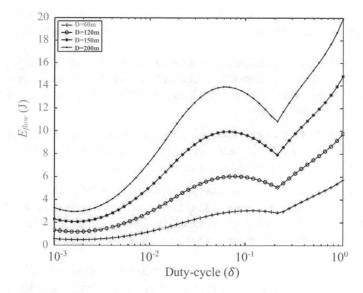

Fig. 6.19 Average energy consumption versus duty cycle for different values of D (Vuran and Akyildiz 2010)

driven simulation engine. The channel errors, packet collisions, and energy consumption at the transceiver are accurately modeled based on ns-2 (Chap. 8). Simulation results are obtained for a sensor topology of 300 nodes randomly deployed in a 100×100 m^2 sensor field. The sink is located at the coordinates (80,80). In each simulation, an event occurs in an event area located at coordinates (20,20) with an event radius of 20 m. Each source node reports its event information to the sink. To investigate the effect of duty cycle, each simulation is performed for duty-cycle values of $\delta \in [0.1,1]$. Each simulation lasts for 300 s and the average of 10 trials for each of 10 different random topologies is shown along with their 95% confidence intervals.

To assess XLP, the performance metrics evaluated are the throughput, goodput, energy efficiency, number of hops, and latency.

- Throughput. It is defined to be the number of bits per second received at the sink. Only unique packets are considered since multiple copies of a packet can be received at the sink for certain protocols.
- Goodput. It is the ratio between the total number of unique packets received at the sink and the total number of packets sent by all the source nodes. As a result, the overall communication reliability is investigated.
- Energy efficiency. The most important metric in WSNs. The average energy consumption per unique packet that is received at the sink is considered in this analysis; it is the inverse of energy efficiency. Hence, a lower value refers to a more energy-efficient communication.

- Number of hops. It is set to be the number of hops each received packet traverses to reach the sink. This metric is used to evaluate the routing performance of each suite.
- Latency. It is the time that passes between the time where a packet is generated at a source node and the time it is received at the sink. This delay accounts for the queuing delay and the contention delay at the nodes, as well as the specific protocol operation overhead.

A multiplicity of factors influences XLP operation; these are the angle-based routing, SNR threshold, ξ_{Th}, and duty-cycle parameter, δ. The effects of these parameters on the XLP performance metrics are laid out:

- The route failure rate versus the duty cycle, δ, with and without angle-based routing. In these experiments, a snapshot of the network is considered and the routes are found considering this topology. The route failure is the ratio of the number of unsuccessful routes between each node in the network and all possible routes.

   The results shown in Fig. 6.20 disclose that route failure rate increases as the duty-cycle parameter δ is decreased. On the other hand, angle-based routing limits the route failure rate to less than 10% for $\delta \geq 0.3$. This leads to up to 70% drop in failure rate. Note that the failure rate of XLP with angle-based routing also rises as δ is further decreased since the probability that at any given time the network is partitioned increases.

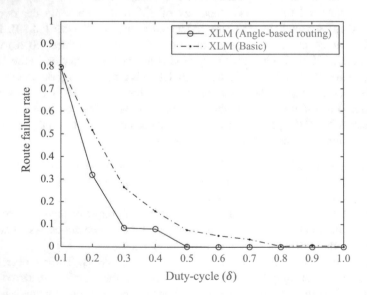

Fig. 6.20 Route failure rate for XLP with/without angle-based routing (Vuran and Akyildiz 2010)

- The total throughput received at the sink versus the duty cycle, δ, for different SNR threshold, ξ_{Th}, values. Figure 6.21a displays the increase in network throughout as the duty cycle, δ, increases. Clearly, the step-up in the duty cycle results in an augmentation in the number of nodes that are active at a given time; consequently, the capacity of the network increases. This is also evident from the buffer occupancy analysis in Eqs. (6.25)–(6.31).

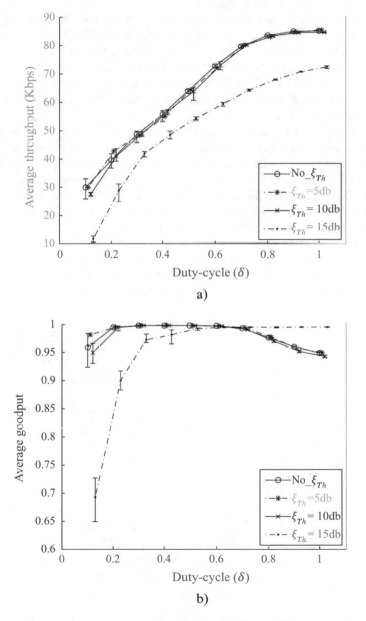

Fig. 6.21 Average throughput/goodput versus duty cycle for different values of the SNR threshold. (**a**) Average throughput. (**b**) Average Goodput (Vuran and Akyildiz 2010)

The effect of the SNR threshold, ξ_{Th}, is also shown in Fig. 6.21a. The No_ξ_{Th} curve is the case where the first condition in Eq. (6.24) is not implemented. In other words, nodes contend for participating in routing irrespective of the received SNR value. It can be observed that increasing the SNR threshold, ξ_{Th}, improves the network throughput up to a certain ξ_{Th}; above this value, the network throughput degrades. This shows that a very conservative operation of XLP leads to performance degradation.

- The goodput versus the duty cycle, δ, for different SNR threshold, ξ_{Th}, values. As Fig. 6.21b clarifies, XLP provides reliability above 90% for $\delta \geq 0.2$ and $\xi_{Th} \leq$ 10 dB. The lessening in goodput at $\delta = 0.1$ is due to the fact that the connectivity of the network cannot be maintained at all times. Moreover, for $\xi_{Th}= 15$ dB, the goodput decreases to 0.7 as the duty cycle is reduced. This is because potential receivers with the desired channel quality cannot be found; hence, the reliability of XLP degrades.

  For high duty cycle $\delta > 0.7$, a slight decrease in goodput is observed for $\xi_{Th}<15$ dB. This accounts for the increased contention in the network since higher number of nodes is active for participation in routing at a given time. Contrarily, for $\xi_{Th}= 15$ dB, a fewer number of nodes are selected for contention participation; thus, collisions are limited and the goodput is not affected.

- End-to-end latency versus the duty cycle, δ, for different SNR threshold, ξ_{Th}, values. In Fig. 6.22, it is obvious that increasing the SNR threshold, ξ_{Th}, improves the end-to-end latency performance up to a certain ξ_{Th} value. Also, $\xi_{Th}= 10$ dB results in the lowest latency. Moreover, there is a suitable operating point for duty cycle, δ, considering end-to-end latency ($\delta \cong 0.6$); above this value, the end-to-end delay starts to increase as a result of the increase in receiver-based contention.

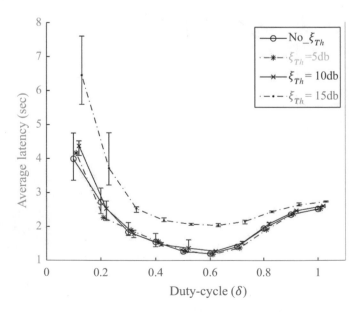

Fig. 6.22 Average latency versus duty cycle for different values of SNR threshold (Vuran and Akyildiz 2010)

Last, throughout this section, it was emphasized that XLP provides the functionalities of medium access, routing, and congestion control. Based on the initiative determination concept, XLP serves as a proof of concept that performs receiver-based contention, initiative-based forwarding, local congestion control, and distributed duty-cycle operation to realize efficient and reliable communication in WSNs. The "initiative determination" concept is the first step in cross-layering that replaces the whole traditional layered WSN protocol architecture, so that both the information and the functionalities of traditional communication layers are blended in a single module. Analytical performance evaluation and simulation experiment results have shown that XLP improves the communication performance and outperforms the traditional layered protocol architectures in terms of both network performance and implementation complexity.

6.3 Cross-Layer Design for WSNs Security

As this book reveals, a wide range of WSN applications have been recognized in oceans and wildlife, manufacturing machinery performance, building safety, earthquake monitoring, and military arenas. The aspects of WSNs have been under intense research, mainly on energy efficiency, network protocols, and distributed databases. However, relatively few works have been reported on security issues, which are also important, especially in battlefield applications, premises security and surveillance, and critical systems such as airports, hospitals, etc. A network cannot perform efficiently or may become useless at the worst, with the absence or lack of a security mechanism that protects the privacy and integrity of data. Although different applications may require different security levels, there are four fundamental security requirements, namely (Xiao et al. 2006b):

- Availability. The service offered by WSN nodes should be available to their users whenever expected.
- Authenticity of origin. The identity of which one interacts with is the expected one.
- Authentication of data (integrity). The received data should be authentic and not tampered.
- Confidentiality (privacy). The information exchanged should be understood by the intended users only. This is often realized by encrypting the messages with a key that is usually made available by the authentication process.

For a WSN, two additional requirements arise:

- Survivability. The ability to provide a minimum level of service in the presence of power loss, failures, or attacks.
- Leveling of security services. The possibility of changing security levels as resource availability changes.

The typical characteristics of WSNs, as clarified all over this book, make providing an efficient and scalable security solution remarkably tricky for the following causes (Perrig et al. 2004; Xiao et al. 2006a; Pathan et al. 2006):

- Vulnerability of channels owing to the shared wireless medium.
- Vulnerability of nodes in open network architecture. Providing open-access architecture contributes to help end users control their content without relying on big Internet companies and to foster innovation by enabling experiments and deployment of innovative functionalities within the network for small players as well.
- Absence of infrastructure that form its backbone.
- Changing network topology due to mobility, or duty cycling that puts some nodes on a standby mode. WSN topology may also change due to power depletion or node failure.
- Hostile deployment environments. A main motive for WSNs is their deployment in hard-to-reach or hostile locations.
- Resource limitations, such as limited processing power and small memory size.
- Large number of nodes densely distributed.

Due to their features, WSNs are vulnerable to unique attacks that may not threaten traditional networks. Various kinds of active and passive attacks have been recognized (Xiao et al. 2006b):

- Denial-of-service (DoS) attack for the purpose of exhausting battery power. For instance, a malicious node could prohibit another node from going back to sleep causing the battery to drain.
- Eavesdropping and invasion. This is fairly easy in wireless communication, if no proper security measures are taken. An adversary could easily extract useful information from conversations between nodes. With this information, a malicious user could join the network undetected by impersonating as a trusted node, to access private data, disrupt the normal network operations, or trace the actions of any node in the network.
- Physical node tampering leading to node failure.
- Battery exhaustion on a node.
- Radio jamming at the physical layer.

In light of these attacks, security techniques that consider the characteristics of WSNs are to be devised. Since large numbers of sensor nodes are distributed in a WSN, low cost and low power are becoming the core design challenges. Low cost constrains the resources that can be implemented on the devices, while the low power requires the operations to be done efficiently. Moreover, because of the large-scale and distributed nature of WSNs, the protocols and algorithms must be scalable. A number of solutions have been proposed specifically for securing WSNs (Wood and Stankovic 2002; Perrig et al. 2002; Liu and Ning 2003; Shi and Perrig 2004; Hu et al. 2005; Liu et al. 2005a). Most of the solutions deal with attacks targeting one protocol layer, yet the layered scheme is inadequate in providing security for WSNs; instead, cross-layer solutions are needed to improve performance.

6.3.1   Challenges of Layered Security Approaches

As WSNs pose unique challenges, security techniques used in traditional networks cannot be applied directly for several reasons (Naeem and Loo 2009; Kumar et al. 2014):

- Economically, to make WSNs viable, sensor devices are limited in their energy, computation, and communication capabilities.
- Unlike traditional networks, sensor nodes may be deployed in accessible areas, thus increasing the risk of physical attacks.
- WSNs interact closely with their physical environments and with people, posing new security problems.

Consequently, existing security mechanisms are inadequate, and new approaches become justifiable. Owing to WSN resource limitations on computation, storage, and bandwidth, the following aspects should be carefully considered when designing a security scheme (Xiao et al. 2006a):

- Power efficiency. Energy supply is scarce and hence energy consumption is a primary metric to be considered.
- Node density and reliability. WSNs may be intended to scale up to large number of nodes, hence instigating more scalable solutions, contrarily to ad hoc networks. Sensor nodes are prone to failures, while existing security designs can address only a small, fixed threshold number of compromised nodes; the security protection breaks down when such threshold is exceeded (Ye et al. 2005).
- Adaptive security. With numerous combinations of sensing, computing, and communication technology, WSNs are deployable with network densities, ranging from extreme sparse to extreme dense deployments. Moreover, WSNs are intended to interact with environments whose traffic patterns are not human-driven; this requires different or at least adaptive security protocols.
- Self-configurability. Like ad hoc networks, WSNs are required to be self-configured. However, factors such as traffic versus energy trade-offs may necessitate innovative solutions; for instance, sensor nodes may have to learn about their geographical position.
- Simplicity. Since sensor nodes are tiny and their energy is limited, the operating and networking software must be kept orders of magnitude simpler as compared to other computing networks.
- Sensor nodes may not have a unique ID like an IP address. This is because the unique ID will generate a significant overhead resulting from the large number of sensors.

To effectively address the above issues, it may be advantageous to break with the conventional layering rules for networking software. The limitations of the layered security approaches are presented in the coming section.

6.3.2   Limitations of Layered Security Approaches

To effectively address the challenges emphasized in the previous section and move
to the cross-layered designs, the limitations of the layered security approaches are to
be laid out and understood (Xiao et al. 2006a):

*   Redundant security provisioning. WSNs are subject to a large number of attacks,
    and each security mechanism consumes some tangible resources, such as battery,
    memory, computation power, and bandwidth. The provision of maximum-
    security services in each sensor node can lead to unnecessary waste of system
    resources and can significantly reduce the network lifetime. Without a systematic
    view, individual security protocols developed for different protocol layers might
    provide redundant security services and hence consume more WSN resource than
    needed. An unorganized design of security provisioning while consuming net-
    work resources may accidentally launch a DoS attack, denoted as security service
    DoS (SSDoS) attack. Generally, there may be several protocol layers within the
    network protocol stack capable of providing security services to the same attack.
    In such a case, when the original data goes downward through the protocol stack
    from the highest layer, some part of the data packets may redundantly go through
    the security provision operations of different layers.
*   Nonadaptive security services. Because attacks on a WSN may come from any
    protocol layer, a counterattack scheme in a protocol layer is unlikely to guarantee
    security all the time. Specifically, link layer security typically addresses confi-
    dentiality provisioning, two-party authentication, and data freshness, but none of
    the security problems of the physical layer. However, an insecure physical layer
    may practically render the entire network insecure; understandably, multilayer
    solutions or cross-layer solutions can achieve better performance. Furthermore,
    self-adaptive security services are flexible in dealing with the dynamic network
    topology as well as with various types of attacks.
*   Power inefficiency. In designing a WSN, energy efficiency is crucial. Several
    causes of power consumption arise, such as idle listening, retransmissions
    resulting from collisions, control packet overhead, and unnecessarily high trans-
    mission power. Correspondingly, different methods were developed for reducing
    power consumption. Some approaches limit the transmission power so as to
    increase the spatial reuse while maintaining network connectivity (Wattenhofer
    et al. 2001; Chen et al. 2002; Santi 2005; Wang 2008). At the network layer,
    power-aware routing protocols result in significant power savings (Aslam et al.
    2003; Chang and Tassiulas 2004). At the MAC layer, the wireless transceivers
    can be turned *off* whenever possible, to reduce the idle listening power as well as
    the number of collisions (Liu et al. 2005b). Depending on the specific applica-
    tions, measures can be taken at the application layer to efficiently improve power
    consumption (Madden et al. 2002, 2003, Madden et al. 2005). In order to reduce
    power consumption, several key management techniques were tailored for WSNs
    (Yu and Guan 2008; Zhang and Varadharajan 2010; Bechkit et al. 2013). Progres-
    sively, it has been conceived that power efficiency design cannot be addressed
    completely at any single layer of the WSN protocol stack (Min et al. 2002).

6.3.3   Guidelines for Securing WSNs

Being aware of the challenges and limitations of layered security approaches, four guiding principles arise as worthy of care for securing WSNs (Jones et al. 2003):

- Security of a network is determined by the security over all protocol layers. Specifically, provisioning confidentiality, two-party authentication, and data freshness address security of the link layer. Referring to Fig. 6.23, it is clear that securing the link layer confers the layers above some security; however, it does not address security problems in the physical layer below, most notably jamming. An insecure physical layer may practically render the entire network insecure, even if the layers above are secure. This is especially true in the WSN environment since basic wireless communication is inherently not secure.
- In a massively distributed network, security measures should be amenable to dynamic reconfiguration and decentralized management. Given the nature of WSNs, a security solution must work without prior knowledge of the network configuration after deployment. Also, the security solution should work with minimal or no involvement of a central node to communicate, globally or regionally, shared information.
- In a given network, at any given time, the cost incurred due to the security measures should not exceed the cost assessed due to the security risks at that time. The sensor network is expected to experience different magnitudes of risk at different times, especially considering the, typically, long-lived nature of a network. In principle, security services should adapt to changes in assessed security risk. This entails that a cost model for both security provisioning and risk must be an integral part of the security model.
- If physical security of nodes in a network is not guaranteed, the security measures must be resilient to physical tampering with nodes in the field of operation. For example, a WSN deployed in a battlefield should exhibit graceful degradation if some network nodes are captured.

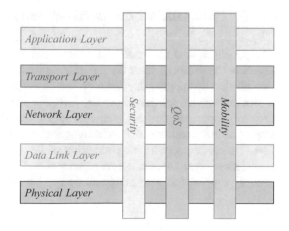

Fig. 6.23 Holistic view of cross-layer design. A cross-layer design normally targets at least one of the three goals, security, QoS, or mobility. (Fu et al. 2014)

6.3.4 Trends in Cross-Layer Design for Security

Several works addressed cross-layer security design at different levels of the protocol stack. A cross-layer design approach was introduced for key management in multicast communications in wireless ad hoc networks (Lazos and Poovendran 2004). Following this approach, cryptographic keys to valid group members are distributed in an energy-efficient manner. It was ascertained that a cross-layer design approach for key distribution incorporating network layer (routing) as well as physical layer (energy) parameters leads to energy savings. Moreover, it was disclosed that heuristics are needed to reduce the computational complexity. Further reduction in energy expenditure is achieved by assigning common keys to nodes in order to receive messages from a sender via a common path (Salido et al. 2007). Based on the Hamming distance between codewords, a computationally viable heuristic called VP3 is developed by using codewords to represent paths and to group nodes based on the length of the common path. Simulation results illustrate the improvements achieved by VP3.

Considering the physical and network layer in combination, an optimization problem was formulated to minimize the energy required for rekeying (Eschenauer and Gligor 2002). In this formulation, a suboptimal cross-layer algorithm that considers the node transmission power (physical layer property) and the multicast routing tree (network layer property) is devised to construct an energy-efficient key distribution scheme (application layer property).

A robust and energy-efficient solution for secure operation of WSNs is proposed (Jones et al. 2003). The approach motivates a new paradigm where security is based on using parameterized frequency hopping and cryptographic keys in a unified framework, to provide differential security services that can be dynamically configured in order to accommodate changing application and network system state in WSNs.

6.3.5 Proposals for Cross-Layer Design for Security

The paradigm is to secure WSNs based on a holistic approach that secures multiple layers in the protocol stack (Fig. 6.23). An important aspect of this paradigm is the exploitation of the interplay between security measures in different layers to provide a security service for the whole network. For security provisioning in WSNs, each protocol layer emphasizes particularly on specific aspects. The physical layer improves information confidentiality using encoding. The MAC layer and network layer are concerned with the encryption of data frames and routing information. The application layer focuses on management and exchange of keys, which provides security support for encryption and decryption of the lower layers.

When considering the security issues in WSNs, the characteristics of each layer should be under focus, and the cross-layer design should compromise between security and network performance with a focus on reducing as much redundancy as possible. For instance, if the objective is to provide energy-efficient security provisioning, the following measures may be integrated:

- At the physical layer, transmission power can be automatically tuned according to the interference strength, which reduces energy consumption and avoids congestion attacks.
- At the MAC layer, the number of retransmissions can be reduced, thus preventing exhaustion attacks while saving energy.
- At the network layer, multipath routing can be adopted, which avoids the routing "hot spot/energy-hole problem"[2] and reduces the energy consumption due to congestion (Li and Mohapatra 2007; Popa et al. 2007).

As discussed throughout, security of WSNs involves all protocol layers. Moreover, at each protocol layer, multiple functional blocks are cross-related to a security solution. Therefore, an effective approach is to develop a cross-layer security scheme individually for each category of security issues, as the following bullets illustrate (Xiao et al. 2006a):

- Cross-layer security for diverse requirements and service types. A WSN may include different types of sensors and perform multiple concurrent applications. Different application scenarios have diverse security requirements. Even within an application, each individual task may have different security concerns.

   Resource limitations and the specific architecture of WSNs call for customized security mechanisms. An approach is presented to classify the types of data existing in WSNs and to identify possible communication security threats according to that classification (Slijepcevic et al. 2002). A communication security scheme defines a security mechanism for each type of data. By employing this multitiered security architecture where each mechanism has different resource requirements, efficient resource management is realized, which is essential for WSNs.

   A link layer security framework, SecureSense, provides energy-efficient secure communication in WSNs (Xue and Ganz 2003). Using runtime security service composition, the proposed SecureSense enables a sensor node to opti-

---

[2]The traffic pattern inherent to WSNs is convergecast, i.e., messages are generated from sensor nodes and are collected by the sink. As a consequence, nodes closer to the sink are more overloaded than others and are subject to premature energy depletion. This issue is known as the funneling effect or the "hot spot/energy-hole problem," since the neighbors of the sink represent the bottleneck of traffic; it is also called the "crowded center effect." Mobile elements can help reduce the funneling effect, as they can visit different regions in the network and spread the energy consumption more uniformly, even in the case of a dense WSN architecture.

mally allocate its resources (CPU cycles, memory consumption, and RF messages) to appropriate security services depending on observed external environments, internal constraints, and application requirements. SecureSense service
composition varies with respect to both security provision strength (how easy it
can be broken) and security concerns (e.g., integrity versus confidentiality).
Compared to a stationary security model with maximum security provision,
SecureSense can significantly prolong the network lifetime without degrading
the security requirements of the applications.

These schemes have considered diverse security concerns for diverse requirements and services. However, they have not taken into account the fact that the
diversity of these services or requirements may also be reflected at different
protocol layers. Different service types require messages to be encrypted differently. Different encryption schemes also consume different amounts of energy.
Security overhead and energy consumption should correspond to the sensitivity
of the encrypted information. Security requirements can span different layers
jointly, so as to minimize the security-related energy consumption.

- Cross-layer for intrusion detection. Detecting intrusion has been under focus on
  routing and MAC protocols; however, the existing secure protocols or intrusion
  detection schemes are proposed for one protocol layer which does not address the
  security concerns that may arise at other protocol layers. Overall security is hence
  clearly lacking, thus imposing the necessity to have a security monitoring tool
  that embodies a cross-layer detection framework which consolidates various
  schemes in different protocol layers.

  Protocol design has the task to consider cross-layer architecture for intrusion
  detection (Xue and Ganz 2003). Noticeably, some existing solutions for one
  protocol layer are also falling short; specifically, the general assumption that
  MAC is for one-hop connectivity may not actually be true in WSNs (Akyildiz
  et al. 2002). Intrusion at the physical layer has mostly been overlooked; however,
  this type of attacks is serious and may go undetected. Indeed, if a channel is
  intentionally jammed by malicious users, security detection schemes based on
  MAC or routing protocols will fail.

- Cross-layer design for power efficiency. It is desirable to consider the energy
  consumption at each design stage and across protocol layers, so as to achieve the
  trade-off between energy consumption, network performance, and complexity
  and maximize lifetime of the whole WSN. A cross-layer approach can conserve
  energy while providing network security provisioning.

  At the MAC layer, the carrier detection is liable to DoS attacks. A malicious
  node can take advantage of the interactions in MAC layer to repeatedly request
  for channel, thus preventing other nodes from connecting with the target node and
  exhausting its battery due to frequent responses. From the information collected
  from other layers, the malicious node can be identified and isolated or restrained.

  At the network layer, a suitable route may be followed using information from
  other layers. From the information of the battery usage, a node with more energy

left is selected to stand more computational load for security or to relay more traffic. While from the authentication information, the choice is for a route far from malicious nodes or attacked areas. The geographical location information can help resist attacks such as sinkhole[3] (Shafiei et al. 2014). The safest and most energy-conserving node is the inactive node, i.e., the node in sleeping mode. Various node-sleeping schemes deserve more consideration.

- Cross-layer design for key management. Due to the limited capacity of a sensor node, it is required to save storage space, decrease computational complexity, and reduce communication overhead required for key management. Key management schemes, such as basic random key (Eschenauer and Gligor 2002) and polynomial poll-based key (Deng et al. 2005), are different in complexity, scalability, and effectiveness in resisting cracking. Adaptive key management schemes have to account for information such as security level, congestion, location, and remaining energy; essentially, it is deriving the overall optimization to extend across multiple protocol layers. Key management schemes based on such optimization embody different interacting components located at multiple layers to really deliver overall optimized performance.
- Cross-layer design for detecting selfish nodes. In a WSN, the connectivity of the network critically relies on the cooperation among nodes. If a node intentionally stops forwarding packets for its neighboring nodes, the network will eventually become out of service; such node is called a selfish node. To avoid this common issue, two techniques are available:

  - Implementing a mechanism in the communication protocols to guarantee that a node has enough interest in forwarding packets to other nodes
  - Developing a scheme for the communication protocols to detect selfish nodes and then warning or penalizing them when selfishness is detected and quickly taking them back to the collaboration mode

Both solutions heavily depend on the cross-layer design methodology, as selfish behavior can emerge at any protocol layer, in particular, MAC and network layers. When a cross-layer design is considered, the solution is more effective in avoiding selfish behavior at one particular protocol layer, as well as at multiple protocol layers. As an approach, a component may be contained in the network layer of a node to monitor packet forwarding by this node successors, while another component contained in the node MAC layer appends two-hop information such as two-hop acknowledgments to the standard MAC packets and forwards them. Such two-hop information will be used by the upper layer component to detect selfish nodes. When detected, actions can be taken by the component in the MAC layer. Such a scheme can detect a selfish node more quickly, due to the faster actions of a

---

[3] In sinkhole attack, a malicious node advertises itself as a best possible route to the basestation, which deceives its neighbors to use the route more frequently. Thus, the malicious node has the opportunity to tamper with the data, damage the regular operation, or even conduct further challenges to the security of the network.

MAC protocol than a networking protocol. This cross-layer architecture also reduces the communication overhead compared with a standard one-layer approach and gives more robustness against selfish behavior.

Table 6.1 Cross-layer design approaches classified

| | Taxonomy | |
	Layer interactions (Sect. 6.2.1)	Single-layer integrated modules (Sect. 6.2.2)
Cross-layer design approaches (Sect. 6.2)	Cross-layering MAC and network layers (Sect. 6.2.1.1): PANEL (the proposed solution combines the network formation procedure defined at the MAC layer by the IEEE 802.15.4 standard with a topology reconfiguration algorithm operating at the network layer. PANEL is devised to self-configure the IEEE 802.15.4/ZigBee WSN by electing, in a distributed way, a suitable PAN coordinator. A protocol implementing this solution in IEEE 802.15.4 is also provided. Performance results show that the proposed cross-layer approach minimizes the average number of hops between the nodes of the network and the PAN coordinator allowing to reduce the data transfer delay and determining significant energy savings compared with the performance of the IEEE 802.15.4 standard) CLB (it is a simple cross-layer routing protocol that enhances the WSN lifetime by balancing the energy consumption in the forwarding task. To do so, the MAC layer informs the network layer about all the overheard communications of the neighboring nodes. According to this information and in order to balance the energy consumption of the forwarding nodes, a node chooses its next hop among the less-used ones. Hence, the choice of the next hop is not probabilistic and leads to better energy consumption balancing. The obtained results have shown that CLB uses all forwarding nodes in an equitable manner; this enables to avoid the network partitioning and to enhance the network lifetime)	XLP (it achieves congestion control, routing, and medium access control in a cross-layer fashion. The design principle is based on the cross-layer concept of "initiative determination," which enables receiver-based contention, initiative-based forwarding, local congestion control, and distributed duty-cycle operation to realize efficient and reliable communication in WSNs. The initiative determination requires simple comparisons against thresholds and thus is very simple to implement, even on computationally impaired devices. XLP is the first protocol that integrates functionalities of all layers from physical to transport into a cross-layer protocol. A cross-layer analytical framework is developed to investigate the performance. XLP significantly improves the communication performance and outperforms the traditional layered protocol architectures in terms of both network performance and implementation complexity)

(continued)

Table 6.1 (continued)

	Taxonomy	
	Layer interactions (Sect. 6.2.1)	Single-layer integrated modules (Sect. 6.2.2)
Cross-layer design approaches (Sect. 6.2)	Cross-layering physical and MAC and network layers (Sect. 6.2.1.2): TPGFPlus (to optimize the system as a whole, this algorithm is designed on the basis of multiple layers interactions, taking into account the following. At the physical layer, sensor nodes are developed to scavenge energy from the environment, i.e., node rechargeable operation. Each node can adjust its transmission power depending on its current energy level (the main object for nodes with energy harvesting is to avoid the routing hole when implementing the routing algorithm). At the MAC layer, where an energy-balanced sleep scheduling scheme, duty cycle, and energy consumption-based connected k-neighborhood are applied to allow sensor nodes to have enough time to recharge energy, which takes nodes' current energy level as the parameter to dynamically schedule nodes to be active or asleep. At the network layer, a forwarding node chooses the next-hop node based on two-hop neighbor information rather than one-hop. Performance of TPGFPlus was evaluated under three forwarding policies. Simulations show that by cross-layer optimization, shorter paths are found, resulting in shorter average path length, without causing much energy consumption. On top of these, a considerable increase of the network sleep rate is achieved)	

6.4   Conclusion for Reality

A book is a whole life full of emotions, visions, and feelings. This chapter lived the eruption of the coronavirus. An outbreak of respiratory disease caused by the new coronavirus was first detected in China and has spread in more than 213 countries. The virus has been named "SARS-CoV-2" and the disease it causes has been named "coronavirus disease 2019" (COVID-19). On January 30, 2020, the World Health

Organization (WHO) declared the outbreak a public health emergency of international concern (PHEIC). Early on, many of the patients at the epicenter of the outbreak in Wuhan, Hubei Province, China, had some link to a large seafood and live animal market, suggesting animal-to-person spread. Later, a growing number of patients reportedly did not have exposure to animal markets, indicating person-to-person spread. Person-to-person spread was subsequently reported outside Hubei and in countries outside China. On Wednesday, March 11, 2020, the coronavirus outbreak has been labeled a pandemic by the WHO chief. A pandemic is defined as a disease that can infect and sicken humans, can transmit easily from one human to another, and has spread worldwide. Pandemics occur when a new form of virus emerges (either a mutated version or a combination with another variation) and is capable of transmitting from person to person. As of submitting the book manuscript, the world has seen a total of 39,388,943 confirmed cases, 29,508,233 recoveries, and 1,105,915 deaths (worldometer 2020). The economic fallout was dangerous; there have been widespread supply shortages of pharmaceuticals and manufactured goods due to factory disruption in China. The technology industry, in particular, has been hit by delaying shipments of electronic goods. A worldwide rolling recession resulted as the disease spread to different areas. The economic impacts of quarantines and travel restrictions were severe. Schools, universities, theatres, and stadiums were shut. The Tokyo 2020 Summer Olympics that were due from July 21 to August 9 were postponed for one year, marking the first time that an entire Olympics has ever been postponed, since 1896, the start of the modern Olympics in Athens.

Back to cross-layering, WSNs differ from other wireless networks in several ways:

- They consist of physically small network nodes, which perform sensing, processing, and then radio communications.
- Each node is configured with the same peer-to-peer networking protocol, thereby allowing a group of sensor nodes to form a self-configuring network.
- The sensor nodes are energy constrained since they are designed to operate in specific areas for years with no maintenance.

Depending on the specific application for which they are used, WSNs can be further divided into event-driven sensor networks or continuous monitoring sensor networks:

- On event-driven sensor networks, the nodes remain in the sleep mode until some event occurs, as in the case of a sensor network devised for sensing forest fires. However, the main problem with this type of networks is being able to switch nodes from a sleeping mode to a listening mode in a defined time.
- On continuous monitoring sensor networks, data are continuously transmitted from source nodes to sink nodes.

One of the main and foremost problems faced by WSNs is that they are energy constrained, due to the fact that WSNs consist of sensor nodes which are battery operated; therefore, it is impossible to recharge them, as they are intended to operate in specific areas for years with no maintenance. Hence, it is important to devise ways by which the energy efficiency of these sensor nodes can be increased so that the overall lifetime of the network is also improved. Cross-layer techniques can achieve the goal of maximizing the energy efficiency in WSNs.

Cross-layer design is a new concept, which has been devised for protocols of wireless networks such as ad hoc networks and sensor networks. A significant number of papers have proposed the use of cross-layer techniques in WSNs in order to achieve different objectives. Furthermore, it has been proved that cross-layer techniques help to improve energy conservation in WSNs.

With cross-layer techniques, the different layers of the conventional open system interconnection (OSI) model interact with each other, irrespective of their positions in the model, to achieve a specific result. The traditional OSI layer architecture is modular in nature and has been implemented successfully in the case of wired networks. In the case of WSNs, which have many constraints in terms of processing, memory, and energy, it becomes difficult to apply only the traditional protocol structure. Cross-layer designs have emerged as an effective approach and have been applied to WSNs. Constraints on energy, memory, storage resources, and low radio transmission capabilities of the wireless sensor nodes make cross-layer support more attractive.

Architecture is important for proliferation of technology, and at a time when wireless networking may be on the verge of a takeoff, its importance needs to be kept in mind. In venturing into the space of cross-layer design, it may, however, be useful to note some adverse possibilities and exercise appropriate attention (Kawadia and Kumar 2005):

- Unbridled cross-layer design can lead to a spaghetti design and stifle further innovations since the number of new interactions introduced can be large.
- Such design can suppress proliferation since every update may require complete redesign and replacement.
- Cross-layer design creates interactions, some intended and others unintended. Dependency relations may need to be examined, and timescale separation may need to be enforced. The consequences of all such interactions need to be well understood, and theorems establishing stability may be needed.

Proposers of cross-layer design must therefore consider the totality of the design, including the interactions with other layers and also what other potential suggestions might be barred because they would interact with the particular proposal being made. They must also consider the long-term architectural value of the suggestion. Cross-layer design proposals must therefore be holistic rather than fragmenting.

Not to be overlooked, cross-layer designs are expected to be the suitable solution to closely examine the trade-off between added security, vulnerability, and network performance. Several interesting issues and open questions arise. To name a few, how to trade off the security level against system performance with minimal power computation? For sake of network survivability, how should be the cross-layer interactions to detect attacks? Also, how to provide intrusion tolerance and ensure graceful degradation designs? How to tolerate the lack of physical security, through redundancy and/or added knowledge about the physical environment?

Cross-layering is a WSN specificity, and it deserves more attention, further exploration.

6.5 Exercises

Note: A technical report follows the template of a peer-reviewed journal paper.

1. Search the literature for more protocols that can be categorized based on Sect. 6.2.1.
2. Explore for more protocols that can be categorized based on Sect. 6.2.2.
3. Write a technical report on the heuristic approach hinted in Sect. 6.1.

Section 6.3 elaborates on the cross-layer design for security. Write your own report on this topic.

References

Abbagnale, A., E. Cipollone, and F. Cuomo. 2008. Constraining the Network Topology in IEEE 802.15.4. In *Advances in Ad Hoc Networking: The 7th IFIP Annual Mediterranean Ad Hoc Networking Workshop*, ed. P. Cuenca, C. Guerrero, R. Puigjaner, and B. Serra, vol. 265, 167–178. Boston: Springer.

Akan, O., and I. Akyildiz. 2005. Event-to-Sink Reliable Transport in Wireless Sensor Networks. *Transactions on Networking* 13 (5): 1003–1016.

Akyildiz, I.F., W. Su, Y. Sankarasubramaniam, and E. Cayirci. 2002. A Survey on Sensor Networks. *Communications Magazine* 40 (8): 102–114.

Akyildiz, I.F., M.C. Vuran, and O.B. Akan. 2006. A Cross-Layer Protocol for Wireless Sensor Networks. In *40th Annual Conference on Information Sciences and Systems*, 1102–1107. Princeton: IEEE.

Aslam, J., Q. Li, and D. Rus. 2003. Three Power-Aware Routing Algorithms for Sensor Networks. *Wireless Networks and Mobile Computing* 3 (2): 187–208.

Bai, Y., S. Liu, M. Sha, Y. Lu, and C. Xu. 2008. An Energy Optimization Protocol Based on Cross-Layer for Wireless Sensor Networks. *Journal of Communications* 3 (6): 27–34.

Baronti, P., P. Pillai, V. Chook, S. Chessa, A. Gotta, and Y. Hu. 2007. Wireless Sensor Networks: A Survey on the State of the Art and the 802.15.4 and ZigBee Standards. *Computer Communications* 30 (7): 1655–1695.

Bechkit, W., Y. Challal, A. Bouabdallah, and V. Tarokh. 2013. A Highly Scalable Key Pre-Distribution Scheme for Wireless Sensor Networks. *Transactions on Wireless Communications* 12 (2): 948–959.

Bouabdallah, F., N. Bouabdallah, and R. Boutaba. 2009. On Balancing Energy Consumption in Wireless Sensor Networks. *Transactions on Vehicular Technology* 58 (6): 2909–2924.

Cardei, M., and D.-Z. Du. 2005. Improving Wireless Sensor Network Lifetime through Power Aware Organization. *Wireless Networks* 11: 333–340.

Casari, P., M. Nati, C. Petrioli, and M. Zorzi. 2006. ALBA: An Adaptive Load-Balanced Algorithm for Geographic Forwarding in Wireless Sensor Networks. In *The Conference on Military Communications (MILCOM)*, 1–9. Washington, DC: IEEE.

Chang, J.-H., and L. Tassiulas. 2004. Maximum Lifetime Routing in Wireless Sensor Networks. *Transactions on Networking* 12 (4): 609–619.

Chen, D., and P.K. Varshney. First Quarter 2007. A Survey of Void Handling Techniques for Geographic Routing in Wireless Networks. *Communications Surveys & Tutorials* 9(1): 50–67.

Chen, B., K. Jamieson, H. Balakrishnan, and R. Morris. 2002. Span: An Energy-Efficient Coordination Algorithm for Topology Maintenance in Ad Hoc Wireless Networks. *Wireless Networks* 8 (5): 481–494.

Chiang, M. 2005. Balancing Transport and Physical Layers in Wireless Multihop Networks: Jointly Optimal Congestion Control and Power Control. *Journal on Selected Areas in Communications* 23 (1): 104–116.

Chilamkurti, N., S. Zeadally, A. Vasilakos, and V. Sharma. 2009. Cross-Layer Support for Energy Efficient Routing in Wireless Sensor Networks. *Journal of Sensors* 2009: 1–9.

Cipollone, E., F. Cuomo, S.D. Luna, U. Monaco, and F. Vacirca. 2007. Topology Characterization and Performance Analysis of IEEE 802.15.4 Multi-Sink Wireless Sensor Networks. In *The 6th Annual Mediterranean Ad Hoc Networking WorkShop*, 196–203. Corfu: Ionian University.

Crossbow. MICA2. January 1, 2002. http://www.eol.ucar.edu/isf/facilities/isa/internal/CrossBow/DataSheets/mica2.pdf. Accessed 3 Feb 2014.

Cuomo, F., S.D. Luna, U. Monaco, and T. Melodia. 2007. Routing in ZigBee: Benefits from Exploiting the IEEE 802.15.4 Association Tree. In *International Conference on Communications (ICC)*, 3271–3276. Glasgow: IEEE.

Cuomo, F., A. Abbagnale, and E. Cipollone. 2013. Cross-Layer Network Formation for Energy-Efficient IEEE 802.15.4/ZigBee Wireless Sensor Networks. *Ad Hoc Networks* 11 (2): 672–686.

Deng, J., Y. Han, P.K. Varshney, J. Katz, and A. Khalili. 2005. A Pairwise Key Predistribution Scheme for Wireless Sensor Networks. *Transactions on Information and System Security* 8 (2): 228–258.

Ding, G., Z. Sahinoglu, P. Orlik, J. Zhang, and B. Bhargava. 2006. Tree-Based Data Broadcast in IEEE 802.15.4 and ZigBee Networks. *Transactions on Mobile Computing* 5 (11): 1561–1574.

Dobson, S.A., et al. 2006. A Survey of Autonomic Communications. *Transactions on Autonomous and Adaptive Systems* 1 (2): 223–259.

Eschenauer, L., and V.D. Gligor. 2002. A Key-Management Scheme for Distributed Sensor Networks. In *The 9th Conference on Computer and Communications Security*, 41–47. Washington, DC: ACM.

Fu, B., Y. Xiao, H. Deng, and H. Zeng. 2014. A Survey of Cross-Layer Designs in Wireless Networks. *Communications Surveys & Tutorials* 16 (1): 110–126.

Gu, Y., G. Hwang, T. He, and D. Du. 2007. uSense: A Unified Asymmetric Sensing Coverage Architecture for Wireless Sensor Networks. In *The 27th International Conference on Distributed Computing Systems (ICDCS)*. Toronto: IEEE.

Haapola, J., Z. Shelby, C.A. Pomalaza-Raez, and P. Mähönen. 2005. Cross-layer Energy Analysis of Multi-hop Wireless Sensor Networks. In *The 2nd International Conference on Embedded Wireless Systems and Networks (EWSN)*, 33–44. Istanbul: ACM/SIGBED.

Han, G., Y. Dong, H. Guo, L. Shu, and D. Wu. 2015. Cross-Layer Optimized Routing in Wireless Sensor Networks with Duty Cycle and Energy Harvesting. *Wireless Communications and Mobile Computing* 15 (16): 1957–1981.

Hefeida, M.S., T. Canli, and A. Khokhar. 2013. CL-MAC: A Cross-Layer MAC protocol for heterogeneous Wireless Sensor Networks. *Ad Hoc Networks* 11 (1): 213–225.

Hill, J., R. Szewczyk, A. Woo, S. Hollar, D. Culler, and K. Pister. 2000. System Architecture Directions for Networked Sensors. In *The 9th International Conference on Architectural Support for Programming Languages and Operating Systems (ASPLOS IX)*, 93–104. Cambridge, MA: ACM.

Holland, M., T. Wang, B. Tavli, A. Seyedi, and W. Heinzelman. 2011. Optimizing Physical Layer Parameters for Wireless Sensor Networks. *ACM Transactions on Sensor Networks (TOSN)* 7 (4): 1–28.

Hu, Y., A. Perrig, and D.B. Johnson. 2005. Ariadne: A Secure On-Demand Routing Protocol for Ad Hoc Networks. *Wireless Networks (Springer Nature)* 11: 21–38.

Jones, K.H., A.H. Wadaa, S. Olariu, L.T. Wilson, and M.Y. Eltoweissy. 2003. Towards a New Paradigm for Securing Wireless Sensor Networks. In *The Workshop on New Security Paradigms (NSPW)*, 115–121. Ascona: ACM.

Karp, B., and H.T. Kung. 2000. GPSR: Greedy Perimeter Stateless Routing for Wireless Networks. In *The 6th Annual International Conference on Mobile Computing and Networking (MobiCom)*, 243–254. Boston: ACM SIGMOBILE.

Kawadia, V., and P.R. Kumar. 2005. A Cautionary Perspective on Cross-Layer Design. *Wireless Communications*: 3–11.

Kim, W.S., I.W. Kim, S.E. Hong, and C.G. Kang. 2003. A Seamless Coordinator Switching (SCS) Scheme for Wireless Personal Area Network. *Transactions on Consumer Electronics* 49 (3): 554–560.

Kim, J., J. Lee, and S. Kim. 2009. An Enhanced Cross-Layer Protocol for Energy Efficiency in Wireless Sensor Networks. In *The 3rd International Conference on Sensor Technologies and Applications*, 657–664. Glyfada: IEEE.

Kumar, V., A. Jain, and P.N. Barwal. 2014. Wireless Sensor Networks: Security Issues, Challenges and Solutions. *International Journal of Information & Computation Technology* 4 (8): 859–868.

Lachenmann, A., P.J. Marrón, D. Minder, and K. Rothermel. 2005. An Analysis of Cross-Layer Interactions in Sensor Network Applications. In *International Conference on Intelligent Sensors, Sensor Networks and Information Processing (ISSNIP)*, 121–126. Melbourne: IEEE.

LAN/MAN Standards Committee. 2006. *Part 15.4: Wireless Medium Access Control (MAC) and Physical Layer (PHY) Specifications for Low-Rate Wireless Personal Area Networks (LR-WPANs)*. Computer Society, Washington, DC: IEEE.

Lazos, L., and R. Poovendran. 2004. Cross-Layer Design for Energy-Efficient Secure Multicast Communications in Ad Hoc Networks. In *International Conference on Communications (ICC)*, 3633–3639. Paris: IEEE.

Leong, B., S. Mitra, and B. Liskov. 2005. Path Vector Face Routing: Geographic Routing with Local Face Information. In *The 13th International Conference on Network Protocols (ICNP)*, 147–158. Boston: IEEE.

Li, J., and P. Mohapatra. 2007. Analytical Modeling and Mitigation Techniques for the Energy Hole Problem in Sensor Networks. *Pervasive and Mobile Computing* 3 (3): 233–254.

Liu, D., and P. Ning. 2003. Location-based Pairwise Key Establishments for Static Sensor Networks (SASN). In *The 1st Workshop on Security of Ad Hoc and Sensor Networks*, 72–82. Fairfax: ACM.

Liu, J., F. Zhao, P. Cheung, and L. Guibas. 2004. Apply Geometric Duality to Energy-Efficient Non-Local Phenomenon Awareness using Sensor Networks. *Wireless Communications* 11 (6).

Liu, D., P. Ning, and R. Li. 2005a. Establishing Pairwise Keys in Distributed Sensor Networks. *Transactions on Information and System Security* 8 (1): 52–61.

Liu, Y., I. Elhanany, and H. Qi. 2005b. An Energy-Efficient QoS-Aware Media Access Control Protocol for Wireless Sensor Networks. In *The International Conference on Mobile Adhoc and Sensor Systems (MASS)*, 189–191. Washington, DC: IEEE.

Liu, S., Y. Bai, M. Sha, Q. Deng, and D. Qian. 2008. CLEEP: A Novel Cross-Layer Energy-Efficient Protocol for Wireless Sensor Networks. In *The 4th International Conference on Wireless Communications, Networking and Mobile Computing (WiCom)*, 1–4. Dalian: IEEE.

Lo Bello, L., and E. Toscano. 2009. Coexistence Issues of Multiple Co-Located IEEE 802.15.4/ZigBee Networks Running on Adjacent Radio Channels in Industrial Environments. *Transactions on Industrial Informatics* 5 (2): 157–167.

Madan, R., S. Cui, S. Lall, and A. Goldsmith. 2006. Cross-Layer Design for Lifetime Maximization in Interference-Limited Wireless Sensor Networks. *Transactions on Wireless Communications* 5 (11): 725–729.

Madden, S., M.J. Franklin, J.M. Hellerstein, and W. Hong. 2002, Winter. Tag: A Tiny Aggregation Service for Ad Hoc Sensor Networks. *Operating Systems Review*: 131–146.

———. 2003. The Design of an Acquisitional Query Processor for Sensor Networks. In *The International Conference on Management of Data (SIGMOD)*, 491–502. San Diego: ACM SIGMOD.

Madden, S.R., M.J. Franklin, J.M. Hellerstein, and W. Hong. 2005. TinyDB: An Acquisitional Query Processing System for Sensor Networks. *Transactions on Database Systems (TODS)* 30 (1): 122–173.

Melodia, T., M.C. Vuran, and D. Pompili. 2005. The State of the Art in Cross-Layer Design for Wireless Sensor Networks. In *Lecture Notes in Computer Science-Wireless Systems and Network Architectures in Next Generation Internet (EuroNGI)*, ed. M. Cesana and L. Fratta, vol. 3883, 78–92. Berlin, Heidelberg: Springer Nature.

Min, R., M. Bhardwaj, N. Ickes, A. Wang, and A. Chandrakasan. 2002. Microsensors, The hardware and the Network: Total-System Strategies for Power Aware Wireless. In *CAS Workshop on Wireless Communications and Networking*. Pasadena: IEEE.

Moore, D.C., J.J. Leonard, D.L. Rus, and S.J. Teller. 2004. Robust Distributed Network Localization with Noisy Range Measurements. In *The 2nd International Conference on Embedded Networked Sensor Systems (SenSys)*, 50–61. Baltimore: ACM.

Naeem, T., and K.K. Loo. 2009. Common Security Issues and Challenges in Wireless Sensor Networks and IEEE 802.11 Wireless Mesh Networks. *International Journal of Digital Content Technology and its Applications (JDCTA)* 3 (1): 88–93.

Nath, S.K., and P.B. Gibbons. 2007. Communicating via Fireflies: Geographic Routing on Duty-Cycled Sensors. In *The 6th international Conference on Information Processing in Sensor Networks (IPSN)*, 440–449. Cambridge, MA: ACM.

Pathan, A.S.K., H.-W. Lee, and C.S. Hong. 2006. Security in Wireless Sensor Networks: Issues and Challenges. In *The 8th International Conference Advanced Communication Technology (ICACT)*, 1043–1048. Phoenix Park: IEEE.

Perrig, A., R. Szewczyk, J.D. Tygar, V. Wen, and D.E. Culler. 2002. SPINS: Security Protocols for Sensor Networks. *Wireless Networks* 8: 521–534.

Perrig, A., J. Stankovic, and D. Wagner. 2004. Security in Wireless Sensor Networks. *Communications* 47 (6): 53–57.

Petrioli, C., M. Nati, P. Casari, M. Zorzi, and S. Basagni. 2014. ALBA-R: Load-Balancing Geographic Routing Around Connectivity Holes in Wireless Sensor Networks. *Transactions on Parallel and Distributed Systems* 25 (3): 529–539.

Pichler, M., S. Schwarzer, A. Stelzer, and M. Vossiek. 2009. Multi-Channel Distance Measurement With IEEE 802.15.4 (ZigBee) Devices. *Journal of Selected Topics in Signal Processing* 3 (5): 845–859.

Pompili, D., and I.F. Akyildiz. 2010. A Multimedia Cross-Layer Protocol for Underwater Acoustic Sensor Networks. *Transactions on Wireless Communications* 9 (9): 2924–2933.

Pompili, D., M.C. Vuran, and T. Melodia. 2006. Cross-layer Design in Wireless Sensor Networks. In *Sensor Network and Configuration: Fundamentals, Techniques, Platforms, and Experiments*, ed. M.P. Mahalik. Berlin, Heidelberg: Springer-Verlag.

Popa, L., A. Rostamizadeh, R. Karp, C. Papadimitriou, and I. Stoica. 2007. Balancing Traffic Load in Wireless Networks with Curveball Routing. In *The 8th International Symposium on Mobile Ad Hoc Networking and Computing (MobiHoc)*, 170–179. Quebec: ACM.

Puccinelli, D., E. Sifakis, and M. Haenggi. 2006. A Cross-Layer Approach to Energy Balancing in Wireless Sensor Networks. In *Lecture Notes in Control and Information: Networked Embedded Sensing and Control*, ed. P. Tabuada, vol. 331, 309–324. Berlin, Heidelberg: Springer Nature.

Salido, J., L. Lazos, and R. Poovendran. 2007. Energy and Bandwidth-Efficient Key Distribution in Wireless Ad Hoc Networks: A Cross-Layer Approach. *Transactions on Networking* 15 (6): 1527–1540.

Santi, P. 2005. Topology Control in Wireless Ad Hoc and Sensor Networks. *Computing Surveys (CSUR)* 37 (2): 164–194.

Seada, K., M.A. Zúñiga, A. Helmy, and B. Krishnamachari. 2004. Energy-Efficient Forwarding Strategies for Geographic Routing in Lossy Wireless Sensor Networks. In *The 2nd International Conference on Embedded Networked Sensor Systems (SenSys)*, 108–121. Baltimore: ACM.

Semchedine, F., L. Bouallouche-Medjkoune, L. Bennacer, N. Aber, and D. Aïssani. 2012. Routing Protocol Based on Tabu Search for Wireless Sensor Networks. *Wireless Personal Communications* 67: 105–112.

Shafiei, H., A. Khonsariab, H. Derakhshia, and P. Mousavia. 2014. Detection and Mitigation of Sinkhole Attacks in Wireless Sensor Networks. *Journal of Computer and System Sciences* 80 (3): 644–653.

Shah, R.C., and J.M. Rabaey. 2002. Energy Aware Routing for Low energy Ad Hoc Sensor Networks. In *Wireless Communications and Networking Conference (WCNC)*, 350–355. Orlando: IEEE.

Shi, E., and A. Perrig. 2004. Designing Secure Sensor Networks. *Wireless Communications* 11 (6): 38–43.

Shu, L., Y. Zhang, L.T. Yang, Y. Wang, M. Hauswirth, and N. Xiong. 2010. TPGF: Geographic Routing in Wireless Multimedia Sensor Networks. *Telecommunication Systems* 44: 79–95.

Slijepcevic, S., M. Potkonjak, V. Tsiatsis, S. Zimbeck, and M.B. Srivastava. 2002. On Communication Security in Wireless Ad-Hoc Sensor Networks. In *The 11th International Workshops on Enabling Technologies: Infrastructure for Collaborative Enterprises (WET ICE)*, 139–144. Los Alamitos: IEEE.

Song, L., and D. Hatzinakos. 2007. A Cross-Layer Architecture of Wireless Sensor Networks for Target Tracking. *Transactions on Networking* 15 (1): 145–158.

Srivastava, V., and M. Motani. 2005. Cross-Layer Design: A Survey and the Road Ahead. *Communications Magazine* 43 (12): 112–119.

Suh, C., Y.-B. Ko, and D.-M. Son. 2006. An Energy Efficient Cross-Layer MAC Protocol for Wireless Sensor Networks. In *Lecture Notes in Computer Science-Advanced Web and Network Technologies, and Applications (APWeb)*, ed. H.T. Shen, J. Li, M. Li, J. Ni, and W. Wang, vol. 3842, 410–419. Berlin, Heidelberg: Springer.

Sulaiman, T.H., K. Sivarajah, and H.S. Al-Raweshidy. 2007. Improved PNC Selection Criteria and Process for IEEE802.15.3. *Communications Magazine* 45 (12): 102–109.

Urrutia, J. 2007. Local Solutions for Global Problems in Wireless Networks. *Journal of Discrete Algorithms* 5 (3): 395–407.

van Dam, T., and K. Langendoen. 2003. An Adaptive Energy-Efficient MAC Protocol for Wireless Sensor Networks. In *The 1st International Conference on Embedded Networked Sensor Systems (SenSys)*, 171–180. Los Angeles: ACM.

van der Schaar, M., and N.S. Shankar. 2005. Cross-Layer Wireless Multimedia Transmission: Challenges, Principles, and New Paradigms. *Wireless Communications* 12 (4): 50–58.

van Hoesel, L., T. Nieberg, J. Wu, and P.J.M. Havinga. 2004. Prolonging the Lifetime of Wireless Sensor Networks by Cross-Layer Interaction. *Wireless Communications* 11 (6): 78–86.

Venkitasubramaniam, P., S. Adireddy, and L. Tong. 2003. Opportunistic ALOHA and Cross Layer Design for Sensor Networks. In *Military Communications Conference (MILCOM)*, 705–710. Boston: IEEE.

Vuran, M.C., and I.F. Akyildiz. 2006. Spatial Correlation-Based Collaborative Medium Access Control in Wireless Sensor Networks. *Transactions on Networking* 14 (2): 316–329.

———. 2009. Error Control in Wireless Sensor Networks: A Cross Layer Analysis. *Transactions on Networking* 17 (4): 1186–1199.

———. 2010. XLP: A Cross-Layer Protocol for Efficient Communication in Wireless Sensor Networks. *Transactions on Mobile Computing* 9 (11): 1578–1591.

Wang, Y. 2008. Topology Control for Wireless Sensor Networks. In *Wireless Sensor Networks and Applications. Signals and Communication Technology*, ed. Y. Li, M.T. Thai, and W. Wu, 113–147. Boston: Springer.

Wang, F., and J. Liu. 2009. Duty-Cycle-Aware Broadcast in Wireless Sensor Networks. In *The 28th International Conference on Computer Communications (INFOCOM)*, 468–476. Rio de Janeiro: IEEE.

Wang, X., G. Xing, Y. Zhang, C. Lu, R.B. Pless, and C. Gill. 2003. Integrated Coverage and Connectivity Configuration in Wireless Sensor Networks. In *The 1st International Conference on Embedded Networked Sensor Systems (SenSys)*, 28–39. Los Angeles: ACM.

Wattenhofer, R., L. Li, P. Bahl, and Y.-M. Wang. 2001. Distributed Topology Control for Power Efficient Operation in Multihop Wireless Ad Hoc Networks. In *The Conference on Computer Communications - The 20th Annual Joint Conference of the IEEE Computer and communications Societies (INFOCOMM)*, 1388–1397. Anchorage: IEEE.

Wood, A.D., and J.A. Stankovic. 2002. Denial of Service in Sensor Networks. *Computer* 35 (10): 54–62.

worldometer. COVID-19 Coronavirus Pandemic. January 1, 2020. https://www.worldometers.info/coronavirus/?. Accessed 8 July 2020.

Xiao, M., X. Wang, and G. Yang. 2006a. Cross-Layer Design for the Security of Wireless Sensor Networks. In *The 6th World Congress on Intelligent Control and Automation*, 104–108. Dalian: IEEE.

———. 2006b. Cross-Layer Design for WSNs Security. In *Security in Sensor Networks*, ed. Y. Xiao, 311–328. Boca Raton: CRC Press.

Xie, P., and J.-H. Cui. 2007. R-MAC: An Energy-Efficient MAC Protocol for Underwater Sensor Networks. In *The 2nd International Conference on Wireless Algorithms, Systems and Applications (WASA)*, 187–198. Chicago: IEEE.

Xue, Q., and A. Ganz. 2003. Runtime Security Composition for Sensor Networks (SecureSense). In *The 58th Vehicular Technology Conference (VTC)*, 2976–2980. Orlando: IEEE.

Yan, T., T. He, and J.A. Stankovic. 2003. Differentiated surveillance for sensor networks. In *The 1st International Conference on Embedded Networked Sensor Systems (SenSys)*, 51–62. Los Angeles: ACM.

Ye, W., J. Heidemann, and D. Estrin. 2002. An Energy-Efficient MAC Protocol for Wireless Sensor Networks. In *The 21st Annual Joint Conference of the IEEE Computer and Communications Societies (INFOCOM)*, 1567–1576. New York: IEEE.

———. 2004. Medium Access Control with Coordinated Adaptive Sleeping for Wireless Sensor Networks. *IEEE/ACM Transactions on Networking* 12 (3): 493–506.

Ye, F., H. Luo, S. Lu, and L. Zhang. 2005. Statistical En-Route Filtering of Injected False Data in Sensor Networks. *Journal on Selected Areas in Communications* 23 (4): 839–850.

Yessad, S., L. Bouallouche, and D. Aissani. 2011. Proposition and Evaluation of a Novel Routing Protocol for Wireless Sensor Networks. In *The 5th International Workshop on Verification and Evaluation of Computer and Communication Systems (VECOS)*, 1–9. Tunis: BCS, The Chartered Institute for IT.

Yessad, S., N. Tazarart, L. Bakli, L. Medjkoune-Bouallouche, and D. Aissani. 2012. Balanced energy efficient routing protocol for WSN. In *International Conference on Communications and Information Technology (ICCIT)*, 326–330. Hammamet: IEEE.

Yessad, S., L. Bouallouche-Medjkoune, and D. Aïssani. 2015. A Cross-Layer Routing Protocol for Balancing Energy Consumption in Wireless Sensor Networks. *Wireless Personal Communications* 81: 1303–1320.

Yetgin, H., K.T.K. Cheung, M. El-Hajjar, and L. Hanzo. 2015. Cross-Layer Network Lifetime Maximization in Interference-Limited WSNs. *Transactions on Vehicular Technology* 64 (8): 3795–3803.

Yu, Z., and Y. Guan. 2008. A Key Management Scheme Using Deployment Knowledge for Wireless Sensor Networks. *Transactions on Parallel and Distributed Systems* 19 (10): 1411–1425.

Yuan, Z., L. Wang, L. Shu, T. Hara, and Z. Qin. 2011. A Balanced Energy Consumption Sleep
    Scheduling Algorithm in Wireless Sensor Networks. In *The 7th International Wireless Com-
    munications and Mobile Computing Conference (IWCMC)*, 831–835. Istanbul: IEEE.
Zamalloa, M.Z., K. Seada, B. Krishnamachari, and A. Helmy. 2008, June. Efficient Geographic
    Routing over Lossy Links in Wireless Sensor Networks. *Transactions on Sensor Networks*.
Zhang, J., and V. Varadharajan. 2010. Wireless Sensor Network Key Management Survey and
    Taxonomy. *Journal of Network and Computer Applications* 33 (2): 63–75.
Zigbee Alliance. Zigbee. Zigbee Alliance. January 1, 2020. https://zigbeealliance.org/solution/
    zigbee/. Accessed 18 April 2020.
Zorzi, M., and R.R. Rao. 2003. Geographic Random Forwarding (GeRaF) for Ad Hoc and Sensor
    Networks: Multihop Performance. *Transactions on Mobile Computing* 2 (4): 337–348.
Zuniga, M., and B. Krishnamachari. 2004. Analyzing the Transitional Region in Low Power
    Wireless Links. In *1st Annual Communications Society Conference on Sensor and Ad Hoc
    Communications and Networks (SECON)*, 517–526. Santa Clara: IEEE.

Part III
WSNs Experimentation and Analysis

Chapter 7
Testbeds for WSNs

Simulation is imagining ... Testbed is sore reality.

7.1 WSN Testbeds Principles

As iterated throughout this book, WSNs are large-scale distributed embedded systems incorporating small, limited energy and resource-constrained sensor nodes communicating over wireless media. Because of their massively distributed nature, the design, implementation, and evaluation of sensor network applications, middleware, and communication protocols are difficult tasks. The first design steps can often be made with the help of simulations; however, they frequently force the designer to make nonrealistic assumptions about traffic, failure patterns, and topologies. The coming after steps of implementation and evaluation of application performance, as well as assessment of error resilience and other nonfunctional properties, require the use of real hardware, realistic environments, and realistic experimental setups.

Practically, real experiments with distributed systems like WSNs become very cumbersome if the number of nodes exceeds a few dozens. In fact, all the phases of the experiment are almost infeasible without a targeted, specialized support, specifically:

- Deployment of the nodes in the desired, maybe heterogeneous or hierarchical configuration
- Making changes in the software of individual nodes
- Conducting experiments that include both data processing and self-reconfiguration of the network

For all but the smallest experiments, a dedicated infrastructure supporting the above listed steps is necessary. This infrastructure, referred to as *testbed*, makes it possible to create, modify, and observe the target configuration, as hardware and software, in its whole complexity including nodes, communication protocols,

© The Author(s), under exclusive license to Springer Nature Switzerland AG 2023
H. M. A. Fahmy, *Concepts, Applications, Experimentation and Analysis of Wireless Sensor Networks*, Signals and Communication Technology,
https://doi.org/10.1007/978-3-031-20709-9_7

middleware, and application. This target configuration is in some literature referred to as system under examination (SUE) (Handziski et al. 2006).

Current surveys and forecasts predict that the number of wireless devices is going to increase tremendously. These wireless devices can be computers of all kinds, notebooks, netbooks, smartphones, and sensor nodes that evolve into real-world scenarios forming in the future a "real-world Internet." In current research of the Future Internet, small battery-driven devices forming the "Internet of Things" are of special focus as the number of wireless devices is going to increase tremendously. A survey of the Wireless World Research Forum predicted that in the year 2017, there will be seven trillion wireless devices for seven billion humans which is equivalent to 1000 devices per human being on average (Gantz et al. 2008).

In recent networking research, testbeds gain more and more attention, especially in the context of Future Internet and WSNs. This development stems from the fact that simulations and even emulations are not considered sufficient for the deployment of new technologies as they often lack in-depth view of the inner minute functioning details. In order to investigate how protocols and algorithms perform in the real world, experimental research is a capable means. In research institutions, testlabs or testbeds are deployed exclusively for all kinds of research experiments. Setup in realistic environment is indispensable to understand large-scale networks enclosing many devices.

Several global initiatives for experimental network research have started. While these large projects include all aspects of the Future Internet, some special projects focus on the Internet of Things and the real-world Internet. Especially in the Internet of Things with WSNs, there is an upcoming need for large heterogeneous testbeds available 24 hours a day that can be automatically used without supervision.

The deployment of testbeds is challenging and user and operator requirements need to be considered carefully. Therefore, the goal is to design an architecture that allows operators of WSN testbeds to offer numerous users access to their testbeds in a standardized flexible way that matches these requirements. In Sect. 7.1.1 of this chapter, these requirements are comprehensively identified. Section 7.1.2 describes Full-scale and Miniaturized Testbeds. Section 7.1.3 illustrates the concepts of Virtualizing and Federating Testbeds. Section 7.2 focuses on the design and applications of the most significant testbeds, regarding concept illustration, hardware, software, and specific deployment and experimentation. Chapter 9 surveys the WSN manufacturing companies, while Chap. 10 focuses on the datasheets of the sensor motes and the multiplicity of components used in typical testbed implementations.

7.1.1   Requirements from Testbed Deployment

This section identifies the main requirements and design goals for a WSN testbed. Precisely, all phases of the sensor network life cycle are to be considered, specifically, design, deployment, test, and experiment implementation and evaluation. A testbed solution supporting these features will offer substantial help in terms of speeding up the preparation and conducting of experiments with different SUEs and tuning their parameters. As such, a testbed, which is expected to replicate an environment and support all research and experimentation activities, must meet a number of requirements (Lundgren et al. 2002; Handziski et al. 2006; Slipp et al. 2008):

- Building different SUE architectures. A number of different WSN architectures have emerged. As previously elaborated in Chap. 3, in the flat architecture the sensor network is composed of homogeneous sensor nodes running the same application and protocol code. In a segmented architecture, a number of flat networks are coupled by gateways. The different flat networks can use incompatible radio technologies. In a multitier architecture or hierarchical architecture, a sensor network application is partitioned such that parts of it run on low-end sensor nodes, whereas other parts run on more capable high-end sensor nodes which have no energy constraints and have better memory and computational resources. An example is the IEEE 802.15.4 protocol, where the full-function devices (FFDs) have much more responsibilities than the reduced-function devices (RFDs) (LAN/MAN Standards Committee 2006). In addition to being connected to the low-end nodes, the high-end nodes can interact among themselves and with entities higher in the hierarchy via a backbone network.
- Instrumentation and experimental control. The testbed must provide instrumentation for generating data and examining network performance. Facilities must also be provided to monitor, program, test, and run the SUE. It is awkward to obtain an adequate understating of the behavior of some networks without comprehensive instrumentation.

  Another requirement in the setup of experiments is to have control over the times when certain actions like the configuration or start of sensor nodes have to be performed. For example, to investigate the influence of interference on the transmission of data between two nodes, the transmitting node and the interfering node should be started at the same time. It is practically simple for an experimenter to describe such scenarios and have them executed in the testbed.

  The testbed must also ensure programmable configuration of network and node parameters such as application selection, data logging, selection of the nodes that will participate, and ability to pause, resume, and stop experiments. Repeatability of experiments is a requirement to account for varying environmental necessities and functional parameters.

  For an industrial environment, the testbed may have to replicate the multipath that exists in radio harsh environment (RHE) and provide for the control of electromagnetic interference (EMI) that would particularly exist. It should also provide control over signal strength to enable testing under varying signal strength conditions; this includes effects due to changes in distance.
- Easy reprogramming and debugging. The implementation and debugging of sensor network applications requires frequent reprogramming of the nodes with new software. This is needed to compare different solutions, but most critically for stepwise debugging and improving of the software. There exist approaches

(and protocols) to distribute new application code within the sensor network application and over the wireless interface (Wan et al. 2002; Hui and Culler 2004), but this can be tedious in case of low-communication bandwidth and sporadic reachability of individual nodes. It is therefore desirable to support the reprogramming of nodes in an "out-of-band" fashion using the testbed infrastructure. To save time, especially in large testbeds, the reprogramming of nodes should be executable in parallel. Fixing bugs on single nodes can be a hard task and it gets considerably more complicated in a distributed system like a WSN, where individual nodes only contribute slightly to the global state, and race conditions can cause serious headache to the programmer. Determining the current state of the system and how it was reached is crucial for distributed debugging.

However, hosting a distributed debugger that allows a complete "happened-before" ordering of the events simply exceeds the computational capacities of the sensor nodes. A testbed should provide some support for distributed debugging. Simply, the programmer of a sensor node should have some *printf( )*-like routine for generating debug messages with additional timestamping at its disposal. When the timestamps have a reasonable resolution and are properly synchronized with the timestamps generated at other nodes, it is possible to figure out the ordering of events. Noteworthy, generating messages with debug information changes the execution timing of the sensor node software and hence possibly its behavior. This cannot be avoided, but a testbed should give the possibility to transport such debug messages in an "out-of-band" fashion, i.e., without using the primary wireless air interface of a sensor node. This way, the debug messages do not influence the protocol behavior or the bandwidth sharing between sensor nodes, nor is any additional congestion created.

- On-the-fly configuration changes. The main purpose of a testbed is the investigation of different solutions in reproducible experiments under controlled conditions. Checking the robustness of applications and protocols against node failures or addition of new nodes is one of the major experiments to be conducted. A WSN testbed should offer support for testing this robustness under realistic circumstances. Specifically, it should be possible to emulate the expiry of sensor nodes due to energy depletion (fault injection) or the addition of new nodes to the network while assuring complete repetitiveness of the experiment across different software solutions.

  Energy consumption is one of the major performance metrics of sensor network protocols and applications. Sensor nodes often have finite energy budgets and sensor networks have highly time-variable topologies due to the energy depletion of nodes and deployment of new nodes.

- Remote accessibility. The testbed must be remotely accessible to promote collaboration between geographically dispersed teams. Working conditions may vary depending on the location, as well as mote types and software, which necessitate exchange of information in order to reach the best possible installation and equipment.

- Testbed management. In WSN applications, having unique node identifiers is usually not overly important. But this does not apply to testbed management. To keep the testbed operational, it is important to correctly identify nodes, for instance, to decide whether experimental results pertain to the operation of the SUE or whether they are the result of misconfiguration and/or malfunctioning of the sensor nodes. Malfunctioning nodes must be identified and replaced before the experiment is repeated. To this end, knowledge of the exact identifier of the nodes and its position is helpful. Since the exchange of nodes in a testbed will be a frequent event (especially in large testbeds), the node identifiers and their positions should be independent of each other, i.e., the node identifier should not encode its position or vice versa. To find malfunctioning nodes, an association between node identifiers and positions should be nonetheless maintained. Ideally, this association is automatically updated every time a node is added or removed from the testbed, and the testbed user should be able to formulate queries about this association. As a secondary benefit of such an association, the localization information can be made available to the SUE. The SUE implementer hence does not necessarily need to implement localization algorithms on his/her own and can concentrate on other aspects of the application. When, on the other hand, the problem happens to be the implementation and test of a localization algorithm, the known positions provide a ground truth against which the new algorithm can be compared.
- Scalability and extensibility. It is not only a general requirement for sensor networks, but also the testbed infrastructure should support nontrivial numbers of sensor nodes. Consider as an example the continuous generation of debug data or the gathering of application data. Data from all sensor nodes are required to obtain full insight into the operation of the network. For a large sensor network, the sheer volume of debug data can be overwhelming and appropriate mechanisms for aggregating, compressing, or filtering this data are needed.

  Easy extensibility means supporting the variety of existing and newly emerging mote platforms. A rigid SUE is definitely limited and doomed to fading out with the introduction of different technologies and techniques, or the need to use other motes.
- The cost of the testbed must be reasonable. A testbed must ensure the support of several SUEs and provide trustworthy outcomes without unnecessary spending due to overestimated network size, or choice of overpowered motes and equipment.

7.1.1.1   Additional Requirements

Furthermore, the testbed operation would be more convenient if the following requirements are mostly supported, depending on the application and environment (Slipp et al. 2008):

- Real-time monitoring. The testbed should provide the ability to monitor, in real time, the progress of the test with typical measured values and specific performance metrics such as time delay and error assessment.
- Malfunction alerts. In the event of a malfunction, there should be an alert for the operators so that the appropriate action can be taken as soon as possible. This will minimize downtime and avoid wasting time running a test for which data may be tainted.
- Collaboration. The testbed should promote collaboration between researchers by providing the means to easily share WSN technologies developed for use in the testbed. For example, it should be easy to share routing and security protocols and power models.
- Support for simultaneous users. But it is to be noted that simultaneous users would create conditions that are not repeatable. This refers mostly to the sharing of the wireless bandwidth and potential interference.
- Mobility. Although mobility is important to some industrial networks, it may not be a concern to petroleum facilities consisting almost entirely of static nodes.

7.1.1.2   User Requirements from a Testbed

A user is responsible of doing research on WSNs based on testbed experimentation, and he investigates protocols and communication concerns on the testbed before typical WSN implementation, modification, or extension. From a user standpoint, several requirements are necessary to make testbeds useful and efficient (Chatzigiannakis et al. 2010):

- Transparent access to the testbed. Transparency in this context means that sensor nodes in a remote testbed can be handled in the same way as local sensor nodes. In order to use the testbed, a robust, fast, and simple reservation is essential, including support for isolated (noninterfering) experiments with other users.
- Heterogeneity and scale of the testbeds. They are significant in order to address challenges that occur in real-world deployment scenarios. Many algorithms and applications do not scale or behave unstably in large scale; therefore, the network size should be as large as possible. As deployment cost is still a dominating factor, the concept of testbed federation to enable experiments at scale can help to reduce overall facility cost. To enable federation standardization, application program interfaces (APIs) is a key concept if not a prerequisite. In the same way, heterogeneity can affect the performance of an application. Heterogeneity might include memory (program and data), wireless transceiver characteristics, or microcontroller. Heterogeneity can be introduced by using different classes of devices from an embedded 8-bit microcontroller platform to a fully equipped PC. A full support of a testbed for heterogeneity even allows performing interoperability tests.
- Reproducibility. It is a property that is very difficult to achieve in a testbed. However, the support to repeat experiments is needed to reach statistical

soundness of the experiment. An additional supporting feature is detailed logging capabilities to enable post-failure analysis and debugging support. User interaction is very helpful to stimulate events or errors in the testbed during experiment runtime in order to investigate reactions of the system under test.

- Mobility. In the future, support for mobility in testbeds will become more important. Therefore, testbeds deployed today must be prepared for a next step extension by mobile devices.

7.1.1.3 Operator Requirements from a Testbed

The operator is the one who checks the testbed access and the functioning of its constituents, obtains the required measurements, and reports malfunctions. The operator requirements additionally include several qualities that make testbeds handy to manage (Chatzigiannakis et al. 2010):

- Robustness of the testbed. It is definitely desirable for the user too. When nodes or the whole WSN crash due to software errors, the testbed must recover from such failure. Therefore, support for remote or even better automatic reset of the experimental facility to a safe and basic state is compulsory.
- Access control system to the testbed. It is a need to monitor the activities and allow for future accounting and even payment per use.
- Easy installation and a fully automated reservation system. A fully integrated testbed management and maintenance system including performance-monitoring capabilities is definitely helpful and can be addressed as a long-term goal when the testbeds grow in size. The testbed software must be ready for future extensions; therefore, open software is preferable over a closed or proprietary system.

7.1.2 Full-Scale and Miniaturized Testbeds

What are the approaches followed for the design and implementation of testbeds? Specifically, two distinct approaches are taken (Maltz et al. 2001; De et al. 2005a):

- Full-scale testbeds that provide propagation over the actual distances where the network is designed to operate
- Miniaturized testbeds that foreshorten actual distances using some electronic means

This dichotomy exists for several reasons based primarily on need and practicality. Based on need, two classifications of users are distinguishable:

- Users involved in the evaluation and development at the physical layer and cannot ignore the effects of multipath on network performance. Researchers interested in the physical layer typically require a full-scale testbed, as pure software simulation is not that practical. Although there are hardware multipath simulators in existence, such as Elektrobit's Propsim line of radio channel emulators (EB 2006), they have a limited number of paths and they are extremely expensive for equipping a wireless lab.

- Users involved in the development of routing protocols that need control over the topology to determine the effects on network performance. However, the full-scale testbeds do not lend themselves easily to topology control due to the relatively large distances between motes and the irregular geometry of their placement. Researchers interested in topology control typically need a miniaturized testbed. As its name implies, in a miniaturized testbed the motes are placed closer together. This placement is usually according to a regular geometry, such as a grid as used in ORBIT (Raychaudhuri et al. 2005) or Kansei (Arora et al. 2006). Such testbeds are typically equipped with some form of attenuator as in EWANT (Sanghani et al. 2003), MiNT (De et al. 2005b), and MeshTest (Clancy and Walker 2007), and they may have their signals routed through an antenna or connected through a hardwired matrix such as in MeshTest (Clancy and Walker 2007). However, by definition, the miniaturized testbed does not provide realistic multipath, for instance, efforts as presented in (Judd and Steenkiste 2004) utilize an FPGA-based solution for emulating a signal propagation model, which are still far from reality (De et al. 2005a).

On the practicality side, researchers must balance the trade-off between available space and cost. Full-scale testbeds take up a lot of space, require considerable infrastructure, and are relatively expensive to operate. Miniaturized testbeds can usually be placed in available laboratory space and are considerably less expensive than their counterparts; they also have the added advantage of providing a more controllable environment, thus supporting more repeatable experiments.

As detailed in Sect. 7.2, falling in the category of full-scale testbeds are MoteLab (Werner-Allen et al. 2005), Trio (Dutta et al. 2006), TWIST (Handziski et al. 2006), SignetLab (Crepaldi et al. 2007), and SenseNeT (Dimitriou et al. 2007). Moreover, large testbeds are accompanied by software that provides users with the facilities to conduct experiments with the testbed nodes, but are generally limited in their adaptability and configurability to the users' needs. Capabilities such as reconfiguring the network topology or federating multiple testbeds to form a larger virtualized facility may be lacking.

In the category of miniaturized testbeds are MiNT (De et al. 2005b), MiNT-m (De, Raniwala and Krishnan, et al. 2006), and MeshTest (Clancy and Walker 2007). MeshTest is particularly interesting because of its ability to control interference and attenuation while supporting mobility and providing users with control over topology. It does this using a unique RF matrix switch. MiNT is a hybrid testbed, providing both real nodes and ns-2 simulation to provide scalability. It focuses on ad hoc networks and uses attenuators to reduce transmission ranges, but its major drawback is the use of manual attenuators. MiNT-m performs similarly using a number of mobile robots for the mobile nodes.

ORBIT (Raychaudhuri et al. 2005) is a distinct wireless testbed in significant respects. It includes both an indoor miniaturized testbed (sandboxes) and an outdoor full-scale testbed, the miniaturized testbed provides a more controlled environment that facilitates reproducibility, while the full-scale testbed supports real-world testing. ORBIT is also one of the few to incorporate electromagnetic interference (EMI), which it does by employing a vector signal generator (VSG).

Before delving into testbed platforms as made available in the literature, the next section built upon (Baumgartner et al. 2010) introduces the concept of virtual links and federated testbeds in light of their importance for testbed design and implementation.

7.1.3 Virtualizing and Federating Testbeds

In recent years, experimentally driven research on WSNs has been contributory in advancing the state of the art towards new sensing applications, network architectures, and protocol stacks optimized to operate over varied radio technologies and restricted resources under specific deployment strategies. The most commonly applied technique is simulation, which allows rapid development, offers debugging tools, and enables easy repeatability. The technical step that follows is to implement the system on real hardware platforms and experiment through tailored testbed environments. This allows researchers to avoid the inherent limitations of simulation regarding typical hardware characteristics (e.g., buffer sizes, available interrupts) and communication technology behavior (e.g., transmission rates, interference patterns).

In most of the cases, due to the costs of hardware, researchers evaluate their solutions in local testbeds of limited size. While small testbeds provide useful insights into the effectiveness of the system in real conditions, they only offer limited support in terms of heterogeneity, scalability, and mobility. Furthermore, in most cases, as much as a tightly coupled network and software architecture are applicable on a testbed, they limit the number of possible configurations of that testbed.

To overcome limitations in scale, a number of testbeds of significant size have been developed in the last few years. Their size currently levels up to 1000 nodes, and there is a trend towards building even larger testbeds as seen by projects such as SENSEI (Presser et al. 2009) and WISEBED (Chatzigiannakis et al. 2010). Acknowledging the clear and continuing need for large open testbeds in WSN research, the use of federated testbeds that unite isolated WSN testbeds via virtual links provide promising solutions to such questions as:

- How to deal with the ever-increasing total number of nodes demand?
- How do large testbeds cope with heterogeneity in available sensors, radios, computational resources, etc.?
- How to maintain a very large WSN testbed efficiently?
- How to provide hybrid simulation approaches, i.e., the combination of real and simulated testbeds in order to produce extremely large-scale WSN testbeds?

- How to utilize the facilities provided by these testbeds and adapt them to each experiment's needs, i.e., how to define and use specific network topologies that fit into the target application domain?

As suggested in (Baumgartner et al. 2010), virtualized network links are visioned in the following ways:

- Between physically distinct testbeds of varying features (location, size, etc.) and between specific nodes of such testbeds, resulting in larger testbeds with customized cross-network edges
- Between nodes inside a single testbed, thus defining a customized network topology
- Between real and simulated nodes, enabling hybrid simulation for massive network sizes

A virtual link basically enables two testbed nodes, which have otherwise no direct physical radio connection, to communicate in a way that is transparent to the user applications; additionally, existing links, which are reachable within one-hop radio range, can be selectively deactivated between neighboring nodes. Both kinds of virtualization are done in a way that is entirely transparent to a deployed application.

7.1.3.1   Virtual Links and Federated Testbeds

This section details the virtualizing testbed approach. A virtualized testbed is defined as follows (Baumgartner et al. 2010):

- A single physical testbed with a virtualized topology
- Two or more physically distinct testbeds federated into a single unified testbed
- A simulated testbed similarly federated with a physical testbed
- Any combination of the above

The key components of the proposed architecture are shown in Fig. 7.1. A testbed server that acts as the Internet-facing gateway represents each testbed, physical or simulated. A testbed is composed of a number of sensor nodes that can communicate with the testbed server, potentially via gateway devices inside a physical testbed.

A virtual link is then a unidirectional connection between two nodes, in the same or in different testbeds, which would not normally be able to communicate. An arbitrary number of virtual links can thus be created to define a virtualized topology and federate distinct testbeds. It is also possible to deactivate existing physical reachability between two nodes by selectively dropping packets to allow complete topology control. Specifically, virtual links are enabled with a special software on each sensor node; it is a virtual radio which contains a routing table of the form {ID, interface}, such that when sending a message to a specific node ID, the radio can decide on which "interface" to send this message, the node's real radio or the virtual interface which forwards the message to the testbed server.

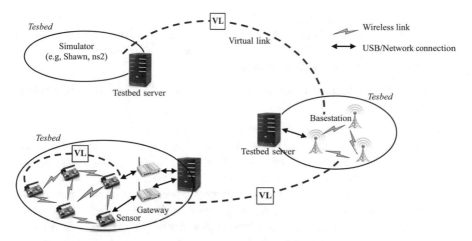

Fig. 7.1 Virtualized testbed architecture. (Based on Baumgartner et al. (2010))

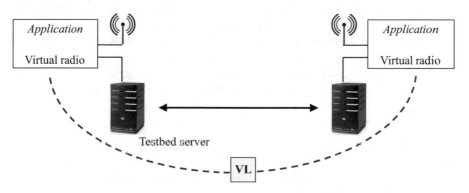

Fig. 7.2 Communication between virtual radio drivers on sensor nodes. (Based on Baumgartner et al. (2010))

7.1.3.2   Topology Virtualization

As illustrated in Fig. 7.2, topology virtualization involves two key elements, virtual radio components and testbed servers.

Experimentation setup is initialized according to a determined arrangement:

- The IDs of virtual radios across the entire virtualized topology are configured by an overall controlling component to ensure uniqueness.
- Virtual topology is configured by each testbed server informing its local sensor nodes of their virtual neighbors, where a virtual neighbor entry in a sensor node's virtual radio simply consists of an ID along with "virtual," meaning any packets to this ID should be sent to the testbed server for further routing.
- How messages reach the testbed server depends on the architecture of the deployed testbed. Routing may be via an out-of-band backbone infrastructure

when sensor nodes are connected 1:1 with gateway devices, or alternatively may reuse the wireless medium of the sensor nodes in testbeds where not every sensor node is directly connected to a gateway device. In either case the procedure is transparent to the application software.

The process of sending a message through the virtualized architecture works as follows:

- Applications on a sensor node send a packet to its virtual radio component. On some operating systems, such as TinyOS, using a virtual radio instead of a real one is simply a matter of component configuration. On others, it may require changing radio function calls in the application's source.
- The virtual radio component then uses its local routing table, as configured by the testbed server to decide on which interface to send this message, whether via the real radio or the virtual topology service thru the testbed server. When broadcasting, a packet is simply sent on both interfaces.
- If the message is sent to the testbed server, the server examines the destination ID of the packet and forwards this either to another testbed server, which is responsible for that node, or to the corresponding node in the local testbed. If the packet is broadcast, the testbed server forwards the message to all virtual neighbors of this node by generating one message for each neighbor.
- Finally, when receiving a message on the real radio interface, the virtual radio component checks its routing table to determine whether or not the sender is configured to be a neighbor in the currently configured topology; if not the packet is dropped and so it never reaches the application.

All components of this procedure are completely transparent to the application, which simply sees a radio component conforming to a common radio API.

Through experimentation, building scalable federated testbeds provided interesting results (Baumgartner et al. 2010):

- The interconnection of two closely located testbeds, i.e., with short delays between testbed servers, works very well in practice, as the virtual links operate significantly faster than physical links. An experiment can be tuned so that applications cannot detect that they are running in a physically separated network.
- When moving to wider-area networks, the existence of a sufficiently fast Internet backbone becomes a must, since latency may degrade and severely affect realism in large intercontinental federations.

Optimistically, topology virtualization is a promising approach to create large federations of physically detached and heterogeneous networks.

The following section details the most widely quoted testbeds, their hardware and software composition, and how they may be categorized. Many of the testbeds were snapshots in time, they satisfied a given experimentation and ceased to exist, and that is why the focus of this chapter is on the experimentation on typically implemented testbeds aside from their current existence.

Tables 7.2 and 7.3 compare the pertinent features and composition of the presented testbeds and focus on the categories that embrace them.

7.2   Testbeds Illustrated

7.2.1   ORBIT

The open-access research testbed for next-generation wireless networks (ORBIT) was first funded in 2003 under the network research testbed (NRT) program and subsequently under follow-on grants (ORBIT 2014). Construction of the 464.5 m^2 (5000 ft^2), 400-node ORBIT radio grid facility at the WINLAB[1] (WINLAB 2010) Tech Center II building in North Brunswick, NJ, was completed in mid-2005, leading to the first community release of testbed services in October 2005. Since then, it has become a widely used community resource for evaluation of emerging wireless network architectures and protocols.

Generally, the development of a general-purpose open-access wireless multiuser experimental facility poses significant technical challenges that do not arise in wired network testbeds such as Emulab (Flux Group 2014). In particular, it is far more difficult to set up a reproducible wireless networking experiment due to random time variations in mobile user location and associated wireless channel models. In addition, wireless systems tend to exhibit complex interactions between the physical, medium-access control and network layers, so that strict layering approaches often used to simplify wired network prototypes cannot be applied here. Some of the basic characteristics of radio channels that need to be incorporated into a viable wireless network testbed are as follows (Raychaudhuri et al. 2005):

- Radio channel properties depend on specific wireless node locations and surroundings.
- Physical layer bit rates and error rates are time varying.
- Shared medium layer-2 protocols on the radio link have a strong impact on network performance.
- There are complex interactions between different layers of the wireless protocol stack, and currently, their mutual interaction cannot be studied easily.
- Users exhibit random mobility; location also plays a role.

The key design goals as adopted in ORBIT are as follows (Raychaudhuri et al. 2005):

- Scalability in terms of the total number of wireless nodes (hundreds)
- Reproducibility of experiments, which can be repeated with similar environments to get similar results
- Open-access flexibility giving the experimenter a high level of control over protocols and software used on the radio nodes

---

[1] Wireless Information Network Laboratory (WINLAB), an industry-university cooperative research center focused on wireless technology, was founded at New Jersey's Rutgers University in 1989. Its research mission is to advance the development of wireless networking technology by combining the resources of government, industry, and academia. The center's educational mission is to train the next generation of wireless technologists via graduate research programs that are especially relevant to industry.

- Extensive measurement capability at radio PHY, MAC, and network levels, with the ability to correlate data across layers in both time and space
- Remote-access testbed capable of unmanned operation and the ability to robustly deal with software and hardware failures

Functionally, ORBIT is a two-tier wireless network emulator/field trial designed to achieve reproducible experimentation while also supporting realistic evaluation of protocols and applications:

- The radio grid testbed. It is central to the ORBIT facility that uses a novel approach based on a 20 × 20 two-dimensional grid of programmable radio nodes, which can be interconnected into specified topologies with reproducible wireless channel models.
- Outdoor ORBIT network. Once the basic protocol or application concepts have been validated on the radio grid emulator, users can migrate their experiments to the outdoor ORBIT network which provides a configurable mix of both high-speed cellular (WiMAX, LTE) and 802.11 wireless access in a real-world setting.

The main ORBIT radio grid and outdoor testbeds have been further supplemented with a number of experimental "sandboxes" which allow researchers to debug and test their code without tying up the resources of the larger radio grid. Further details are provided in Sect. 7.2.1.1.

As of 2014, there are over 1000 registered ORBIT users who have conducted a total of over 200,000 experiment-hours on the radio grid testbed to date, with 55,701 experiment-hours served during 2013. The ORBIT testbed is also being used to support wireless aspects of the global environment for network innovation (GENI) future Internet testbed (Sect. 7.2.12), and the ORBIT management framework (OMF) is being used as one of the core control frameworks in GENI. Examples of specific experiments that have been run on the ORBIT testbed include multi-radio spectrum coordination, cognitive radio networks, dense Wi-Fi networks, cellular/ Wi-Fi multi-homing, vehicular and ad hoc network routing, storage-aware/delay-tolerant networks, mobile content delivery, location-aware protocols, interlayer wireless security, future Internet architecture, and mobile cloud computing.

ORBIT is available for remote or on-site access by academic researchers both in the United States and internationally (prospective users should first send in an account signup request using a registration form). Users will have access to the following resources (ORBIT 2014):

- Range of radio resources including Wi-Fi 802.11a/b/g 802.11n 802.11 ac, Bluetooth (BLE), ZigBee, software-defined radio (SDR) platforms such as universal software-defined radio (USRP)[2] (National Instruments 2014), USRP N210 (Ettus Research 2012), USRP X310 (Ettus Research 2014), wireless open-access

---

[2]Universal Software Radio Peripheral (USRP) is a range of software-defined radios designed and sold by Ettus Research and its parent company, National Instruments. Developed by a team led by Matt Ettus, the USRP product family is intended to be a comparatively inexpensive hardware platform for software radio and is commonly used by research labs, universities, and hobbyists (Ettus 2014).

research platform (WARP) (WARP 2014), and RTL-SDR (RTL-SDR.com 2014).

- Software-defined networking (SDN) resources, NEC (NEC 2014) and Pronto[3] switches (Pica8 2014), and NetFPGA 1G (NetFPGA 2014b) and NetFPGA 10G (NetFPGA 2014a) platforms.
- WiMAX and LTE basestations and clients.

7.2.1.1 Hardware

The ORBIT laboratory is comprised of three test domains in which experimenters can use the hardware provided by WINLAB for use in wireless experimentation. As emphasized in the coming sections these domains are as follows (ORBIT 2014):

- ORBIT grid
- Outdoor testbed
- Sandboxes

Moreover, a Chassis Manager (CM) is a simple, reliable, platform-independent subsystem for managing and autonomously monitoring the status of each node in the ORBIT network testbed.

7.2.1.1.1 ORBIT Grid

The main domain is the ORBIT grid. As illustrated in Fig. 7.3, it consists of a multiply interconnected, 20 × 20 grid of ORBIT radio nodes, some non-grid nodes to control RF spectrum measurements and interference sources, front-end servers, application servers, and back-end servers that support various ORBIT services. The back-end ORBIT infrastructure delivers overall services and control to the test facility. Each node of the 20 × 20 grid has at least two wireless interfaces.

A radio node is a PC typically equipped with two 100BaseT Ethernet ports, radio cards, and a CM to control the node. The two Ethernet ports are connected to data and control subnets. While control is used primarily for node access, experiment management, and measurement collection, the data subnet is exclusively available to the experimenter. In addition to the standard radio cards on most of the nodes, there are many other devices throughout including devices for various communication protocols as well as software radio platforms such as USRP.

A recent LV-67J motherboard, which features Intel third-generation core i7/ i5/ i3 equipped with a two DDR3[4] 1066/1333 MHz up to 16-GByte SDRAM

---

[3] Pronto and PICA8 were merged in February 1, 2012; the merged company is named PICA8, Inc. Pronto became PICA8's brand name (Pronto Systems 2012).

[4] Double data rate type three (DDR3) is the current standard for the fast SDRAM system memory and predecessor to DDR4 (Murray 2012).

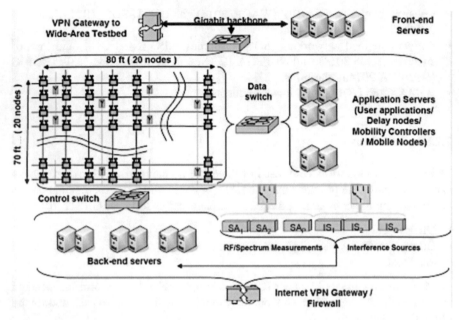

Fig. 7.3 ORBIT hardware (ORBIT 2014)

(COMMELL 2014e), is intended to replace the LV-67C and the LV-67G still being used in the grid as well as some of the sandboxes. LV-67C is provided with an Intel core i7/ i5/ i3 Pentium desktop and two up to 8-GByte DDR3 1066/1333 MT/sec DIMM[5] (COMMELL 2014b). The LV-67G is a second-generation Intel core i7/ i5/ i3/Pentium/Xeon desktop with two up to 16-GByte DDR3 1066/1333 MHz DIMM (COMMELL 2014d).

Experimenters can access the ORBIT radio grid via an Internet portal, which provides a variety of services to assist users with setting up a network topology, programming the radio nodes, executing the experimental code, and collecting measurements.

7.2.1.1.2 Outdoor Testbed

The secondary large ORBIT domain is the outdoor testbed consisting of 22 nodes:

- Ten fixed nodes placed at five different locations within ORBIT/WINLAB facility. Each location features two ORBIT radio nodes.

---

[5] Dual in-line memory module (DIMM) is a double SIMM (single in-line memory module). Like a SIMM, it is a module containing one or several RAM chips on a small circuit board with pins that connect it to the computer motherboard. Transfer speed is computed in million transfers/sec (MT/sec) (WhatIs.com 2014).

- Twelve mobile nodes that can be installed in ORBIT vehicular facilities. There have been three versions of mobile nodes to date; however, there are very limited numbers that have actually been deployed from the latest version, LV-67K, which is a third-generation Intel core i7/i5/i3 mobile processor with two up to 16-GByte DDR3 1066/1333/1600 MHz SO-DIMM[6] (COMMELL 2014f). The two versions being used are the LV-67B that features an Intel Penryn processor with two up to 8-GByte DDR3 800/1066 MHz SO-DIMM (COMMELL 2014a) and the LV-67F provided with Intel core i7/i5/i3, Celeron, Pentium Mobile processor and 2- up to 8-GByte DDR3 800/1066 MHz SO-DIMM (COMMELL 2014c).

7.2.1.1.3   Sandboxes

In addition to the main grid, there are also nine additional smaller test grids (sandboxes). The LV-67C and the LV-67G used in the ORBIT grid are also adopted in the sandboxes. Available sandboxes include Wi-Fi, WiMAX, OpenFlow[7] (McKeown et al. 2008), and USRP2.

7.2.1.1.4   Chassis Manager

The ORBIT Chassis Manager (CM) is a simple, reliable, platform-independent subsystem for managing and autonomously monitoring the status of each node in the ORBIT network testbed. Each ORBIT grid node consists of one radio node with two radio interfaces, two Ethernet interfaces for experiment control and data, and one CM with a separate Ethernet network interface. The radio nodes are positioned about 1 m apart in a rectangular grid. Each CM is tightly coupled with its radio node host. CM subsystems are also used with non-grid support nodes. There are basic requirements the CMs must perform:

- Issue a system reset to the radio node.
- Control the power state of the radio node.
- Obtain chassis status.
- Provide a pass-through *Telnet*[8] (TechTarget 2014) session to the radio node.

---

[6] Small outline dual in-line memory module (SO-DIMM) used in laptops is about half the length of a regular-size DIMM. Most desktop computers have plenty of space for RAM chips, so the size of the memory modules is not a concern. However, with laptops, the size of components and memory modules, as well, matters significantly (PC Glossary 2014).

[7] OpenFlow is an open standard that enables researchers to run experimental protocols in the campus networks used every day. It is added as a feature to commercial Ethernet switches, routers, and wireless access points and provides a standardized hook to allow researchers to run experiments, without requiring vendors to expose the internal workings of their network devices. Major vendors, with OpenFlow-enabled switches now commercially available, are currently implementing OpenFlow.

[8] *Telnet* is a user command and an underlying TCP/IP protocol for accessing remote computers.

- Provide CM diagnostics. These include the following:
  - Provide a means to locally interrogate the grid position of the node. The grid position is either the x,y coordinate of the node or the ID number of a non-grid node. The grid position is used to create the static IP address of the CM.
  - Provide local visual indication of a node operational status.
  - Provide a control API to the experiment controller. An experiment controller (EC), also referred to as the "node handler," is the ORBIT system component that configures the grid of radio nodes for each experiment.

7.2.1.2 Software

The ORBIT radio grid testbed is operated as a shared service to allow a number of projects to conduct wireless network experiments on-site or remotely. Although only one experiment can run on the testbed at a time, automating the use of the testbed allows each one to run quickly, saving the results to a database for later analysis.

Precisely, as displayed in Fig. 7.4, ORBIT may be viewed as a set of services into which one inputs an experimental definition and receives the experimental results as output. The experimental definition is a script that interfaces to the ORBIT services. These services can reboot each of the nodes in the 20×20 grid; then load an operating system, any modified system software, and application software on each node; and then set the relevant parameters for the experiment in each grid node and in each non-grid node needed to add controlled interference or monitor traffic and interference. The script also specifies the filtering and collection of the experimental data and generates a database schema to support subsequent analysis of that data.

7.2.1.2.1 Experiment Control

The main component of the experiment management service is the node handler that functions as an experiment controller. It multicasts commands to the nodes at the

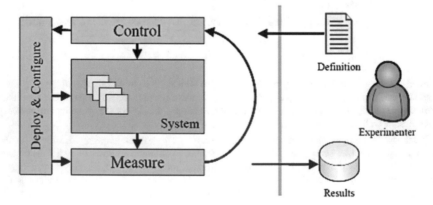

Fig. 7.4 ORBIT software (ORBIT 2014)

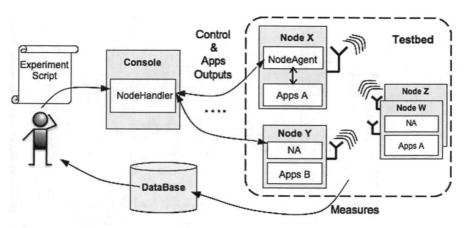

Fig. 7.5 Execution of an experiment from a user point of view (ORBIT 2014)

appropriate time and keeps track of their execution. The node agent software component resides on each node, where it listens and executes the commands from the node handler; it also reports information back to the node handler. The combination of these two components gives the user the controls over the testbed and enables the automated collection of experimental results. Because the node handler uses a rule-based approach to monitoring and controlling experiments, occasional feedback from experimenters may be required to fine-tune its operation. Figure 7.5 illustrates the execution of an experiment from the user's point of view.

Finally, using the node handler, via a dedicated image node experiment, the user can quickly load hard disk images onto the nodes of his/her experiment. This imaging process allows different groups of nodes to run different OS images; it relies on a scalable multicast protocol and the frisbee system for saving, transferring, and installing entire disk images (Hibler et al. 2003). Similarly, the user can also use the node handler to save the image of a node's disk into an archive file.

7.2.1.2.2 Measurement and Result Collection

The ORBIT measurement framework and library (OML) are responsible for collecting the experimental results. It is based on a client/server architecture as depicted in Fig. 7.6. The node handler for a particular experiment execution starts one instance of an OML collection server. This server listens and collects experimental results from the various nodes involved in the experiment; it uses an SQL database for persistent data archiving of results.

On each experimental node, one OML collection client is associated with each experimental application. An application forwards any required measurements or outputs to the OML collection client. The OML client will optionally apply some filter/processing to the measurements/outputs and then sends them to the OML collection server.

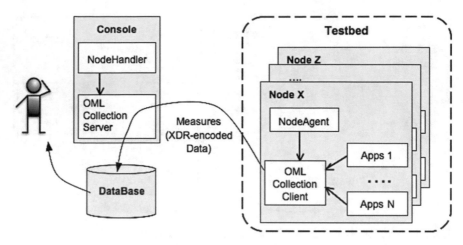

Fig. 7.6 OML component architecture (ORBIT 2014)

Finally, the ORBIT platform also provides the Libmac library. Libmac is a user-space C library that allows applications to inject and capture MAC layer frames, manipulate wireless interface parameters at both aggregate and per-frame levels, and communicate wireless interface parameters over the air on a per-frame level. Users can interface their experimental applications with Libmac to collect MAC layer measurements from their experiments.

7.2.2 MoteLab

Manually reprogramming dozens of WSN nodes, deploying and locating them on the physical environment, and instrumenting them to gather, extract, and debug performance data are tedious and time-consuming. To address this need, MoteLab (Werner-Allen et al. 2005), one of the early testbeds, has been developed and deployed at Harvard University; it is a Web-based sensor network testbed. MoteLab has several features:

- It consists of a set of permanently deployed sensor network nodes connected to a central server in charge of handling reprogramming and data logging while providing a Web interface for creating and scheduling jobs on the testbed.
- It accelerates application deployment by streamlining access to a large, fixed network of real sensor network devices; it also speeds up debugging and development by automating data logging, thus allowing the performance of sensor network software to be evaluated offline.
- Additionally, by providing a Web interface, MoteLab permits both local and remote users access to the testbed; also, its scheduling and quota system ensure fair sharing.

- A node is equipped with a network-connected digital multimeter, allowing the MoteLab back end to continuously monitor the energy usage of the node. Current consumption data is logged and returned with other data generated during the experiment. Gathering energy usage data simply requires checking a box in the Web interface.

MoteLab proved to be invaluable for both research and teaching; its source is freely available, easy to install, and already in use at several other research institutions.

7.2.2.1 Technical Details

MoteLab is a set of software tools for managing a testbed of Ethernet-connected sensor network nodes. A central server handles scheduling, reprogramming nodes, logging data, and providing a Web interface to users. Users access the testbed using a Web browser to set up or schedule jobs and download data.

MoteLab hardware and software are detailed in the sections to come. Figure 7.7 displays MoteLab software components and illustrates how they communicate. From the figure, three external users are using the lab. User A is setting up a job to be run later. User B is accessing the MySQL tables directly to process data collected during a previous experiment. User C has a job running and has made a direct connection to a serial forwarder (SF) to receive or send messages to the attached node.

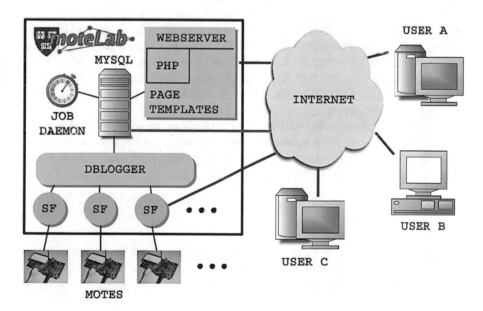

Fig. 7.7 Interaction between MoteLab components and users (Werner-Allen et al. 2005)

7.2.2.1.1   MoteLab Hardware

MoteLab software manages a fixed array of WSN nodes equipped with Ethernet interface backchannel boards allowing remote reprogramming and data logging. The original testbed comprised 26 Mica2 motes (Crossbow 2002). A Mica2 mote consists of a 7.3-MHz ATmega128L processor, 128 KByte of code memory, 4 KByte of data memory, and a Chipcon CC1000 radio operating at 433 MHz with a data rate of approximately 34 Kbps (Chap. 10). These were attached to 26 Ethernet interface boards, specifically, 6 Ethernet PRogramming Boards (EPRBs) developed at Intel research by Phil Buonadonna and 20 Crossbow MIB 600 emotes (Crossbow 2004a). Both provide one TCP port for reprogramming and another for data logging. MoteLab was later upgraded to embody 30 MicaZ motes (Crossbow 2006), which upgrade the Mica2 with Chipcon CC2420 IEEE 802.15.4-compliant radios (Chap. 10).

7.2.2.1.2   MySQL Database Back End

MoteLab software runs on a central server running Linux with Apache, MySQL, and PHP. A MoteLab job consists of some number of executables and testbed nodes, a description mapping each node used to an executable, several Java class files used for data logging, and other configuration parameters, such as whether or not to perform power profiling during the experiment. Once a job is created, MoteLab stores the configuration information allowing the same job to be run multiple times, for different amounts of time or at different times of day.

   MoteLab uses a MySQL database to store all information needed for a testbed operation. This information is divided into two categories, job-generated data and testbed state. When a user account is created, a MySQL database is created for that user to hold all his job-generated data. A new set of tables is created for each instance of a job run, one table for each message type associated with the job. The user is given access rights to his database, allowing him to leverage the MySQL query language for post-processing.

   A separate database holds all lab state information, including user information and access rights, node state, information about uploaded executables and class files, job properties, and a representation of the lab schedule. Such state information is provided to and modified by all of the other main MoteLab components.

7.2.2.1.3   Web Interface

MoteLab uses PHP to generate dynamic Web content and JavaScript to provide an interactive user experience. This allows users to access the lab in a platform-independent way. After logging in, a normal user has access to several Web interface functionalities:

- Home Page. It provides a summary of pending, running, and completed jobs and the ability to download data logged by the lab during past experiments.
- User Info. Such as the instructions for database access, serial forwarder (SF) access, and the ability to change lab passwords.
- Create Job. Users can upload executables, choose which executables will run on which testbed nodes, upload class files for message parsing, and choose from among various options including whether to run power profiling during the experiment. Administrators can also choose to run a job as a daemon for a given period at specified intervals and choose programs to run on MoteLab during and after job execution.
- Edit Job. It permits the same abilities as the job creation page, but reloads it with information from a stored job.
- Schedule. It presents a view of the current state of the lab, including finished, running, and pending jobs, and the ability to schedule a job at various degrees of granularity. Users can delete their own pending jobs; administrators can delete any.

Two additional pages are provided for administrators. The first allows new user accounts to be created and modified, and the second allows lab partitioning to be configured.

7.2.2.1.4   DBLogger

DBLogger is a Java program to be started at the beginning of every job. It connects to each node via the data logging TCP port and uses class introspection to parse messages sent over its serial port and inserts them into the appropriate MySQL database. The individual fields of each message sent are parsed and their values extracted into the database. The resulting table structure is identical to the message structure, with the addition of fields identifying which testbed node originated the message, the time the message was inserted into the database, and a global sequence number.

7.2.2.1.5   Job Daemon

The Job Daemon is a Perl script[9] (Perl.org 2002) run as a cron job[10] (HostGator 2002). The Job Daemon sets up experiments that involves reprogramming nodes and

---

[9] Perl is a family of high-level, general-purpose, interpreted, dynamic programming languages. The languages in this family include Perl 5 and Perl 6.

[10] A cron job is a Linux command for scheduling a command or script on a server to automatically complete repetitive tasks. Scripts executed as a cron job are typically used to modify files or databases; however, they can perform other tasks that do not modify data on the server, like sending out email notifications.

starting other necessary system components (including the DBLogger and serial forwarders) and tears them down when finished, that is, stopping node activity, killing processes necessary during the job, and dumping the data from the MySQL database into a format suitable for download.

7.2.2.1.6   User Quotas, Direct Node Access, Power Measurement

Several practices in MoteLab operation merit attention:

- User quotas. They facilitate sharing the lab between multiple users. The quota does not control how much total access a user can have; rather, it limits the number of outstanding jobs that the user can post to the lab at once.
- Direct node access. In addition to logging data to a database through DBLogger, users have direct access to each node's serial port over a TCP/IP connection. This permits the use of custom programs for monitoring and injecting data into the running application. Because the interface boards allow a single TCP connection to the node, the TinyOS SerialForwarder program (TinyOS Wiki 2012), which acts as a TCP multiplexer, is used.
- In situ power measurement. One node on the network is connected to a networked Keithley 2701 digital multimeter (Keithley Instruments 2002); the use of this device is displayed on the Create Job page. The Keithley multimeter can sample continuously at 250 Hz, and it bursts at 3000 Hz; actually, only continuous operation was supported. Time-stamped current data is included in the download archive if the user has selected this option.

7.2.2.2   Use Models

There are two different ways to use MoteLab, batch mode and real-time access. Users can schedule a large number of jobs to be run unattended in a batch fashion, or they can interact directly with their running job by attaching to the exposed per-node serial forwarders or by exploiting real-time access to the MySQL database.

7.2.2.2.1   Batch Use

Sensor network experimentation begins on the desktop. Following local testing to verify that his application produces data, a user is ready to use MoteLab. After logging on, users proceed to the Create Job page. Subsequently, uploading the necessary files and specifying job parameters, they go to the Schedule page and schedule their job some time later in the future.

When the job is ready to run, the Job Daemon reprograms the network and starts the DBLogger with the user-uploaded class files. The job is now live, and data sent to the serial port of any node will be parsed and inserted into the appropriate MySQL tables created for this job. When the job completes, the Job Daemon removes the executable from the lab and archives job data. After a job completion, post-processing can be done by parsing the data dump files in the job download or by directly accessing the MySQL database.

7.2.2.2.2   Real-Time Access

MoteLab allows researchers to connect directly to the serial forwarder running during their job via a set of dedicated ports on the MoteLab machine; this permits several ways of interacting with a running job:

- A researcher may have a data set to inject for simulation, either because the data collected is not of the type that could be collected on MoteLab or to make experiments reproducible.
- Another use of the direct serial forwarder access is to analyze real-time data. Real-time data processing is possible either by connecting to the serial forwarders providing a data stream for each node or by accessing the MySQL database during the job.

The Connectivity Daemon is an example of the applications that use real-time access, as well as almost every other feature available on MoteLab. The Connectivity Daemon is a job used to collect information eventually used to graphically illustrate connectivity between lab nodes on the Maps page.

7.2.2.3   MoteLab Applications

Several applications have benefited from MoteLab for research and teaching:

- MoteTrack (Lorincz and Welsh 2004). It is an RF-based location tracking system developed for TinyOS-based motes. MoteTrack represents a case where MoteLab is used not just to develop a complete sensor network application in an arbitrary environment, but as a valuable infrastructure in its own right. The distribution of MoteLab nodes around the University of Harvard campus allowed to achieve good coverage and develop a building-wide location tracking system, which would not have been possible in a single, smaller lab.
- The Harvard CodeBlue project (Lorincz et al. 2004) developed robust protocols and services for integrating wireless devices into a range of medical care settings. MoteLab was an essential resource for developing the CodeBlue system, permitting tests and experimentation in a realistic setting on real motes.
- Instructional use. MoteLab is a valuable tool for teaching sensor network concepts, allowing students to experiment with a real testbed. The Web-based interface simplifies the mechanical aspects of programming and debugging the network.
- External users and external MoteLabs. MoteLab accounts were made available to external researchers upon request.

7.2.3 Meerkats

Meerkats as developed at the University of California, Santa Cruz (UCSC), is a wireless network of battery-operated camera nodes that can be used for monitoring and surveillance of wide areas (Boice et al. 2005). An important feature of Meerkats when compared with systems like Cyclops (Rahimi et al. 2005) is that nodes are equipped with sufficient processing and storage capabilities to be able to run relatively sophisticated vision algorithms, e.g., motion estimation locally and/or collaboratively.

Meerkats' main contributions include the following:

- Application-level visual sensor acquisition and processing techniques such as image acquisition policies including cooperative, event-driven policies, and visual analysis for event detection, parameter estimation, and hierarchical representation.
- Resource management strategies that dynamically assess the power versus application-level requirements to make decisions on the tasks to be performed by the system, e.g., what data representation level to use in transmitting data at a given point in time.
- Network-level techniques for bandwidth and power adaptive routing as well as media scaling.

The coming section focuses on Meerkats' hardware and software composition.

7.2.3.1 Hardware

Meerkats consists of eight visual sensor nodes and one gateway also used as the information sink. The laptop is a Dell Inspiron 4000 with PIII CPU, 512-MByte memory, and 20-GByte hard disk, and it is used as the sink and as a gateway. It runs Linux (kernel 2.4.20) and uses an Orinoco Gold 802.11b wireless card for communication (Lucent Technologies 2015).

Visual sensor nodes use Crossbow's Stargate (Crossbow 2004b). In a Meerkats node (Fig. 7.8), the Crossbow's Stargate platform has an XScale PXA255 CPU (400 MHz) with 32-MByte flash memory and 64-MByte SDRAM. PCMCIA and CompactFlash connectors are available on the main board. The Stargate also has a daughter board with Ethernet, USB, and serial connectors. Each Stargate is connected with an Orinoco Gold 802.11b PCMCIA wireless card and a Logitech QuickCam Pro 4000 webcam (PC 2015) connected via USB. The QuickCam can capture video with a resolution up to 640 × 480 pixels. A customized 7.4-volt, 1-Ah, 2-cell lithium-ion (Li-Ion) battery is used and an external DC-DC switching regulator with efficiency of about 80%. The operating system is Stargate version 7.3, which is an embedded Linux (kernel 2.4.19) system.

The choice of Crossbow's Stargate as the Meerkats node main component was based on several considerations:

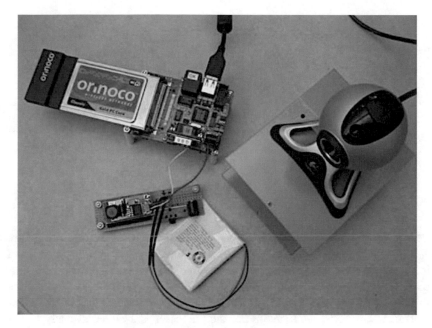

Fig. 7.8 Visual sensing node in Meerkats (Boice et al. 2005)

- Design focus is not on hardware, so off-the-shelf components are picked up.
- A platform that runs an open-source operating system is chosen.
- Since the webcam is the visual sensor, there is a need for a board with a USB connector.
- A platform is required to provide reasonable processing and storage capabilities.

An important feature provided by the Stargate is its battery monitoring capability. This is achieved through a specialized DS2438 chip (Maxim Integrated 2005) on the main board. Two kernel modules provide access to the battery monitor chip and retrieve information about the battery's current state.

7.2.3.2 Software

The Meerkat's node software organization, shown in Fig. 7.9, consists of three main components, namely, the Resource Manager, Visual Processing, and Communication modules.

7.2.3.2.1 Resource Manager

The Resource Manager is the main thread of control running on the Meerkats' node. It controls the activation of the webcam and wireless network card in order to

Fig. 7.9 Meerkats'
software organization
(Boice et al. 2005)

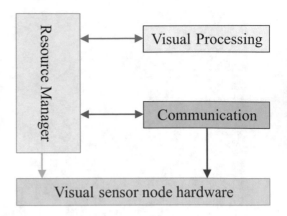

perform image acquisition/processing and communication-related tasks (e.g., transmit an image), as needed. For energy conservation, the Resource Manager has the Meerkats' sensor node operating on a duty-cycle basis, i.e., the node periodically wakes up, performs some tasks as needed, and goes back to either idle or sleep mode. While sleep is the mode with the lowest power requirements, idle mode has a number of variations. At a minimum, the processor is awake and ready to work, even though there are no active processes running. Other variations of idle are as follows:

- Processor and wireless network card ready
- Processor and webcam ready
- Processor, wireless network card, and webcam ready

These variations correspond to the cases where the node is ready to engage in communication-related tasks, image acquisition/processing tasks, or both. More on energy consumption is presented in Sect. 7.2.3.3. An accurate power consumption analysis for the different elementary tasks forming a duty cycle, along with a number of different duty cycle configurations and related energy measurements, was presented in (Margi et al. 2006).

7.2.3.2.2   Visual Processing

The Visual Processing module performs all vision-related tasks, including image acquisition, compression, and processing; it is invoked by the Resource Manager after activating the webcam. The goal is to detect events, in the form of moving images. Upon completion, Visual Processing returns control to the Resource Manager with a parameter flagging whether an event has been detected and, as well, a set of parameters including the number of moving blobs in the image and the velocity of each blob. If an event is detected, the relevant portion of the image is JPEG compressed and transmitted to the sink.

Moving blobs in the image are detected using a fast motion analysis algorithm described in (Lu and Manduchi 2006). The algorithm is comprised of three stages:

- First, local differential measurements are used to determine an initial labeling of image blocks and exercising a total least squares approach with fast implementation.
- Then, belief propagation is used to impose spatial coherence and resolve aperture effect inherent in textureless areas.
- Finally, the velocity of the resulting blobs is estimated via least squares regression. On the Meerkats' node, the motion analysis algorithm, applied on a pair of consecutive images, takes about 0.9 s and consumes 0.16 coulomb.

7.2.3.2.3   Communication

Communication between nodes and the sink is based on 802.11b links. Multihop routing is performed using the dynamic source routing (DSR) protocol (Johnson and Maltz 1996). This is an on-demand routing mechanism especially designed for multihop wireless ad hoc networks. The version of DSR running on the Meerkats' nodes was ported from the DSR kernel module available for the PocketPC Linux (Song 2001).

Two types of data are handled by the application layer, specifically, control packets exchanged between nodes via UDP for synchronization and alerting and image data transmitted from nodes to the sink via TCP. The sink runs a multithreaded server program that listens for connection requests from sensor nodes, opens a connection, receives image files, and renders images on the sink's console.

Experiments revealed sporadic instability problems using the 802.11b links. In order to minimize the effect of this instability, a simple fault recovery procedure is implemented:

- When control packets are transferred via UDP, the receiver is required to send an ACK back to the sender. If within a fixed period of time the sender does not receive an ACK from the receiver, it resends the same control message during the next duty cycle.
- In the case of image data being sent from a camera node to the sink via TCP, a timer is set up at the sender to monitor the establishment of a TCP connection. If the TCP connection is not built within a fixed period of time, the sender considers that transmission failed and tries to set up a TCP connection again in the next duty cycle.

Two nodes may coordinate when tracking a moving object in the scene. In an adopted master-slave scenario, the master node acquires and processes images on a regular basis. If it detects an event, it sends a short alert packet to the slave node. Meerkats nodes are not interruptible while in sleep more and, therefore, the slave node needs to periodically wake up and listen for messages from the master; if it receives an alert packet, it takes and compresses an image.

7.2.3.3   Energy Consumption Characterization Benchmark

Energy consumption characterization benchmark consists of a set of basic operations that are representative of activities performed by visual sensor nodes. Meerkats benchmark consists of five main task categories (Margi et al. 2006):

- Idle. The idle or baseline benchmark captures the energy consumption behavior of the node when only basic operating system tasks are running. This benchmark characterizes energy consumption when the system is idle and also serves as baseline for all other tasks.
- Processing-intensive. The characterization of processing-intensive tasks is performed using the FFT benchmark (Frigo and Johnson 2015), which is part of SPEC CPU2000 (Henning 2000), an industry-standardized CPU-intensive benchmark suite.
- Storage-intensive. The storage media available on the Stargate is flash memory. In order to understand its energy consumption behavior, a program is used to read and write files with random data.
- Communication-intensive. To characterize the energy consumed by communication-related tasks, a set of UDP client/server programs is used. The client program transmits a certain amount of random bytes, provided as an argument, to the server. To obtain the energy cost of transmission, the client program is run on the Stargate being monitored, and then the Stargate is monitored when running the server program to obtain the energy cost of reception.
- Visual sensing. Power consumed by the webcam is characterized using the *videotime* program available on the Stargate 7.3, to acquire a sequence of frames.

At steady state, Meerkats nodes running the different benchmarks with different combinations of active hardware subsystems highlighted a number of interesting observations:

- There is a considerable difference in power consumption when comparing results from the "sleep" and "idle" benchmarks.
- Communication-related tasks (i.e., receive and transmit) are less expensive than intensive processing and flash access when the radio modules are loaded.
- The processing-intensive benchmark results in the highest current requirement.
- Flash reads and writes cost about the same.
- Transmit is only about 5% more expensive than receive.

7.2.3.4   Image Acquisition Analysis

The goal of Meerkats is to detect and track moving bodies within the area covered. Ideally, when a body enters the field of view (FOV) of a camera, the camera would take one or more images of it. The visual data is used for event detection, data transmission in the chosen representation, and activation of nearby nodes, which are likely to see the body next. However, due to the finite acquisition rate of the cameras, it is possible that a moving body traverses a camera's FOV without being detected, and it is therefore important to assess the probability of this occurrence.

Let the presence of a moving body in the network be denoted by the event X1. If a body enters the ith camera FOV (FOVi), an event F_i^1 occurs. Every time a body circulating in the area covered by the network enters the ith camera's FOV and is not detected, there is a "miss" event M_i^1 for camera i. Moreover, in general, one may consider the case of n bodies circulating in the network (event X^n), r, which enter the FOVi at some point (event F_i^r), with the ith camera missing k of the body in its FOV (event M_i^k). It can be assumed that M_i^k is independent of X^n given F_i^r (since objects outside the camera's FOV cannot be detected):

$$P(M_i^k|F_i^r, X^n) = P(M_i^k|F_i^r) \qquad (7.1)$$

Further, assuming that $P(M_i^k|F_i^r)$ is binomial, it means that each "miss" event is independent from the others. This makes sense if the case of "rare events," that is, when two bodies are unlikely to appear at the same time in the same FOV. It is also possible postulate that $P(F_i^r|X^n)$ is binomial, a reasonable assumption in the case of independently moving bodies.

A possible measure of performance of a camera node is the *miss rate* MR_i, the ratio of the expected numbers of "miss" events to the expected number of bodies in the network:

$$MR_i = E[M_i]/E[X] \qquad (7.2)$$

where $E[.]$ is the expectation operator.

Let $P_{(M|F)} = P(M_i^1|F_i^1)$ and $P_{(F|X)} = P(F_i^1|X^1)$. Using the total probability theorem and remembering that the conditional distributions of interest are binomial:

$$
\begin{aligned}
E[M_i] &= \sum_k kP(M_i^k) \\
&= \sum_n \sum_r \sum_k P(M_1^k|F_i^r) \, P(F_i^\tau|X^n)P(X^n) \\
&= \sum_n \sum_r rP_{(M|F)}P(F_i^r|X^n) \, P(X^n) \\
&= P_{(M|F)} \sum_n E(F|X^n) \, P(X^n) = P_{(M|F)}P_{(F|X)}E(X)
\end{aligned}
\qquad (7.3)
$$

Hence, from Eq. 7.2

$$MR_i = P_{(M|F)}P_{(F|X)} \qquad (7.4)$$

Farther results are made available in (Boice et al. 2005).

7.2.4   MiNT

A miniaturized network testbed for mobile wireless research (MiNT) (De et al. 2005b) serves as a platform for evaluating mobile wireless network protocols and their implementations. Like a generic wireless network testbed, MiNT consists of a set of wireless network nodes that communicate over one or multiple hops with one another using wireless network interfaces. A prime feature of MiNT is that it dramatically reduces the physical space requirement for a wireless testbed while providing the fidelity of experimenting on a large-scale testbed. For example, using MiNT, it is possible to set up an IEEE 802.11b-based three-hop wireless network with up to eight nodes on a 3.66 m × 1.83 m (12 ft ×6 ft) table. This space reduction is achieved by attenuating the radio signals on the transmitter and the receiver. Through this miniaturization, it is possible to substantially reduce setup, fine-tuning, and management efforts required for a wireless network testbed. Additionally, attenuation on the transmitters reduces the interference of the testbed with the production wireless networks operating in its vicinity.

Specifically, MiNT makes several contributions:

- It introduces the architecture and implementation of a miniaturized wireless network testbed that features mobile multihop ad hoc networking on a tabletop. The testbed additionally incorporates comprehensive remote management, traffic monitoring, and fault injection facilities.
- It introduces pioneering hybrid simulation platforms that can run unmodified ns-2 simulations while its link, MAC, and physical layers are replaced by real hardware and driver implementations. The large number of wireless network protocols and traffic models already coded for ns-2 can thus be directly used on MiNT. MiNT allows unmodified ns-2 scripts to be executed on a set of physical nodes. Since the effects of radio signal propagation, like multipath fading and interference, are better captured while executing simulations in the hybrid mode, it produces much more realistic results for simulation experiments.
- It verifies the fidelity of the miniaturization approach and points out its limitations through extensive experimentation on an operational prototype.

7.2.4.1   MiNT Architecture

MiNT consists of a collection of core nodes managed remotely by a central controller node, as shown in Fig. 7.10. A core node communicates with its peers in the testbed using an IEEE 802.11b wireless NIC that is connected to a low-gain external antenna through radio signal attenuators. The antenna is mounted on a mobile robot to enable mobility. Each core node has another optional wireless interface for the purpose of sniffing traffic and collecting packet trace. The controller node oversees the operations of all the core nodes. A core node communicates with the controller node through a dedicated network interface that can be either wired Ethernet or any other wireless technology that does not interfere with the 802.11b transmissions in the testbed.

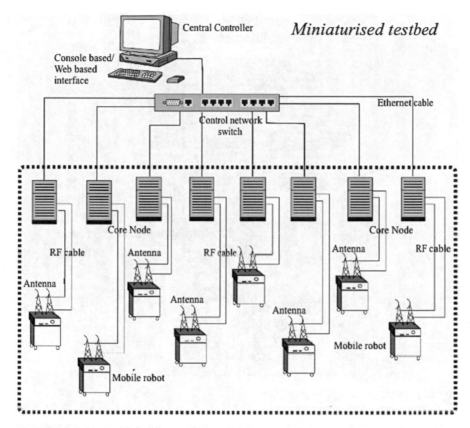

Fig. 7.10 MiNT architecture (De et al. 2005b)

7.2.4.1.1   Core Nodes

A collection of core nodes constitutes a MiNT testbed. As the MiNT goal is to build a multihop wireless testbed, the design of a core node is at the heart of the overall testbed design. A typical wireless testbed spans a large geographical area because the radio signal can be received over a large radius of the order of a few hundred meters. In order to build a testbed that can fit on a tabletop, it is crucial to restrict the radio signal within a small space; this enables to set up several nodes on a table and still establish multiple collision domains. Mobility is one more issue in the design of core nodes. The design of the core nodes with respect to miniaturization of the overall testbed and mobility of the core nodes is emphasized in what follows:

- Miniaturization. The key to miniaturization of the testbed lies in limiting the radio signals within a small space. The simplest technique is to adjust the transmit power on the wireless interface card.

    Fixed radio signal attenuators are used to limit the transmit power to a miniaturized tabletop setup. A desktop PC equipped with a NETGEAR MA311

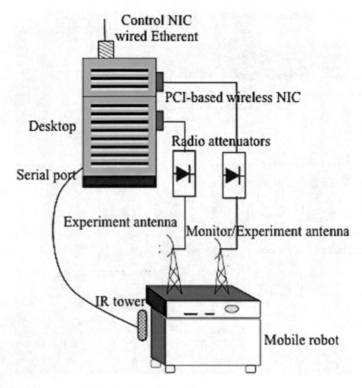

Fig. 7.11 MiNT core node (De et al. 2005b)

wireless PCI card (NETGEAR 2002) that does not have an internal antenna is
used for the core nodes. The PCI wireless NIC is connected to a radio signal
attenuator that in turn connects to an external antenna using an RF cable.
* Mobility. Node mobility is implemented using mobile robots. As the desktop
  node itself is not mobile, the external antenna is placed on the mobile robot. This
  limits the robot movement by the length of the cable connecting the external
  antenna to the wireless card. A LEGO robot from Mindstorms (LEGO 2014) is
  controlled from the desktop PC using an infrared (IR) tower. The IR tower is
  attached to the robot so that infrared signal from one tower does not interfere with
  another robot's movements (Fig. 7.11).

7.2.4.1.2   Controller Node

The controller node enables centralized control and management of the testbed
through a console-based/Web-based remote access. The functionalities provided
by the controller node are used by the administrator and the users (experimenters):

- The administrator is primarily concerned with status monitoring and routine maintenance, e.g., software upgrades, of the testbed nodes.
- A user accessing a shared MiNT testbed deployment requires other functionalities that let him configure each node, monitor the status of individual links, set up scripts on different nodes, and control experiment execution on the testbed.

A remote management system is the underlying mechanism to enable this remote operability of MiNT; it is based on the simple network management protocol (SNMP)[11] (Ranjan et al. 2005), where each testbed node is treated as a managed device.

In order to collect management data while an experiment is in progress, a control network is installed to be separate from the wireless network used for experiments. This control network operates on a noninterfering channel; in the current MiNT prototype, it is over wired Ethernet. One can also use 802.11a for a control channel since it does not interfere with 802.11b channels used for the experiments. The wireless control interface is not attenuated, enabling each node to communicate with the controller node over a single hop. This is unlike a full-scale testbed, where the control network also needs to operate over multiple hops (Chambers 2002).

7.2.4.2 Experimentation on MiNT

This section clarifies how experiments are controlled and analyzed on MiNT.

7.2.4.2.1 Experiment Control

Defining an experiment on any testbed involves several steps, exclusively, configuring network topology, setting up applications, defining mobility patterns, and setting the required per-node parameters. MiNT facilitates this configuration through a graphical user interface (GUI) that can be used by an experimenter to set up and manage experiments. The following steps illustrate experiment setup:

- Topology configuration. In configuring a wireless network topology, an experimenter is primarily interested in the radio connectivity between different node pairs. This is achieved by placing the nodes in such a way that each node pair satisfies specified link properties, like signal-to-noise ratio (SNR) or link error rate. In manual topology configuration as adopted in MiNT, the user determines the correct location of all the nodes to satisfy the link properties. However, with a large number of nodes, this method quickly becomes tedious. Ideally, the user

---

[11] Simple network management protocol (SNMP) is the protocol governing network management and the monitoring of network devices and their functions; it uses the User Datagram Protocol (UDP) and is not necessarily limited to TCP/IP networks. SNMP is described formally in the Internet Engineering Task Force (IETF) Request for Comment (RFC) 1157 and in a number of other related RFCs.

should declaratively specify the topology constraints, and the node positions should be automatically calculated based on a priori measurements done on the testbed. For automated topology configuration, one can start by calculating approximate node positions from relative signal strength using multihop trilateration[12] (Encyclopædia Britannica 2014). Iteratively changing the node locations and measuring the signal quality to achieve the desired pairwise configuration can then improve the initial positions.

- Application configuration. This involves setting up the traffic generators and traffic sinks and can be done in two ways:

  - The user can write his applications.
  - He can choose from a MiNT-supported library of ready-made applications, similar to the traffic sources/sinks provided by the ns-2 simulator.

- Mobility configuration. A user can configure node mobility by specifying the following:

  - Node trajectories
  - Target locations
  - Mobility models, such as the random waypoint model and the random walk model

  Mobility scripts are installed on each node using the GUI. Multiple nodes moving at the same time could collide; a script for full mobility must thus be validated to avoid such node collisions.

- Setting node/card properties. Changing node/card configurations and installing kernel modifications are common user requirements that are provided via the GUI. Users are also granted privileged access, which is required for accessing many of the functionalities such as raw socket and broadcast socket. An alternate approach to providing privileged access is to support limited access programming interfaces providing similar functionalities.

- Experiment execution. The next step in experiment control is providing the user with ways to fine-tune an experiment by observing the results during execution. In addition to simultaneous start/stop of an experiment on all the testbed nodes, an ability to pause the experiment, modify parameters on the fly, and then continue the experiment could substantially reduce experimentation time.

- Application/protocol debugging. MiNT is a distributed experimentation platform; hence, an experimenter faces all the difficulties of debugging distributed applications and protocols. To address this problem, MiNT incorporates a fault injection and analysis tool, which was earlier implemented for wired network protocol testing (De et al. 2003). The tool helps a developer generate realistic

---

[12]Trilateration is the method of surveying in which the lengths of the sides of a triangle are measured, usually by electronic means, and, from this information, angles are computed. By constructing a series of triangles adjacent to one another, a surveyor can obtain other distances and angles that would not otherwise be measurable.

network faults, like dropping, delaying, or corrupting of specific packets, using a simple scripting language. It is also possible to check for violations of protocol conditions and thus catch implementation bugs. Such facility is also useful in understanding the behavior of wireless protocols like AODV in the presence of multiple errors such as control packet losses.

Once an experimental configuration is finalized, the user can save all the configuration settings, such as node coordinates, applications, and mobility scripts. A saved configuration can be then used to quickly and automatically set up the experiment next time onwards.

7.2.4.2.2 Experiment Analysis

A crucial component in an experiment life cycle is its analysis stage. A network application/protocol is usually analyzed by looking at various packet dynamics during the experiment execution. MiNT incorporates a full-scale packet trace collection, aggregation, and visualization facility to aid such analysis:

- Trace collection. Network sniffers, such as *tcpdump*[13] (Nguyen 2004) and *ethereal*[14] (Nguyen 2004), are standard tools for Ethernet-layer packet capture. One can additionally switch a wireless card to the RF monitor mode, where it can capture all 802.11 link-level transmissions including 802.11 protocol headers and control frames. In a distributed environment, multiple monitor nodes are required to collect the entire network trace (Yeo et al. 2004). In MiNT, each core node also performs the monitor function using an additional wireless interface. This approach is most accurate in reconstructing each testbed node's view of the wireless channel during an experiment. It is also possible to separate the monitoring facility from experiment nodes. This requires strategically placing the nodes to completely cover the signal space of all the nodes. Additionally, the packets observed by a monitor node could be different from those seen by an experiment node.
- Trace aggregation. The trace collected on each node is sent to the controller node over the control network. Here all the traces are merged based on timestamps. This merge step requires that all nodes be synchronized at the beginning of any experiment. It is possible for the same packet to be captured by multiple monitor nodes. The duplicate packets are eliminated to create the final trace.
- Trace visualization. Trace visualization shows the transition of packets with respect to time. Visualization could be real time or offline, depending on whether

---

[13] *Tcpdump* is a command-line tool for monitoring network traffic. *Tcpdump* can capture and display the packet headers on a particular network interface or on all interfaces. *Tcpdump* can display all of the packet headers, or just the ones that match particular criteria.

[14] Network traffic analyzer (console) *Ethereal* is a network traffic analyzer, or "sniffer," for Unix and Unix-like operating systems. A sniffer is a tool used to capture packets off the wire. *Ethereal* decodes numerous protocols.

the collected trace on individual nodes are transported and aggregated, while the experiment is running, or at the end of the experiment. Real-time visualization requires that parse, collate, and display operations be done in real time. Display of the network-wide packet dynamics must show the packet exchanges over time for each node. Also, different frames, like control, management, and data frames, must be highlighted separately for ease of understanding. The MiNT prototype supports offline analysis and uses Ethereal for visualizing the aggregated trace.

• Data filtering. Another useful element of experiment analysis is the set of filters used to reduce the amount of trace collected on each node. This aids the online visualization of trace by reducing the amount of traffic that must be transferred in real time. The user could not only specify the network layer at which the packets are collected, but also the types of packets, e.g., HELLO packets, that are collected at each node. A similar filter is available with the visualization tool to further aid the trace analysis.

7.2.4.2.3   Fidelity of MiNT

It was shown by experimentation that the miniaturization technique based on attenuators does not affect the fidelity of the results. Results of experiments conducted on the testbed are compared with and without the use of attenuators on the signal path. Comparisons verify that the miniaturization technique does not alter the behavior of any layer in the network stack; it only shrinks the physical space used by the testbed:

• Physical layer. As signal propagation is a key aspect of the wireless physical layer, the impact of attenuation on signal propagation characteristics is studied. In this experiment, two nodes are connected in ad hoc mode and different levels of attenuation are applied. The resulting spatial distribution of signal quality (SNR) is compared with that of the non-attenuated case as listed below:

  – The obtained results reveal that when the attenuation is removed completely, the signal quality improves, but the nature of its variation is preserved compared to the attenuation cases.
  – To configure a topology in MiNT, it is required to reduce signal attenuation and keep the space unchanged, which makes the entire space better connected. By adjusting attenuation level to a specific research task's needs, one can trade off the minimum signal quality with the physical space requirement of the setup.

• MAC layer. This experiment studies the impact of attenuation on fairness property of the channel access algorithm. A string topology of four nodes is set, as shown in Fig. 7.12a. Node N2 is sending unicast traffic to node N1 and node N3 to node N4. Since N2 and N3 are in the interference range of each other, they contend for access to the shared wireless medium. Two different setups are

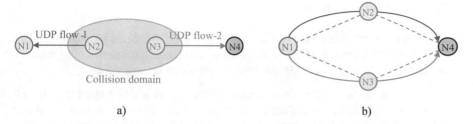

Fig. 7.12 MiNT fidelity experimentation. (**a**) String topology where N2 and N3 contend (**b**) Two-hop topology to run AODV (De et al. 2005b)

compared, one with attenuators and the other without attenuators, while keeping the same link quality across both setups. The instantaneous throughput of the two UDP flows for both cases is measured. It was shown that the channel is shared equally between the two contending flows, and the bandwidth sharing behavior is same in the attenuated and the non-attenuated case.

- Routing layer. This experiment shows that the behavior of the routing layer protocols is not affected by introducing attenuators on the signal path. A four-node network topology is devised, where the end nodes are connected over two hops, as shown in Fig. 7.12b. This experiment resorts to the AODV-UU[15] protocol (SourceForge 2013) to route packets between N1 and N4. The link quality is maintained alike across the attenuated and the non-attenuated runs.

  In each experiment, the route between node N1 and node N4, as chosen by AODV-UU, is made to fail by artificially failing the intermediate hop. The obtained results depict the time taken for new route discovery when such a failure occurs; it was shown that the time taken in both attenuated and non-attenuated cases holds similar values.

- Transport layer. To prove that the transport layer is unaffected by attenuators, a one-hop TCP experiment is set. Two nodes are connected in ad hoc mode and the throughput of a TCP connection between them is measured. The link quality is again maintained same across the attenuated and the non-attenuated setup.

The average TCP throughput is exhibited to be identical for the non-attenuated and the attenuated cases for a period of 120 seconds. Even the instantaneous variations have similar comportment, suggesting that the transport layer behavior is not affected by the use of attenuation.

---

[15] AODV-UU is an implementation of the Ad hoc On-demand Distance Vector routing protocol (IETF RFC 3561). It runs in Linux and ns-2 and was initially created at Uppsala University, hence the UU suffix.

7.2.4.2.4   MiNT Limitations

The key feature of a MiNT testbed is its ability to limit the signal propagation range between two nodes to within a few feet through the use of attenuators. However, the attenuation approach has certain limitations that are to be under focus:

- Selective attenuation. The most obvious difference in MiNT from a typical full-scale testbed is that in MiNT the radio signals are attenuated at the transmitter and the receiver ends. As the core nodes are placed in a noisy environment, the nodes operate in the presence of external noise sources, like microwave oven, cordless phones, and other interfering channels. The RF signals from these noise sources are attenuated only at the receivers. Also, the thermal noise at the receiver is unattenuated because it does not go through the receiver antenna. Since the attenuation of signal is more than that of the noise, one might suspect that the signal-to-noise ratio (SNR) for a link in MiNT is lower than that of an unattenuated testbed. However, this effect can be overcome by reducing either the attenuation level or the distance between the nodes.
- Near-field effect. Typically in MiNT trimmed space, since the nodes, and hence the antennas, are placed in proximity of each other, the receiver is in the near-field zone of the sender. This is unlike a full-scale testbed, where the nodes are typically placed far from each other; hence, the receiver is usually in the far-field zone of the sender.
- Spatial variation of signals. Multipath effects in signal propagation lead to small-scale variation in the signal strength. On the other hand, constructive and destructive interference resulting from the multipath effects are dependent only on the frequency of the signals. Hence, a solution to this problem is to scale down the frequency of the signals which would make the number of crests and troughs the same. However, changing the frequency would change the properties of the wireless medium under test and hence is not a viable solution.

  This limitation impacts the mobility-related experiments where the extent of signal quality variation encountered by a mobile node in MiNT will differ from that of a full-scale testbed.
- Non-repeatability. Finally like any other testbed, experiments on MiNT are not exactly repeatable because the external factors affecting signal propagation cannot be fully controlled across experiments.

7.2.4.3   Hybrid Simulation

Raising doubts about the veracity of simulation results is common. The drawback is mostly attributed to the lack of detailed models for the physical layer properties such as signal propagation and error characteristics. A usual practice in many academic researches is to use simplistic physical layer models. This is one of the prime reasons for the lack of simulation fidelity. With growing interest in cross-layer designs of protocols, it becomes imperative to provide accurate results at different layers in the protocol stack. Hybrid simulation alleviates some of these problems faced by pure simulation.

Hybrid simulation is a technique where some layers of the simulator's protocol stack are replaced with their real implementations. It is clear that majority of the inaccuracies in simulations stem from incomplete physical layer models. In MiNT, the link layer, the MAC layer, and physical layer of the ns-2 simulator are replaced with a wireless card driver, firmware, and real wireless channel, respectively.

The benefit of the hybrid simulation approach is that it requires minimal change to the already existing simulation code and scripts. The same simulation experiment can be used to obtain results in a realistic setting. The questionable effects of the physical layer models in simulation are removed through the use of a real wireless channel.

7.2.4.3.1 Implementation Issues

MiNT provides a way to conduct simulations in realistic settings to test, debug, and evaluate protocol implementations before going for their larger-scale deployment; ns-2 simulator is accordingly modified to support hybrid simulations. The challenges involved in implementing hybrid simulation capability into a standard discrete-event simulator and the techniques used to overcome these challenges for the ns-2 simulator are underlined below:

- Event scheduler. Two key design components in a simulator are as follows:

  - The way to model execution logic of different entities based on events, activities, or processes.

    The way the simulation time is advanced. ns-2 is a discrete-event simulator, where the execution logic is based on events, and the time is advanced at the pace of event execution time using a global virtual clock.

  In a hybrid simulation, all packet communication is carried over real wireless medium. This leads to inconsistency between the virtual clock that determines the dispatch rate of simulation events and the real-world clock that determines the transmission rate of packets over an actual wireless channel.

  To overcome such issues, MiNT uses the system clock on all the nodes that are synchronized at the beginning of each experiment, to update the simulator's virtual clock. Events are thus dispatched according to their real execution time instead of being executed as soon as the previous event has finished execution. ns-2's built-in RealTime Scheduler is modified accordingly.
- Limiting the number of events. The correctness of hybrid simulation requires that events should not be scheduled "in the past." For instance, if the amount of time spent in processing the simulator's execution logic is too large, then an event dispatching a packet to another node could be delayed and may be dequeued by the scheduler when the real time has advanced past its scheduled execution time. Such delayed event execution is prevented by reducing the number of events that the scheduler needs to process.

- Transmission/reception of packets. The internal packet format used in a simulator does not conform to the exact specifications of the real protocols. Hence, a packet from the simulator needs to be modified before it can be sent over the wireless medium. In order to transmit an ns-2 packet sent from the routing layer onto the link layer, a wrapper is implemented to encapsulate an ns-2 packet in a UDP packet payload and delivers it to the destination node using the standard socket layer.
- Changes to the ns-2 script. The goal is to minimize changes to the existing ns-2 scripts to have them executed on the hybrid simulation platform. To provide a single-script abstraction, the required changes are kept independent of the individual core nodes. All changes are composed at the central distribution node, and the same script is loaded on all the core nodes.

The changes to an existing script are as follows:

- The script must point to the MiNT link layer implementation instead of the ns-2 link layer.
- Each testbed node is assigned a physical node-id that is used in the ns-2 script. The physical node-id for each node is preassigned and the ns-2 script reads it from an environment variable local to each node.

In MiNT, only one virtual node is mapped onto a physical node. This might limit the size of the network that can be tested in hybrid simulation by the number of physical nodes available.

7.2.4.3.2 Hybrid Simulation vs. Pure Simulation

A study of the impact of physical layer characteristics is conducted; specifically, signal propagation and error characteristics influence the data transfer rates for both the platforms. Significant findings are obtained and elucidated.

7.2.4.3.2.1 Signal Propagation

This experiment tracks the impact of signal propagation on experimental results in pure simulation and hybrid simulation. Two unicast flows are used, between nodes N1–N2 and N3–N4, as shown in Fig. 7.13. The MAC layer on the senders N1 and N3 senses the channel before transmitting. A channel is perceived busy if the signal from one active sender, say N1, reaches the other sender, say N3. If N1 cannot sense N3, then there will be no interference, and the two flows will be active simultaneously, giving higher throughput to both flows.

In ns-2, the two-ray ground propagation model[16] (Henderson 2011) is used, with a ratio of 1:2 for hearing and sense ranges, that is, 6.71 m:13.41 m (22 ft:44 ft). In

---

[16] A single line-of-sight path between two mobile nodes is seldom the only means of propagation. The two-ray ground reflection model considers both the direct path and a ground reflection path.

Fig. 7.13 Topology to
study the impact of signal
propagation on channel
access (De et al. 2005b)

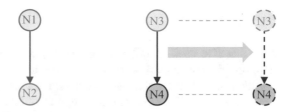

MiNT, the signal propagation is dependent on the environment, and this determines whether one node can hear/sense another node's activity. In ns-2, the channel capacity is set to 2 Mbps; in MiNT, the card transmission rate is set to 2 Mbps. For both cases, a constant bit rate (CBR)[17] traffic source (Cisco 2005) is used on N1 and N3 to pump packets of size 1000 Bytes at 2 Mbps that ensures that both senders are constantly trying to access the channel.

The obtained results reveal the impact of signal propagation characteristics on the behavior of the MAC layer. Use of the two-ray ground propagation model in pure simulation leads to the MAC layers of the senders perceiving the other sender's transmission till they are out of "sense range"; throughput variation is uniform till the senders are out of each other's range. In hybrid simulation, the signal quality variation is nonuniform, and the senders move in and out of the sense threshold, and hence, there is a nonuniform throughput variation in hybrid simulation.

It is clear that the interference is initially higher leading to channel contention between the senders, but later the interference fades, and both flows can pump data simultaneously. Pure simulation fails to capture this nonuniform spatial and temporal variation of throughput, which is an artifact of signal propagation characteristics.

7.2.4.3.2.2 Error Characteristics

The conducted experiments illustrate the difference in error characteristics when using pure ns-2 simulation and hybrid simulation running on MiNT. A CBR traffic source to pump data from one node to another is used. In pure ns-2-based simulation studies, each packet is corrupted according to a uniform random variable and prespecified error probability. On the other hand, errors in hybrid simulation occur due to the ambient noise in the environment.

The obtained results show that simple bit error models in simulation could produce qualitatively different behavior than those observed in real radio channels as seen on MiNT. Therefore, testing wireless protocols that depend on accurate bit error characteristics becomes much easier and produces realistic results with the use of a hybrid simulation technique.

---

[17]The CBR service class is designed for ATM virtual circuits (VCs) needing a static amount of bandwidth that is continuously available for the duration of the active connection. An ATM VC configured as CBR can send cells at peak cell rate (PCR) at any time and for any duration. It can also send cells at a rate less than the PCR or even emit no cells.

7.2.5 MiNT-m

MiNT-m (De et al. 2006) is an experimentation platform devised specifically to support arbitrary experiments for mobile multihop wireless network protocols. In addition to inheriting the miniaturization feature and hybrid simulation from its predecessor MiNT, MiNT-m has several additional features:

- It enables flexible testbed reconfiguration on an experiment-by-experiment basis by putting each testbed node on a centrally controlled untethered mobile robot.
- To support mobility and reconfiguration of testbed nodes, MiNT-m includes a scalable mobile robot navigation control subsystem, which in turn consists of a vision-based robot positioning module and a collision avoidance-based trajectory planning module.
- Further, MiNT-m provides a comprehensive network/experiment management subsystem that affords a user full interactive control over the testbed as well as real-time visualization of the testbed activities.
- Finally, because MiNT-m is designed to be a shared research infrastructure that supports 24×7 operation, it incorporates an innovative automatic battery recharging capability that enables testbed robots to operate without human intervention for weeks.

To support autonomous node mobility and topology reconfiguration in a wireless network testbed, it is necessary to mount each testbed node on a mobile robot. Though conceptually simple, there are several technical challenges in designing and implementing such a wireless testbed:

- Each testbed node must be battery-operated and self-rechargeable. The key design issue is how to build completely untethered mobile robots that can operate autonomously, thereby far exceeding in usability the ones that are simply battery operated and thus requiring frequent management.
- To set up a given initial topology or to enact a particular runtime node movement pattern, an accurate positioning mechanism is required to track and control the position of each wireless network testbed node.
- To grow a mobile wireless network testbed to a significant size of about 100 nodes, the targeted size of the MiNT-m project, the cost of each testbed node must be low, and the design of various testbed control functions, such as node movement and position tracking, must be scalable.

As detailed in Sect. 7.2.5.1, a MiNT-m node is built using a low-cost commodity robotic vacuum cleaner called Roomba (iRobot 2014), which supports a limited number of externally controllable movements and is able to carry a large payload up to 13.6 kg (30 pounds), as well comes with an effective auto-recharging capability. Mounted on each Roomba is a wireless network node supporting four 802.11 interfaces, each of which is attached to an antenna through a radio signal attenuator to reduce its signal coverage.

A network/experiment management system is essential to the robustness and usability of any wireless network testbed. The network/experiment management system designed for MiNT, called MOVIE (Mint-m cOntrol and Visualization InterfacE), has a multiplicity of functions:

- Providing real-time display of network traffic load distribution, pairwise end-to-end routes, node/link liveliness, protocol-specific state variables, positions of individual nodes, and inter-node signal-to-noise ratios.
- Allowing users to control a simulation run dynamically, including pausing a simulation run at a user-specified breakpoint, inspecting its internal states and/or network conditions, modifying different simulation parameters, and resuming the run.
- Supporting a rollback mechanism that allows one to go back to a previous state of a long-running simulation and resume from there with a different set of simulation parameters.

7.2.5.1   MiNT-m Architecture

MiNT-m derives from MiNT (De et al. 2005b) the feature of using radio signal attenuation to shrink physical space. The improvement is in designing completely untethered nodes that was lacking in MiNT due to use of desktop PCs as testbed nodes. More specifically, MiNT-m mounts a battery-powered small form-factor RouterBOARD (MikroTik 2002) on the Roomba robotic vacuum cleaner.

7.2.5.1.1   Hardware Components

MiNT-m hierarchy is built upon the control server, the tracking server, and the mobile nodes (Fig. 7.14):

- Logically, each MiNT-m testbed node is a wireless networking device mounted on a mobile robot. A testbed node design entails more than a factor. Cost is first, since MiNT-m was planned to scale to a size in the order of 100 nodes. Next, for mobility, the wireless networking device should have a small form factor so that it can be easily mounted on a simple robot; also it should be power efficient to maximize its runtime even on a small battery.

  A mobile node comprises of a wireless computing device and a mobile robot for physical movement. In MiNT-m, the wireless device is RouterBOARD 230 (MikroTik 2005), a small-form-factor PC with a 266-MHz processor and runs on an external laptop battery. It also comes with a PCI extension board (RB-14), which allows connecting four Qualcomm Atheros-based 802.11 a/b/g mini-PCI cards (Qualcomm Atheros 2014). Each of these cards is connected to a 2-dBi external antenna through a 22-dB attenuator. This adds a total of 44-dB attenuation on the signal path from transmitter to receiver and thus makes it possible to deploy a 12-node MiNT-m prototype within a space of 3.37 m × 4.29 m

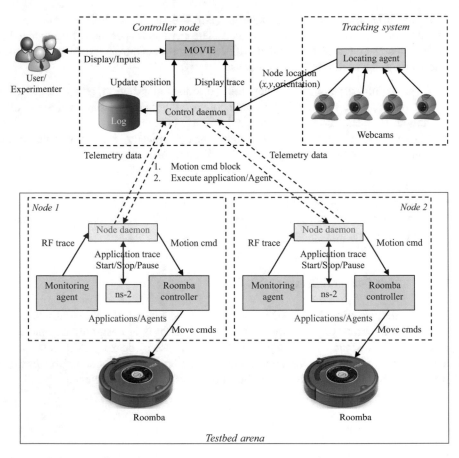

Fig. 7.14 MiNT-m hardware and software components. (Based on De et al. (2006))

(132.75″ × 168.75″) (Fig. 7.15). In addition to the fixed attenuation, the transmit power on the mini-PCI cards can be altered by 20 dBm to provide additional flexibility in tuning inter-node signal-to-noise ratio.

Roomba IR-based remote control facility permits two primitives for arbitrary movements:

- Move the mobile robot forward.
- Turn the robot by a specified angle.

Roomba's remote control codes are learnt using a Spitfire programmable remote controller (Innotech Systems 2000). The central control server moves a testbed node by sending a movement command to the testbed node's RouterBOARD, which then sends a corresponding command to Spitfire over its serial port. Eventually, Spitfire issues the associated infrared code to instruct the node's Roomba to move accordingly. The RouterBOARD is equipped with four wireless NICs each connected to a separate omnidirectional antenna via a radio signal attenuator (Fig. 7.16).

Fig. 7.15 MiNT-m prototype with 12 nodes and charging stations (top left corner) (De et al. 2006)

Roomba's auto-recharging circuitry is modified to power up both Roomba and the wireless network node. Moreover, a residual power estimation and a recharge scheduling algorithm is designed to keep track of the battery status of each node and determine the next recharge time for a node. A vision-based positioning system is designed to track the position of each mobile node in the testbed. The positioning system robustly tracks the nodes with zero false positives and requires only commercial off-the-shelf webcams. The resulting node position estimates are used for monitoring and for planning trajectories for collision-free node movement.

- The control server is a PC equipped with three wireless network interfaces. All control traffic is transported on an IEEE 802.11g channel and thus does not interfere with IEEE 802.11a channels that are used in actual experiments. Multiple NICs give the flexibility to scale the testbed to the increasing number of testbed nodes.
- The tracking server is a cluster of three PCs that periodically receives snapshots of the entire testbed, as captured by a (3 × 2) grid of commodity web cameras and uses them for testbed node identification and positioning. Smaller physical space requirement also reduces the number of cameras needed.

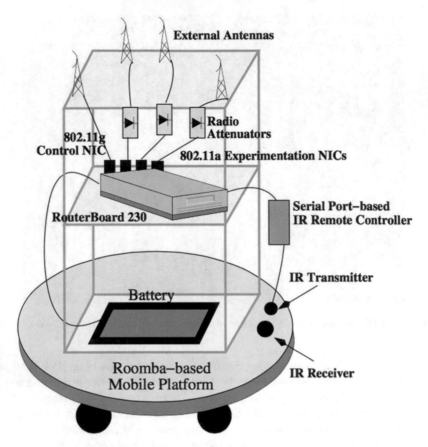

Fig. 7.16 MiNT-m testbed node (De et al. 2006)

7.2.5.1.2   Software Components

The key software components in MiNT-m are as follows:

- The control daemon running on the central control server.
- The node daemon residing on each testbed node
- MOVIE, the network monitor, and control interface

The control daemon running on the control server collects position updates of testbed nodes from the tracking server and event traces from experiment nodes and correspondingly updates the MOVIE display. It also communicates user-issued control commands, regarding node position or configuration changes, to individual node daemons that in turn control the movement of mobile robots. Because all event messages from the testbed nodes pass through the control server, the control daemon also maintains a complete log of activities in the testbed.

The node daemons on the testbed nodes communicate with the central control daemon over an IEEE 802.11g channel that is determined at start-up time. The messages that are communicated are either movement commands from the central control daemon or simulation events reported by testbed nodes back to the central control server. Other programs running on testbed nodes, for example, an ns-2 simulator, a TCP sender, or an RF monitoring agent, rely on the node daemon for any communications with the central control server. For example, critical events in the event trace that an ns-2 simulation run generates are passed in real time through the node daemon to the controller node for display.

MOVIE provides a comprehensive monitor and control interface that offers real-time visibility into the testbed activity and supports full interactive control over testbed configuration and hybrid simulation runs. MOVIE is derived from a Network Animator (NAM), an offline visualization tool for ns-2 traces, but introduces several powerful features for real-time monitoring and controlling simulation runs and for interactively debugging simulation results such as protocol-specific breakpoints and state rollback (Sect. 7.2.5.4).

7.2.5.2 Using MiNT-m

Running a hybrid simulation on MiNT-m involves three steps: experiment configuration, experiment execution, and experiment analysis.

7.2.5.2.1 Experiment Configuration

To configure an experiment running on MiNT-m, a user could specify the testbed topology, the applications to run on the testbed nodes, the mobility patterns of testbed nodes, and the network interface card parameters such as radio channel, transmission power, etc. MOVIE allows configuring the network topology through a simple drag of a node icon in the canvas. Accordingly, the control daemon triggers physical movement of the chosen node, followed by update of the pairwise signal strengths in MOVIE.

When the user runs an ns-2 simulation on the MiNT-m testbed, an ns-2 instance runs on each testbed node. In order to use MiNT-m as a protocol development platform, Linux implementations of the protocol can be installed and executed on each testbed node. To describe the node mobility pattern, the user specifies the intermediate positions and final destinations, along with their relative temporal offsets with respect to the beginning of the simulation run. From this information, instead of statically computing a global trajectory for each moving testbed node, MiNT-m relies on a runtime collision avoidance algorithm that dynamically resolves possible collisions among testbed nodes by halting some of them when collisions become imminent.

The user can also configure individual testbed nodes. One can first gain a root shell on individual nodes and then deploy applications or kernel modules and then change their wireless network card parameters, such as transmit power and retry count.

7.2.5.2.2   Experiment Execution

Executing an experiment in MiNT-m has several facets:

- Through MOVIE the user initiates the experiment and controls execution by observing its progress and intermediate results.
- MiNT-m enables starting, stopping, temporarily pausing an experiment, modifying simulation parameters on the fly, and then resuming the experiment.
- MiNT-m supports the ability to roll back an experiment back to a previous specified time, modify some simulation parameters, and restart the simulation run from the restored state.
- MiNT-m duplicates from VirtualWire (De et al. 2003) a facility to introduce controlled faults that are designed to expose potential bugs in protocol implementations.

7.2.5.2.3   Experiment Analysis

MiNT-m allows the user to specify simulation events of interest and to request the associated values to be displayed in real time. In addition, MOVIE supports real-time display of several wireless network parameters that are generally useful across all wireless protocols, such as the inter-node signal-to-noise ratio, the throughput on each wireless link, and the route between a pair of nodes.

7.2.5.3   Autonomous Node Mobility

7.2.5.3.1   Position and Orientation Tracking

To enable autonomous robot movement, the central control daemon must keep track of the current position and orientation of each testbed node. An optical or vision-based position/orientation tracking system is adopted; it only requires off-the-shelf webcams and color patches mounted on testbed nodes. The resulting tracking system is able to uniquely identify each testbed node and accurately pinpoint its (X, Y) position and orientation (θ). Moreover, it can scale to over 100 nodes, which is the target size of MiNT-m design.

MiNT-m testbed covers a floor space of 3.37 m × 4.29 m (11.06′ × 14.06′). A Logitech QuickCam 4000 is used; its image resolution is 320 × 240 pixels. Each webcam is placed at a height of 2.77 m (9.1 ft) from the ground and is able to cover a floor region of approximately 2.21 m × 1.68 m (7.25′ × 5.5′) which means each pixel corresponds to 0.48 cm² (0.075 inch²). To cover the entire testbed arena, the prototype uses six webcams. These webcams are placed such that they overlap with one another and the overlap area is large enough to completely hold a Roomba. As a result, every Roomba is fully captured by at least one webcam and the image streams from the six webcams can be processed independently.

Colors represented in the HSV model are used to identify each testbed node and its position/orientation. The HSV (hue, saturation, value) model, also called HSB (hue, saturation, brightness), defines a color space in terms of three components (ACA Systems 2011):

- Hue (H), the color type (such as red, green). It ranges from 0° to 360°, with red at 0°, green at 120°, blue at 240°, and so on.
- Saturation (S), of the color, ranges from 0% to 100%. Sometimes it is called the "purity." The lower the saturation of a color, the more "grayness" is present and the more faded the color will appear.
- Value (V), the brightness (B), of the color ranges from 0% to 100%. It is a nonlinear transformation of the RGB color space.

MiNT-m associates a four-color pattern with each testbed node, as shown in Fig. 7.17. The head and tail color patches are the same for all testbed nodes. Only the center patch, which consists of two colors, is used in node identification. The location of a testbed node is the centroid of the ID patch. The orientation is determined based on its direction, computed as the vector connecting the centroid of the tail patch to that of the head patch. Using the same colors for head and tail

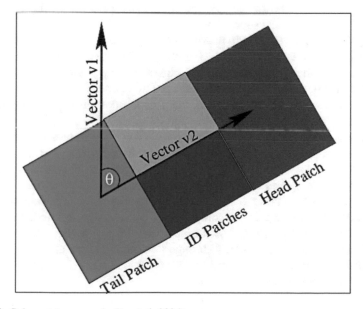

Fig. 7.17   Color patch on a node (De et al. 2006)

patches introduces redundancies that guard against noises and simplifies the determination of robot orientation. The vector from the centroid of the tail patch to the centroid of the head patch is used to determine the Roomba's direction, thereby computing a node's orientation in the testbed arena. The node location and identification are done using the center ID patches.

The color recognition algorithm used in MiNT-m uses standard image processing techniques for edge detection.

7.2.5.3.2   Node Trajectory Determination

The MiNT-m trajectory computation is based on a static trajectory planning algorithm, which computes a robot's path assuming the world is static, and a dynamic collision avoidance algorithm, which detects and resolves collision by fine-tuning precomputed trajectories.

Given the current position and the target destination of a testbed node, the control server takes a snapshot of the positions of other testbed nodes and treats them as obstacles in the calculation of the testbed node trajectory. The static trajectory planning algorithm first checks if there is a direct path between the testbed node current position and its destination. If such path does not exist, the algorithm identifies the obstacle closest to the source position and finds a set of intermediate points that lie on the line which passes through the obstacle and is perpendicular to the line adjoining the source and destination and have a direct path to both the source and destination. If no such intermediate points exist, the algorithm finds a random intermediate point that is δ steps away from the obstacle closest to the source and is directly connected to the source, and the algorithm repeats from this new intermediate point as if it is a new source.

In Fig. 7.18, node N1 is set to move from $A_{initial}$ to A_{final}. However, N2, N3, and N4 block the direct path between $A_{initial}$ and A_{final}. The trajectory planning algorithm first figures out that N3 is the obstacle closest to $A_{initial}$ and then computes the intermediate points P1, P2, ..., P6 to search for two-hop paths to A_{final}. Because the paths L1 and L2 are partially blocked, the algorithm eventually chooses path L3, which passes through the intermediate point P3.

In addition to static trajectory planning, MiNT-m also requires a dynamic collision avoidance algorithm, because while testbed nodes are moving, the robot movement may not be perfect. Given a snapshot of the testbed, which appears once every 1/4 second in the MiNT-m prototype. MiNT-m performs a proximity check for each testbed node; if any two nodes are closer than a threshold distance, both of them stop, a new path is recomputed for each, and the algorithm moves them on their new trajectory one by one. In the event that two nodes collide with each other, the algorithm again detects it through a proximity check and stops the nodes immediately. In this case, the algorithm also recomputes a new path for each of the two nodes and moves them one by one.

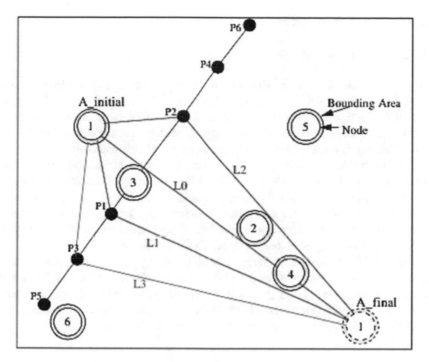

Fig. 7.18 Finding the trajectory from the N1 current position (De et al. 2006)

7.2.5.3.3 24 × 7 Autonomous Operations and Auto-recharging

To render the testbed self-manageable and providing uninterrupted 24 × 7 continuous operation, each testbed node is powered with batteries that are recharged periodically. MiNT-m supports automatic recharging of the nodes' batteries.

Roomba provides a docking station to charge its batteries. The Roomba docking station emits an IR beacon that is received by a Roomba over a distance of around 1.52 m (5 ft). When a Roomba's battery power drops below a threshold, it starts looking for a beacon emitted by the docking station and uses the signal to home into the docking station for recharge. On the other hand, Roomba's built-in battery cannot be used to directly power the RouterBOARD; hence, a separate universal laptop battery is used to power the RouterBOARD. To recharge the RouterBOARD battery along with the Roomba battery, the RouterBOARD battery is connected to the charging tip of the Roomba battery. This allows both batteries to be charged simultaneously from the same docking station.

7.2.5.4   Hybrid Simulation

7.2.5.4.1   Pause/Breakpointing

In hybrid simulation mode, the simulator is running in a distributed manner across all the nodes in the testbed. Debugging such a distributed application is a challenging task. In addition to the simultaneous start and stop of an experiment on all nodes, MiNT-m simplifies protocol debugging by introducing other standard features of a typical debugger, namely, pause and breakpointing of the experiment.

The implementation of the pause feature in MiNT-m requires modification to the RealTime scheduler in the hybrid ns-2. When the simulation is paused, the execution of events pending in the event queue as well as those in transit to other nodes is stalled. In the pause state, the user is allowed to change the physical configuration of the testbed, or alter any physical parameters of the nodes in the testbed, like node positions or transmit power, before resuming the execution.

The breakpoint feature is implemented by using the pause mechanism. In breakpointing, the user specifies a watch on ns-2 packet header fields. Each node matches the outgoing/incoming packet headers for prespecified values, and when a match occurs, a breakpoint signal is sent to the controller node. The controller node then informs all the nodes to pause their experiment execution.

7.2.5.4.2   Rollback Execution

The rollback feature for an experiment running in hybrid simulation mode gives the flexibility to a user to repeat the experiment from a snapshot time in the past with modified parameters fed to the experiment. This saves on experimentation time as the entire simulation experiment need not be repeated from the beginning.

7.2.5.4.3   Performance

The core computing platform used is a processor-limited RouterBOARD-230 that has a 266-MHz CPU. As more processing overhead on the system is added, the maximum achievable throughput goes down. The throughput degradation of a single hop is measured with different features, specifically, remote tracing, per-packet local tracing, experiment breakpointing, and experiment rollback:

- Tracing. A study is conducted for the impact of different forms of tracing on the maximum throughput achieved between two communicating MiNT-m nodes:

    - Without tracing, ns-2 application agents could only achieve 20.5 Mbps as compared to 33 Mbps achievable by a simple UDP flow running between the same nodes. This is because of the additional processing overhead introduced by ns-2, in contrast with a simple UDP sender that needs almost no processing to prepare a packet.

- A throughput degradation results from any form of tracing that introduces further CPU processing overhead due to string operations done by ns-2. For instance, online remote tracing as done for selected events results in a 17.043-Mbps throughput.

- Breakpointing of experiments. This feature requires matching expressions to trigger the breakpoints when the event occurs. Since matching different fields incurs overhead, the throughput reduces. Despite the CPU bottleneck, the throughput overhead increases only slightly with increasing number of expressions. This is because the expressions are only checked once for each packet, limiting the extra processing burden introduced by breakpointing.

- Rollback. This feature requires regular snapshot (using *fork()* system call) of the ns-2 process running on every MiNT-m node. Linux kernel's *fork()* system call automatically uses copy-on-write technique to avoid copying of all the pages at the fork time. This spreads out the throughput degradation to a few seconds after the *fork()* system call. Even with a 1-minute snapshot granularity, the overall throughput degradation was less than 0.25 Mbps.

7.2.6 Kansei

The Kansei testbed at the Ohio State University is designed to facilitate research on networked sensing applications at scale (Arora et al. 2006). The basic idea is to couple one or more generic platform arrays that support a broad set of users, with multiple domain-specific sensing platform arrays. Based on this concept, Kansei needs to be extensible to readily add new platforms, particularly domain-specific ones. To address the scaling challenge, the idea is to use arrays that are large enough so as to mirror the deployment scale, and, if they are not large enough, they should be capable to high-fidelity capture radio phenomena at a resolution that enables their scaling via software. The Kansei facility has been developed since Spring 2004, partly through equipment support obtained from the Defense Advanced Research Projects Agency (DARPA) for the ExScal project (Arora et al. 2005) as well as Intel Corporation and the Ohio State University. While a basic purpose for developing Kansei was to shorten the long cycle time of ExScal field testing in multiple outdoor settings, Kansei was supporting a significant number of diverse use cases and users.

Since Kansei has been made openly available, it has been used for research projects at Ohio State University and elsewhere, at project-based graduate and undergraduate courses, as well as in short classes for training XSM and Stargate users. Kansei has also assisted in transitioning software to industry partners, in part by getting them to execute validation tests on components being transitioned.

By its design focus on sensing and scaling, Kansei embodies a combination of characteristics:

- Heterogeneous hardware infrastructure with dedicated node resources for local computation, storage, data exfiltration,[18] and back-channel communication to support complex experimentation.
- Time-accurate hybrid simulation engine for simulating substantially larger arrays using testbed hardware resources.
- High-fidelity sensor data generation and real-time data and event injection.
- Software components and associated job control language to support complex multitier experiments utilizing real hardware resources and data generation and simulation engines.
- Kansei exports a Web interface on which experiments can be scheduled and the results retrieved.

7.2.6.1 Kansei Composition

Kansei consists of a set of hardware platforms, access to which is managed by a remotely accessible Director framework. The composition supports several tools, for high-fidelity sensor data generation (Sect. 7.2.6.2) and hybrid simulation (Sect. 7.2.6.3).

7.2.6.1.1 Hardware Infrastructure

Kansei's hardware infrastructure consists of three components: stationary array, portable array, and mobile array:

7.2.6.1.1.1 The Stationary Array

The stationary array consists of 210 sensor nodes placed on a 15×14 rectangular grid benchwork with 91.4-cm (3-ft) spacing. Each node in the stationary array consists of two hardware platforms, eXtreme Scale Motes (XSMs) (Arora et al. 2005), and Stargates (Crossbow 2004b) (Fig. 7.19). XSM is a derivative of Berkeley prototype sensor nodes and was developed by Crossbow and DARPA Network Embedded System Technology (NEST) team at Ohio State University, for use in the ExScal Project. Each XSM is equipped with a 7.3-MHz 8-bit CPU, 128-KByte instruction memory, and 4-KByte RAM. For communication, the mote uses a 433-MHz low-power radio. The radio's reliable communication range is between 15 and 30 m when placed on ground level. Each mote accommodates a variety of sensors, such as a photocell, a passive infrared (PIR), a temperature and a magnetometer sensor, and a microphone. The motes run TinyOS (TinyOS 2012),

---

[18] Data exfiltration, also called data extrusion, is the unauthorized transfer of data from a computer. Such a transfer may be manual and carried out by someone with physical access to a computer, or it may be automated and carried out through malicious programming over a network (TechTarget 1999).

a)   b)

Fig. 7.19 Stationary array. (**a**) eXtreme Scale Motes (XSMs). (**b**) Stargate (Arora et al. 2006)

a lightweight event-based operating system that implements the networking stack
and communication with the sensors, and provides the programming environment
for this platform.

Stargate (Crossbow 2004b) is an expandable single-board computer with Intel's
400-MHz PXA255 CPU running the Linux operating system. It also has a daughter
card, which contains an interface to a mote and a number of other interfaces
including RS-232 serial, 10/100 Ethernet, and USB. Its in-band[19] communication
(The Free Dictionary 2014) is via an 802.11b wireless NIC card. The characteristics
at outdoor environments for these specific 802.11b radios were extensively
measured.

The XSM is connected to the Stargate through a dedicated 51-pin connector. The
Stargate devices serve as integration points for the mote-level devices, providing
them with channels for data collection, data analysis, and local sensor data generation and injection. These devices are connected through high-speed network
switches to an Ethernet back-channel network, which provides high-bandwidth
connectivity to and from the nodes for management commands and data injection
and extraction.

The Ethernet back channel of the stationary array connects to a cluster of PCs.
One PC serves as the primary server node for the Kansei Director platform as well as
for the remote access to Kansei. Other PCs are used for running visualizations,
compute-intensive analysis, high-fidelity sensor data generation, hybrid simulation,
and diagnostic analysis. Lately, 150 nodes were upgraded to contain Tmote Sky
nodes as a third hardware platform, which features an IEEE 802.15.4 radio operating
at 2.4 GHz and an integrated onboard antenna (Moteiv 2006b).

---

[19] In-band signaling or CAS, channel-associated signaling. Transmission of control signals in the
same channel as data. This is commonly used in the public switched telephone network (PSTN)
where the same pair of wires carries both voice and control signals, e.g., dialing or ringing.

Fig. 7.20 Mobile node on the stationary array (Arora et al. 2006)

The nodes are placed on customized benchwork, with a Plexiglas[20] plane (Arkema 2013) layered on top to support the mobile nodes. Four high-resolution Sony SNC-RZ30N cameras (Sony Corporation 2002) with pan-tilt-zoom and wireless as well as networked programmability provide slew-to-cue capability for configurable image feeds of indoor testbed operation. These image feeds will serve sensing, visualization, and, in some experiments, ground truth purposes.

7.2.6.1.1.2 Portable Array

The stationary array infrastructure is designed to be coupled with one or more portable arrays for in situ recording of sensor data and field testing of sensor network applications. Each portable array consists of domain-specific sensors and generic software services for data storage, compression, exfiltration, time synchronization, and management.

Kansei currently includes a portable array of 50 Trio motes. The UC-Berkeley-designed Trio integrates the XSM sensor board (acoustic, passive infrared, two-axis magnetometer, and temperature) with Tmote Sky nodes and a solar power charging system. The Tmote Sky features an IEEE 802.15.4 radio operating at 2.4 GHz and an integrated onboard antenna. This particular array duplicates the sensors in the stationary array for at-scale high-fidelity sensing validation studies. The solar-powered charging makes it suitable for long-term deployments.

---

[20]Plexiglas MC diffusion acrylic sheet designed specifically for the lighting industry; the sheet diffuses light from LED sources without sacrificing significant light transmission.

7.2.6.1.1.3 Mobile Array

This platform consists of five robotic mobile nodes that operate on the transparent Plexiglas mobility plane (Fig. 7.20). The transparency of the plane allows light sources mounted under the robots to activate the photo sensors of the nodes in the stationary array. Robots from Acroname, Inc. (Acroname Inc. 1994), with built-in motor boards and a Stargate interface are used. A Stargate on each robot features an 802.11b radio with the optional attenuated antenna, as in the stationary array. In addition, each robot contains an XSM and Tmote Sky node to communicate with the stationary array as well as to run native code for the XSM and Tmote platforms.

7.2.6.1.2 Director: A Uniform Remotely Accessible Framework for Multitier WSN Applications

The Kansei Director is an extensible software platform that enables integrated experimentation on the stationary array, portable arrays, and mobile array. It provides basic services:

- Experiment scheduling, deployment, monitoring, and management for all array platforms.
- Creation and management of testbed configurations in support of multiuser and multiple-use scenarios, such as for allocation to experiments. Thus, "jobs" potentially consisting of multiple WSN executables, scripts, and data files can be programmed to run on a specific configuration of the testbed for a specified length of time. The status of these jobs may be monitored during their execution.
- Gathering the state of the Kansei testbed to optimize resource utilization. The complexity of network-embedded applications is growing rapidly, yielding for applications that are multiphase and that are reconfigured from time to time. WSN resources, however, are not growing at a rate that significantly exceeds application needs. Hence, unlike traditional network-based systems, network-embedded computing continues to involve operating networks "on the edge," as opposed to well within network capacity. Thus, application-dependent optimization of resource utilization is an important integration challenge. This implies awareness of network resources and rapid configuration of applications in accordance to available resources, which is done by collecting state information.
- Reconfiguration at runtime. A core integration challenge is to support application management neatly, both for human users and for mechanization. Thus, the Director also supports the orchestration of an experiment consisting of multiple phases. In ExScal, for example, a "localization" phase calculates and disseminates to each mote its (x,y) grid position, which is stored in flash memory. The mote is then rebooted to a "sensing" phase that initializes itself by reading this localization information. Complex multiphase experiments especially occur when iteratively tuning the application to the environment in conjunction with tuning the middleware to the application.

- Implementing a core set of system-level utilities and runtime components. Examples include tools for data injection, for instance, when an experiment requires the injection of the output data from a previous phase as input for the next one, health monitoring, and logging for all array platforms. Components could simply be specific middleware such as for routing, or runtime components for implementing "reflective" applications that, for example, monitor resource utilization on the node and reconfigure themselves appropriately. It is to be noted that the development of applications and system utilities and components is done outside the Director.
- System administration services. These services include user management, such as creation and deletion of users, and the assignment of access rights, as well as platform administration, such as the restarting or setting the network configuration of a node.
- Remotely exposing services for experiment scheduling, configuration, deployment and management, system utility deployment and configuration, and system administration. End users access these services through a Web interface and programs through Web services.
- Supporting as a "plug-and-play" both the current and the future variety of hardware and operating system platforms across all tiers. Each platform exposes its tailored uniform set of services. This uniformity allows the end user of Kansei to essentially be platform skeptical in the specification of the experiment, including in its orchestration.

7.2.6.1.2.1 Director Architecture

In the main Director, several subsystems are contained (Fig. 7.21):

- The Configuration subsystem that manages testbed configurations, such as a topology and its nodes on the stationary array and on the portable arrays used in an experiment.
- The Access management subsystem that manages the levels of users and their access rights.
- The Platform management subsystem that abstracts the services of the arrays to enable platform plug-and-play, through platform manifest files which are installed on Kansei when a new platform is incorporated.
- The Experiment management subsystem is responsible for experiment scheduling, configuration, deployment, and monitoring.

The Director uses orchestration services for the sequencing of steps within a multiphase job.

Director services are implemented not just by components that run on the top level of the arrays, but also on nodes within arrays. For example, deploying an application on a mote involves invoking a director component on the Director server, which in turn invokes a component on the Stargate that serves as a gateway to the mote. Just as the Director is hierarchical, so are several of the Kansei utilities.

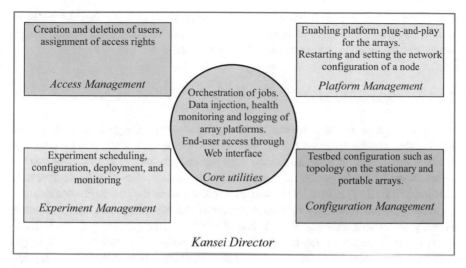

Fig. 7.21 Kansei Director architecture

Software implementation is performed as follows:

- The main Director runs on a Linux server, which also runs the Web server for the Web interface.
- Testbed scheduling, administration, management, and experimentation are implemented in a multi-threaded daemon that uses scripts and utilities written as Perl (Perl.org 2002) modules and which encapsulate testbed services, such as UISP[21] (UISP 2014) for XSM programming via the serial port of the Stargate.
- PHP modules implement the Web-accessible testbed services, such as job creation, storage of experiment data, and a testbed health-monitoring page.
- A MySQL database provides persistence for storing job configurations and user reservations. Data generated by jobs are stored on the server file system and may be retrieved by links on the Web interface.

7.2.6.2   High-Fidelity Sensor Data Generation Tools

Sensor data generation is a key component of high-fidelity design and testing of applications at scale. In addition to its utility in validation of applications and network services, it provides a theoretical basis for the design of algorithms for efficient sampling, compression, and exfiltration of the sensor readings. Kansei users

---

[21] UISP is a tool for AVR (and AT89S) microcontrollers, which can interface to many hardware in-system programmers. UISP was written to work in a GNU/Linux environment, but can also run inside Microsoft Windows systems, by using *Cygwin*.

can generate sensor data fields of arbitrary size at high fidelity using the methods detailed below, specifically sample-based modeling tools and synthetic data generation from parametric and probabilistic sensor models.

7.2.6.2.1   Sample-Based Modeling Tools

For many sensor modalities, the physical phenomena of signal generation and propagation are too complex for accurate parametric modeling and computationally feasible simulation. In these instances, a generic sample-based model can be used to simulate sensor readings at a large scale. The model maintains a database of sensor snippets indexed by ground truth parameters collected for the source phenomena. To capture spatial correlations in sensor readings, the recordings are made simultaneously on an appropriately sized patch of sensors. Examples are passive infrared energy recordings on a small mesh of sensors as a personnel intruder passes through a tile of sensors, acoustic energy recorded on a small mesh of sensors for a windy day indexed by the wind speed at that time and location, and signal energy, time, and direction of arrival recordings for all neighboring sensor locations for a buzzer node indexed by their relative location to the source. Generation of the sensor data at the desired scale is accomplished by replaying the snippets with appropriate time and spatial shifts.

7.2.6.2.2   Synthetic Data Generation Using Parametric Models

For many sensor modalities the physical relationship between the sensor reading and the underlying natural phenomena is well understood and the sensor readings are dominated by the foreground signal. Consequently, sensor readings can be generated from a parametric model of the underlying phenomena. In Kansei, physics-based parametric sensor models are developed for a variety of sensing modalities including models of passive acoustic, seismic, infrared, and magnetic sensors.

7.2.6.2.3   Probabilistic Modeling Tools

An alternative modeling strategy relies on accurate estimation of the spatial and temporal correlation of the sensor readings. Many sensor modalities can be modeled as time-varying random Markov fields. Examples are temperature, gas, humidity, and turbulent wind energy distribution.

7.2.6.3   Hybrid Simulation

For simulation to be an effective tool in evaluating sensor network algorithms, it has to correctly model the physical environment for radio signal propagation as well as adequately represent the application being run by the sensor network. Kansei features a high-fidelity hybrid simulation capability where a PC simulation server is connected to the stationary or the portable array.

A hybrid simulator has to coordinate both real and simulated events. Thus, the problem of reconciling the real and simulated time arises. One approach is to allow the real events to occur at their own speed and periodically resynchronize the simulated part with real events. However, this approach has potential scalability and fidelity problems. In hybrid simulation, hundreds of virtual sensor nodes can be simulated using real radio hardware to communicate messages. This ensures fidelity of the simulator with respect to the radio propagation in realistic deployment environments.

In Kansei, the hybrid simulator is applicable to the Berkeley motes running TinyOS (TinyOS 2012) applications. A part of the simulator is TOSSIM (TinyOS Wiki 2013), a TinyOS simulator. The main simulator component is running on the PC. For hybrid modeling, the simulator utilizes the out-of-band access to the physical sensor nodes on the stationary array. The simulator allows TOSSIM to run the application on the host PC but relays the communication and sensing requests to the physical motes connected to the PC. This is done by replacing the components that simulate communication and sensing in TOSSIM with components that handle the interaction with the motes.

7.2.7 Trio

Trio is one of the largest solar-powered outdoor sensor networks; it offers a unique platform on which both systems and application software can be tested safely at scale (Dutta et al. 2006). The testbed is based on Trio, a new mote platform that provides sustainable operation, enables efficient in situ interaction, and supports fail-safe programming. The motivation behind this testbed was to evaluate robust multi-target tracking algorithms at scale.

Outdoor sensor network deployments like ZebraNet (Liu et al. 2004), GDI (Szewczyk et al. 2004), Redwoods (Tolle et al. 2005), VigilNet (He et al. 2006), and ExScal (Arora et al. 2005) provide unmatched realism, but these networks have achieved either large scale or long life, but usually not both. Contrarily, indoor testbeds like MoteLab (Werner-Allen et al. 2005) and Mirage (Brent et al. 2005) use real radio hardware that provides much greater communication realism but does not capture the nuances of outdoor environments. Not to be sidestepped from consideration, portable testbeds like EmStar (Elson et al. 2003) allow realistic outdoor experimentation, but their wired backchannels have two sides of a coin; on one side, they provide great visibility, but on the other side, they limit the scale of the deployment. Testing at realistic scales is imperative because each order of magnitude increase in network size ushers in a new set of unforeseen challenges.

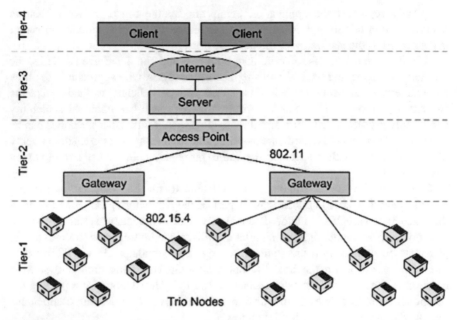

Fig. 7.22 The four tiers of Trio (Dutta et al. 2006)

Trio is an open four-tier experimental platform designed to better address the requirements of a large-scale, long-lived, outdoor testbed as illustrated in Fig. 7.22 (Dutta et al. 2006):

- Trio nodes reside in the lowest tier (tier 1) of the architecture. Trio provides support for application-level experimentation through a sensor suite optimized for detection and classification of humans and vehicles as well as support for system-level experimentation through hardware and firmware for fault-tolerant operation. Trio integrates Telos (Moteiv 2004), the eXtreme Scale Mote (XSM) (Arora et al. 2005), and Prometheus (Jiang et al. 2005); it improves upon their designs in several ways. Trio addresses a multiplicity of issues:

  – Sustainable operation through a solar-power-based renewable energy supply with super capacitor and lithium-ion battery storage elements.
  – Support for efficient in situ maintenance and fail-safe operation and environmentally hardened package design.
  – Scalability by reducing the cost of human-in-the-loop operations, such operations that require human interaction.
  – Fail-safe flexibility is addressed through the use of the Deluge network reprogramming system (Hui and Culler 2004), watchdog[22] (Dien and Ghing-

---

[22] A watchdog timer (WDT), also known as a computer operating properly (COP) timer, is an embedded timing device that automatically prompts corrective action upon system malfunction detection. If software hangs or is lost, a WDT resets the system microcontroller via a 16-bit counter.

Hsin 2000) and grenade[23] (Stajano and Anderson 2000) timers, and one-touch recovery.

The 557 Trio nodes in the testbed are organized into multiple routing trees, with each tree rooted at a gateway (Fig. 7.22). Gateways forward traffic between the 802.15.4 Trio network and an 802.11 wireless backbone network. Each Trio node dynamically associates with the closest gateway based on routing cost. Gateways are physically distributed throughout the mote tier and support network scalability by adding gateways as the number of nodes increases. Gateways support sustainable operation through solar power, and since they simply forward traffic statelessly, flexibility is not required.

- Seven gateways sit in tier 2 of the system architecture along with 802.11 repeaters and access points which bridge the 802.11 network to an 802.3 Ethernet network.
- A single-root server resides in tier 3 and connects to all of the gateways. The server maintains a TCP session with each gateway, while a daemon on the server multiplexes these TCP sessions using gather-scatter communications and exposes them as a single TCP session. This approach simplifies client interaction by presenting a unified view of the network and abstracting superfluous implementation details. The server also runs network monitoring and management software that allows active querying or passive monitoring of the network and stores the resulting data for online or later offline analysis. The monitoring software supports scalability by aggregating large amounts of information from several sources into simple but informative graphics and tables. The management software supports flexibility by allowing network programming and other control functions. Section 7.2.7.1.3 considers the software that runs on the server.
- The clients in tier 4 consists of one or more desktop computers that run client-side applications. These applications access the network via the tier 3 server, which forwards traffic to and from the gateways in tier 2. The multi-target tracking (MTT) was the first application to use the Trio testbed.

More elaborate Trio details are provided in the sections to come.

7.2.7.1 Trio Architecture

7.2.7.1.1 Tier 1: The Trio Node

The Trio node, shown in Fig. 7.23, is designed for long-lived operation with minimal physical maintenance. Each node is based on three components:

---

[23]The grenade timer is an evolution of the watchdog timer that can impose a hard limit on the CPU time that a guest program may consume, in the absence of a protected mode on the host processor. Unlike its predecessor, it is resistant to malicious attacks from the software it controls, but its structure remains extremely simple and maps to very frugal hardware resources.

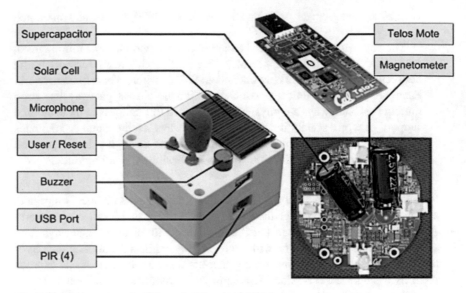

Fig. 7.23 Components of a Trio node (Dutta et al. 2006)

- Telos mote (Moteiv 2004), which provisions for low-power operation and remote reprogrammability, a necessity for flexible, long-lived applications.
- XSM mote (Arora et al. 2005) provides a Trio grenade timer for fail-safe operation and includes a sensor suite of passive infrared (PIR) motion sensors, a magnetometer, and a microphone.
- The Prometheus solar charging system (Jiang et al. 2005) ensures sustainability via a renewable energy supply.

A Trio node supports sustainable operation, efficient physical interaction, and fail-safe flexibility as detailed below. Software residing in tier 1 is illustrated in Fig. 7.25.

7.2.7.1.1.1   Sustainable Operation

Sustainable operation is supported in two ways, through a renewable energy supply and by environmentally hardening the Trio enclosure:

- Renewable energy. Trio circumvents the typical lifetime limitation resulting from a non-rechargeable battery by including a renewable energy supply based on the Prometheus solar charging system (Jiang et al. 2005) for maintenance-free self-charging. The original Prometheus design is modified to improve its performance and ensure fail-safe operation. A Trio node with a depleted capacitor or battery starts to wake up after solar energy charges the supercapacitor enough to produce a supply voltage of 1.8 V, the minimum operating voltage for the MSP430 processor and CC2420 radio. However, initializing sensor modules and writing to flash requires a higher supply voltage. A component of the modified

Prometheus driver enforces hysteresis and waits to wake up the rest of the system until the supply voltage rises past 2.75 V, which is enough to power the sensors and write to the flash memory. The application program is only started once the system voltage exceeds 2.75 V.
- Environmental hardening. One of the key design challenges was to harden the Trio enclosure for an outdoor environment without hampering sensor performance or node maintainability. Several components of Trio, e.g., solar cell, PIR sensor, microphone, buzzer, and user/reset switches, are exposed to the environment for sensing, solar energy harvesting, and user input. These components are made weather-resistant so that Trio nodes could operate under varying weather conditions.

7.2.7.1.1.2 Efficient Physical Interaction

The Trio node supports scalable operation through efficient physical interaction. The node also provides audible feedback about certain states and state transitions. When a node's capacitor voltage drops below a safe operating voltage, the node chirps every few seconds. These audible cues allow operators to passively gauge system status.

7.2.7.1.1.3 Fail-Safe Flexibility

Since Trio can be programmed wirelessly using the Deluge network programming system, it is possible to program Trio with a buggy or even Byzantine program[24] (Lamport et al. 1982). Deluge is included in the Trio platform software, so network programming is automatically compiled into every application that uses the Trio libraries. The external flash can be used to store up to seven programs and simple Deluge commands can be issued to switch between the programs.

Several mechanisms are used to support fail-safe operation and recover from buggy or Byzantine programs:

- Watchdog timer to ensure that software is making progress, tasks are executing, and interrupts are being handled
- Grenade timer to guarantee that a node can recover from Byzantine applications by periodically transferring control to a trusted kernel
- USB override that allows even the trusted code to be reprogrammed if necessary
- Hardware override on the power system to guarantee that the system always reverts to the solar power supply in the event the battery dies during operation

---

[24]The Byzantine problem is built around an imaginary general who makes a decision to attack or retreat and must communicate the decision to his lieutenants. A given number of these actors are traitors (possibly including the general). Traitors cannot be relied upon to properly communicate orders; worse yet, they may actively alter messages in an attempt to subvert the process.

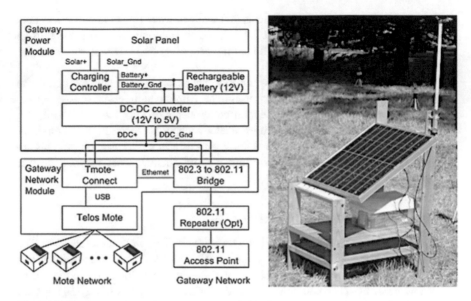

Fig. 7.24 Gateway node architecture (Dutta et al. 2006)

7.2.7.1.2   Tier 2: A Network of Gateways

In a large-scale deployment, a wireless high-bandwidth gateway backbone spread throughout the network can serve several purposes. Specifically, it can partition the traffic to lessen the overall network utilization, provide points for traffic observation, and support scalability through hierarchy. Hierarchy allows a large sensor network to be partitioned into multiple smaller networks that operate in parallel.

Figure 7.24 shows a gateway node and the backbone network architecture. The backbone network consists of a gateway node that forwards mote traffic to and from the 802.11 backbone network, optional 802.11 repeaters, and an 802.11 access point that connects this network to the root server.

Seven gateway nodes are available in tier 2. A gateway node includes three major components:

- Telos mote
- Tmote Connect software (Moteiv 2006a) to be installed on a Telos-to-Ethernet gateway such as NSLU2 (Linksys 2008a) to forward messages from the attached Telos mote to the Ethernet interface
- An 802.3-to-802.11 bridge that forwards messages from the Tmote Connect to the 802.11 network.

A 9-dBi omnidirectional antenna extends the gateway radio range. To attain sustainable operation, the gateway nodes are designed to operate on solar power. The power supply for a gateway node consists of a solar panel, a charging controller, a gel cell battery, and a DC-DC converter.

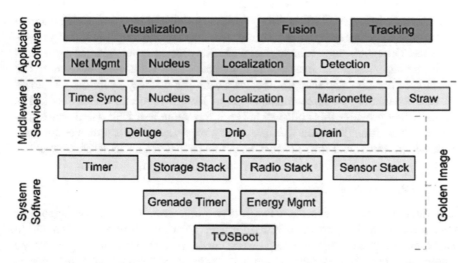

Fig. 7.25 Trio software. Each color matches a tier in Fig. 7.22 (Dutta et al. 2006)

Scalability is supported through hierarchy. Seven gateway nodes are deployed to support 557 Trio sensor nodes. Because the diameter of the network was larger than the 200-m range of the gateway nodes, 802.11 repeaters with higher-gain antennas were placed at key locations in the field. The backbone network required basic IP routing, and management was performed through the web consoles of both the Tmote Connect and the 802.3-to-802.11 bridge. Each gateway node was assigned an IP address on the same subnet as the access point. Figure 7.25 displays the software that runs the gateway functions.

7.2.7.1.3 Tier 3: The Root Server

As iterated, the Trio testbed consists of a total of 557 Trio nodes distributed over an area of approximately 50,000 m². The testbeds' large-scale and remote location makes it difficult to monitor the nodes directly and raises the need for remotely accessible tools to manage the network. Thus, the Golden Image (Techopedia 2015a) and the management framework that runs alongside testbed applications on the mote include the Nucleus network management system, a second-generation version of simple network management system (SNMS) (Tolle and Culler 2005). Figure 7.25 displays the used software.

7.2.7.1.3.1 Network Health Monitoring

The Nucleus query system enables a testbed user to determine which nodes are running at any particular time. The Nucleus query server that runs on the root server provides an XML-RPC[25] (Kidd 2001) interface to be used by a monitoring daemon that periodically injects queries into the network, collects responses, and records statistics. The monitoring daemon tracks which nodes are running, which nodes had been running but have stopped responding, and which nodes have never run. The monitoring daemon also marks a node as awake if a gateway overhears a protocol message containing a source address such as Deluge (Hui and Culler 2004). The daemon then provides this collected health information to a PHP-based web application, which fuses this data with previously measured GPS coordinates for each node and produces real-time network health maps that can be accessed remotely.

7.2.7.1.3.2 Power Monitoring

In addition to ensuring that the network is running, a user of the testbed should be able to verify that it is running sustainably. The monitoring daemon uses Nucleus to query the Prometheus logic running on each node, periodically collecting measured battery and capacitor voltage, along with flags indicating whether the node is charging its battery or running on it. This information is also displayed on the map provided by the web management console for online viewing and is logged on the server for offline charting and analysis.

7.2.7.1.3.3 Monitoring Network Programming

The monitoring daemon also collects information from Deluge, which permits tracking the progress of an image through the network and visually identifying nodes that are unable to acquire the image. Low battery voltages can prevent Deluge from writing data to the flash storage, which leads to low-voltage nodes requesting new data but never saving it. This Deluge "tension" can create hotspots of traffic within the network that impede the flow of application and management data. The acquired health maps enabled identifying the tense nodes and rebooting them or simply shutting them off.

7.2.7.1.3.4 Monitoring and Control of Applications

The Nucleus management framework, or an alternate visibility and debugging system called PyTOS (Whitehouse et al. 2006), can provide remote monitoring and control of an application running on the testbed. Even though executing a new application stops the Golden Image[26] (Techopedia 2015a) from running, nonetheless

---

[25] XML-RPC is a quick-and-easy way to make procedure calls over the Internet. It converts the procedure call into an XML document, sends it to a remote server using HTTP, and gets back the response as XML.

[26] In network virtualization, a Golden Image is an archetypal version of a cloned disk that can be used as a template for various kinds of virtual network hardware. Using golden images as templates, managers can create consistent environments where the end user does not have to know a lot about the technology.

maintaining the ability to query nodes and build health maps is highly desirable. Though management traffic can conflict with an application being tested, perpetually available management is fundamental to the successful operation of a long-lived outdoor testbed.

7.2.7.1.4   Tier 4: Client Applications

Trio is created for a large-scale study of multi-target tracking algorithms developed at UC Berkeley. In (Oh et al. 2005), MTT algorithms are ported to receive detection events via the root server. Using a 144-node subset of Trio, they successfully demonstrated in front of a large audience, real-time tracking of three people crossing paths through the center of the Trio field, as shown in Fig. 7.25. The use of Trio for this application has highlighted problems with the system software and has raised new challenges that would not have been discovered in a small-scale or indoor setting.

7.2.7.2   Experimenting with Trio

7.2.7.2.1   Familiarities with Renewable Energy

Renewable energy, in the form of a solar power supply, has been both the benediction and nuisance of this experience. The most promising discovery was how renewable energy fundamentally simplifies system operation, management, and maintenance, enabling the familiar "deploy first, develop later" approach used with wired testbeds. The tricky side stemmed from the dynamics of solar power and the logistics of node initialization that raised many new concerns and exposed several unknown weaknesses in the network protocols and management strategies.

7.2.7.2.2   Limited Availability

Trio can be operated at 100% duty cycle during only a few hours in the middle of the day when direct sunlight is present. However, a duty cycle ranging from 20% to 40%, depending on the time of year, allows continuous operation. This availability limitation stems from several factors:

- Limited awareness of the subtleties of solar energy harvesting. Specifically, seasonal and daily variation in solar power, the angle of inclination of the solar cell, the effect of dirt and bird droppings on the solar cell, the importance of maximizing power transfer from the solar cell, and the policy surrounding energy transfer between the primary and secondary energy stores

- High-power draw and lack of a low-power TinyOS MAC layer for the CC2420 radio
- Several inaccuracies in the design of the Prometheus solar energy harvesting system

7.2.7.2.3 Emergency Battery Daemon

As a consequence of limited availability that is operating with a power deficit, it is impractical to rely on the battery to supply enough power at all times. Thus, Prometheus is prevented from automatically switching to the battery in times of low energy availability. Because Prometheus no longer switches automatically from capacitor to battery, a module called the Battery Daemon is added to permit manually managing this switchover. The Battery Daemon uses Drip (Tolle and Culler 2005) to disseminate a command that directs each node to acquire a short lease. While holding the lease, a node can switch to battery when either the capacitor voltage runs low or with a different command, until the lease expires.

7.2.7.2.4 Epidemic Protocol Failures

The Golden Image includes Deluge (Hui and Culler 2004) and Drip (Tolle and Culler 2005), both of which use the Trickle algorithm (Levis et al. 2004). In these protocols, one node can send an advertisement message that contains out-of-date metadata, which causes neighboring nodes to generate traffic in order to update the advertising node. When exercising the protocols at scale, unstable solar power supply led to nodes powering down, losing their saved metadata, and sending out-of-date advertisements when power was restored. During times of low or occluded sunlight, these reboots happened frequently enough that the excess update traffic created network hotspots. Such excess traffic noticeably slowed down network programming time and disrupted network monitoring and management operations. The power instability present in Trio has exposed several problems in network and transport protocols that are unlikely on a stable indoor testbed.

7.2.7.2.5 Variability at Scale

Non-justified significant variance is revealed across the nodes in their solar energy harvesting and an almost linear growth in the percentage of nodes using the battery in the afternoon from 0% at 13:30 hours to just below 70% at 16:30 hours. These results are surprising because all nodes in Trio run the same software and are oriented in the same way with the solar cells facing south.

7.2.8   TWIST

TKN wireless indoor sensor network testbed (TWIST) is a scalable and flexible testbed architecture for indoor deployment of WSNs; it is developed and experimented at TKN (Telecommunication Networks Group at Technical University of Berlin) (Handziski et al. 2006). TWIST is based on cheap off-the-shelf hardware and uses open-source software. It is thus cost-effective and open for solutions that can be reproduced by others. The design of TWIST is based on an analysis of typical and desirable use cases and is thus capable of supporting a multiplicity of features:

- It provides basic services like node configuration, network-wide programming, out-of-band extraction of debug data, and gathering of application data.
- It supports experiments with heterogeneous node platforms.
- It provides active power supply control of the nodes. This enables easy transition between USB-powered and battery-powered experiments, dynamic selection of topologies, and controlled injection of node failures into the system.
- It permits creation of both flat and hierarchical sensor networks. For this a layer of "super nodes" is introduced, that not only forms a part of the testbed infrastructure but can also play a role as element of the sensor network.

The self-configuration capability, the use of hardware with standardized interfaces, and open-source software make the TWIST architecture scalable, affordable, and easily replicable. A specific realization of TWIST spans three floors of an office building and supports over one hundred sensor nodes.

7.2.8.1   TWIST Architecture

The following sections elaborate a description of the individual testbed entities, starting from the lowest layer, the sensor nodes, and moving up to the testbed backbone with the attached server and control station (Fig. 7.26).

7.2.8.1.1   Sensor Nodes

The sensor nodes need a set of hardware capabilities facilitating their smooth integration with the components of the testbed infrastructure. The overall architecture of the TWIST is remarkably centered on the use of the USB interface. A heterogeneous mixture of WSN platforms is supported, as long as they disseminate capabilities such as power supply, programming, and communication via a standard-compliant USB interface. Generally, any platform having a USB 1.1 interface can be used. This feature is supported by Telos (Moteiv 2004) and EyesIFX (Handziski et al. 2004) mote families, which have been successfully interfaced with TWIST.

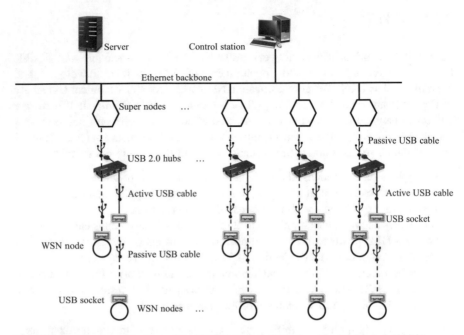

Fig. 7.26 Hardware architecture of the TWIST testbed. (Based on Handziski et al. (2006))

The operating system running on the sensor nodes has to satisfy several basic requirements:

- It has to provide a suitable execution environment for the application logic of the SUE.
- It should support node configuration and instrumentation of the application code and allow for out-of-band communication with the super nodes over the USB infrastructure.

TinyOS (2012) is chosen as it satisfies these requirements and runs on both the Telos (Moteiv 2004) and the EyesIFX (Handziski et al. 2004) platforms.

7.2.8.1.2   Testbed Sockets and USB Cabling

A testbed socket is the point where the USB interface of the sensor node attaches to the USB infrastructure of the testbed. The sockets have unique identifiers, and their geographical position is known and does not change over time. The node identifiers are associated to the socket identifiers and hence to the geographic position of the sockets. The sockets are connected to other testbed components using a combination of passive and active USB cables, depending on the distance between the socket and the next element of the infrastructure, the USB hubs. Using passive cables, a maximum distance of 5 m can be bridged. For greater distances, active USB cables can be used (single-port USB hubs with fixed cable), or several USB hubs can be daisy-chained together.

7.2.8.1.3   USB Hubs

The hubs are the central element of the TWIST USB infrastructure and provide support for some of the most important features of TWIST:

- At the most basic level, the USB hub is a multiplexing device that enables to break the one-to-one correspondence between the sensor nodes and the second-level testbed devices that can be found in many of the existing WSN testbeds. This enables significant cost savings without compromising the testbed functionality.
- Even more, the USB hubs give TWIST one of its most powerful capability, the binary power-control over the sensor nodes in the testbed.

The USB hub specification 2.0 requires that self-powered hubs support port power switching. By sending a suitable USB control message, the software can control the power state of a given port on the hub, effectively enabling or disabling the power supply for any attached downstream device. In the case of TWIST, these downstream devices are the sensor nodes plugged into the testbed sockets. Depending on whether the sensor node attached to the socket is battery equipped or not, four different state transitions are enabled:

- From "USB-powered" to "OFF"
- From "OFF" to "USB-powered"
- From "USB-powered" to "battery-powered"
- From "battery-powered" to "USB-powered"

7.2.8.1.4   Super Nodes

If TWIST only relied on the USB infrastructure, it would have been limited to 127 USB devices (both hubs and sensor nodes) with a maximum distance of 30 m between the control station and the sensor nodes (achieved by daisy-chaining of up to 5 USB hubs). While suitable for small- to medium-size testbeds, these limitations do not allow for scalability of the architecture and support for deployments over large geographical areas.

To tackle the scalability problem, several requirements must be met:

- There is a need for a distributed solution that spreads the testbed functionality among multiple entities.

- The super nodes must have the ability to interface with the earlier described USB infrastructure. In addition, they have to support a secondary communication technology that does not have the size and cable length limits of the USB standard and that forms the testbed backbone to which the server and control stations can be attached.
- Adequate computational, memory, and energy resources are needed.

To satisfy these requirements, while keeping reasonable expenses for a medium- to large-scale testbed, the class of 32-bit embedded devices is used for attaching networked storage. At the same time, these devices have similar capabilities as the so-called high-end wireless sensor nodes or microservers, enabling dual use of the super nodes as parts of the testbed and as parts of the SUE. For TWIST, the Linksys Network Storage Link for USB 2.0 (NSLU2) (Linksys 2008a) depicted as a super node has two USB 2.0 ports; it uses an IXP420 processor from Intel's XScale family (clocked at 133 MHz), with 32-MByte SDRAM and 8-MByte Flash as persistent storage. One particular feature of the IXP4xx family is the two integrated network processor engines (NPEs) that implement, among else, two full Ethernet MAC and physical layer units along with the related packet-processing functionality. The Linksys-supplied firmware for NSLU2 is a customized Linux OpenSlug (Linksys 2008b), now called SlugOS/BE (Linksys 2009).

7.2.8.1.5   Server

The server and the control stations must interact with the super nodes using the testbed backbone, so they have to support the same communication technology. The role of the server is critical; it contains the testbed database, provides persistent storage for debug and application data from the SUE, runs the daemons that support the system services in the network, etc.; its hardware resources should thus be adequately dimensioned to guarantee high levels of availability.

The operating system support on the server is also based on Linux, a standard Fedora Core 3 server installation (Red Hat 2004). Fedora Core 3 was codenamed Heidelberg and was over in January 16, 2006, for a newer version. The current version is Fedora 20, codenamed Heisenbug, and was released on December 17, 2013 (Fedora 2014).

For the management of the super node network, the server runs the DHCP, DNS, NTP, and NFS daemons, as well as the UnionFS kernel module. At the heart of the server is the PostgreSQL database (PostgreSQL 2013) that stores a number of tables including configuration data like the registered nodes (identified as NodeIDs) and the sockets and their geographical positions (identified as SocketIDs) as well as the dynamic bindings between the SocketIDs and NodeIDs. The database is also used for recording debug and application data from the SUE. A significant motive for choosing PostgreSQL is the availability of the PostGIS extension that permits to represent the locations of the sockets in a natural 3D coordinate system and provides support for spatial queries and experimentation with location-based services.

7.2.8.1.6   Control Station

The control station hardware can be any workstation that is attached to the backbone, though the ability to run Linux eases its integration into the testbed. Software functions are two sided:

- A number of developed Python scripts that run locally on the super nodes provide functionalities like sensor node programming, executing power control, collecting debug and application data, and more.
- The actual invocation of these scripts is done by the control station using *ssh* remote command execution[27] (Red Hat 2008).

Without further optimization, the control station would have to log onto the super nodes serially and invoke the Python scripts. Clearly, this would require a lot of time when activities involving all the nodes (like reprogramming) have to be executed. To speed up such tasks, a hierarchical threading approach is adopted to exploit parallelism; the control station first creates a separate thread of control for each of the super nodes. Every such thread starts the Python scripts on its associated super node via the *ssh* remote command execution. On the super node, each of these Python scripts in turn creates separate threads for quasi-parallel reprogramming of all the attached sensor nodes. In this way, by utilizing the natural parallelism in the system, it is possible to execute network-wide tasks in approximately the same amount of time as it would have taken on a single sensor node.

7.2.8.2   TWIST Installation

This section illustrates how TWIST is installed to satisfy the multiplicity of features, laid out earlier, using cheap off-the-shelf hardware and open-source software.

7.2.8.2.1   Matching SUE and TWIST Architectures

The flat, segmented, and multitier sensor network architectures can be easily realized with TWIST due to the flexible boundary between the SUE and testbed functionalities:

- Flat sensor networks. Under this scenario, the boundary between the SUE and the testbed is just the USB interface on the sensor nodes. The super node, the server, and the control station exclusively perform testbed functions. The testbed is used to program the sensor network and to extract debug data or application-related data from the WSN. The extracted data can be preprocessed, compressed, filtered, or aggregated already in the super nodes in a distributed fashion. The debug data

---

[27] The *ssh* command is a secure replacement for the *rlogin*, *rsh*, and *telnet* commands. It allows to log in to a remote machine, as well as execute commands on that machine.

is transferred out-of-band and does not consume any wireless bandwidth. In cases where even the instrumentation of the SUE code with the debug code causes unwanted interactions, the sensor node population can be partitioned into SUE nodes and nodes doing only packet sniffing on the wireless medium.

- Hierarchical sensor networks. In such a case, the super nodes can play a role both in the SUE and in the testbed. They do this in two ways. Either to let some of the super nodes act as high-level nodes in the sensor network application and others as testbed nodes or to execute both roles at the same time on a single super node. Hence, the super nodes are dual-use devices. Again, debug data originating in sensor nodes is not transmitted over the wireless channel. But, for time-sharing, the super nodes have to split their computational resources and bandwidth between SUE and testbed-related functionalities.
- Segmented sensor networks. To implement this scenario, super nodes can be used as gateways between different flat segments. The selected super node hardware offers the possibility to equip them with WLAN, Bluetooth, or other communication technology for this purpose.

The communication between the super nodes, the server, and the control station is carried out using TCP/IP, making it easy to export the testbed services to authorized remote users. The super nodes also play a key role in addressing the scalability requirement. They can be used to filter, aggregate, or compress the generated data, thus pushing the "congestion barrier" on the backbone network towards higher numbers of nodes. The server and control station can be used to store the aggregated data and present online and offline evaluations to the user.

7.2.8.2.2   Programming and Time Synchronization

Reprogramming is supported by TWIST over the USB interface, and it is parallelized as possible by letting the super nodes reprogram their nodes in parallel. The current approach of TWIST towards distributed debugging follows the *printf()* approach. Specifically, when the application on the sensor node dumps debug data, this data is transported over the USB interface to the super node and timestamped there. The super nodes receive their timing information via the network time protocol (NTP) protocol (Authors 2011) from the server. The precision of NTP in local networks is in the range of hundreds of microseconds to few milliseconds. The time resolution achievable by this approach is sufficient for many WSN applications.

7.2.8.2.3   Power Supply Control

A key facility of TWIST is binary power supply control. By switching off the USB connection of a sensor node and thus its power, the extinction of nodes can be emulated. Conversely, by repowering the USB connection, the deployment of new nodes is mimicked. Importantly, this "life control" does not require any cooperation from the sensor nodes. Thus, it is possible to observe, under controlled, precisely repeatable conditions, the response of self-configuration algorithms of the SUE, such as routing, to such configuration changes.

When an experiment requires battery-driven nodes, for example, to obtain lifetime results, this can also be achievable with TWIST; when both a battery and the USB power source are available, the node is powered from the USB port. When the USB port is switched off, the nodes run on battery power only.

7.2.8.2.4 Management

Management in TWIST has several characteristics:

- Powering the nodes via USB cables has a major benefit in alleviating the need for frequent battery changing, which in a larger testbed could create significant additional work and costs.
- As already discussed in Sect. 7.2.8.1, the association between node identifiers and geographic positions is created via the USB sockets.
- The super nodes play a key role in the automatic maintenance of this association:

  - The USB interface on the super nodes detects when a node is plugged into a socket or when it is removed. As a result, a software event is triggered.
  - On receiving such an event, the super node extracts the node's manufacturer serial number from the event data and determines the unique node identification (nodeID) from a database on the server.
  - After that, the super node registers the binding between the node and the socket identifier in the database. This database also contains the geographical position of each socket. It is thus possible to put nodes into arbitrary sockets and to automatically keep the database in a consistent state. Furthermore, it is an easy task to figure out the precise position of a sensor node given its identification, to determine all sensor node identifications pertaining to a given geographical area and so forth.

7.2.8.3 TWIST Deployment

The local instance of TWIST spans three floors at the TKN building. Specifically, 90 locations are fixed for nodes with known positions and there are additional 90 free slots on the USB hubs. Also, 37 NSLU2s are used, 53 USB hubs and about 600 m of USB cables. The NSLU2s communicate over Ethernet. An alternative is a USB-to-WLAN adapter that can be attached to the free port of the NSLU2 and to establish a wireless backbone network.

One of the assets of TWIST is the possibility to experiment with different node densities, network sizes, and node dynamics. For instance, such parameters are changed to test the DRAIN routing protocol (Tolle and Culler 2005).

7.2.9   *SignetLab*

SignetLab is a sensor network testbed deployed at the University of Padova, Italy (Crepaldi et al. 2007). In its design, a twofold approach is adopted, exactly, the design of the physical deployment and design of the software tool. The software tool is made as independent from the physical deployment as possible; this allows the testbed to grow and change without the need to reimplement the software. It also permits other laboratories to easily make use of this tool without the need to replicate the hardware used in SignetLab. The software tool was freely available on the SignetLab group website.

The works that are most relevant to SignetLab are MoteLab (Werner-Allen et al. 2005), Mobile Emulab (Johnson et al. 2006), and TWIST (Handziski et al. 2006). They are aimed at maximizing testbed utilization among different users by providing a web interface through which users can schedule jobs.

7.2.9.1   Hardware

The choice of hardware for SignetLab supports a number of goals:

- The radio should provide sufficient range and power settings to allow the testing of a variety of protocols.
- The nodes must provide a means to alter their sensing capability in order to provide support for a variety of applications.
- The processor on the nodes should provide sufficient computational resources to allow the execution of interesting protocols and applications while still being realistic for a sensor node.
- There should be a reasonable way to get real-time status and debugging information from the testbed without interfering with the execution of the main application.

7.2.9.1.1   Deployment Space

SignetLab is deployed in a 10 m × 11 m laboratory due to space limitations at the University of Padova. The deployment is on a grid suspended 60 cm from the ceiling and 2.4 m above the floor. In this way, the laboratory is not overtaken by the sensor network deployment. The network is made up of 48 EyesIFXv2 nodes (Handziski and Lentsch 2005), separated by 160 cm in one direction and 120 cm in the other direction. These distances were chosen to provide a uniform distribution in the laboratory.

7.2.9.1.2   Sensor Nodes

The EyesIFXv2 nodes were developed during a 3-year European research project on self-organizing energy-efficient sensor networks (Handziski et al. 2004). The nodes use an ultralow-power MSP430F1611 processor with 10 KB on chip RAM, 48-KByte Flash, and an additional 512-KByte serial EPROM. The radio chip is a low-power FSK/ASK transceiver, providing half-duplex, low data rate communication in the 868-MHz ISM band. It operates using FSK modulation, with $< -$ 109 dBm sensitivity, enabling up to 64 Kbps, half-duplex, wireless connectivity.

The platform is also equipped with an onboard stripline antenna and a subminiature version A (SMA)[28] (Wellshow 2015) connector for external antenna. The external antenna is the default. The onboard antenna can be selected by soldering a resistor into the correct location on the chipboard. However, using either of the available antennae created a radio range, which reduced the testbed to a one-hop network. One option was to use a low-gain setting at the receiver; however, this does not decrease the interference range of the transmitters. As this option is not required, SignetLab uses homegrown, low-gain antennae inserted into the external antenna plug to provide multihop transmission ranges.

The nodes are equipped with onboard temperature and light sensors as well as a serial peripheral interface (SPI) expansion port that can be used for additional sensing capabilities. The SPI bus is shared between the expansion port, the radio, and the processor. Therefore, there is a hard restriction on the amount of resources used at a time.

The nodes can be powered either by batteries with a capacity 1000 mAh or through a power supply connected via an external polarized connector or a USB connection.

7.2.9.1.3   Backplane Connection

To avoid interference between debugging and data gathering with the operation of the testbed, a backplane using USB connections is made available. These same USB connections are used to supply power to the nodes; therefore, only a single cable is required to connect each node. Figure 7.27 depicts the backplane architecture, which is composed of two tiers of hubs. Each of the used 15 hubs has its own power supply. The gray squares represent the 12 second-tier hubs, each of which connects four sensor nodes. The solid rectangles represent the three first-tier hubs, each connecting four second-tier hubs. The first-tier hubs are connected directly to the controlling

---

[28] SMA is a coaxial RF connector with a 50-ohm impedance, 1/4–36 thread-type coupling mechanism. SMA offers excellent electrical performance from 0 to 18 GHz.

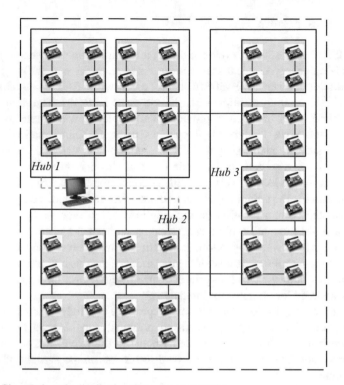

Fig. 7.27 SignetLab node distribution (Crepaldi et al. 2007)

PC. One of the driving factors in this layout was the fact that USB cable lengths could not be greater than 5 m, in order to keep transmission error rates sufficiently low. This is due to insufficient power at the hubs to transmit signals that can be accurately decoded over long distances.

7.2.9.2   Software Tool

The software tool of SignetLab was designed to support a number of goals:

- Providing a single intuitive to use programming interface to all users.
- Being supported on multiple operating systems to allow easy integration into users' work environment.
- Supporting multiple physical sensor network testbeds, i.e., different node technologies, different node layouts, etc.
- Programming nodes, either all or some subset, including compiling and uploading code, should be simple and automated, giving the users as much control as possible during their use of the testbed.
- It should be easy for users to add functionality to the tool.

The SignetLab software tool is a Java application and a set of configuration files that set up the environment; it does not have a component installed on each node and does not rely on TinyOS. A number of example plugins for TinyOS was offered to demonstrate the tool use.

The GUI node selection pane reproduces the topology of the network as specified in the topology configuration file. The user has the ability to select the entire set of nodes or any subset of nodes either by clicking on the nodes and moving them or by using the selection menu. Once nodes are selected, various plugins can be used to program the nodes and begin code execution.

7.2.9.3 Analysis of SignetLab

Considering the signal propagation from a single sensor node for a given transmit power level, two metrics are defined to analyze the testbed. Theoretically, in the absence of any interference or reflections, the area where the signal is received at greater than some strength, x, would define a circle. Practically, a horizontal slice of the transmission pattern at a given signal strength, x, does not describe a perfect circle. For an indoor environment, the contour resulting from such a slice would, in general, be very different.

The two metrics are defined in terms of *inscribing* and *circumscribing* circles for the signal strength slices:

- The *greatest continuous distance reached* is defined as the radius of the inscribed circle, which is the distance inside which the average received signal strength is guaranteed to be greater than x.
- The *farthest distance reached* is defined as the radius of the circumscribed circle, which is the distance outside which the average received signal strength is guaranteed to be less than x.

Instead of the received signal strength as the metric to a slice, the percentage of packets received is used, which is essentially the same, as received signal strength can always be translated to a probability of packet error. Obtained experimentation results show the following:

- The performance of the network with respect to the farthest distance is not very sensitive to the definition of reachability in terms of percentage of packets received.
- The continuous distance is more sensitive because a single node in a region of poor signal quality will reduce the defined radius of the inscribed circle.

These metrics are key to the design of SignetLab to provide the ability to fit the nodes within the space of the lab while still providing an adequate multihop environment to test routing protocols. More results are offered in (Crepaldi et al. 2007).

7.2.10   WISEBED

The WISEBED project is a joint effort of nine academic and research institutes across Europe. The project took place between June 2008 and May 2011. It was funded by the European Commission under the Information Communication Technologies program as part of the Seventh Framework, project number 224460. The WSN testbed, WISEBED, architecture (Coulson et al. 2012; FIRE 2014) was designed with a focus on generality through creating a set of standardized APIs by which a testbed is accessed from a user's perspective (Chatzigiannakis et al. 2010). Users can thus access compatible testbeds using the same clients, no matter if the testbed comprises only a single node connected to a laptop or a full-blown testbed with thousands of nodes. Using the same API and hence the same client software, researchers can automatically deploy the same experiment to a number of testbeds and compare results.

Implementations of the APIs are up to the operators of testbeds. Also, a number of back-end implementations of the WISEBED APIs are available under open-source licenses and individuals or research organizations can easily download and deploy the software to run WISEBED-compatible testbeds. Thus, a growing ecosystem, of testbed client/back-end software including documentation, is built and it is possible to put online descriptions of experiments for use by others for repeating experiments and verifying the results.

WISEBED provides solutions for small- and large-scale experiments. The main benefit is that these experiments can be performed across testbed platforms at different sites spread all over the world even with heterogeneous sensor node hardware. All testbed sites implement the same interfaces and therefore provide a consistent API to users. Thereby, testbed users gain complete remote access to the sensor nodes including program memory and a stream-oriented debugging interface achieving the same flexibility like doing experiments in a local environment. WISEBED operators, on the other hand, keep control of their testbed site by means of authentication and reservation mechanisms that provide security features and allow only those registered with appropriate user rights to perform experiments.

A significant opportunity is to use the WISEBED software components to deploy a private testbed that is not part of the WISEBED federation but that is API compatible. The WISEBED APIs are designed in a way that they are technology independent. This allows deploying different back-end implementations ranging from small-scale deployments on a single PC to a full-blown testbed federation. In the simple case, no authentication or reservation may be required and dummy implementations for both APIs can be used. In the case of a large federation, more complex implementations are required.

Currently, some but not all of the testbeds are still operational and some but not all of the software solutions are still being developed (FIRE 2014).

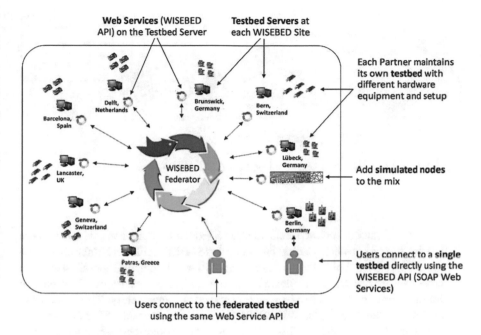

Web Services (WISEBED API) on the Testbed Server

Testbed Servers at each WISEBED Site

Each Partner maintains its own **testbed** with different hardware equipment and setup

Add **simulated nodes** to the mix

Users connect to a **single testbed** directly using the WISEBED API (SOAP Web Services)

Users connect to the **federated testbed** using the same Web Service API

Fig. 7.28 Testbed federation architecture (Hellbrück et al. 2011)

7.2.10.1 Architecture

WISEBED architecture as embodying the following components is detailed in Fig. 7.28 (Hellbrück et al. 2011):

- Testbed server (TS). Each testbed comprises a number of sensor nodes that are managed and controlled by TS. The TS exposes the functionalities of a testbed to users in the Internet by running a software that provides WISEBED API-compatible Web services. Using these Web services, users can run experiments on single testbeds. The WISEBED APIs that users call to interact with the testbed consist of the following:

  - Sensor network authentication and authorization (SNAA) API
  - Reservation system (RS) API

  The above two APIs provide interfaces for authentication and authorization of users as well as resource reservation.

  - Wireless sensor network (WSN) API. This API describes the main entry point for conducting experiments and allows users to interact with the nodes (e.g., program/reprogram and send messages).
  - Controller API. Users that conduct experiments start a Web service endpoint implementing the controller API where the Web service listens for output generated by the nodes (e.g., using the serial interface).

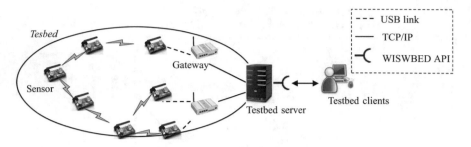

Fig. 7.29 Testbed architecture. (Based on Hellbrück et al. (2011))

- Federator. Testbeds are interconnected by a federator component, which exposes a federation of testbeds using the same WISEBED APIs and thus creating a virtual large-scale unified testbed. From an API standpoint, the federation appears to be a single testbed. This allows users to use all sensor nodes of all testbeds at the same time and as part of a single experiment transparently. To interconnect spatially divided testbeds, WISEBED employs the concept of virtual links as enlightened in Sect. 7.1.3. Virtual links emulate broadcast connectivity between arbitrary nodes by tunneling messages between the communication partners using the WISEBED APIs. Additional nodes can be added to the federation by integrating simulated nodes into experiments. The Shawn network simulator (Kröller et al. 2005; Fekete et al. 2007) has been extended to support the WISEBED APIs and can hence be part of a federated testbed.

The architecture of a typical testbed is detailed in Fig. 7.29. Two different types of testbed architectures are distinguished, wired and wireless:

- Wired testbed where every sensor node is attached to a host system (gateway) via a serial USB connection. The use of additional gateway components may be required if not all sensor nodes can be directly attached to the testbed server, e.g., because they are spread over multiple rooms. Typically, such a gateway is an embedded PC or a netbook-class computer.
- Wireless testbeds that have at least one sensor node that is not connected to any kind of wired communication backbone, i.e., communication with and reprogramming of such nodes can only be done wirelessly.

The testbed server and the gateways habitually communicate through a wired infrastructure, such as Ethernet to provide a reliable backbone. Additional sensor nodes can be attached wirelessly as described above. Programming and reprogramming of these nodes are based on an over-the-air programming (OTAP) mechanism which must be present on the sensor nodes at any time.

7.2.10.2 WISEBED-Compatible Testbeds

The nine partners (Fig. 7.28) provided testbeds comprising a total of 700+ hetero-geneous wireless sensor nodes of various architectures and vendors. Table 7.1 details as possible the constituents of each testbed. The listed data depend on the available information that is vastly diversified in the literature; a first priority is accorded for (FIRE 2014) and then to a recent publication (Hellbrück et al. 2011). All these testbeds are permanent installations with wired backbones. The wired backbone enables out-of-band interaction with the nodes to collect traces and debug informa-tion or send commands to the nodes. Apart from the permanently installed wired testbeds, several sites offer on-demand extensions such as the possibility to intro-duce mobility into experimentation by using robots that piggyback wireless sensor nodes and move autonomously through the testbeds.

In addition to the WISEBED testbeds provided by the nine partners, a number of WISEBED-compatible testbeds have been recently deployed (Hellbrück et al. 2011):

- Wireless solar-powered outdoor testbeds are available at the University of Braun-schweig and the University of Lübeck, Germany. They can be used to perform experiments under realistic, outdoor conditions.
- The University of Applied Sciences in Lübeck, Germany, extended one of the WISEBED API implementations to support their custom-made TriSOS (3-SOS) sensor node platform (Hellbrück 2012). The troika TriOS stands on smart object systems, self-organizing systems, and service-oriented systems. The TriSOS platform is based on the AVR Raven evaluation board. Their testbed is now run using WISEBED software and is accessible using the WISEBED APIs and clients.
- The upcoming testbed of the citywide deployment of the SmartSantander project (SmartSantander 2014) is based on WISEBED technology and will contribute actively in the further evolution of APIs and client and back-end implementations.
- The project MOVEDETECT (Langmann et al. 2013), which is funded by the Federal Office for Information Security (BSI), based its deployment architecture on the WISEBED APIs and the testbed runtime implementation.

The above-listed activities reveal the WISEBED ecosystem as made available in the WISEBED website (FIRE 2014).

7.2.11 Indriya

Indriya (Doddavenkatappa et al. 2012) is a WSN testbed deployed at the National University of Singapore (NUS) across three different floors of the main School of Computing building. The deployment over three floors covers spaces used for different purposes, including laboratories, tutorial rooms, seminar rooms, study

Table 7.1 WISEBED testbed sites

Partner	Operational	Nodes	Sensors	Wireless interface	Amount
Lübeck, Germany	Yes	Pacemate by ITM[a]	Heart rate	Xemics[c] RF (868 MHz)	60
		iSense[b] by Coalesenses[a]	Temperature, light, PIR, acceleration	IEEE 802.15.4 (2.4 GHz)	60
		TelosB[b]	Temperature, humidity, light	IEEE 802.15.4 CC2420[b] (2.4 GHz)	60
Lancaster, UK	No	TelosB[b]	Temperature, humidity, light	IEEE 802.15.4 CC2420 (2.4 GHz)	16
Bern, Switzerland	No	TelosB	Temperature, humidity, light	IEEE 802.15.4 CC2420 (2.4 GHz)	40
		MSB-430	Temperature, humidity, acceleration	CC1020[b] (402,424,426,429,433,447,449,469,868,915 MHz)	7
Patras, Greece	No	TelosB	Temperature, humidity, light	IEEE 802.15.4 CC2420 (2.4 GHz)	50
		iSense	Temperature, light, PIR, acceleration	IEEE 802.15.4 (2.4 GHz)	100
		iSense	Temperature, humidity, PIR	IEEE 802.15.4 (2.4 GHz)	10
Geneva, Switzerland	No	iSense	Temperature, light, PIR, AMR, acceleration	IEEE 802.15.4 (2.4 GHz)	28
		MicaZ[b]	Temperature, light, acceleration, acoustic, and magnetic fields	IEEE 802.15.4 (2.4 GHz)	2
Delft, The Netherlands	No	SOWNet[a] T-Node[b]	Alcohol, temperature, PIR, humidity, heart-beat, light, magnetic	CC1000 (868 MHz)[b]	24
		Tmote Sky[b]	Temperature, humidity, light	IEEE 802.15.4 CC2420 (2.4 GHz)	8
		SOWNet G-Node G301[b]	Temperature	CC1101 low-power sub-1-GHz transceiver (315/433/868/915 MHz)[b]	108
Berlin, Germany	Yes	MSB-A2[b]	Temperature, humidity	CC1100 (864–970 MHz)[b]	100
Brunswick, Germany	No	iSense	Pressure, PIR	IEEE 802.15.4 (2.4 GHz)	30
				Total	**703**

[a]Manufacturer information available in Chap. 9
[b]Datasheets are available in Chap. 10
[c]Xemics, Inc., based in Switzerland, was acquired by Semtech Corporation in June 2005. Xemics, Inc., is a fabless developer of ultralow-power analog, radio-frequency (RF), and digital integrated circuits, including the CoolRISC microcontroller family (EE Times 2005)

areas, and walkways. The network has several inter-floor links providing three-dimensional connectivity. It has been available for internal use since April 2009 and is publicly available since December 2009. Users from more than 35 universities use the testbed for research. It is also used for teaching within NUS. Indriya is installed with 127 TelosB motes. More than 50% of the motes are equipped with different sensor modules, including passive infrared (PIR), magnetometer, accelerometer, etc. It is built on a reliable active-USB infrastructure that employs special active USB cables. The infrastructure provides a remote programming back channel and it also supplies electric power to the sensor devices.

Indriya's design has the following three advantages:

- It is designed to reduce the costs of both deployment and maintenance of a large-scale testbed. The average installation cost per node in Indriya is substantially less compared to the costs in MoteLab (Werner-Allen et al. 2005) and Kansei (Arora et al. 2006) testbeds. When compared to TWIST (Handziski et al. 2006), which is also centered on an active-USB infrastructure, Indriya avoids the costs and difficulties involved in setting up and maintaining a large number of single-board computers like NSLU2 (Linksys 2008a, 2009). Indriya has been in use by beyond 100 users. The total maintenance cost so far is less than US $500 plus a recurring cost of 1–2 hours per week of time spent by one PhD student. Most of the cost is spent on replacing failed AC-to-DC adapters, as these devices are not designed for long-term usage.
- As deployment of Indriya is over three floors, wireless connectivity among nodes is three-dimensional. This allows experimentation of protocols that are sensitive to placement and connectivity, such as geographical routing protocols.
- Unlike most of the existing testbeds that provide only a wireless infrastructure, Indriya is equipped with different types of sensor boards, thus allowing evaluation of numerous WSN applications.

7.2.11.1 Indriya Composition

7.2.11.1.1 Motes

It is not feasible to use batteries to power motes for sustained and long-term experimentation, in particular for a large-scale testbed. On the other hand, wall-powering individual nodes suffers significant equipment and labor cost for installing power points and electric cables. In order to avoid these costs, USB-based motes are selected so that they can be powered by the remote programming back channel which is built over USB active cables. TelosB devices (Moteiv 2004) are chosen, as they are the most popularly used USB motes in the WSN community. TelosB has a Texas Instruments MSP430 microcontroller with 10 KByte of RAM used for storing program data only (Texas Instruments 2011). The program code is stored in an internal Flash of size 48 KByte. TelosB has a Chipcon CC2420 radio transceiver operating at 2.4 GHz with indoor range of approximately 20–30 m.

7.2.11.1.2 Sensors

More than 50% of the motes in Indriya are installed with different types of sensors, thus allowing experimentation of diverse WSN applications. The main types of sensors deployed are WiEye (EasySen LLC 2008c), SBT80 (EasySen LLC 2008b), and SBT30EDU sensor boards (EasySen LLC 2008a). The WiEye board is commonly used to detect the presence of objects that emit invisible infrared rays, particularly, human beings. The SBT30EDU board includes visual light, acoustic, and infrared sensors. In addition to these sensors, SBT80 contains temperature, two-axis acceleration, and two-axis magnetic sensors.

Sensor manufacturers are listed in Chap. 9, and datasheets are obtainable in Chap. 10.

7.2.11.1.3 USB Active Cables

While using a normal USB cable, the maximum distance between a host and a TelosB mote is limited to 5 m; this limitation is overcome by employing USB active cables. As stated in Sect. 7.2.8, an active cable is a special USB cable that incorporates electronics to sustain data signals so that five of them can be daisy-chained to cover a maximum distance of 25 m. It is crucial to use high-quality active cables; otherwise, a host computer often loses USB connection with the sensor devices.

7.2.11.1.4 Design of a Back Channel for Remote Programming

Existing testbed deployments either individually attach testbed nodes to single-board computers such as Stargate NetBridge (Crossbow 2007b) or use such computers as super devices with each controlling a group of nodes. In both cases, the single-board computers are accessed over Ethernet. Although the latter design is relatively cost-effective, the required number of super devices is still quite large. Particularly, TWIST is based on such a design and it uses 46 NSLU2 single-board computers to control 204 sensor nodes.

Unlike these deployments, Indriya does not employ single-board computers. Instead, it is based on an efficient cluster-based design with each cluster consisting of a single cluster head that can accommodate up to 127 sensor devices. Moreover, an individual cluster can geographically span a circle of diameter of up to 50 m. Three floors of the large building are covered with only six clusters. Compared to existing 3D testbeds, Indriya is the largest in terms of geographical size as measuring 23,500 m^3, when compared to the 12,000 m^3 MoteLab and 6630 m^3 TWIST (Gnawali et al. 2009).

Figure 7.30 depicts the design of a cluster of Indriya. A Mac Mini PC (Apple Inc. 2014) constitutes the cluster head. This PC is a very small footprint computer (19.7 cm × 19.7 cm × 3.6 cm) but as resourceful as a desktop PC. The motes are connected to the cluster head using USB hubs and active cables. Belkin seven-port

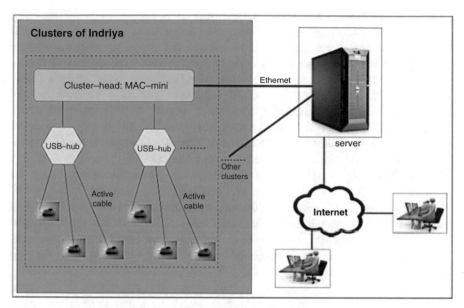

Fig. 7.30 Cluster-based structure in Indriya (Doddavenkatappa et al. 2012)

USB hubs (Belkin International, Inc. 2014) and a mix of locally supplied and ATEN USB active cables are used (ATEN 2014). Cluster heads are connected to the server via Ethernet. The server manages the testbed and provides the user interface.

7.2.11.1.5 User Interface

Indriya uses MoteLab's user interface software that provides Web-based access to the testbed nodes. As the design of Indriya is cluster-based that differs from that of MoteLab, changes are required, particularly in the code that is responsible for communicating with testbed nodes. However, from the perspective of users, clusters in Indriya are transparent and the testbed is simply a wireless network of 127 nodes. The interface allows users to evaluate WSN systems implemented over TinyOS (TinyOS 2012), the de facto standard operating system for WSNs. Users can upload, monitor, and control their jobs remotely and in real time.

7.2.11.2 Indriya Compared

This section compares Indriya against three existing testbed deployments, MoteLab, Kansei, and TWIST, as presented in Sects. 7.2.2, 7.2.6, and 7.2.8. The comparison is basically from the perspective of deployment cost and difficulties involved in setting up and maintaining a large-scale testbed:

- MoteLab as deployed at Harvard University is composed of 190 Tmote Sky sensor nodes with currently 85 of them being active (as of February 2011). Back-channel support is Ethernet-based with individual motes attached to separate Stargate NetBridge single-board computers. Like MoteLab, devices in Kansei are also individually coupled to devices called eXtreme Scale Stargate (XSS) (Arora et al. 2005) which are, configuration-wise, similar to NetBridge. Each XSS has an Intel 400-MHz PXA255 XScale processor with 64-MByte SDRAM, 32-MByte Flash, type II PCMCIA slot, USB port, and 51-pin mote connector; packaging is watertight (Crossbow 2004b). Kansei is deployed at the Ohio State University (OSU) with 210 eXtreme Scale Motes (XSMs) (Arora et al. 2005). The XSM circuit board has a $3'' \times 3''$ footprint and the enclosure has dimensions of $3.5'' \times 3.5'' \times 2.5''$. The XSM platform integrates an Atmel ATmega128L microcontroller, a Chipcon CC1000 radio operating at 433 MHz, a 4-Mbit serial Flash memory, quad infrared, dual-axis magnetic and acoustic sensors, and weatherproof packaging. Datasheets are embodied in Chap. 10. XSMs are commercially available under the trade name of MSP410CA Mote Security Package (Willow Technologies 2013).
- Contrary to MoteLab and Kansei, Indriya incorporates an efficient cluster-based design eliminating the need for coupling individual motes to separate single-board computers. Instead, Indriya uses a MAC Mini capable of controlling 127 sensor devices. This significantly reduces the cost per node in addition to avoiding painstaking difficulties involved in setting up and maintaining single-board computers. Moreover, single-board devices are typically wall-powered, thus incurring both labor and equipment costs for installing power points and electric cables, whereas nodes in Indriya are powered over USB.
- The average cost per node in Indriya is US $158, which is considerably less, compared to the costs in MoteLab and Kansei. The cost per node in MoteLab and Kansei is almost the same as they incorporate similar designs and devices. The cost is approximately US $548 (NetBridge/XSS: US $449 + Tmote/XSM: US $99) plus the cost of providing wall power and Ethernet connectivity. Furthermore, extending Indriya with additional motes is comparatively less pricy as each MAC Mini can accommodate up to 127 USB devices, and currently, there is an average of only 22 nodes per cluster.
- Since both TWIST and Indriya incorporate similar USB backbones, the average cost per node in these deployments is almost the same. But this is true only to the existing deployment instance of TWIST; adopting its design is actually expensive. This is because of the fact that NSLU2 is a discontinued product since 2008 and is now called SlugOS/BE (Linksys 2009). A natural replacement to NSLU2 is either Stargate NetBridge or WRT600N (Linksys 2007). Stargate NetBridge is a Crossbow (Crossbow 2007b) modified and expensive version of the original NSLU2. On using either of these latest devices, the cost of replicating TWIST will rise by at least US $70 per node compared to Indriya.
- Moreover, comparing Indriya versus TWIST, which is also centered on a USB infrastructure, TWIST uses 46 single-board and wall-powered NSLU2 computers to manage 204 testbed nodes, whereas in Indriya, there are only 6 Mac Mini PCs

to manage 127 nodes, with each Mac Mini capable of accommodating more than 100 USB devices. Both testbeds span three floors of a building but Indriya covers almost three times larger geographical volume, as per the dimensions provided in (Gnawali et al. 2009).

- In TWIST, NSLU2 devices use the OpenSlug distribution of Linux customized specifically for testbed usage (Linksys 2008b). On the other hand, Mac Mini devices in Indriya can use any desktop OS without any specific changes. Currently, Ubuntu Linux running the 2.6.12 kernel is used without any modification (Ubuntu 2014). This also allows employing standard tools available for programming and managing sensor motes, while NSLU2-like single-board devices demand significant changes or a new set of tools. Another important issue with NSLU2 devices is that they are very resource-constrained, particularly, their limited flash memory. Requirements such as file system over network using NFS[29] (Indiana University 2014)-like protocols are memory greedy. All these issues significantly add to the difficulties involved in setting up and maintaining a large-scale testbed.

7.2.12   GENI

The Global Environment for Network Innovations (GENI) project (The GENI Project Office 2008; Berman et al. 2014) concretely illustrates an architecture where WSN fabrics are key components (Sridharan et al. 2011). GENI is an innovated experimental network research infrastructure. It includes support for control and programming of resources that span facilities equipped with fiber optics and switches, high-speed routers, citywide experimental urban radio networks, high-end computational clusters, and sensor grids. GENI is intended to support sizeable numbers of users and large simultaneous experiments with extensive instrumentation designed to make it easy to collect, analyze, and share real measurements and to test load conditions that match those of current or projected Internet usage.

GENI provides a virtual laboratory for education, networking, and distributed systems research; it is well suited for exploring networks at scale, thereby promoting innovations in network science, security, services, and applications (GENI 2014). GENI allows experimenters to the following:

- Obtain compute resources from locations around the United States
- Connect compute resources using layer 2 networks in topologies best suited to their experiments
- Install custom software or even custom operating systems on these compute resources

---

[29]NFS stands for network file system, a file system developed by Sun Microsystems, Inc. It is a client/server system that allows users to access files across a network and treat them as if they resided in a local file directory.

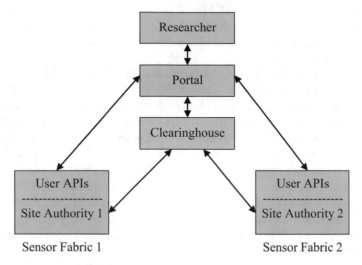

Fig. 7.31 Federated fabric/GENI model (Sridharan et al. 2011)

- Control how network switches in their experiment handle traffic flows
- Run their own layer 3 and above protocols by installing protocol software in their compute resources and by providing flow controllers for their switches

Figure 7.31 depicts the GENI architecture from a usage perspective. In a nutshell, GENI consists of three entities, namely, researchers, clearinghouses, and sites also known as resource aggregates. Typically, a researcher interacting, via specially designed portal, queries a clearinghouse for the set of available resources at one or more sites and requests reservations for the resources he requires. To run an experiment, a researcher configures the resources allocated to his slice, which is a virtual container for the reserved resource, and controls his slice through well-defined interfaces.

By programmable WSN fabrics (Sridharan et al. 2011), it is meant that individual sensor arrays offer not just resources on which programs can be executed; they also provide network abstractions for simplifying WSN application development and operation. Examples include APIs for scheduling tasks, monitoring system health, and in-the-field programming and upgrade of applications, network constituents, and sensing components. Fabrics can also support and manage the concurrent operation of multiple applications. Figure 7.32 compares the traditional WSN model with the emerging fabric model of WSNs.

Through federating WSN testbeds (Sect. 7.1.3), multiple WSN testbeds are loosely coordinated to support geographically and logically distinct resource sharing. A federation provides users with a convenient, uniform way of discovering and tasking desired WSN resources. Experiments can simultaneously use resources in multiple testbeds, for applications ranging from regression testing, producer-consumer, and parallel processing to enterprise-edge cooperation.

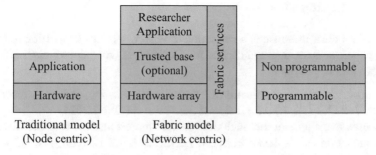

Fig. 7.32 Traditional and fabric models (Sridharan et al. 2011)

Researchers and sites in GENI establish trust relationships and authenticate each other via GENI clearinghouses (Sridharan et al. 2011). The clearinghouse keeps track of the authenticated users, resource aggregates, slices, and reservations. Each resource provider may be associated with its own clearinghouse but there are also central GENI clearinghouses for federated discovery and management of all resources owned by participating organizations. GENI also relies on all entities to describe their underlying resources. Resource descriptions serve as the glue for the three entities because all interactions involve some description of resource, be it a physical resource, such as a router and a cluster, or a logical resource, such as CPU time or wireless frequency.

GENI is not included in Table 7.2 due to lack of information regarding typical WSN deployment.

7.2.12.1   Federated WSN Fabrics

As previously illustrated (Fig. 7.31), the federated WSN fabric model distinguishes three actors:

- The clearinghouse that enables discovery and manages resource inventory and allocation
- The site that owns and maintains WSN aggregate resources
- The researcher who deploys/tests applications via a portal and who is not necessarily a WSN expert

In what follows the roles and requirements of each of these actors will be laid out. Noteworthy, the main goal of the federated WSN fabric model is to make user experimentation easy, repeatable, verifiable, and secure while maximizing resource utilization.

7.2.12.1.1   Clearinghouse Tasks

As the GENI clearinghouse is a collection of related services supporting federation among experimenters, aggregates, and the GENI Meta Operations Center (GMOC), it has two broad functions:

- Identification and authentication of various actors in the system as revealed in Sects. 7.2.12.1.1.1, 7.2.12.1.1.2, and 7.2.12.1.1.3 (GENI 2014)
- Resource management that includes resource representation, resource discovery, and allocation as displayed in Sects. 7.2.12.1.1.4, 7.2.12.1.1.5, and 7.2.12.1.1.6 (Sridharan et al. 2011)

7.2.12.1.1.1   Federation Services

The clearinghouse represents a trust anchor for all software entities (tools, aggregates, services) in the GENI federation. Any member of the GENI federation trusts anything trusted by the GENI federation. The installation of the GENI certificate as a trust root at any GENI service allows for federated trust across all users, aggregates, and services. In this way, there is no need for each entity to explicitly trust each other's entity to allow for federation-wide trust; each entity needs only to trust GENI.

The clearinghouse provides a series of services for managing and asserting the credentials of entities trusted by GENI:

- An Identity Provider (IdP) that provides certificates and public key infrastructure (PKI) materials to human users, registering them with the GENI federation as GENI users.
- A Project Authority to assert the existence of projects and the roles of members, such as principal investigator (PI) and experimenter.
- A Slice Authority that offers experimenters with slice credentials by which to invoke aggregate manager (AM) API calls on federation aggregates.
- A Service Registry that delivers to experimenters with "yellow pages" of URLs of all trusted services of different types. In particular, the list of all available aggregate managers trusted by GENI, possibly satisfying particular criteria, is provided.
- A single sign-on portal, which provides Web-based authentication and access to the authorized clearinghouse services and other GENI user tools.

7.2.12.1.1.2   Authorization Services

The clearinghouse provides services to determine whether particular actions (within the clearinghouse or with respect to a particular aggregate) are permitted by federation policy. There are two essential types of authorization policy, specifically, Trust Policy and Resource Allocation Policy:

- Trust Policy is a statement or sequence of statements from which allowable actions may be inferred from the attributes of a principal. The GENI software architecture recognizes two types of credentials:
  - Attribute-based access control (ABAC) provides a representation for trust delegation statements and a reasoning engine that proves that a given entity is trusted to take a particular action based on the set of ABAC statements provided.
  - Slice federation architecture (SFA) credentials use a table-driven mechanism to map attributes into allowable actions.

- Resource Allocation Policy is a statement limiting the resource allocations or allocation behaviors associated with a given project, slice, or experimenter. For example, it may be required to limit the number of compute nodes (computers or VMs) allocated to a given project at any given time.

The clearinghouse authorization service determines whether a given action is permitted by policy. It contains a series of guards, each of which may veto a given action, i.e., an act is authorized if and only if any guard does not prohibit it.

The clearinghouse provides a credential store that provides authorized read/write access to all credentials for all GENI-trusted entities. This store allows for federation or local authorization services or other policy decision or enforcement points to have access to the appropriate credentials without needing to carry or compute these at the time of each customer request.

7.2.12.1.1.3   Accountability Services

The clearinghouse provides services that log transactions (successful or failed) between user tools and aggregates to support real-time and ex post facto forensics analytics. By maintaining logs and databases of transaction callers and arguments, of projects and their slices and slivers, the GMOC can have critical timely traceback to find the identities of possibly misbehaving users or responsible project leads. They can then, depending on the situation, contact the project lead and shut down all or some slivers associated with a misbehaving aggregate or user or some combination thereof.

The logging service provided by the clearinghouse fronts a store for writing and querying data associated with transactions, allowing for determining what entity made what requests and got what results. The logging service provides the traceability between slivers and slices. The Slice Authority provides the link of slices to projects, while the Project Authority provides the link of projects to investigators. Together, these provide the ability to find the responsible party to contact in case of problematic behavior on the part of an experimenter.

Resource management is emphasized in the upcoming sections (Sridharan et al. 2011).

7.2.12.1.1.4   Resource Representation

A basic issue for federated WSN fabrics is how to represent a resource in a way that allows multiple sites with different types of fabrics to publish resources to the same clearinghouse while allowing sites, portals, and clearinghouses to evolve over time. The choice of this representation potentially affects all actors; specifically, sites need to advertise the resources, portals need to request the resources, and clearinghouses need to match portal requests to resources available at sites. Also, clearinghouses may need to communicate with each other for federated resource discovery and allocation. All of the above requires a language that can be used to precisely specify information about the resource.

It is to be noted that the need for a resource description language does not mean that the same type of device/network must be defined by all fabrics in a globally unique approach. Given the vast heterogeneity of sensor devices, aggregate architectures, fabric service abstractions/semantics, and administrative domains, each fabric may locally define its resources in a unique way. For the sake of illustration, even the use of IP addressing for WSN devices remains a controversial issue, some fabrics may use this choice while others may not. Likewise, each fabric may choose to associate only locally unique identifiers with devices in its namespace, while others might insist on globally unique identifiers.

The WSN device/network specifications, RSpecs, tend to be declarative rather than descriptive. In other words, they concisely define what the resource is and eschew details about how the resource is used. RSpecs need not necessarily be human-readable because most researchers are expected to interact with portals in appropriate ways, e.g., with graphical interfaces or library support to help and automate the composition of resource requests.

7.2.12.1.1.5   Resource Discovery

Sites advertise their resources to well-known clearinghouses, so that researchers can discover their resources. Clearinghouses of different levels may discover resources directly from the fabric provider through their advertisements or indirectly through other clearinghouses. A hybrid clearinghouse architecture is envisioned where hierarchical, as well as peer-to-peer, communication is possible. Figure 7.33 shows the communications in such a (hypothetical) federation. Push and pull models of resource discovery are conceived. In the push model, a clearinghouse periodically announces to its peers or upper-level clearinghouses the available resources at its associated fabrics that can be shared. In the pull model, a clearinghouse requests from its peers or upper-level clearinghouses their latest resource availability. The pull model is likely to be used in an on-demand manner when a clearinghouse cannot find enough resources to satisfy a user request.

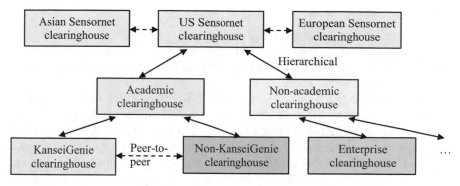

Fig. 7.33 Clearinghouse to clearinghouse interaction architecture (Sridharan et al. 2011)

7.2.12.1.1.6   Resource Allocation

In a federated experiment, a researcher might want to request, via one or more clearinghouses, resources from multiple sites into a slice. It is possible that not all requested resources are available at the same clearinghouse, so portals may have to coordinate requests. Broadly speaking, there are two approaches for federated resource allocation depending on whether the clearinghouse or the portal will own the responsibility of getting all of the requested resources:

- The portal will directly request the resource from multiple clearinghouses, though this approach lacks scalability. This approach is adopted by KanseiGenie (Sridharan et al. 2011).
- The portal communicates with a single clearinghouse, which in turn communicates with other clearinghouses to get the requested resources. This approach requires clearinghouse-to-clearinghouse resource delegation, as shown in Fig. 7.33.

7.2.12.1.2   Site Requirements

7.2.12.1.2.1   Sliceability

In the GENI model of experimentation, each researcher owns a virtual container, a slice, to which he can deploy/execute experiments and add/remove resources. This view fundamentally decouples the physical location of the resource from its reuse. It follows that all resources leased to a researcher should be able to communicate with each other and only with each other. Sliceability is also fundamental for federated experimentation where a researcher selects resources from multiple sites and adds them to a federated slice and runs experiments on this federated slice.

Sliceability may be fine-grained. To share memory, processing, or links between slices in a transparent manner, it is necessary to achieve node/network virtualization of resources. A fabric model suits virtualization since it allows users to interact with resources only through well-defined APIs.

7.2.12.1.2.2 Virtualization

The requirement that sites allow WSN resources to be sliced finely enables multiple slices to coexist. The challenge in virtualization is to provide as much control to the users (as low in the network stack as is possible) while retaining the ability to share and safely recover the resource.

Virtualization in WSN fabrics is nontrivial. Not only do sites have to virtualize the hardware, but also the network. Recall that WSN fabrics may span multiple arrays of sensors, and multiple researchers may run their experiments concurrently on subsets of one or more arrays. Usually, sensors are densely deployed over space and share with different degrees the same geographical space; in addition, they are subject to similar, if not statistically identical, physical phenomena/environments. Wireless interference between slices is thus an inherent problem due to the broadcast nature of the wireless communications. Virtualization has to thus isolate the communications of an experiment running on a slice, to enable repeatable performance. For instance, channel properties such as signal-to-noise ratios among wireless nodes may need to be (statistically) similar across repeated experiments.

7.2.12.1.2.3 Programmability

WSN fabrics are expected to provide the hardware and software infrastructure for an end-to-end reprogramming service, which reliably deploys the sensing applications composed by researchers on the corresponding slice. Sites should also provide monitoring and logging services. In particular, they should also provide feedback to the researcher about the environment and any failures that occur during programming or execution of an experiment. When sites are situated in environments that are not representative of sensing phenomena, it is desirable that they provide services for external sensor data injection. Finally, sites and/or portals should support workflow services that will allow staging and complex experimentation.

7.2.12.1.3 Researcher Requirements

7.2.12.1.3.1 Resource Utilization

To simplify the researcher's task in using federated resources, a portal needs to provide a uniform resource utilization or experimentation framework. This is challenging since the federation may consist of fabrics with a great variety of available platforms, sensors, radios, operating systems, and libraries. For instance, while for some platforms such as XSM (Arora et al. 2005) and TelosB (Moteiv 2004) the default is to program on bare metal, others such as iMote2 (Crossbow 2005), SunSPOT (Ritter 2007; Tantisureeporn and Armstrong 2008), and Stargate (Crossbow 2004b) host their own operating systems. Moreover, the execution environments in these platforms vary from a simple file download and programming the Flash to command line interfaces and virtual machines.

All of the above necessitates an experiment specification language that enables researchers to configure slices in a generic manner. Intuitively, an experiment specification should include the resource description that the experiment is to run on. It also includes a selection of user services that is relevant to the experiment. In addition to these declarative elements, the experiment specification language includes procedural descriptions (or workflow elements). Unlike resource specifications where readability is not important, experiment specifications should provide good readability since a researcher might want to script his experiments to iterate through a bunch of test parameters. Also, he might need to reuse the same experiment specification on different slices, which makes the experiments repeatable.

7.2.12.1.3.2   Resource Translation

It is often more convenient for a researcher to request a networked resource in an abstract manner. For instance, requesting a 5×5 connected grid or a linear array of 10 nodes with 90% link delivery radio is much easier than identifying specific sensor devices which match the required topology. Since the resources published at the clearinghouses are specified concretely, a portal needs to translate the abstract specification to embed it into site resources, although it is possible that this be realized at the clearinghouse as well.

In a federated setting where resources are variously represented by different sites, a service is required to provide a mapping between the researcher's resource need and a resource request that can be processed by different clearinghouses. This service is likely to be implemented at the portal, if not in a clearinghouse.

7.2.12.2   Why Use GENI?

GENI might be fitting for experimenters eager for the following requirements (GENI 2014):

- A large-scale experiment infrastructure. GENI can potentially provide the experimenter with more resources than is typically found in any one laboratory. GENI gives access to hundreds of widely distributed resources including compute resources such as virtual machines and "bare machines" and network resources such as links, switches, and WiMax basestations.
- Non-IP connectivity across resources. GENI allows to set up layer 2 connections between compute resources and run one's own layer 3 and above protocols connecting these resources.
- Deep programmability. With GENI, the experimenter can program not only the end hosts of his experimental network but also the switches in the core of his network. This allows to experiment with novel network layer protocols or with novel IP-routing algorithms.
- Reproducibility. The experimenter can get exclusive access to certain GENI resources including CPU resources and network resources. This gives control

over the experiment's environment and hence the ability to repeat experiments under identical or very similar conditions.

- Instrumentation and measurement tools. GENI has two instrumentation and measurement systems that the experimenter can use to instrument his experiments. These systems provide probes for active and passive measurements, measurement data storage, and tools for visualizing and analyzing measurement data.

7.2.12.3   Key GENI Concepts

Based on (GENI 2014), several GENI concepts are identified and clarified in the coming sections.

7.2.12.3.1   Project

A project organizes research in GENI; it contains both people and their experiments. A project is created and led by a single responsible individual, the project lead. A project may have many experimenters as its members and an experimenter may be a member of many projects. The project lead is ultimately accountable for all actions by project members in the context of the project. GENI experimenters must have project lead privileges to create projects. Only faculty and senior members of an organization can be project leads (e.g., students cannot be project leads).

Figure 7.34 illustrates a situation where a professor is the lead for two GENI projects, one that he uses for his research project Hactar and the other for the networking class CS404 he is teaching. Members of the project Hactar are the

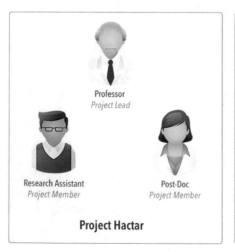

Fig. 7.34   GENI projects (GENI 2014)

professor's research assistant and his postdoc. Members of the project CS404 are the teaching assistant for CS404 and all the students in the class. The professor gives his teaching assistant administrative privileges to project CS404 to permit him adding students to the project or removing them.

7.2.12.3.2   Slice

GENI is a shared testbed, i.e., multiple experimenters may be running multiple experiments at the same time. This is possible because of the concept of a slice. A GENI slice is as follows:

- The unit of isolation for experiments. A GENI experiment lives in a slice. Only experimenters who are members of a slice can make changes to experiments in that slice.
- A container for resources used in an experiment. GENI experimenters add GENI resources (compute resources, network links, etc.) to slices and run experiments that use these resources. An experiment can only use resources in its slice.
- A unit of access control. The experimenter that creates a slice can determine which project members have access to the slice, i.e., are members of the slice. The project lead is automatically a member of all slices created in a project.

Figure 7.35 shows two slices created by the research assistant in project Hactar. He has added to slice 1 three compute resources connected by three network links. He has also added the postdoc associated with his project as a member of the slice. His professor was automatically added to the slice as he is the project lead. Slice

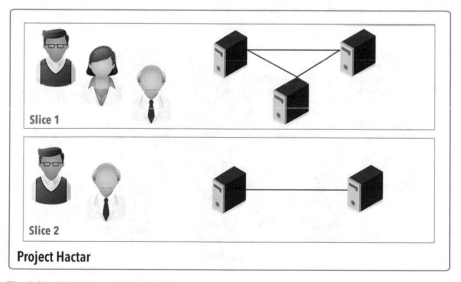

Fig. 7.35   GENI slices (GENI 2014)

2 has two compute resources connected by a link. He has not added the postdoc as a member of this slice and so she cannot perform any actions on this slice or even view the resources in this slice. An experiment in slice 1 can only use resources in slice 1 and an experiment in slice 2 can only use resources in slice 2.

7.2.12.3.3   Aggregates

A GENI aggregate provides resources to GENI experimenters. For example, a GENI rack at a university is an aggregate (Flux Research Group 2015); experimenters may request resources from this aggregate and add them to their slice. Different aggregates provide different kinds of resources. Some aggregates provide compute resources, virtual machines, or bare machines or both. Some aggregates provide networking resources that experimenters can use to connect compute resources from multiple aggregates. Figure 7.36 shows a GENI slice with resources from multiple aggregates.

7.2.12.3.4   The GENI AM API and GENI RSpecs

Experimenters request resources from aggregates using a standard API called the GENI aggregate manager API or GENI AM API. The AM API allows experimenters to, among other things, the following:

- List the resources available at an aggregate
- Request specific resources from the aggregate be allocated to their slices
- Find the status of resources from the aggregate that are allocated to their slices
- Delete resources from their slices

Fig. 7.36 GENI slice and aggregates (GENI 2014)

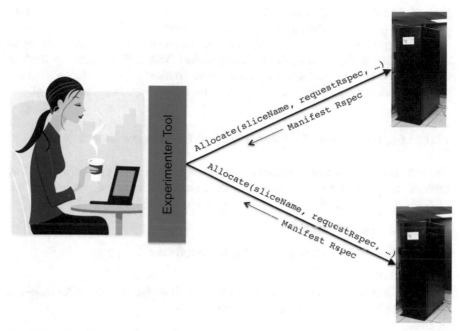

Fig. 7.37 GENI slice, resources, and aggregates (GENI 2014)

The AM API uses resource specification documents, commonly referred to as GENI RSpecs, to describe resources. RSpecs are just XML documents in a prescribed format. Experimenters send to aggregates a "request" RSpec that describes the resources they want and get back from the aggregates a "manifest" RSpec that describes the resources they got. The manifest includes information the experimenters will need to use these resources such as the names and IP addresses of compute resources (e.g., virtual machines), user accounts created on the resources, and VLAN tags assigned to network links. Most experimenters will not need to learn details of the AM API or read/write RSpec files; GENI experimenter tools hide much of this complexity.

There is a third type of RSpec called an "advertisement" RSpec. This is the RSpec returned by an aggregate when an experimenter lists the resources available at the aggregate. It describes all the resources available at the aggregate.

Figure 7.37 displays an experimenter adding resources from two different aggregates to his slice using the allocate call of the GENI AM API.

7.2.12.3.5   Getting Access to GENI and GENI Resources

Experimenters need an account to use GENI and can get an account from any one of GENI's federated authorities called clearinghouses. Commonly used clearinghouses include the GENI Project Office (GPO), Emulab (Sect. 7.2.13.1), and PlanetLab

(Sect. 7.2.13.2). The GPO provides system engineering and project management expertise to guide the planning and prototyping efforts of GENI.

GENI aggregates federate with one or more clearinghouses, i.e., they choose to trust GENI accounts issued by these clearinghouses. Most GENI aggregates federate with all three of the clearinghouses listed above so most experimenters should not have to concern themselves with which clearinghouse to use to get an account.

7.2.12.3.6 Tying Up All Together: The GENI Experimenter Workflow

The succeeding is a sample workflow for a typical GENI experiment. Without being exhaustive, the objective is to show how the concepts described above tie together.

7.2.12.3.6.1 Experiment Setup

An experiment can thus be set according to several steps:

- Get a GENI account.
- Join an existing project or create a new project. Only faculty and senior technical staff with project-lead privileges can create projects.
- Create a slice.
- Use an experimenter tool to:
    - Craft a Request RSpec that specifies the resources needed
    - Make the appropriate GENI AM API calls on the aggregates from where the resources are being requested

7.2.12.3.6.2 Experiment Execution

Use information in the manifest RSpec returned by the aggregates to log into compute resources, install software, send traffic on the network links, etc.

7.2.12.3.6.3 Finishing Up

Delete resources obtained from the aggregates by using an experimenter tool to make the appropriate GENI AM API calls on these aggregates.

7.2.13 Further Testbeds

This section spotlights on more testbeds for abundant consideration of what has been presented in the literature. Noteworthy surveys are accessible in (Jiménez-González et al. 2013) and (Horneber and Hergenroder 2014).

7.2.13.1 Emulab

Emulab is a network testbed, incorporating a number of centrally managed general-purpose computers and network resources (White et al. 2002). The initial and largest Emulab, which includes several hundred nodes, is located at the University of Utah, where it is managed by the Flux Research Group (Flux Group 2014). Emulab software is available on an open-source basis and has been used to stand up dozens of Emulabs worldwide. There are also installations of the Emulab software at more than two dozen sites around the world, ranging from testbeds with a handful of nodes up to testbeds with hundreds of nodes. Emulab is widely used by computer science researchers in the fields of networking and distributed systems. It is also designed to support education and has been used to teach classes in those fields (Flux Research Group 2015).

Emulab is a public facility, available without charge to most researchers worldwide; it provides integrated access to a wide range of experimental environments:

- Emulation. An emulated experiment allows specifying an arbitrary network topology, giving experimenters a controllable, predictable, and repeatable environment, including PC nodes on which full "root" access is granted and running any chosen operating system.
- Live-Internet experimentation. Using the RON (MIT 2015) and PlanetLab (Bavier et al. 2004; PlanetLab 2007) testbeds. Emulab provides a full-featured environment for deploying, running, and controlling experimenter application at hundreds of sites around the world.
- 802.11 wireless. Emulab's 802.11a/b/g testbed is deployed on multiple floors of an office building. Nodes are under the experimenter's full control and may act as access points, clients, or in ad hoc mode. All nodes have two wireless interfaces, plus a wired control network.
- Software-defined radio. USRP devices from the GNU Radio project give the experimenter control over layer 1 of a wireless network; everything from signal processing up is done in software.

Emulab unifies all of these environments under a common user interface and integrates them into a common framework. This framework provides abstractions, services, and namespaces common to all, such as allocation and naming of nodes and links. By mapping the abstractions into domain-specific mechanisms and internal names, Emulab masks much of the heterogeneity of the different resources.

The Emulab software also forms the basis for ProtoGENI, a GENI prototype and control framework, which extends the Emulab model to bridge multiple physical sites and support the GENI API (Ricci et al. 2012).

The fundamental GENI abstraction of multiple general-purpose computers, interconnected by layer two networks in experimenter-specified topologies, draws directly on the basic capabilities of Emulab. In addition, several of the specifics of the GENI API derive basic ideas and/or implementation from Emulab and ProtoGENI. Notable among these are the content and format of GENI Rspecs and portions of the experiment life cycle. In addition, the ProtoGENI implementation of multiple sites interconnected at layer two over Internet2 links is a major influence on GENI stitching.

The primary ProtoGENI site at Utah and several other ProtoGENI sites also share resources with GENI by exporting aggregates that speak the GENI API. The primary resources exported are bare metal computers connected by layer two and layer three networks, although virtual machines are also available. One of the GENI rack implementations, InstaGENI, is built on Emulab/ProtoGENI software.

7.2.13.2   PlanetLab

PlanetLab is a global research network that supports the development of new network services (Bavier et al. 2004). A consortium of academic, industry, and government institutions for the benefit of the research community manage PlanetLab (PlanetLab 2007). PlanetLab embodies a collection of machines distributed over the globe; most of the machines are hosted by research institutions and are connected to the Internet. The goal for PlanetLab is to grow to 1000 widely distributed nodes that peer with the majority of the Internet's regional and long-haul backbones.

Many key GENI (Sect. 7.2.12) components trace their origins, either in code or in concept, back to roots in PlanetLab. The slice-based federation architecture (SFA), an approach to federation growing primarily out of the PlanetLab experience, is the basis for much of the GENI API (Peterson et al. 2009). In particular, the aggregate manager API is adapted from the SFA.

Other ideas pioneered in PlanetLab and prominent in GENI include the following:

- Use of a common overlay network infrastructure as a mechanism for validating and deploying novel network services
- Providing a widely available leasing service for networked compute resources
- The slice concept
- Use of lightweight host virtualization for efficient support of many long-lived services

As an overlay network deployed over the global Internet, PlanetLab is intended to support experiments that run at or above layer three. This design decision permits PlanetLab to present a simple, well-understood, and powerful network abstraction to potential experimenters. More recently, some PlanetLab variants, such as Virtual Network Infrastructure (VINI) (Bavier et al. 2006), VICCI (Peterson et al. 2011), and Great Plains Environment for Network Innovation (GpENI) (Sterbenz et al. 2011; Medhi et al. 2014) are reaching into lower layers of the network stack.

In addition to architectural contributions, PlanetLab also shares resources with GENI by supporting the AM API and exporting a GENI aggregate. Via this interface, experimenters have access to PlanetLab's worldwide network of virtualized compute resources. Some experimenter tools originally developed with

PlanetLab in mind were readily generalized into a GENI context. Examples include the GENI user shell (Gush) (Albrecht and Huang 2011), Stork (Hartman et al. 2014), and Raven (Hartman and Baker 2014) experiment management tools.

7.2.13.3 Mobile Emulab

Mobile Emulab, developed at the University of Utah, is a general-purpose testbed that uses robots to provide mobility to WSN nodes; it has been open for public use until 2008 and was popular for research on mobile WSNs (Johnson et al. 2006). Mobile Emulab is designed to provide unified access to a variety of experimental environments:

- A Web-based front end, through which users create and manage experiments.
- A core that manages the physical resources within a testbed. It consists of a database and a wide variety of programs that allocate, configure, and operate testbed equipment.
- Numerous back ends which interface to various hardware resources. Back ends include interfaces to locally managed clusters of nodes, virtual and simulated nodes, and a PlanetLab interface (Sect. 7.2.13.2).

Mobile Emulab uses a modular and distributed architecture and is designed for open use. It provides a high degree of interaction with the user allowing remote operation through a GUI. The user can position the robots, run programs, and configure data logging. The GUI also shows live maps and images of the experiment. Emulab users create experiments, which are essentially collections of resources that are allocated to a user by the testbed management software and act as a container for control operations by the user and system.

Mobile Emulab is composed of robots and fixed motes. It is composed of six Acroname Garcia robots and a static WSN composed of 25 Mica2 nodes deployed in an L-shaped controlled indoor area of 60 m^2 and 2.5 m high. Overlooking this area are six cameras used by the robot tracking system and three webcams that provide live feedback to testbed users.

The robots operate completely wirelessly through the IEEE 802.11b card and a battery that provides 2–3 hours of use to drive the robot and power the onboard computer and mote. Precisely, Garcia robots are equipped with an XScale-based Stargate (Crossbow 2004b) small computer running Linux and connected to a 900-MHz Mica2 mote (Crossbow 2002) (Fig. 7.38). To the Stargate is attached an IEEE 802.11b card that acts as a separate "control network," connecting the robot to the main testbed and the Internet. The Stargate serves as a gateway for both Emulab and the experimenter to control and interact with the mote and for the user to run arbitrary code. Users can log in to the Stargate and will find their Emulab home directory NFS-mounted.

The 25 stationary motes are arranged on the ceiling in a 2-meter grid and on the walls near the floor. All of the fixed motes are attached to MIB500CA serial programming boards (Crossbow 2004a) to allow for programming and

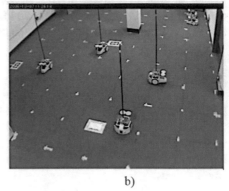

a)   b)

Fig. 7.38 Mobile Emulab. (**a**) Acroname Garcia robot with Stargate, Wi-Fi, and Mica2. (**b**) GUI to track robots and control them with click and drag (Johnson et al. 2006)

communication. The ten near-floor motes also feature an MTS310 full multisensor board (Crossbow 2007a) with magnetometers that can be used to detect the robot as it approaches. These motes are completely integrated with the Emulab software, making it trivial to load new kernels onto motes, remotely interact with running mote kernels via their serial interfaces, or access serial logs from experiments.

It is to be noticed that in Emulab two wireless networks are included, one for the robots and one for the WSN. Robots are localized using the overhead cameras.

7.2.13.4   SenseNet

SenseNet (Dimitriou et al. 2007) is a low-cost sensor network testbed that exploits only the wireless channel to transfer data and offers benefits to users and adminis-trators like ease of deployment, ease of use, no need of a wired infrastructure, coping with multiple users at the same time, and most importantly scalability.

SenseNet satisfies several design features:

- Absence of a wired backchannel. The common fact in other testbed approaches is the existence of a wired backchannel that is used for reprogramming, data logging, and network monitoring purposes. This implies that an Ethernet or USB channel exists through which newer versions of code are downloaded to the mote; also, packets originating from that mote are uploaded to a sink. The benefit of this approach is that the wireless link is left free for purely application purposes at the expense of increasing cost, maintenance problems, and reduced scalability.
- Absence of super nodes. In some implementations (Dutta et al. 2006; Handziski et al. 2006), as illustrated in Sects. 7.2.7 and 7.2.8, there exist some nodes with advanced capabilities compared to those of the motes. Their existence offers

solutions to problems like addressing a remote node, aggregating packets, forwarding traffic between different types of networks, or acquiring the role of a cluster head in a routing protocol. However, such approach might reduce scalability as the administrators, and sometimes users, have to guarantee the proper functionality of these nodes to guarantee that tasks can be performed.

- Existence of multiple users at the same time. The existence of multiple users at the same time offers the capability of simultaneously hosting more than one experiment, which prevents one user from blocking all other potential users of the infrastructure.

- Scalability. This means that, transparently to users or existing tasks, adding new nodes in the network is straightforward and without extra hardware. Moreover, from an algorithmic point of view, adding a new node should only affect the new node's neighborhood and not the entire network. Reliance of SenseNet only on software that utilizes the wireless channel in order to achieve the necessary functionality helped in boosting scalability. More specifically, deployment does not demand cables, special hardware, and specific drivers. The only requirement is the inclusion of software in the motes.

SenseNet architecture embodies several components:

- Web Server. Communication between a user and SenseNet is carried out through a Web interface. When assigned a username and password, users from everywhere can access the services provided by the testbed. The web pages are written in HTML/JSP[30] (Tutorialspoint 2014) and provide a simple and practical user interface. The web server used is Apache Tomcat, which is free to download and use.

- Application Server. When a new task arrives to the system, the application server is triggered through a remote method invocation (RMI)[31] call (Techopedia 2015b). All information about such task is retrieved from the database as stored by the Web server. The application server sends to the appropriate motes, using Deluge (Hui and Culler 2004), the executables of the task and keeps motes' status till task termination.

- MySQL Database Server. It is the repository where the Web server places information regarding user tasks and where the application server finds what to retrieve and exploit appropriately. The database holds the most recent status of the sensor network and the motes, needed for both the application and Web interface.

- Serial Forwarder. It is a tool that comes with the used TinyOS. It is mainly used to provide a communication link between the serial port, where the basestation is

---

[30] JavaServer Pages (JSP) is a technology for developing Web pages that support dynamic content that helps developers insert a Java code in HTML pages by making use of special JSP tags.

[31] RMI is a distributed object technology developed by Sun for the Java programming language. RMI permits Java methods to refer to a remote object and invoke methods of the remote object.

connected, and the Ethernet (TCP/IP). Hence, the SerialForwarder (TinyOS Wiki 2012) binds the serial port and waits for connections from applications that need to send or receive packets.

SenseNet was experimented on several topologies that involved eight Mica2 motes and provided satisfactory performance.

7.2.13.5 Ubiquitous Robotics

Ubiquitous robotics integrates a wide variety of heterogeneous technologies including networked mobile robots, WSN and RFID networks, camera networks, and networks of personal mobile computing devices. These technologies have been divided in two groups (Jiménez-González et al. 2013):

- Ubiquitous systems with physical actuation capabilities. These are the robots, which can move, carry sensors or other ubiquitous systems, and can interact with the environment.
- Ubiquitous systems without physical actuation capabilities. They include technologies based on nodes with sensing, computational, and communication capabilities that organize autonomously into networks. They also group WSN and RFID networks, camera networks, and personal mobile computing networks. These devices can sense the environment, can interact with humans, and can perform actions such as turning the lights on, but they cannot perform physical actions and are static unless mounted on robots or carried by humans.

In ubiquitous robotics testbeds, multirobot (MR) systems can be comprised of ground, aerial, or marine robots (Reich et al. 2008; Kitts and Mas 2009):

- Ground robots typically use small- or medium-sized platforms and include sensors such as cameras, RGB-D sensors, laser-range finders, ultrasound sensors, bumpers, GPS receivers, and inertial navigation systems.
- Aerial robots, although limited in payload and, thus, in onboard sensing and processing, can move in 3D. Vertical takeoff and landing quad-rotors are the most commonly used, although blimps, fixed-wing platforms, and helicopters are also found in outdoor testbeds.
- Underwater or surface vehicles are rarely found in ubiquitous robotics testbeds.

While low cost, low size, and low energy constrain the features of sensors integrated in WSN platforms, robots can carry and provide mobility to sensors with higher performance. Moreover, WSNs were designed for low-rate and low-range communications, whereas Wi-Fi networks, typically used by multirobot systems, can provide up to 36 Mbps experimental bound at greater distances. As camera networks are frequent in ubiquitous robotics testbeds, they mostly adopt schemes with decentralized image processing for scalability and bandwidth efficiency. Also, the popularization and improvement of performance of smartphones and PDAs have boosted their use in testbeds.

Basically, testbeds need an architecture that integrates heterogeneous components. A high percentage of the testbed flexibility, extensibility, and scalability
depends on its architecture. If the testbed is designed to solely serve one experiment
or functionality, its architecture tends to be monolithic. In contrast, the architecture
of general-purpose testbeds with open public access tends to be modular and use
standard interfaces and open-source software. Some testbeds include usability tools
such as simulators and tools for experiment programming, logging, and monitoring.

In general, the purpose of testbeds as experimental tools is twofold:

- In some cases, as placed in indoor laboratories, they provide a controlled environment to allow algorithm testing and debugging with simulation-like
conditions.
- In other cases they are used to fill the gap between research and market, enabling
testing in conditions close to the final application. These are typically deployed in
settings, which can range from office buildings to an entire city.

Figure 7.39 shows a typical ubiquitous robotic testbed.

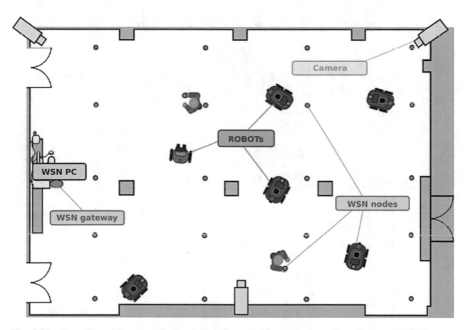

Fig. 7.39 General architecture of ubiquitous robotic testbed (Jiménez-González et al. 2011)

Table 7.2 Testbed hardware/software compared

Testbed	Platform[a]	Processor[b]	Memory		OS	Sensors[b]	Communication
			EEPROM/RAM	Flash			
ORBIT (Sect. 7.2.1)	Grid (400 nodes)	Intel third generation core i7/i5/i3	Two DDR3 1066/1333 MHz up to 16-GB SDRAM	Phoenix 64-Mbit SPI flash BIOS	All popular OS: Linux, Windows XP/Vista/7	N/A	Intel 82574L Gigabit Ethernet Controller Intel 82579LM Gigabit Ethernet Controller
	LV-67 J motherboard						
	Outdoor (22 nodes)	Intel third generation core i7/i5/i3 mobile processor	Two up to 16-GB DDR3 1066/1333/1600 MHz SO-DIMM	Phoenix 16-Mbit SPI flash BIOS	All popular OS: Linux, Windows XP/Vista/7	N/A	Intel 82579LM Gigabit Ethernet Controller Intel 82574L Gigabit Ethernet Controller
	LV-67 K motherboard (recent)						
	Sandboxes (nine nodes)	Intel core i7/i5/i3 Pentium desktop	Two up to 8-GB DDR3 1066/1333 MT/sec DIMM	Phoenix 16 Mbit SPI flash BIOS	All popular OS: Linux, Windows XP/Vista/7	N/A	Two Intel 82574L Gigabit Ethernet Controllers
	LV-67C motherboard						
	LV-67G motherboard	Intel second-generation core i7/i5/i3/Pentium/Xeon desktop	Two up to 16-GB DDR3 1066/1333 MHz DIMM	Phoenix 16-Mbit SPI flash BIOS	All popular OS: Linux, Windows XP/Vista/7	N/A	Two Intel 82574L Gigabit Ethernet Controllers
MoteLab (Sect. 7.2.2)	(184 nodes)	7.37-MHz 8-bit Atmel ATmega 128L	4 KB (EEPROM)	128 KB (instruction memory)	TinyOS	Crossbow MTS/MDA sensor boards	CC2420 operating at 2.4 GHz
	MicaZ						

Meerkats (Sect. 7.2.3)	(One node)	Laptop (gateway and sink)	Dell Inspiron 4000 PIII CPU 750-MHz 20-GB hard disk	512 MB (RAM)	–	Linux	–	Orinoco Gold IEEE 802.11b wireless card
	(8 Eight nodes)	Stargate	400-MHz 32-bit PXA255 XScale	64 MB (SDRAM)	32 MB	Linux	Logitech QuickCam Pro 4000 webcam	Orinoco Gold IEEE 802.11b wireless card, USB
MiNT (Sect. 7.2.4)	Multiple core nodes		Desktop PC and LEGO Mindstorms robot	N/A	N/A	N/A	N/A	IEEE 802.11b NETGEAR MA311 wireless PCI card
	Central node		Desktop PC	N/A	N/A	N/A	N/A	Wired Ethernet or IEEE 802.11a
MiNT-m (Sect. 7.2.5)	Control server		Desktop PC	N/A	N/A	N/A	N/A	3 IEEE 802.11 a/b/g PCI cards
	Tracking server (three PCs)		Desktop PC	N/A	N/A	N/A	N/A	N/A
	Testbed node (100 nodes planned)	RouterBOARD 230	Small form-factor PC with 266-MHz processor (Intel Pentium MMX architecture)	512 MB (SDRAM)	2-Mbit Flash BIOS	Linux, MS DOS 6.22, OpenBSD[d], FreeBSD[d]	N/A	4 Atheros IEEE 802.11 a/b/g PCI cards
Kansei (Sect. 7.2.6)	Stationary array (210 sensor nodes)	XSM	7.37-MHz 8-bit Atmel ATmega 128L	4 KB (EEPROM)	128 KB (instruction memory)	TinyOS	Photocell, PIR, temperature, magnetometer, and microphone	Chipcon CC1000 radio operating at 433 MHz

(continued)

Table 7.2 (continued)

Testbed	Platform[a]	Processor[b]	Memory EEPROM/RAM	Flash	OS	Sensors[b]	Communication
	Stargate	400-MHz 32-bit PXA255 XScale	64 MB (SDRAM)	32 MB	Linux	–	PCMCIA slot, compact Flash slot, 51-pin expansion connector for MICA2 motes and other peripherals. Ethernet, serial, JTAG, USB connectors via 51-pin daughter card interface
	Tmote Sky	8-MHz 16-bit MSP430 F1611	10 KB (RAM)	48 KB	TinyOS	Integrated humidity, temperature, and light sensors	CC2420 operating at 2.4 GHz
Portable array (50 trio motes)	XSM	7.37-MHz 8-bit Atmel ATmega 128L	4 KB (EEPROM)	128 KB (instruction memory)	TinyOS	Acoustic, PIR, two-axis magnetometer, and temperature	Chipcon CC1000 radio operating at 433 MHz
	Tmote Sky	8-MHz 16-bit MSP430F1611	10 KB (RAM)	48 KB	TinyOS	Integrated humidity, temperature, and light sensors	CC2420 operating at 2.4 GHz
Mobile array (five robotic nodes)	XSM	7.37-MHz 8-bit Atmel ATmega 128L	4 KB (EEPROM)	128 KB (instruction memory)	TinyOS	Acoustic, PIR, two-axis magnetometer, and temperature	Chipcon CC1000 radio operating at 433 MHz
	Tmote Sky	8-MHz 16-bit MSP430F1611	10 KB (RAM)	48 KB	TinyOS	Integrated humidity, temperature, and light sensors	CC2420 operating at 2.4 GHz

Trio (Sect. 7.2.7)	Tier 1 (557 mote nodes)	Telos	8-MHz 16-bit MSP430F1611	10 KB (RAM)	48 KB	TinyOS	—	CC2420 operating at 2.4 GHz
		XSM	7.37-MHz 8-bit Atmel ATmega 128L	4 KB (EEPROM)	128 KB (instruction memory)	TinyOS	Acoustic, PIR, two-axis magnetometer, and temperature	Chipcon CC1000 radio operating at 433 MHz
		Prometheus	—	—	—	—	—	—
	Tier 2 (seven gateway nodes)	Telos	8-MHz 16-bit MSP430F1611	10 KB (RAM)	48 KB	TinyOS	—	CC2420 operating at 2.4 GHz
		NSLU2 with Tmote Connect	133-MHz 32-bit IXP420	32 MB (SDRAM)	8 MB	Linux	—	USB 1.0/1.1/2.0, IEEE 802.3, IEEE 802.3u
		802.3-to-802.11 bridge	—	—	—	—	—	IEEE 802.3, IEEE 802.3u, IEEE 802.11
	Tier 3 (one server node)	Server computer	N/A	N/A	N/A	N/A	N/A	N/A
	Tier 4 (clients)	Desktop PC	N/A	N/A	N/A	N/A	N/A	N/A
TWIST (Sect. 7.2.8)	(37 nodes)	NSLU2	133-MHz 32-bit IXP420	32 MB (SDRAM)	8 MB	Linux	—	Ethernet (IEEE 802.3, IEEE 802.3U), USB
	Nodes (90 fixed locations+90 free slots on the USB hubs, 57 USB hubs)	EyesIFX	8-MHz 16-bit MSP430F149	2 KB (SRAM)	60 kB (instruction memory) TinyOS	TinyOS	Light, temperature	TDA5250 radio transceiver with speed 64 Kbps
		Telos	8-MHz 16-bit MSP430F1611	10 KB (RAM)	48 KB	TinyOS	Humidity, light, and temperature	CC2420 operating at 2.4 GHz, USB

(continued)

Table 7.2 (continued)

Testbed	Platform[a]	Processor[b]	Memory		OS	Sensors[b]	Communication
			EEPROM/RAM	Flash			
SignetLab (Sect. 7.2.9)	EyesIFXv2 (48 nodes)	8-MHz 16-bit MSP430F1611	10 KB (RAM)	48 KB (ROM) 512 KB SPI (EPROM)	TinyOS	Light, temperature	64- Kbps 868-MHz ASK/FSK Infineon Wireless Transceiver TDA5250, USB
Indriya (Sect. 7.2.11)	TelosB (127 motes)	8-MHz 16-bit TI MSP430	10 KB (RAM)	1 MB (External Flash)	TinyOS	PIR, light, acoustic, temperature, two-axis acceleration, and two-axis magnetic sensors	CC2420 operating at 2.4 GHz, USB
WISEBED (Sect. 7.2.10)	G-Node (108 nodes)	16-MHz 16-bit MSP430F2418	8 KB (RAM)/ 116 KB (ROM)	256 B internal +8 Mbit external	TinyOS	Temperature	CC1101 low-power sub-1-GHz transceiver, USB
	iSense (228 nodes)	–	–	–	–	Temperature, light, PIR, AMR, accelerometer	IEEE 802.15.4 (2.4 GHz)
	MicaZ (2 nodes)	7.37-MHz 8-bit Atmel ATmega 128 L	4 KB (EEPROM)	128 KB (instruction memory)	TinyOS	Crossbow MTS/MDA sensor boards	CC2420 operating at 2.4 GHz
	Pacemate (60 nodes)	–	–	–	–	Heart rate	Xemics[e] RF (868 MHz)
	MSB-A2 (100 nodes)	LPC2387 based on 72-MHz 32-bit ARM7TDMI-S	98 KB (SRAM) /512 KB (ROM)	512 KB	Contiki[f]	Temperature, humidity	CC1100 (864–970 MHz)[a]

(7 nodes)	MSB-430	8-MHz 16-bit RISC TI MSP430F1612	5 KB	55 KB	Contiki, ScatterWeb[g]	Temperature, humidity, acceleration	CC1020 (804–940 MHz, 863–870 MHz)[a]
(24 nodes)	T-Node	7.37-MHz 8-bit Atmel ATmega 128L	4 KB (SRAM)	128 KB (program memory) 512 KB (data memory)	TinyOS	Alcohol, temperature, PIR, humidity, heartbeat, light, magnetic	CC1000 (868 MHz)
(166 nodes)	TelosB	8-MHz 16-bit TI MSP430	10 KB (RAM)	1 MB (external Flash)	TinyOS	PIR, light, acoustic, temperature, two-axis-acceleration, and two-axis magnetic sensors	CC2420 operating at 2.4 GHz, USB
(8 nodes)	Tmote Sky	8-MHz 16-bit MSP430 F1611	10 KB (RAM)	48 KB	TinyOS	Integrated humidity, temperature, and light sensors	CC2420 operating at 2.4 GHz

[a]Values and units are as reported in the source paper

[b]Datasheet is available in Chap. 10

[c]The OpenBSD project produces a FREE, multiplatform 4.4BSD-based UNIX-like operating system (OpenBSD 2014)

[d]FreeBSD is an advanced computer operating system used to power modern servers, desktops, and embedded platforms (FreeBSD 2014)

[e]Xemics, Inc., based in Switzerland, was acquired by Semtech Corporation in June 2005. Xemics, Inc., is a fabless developer of ultralow-power analog, radio-frequency (RF), and digital integrated circuits, including the CoolRISC microcontroller family (EE Times 2005)

[f]Contiki is an open-source, highly portable, multitasking operating system for memory-efficient networked embedded systems and wireless sensor networks. Contiki is designed for microcontrollers with small amounts of memory (Dunkels et al. 2004; Contiki-Developers 2016)

[g]ScatterWeb OS for WSN (ScatterWeb 2015)

7.3   Conclusion for Extension

Testbeds are representative of WSNs, they support the diversity of their hardware and software constituents, they are deployed in the same conditions and would-be environment, and they make use of the protocols to be used at a larger scale. Testbeds are intended to safeguard would-be implemented WSNs from malfunctions that may not be seen in theoretical simulations. Malfunctions may be in inconvenient hardware, buggy software, and deployment prone to energy depletion and radio interferences. By momentarily tolerating faults that cannot be accepted in everyday actual WSNs, testbeds find the curing solutions.

In the literature many testbeds are reported, not all are typically implemented, and not all are available now. Knowledge is to be acquired from those who got it by researching, trying, and experimenting; this chapter considers testbeds with authentic information even if they ceased to subsist. Pioneering testbeds, as fully illustrated, continue to offer models in concepts, implementation, and applications. Some of the testbeds are built for general use, while others are meant for typical applications such as visual surveillance.

As fully detailed in this chapter, based on the researchers and practitioners' interests, testbeds can be classified under several categories. They may be full scale or miniaturized, deployed on a 2D or 3D pattern, mobile or static, provide Web services or are just accessible from the deployment location, limited to homogeneous platforms or they are extended to support heterogeneity, provide for hybrid simulation as a tool for enhanced analysis or are contented with experimentation analysis. As elucidated, a testbed is not confined to a single category; it may be static, deployed on a 2D field, and provide Web access (Table 7.3).

Table 7.3   Testbed size/simulation/homogeneity/deployment/mobility/Web compared

Testbed	Full-scale/ miniaturized	Hybrid simulation	Homogeneous/ heterogeneous platforms	Deployment	Mobile	Web interface
ORBIT (Sect. 7.2.1)	Full-scale/ miniaturized	N/A	Heterogeneous	Indoor/ outdoor	No	No
MoteLab (Sect. 7.2.2)	Full-scale	No	Homogeneous	3D-indoor	No	Yes
Meerkats (Sect. 7.2.3)	Miniaturized	No	Homogeneous	Indoor	No	No
MiNT (Sect. 7.2.4)	Miniaturized	Yes	N/A	Indoor	Yes	No
MiNT-m (Sect. 7.2.5)	Miniaturized	Yes	N/A	Indoor	Yes	No
Kansei (Sect. 7.2.6)	Full-scale	Yes	Homogeneous	Indoor	Yes	Yes
Trio (Sect. 7.2.7)	Full-scale	No	Homogeneous	Outdoor	No	Yes

(continued)

Table 7.3 (continued)

Testbed	Full-scale/ miniaturized	Hybrid simulation	Homogeneous/ heterogeneous platforms	Deployment	Mobile	Web interface
TWIST (Sect. 7.2.8)	Full-scale	No	Heterogeneous	3D-indoor	No	Yes
SignetLab (Sect. 7.2.9)	Full-scale	No	Homogeneous	Indoor	No	Yes
WISEBED (Sect. 7.2.10)	Full-scale	No	Heterogeneous	Indoor	No	No
Indriya (Sect. 7.2.11)	Full-scale	No	Homogeneous	3D-indoor	No	No
GENI (Sect. 7.2.12)	N/A	N/A	N/A	N/A	N/A	Yes
Emulab (Sect. 7.2.13.1)	Full-scale	No	Heterogeneous	Indoor	No	Yes
PlanetLab (Sect. 7.2.13.2)	Full-scale	No	Heterogeneous	Indoor	No	Yes
Mobile Emulab (Sect. 7.2.13.3)	Miniaturized	No	Homogeneous	3D-indoor	Yes	Yes
SenseNet (Sect. 7.2.13.4)	Miniaturized	No	Heterogeneous	Indoor	No	Yes

Testbeds and simulators are complementary; ideally getting benefits from both of them is the best option. Theoretical simulation studies provide numerical metrics that are truly needed for practical testbed implementation and deployment. But is the topmost approach always possible? Not all the wishes are usually attainable. Testbeds are the expensive choice, both in money and effort; simulation is realistically the less risky resort when budgets and time are short and when typical deployment is not insisting.

Simulators are the inevitable tools for analysis; they help in previewing the performance metrics needed for proper testbed deployment. The next chapter considers in full detail the most common WSN simulators.

7.4 Exercises

1. What is a testbed? Why are testbeds needed?
2. Identify the requirements from testbeds.
3. Explain virtualization, federation.

4. How is topology virtualized?
5. Compare full-scale and miniaturized testbeds.
6. Why is TinyOS widely used in WSNs?
7. Discuss the homogeneity of the testbeds involved in this chapter.
8. How do Web services vary in the offering testbeds?
9. Which of the laid-out testbeds support federation? How?
10. For the testbeds presented in this chapter, compare:

- The outdoor testbeds
- The indoor testbeds
- The full-scale testbeds
- The miniaturized testbeds
- The 3D testbeds

11. Search the literature for more outdoor, indoor, full-scale, miniaturized, visual, 3D, and Web interfaced wireless sensor testbeds.

References

ACA Systems. 2011. *What is HSB/HSV Color Spaces?* ACA Systems. January 1, 2011. http://www.acasystems.com/en/color-picker/faq-hsb-hsv-color.htm. Accessed 1 Dec 2014.

Acroname Inc. 1994. *About Acroname.* Acroname Inc. January 1, 1994. http://www.acroname.com/about.html. Accessed 31 Aug 2014.

Albrecht, J., and D.Y. Huang. 2011. Managing Distributed Applications Using Gush. In *Testbeds and Research Infrastructures. Development of Networks and Communities*, ed. T. Magedanz, A. Gavras, N.H. Thanh, and J.S. Chase, vol. 46, 401–411. Berlin, Heidelberg: Springer.

Apple Inc. 2014. *Mac Mini.* Apple Inc. January 1, 2014. https://www.apple.com/mac-mini/. Accessed 14 Aug 2014.

Arkema. 2013. *Plexiglas MC diffusion.* Arkema. July 1, 2013. http://www.plexiglas.com/export/sites/plexiglas/.content/medias/downloads/sheet-docs/plexiglas-mc-diffusion.pdf.   Accessed 30 Aug 2014.

Arora, A., et al. 2005. ExScal: Elements of an Extreme Scale Wireless Sensor Network. In *11th IEEE International Conference on Embedded and Real-Time Computing Systems and Applications*, 102–108. Hong Kong: IEEE.

Arora, A., E. Ertin, R. Ramnath, M. Nesterenko, and W. Leal. 2006. Kansei: A High-Fidelity Sensing Testbed. *Internet Computing* 10 (2): 35–47.

ATEN. 2014. *USB Extenders.* ATEN. January 1, 2014. http://www.aten.com/products/productList.php?prdClassInput=By%20Function&prdSubClassInput=2011011816087001&prdInput=20140402113624001. Accessed 13 Aug 2014.

Authors, Contributing. December 24, 2011. What is NTP? http://www.ntp.org/ntpfaq/NTP-s-def.htm. Accessed 12 July 2014.

Baumgartner, T., et al. 2010. Virtualising Testbeds to Support Large-Scale Reconfigurable Experimental Facilities. In *Wireless Sensor Networks*, ed. J.S. Silva, B. Krishnamachari, and F. Boavida, vol. 5970, 210–223. Berlin, Heidelberg: Springer.

Bavier, A., et al. 2004. Operating System Support for Planetary-Scale Network Services. In *First Symposium on Networked Systems Design and Implementation (NSDI)*, 253–266. San Francisco: USENIX.

Bavier, A., N. Feamster, M. Huang, L. Peterson, and J. Rexford. 2006. In VINI Veritas: Realistic and Controlled Network Experimentation. In *Conference on Applications, Technologies, Architectures, and Protocols for Computer Communications ()SIGCOMM*, 3–14. Pisa: ACM.

Belkin International, Inc. 2014. *7-Port Powered Hub*. Belkin International, Inc. January 1, 2014. http://www.belkin.com/us/Products/Macbook-%26-PC/Hubs-%26-Docks/c/ WSMACPCHUBS7PPH/. Accessed 14 Aug 2014.

Berman, M., et al. 2014. GENI: A Federated Testbed for Innovative Network Experiments. *Computer Networks* 61: 5–23.

Boice, J., et al. 2005. *Meerkats: A Power–Aware, Self–Managing Wireless Camera Network for Wide Area Monitoring*. Technical, Department of Computer Engineering, University of California. Santa Cruz: University of California.

Brent, N.C., et al. 2005. Mirage: A Microeconomic Resource Allocation System for Sensornet Testbeds. In *Second IEEE Workshop on Embedded Networked Sensors (EmNetS-II)*, 19–28. Sydney: IEEE.

Chambers, B.A. 2002. The Grid Roofnet: A Rooftop Ad Hoc Wireless Network. MSc Thesis, Massachusetts Institute of Technology.

Chatzigiannakis, I., S. Fischer, C. Koninis, G. Mylonas, and D. Pfisterer. 2010. WISEBED: An Open Large-Scale Wireless Sensor Network Testbed. In *Lecture Notes of the Institute for Computer Sciences, Social Informatics and Telecommunications Engineering-Sensor Applications, Experimentation, and Logistics*, ed. N. Komninos, vol. 29, 68–87. Berlin, Heidelberg: Springer.

Cisco. 2005. *What Is Constant Bit Rate?* Cisco. December 12, 2005. http://www.cisco.com/c/en/us/ support/docs/asynchronous-transfer-mode-atm/atm-traffic-management/10422-cbr.html#what. Accessed 25 Sept 2014.

Clancy, T.C., and B.D. Walker. 2007. MeshTest: Laboratory-Based Wireless Testbed for Large Topologies. In *3rd International Conference on Testbeds and Research Infrastructure for the Development of Networks and Communities (TridentCom)*, 1–6. Lake Buena Vista: IEEE.

COMMELL. 2014a. *LV-67B User's Manual*. COMMELL. January 1, 2014. http://www.commell. com.tw/download/manual/lv-67b_manual_v15.pdf. Accessed 25 Dec 2014.

———. 2014b. *LV-67C*. COMMELL. January 1, 2014. http://www.commell.com.tw/product/ SBC/LV-67C.HTM#Overview. Accessed 25 Dec 2014.

———. 2014c. *LV-67F User's Manual*. COMMELL. January 1, 2014. http://www.spectra.de/ produkte/K123595/web/Handbuch-LV-67F.pdf. Accessed 25 Dec 2014.

———. 2014d. *LV-67G*. COMMELL. January 1, 2014. http://www.commell.com.tw/Download/ Datasheet/LV-67G_Datasheet.pdf. Accessed 25 Dec 2014.

———. 2014e. *LV-67J*. COMMELL. January 1, 2014. http://www.commell.com.tw/Product/SBC/ LV-67J.HTM. Accessed 25 Dec 2014.

———. 2014f. *LV-67K User's Manual*. COMMELL. January 1, 2014. http://www.commell.com. tw/Download/Manual/LV-67K_Manual_V14.pdf. Accessed 25 Dec 2014.

Contiki-Developers. 2016. *Contiki: The Open Source OS for the Internet of Things*. Contiki-Developers. January 1, 2016. http://www.contiki-os.org/index.html. Accessed 14 Sept 2016.

Coulson, G., et al. 2012. Flexible Experimentation in Wireless Sensor Networks. *Communications of the ACM* 55 (1): 82–90.

Crepaldi, R., et al. 2007. The Design, Deployment, and Analysis of SignetLab: A Sensor Network Testbed and Interactive Management Tool. In *3rd International Conference on Testbeds and Research Infrastructure for the Development of Networks and Communities (TridentCom)*, 1–10. Lake Buena Vista: IEEE.

Crossbow. 2002. *MICA2*. January 1, 2002. http://www.eol.ucar.edu/isf/facilities/isa/internal/ CrossBow/DataSheets/mica2.pdf. Accessed 3 Feb 2014.

———. 2004a. *MPR/MIB User's Manual*. August 1, 2004. http://www-db.ics.uci.edu/pages/ research/quasar/MPR-MIB%20Series%20User%20Manual%207430-0021-06_A.pdf. Accessed 7 Jan 2014.

———. 2004b. *Stargate: X-Scale, Processor Platform*. Crossbow. January 1, 2004. http://www.eol.ucar.edu/isf/facilities/isa/internal/CrossBow/DataSheets/stargate.pdf.   Accessed 19 Mar 2014.

———. 2005. *Imote2: High-performance Wireless Sensor Network Node*. Crossbow. January 1, 2005. http://web.univ-pau.fr/~cpham/ENSEIGNEMENT/PAU-UPPA/RESA-M2/DOC/Imote2_Datasheet.pdf. Accessed 19 Mar 2014.

———. 2006. *MICAz*. January 1, 2006. http://www.openautomation.net/uploadsproductos/micaz_datasheet.pdf. Accessed 5 Feb 2014.

———. 2007a. *MTS/MDA Sensor Board Users Manual*. August 1, 2007. http://www.investigacion.frc.utn.edu.ar/sensores/Equipamiento/Wireless/MTS-MDA_Series_Users_Manual.pdf. Accessed 7 Jan 2014.

———. 2007b. *Stargate NetBridge: Embedded Sensor Network Gateway*. Crossbow. January 1, 2007. http://www.openautomation.net/uploadsproductos/stargate_netbridge_datasheet.pdf. Accessed 12 Aug 2014.

De, P., A. Neogi, and T.-C. Chiueh. 2003. VirtualWire: A Fault Injection and Analysis Tool for Network Protocols. In *23rd International Conference on Distributed Computing Systems (ICDCS)*, 214–221. Providence: IEEE.

De, P., A. Raniwala, S. Sharma, and T.-C. Chiueh. 2005a. Design Considerations for a Multihop Wireless Network Testbed. *Communications Magazine* 43 (10): 102–109.

———. 2005b. MiNT: A Miniaturized Network Testbed for Mobile Wireless Research. In *24th Annual Joint Conference of the IEEE Computer and Communications Societies (INFOCOM)*, 2731–2742. Miami: IEEE.

De, P., et al. 2006. MiNT-m: An Autonomous Mobile Wireless Experimentation Platform. In *The 4th International Conference on Mobile Systems, Applications and Services (MobiSys)*, 124–137. Uppsala: ACM.

Dien, and Ghing-Hsin. 2000, August 29. Computer Watchdog Timer. US Patent US6112320 A.

Dimitriou, T., J. Kolokouris, and N. Zarokostas. 2007. Sensenet: A Wireless Sensor Network Testbed. In *The 10th ACM Symposium on Modeling, Analysis, and Simulation of Wireless and Mobile Systems (MSWiM)*, 143–150. Chania: ACM.

Doddavenkatappa, M., M.C. Chan, and A.L. Ananda. 2012. Indriya: A Low-Cost, 3D Wireless Sensor Network Testbed. In *Lecture Notes of the Institute for Computer Sciences, Social Informatics and Telecommunications Engineering-Testbeds and Research Infrastructure. Development of Networks and Communities*, ed. T. Korakis, H. Li, P. Tran-Gia, and H.-S. Park, vol. 90, 302–316. Berlin, Heidelberg: Springer.

Dunkels, A., B. Gronvall, and T. Voigt. 2004. Contiki-A Lightweight and Flexible Operating System for Tiny Networked Sensors. In *29th Annual IEEE International Conference on Local Computer Networks (LCN)*, 455–462. Tampa: IEEE.

Dutta, P., et al. 2006. Trio: Enabling Sustainable and Scalable Outdoor Wireless Sensor Network Deployments. In *The 5th International Conference on Information Processing in Sensor Networks (IPSN)*, 407–415. Nashville: ACM/IEEE.

EasySen LLC. 2008a. *SBT30EDU: Sensor and Prototyping Board*. EasySen LLC. January 29, 2008. http://www.easysen.com/support/SBT30EDU/DatasheetSBT30EDU.pdf. Accessed 12 Aug 2014.

———. 2008b. *SBT80: Multi-Modality Sensor Board for TelosB Wireless Motes*. EasySen LLC. January 29, 2008. http://www.easysen.com/support/SBT80v2/DatasheetSBT80v2.pdf. Accessed 12 Aug 2014.

———. 2008c. *Sensor board for wireless surveillance and security applications*. EasySen LLC. January 29, 2008. http://www.easysen.com/support/WiEye/DatasheetWiEye.pdf. Accessed 12 Aug 2014.

EB. 2006. *A New Version of Propsim C8 for the Testing of 4X4 MIMO Systems*. EB. January 16, 2006. https://www.elektrobit.com/news-988-248-a_new_version_of_propsim_c8_for_the_testing_of_4x4mimo_systems. Accessed 2 July 2014.

EE Times. 2005. *Semtech to Acquire Xemics for $43 Million*. EE Times. June 20, 2005. http://www.eetimes.com/document.asp?doc_id=1154789. Accessed 17 July 2014.

Elson, J., et al. 2003. *Emstar: An Environment for Developing Wireless Embedded Systems Software*. Technical, Center for Embedded Network Sensing, University of California. Los Angeles: University of California.

Encyclopædia Britannica. 2014. *Trilateration*. Encyclopædia Britannica, Inc. January 1, 2014. http://www.britannica.com/EBchecked/topic/605329/trilateration. Accessed 24 Sept 2014.

Ettus. 2014. *USRP2: The Next Generation of Software Radio Systems*. Ettus. January 1, 2014. http://www.ece.umn.edu/users/ravi0022/class/ee4505/ettus_ds_usrp2_v5.pdf. Accessed 22 Dec 2014.

Ettus Research. 2012. *USRP N200/N210 Networked Series*. Ettus Research. September 14, 2012. https://www.ettus.com/content/files/07495_Ettus_N200-210_DS_Flyer_HR_1.pdf. Accessed 24 Dec 2014.

———. 2014. *USRP X300 and X310 X Series*. Ettus Research. January 1, 2014. http://www.ettus.com/content/files/X300_X310_Spec_Sheet.pdf. Accessed 24 Dec 2014.

Fedora. 2014. *Releases/HistoricalSchedules*. Fedora. January 14, 2014. http://fedoraproject.org/wiki/Releases/HistoricalSchedules. Accessed 10 July 2014.

Fekete, S.P., A. Kröller, S. Fischer, and D. Pfisterer. 2007. Shawn: The Fast, Highly Customizable Sensor Network Simulator. In *Fourth International Conference on Networked Sensing Systems (INSS)*. Braunschweig: Transducer Research Foundation and the IEEE Sensors Council.

FIRE. 2014. *WISEBED. Future Internet Research & Experimentation (FIRE)*. January 1, 2014. http://www.wisebed.eu/#home. Accessed 16 July 2014.

Flux Group. 2014. *Emulab - Network Emulation Testbed Home*. School of Computing - University of Utah. January 1, 2014. http://www.emulab.net. Accessed 20 Dec 2014.

Flux Research Group. 2015. *Emulab - Network Emulation Testbed Home*. Flux Research Group, University of Utah. January 1, 2015. http://www.emulab.net. Accessed 20 Feb 2015.

FreeBSD. 2014. *FreeBSD*. January 1, 2014. https://www.freebsd.org. Accessed 20 Dec 2014.

Frigo, M., and S.G. Johnson. 2015. *BenchFFT*. Janaury 1, 2015. http://www.fftw.org/benchfft/. Accessed 12 Feb 2015.

Gantz, J.F., et al. 2008. *The Diverse and Exploding Digital Universe: An Updated Forecast of Worldwide Information Growth Through 2011*. EMC. March 1, 2008. http://www.emc.com/collateral/analyst-reports/diverse-exploding-digital-universe.pdf. Accessed 30 June 2014.

GENI. 2014. *Experimentation with GENI*. GENI. January 1, 2014. http://groups.geni.net/geni/wiki/GeniNewcomersWelcome. Accessed 23 Nov 2014.

Gnawali, O., R. Fonseca, K. Jamieson, D. Moss, and P. Levis. 2009. Collection Tree Protocol. In *7th ACM Conference on Embedded Networked Sensor Systems (SenSys)*, 1–14. Berkeley: ACM.

Handziski, V., and T. Lentsch. 2005. *The EyesIFX Platform. Infineon Technologies*. February 2, 2005. www.tinyos.net/ttx-02-2005/platforms/ttx2005-eyesIFX.ppt. Accessed 18 Feb 2015.

Handziski, V., J. Polastre, J.-H. Hauer, and C. Sharp. 2004. Flexible Hardware Abstraction of the TI MSP430 Microcontroller in TinyOS. In *The 2nd International Conference on Embedded Networked Sensor Systems (SenSys)*, 277–278. ACM.

Handziski, V., A. Köpke, A. Willig, and A. Wolisz. 2006. TWIST: A Scalable and Reconfigurable Testbed for Wireless Indoor Experiments with Sensor Networks. In *The 2nd International Workshop on Multi-hop Ad hoc Networks: From Theory to Reality (REALMAN)*, 63–70. Florence: ACM.

Hartman, J., and S. Baker. 2014. *A Provisioning Service for Long-Term GENI Experiments*. January 1, 2014. http://groups.geni.net/geni/wiki/ProvisioningService. Accessed 27 Nov 2014.

Hartman, J.H., J. Cappos, and S. Baker. January 1, 2014. *Stork: An Overview*. January 1, 2014. http://www.cs.arizona.edu/stork/. Accessed 27 Nov 2014.

He, T., et al. 2006. VigilNet: An Integrated Sensor Network System for Energy-Efficient Surveillance. In *ACM Transactions on Sensor Networks (TOSN)*, vol. 2, no. 1, 1–38. ACM.

Hellbrück, H. 2012. *3-SOS. CoSA.* December 31, 2012. http://cosa.fh-luebeck.de/en/research/projects/trisos. Accessed 17 July 2014.

Hellbrück, H., M. Pagel, A. Kröller, D. Bimschas, D. Pfisterer, and S. Fischer. 2011. Using and Operating Wireless Sensor Network Testbeds with WISEBED. In *The 10th IFIP Annual Mediterranean Ad Hoc Networking Workshop*, 171–178. Favignana Island: IEEE/IFIP.

Henderson, T. 2011. *Two-ray Ground Reflection Model.* November 5, 2011. http://www.isi.edu/nsnam/ns/doc/node218.html. Accessed 25 Sept 2014.

Henning, J.L. 2000. SPEC CPU2000: Measuring CPU Performance in the New Millennium. *Computer* 33 (7): 28–35.

Hibler, M., L. Stoller, J. Lepreau, R. Ricci, and C. Barb. 2003. Fast, Scalable Disk Imaging with Frisbee. In *Annual Technical Conference*, 283–296. San Antonio: USENIX.

Horneber, J., and A. Hergenroder. 2014. A Survey on Testbeds and Experimentation Environments for Wireless Sensor Networks. *IEEE Communications Surveys & Tutorials* 16 (4): 1820–1838.

HostGator. 2002. *What Are Cron Jobs? HostGator.* January 1, 2002. http://support.hostgator.com/articles/cpanel/what-are-cron-jobs. Accessed 6 Aug 2014.

Hui, J.W., and D. Culler. 2004. The Dynamic Behavior of a Data Dissemination Protocol for Network Programming at Scale. In *The 2nd International Conference on Embedded Networked Sensor Systems (SenSys)*, 81–94. Baltimore: ACM.

Indiana University. 2014. *What is NFS?* Indiana University. January 1, 2014. https://kb.iu.edu/d/adux. Accessed 14 Aug 2014.

Innotech Systems. 2000. *SpitFIRE II: Universal Infrared Remote Control From Any PC.* Innotech Systems. January 1, 2000. http://www.innotech.com/spitfire6001.htm. Accessed 25 Oct 2014.

iRobot. 2014. *Roomba Vacuum Cleaning Robot.* iRobot. January 1, 2014. http://www.irobot.com/For-the-Home/Vacuum-Cleaning/Roomba. Accessed 25 Oct 2014.

Jiang, X., J. Polastre, and D. Culler. 2005. Perpetual Environmentally Powered Sensor Networks. In *Fourth International Symposium on Information Processing in Sensor Networks (IPSN)*, 463–468. Los Angeles: ACM/IEEE.

Jiménez-González, A., J.R. Martínez-De Dios, and A. Ollero. 2011. An Integrated Testbed for Cooperative Perception with Heterogeneous Mobile and Static Sensors. *Sensors* 11 (12): 11516–11543.

Jiménez-González, A., J.R. Martinez-de Dios, and A. Ollero. 2013. Testbeds for Ubiquitous Robotics: A Survey. *Robotics and Autonomous Systems* 61 (12): 1487–1501.

Johnson, D.B., and D.A. Maltz. 1996. Dynamic Source Routing in Ad Hoc Wireless Networks. In *Mobile Computing*, ed. T. Imielinski and H.F. Korth, vol. 353, 153–181. Springer.

Johnson, D., R. Fish, D.M. Flickinger, L. Stoller, R. Ricci, and J. Lepreau. 2006. Mobile Emulab: A Robotic Wireless and Sensor Network Testbed. In *The 25th Conference on Computer Communications (INFOCOM)*. Barcelona: IEEE.

Judd, G., and P. Steenkiste. 2004. Repeatable and Realistic Wireless Experimentation through Physical Emulation. *SIGCOMM Computer Communication Review* 34 (1): 63–68.

Keithley Instruments. 2002. *Integra Series–Model 2701 Ethernet-based Multimeter/Data Acquisition System.* Keithley Instruments. January 1, 2002. http://www.testequity.com/documents/pdf/keithley/2701-brochure.pdf. Accessed 6 Aug 2014.

Kidd, Eric. 2001. *XML-RPC for C and C++.* January 1, 2001. http://xmlrpc-c.sourceforge.net. Accessed 24 Jan 2015.

Kitts, C.A., and I. Mas. 2009. Cluster Space Specification and Control of Mobile Multirobot Systems. *IEEE/ASME Transactions on Mechatronics* 14 (2): 207–218.

Kröller, A., D. Pfisterer, C. Buschmann, S.P. Fekete, and S. Fischer. 2005. Shawn: A New Approach to Simulating Wireless Sensor Networks. In *Design, Analysis, and Simulation of Distributed Systems (DASD)*, 117–124. San Diego: Welcome to the Society for Modeling & Simulation International (SCS).

Lamport, L., R. Shostak, and M. Pease. 1982. The Byzantine Generals Problem. In *ACM Transactions on Programming Languages and Systems (TOPLAS)*, vol. 4, no. 3, 382–401. ACM.

LAN/MAN Standards Committee. 2006. Part 15.4b: Wireless Medium Access Control (MAC) and Physical Layer (PHY) Specifications for Low Rate Wireless Personal Area Networks (WPANs) (Amendment of IEEE Std 802.15.4-2003). In *Approved Draft Revision for IEEE Standard for Information Technology-Telecommunications and Information Exchange between Systems-Local and Metropolitan Area Networks-Specific Requirements*. New York: IEEE Computer Society.

Langmann, B., et al. 2013. MOVEDETECT – Secure Detection, Localization and Classification in Wireless Sensor Networks. In *Lecture Notes in Computer Science-Internet of Things, Smart Spaces, and Next Generation Networking*, ed. S. Balandin, S. Andreev, and Y. Koucheryavy, vol. 2, 284–297. Berlin, Heidelberg: Springer.

LEGO. 2014. *31313 LEGO MINDSTORMS EV3*. LEGO. January 1, 2014. http://www.lego.com/en-us/mindstorms/products/ev3/31313-mindstorms-ev3/. Accessed 24 Sept 2014.

Levis, P., Neil Patel, David Culler, and S. Shenker. 2004. Trickle: A Self-Regulating Algorithm for Code Propagation and Maintenance in Wireless Sensor Networks. In *First Symposium on Networked Systems Design and Implemention (NSDI)*. San Francisco: USENIX.

Linksys. 2007. *WRT600N: Dual-Band Wireless-N Gigabit Router with Storage Link*. Linksys. January 1, 2007. http://static.highspeedbackbone.net/pdf/Linksys-WRT600N Datasheet.pdf. Accessed 14 Aug 2014.

———. 2008a. *Network Storage Link for USB 2.0 Disk Drives*. Cisco Systems. January 1, 2008. http://support.you.gr/catalog/33/3330F2E4FC0C764CB803BE02D5F4B99F.pdf. Accessed 10 July 2014.

———. 2008b. *OpenSlug*. Linksys. February 27, 2008b. http://www.nslu2-linux.org/wiki/OpenSlug/HomePage. Accessed 10 July 2014.

———. 2009. *SlugOS*. Linksys. March 27, 2009. http://www.nslu2-linux.org/wiki/SlugOS/HomePage. Accessed 10 July 2014.

Liu, T., C.M. Sadler, P. Zhang, and M. Martonosi. 2004. Implementing Software on Resource-Constrained Mobile Sensors: Experiences with Impala and ZebraNet. In *2nd International Conference on Mobile Systems, Applications, and Services (MobiSys)*, 256–269. Boston: ACM.

Lorincz, K., and M. Welsh. 2004. *MoteTrack: A Robust, Decentralized Location Tracking System for Disaster Response*. Technical, Computer Science Group, Harvard University. Cambridge, MA: Harvard University.

Lorincz, K., et al. 2004. Sensor Networks for Emergency Response: Challenges and Opportunities. *Pervasive Computing* 3 (4): 16–23.

Lu, X., and R. Manduchi. 2006. Fast Image Motion Computation on an Embedded Computer. In *Conference on Computer Vision and Pattern Recognition Workshop (CVPRW)*, 120. New York City: IEEE.

Lucent Technologies. 2015. *Orinoco Gold PC Card User's Guide*. Lucent Technologies. January 1, 2015. https://www.mikrotik.com/documentation/manual_2.3/Wavelan/ug_pc.pdf. Accessed 12 Feb 2015.

Lundgren, H., D. Lundberg, J. Nielsen, E. Nordström, and C. Tschudin. 2002. A Large-Scale Testbed for Reproducible Ad Hoc Protocol Evaluations. In *Wireless Communications and Networking Conference (WCNC)*, 412–418. Orlando: IEEE.

Maltz, D.A., J. Broch, and D.B. Johnson. 2001. Lessons from a Full-Scale Multihop Wireless Ad Hoc Network Testbed. *Personal Communications* 8 (1): 8–15.

Margi, C.B., V. Petkov, K. Obraczka, and R. Manduchi. 2006. Characterizing Energy Consumption in a Visual Sensor Network Testbed. In *2nd International Conference on Testbeds and Research Infrastructures for the Development of Networks and Communities (TRIDENTCOM)*. Barcelona: IEEE.

Maxim Integrated. 2005. *DS2438: Smart Battery Monitor*. Maxim Integrated. January 1, 2005. http://datasheets.maximintegrated.com/en/ds/DS2438.pdf. Accessed 12 Feb 2015.

McKeown, N., et al. 2008. OpenFlow: Enabling Innovation in Campus Networks. *ACM SIGCOMM Computer Communication Review* 38 (2): 69–74.

Medhi, D., et al. 2014. The GpENI Testbed: Network Infrastructure, Implementation Experience, and Experimentation. *Computer Networks* 61: 51–74.

MikroTik. 2002. *RouterBOARD*. MikroTik. January 1, 2002. http://routerboard.com/about. Accessed 25 Oct 2014.

———. 2005. *RouterBOARD 200 Series*. MikroTik. November 22, 2005. http://www.routerboard.sk/files/pdf/rb200_manual.pdf. Accessed 25 Oct 2014.

MIT. 2015. *RON (Resilient Overlay Networks)*. MIT. January 1, 2015. http://nms.csail.mit.edu/ron/. Accessed 20 Feb 2015.

Moteiv. 2004. *Telos: Ultra low power IEEE 802.15.4 compliant wireless sensor module*. Moteiv. May 12, 2004. http://www2.ece.ohio-state.edu/~bibyk/ee582/telosMote.pdf. Accessed 27 Mar 2014.

———. 2006a. *Tmote Connect: Wireless Gateway Appliance Software*. Moteiv. December 12, 2006. http://automatica.dei.unipd.it/public/Schenato/PSC/2010_2011/gruppo4-Building_termo_identification/Bibliografia%20Casuale/tmote-connect-datasheet.pdf. Accessed 25 Jan 2015.

———. 2006b. *Tmote sky Data Sheet: Ultra Low Power IEEE 802.15.4 Compliant Wireless Sensor Module*. June 2, 2006. http://www.eecs.harvard.edu/~konrad/projects/shimmer/references/tmote-sky-datasheet.pdf. Accessed 3 Jan 2014.

Murray, M. 2012. *DDR vs. DDR2 vs. DDR3: Types of RAM Explained*. PC Magazine. February 28, 2012. http://www.pcmag.com/article2/0,2817,2400801,00.asp. Accessed 25 Dec 2014.

National Instruments. 2014. *USRP. National Instruments*. January 1, 2014. http://www.ni.com/sdr/usrp/. Accessed 24 Dec 2014.

NEC. 2014. *Switches*. NEC. January 1, 2014. http://www.nec.com/en/global/prod/storage/product/san/switches/index.html. Accessed 24 Dec 2014.

NetFPGA. 2014a. *NetFPGA 10G*. NetFPGA. January 1, 2014. http://netfpga.org/2014/#/systems/3netfpga-10g/details/http://netfpga.org/2014/#/systems/3netfpga-10g/details/. Accessed 25 Dec 2014.

———. 2014b. *NetFPGA 1G*. NetFPGA. January 1, 2014. http://netfpga.org/2014/#/systems/4netfpga-1g/details/. Accessed 25 Dec 2014.

NETGEAR. 2002. *IEEE 802.11b Wireless PCI Adapter - Model MA311*. NETGEAR. January 1, 2002. http://www.downloads.netgear.com/files/ma311_user_guide.pdf. Accessed 24 Sept 2014.

Nguyen, B.T. 2004. http://www.tldp.org/LDP/Linux-Dictionary/html/t.html. August 16, 2004. Accessed 23 Sept 2014.

Oh, S., S. Sastry, and L. Schenato. 2005. A Hierarchical Multiple-Target Tracking Algorithm for Sensor Networks. In *The IEEE International Conference on Robotics and Automation (ICRA)*, 2197–2202. Barcelona: IEEE.

OpenBSD. 2014. *OpenBSD*. January 1, 2014. http://www.openbsd.org. Accessed 3 Dec 2014.

ORBIT. 2014. *Open-Access Research Testbed for Next-Generation Wireless Networks (ORBIT)*. January 1, 2014. http://www.orbit-lab.org/wiki. Accessed 20 Dec 2014.

PC. 2015. *Logitech QuickCam Pro 4000*. PCMag Digital Group. January 1, 2015. http://www.pcmag.com/article2/0,2817,1383352,00.asp. Accessed 12 Feb 2015.

PC Glossary. 2014. *SO-DIMM*. PC Glossary. January 1, 2014. http://pc.net/glossary/definition/sodimm. Accessed 25 Dec 2014.

Perl.org. 2002. *About Perl*. Perl.org. January 1, 2002. http://www.perl.org/about.html. Accessed 7 Aug 2014.

Peterson, L., et al. 2009. *Slice-Based Facility Architecture, Draft Version*. April 7, 2009. http://svn.planet-lab.org/attachment/wiki/GeniWrapper/sfa.pdf. Accessed 25 Nov 2014.

Peterson, L., A. Bavier, and S. Bhatia. 2011. *VICCI: A Programmable Cloud-Computing Research Testbed*. Technical, Computer Science, Princeton University. Princeton: Princeton University.

Pica8. 2014. *First Switching, Routing and SDN Linux Network Operating System*. Pica8. January 1, 2014. http://www.pica8.com/open-switching/open-switching-overview.php. Accessed 24 Dec 2014.

PlanetLab. 2007. *About PlanetLab*. The Trustees of Princeton University. January 1, 2007. http://www.planet-lab.org/about. Accessed 20 Feb 2015.

PostgreSQL. 2013. *PostgreSQL 2013-12-05 Update Release*. December 5, 2013. http://www.postgresql.org. Accessed 7 Jan 2014.

Presser, M., P.M. Barnaghi, M. Eurich, and C. Villalonga. 2009. The SENSEI Project: Integrating the Physical World with the Digital World of the Network of the Future. *Communications Magazine* 47 (4): 1–4.

Pronto Systems. 2012. *Pica8+Pronto= Open Network Platform*. Pronto Systems. February 1, 2012. http://prontosystems.wordpress.com. Accessed 20 Dec 2014.

Qualcomm Atheros. 2014. *Products*. Qualcomm Atheros i. January 1, 2014. http://www.qca.qualcomm.com/products/. Accessed 25 Oct 2014.

Rahimi, M., et al. 2005. Cyclops: In Situ Image Sensing and Interpretation in Wireless Sensor Networks. In *The 3rd International Conference on Embedded Networked Sensor Systems (SenSys)*, 192–204. San Diego: ACM.

Ranjan, A., C. Karbinski, and J. Mathew. 2005. *Simple Network Management Protocol (SNMP)*. WhatIs.com. September 1, 2005. http://whatis.techtarget.com/definition/Simple-Network-Management-Protocol-SNMP. Accessed 24 Sept 2014.

Raychaudhuri, D., et al. 2005. Overview of the ORBIT Radio Grid Testbed for Evaluation of Next-Generation Wireless Network Protocols. In *Wireless Communications and Networking Conference*, 1664–1669. New Orleans: IEEE.

Red Hat. 2004. *Fedora*. Red Hat. November 8, 2004. http://fedoraproject.org/en/about-fedora. Accessed 10 July 2014.

———. 2008. *Using the ssh Command*. Red Hat. March 15, 2008. https://www.centos.org/docs/5/html/5.2/Deployment_Guide/s2-openssh-using-ssh.html. Accessed 12 July 2014.

Reich, J., V. Misra, and D. Rubenstein. 2008. Roomba MADNeT: A Mobile Ad-Hoc Delay Tolerant Network Testbed. *ACM SIGMOBILE Mobile Computing and Communications Review* 12 (1): 68–70.

Ricci, R., J. Duerig, L. Stoller, G. Wong, S. Chikkulapelly, and W. Seok. 2012. Designing a Federated Testbed as a Distributed System. In *Testbeds and Research Infrastructure. Development of Networks and Communities*, ed. T. Korakis, M. Zink, and M. Ott, vol. 44, 321–337. Berlin, Heidelberg: Springer.

Ritter, S. 2007. *Sun SPOTs In Action*. Internet Archive: PayBack Machine. September 24, 2007. http://web.archive.org/web/20080420040447/http://parleys.com/display/PARLEYS/Sun+SPOTs+In+Action. Accessed 17 July 2014.

RTL-SDR.com. 2014. *Chrome RTL-SDR Radio Receiver Updated*. December 23, 2014. http://www.rtl-sdr.com. Accessed 24 Dec 2014.

Sanghani, S., T.X. Brown, S. Bhandare, and S. Doshi. 2003. EWANT: The Emulated Wireless Ad Hoc Network Testbed. In *IEEE Wireless Communications and Networking (WCNC)*, 1844–1849. New Orleans: IEEE.

ScatterWeb. 2015. *ScatterWeb Operating System for WSN*. ScatterWeb. January 1, 2015. http://scatterweb.sourceforge.net. Accessed 3 Jan 2015.

Slipp, J., C. Ma, N. Polu, J. Nicholson, M. Murillo, and S. Hussain. 2008. WINTeR: Architecture and Applications of a Wireless Industrial Sensor Network Testbed for Radio-Harsh Environments. In *6th Annual Communication Networks and Services Research Conference (CNSR)*, 422–431. Halifax: IEEE.

SmartSantander. 2014. *Santander Facility*. SmartSantander. January 1, 2014. http://smartsantander.eu/index.php/testbeds/item/132-santander-summary. Accessed 19 July 2014.

Song, A. 2001. *Piconet II - A Wireless Ad Hoc Network*. School of Information Technology and Electrical Engineering, the University of Queensland, Australia. November 10, 2001. http://piconet.sourceforge.net/thesis/index.html. Accessed 12 Feb 2015.

Sony Corporation. 2002. *Sony IPELA SNC-RZ30N User's Manual*. Sony Corporation. January 1, 2002. http://www.manualslib.com/download/321518/Sony-Ipela-Snc-Rz30n.html. Accessed 30 Aug 2014.

SourceForge. 2013. *AODV-UU*. SourceForge. Febraury 23, 2013. http://sourceforge.net/projects/aodvuu/. Accessed 23 Aug 2015.

Sridharan, M., et al. 2011. From Kansei to KanseiGenie: Architecture of Federated, Programmable Wireless Sensor Fabrics. In *Testbeds and Research Infrastructures. Development of Networks and Communities*, ed. T. Magedanz, A. Gavras, N.H. Thanh, and J.S. Chase, vol. 46, 155–165. Berlin, Heidelberg: Springer.

Stajano, F., and R. Anderson. 2000. The Grenade Timer: Fortifying the Watchdog Timer Against Malicious Mobile Code. In *7th International Workshop on Mobile Multimedia Communications (MoMuC)*. Tokyo: IEEE.

Sterbenz, J.P., et al. 2011. The Great Plains Environment for Network Innovation (GpENI): A Programmable Testbed for Future Internet Architecture Research. In *Testbeds and Research Infrastructures. Development of Networks and Communities*, ed. T. Magedanz, A. Gavras, N.H. Thanh, and J.S. Chase, vol. 46, 428–441. Berlin, Heidelberg: Springer.

Szewczyk, R., A. Mainwaring, J. Polastre, J. Anderson, and C. David. 2004. An Analysis of a Large Scale Habitat Monitoring Application. In *2nd International Conference on Embedded Networked Sensor Systems (SenSys)*, 214–226. Baltimore: ACM.

Tantisureeporn, S., and L.J. Armstrong. 2008. Introducing a New Technology to Enhance Community Sustainability: An Investigation of the Possibilities of Sun Spots. In *The EDU-COM 2008 International Conference on Sustainability in Higher Education: Directions for Change*, 488–495. Perth: Edith Cowan University, Western Australia in association with Khon Kaen University, Thailand and Bansomdejchaopraya Rajabhat University, Thailand.

Techopedia. 2015a. *Golden Image*. Techopedia. January 1, 2015a. http://www.techopedia.com/definition/29456/golden-image. Accessed 24 Jan 2015.

———. 2015b. *Remote Method Invocation (RMI)*. Techopedia. January 1, 2015. http://www.techopedia.com/definition/1311/remote-method-invocation-rmi. Accessed 6 Feb 2015.

TechTarget. 1999. *Data exfiltration (data extrusion)*. TechTarget. January 1, 1999. http://whatis.techtarget.com/definition/data-exfiltration-data-extrusion. Accessed 1 Sept 2014.

———. 2014. *Telnet*. TechTarget. January 1, 2014. http://searchnetworking.techtarget.com/definition/Telnet. Accessed 25 Dec 2014.

Texas Instruments. 2011. *MSP430C11x1, MSP430F11x1A Mixed Signal Microcontroller*. Texas Instruments. January 1, 2011. http://www.ti.com/lit/ds/symlink/msp430f1611.pdf. Accessed 1 Feb 2014.

The Free Dictionary. 2014. *In-band Signalling*. Farlex. January 1, 2014. http://encyclopedia2.thefreedictionary.com/in-band+signalling. Accessed 30 Aug 2014.

The GENI Project Office. 2008. *GENI System Overview*. September 29, 2008. http://www.cra.org/ccc/files/docs/GENISysOvrvw092908.pdf. Accessed 26 Nov 2014.

TinyOS. 2012. *TinyOS*. August 20, 2012. http://www.tinyos.net. Accessed 9 Oct 2013.

TinyOS Wiki. 2012. *Mote-PC serial communication and Serial Forwarder*. TinyOS Wiki. October 21, 2012. http://tinyos.stanford.edu/tinyos-wiki/index.php/Mote-PC_serial_communication_and_SerialForwarder_(TOS_2.1.1_and_later). Accessed 7 Aug 2014.

———. 2013. *TOSSIM*. TinyOS Wiki. May 10, 2013. http://tinyos.stanford.edu/tinyos-wiki/index.php/TOSSIM. Accessed 1 Sept 2014.

Tolle, G., and D. Culler. 2005. Design of an Application-Cooperative Management System for Wireless Sensor Networks. In *The Second European Workshop on Wireless Sensor Networks (EWSN)*, 121–132. Istanbul: IEEE.

Tolle, G., et al. 2005. A Macroscope in the Redwoods. In *The 3rd International Conference on Embedded Networked Sensor Systems (SenSys)*, 51–63. San Diego: ACM.

Tutorialspoint. 2014. *JSP - Overview*. Tutorialspoint. January 1, 2014. http://www.tutorialspoint.com/jsp/jsp_overview.htm. Accessed 6 Feb 2015.

Ubuntu. 2014. *Overview*. Ubuntu. January 1, 2014. http://www.ubuntu.com/desktop/. Accessed 13 Aug 2014.

UISP. 2014. *UISP: AVR In-System Programme*. January 1, 2014. http://www.nongnu.org/uisp/. Accessed 30 Aug 2014.

Wan, Chieh-Yih, A.T. Campbell, and L. Krishnamurthy. 2002. PSFQ: A Reliable Transport Protocol for Wireless Sensor Networks. In *First ACM International Workshop on Wireless Sensor Networks and Applications (WSNA)*, 1–11. ACM.

WARP. 2014. *WARP: Wireless Open Access Research Platform*. January 1, 2014. http://warpproject.org/trac/wiki/WikiStart. Accessed 24 Dec 2014.

Wellshow. 2015. *What is an SMA Connector?* Wellshow Technology Co., Ltd. January 1, 2015. http://www.wellshow.com/technical-support/connector-support/what-is-an-sma-connector/. Accessed 17 Feb 2015.

Werner-Allen, G., P. Swieskowski, and M. Welsh. 2005. MoteLab: A Wireless Sensor Network Testbed. In *The 4th International Symposium on Information Processing in Sensor Networks (IPSN)*. Los Angeles: ACM/IEEE.

WhatIs.com. 2014. *DIMM (Dual In-Line Memory Module)*. WhatIs.com. January 1, 2014. http://whatis.techtarget.com/definition/DIMM-dual-in-line-memory-module. Accessed 25 Dec 2014.

White, B., et al. 2002. An Integrated Experimental Environment for Distributed Systems and Networks. In *The 5th Symposium on Operating Systems Design and Implementation (OSDI)*, 255–270. Boston: ACM.

Whitehouse, K., et al. 2006. Marionette: Using RPC for Interactive Development and Debugging of Wireless Embedded Networks. In *5th International Conference on Information Processing in Sensor Networks (IPSN)*, 416–423. Nashville: ACM.

Willow Technologies. 2013. *Wireless Security Systems*. Willow Technologies. January 1, 2013. http://www.willow.co.uk/html/wireless_security_systems.php. Accessed 13 Aug 2014.

WINLAB. 2010. *About WINLAB*. January 1, 2010. http://www.winlab.rutgers.edu/docs/about/about.html. Accessed 20 Dec 2014.

Yeo, J., M. Youssef, and A. Agrawala. 2004. A Framework for Wireless LAN Monitoring and its Applications. In *3rd ACM Workshop on Wireless Security (WISe)*, 70–79. Philadelphia: ACM.

Chapter 8
Simulators and Emulators for WSNs

Simulation is acting ... Acting is not typical of real-life.

8.1 WSN Testbeds, Simulators, and Emulators

Testbeds, simulators, and emulators are effective tools to evaluate algorithms and protocols at design, development, and implementation stages. Many of these tools are available, each with different features, characteristics, models, and architectures for the performance testing of WSNs. They are the focus of this book, as discussed in the previous chapter and accomplished in this one. It is arduous for a researcher or a practitioner to choose an appropriate tool for performance evaluation without the knowledge of the available tools, their features, as well as their pros and cons. Similarly, efforts to improve existing simulators or design a new one require a detailed understanding of the useful tools readily available. This necessitates comprehensive horizontal and vertical analysis among competitors at different stages. For vertical validation of experiments, a singular approach for conducting simulation, emulation, and real-world experiments in mobile ad hoc networks using a single tool is proposed in (Krop et al. 2007). As such, vertical analysis complements horizontal analysis that is performed by investigating different testbeds at almost the same level of abstraction. Recognizing the immense and diverse literature available, this chapter aims at clarifying in full detail all concepts and features related to modeling and simulation, simulators, and emulators.

Studying and analyzing WSNs goes through different phases and approaches, starting theoretical by simulation, ending up practical via testbeds, or working midway as in emulation. The basic differentiation, as shown in Fig. 8.1, can be outlined in what follows (Coulson et al. 2012):

- Physical testbeds as detailed in the previous chapter excel at high-fidelity evaluation of mature WSN designs, as well as detailed planning for real-world deployments. However, physical testbeds for WSN systems tend to be small in scale, expensive to maintain, and time-consuming to set up. They also lack flexibility, often offering a single, fixed, connectivity topology and restricted heterogeneity,

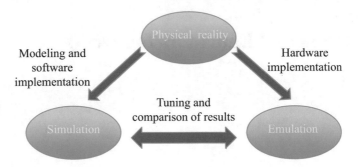

Fig. 8.1 Physical reality, simulation, emulation interrelated

such as only a single type of sensor node, radio, operating system, or programming language. They also tend to be limited in their programmability at lower
levels of the system; for instance, many use fixed operating systems and networking stacks. They are also often unfitted to experimentation scenarios requiring repeatability of experiments since many relevant operating parameters are
beyond user control, such as local radio interference due to infrastructure and
other experiments.

- Simulation is useful for rapidly trying out new ideas and examining the behavior
  of new protocols and mechanisms in varied topologies on a large scale and in a
  repeatable manner (Levis et al. 2003; Fekete et al. 2007). The most notable
  drawback is a lack of fidelity, often making it impractical to simulate fully at
  the instruction-execution level and with high-fidelity radio or power-consumption
  characteristics. While such restrictions are not necessarily problematic in traditional network environments where simulators such as ns-2 are prominent, they
  represent significant drawbacks in WSN environments where resource scarcity
  and incidental physical characteristics are of the essence. Simulation alone is
  therefore of limited use in planning for real-world WSN systems and
  deployments.
- Emulation is situated between physical reality and simulation (Girod et al. 2004;
  Wu et al. 2004; Judd and Steenkiste 2004). Whereas simulation abstractly models
  target systems, emulation duplicates the functionality of one system in terms of
  another system and is therefore capable of much greater fidelity than simulation
  while potentially offering greater flexibility than a purely physical testbed. Emulation is a much less exploited approach in the WSN testbed context despite its
  potential; more precisely, emulation in the form of network overlay technology
  could be used to support different inter-node connectivity patterns in a physical
  testbed. Alternatively, a battery-based power supply on a physical node could be
  emulated by interposing an electricity-powered hardware module with degrading
  power over time.

8.2 Modeling and Simulation

Modeling and simulation are two complementary procedures. Without simulation, models are just paperwork, theories that may not come true. Simulation lacking theoretical study is non-founded and misses a matching vision of a system that should come to life. In this section, basic concepts of modeling and simulation are illustrated with a pinpoint to the essential literature.

8.2.1 Basic Definitions

A model is the first representation of the system to be studied for the purpose of collecting specific metrics, either for the sake of knowledge or for subsequent implementation. Simulation is the upper block of a hierarchy that has to be stepped up (Sargent 2005):

- The problem entity is the system, real or proposed, to be modeled.
- The conceptual model is the mathematical, or logical, or verbal representation of the problem entity developed for a particular study; it is developed through an analysis and modeling phase.
- The computerized model, or simulation model, is the conceptual model implemented on a computer. It is developed through a computer programming and implementation phase. The simulation model specification is a written detailed description of the software design and programming for the conceptual model implementation on a particular computer system. Inferences about the problem entity are obtained by conducting computer experiments on the computerized model in the experimentation phase.

The abovementioned models must undergo several procedures that detect and correct errors in their representation of the required entity and in the true matching of their outcome with the must-have data. For clear understanding, several definitions draw borderlines between a multiplicity of concepts:

- Model verification is ensuring that the computer program of the model and its implementation are correct (Sargent 2005). Verification is also defined as the process of determining that a model implementation accurately represents the developer's conceptual description of the model and its solution (Thacker et al. 2004).
- Model validation is the substantiation that a computerized model within its domain of applicability possesses a satisfactory range of accuracy consistent with the intended application of the model (Schlesinger et al. 1979). Validation is also defined as the process of determining the degree to which a model is an accurate representation of the real world from the perspective of the intended usage patterns of the model (Thacker et al. 2004).

- Operational validation is determining that the model's output behavior has sufficient accuracy for the model's intended purpose over the domain of the model's intended applicability (Sargent 2005).
- Data validity is guaranteeing that the data necessary for model building, model evaluation and testing, and conducting the model experiments to solve the problem are adequate and correct (Sargent 2005).
- Model accreditation is the determination that a model satisfies defined model accreditation criteria according to a specified process (Sargent 2005).
- Model credibility is giving potential users the confidence they need to use a model and to believe the information derived from that model (Sargent 2005).

8.2.2   Validation and Verification

A model should be developed for a specific purpose or application and its validity determined with respect to that purpose. If the purpose of a model is to answer a variety of questions, the validity of the model needs to be determined with respect to each question. Numerous sets of experimental conditions are usually required to define the domain of a model's intended applicability. A model may be valid for one set of experimental conditions but invalid for another. A model is considered valid for a set of experimental conditions if the model accuracy is within its acceptable range, which is the amount of accuracy required for the model's intended purpose. This requires identifying the model output variables and determining their required accuracy. The amount of accuracy required should be specified prior to starting the development of the model or in the early stages of the model development process. If the variables of interest are random variables, then properties and functions of the random variables, such as means and variances, are of interest and are used in determining model validity. Several versions of a model are to be developed prior to obtaining a satisfactory valid model.

The substantiation that a model is valid, i.e., performing model verification and validation is often costly and time-consuming. Tests and evaluations are to be conducted until there is acceptable confidence that a model can be considered valid for its intended application (Sargent 2005). If a test determines that a model does not have sufficient accuracy for any one of the sets of experimental conditions, then the model is invalid. However, determining that a model has enough accuracy for numerous experimental conditions does not guarantee that a model is valid everywhere in its applicable domain.

Verification is concerned with identifying and removing errors in the model by comparing numerical solutions to analytical or highly accurate benchmark solutions. Validation, on the other hand, is concerned with quantifying the accuracy of the model by comparing numerical solutions to experimental data. Thus, verification deals with the mathematics associated with the model, whereas validation deals with the physics associated with the model (Roache 1998). Mathematical errors can eliminate the impression of correctness by giving the right answer for the wrong

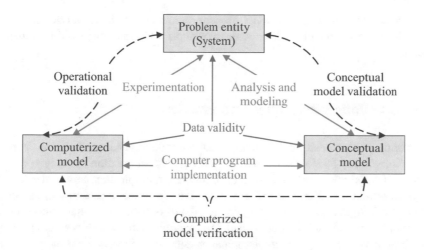

Fig. 8.2 The modeling process (Sargent 2005)

reason; thus, verification should be performed to a sufficient level before the validation activity begins.

The verification activity can be divided into code verification and calculation verification. When performing code verification, problems are devised to ensure that the code can compute an accurate solution. Code verification problems are constructed to verify code correctness, robustness, and specific code algorithms. When performing calculation verification, a model that is to be validated is exercised to ascertain that it is computing a sufficiently accurate solution.

Fundamentally, software validation and verification are different from model validation and verification. Software validation and verification are required when a computer program or code is the end product. Model validation and verification are required when a predictive model is the end product. A code is the computer implementation of algorithms developed to facilitate the formulation and approximate solution of a class of models. A model is the conceptual, mathematical, and numerical description of a specific physical scenario, including geometrical, material, initial, and boundary data.

As shown in Fig. 8.2, an iterative process is used to develop a valid simulation model:

- A conceptual model is developed and validated; this process is repeated until the conceptual model is satisfactory.
- Next the computerized model is built from the conceptual model and is verified; this process is repeated until the computerized model is satisfactory.
- Then, operational validity is conducted on the computerized model. Model changes required by conducting operational validity can be in either the conceptual model or the computerized model.

Verification and validation must be performed anew with any model modification. Commonly, several models are developed prior to obtaining a valid simulation model.

8.3   Simulation Principles and Practice

Simulation is a valuable tool in many areas where analytical methods and experimentation are not feasible. Researchers generally use simulation to analyze system performance prior to physical design or to compare multiple alternatives over a wide range of conditions. However, if simulation fails to reflect a meaningful aspect of reality, the insight into the operating characteristics of the system under study is lost. Generalization and lack of depth lead to inaccurate data, which results in wrong conclusions or inappropriate implementation decisions. Simulation principles, concepts, and practice are found in a wealthy literature that includes among many (Maria 1997), in addition to those referenced in this section.

Most research on MANET routing protocols includes a simulation to study a proposed solution performance. Network simulation packages are complex and require sufficient time to learn and master. Moreover, conducting a study using more than one simulation package might lead to different results, which highlights the inherent simulator technique and its view of the system under study that may be closer or further from the typical reality involved. Factually, matching results among multiple simulators does not mean they are technically credible unless investigators independently validate the simulators and confirm their application for each instance (Andel and Yasinsac 2006).

In the literature, many significant contributions focused on how to make credible simulations; not to be unobserved (Jain 1991; Kurkowski et al. 2005; Andel and Yasinsac 2006). Particularly, in (Kurkowski et al. 2005), there is an interesting consideration of MANET simulation studies as published in (ACM SIGMOBILE 2000). Findings are displeasing; in general, the outcomes published in MANET simulation studies lack believability. The work in (Andel and Yasinsac 2006) questioned the validity of simulation and illustrated how it can produce misleading outcomes; it also presented cautionary advice to enhance simulation credibility.

Compulsorily, to qualify for trustworthiness and believability, a simulation-based research, must be:

- Repeatable. A properly published paper must discuss or reference all details that satisfy the repeatability criteria. A researcher should be able to repeat the simulation results for his own contentment, for future reviews, and for further development. Several factors jeopardize repeatability:

  - Overlooking the identification of the simulator and its version, the operating system, and all variable settings.

- Unavailability to the community of the code and configuration files; in the conducted survey, no paper with a simulation study made a statement about code availability.
- Leaving out the adopted scenarios settings, such as transmission distance and bit rate, the techniques used to avoid initialization bias (influence of empty queues, etc., at the start), and the methods used to analyze the results.

- Rigorous. The scenarios and conditions used to test the experiment must truly exercise the aspect of MANETs being studied. For a rigorous study, factors such as node density, node footprint, coverage, speed, and transmission range must be set correctly to exercise the protocol under test. Explicitly, a study that uses scenarios with average hop counts, between source and destination, below two are solely testing neighbor communication not true routing.
- Statistically sound. The execution and analysis of an experiment must be based on mathematical principles. For a statistically sound study, a careful task must:

  - Account for initialization bias.
  - Execute a number of independent simulation runs, and address sources of randomness such as pseudorandom number generator (PRNG) to ensure the independence of runs.
  - Provide the confidence levels that exist in the results.
  - List all embraced statistical assumptions.
  - Collect data after deleting transient values or eliminating them by preloading tables and queues.

- Unbiased. The results must not be confined to the exclusive scenario used in the experiment. For a study to be unbiased, a project must address initialization bias and random number issues and use a variety of scenarios. A single scenario use is just to prove a limitation or counter a generalization.
- Multiply run. To ensure the credibility of a simulation study multiple runs, up to 30 times, are compulsory. Specifically, executing with different PRNG seeds to account for convergence, deviation, and modal values provides statistical validity.
- Precise. Mathematical models measure results by orders of magnitude, while simulation models show percentage improvements between MANET protocols. However, simulation is functionally imprecise and is subject to errors injected by inaccurate parameters or false assumptions:

  - Transmission range is generally represented as a circle radius.
  - Node distribution is modeled as uniform or random; in fact, roads, trees, water, and other obstacles affect node distribution.
  - Interference models are typically based on signal to noise ratio (SNR) or bit error rate (BER), which disregards interference based on increasing traffic or unpredictable background noise.
  - Use of unfitting radio models, for instance, free-space radio models, is enough during early development, but two-ray and shadow models are more realistic throughout data collection and analysis.

- Unrealistic traffic assumption, such as CBR, which may not be representative of the application under study.
- Node communication is assumed to be bidirectional; however, in wireless communication, there is no guarantee that signal transmission and reception distances are similar. MANET nodes might have different power levels available for transmission.
- Node mobility is modeled as random, but it rarely is. There is usually a pattern to follow.
- Area is assumed to be a square or rectangular network area, which does not reflect reality.

- Validated. The simulation must be validated in all its details, specifically, protocol design, traffic, radio model, against a real-world implementation; or when being at the early phase of concept development, against analytical models or protocol specifications. The later handling is less precise, but it can be refined later when implementation is achieved.

Generally, an alarming lack of believability in MANET simulation research is revealed, even though using simulation to test performance is almost a common factor; that is adopted in 114 out of the 151 (75.5%) published MobiHoc papers (Kurkowski et al. 2005).

8.3.1   Simulating the Advance of Time

For representative simulation, continuous systems and discrete systems are to be comprehensibly differentiated. In continuous systems, the state variables change continuously with respect to time, whereas in discrete systems, the state variables change instantaneously at separate points in time. Unfortunately, for computational experimentation, there are but a few systems that are either completely discrete or completely continuous, although frequently one type dominates the other in such hybrid systems. The real challenge is to find a computational model that mimics closely the behavior of the system under study; specifically the simulation of time advance is tricky. There are a number of means for modeling the progress of time: specifically, time-slicing, discrete-event, and continuous time approaches. The time-slicing method is useful for understanding the basics of the simulation approach. Discrete-event simulation and continuous time simulation are the most commonly adopted approaches, as illustrated below.

8.3.1.1   The Time-Slicing Approach

The simplest method for modeling the progress of time is the time-slicing approach in which a constant time step (Δt) is adopted. It is relatively simple to set up a time-

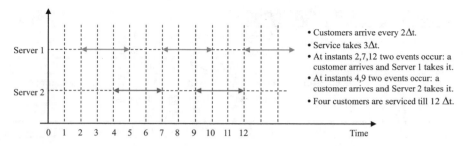

Fig. 8.3 Time-slicing approach

slicing simulation for this situation. There are two main problems with the time-slicing approach:

- First, it is very inefficient. During many of the time steps, there is no change in the system state, and as a result, many computations are unnecessary. This problem is only likely to be exacerbated the larger the simulation becomes.
- A second problem is determining the value of Δt. In most simulations, the duration of activities cannot be counted in whole numbers. Also, there is often a wide variation in activity times within a model from possibly seconds (or less) to hours, days, weeks, or more. The discrete-event simulation approach addresses both of these issues.

As shown in Fig. 8.3, the points of interest are those when a customer arrives, when a server receives it, and when the server completes (five arrivals, five services, four completions); computations are made on all time steps even when there is no state change.

8.3.1.2   The Discrete-Event Simulation Approach

In discrete-event simulation, only the points in time at which the state of the system changes are represented. In other words, the system is modeled as a series of events, that is, instants in time when a state change occurs. Examples of events are as follows: a customer arrives, a server starts, and a server completes service. Each of these occurs at an instant in time. A number of mechanisms have been proposed for carrying out discrete-event simulation, among them are the event-based, activity-based, process-based, and three-phase approaches used by a number of commercial simulation software packages (Pidd 1998).

Figure 8.4 illustrates the seven points of time that are of interest based on the example of Fig. 8.3.

Point in time	Event
2	Customer arrival + service at Server 1
4	Customer arrival + service at Server 2
5	Server 1 completes
7	Customer arrival + service at Server 1 + Server 2 completes
9	Customer arrival + service at Server 2
10	Server 1 completes
12	Customer arrival + service at Server 1+ Server 2 completes

Fig. 8.4 Discrete-event simulation approach

8.3.1.3 The Three-Phase Simulation Approach

In the three-phase simulation approach, events are classified into two types:

- B (bound or booked) events. These are state changes that are scheduled to occur at a point in time. For instance, phone call arrivals in a call center model occur every 3 min. Once a call has been taken by an operator, it can be scheduled to finish 5 min later. This principle applies even when there is variability in the model, by predicting in advance how long a particular activity will take. In general, B-events relate to arrivals or the completion of an activity.
- C (conditional) events. These are state changes that are dependent on the conditions in the model. For instance, an operator can only start serving a customer if there is a customer waiting to be served and the operator is not busy. In general, C-events relate to the start of some activity.

8.3.1.4 The Continuous Simulation Approach

In a whole range of situations, operations are not subject to discrete changes in state, but the state of the system changes continuously through time. The most obvious of these is in operations involving the movement of fluids, for instance, chemical plants and oil refineries. In these systems, tanks of fluid are subject to continuously changing volumes. Systems that involve a high volume of fast-moving items may also be thought of as continuous. In these situations, the level of granularity with which the system is to be analyzed determines whether it is seen as discrete or continuous.

Digital computers cannot model continuous changes in state. Therefore, the continuous simulation approach approximates continuous change by taking small discrete time steps (Δt). The smaller the time step, the more accurate the approximation, but the slower the simulation runs, because of the more recalculations with the more simulated time units. This approach is clearly the same as the time-slicing method described above.

Some discrete-event simulation packages also have facilities for continuous simulation, while it is always possible to imitate a continuous simulation in a discrete-event package by including a regular event that mimics a time step (Δt). This is useful in circumstances where discrete and continuous changes need to be combined, for instance, process failures (discrete change) in a chemical plant (continuous change). In general, discrete-event simulation is more appropriate when a system needs to be modeled in detail, particularly when individual items need to be tracked through the system.

8.3.2 Proof of Concept

A proof of concept (POC) is a demonstration to verify that certain concepts or theories have the potential for real-world application. POC is therefore a prototype that is designed to determine feasibility, but does not represent deliverables. POC is also known as proof of principle (Techopedia 2015b).

According to (Andel and Yasinsac 2006), a proof of concept is possible only after thorough simulation and when testbeds may be available for added experimental information or data, omitting details or oversimplifying the model may lead to erroneous outcomes. There is no necessity to compare protocols against each other. A different understanding of the meaning of proof of concept means a basic simulation (Stojmenovic 2008). Such simulation is built for a very simple model and scenario that matches the model and assumptions needed to design the protocol, with the simple purpose of demonstrating that what is expected about its very basic performance is true. Simulations simply replace theoretical proofs of performance, as they are almost impossible to derive; thus, basic claims and expectations are confirmed.

It is further argued in (Stojmenovic 2008) that a rigorous proof of concept may produce a "lack of concept" when too much realism is introduced, or "defeat" because of introducing too much complexity in the search for realism. Alternatively, it may not be clear which concept is under simulation, or the claim made for a particular concept may end up being valid/invalid for another concept.

It is sometimes advisable, in the pursuit of simplicity, to limit analysis and advances to one parameter or variable at a time, test it on the simplest possible simulator, and understand it fully before making a next step that involves more parameters or variables. For instance, simulations done with realistic physical layers normally lead to investigating phenomena with many variables and puzzles, which leads to few explanations and insufficient hints for future progress. Therefore, simulations should proceed stepwise from the simplified unit disk graph (UDG) toward realistic physical layers (Stojmenovic 2008).

8.3.3   *Common Simulation Shortcomings*

Several simulation deficiencies impact the reliability of a simulation-based study. These defects, as detailed below, are grouped into four categories that match the simulation phases, namely, simulation setup, simulation execution, output analysis, and publishing (Kurkowski et al. 2005).

8.3.3.1   Simulation Setup

Setup begins with determining the simulation type, validating the model, validating the pseudorandom number generator (PRNG), defining variables, and developing scenarios. Though important, simulation setup is a phase of the MANET research job that is frequently disregarded, which blemishes simulation credibility.

8.3.3.1.1   Simulation Type

Identifying the type of simulation, terminating or steady state, is a commonly unnoticed step. As uncovered in the MobiHoc survey, 66 out of the 114 (57.9%) simulation papers did not state whether the simulation was terminating or steady state. It is believed that most simulations are steady state because MANET studies are typically interested in the long-term average behavior of an ad hoc network.

If the simulation type is not defined, unsound statistical results are an unpleasant consequence. Executing one type of simulation and reporting results on the other types of simulation is a common error. Precisely, executing a terminating simulation for a specified number of seconds and claiming the results represent the steady-state performance.

A MobiHoc paper identifying the simulation type used is presented in (Melodia et al. 2005).

8.3.3.1.2   Model Validation and Verification

The step that follows determining the type of simulation is to prepare the simulation model; a model must be validated as a baseline to start any experimentation. Several considerations are to be regarded in this context:

- A common error is to download the ns-2 simulator, compile it, and execute simulations with a model that has not been validated for the environment.
- Making changes to ns-2 without validating these modifications or enhancements.
- A protocol that is being evaluated must be verified to ensure that it has been coded correctly and is operating in accordance with the protocol specifications (Balci 1994).

- Missing to validate the model or verifying code, when software changes, is a usual flaw. Precisely, upgrading to a new compiler may implement a broadcast function in a protocol differently than earlier executions, which might impact protocol performance.

A MobiHoc paper discussing validation prior to evaluation is obtainable in (Zheng et al. 2003).

8.3.3.1.3 PRNG Validation and Verification

With the high computing power obtainable and the complexity of the ns-2 model, MANET researchers need to ensure that PRNG is appropriate for the study. Specifically, the ns-2 PRNG does not allow a separate request stream for each dimension (i.e., a unique request stream) that exists in a simulation study. A three-dimension instance is when a simulation has three different random parameters, such as jitter, noise, and delay. It is required that all three of these parameters (dimensions) be uniformly distributed with each other and within each stream (e.g., the jitter stream needs to be uniformly distributed). In (L'Ecuyer and Simard 2001; Pawlikowski et al. 2002), it is shown that a two-dimensional request on a PRNG is valid for approximately $8\sqrt[3]{L}$, where L is the cycle length. In ns-2, the cycle length is $2^{31} - 1$, which means that only around 10,000 numbers are available in a two-dimensional simulation study. Thus, Pawlikowski et al. (2002) estimate that the ns-2 PRNG is valid just for several thousand numbers before the potential nonuniformity of numbers or the cycling of numbers. This cycling time occurrence is obviously dependent on the number of PRNG calls made during a simulation. Notably, most network simulations spent as much as 50% of the CPU cycles generating random numbers.

On the other hand, testing of PRNG cycling shows that cycling impact is minimal because the repetition of numbers does not occur within the simulator in the exact same state as the previous time (Kurkowski et al. 2005). However, based on (Pawlikowski et al. 2002), the dimensionality of the numbers is likely to cause a correlation hitch. So, before publishing results, a researcher should validate the PRNG to ensure it did not cause any correlation in the results. If the cycle length is an issue with ns-2, Akaroa-2 (McNickle et al. 2010) offers an ns-2 compatible PRNG with a cycle of $2^{191} - 1$. It provides several orders of magnitude more numbers and is valid to 82 dimensions.

8.3.3.1.4 Variable Definition

Variable definition has to be considered in more than a direction:

- ns-2 uses hundreds of configurable variables during an execution so as to satisfy general wired and wireless network simulation requirements. Specifically, there are 538 variables defined in the *ns-default.tcl* file of ns-2.1b7a, and 674 variables defined in the *ns-default.tcl* file of ns-2.27. Such large number of variables makes it difficult to track each variable's default setting.

- The review of the *tcl* driver files, as well as the simulation instances provided by ns-2, shows that many simulation driver files leave key parameters undefined (Kurkowski et al. 2005). Typically, three out of 12 (25%) of the ns-2 MobiHoc simulations did not define the transmission range of a node (Kurkowski et al. 2005). If the transmission range default is changed from an ns-2 version to the next, the simulation outcome would be considerably different. A researcher should define all variables by using his own configuration file or *tcl* driver file (Perrone 2003).

8.3.3.1.5   Scenario Development

The conducted MobiHoc survey highlighted the importance and extent of developing a simulation scenario:

- A simulation scenario should involve the number of nodes, the size of the simulation area, and the transmission range of nodes. Just 48 of the 109 (44%) MANET protocol simulation papers provided all three of these input parameters, itemizing 61 simulation scenarios with a wide range of values. The number of nodes in these scenarios ranged from 10 nodes to 30,000 nodes, the simulation area varied from 25 meter × 25 meter to 5000 meter × 5000 meter, and the transmission ranges were from 3 meter to 1061 meter. The survey highlights the wide range of simulation scenarios used to conduct MANET research and the lack of uniformity in rigorous testing of MANET protocols.
- The derived parameters aggregate multiple input parameters to further characterize a scenario and provide a common basis for comparison across scenarios. For instance, for input parameters width (w) and height (h), several parameters may be derived, such as simulation area ($w{\times}h$), node density ($n/(w \times h)$ where n is the number of nodes, and maximum path ($\sqrt{(w^2 + h^2)}$).
- There is a lack of independence between parameters, such as node density ($n/(w \times h)$) and node coverage ($\pi \times r^2$), where r is the transmission range.
- If there were benchmark scenarios for small-, medium-, and large-sized simulations, then there would be three groupings of values for each simulation area.
- MANET research lacks consistent rigorous scenarios to validate and test solutions to the confronted issues.

8.3.3.2 Simulation Execution

Well-implemented simulation is essential to save execution time cost. Several execution defects that impact data output, analysis, and ultimately results are laid out in what follows.

8.3.3.2.1 Setting the PRNG Seed

In ns-2-based simulation studies, it is essential to set the seed of the PRNG properly for several causes:

- ns-2 uses a default seed of 12,345 for each simulation run; thus, if an ns-2 user does not set the seed, each simulation will produce identical results.
- If the seed is not set or is poorly set, it can revoke the independent replication method typically used in the analysis. Introducing correlation in the replications invalidates the common statistical analysis techniques and the results.

The MobiHoc survey (Kurkowski et al. 2005) reported that none of the 84 simulation papers addressed PRNG issues. The researcher should set the seed correctly in his *tcl* driver file and use the ns-2 *Random Class* for all random variables.

8.3.3.2.2 Scenario Initialization

Such pitfall usually occurs from a lack of understanding of the two types of simulation, terminating or steady state:

- In terminating simulations, the network is usually started in a certain configuration that represents the start of the simulation window. Specifically, if the researcher is trying to simulate a protocol's response to a failure event, he needs to have the failure as the initialization of his analysis.
- The simulation fills the caches, queues, and tables that were initially empty until a steady-state activity is reached. Determining and reaching the steady-state level of activity belongs to initialization. Data generated prior to reaching a steady state is biased by the initial conditions of the simulation and cannot be used in the analysis. Steady-state simulations require that the researcher address initialization bias (Schruben 1982). Typically, in protocols that maintain neighbor information, the size of the neighbor table should be monitored to determine when the table entries stabilize, because the protocol will perform differently with empty routing tables. Akaroa-2 (McNickle et al. 2010) monitors variables during execution to determine a steady state.

Only 8 of the 114 (7.0%) simulation papers in the MobiHoc survey addressed initialization bias, and all 8 used the unreliable method of arbitrarily deleting data.

More information on statistically sound methods of addressing initialization is available in a MobiHoc'2001 paper (Dyer and Boppana 2001).

8.3.3.2.3 Metric Collection

The metrics collected via simulation are of significance; there is no point in correctly running a simulation without obtaining the required data (Pawlikowski et al. 2002). Appropriate output is critical, especially, if it has to be categorized. Unambiguously, if the researcher is trying to track the delivery ratio for data packets and control packets, each type of packet must be identified along with the source and destination to determine the number of each type of packet sent and successfully received. Outputting only the number of packets sent and the number of packets received will not provide the granularity required in the measures.

In Lee and Kim (2000), a MobiHoc paper describes and defines the statistics used in calculating results.

8.3.3.3 Output Analysis

Output analysis is the weak point of many simulation studies. When the preceding steps take longer than planned, enough time is not provided for output analysis at the end of the schedule. Whether it is the publication deadline or the thesis defense date, the correct analysis is habitually conceded as detailed below.

8.3.3.3.1 Single Set of Data

Taking the first set of results from a simulation and accepting them as "correct" is a fault to be avoided. With a lone result, there is a high probability that the single point estimate is not descriptive of the population statistics. A single execution of a discrete-event simulation cannot account for the inherent randomness of the experiment; it may produce good results. However, the single-point estimate produced will not give enough confidence in the unknown population mean. The researcher has to determine the number of runs necessary to produce the confidence levels required for a trustworthy study (Law and Kelton 2000).

In the MobiHoc survey (Kurkowski et al. 2005), only 39 of the 109 (35.8%) MANET protocol simulation papers stated the number of simulation runs. A MobiHoc paper using multiple replications to achieve high confidence is given in (Hu and Johnson 2001), and a proper documentation of the number of replications used and how the quantity was chosen is presented in (Dyer and Boppana 2001).

8.3.3.3.2 Statistical Analysis

Failing to use the correct statistical formulas with the different forms of output is a common imperfection; for instance, using the standard formulas for mean and variance without ensuring the data is independent and identically distributed (IID). Use of IID-based formulas on correlated data produces biased results and hence compromises simulation reliability. To ensure IID and prevent correlated results, a researcher has to use batch means or independent replications of data (Goldsman and Tokol 2000).

From the survey in (Pawlikowski et al. 2002), the statistical methods used in the analysis are overlooked in 76.5% of the papers. A description of the analysis and data used for calculations is presented in (Sadagopan et al. 2003).

8.3.3.3.3 Confidence Intervals

This defect is a wrap-up of several of the previous analysis issues. Confidence intervals provide a range where the population mean is thought to be located relative to the point estimate (Brakmo and Peterson 1996). Confidence intervals account for the randomness and varied output of a stochastic simulation.

In Kurkowski et al. (2005), 98 of the 112 (87.5%) simulation papers using plots did not show confidence intervals on the plots. A MobiHoc paper that used confidence intervals is given in Zheng et al. (2003).

8.3.3.4 Publishing

From Kurkowski et al. (2005), imperfect publishing prevents researchers in the MANET community from benefiting in several ways:

- The lack of consistency in publishing simulation-based results directly impacts the trustworthiness of the studies and inhibits the direct comparison of results.
- A new researcher cannot repeat the studies to start his follow-on work. As previously stated, in the *ns-default.tcl* file of ns-2.27, there are 674 defined variables. To ensure repeatability, a researcher must document the *ns-default.tcl* file used and any changes made to the variable settings in the file. Also, it is prime to state if the code is available and how to obtain it (Perrone 2003). There should be a code statement even if the code's release is restricted by copyright or third-party ownership.
- The lack of labels and units can cause readers of the papers to misinterpret or misunderstand the results:
  - Plots of simulation results are common, i.e., 112 of the 114 (98.2%) simulation papers used plots to describe results.
  - However, 12 of the 112 (10.7%) simulation papers with plots did not provide legends or labels on their charts.

- Additionally, 28 of the 112 (25.0%) simulation papers with plots did not provide units for the data being shown.

• Overlooking documentation and referencing of the parameters set to execute the simulation hamper repeatability and comparisons. Example missed data are statistically detected:

  - Noticeably, 47 of the 109 (43.1%) MANET protocol simulation papers did not state the transmission range of the nodes.
  - Also, 78 of the 109 (71.6%) MANET protocol simulation papers did not mention the packet traffic type used in the simulation.

• Overlooking charts discussion in the text or the text failing to reference charts as supportive leads to ambiguity and mistrust in the published work as a whole.

As exhaustively elucidated throughout Sect. 8.3, trustworthy studies can be barred due to unrepeatable, biased, non-rigorous, and non-statistically sound simulations resulting from deficient simulation setup, faulty execution, erroneous output analysis, and unsound documentation describing the work.

8.3.4 Unreliable Simulation Revealed

Lack of credibility is an inevitable outcome due to imperfect and inaccurate simulation setup, simulation execution, output analysis, and publishing. It is unveiled in Kurkowski et al. (2005) that:

• Less than 15% of the published MobiHoc papers are repeatable.
• It is impractical to repeat a simulation study when the version of a publicly available simulator is unknown. Only seven of the 58 (12.1%) MobiHoc simulation papers that use a public simulator mention the simulator version used.
• It is unthinkable to repeat a simulation study when the simulator is self-developed and the code is unavailable.
• Only eight of the 114 (7.0%) simulation papers addressed initialization bias, and none of the 84 simulation papers tackled random number generator issues. Thus, over 90% of the MobiHoc published simulation results may include bias.
• With regard to compromising statistical soundness, 70 of the 109 (64.2%) MANET protocol simulation papers did not identify the number of simulation iterations used, and 98 of the 112 (87.5%) papers that used plots to present simulation results did not include confidence intervals. Hence, only approximately 12% of the MobiHoc simulation results appear to be based on sound statistical techniques.

As obviously disclosed throughout Sect. 8.3, making a simulation credible is a serious task. Nevertheless, based on incomplete literature reviews and true protocol comparisons, several simulation practices adopt an "I am the best" approach, evading the mentioning of maybe better existing solutions. Moreover, it becomes a harder task to criticize or evaluate such types of work due to poor documentation such as incomplete algorithmic descriptions, vague pseudo-code, and unclear concise descriptions of new ideas and meaningful case studies. Additionally, comparing a proposed solution to some existing ones that are clearly inferior leads nowhere. Even a comparison with a better protocol using different assumptions or metrics gives an unfair advantage. For instance, many papers allege being superior to ad hoc on-demand distance vector routing (AODV) and dynamic source routing (DSR) routing in terms of delay, power consumption, or other metrics. However, AODV and DSR route discovery protocols mainly use hop count as a metric without addressing congestion or power consumption issues (Stojmenovic 2008).

8.3.5   The Price of Simulation

Deciding to simulate may be a necessary choice; it is certainly demanding and definitely challenging, but it is worth perseverance and being up to the price that is not limited to money only. Simulation hurdles are evoked to be (Robinson 2004):

- Expensive. Simulation software is not necessarily low-priced and the cost of model development and use may be considerable, particularly if consultants have to be employed.
- Time-consuming. Simulation is a time-consuming approach. This adds to the cost of its use and signifies that the benefits are not immediate.
- Data hungry. Most simulation models require a significant amount of data. This data is not always immediately available and, where it is, much analysis may be required to put it in a form suitable for the simulation.
- Expertise demanding. Simulation is more than the development of a computer program or the use of a software package. It requires, among other things, skills in conceptual modeling, validation, and statistics, as well as skills in working with people and project management. This expertise is not always readily available.
- Overconfidence insinuative. There is a risk that whatever is produced on a computer is considered right. Through simulation, this is further intensified with the use of an animated display that gives a maybe deceitful appearance of reality. When interpreting the results from a simulation, consideration must be given to the validity of the underlying model and the assumptions and simplifications that have been made.

8.4   Simulators and Emulators

Simulators are knowledge-rich, in networks, hardware, software, and mathematics; studying a simulator urges understanding the underneath foundations in such a multiplicity of areas. A simulator, thus, is not meant to be a black box that produces output for some input. In the quest to master a simulator, this section provides an in-depth layout of its building blocks with a care as to how it is efficient as compared to peer simulators.

Diverse network simulators and emulators are presented in this section. Some are for general networking; others are dedicated to wireless networking with or without a focus on WSNs. Simulators also differ in their time approach, whether it is discrete-event-driven or continuous; furthermore, they are built in a diversity of languages that are different in their techniques, such as the support of object-oriented programming, which necessarily impacts their scalability, or the support of component-based models. Open-source simulators extend their use and lifetime by providing researchers with the ability to add extra insights and scenarios into the main framework. On the contrary, emulators are hardware-dependent, and which one to use is a straightforward choice. Notably, some simulators ceased to exist due to lack of development and support or due to advances in networking that went beyond their intended design goals.

With the wide diversity of simulators, it becomes such a hard task to make a choice. This section provides a thorough analysis of the most widely known simulators, with a focus on their intended use, be it for general networking or for WSNs, and on how the underlying theory and model implementation languages and techniques are manageable. Could someone buy any car? How to make the decision? Which comes first, the brand name or cost? Which engine technology? Manual or automatic? It is a complex procedure that should involve an answer to each question and a priority to which answer should come first. In general, few go for the unknown, a No for taking risks. Brands usually come first. They are trustworthy by history and production. A lower ranked "brand" though may be more efficient for a typical purpose. Then comes the intended use, whether it is for urban driving or for highways, and so on till reaching a narrow selection margin. Nonetheless, the differences between the same classes of cars remain minimal. Simulators are cars. Are the widely used simulators convenient for the task? If none of them is, what are the alternatives? Is it a right decision to start by looking at less-known simulators? Would free license simulators be efficient? Then, how easy are they to learn and use? Nonetheless, simulators are intelligent tools that require up to the level users. This section provides an analysis that makes the tricky decision easier.

Without being exhaustive, this section lays out the simulator and emulator frameworks, chronologically as possible, without sidestepping those deserted, aiming so at introducing the goals and techniques beyond a simulator and making a newly born simulator more familiar and linked to roots in theory and practice. A wide variety of simulators and emulators are presented. A lot more are available in the literature, among many (Egea-López et al. 2005; Singh et al. 2008; Rahman et al. 2009; Imran et al. 2010; Musznicki and Zwierzykowski 2012).

For collective picturing, Table 8.1 assembles, zooms in, and compares the simulators and emulators as a recapitulation of the details exposed in the coming subsections.

8.4.1   The Network Simulator (ns-2)

ns-2 is a discrete-event simulator targeted at networking research. ns-2 provides substantial support for the simulation of TCP, routing, and multicast protocols over wired and wireless (local and satellite) networks (ISI 2011). ns-2 began as a variant of the REAL network simulator (Keshav 1997) in 1989 and has evolved substantially since then. In 1995 ns-2 development was supported by the Defense Advanced Research Projects Agency (DARPA 2015), through the VINT project (Helmy and Kumar 1997) at Lawrence Berkeley National Laboratory (LBNL 2009), in collaboration with Xerox PARC (Xerox PARC 2015), the University of California Berkeley (UCB 2015), and the University of Southern California/Information Sciences Institute (USC/ISI 2015). Later, ns-2 development was supported through DARPA with SAMAN (Lan 2001) and through NSF with CONSER (Chen 2002), both in collaboration with other researchers including ACIRI (The ICSI Networking and Security Group 2015). ns-2 has always included significant contributions from other researchers including wireless code from the Daedalus group at UCB and Monarch projects at Carnegie Mellon University (Johnson 1996) and Sun Microsystems (Oracle 2015).

ns-2 is a discrete-event simulator targeted at networking research. ns-2 provides extensive support for the simulation of wired and wireless networks. It considers TCP, UDP, etc., at the transport layer level, and unicast, multicast, and a multiplicity of other routing protocols at the network layer, as well as traffic sources such as CBR, FTP, HTTP, telnet, etc. (ISI 2011).

ns-2 is object-oriented, and it uses a Tcl/Otcl (Tool command language/Object oriented Tcl) (Heidemann et al. 2015) as a command and configuration interface. Four types of files are related to ns-2:

- Models are described in .tcl or .ns files, which have some common commands without being fully compatible.
- Simulation trace files .tr are created during the session.
- Network Animator (Nam) is a Tcl/TK-based animation tool for viewing network simulation traces and real-world packet traces. Nam supports topology layout, packet-level animation, and various data inspection tools (Buchheim 2002). It is mainly intended as a companion animator to the ns-2 simulator. .nam files are created to visualize the behavior of the network protocols as well as the traffic of the model. Once created, users can operate the .nam file like a media player and repeatedly check the model behavior.

ns-2 is widely used due to the many facilities it provides:

- It is configurable and permits simulation using two languages, OTcl and C++. C++ is used for implementing protocols and extending the ns-2 library. OTcl is used to create and control the simulation environment itself, including the selection of output data. Simulation is run at the packet level, allowing for detailed results.
- Its open-source and modular approach has effectively made it extensible. The object-oriented design of ns-2 allows for the straightforward creation and use of new protocols. The combination of easiness in protocol development and ns-2's popularity has ensured that a high number of different protocols are publicly available, despite not being included as part of the initial simulator's release. Its status as the most used sensor network simulator has also encouraged further popularity, as developers would prefer to compare their work to results from the same simulator.
- Dynamic behavior can be traced using Nam.

The many benefits of ns-2 are attainable, but not with full ease. ns-2 has its own roadblocks that should be located for smooth experimentation:

- It is not highly scalable, which means that large-scale networks may not be fully simulated. Particularly, ns-2 does not scale well for sensor networks; this is partly due to its object-oriented design. While this is beneficial in terms of extensibility and organization, it is a holdback on performance in environments with large numbers of nodes. Every node is its own object and can interact with every other node in the simulation, creating a large number of dependencies to be checked at every simulation interval, leading to an n^2 complexity.
- It is not customizable for WSNs. Packet formats, energy models, MAC protocols, and sensing hardware models all differ from those found in most sensors. Moreover, there is no direct support for mobility and sharing wireless radio channels.
- It does not provide an application model. In many network environments, this may not be a serious setback, but sensor networks often involve interactions between the application level and the network protocol level.
- The APIs are not all complete.
- Real-time simulation is not supported.
- A non-short time period is required for acquaintance and perfection, besides a non-clear source code documentation that could make the get-to-use task harder.

Version 35 of ns-2 was released on November 4, 2011 (Table 8.1).
Simulators built on the ns-2 environment are presented in Sects. 8.4.10 and 8.4.11.

8.4.2   The Network Simulator (ns-3)

The ns-3 consortium is a collection of organizations cooperating to support and develop the ns-3 software. The consortium is governed by an agreement established between the founding members, INRIA and the University of Washington (ns-3 Consortium 2015). Development of ns-3 began in July 2006, it is written from scratch using the C++ programming language. The first release, ns-3.1 was made in June 2008, and afterwards the project continued making quarterly software releases and more recently has moved to three releases per year. ns-3 made its twenty-first release (ns-3.21) in September 2014.

ns-3 is a discrete-event network simulator targeting research and educational use (ns-3 Consortium 2015). ns-3 is a free licensed software that is publicly available for research, development, and use. The goal of the ns-3 project is to develop a preferred, open simulation environment for networking research; it is conforming to the simulation needs of modern networking research and encourages community contribution, peer review, and validation of the software.

The ns-3 software infrastructure encourages the development of simulation models which are sufficiently realistic to allow its use as a real-time network emulator, interconnected with the real world, and which allows many existing real-world protocol implementations to be reused within ns-3. The ns-3 simulation core supports research on both IP and non-IP-based networks. The large majority of its users focus on wireless/IP simulations, which involve models for Wi-Fi, WiMAX, or LTE for layers 1 and 2 and a variety of static or dynamic routing protocols such as OLSR and AODV for IP-based applications.

ns-3 also supports a real-time scheduler that facilitates a number of "simulation-in-the-loop" use cases for interacting with real systems. For instance, users can emit and receive ns-3-generated packets on real network devices, and ns-3 can serve as an interconnection framework to add link effects between virtual machines. Another emphasis of the simulator is on the reuse of real applications and kernel code. Frameworks for running unmodified applications or the entire Linux kernel-networking stack within ns-3 are presently being tested and evaluated.

When approaching ns-3, it is worth noting that:

- ns-3 is open-source, and the project strives to maintain an open environment for researchers to contribute and share their software.
- ns-3 is not a backward-compatible extension of ns-2; it is a new simulator. The two simulators are both written in C++ but ns-3 is a new simulator that does not support the ns-2 APIs. Some models from ns-2 have already been ported from ns-2 to ns-3. The project continues to maintain ns-2 while ns-3 is being built, and will study transition and integration mechanisms.
- ns-3 is lacking the support for protocols, like WSN and MANET, which were supported in ns-2.

Many simulation tools exist for network simulation studies; ns-3 though has several distinguishing features in contrast to other tools:

- ns-3 is designed as a set of libraries that can be combined together and also with other external software libraries. While some simulation platforms provide users with a single, integrated graphical user interface environment in which all tasks are carried out, ns-3 is more modular in this regard. Several external animators and data analysis and visualization tools can be used with ns-3. However, users should expect to work at the command line and with C++ and/or Python software development tools.
- ns-3 is supported on the following primary platforms:

  - Linux x86 and x86_64: gcc versions 4.2 through 4.8.
  - FreeBSD x86 and x86_64: clang version 3.3, gcc version 4.2.
  - Mac OS X Intel: clang-500.2.79, based on LLVM 3.3svn (OS X Mavericks and Xcode 5.0.1), and gcc-4.2 (available with Xcode version 4 or earlier).

- The following platforms are lightly supported:

  - Windows Visual Studio 2012.
  - Windows Cygwin 1.7.

8.4.3   GloMoSim

Global mobile information system simulator (GloMoSim) is a scalable simulation environment that was intended for large wireless and wired communication networks (Takai et al. 1999); however, only wireless networks were considered. Under funding from DARPA (DARPA 2015), a scalable simulation facility has been developed with the objective of simulating networks with up to a hundred thousand nodes linked by a heterogeneous communications capability that includes multicast, asymmetric communications using direct satellite broadcasts, multihop wireless communications using ad hoc networking, and traditional Internet protocols. The scalability of the simulator to very large networks is achieved primarily by exploiting parallelism on state-of-the-art parallel computers. Moreover, parallel model execution achieves dramatic reductions in execution times. A detailed simulation of a large wireless network with 10,000 mobile radios has been implemented. Using parallel execution, it was possible to reduce the execution time sufficiently such that a model with 10,000 wireless nodes could be simulated on a 6-processor symmetric multiprocessor in less time than a network with half as many nodes using purely sequential execution.

8.4.3.1   PARSEC

GloMoSim uses a parallel discrete-event simulation capability provided by PARSEC. PARSEC (for PARallel Simulation Environment for Complex Systems) is a C-based simulation language developed by the Parallel Computing Laboratory at UCLA for the sequential and parallel execution of discrete-event simulation models

(Bagrodia et al. 1998). Moreover, PARSEC can be used as a parallel programming language with the capability of running on several platforms, including UNIX variants and Windows.

PARSEC adopts the process interaction approach to discrete-event simulation. An object, also referred to as a physical process, or set of objects in the physical system, is represented by a logical process. Interactions among physical processes (events) are modeled by time-stamped message exchanges among the corresponding logical processes. One of the important features of PARSEC is its ability to execute a discrete-event simulation model using several different asynchronous parallel simulation protocols on a variety of parallel architectures.

PARSEC is designed to neatly separate the description of a simulation model from the underlying simulation protocol, sequential or parallel, used to execute it. Thus, with few modifications, a PARSEC program may be executed using the traditional sequential (Global Event List) simulation protocol or one of many parallel optimistic or conservative protocols. In addition, PARSEC provides powerful message-receiving constructs that result in shorter and more natural simulation programs.

8.4.3.2 Visualization Tool

GloMoSim has a Visualization Tool that is platform-independent because it is coded in Java (Nuevo 2004). To initialize the Visualization Tool, Java GloMoMain is to be executed from the Java GUI directory. This tool allows to debug and verify models and scenarios; stop, resume, and step execution, show packet transmissions, show mobility groups in different colors, and show statistics.

8.4.3.3 GloMoSim Library

GloMoSim is a scalable simulation library for wireless network systems built using the PARSEC simulation environment (Bagrodia et al. 1998). The protocol stack includes models for the channel, radio, MAC, network, transport, and higher layers. It also supports TCP, IEEE 802.11 CSMA/CA, MAC, UDP, HTTP, FTP, CBR, Telnet, AODV, etc. GloMoSim also supports two different node mobility models. Nodes can move according to the "random waypoint" model (Johnson and Maltz 1996) and the "random drunken" model (Jardosh et al. 2003). In the random waypoint model, a node chooses a random destination within the simulated terrain and moves to that location based on the speed specified in the configuration file. After reaching its destination, the node pauses for a duration that is also specified in the configuration file. In the random drunken model, a node periodically moves to a position chosen randomly from its immediate neighboring positions. The frequency of the change in node position is based on a parameter specified in the configuration file.

As most network systems adopt a layered architecture, GloMoSim is designed using a layered approach similar to the OSI seven-layer network architecture. Simple APIs are defined between different simulation layers to allow the rapid integration of models developed at different layers by different researchers. Actual operational code can also be easily integrated into GloMoSim with this layered design, which is ideal for a simulation model as it has already been validated in real life and no abstraction is introduced. For instance, a TCP model was implemented in GloMoSim by extracting actual code from the FreeBSD operating system. This also reduces the amount of coding required to develop the model (Takai et al. 1999).

8.4.3.4 Aggregation

In contrast to network simulators such as OPNET and ns-2, GloMoSim has been designed and built with the primary goal of simulating very large network models that can scale up to a million nodes using parallel simulation to significantly reduce execution times of the simulation model. It is open-source and uses an object-oriented approach, like ns-2, but to avoid the resulting limitation on scalability, GloMoSim partitions the nodes such that each object is responsible for running one layer in the protocol stack of every node in its given partition, which helps in reducing the overhead of large networks. The coming subsections explore the techniques of node and layer aggregation that are used to achieve this scalability (Takai et al. 1999).

8.4.3.4.1 Node Aggregation

The node aggregation technique is introduced into GloMoSim to give significant benefits to the simulation performance. Initializing each node as a separate entity inherently limits the scalability because the memory requirements increase dramatically for a model with a large number of nodes. With node aggregation, a single entity can simulate several network nodes in the system. Node aggregation technique implies that the number of nodes in the system can be increased while maintaining the same number of entities in the simulation. In GloMoSim, each entity represents a geographical area of the simulation. Hence the network nodes, which a particular entity represents, are determined by the physical position of the nodes.

8.4.3.4.2 Layer Aggregation

For ease of implementation, the various GloMoSim layers are integrated into a single entity. Each entity encompasses all the layers of a simulation. Every layer is implemented as three function calls by the protocol modeler:

- The researcher (developer) has to provide an initialization function that will be called for each layer on every node at the beginning of the simulation.
- The next function call provided by the researcher is automatically invoked when a particular layer of a particular node receives an incoming packet/event. Based on the contents of the message, the appropriate instructions will be executed. Function calls are also provided for a layer to send messages to its lower or upper layer in the simulation.
- At the end of the simulation, another researcher-provided function call is invoked. This can be used to collect any relevant statistics for that layer.

Mostly, the researcher writes pure C code. The presence of the PARSEC runtime and interactions with the runtime are completely hidden from the user.

Actually, GloMoSim is limited to IP networks because of the low-level design assumptions, which makes it similar to ns-2 (Sect. 8.4.1) in its limitations with regard to packet formats, lack of energy models, and the MAC protocols that are not representative of WSNs. Additionally, GloMoSim does not support phenomena occurring outside of the simulation environment. All events must be generated from a node within the network.

GloMoSim stopped releasing updates in 2000 and is limited to some educational institutions; it is now updated as a commercial product called QualNet (SCALABLE Network Technologies 2014).

8.4.4 OPNET

OPNET (Optimized Network Engineering Tool) was launched in 1987 as the first commercially available simulation tool for communication networks; it provides a comprehensive development environment for the specification, simulation, and performance analysis of communication networks[1] (Riverbed Technology 2015). A large range of communication systems from a single LAN to global satellite networks can be supported. Discrete-event simulations are used as the means of analyzing system performance and their behavior.

For maximum effectiveness, a simulation environment should be modular and hierarchical, and take advantage of the graphical capabilities of today's workstations. OPNET is an object-oriented simulation environment that meets all these requirements and is a powerful general-purpose network simulator. OPNET's comprehensive analysis tool is suitable for interpreting and synthesizing output data. A discrete-event simulation of the call and routing signaling was developed using a number of OPNET's features such as the dynamic allocation of processes to model

---

[1] On October 29, 2012, Riverbed acquired OPNET Technologies to build on Riverbed's strong heritage and experience in delivering solutions that improve the performance of technology for business. OPNET Technologies has built its success on application performance management (APM) and is recognized by a leading analyst firm as a leader in the magic quadrant for APM.

virtual circuits transiting through an ATM switch. Moreover, its built-in protoc language support[2] (Ubuntu 2010) provides it with the ability to realize almost any function and protocol (Chang 1999).

OPNET is extensively used for the study of TCP transport across different types of ATM bearer capabilities and DiffServ per-hop behavior. Moreover, it supports routing protocols such as OSPF, RIPEIGRP, BGP, IGRP, DSR, TORA, and PNNI. Also included were MAC, node mobility, ad hoc connectivity, different application models, node failure models, and modeling of power consumption.

For detailed elaboration, the features included in OPNET are as laid out below:

- Modeling and simulation phases. OPNET provides powerful tools to assist users with the building of models, the execution of a simulation, and the analysis of the output data.
- Hierarchical modeling. OPNET employs a hierarchical structure for modeling. Each level of the hierarchy describes different aspects of the complete model being simulated.
- Tailored for communication networks. Detailed library models provide support for several protocols and allow researchers and developers to either modify these existing models or develop new models of their own.
- Automatic simulation generation. OPNET models can be compiled into executable code. An executable discrete-event simulation can be debugged or simply executed, resulting in output data.

OPNET supports C and Java languages. As a commercial package, it has several advantages, such as extensibility, large customer base, provision for professional support, extensive documentation, and inclusion of a large number of built-in protocols. It is though costly and requires considerable time to learn and use. OPNET suffers from the same object-oriented scalability problems as ns-2. Notably, it is not as popular as ns-2 or GloMoSim, at least in research being made publicly available, and thus does not have the high number of protocols available to those simulators.

OPNET IT Guru is an academic, one license per user, free edition that provides pre-built models of protocols and devices. It allows the creation and simulation of different network topologies. Yet, since the set of protocols and devices is fixed, it is not possible to create new protocols nor modify the behavior of existing ones. However common parameter values can be modified. For example, IT Guru includes a model of a wireless LAN device (a laptop using IEEE 802.11b). It is conceivable to create a topology with multiple wireless LAN devices, set parameters such as RTS threshold, packet arrival rate, and data rate, and simulate to measure the network performance. The behavior of IEEE 802.11 DCF cannot be changed, nor can the simulation of IEE 802.11g, n, or e. The academic edition of IT Guru is limited to simulating 50 million events (typical events are receiving a packet, timeout occurring). For example, a wireless LAN network with 10 devices all generating a high

---

[2] protoc is a compiler for protocol buffers definitions files. It can generate C++, Java and Python source code for the classes defined in PROTO_FILE.

load will reach 50 million events in about 5 min of simulation time. The academic edition also limits the number of devices in particular topologies.

The commercial OPNET Modeler provides not only the same ability to create and simulate network topologies as IT Guru (without the limitations of the academic edition) but also access to the models of protocols and devices, that is, the possibility of editing the source code of the IEEE 802.11 DCF model to experiment with variations of the access scheme, including the user-contributed models, Modeler allows for analysis of the latest network protocols and algorithms in use and being researched today. Unlike ns-2 and GloMoSim, OPNET supports modeling different sensor-specific hardware, such as physical-link transceivers and antennas. It can also be used to define custom packet formats.

8.4.4.1 Hierarchical Modeling

OPNET provides four tools called editors to develop a representation of a system being modeled. These editors are Network (upper layer), Node, Process, and Parameter (lower layer); they are organized in a hierarchical fashion to support the concept of model-level reuse (Chang 1999). Models developed at one layer can be used by another model at a higher layer. The Parameter Editor is a utility editor, and is not considered a modeling domain.

8.4.4.1.1 Network Model

Network Editor is used to specify the physical topology of a communications network; it defines the position and interconnection of communicating entities, i.e., nodes and links. The specific capabilities of each node are realized in the Node Editor. A set of parameters or characteristics is attached to each model to customize the node's behavior. A node can be fixed, mobile, or satellite. Simplex (unidirectional) or duplex (bidirectional) point-to-point links connect pairs of nodes. A bus link provides a broadcast medium for an arbitrary number of attached devices. Mobile communication is supported by radio links. Links can also be customized to simulate the actual communication channels.

The complexity of a network model is unmanageable when numerous networks are modeled as part of a single system. Complexity is alleviated by an abstraction known as a subnetwork. A subnetwork may contain many subnetworks. At the lowest level, a subnetwork is composed only of nodes and links. Communication links facilitate communication between subnetworks.

8.4.4.1.2   Node Model

Communication devices created and interconnected at the network level need to be specified in the node domain using the Node Editor. Node models are expressed as interconnected modules; they are grouped into two distinct categories:

- Modules that have predefined characteristics and a set of built-in parameters. Examples are packet generators, point-to-point transmitters, and radio receivers.
- Highly programmable modules referred to as processors and queues; they rely on process model specifications.

8.4.4.1.3   Process Model

Process models, created using the Process Editor, are used to describe the logic flow and behavior of processor and queue modules. Communication between processes is supported by interrupts. Process models are expressed in a language called Proto-C, which consists of state transition diagrams (STDs), a library of kernel procedures, and the standard C programming language.

8.4.4.2   Data Generation

8.4.4.2.1   Probe Editor

Most OPNET models contain objects that are capable of generating vast amounts of output data during simulations. The sources of output data include predefined and user-defined statistics, automatic animation, and custom-programmed animation. Users can use Probe Editor to specify which data to collect. A probe is defined for each source of data that the user wishes to enable.

8.4.4.2.2   Analysis Tool

As previously stated, simulations can be used to generate a number of different forms of output. These forms include several types of numerical data, animation, and detailed traces provided by the OPNET debugger. Moreover, since OPNET simulations support open interfaces to the C language and the host computer's operating system, simulation developers may generate proprietary forms of output ranging from messages printed in the console window, to generation of ASCII or binary files, and to live interactions with other programs.

The service provided by the analysis tool is to display information in the form of graphs. Graphs are presented within rectangular areas called analysis panels. An analysis panel consists of a plotting area with two numbered axes, horizontal and vertical. The plotting area can contain one or more graphs describing relationships between variables mapped to the two axes. The analysis tool can generate error rates and throughputs, as well as delay queue sizes. Packet tracing may be done. Output can be plotted in a graph, such as end-to-end delay vs. queue buffer capacity or loss ratio vs. queue buffer capacity. Probability distribution function, cumulative distribution function, as well as a histogram can be plotted for several data sets.

8.4.4.2.3   Filter Tool

Numeric filters may also operate on the data presented in the Analysis Tool. These are constructed from a predefined set of filter elements in the Filter Editor. Filter models are represented as block diagrams consisting of interconnected filter elements. Filter elements may be either built-in numeric processing elements or references to other filter models.

8.4.5   OMNeT++

OMNeT++ (Objective Modular Network Testbed in C++) simulation environment has been made publicly available since 1997; it is an object-oriented C++ based simulator for modeling communication networks, multiprocessors, and other distributed or parallel systems (Mallanda et al. 2005; Varga and Hornig 2008; OpenSim 2015c). OMNeT++ is a public source and can be used under the Academic Free License (Open Source Initiative 2015), which makes the software free for nonprofit use. The motivation for developing OMNeT++ was to produce a powerful opensource discrete event simulation tool that can be used by academic, educational, and research-oriented commercial institutions for the simulation of computer networks and distributed or parallel systems. OMNeT++ attempts to fill the gap between opensource, research-oriented simulation software such as ns-2 (Sect. 8.4.1) and expensive commercial alternatives like OPNET (Sect. 8.4.4). OMNeT++ is available on all common platforms, including Linux, Mac OS/X, and Windows, using the GCC[3] tool chain (GCC Team 2015) or the Microsoft Visual C++ compiler.

OMNeT++ represents a framework approach. Instead of directly providing simulation components for computer networks, queuing networks, or other domains, it provides the basic machinery and tools to write such simulations. For many areas of application, ready-to-use components already exist. For the simulation of TCP/IP networks, the INET framework contains modules for protocols like UDP, TCP, IP, IPv6, ARP, and Ethernet. Among many published applications, a work extends the

---

[3]The GNU Compiler Collection includes front ends for C, C++, Objective-C, Fortran, Java, Ada, and Go, as well as libraries for these languages (libstdc++, libgcj, . . .). GCC was originally written as the compiler for the GNU operating system. The GNU system was developed to be 100% free software, free in the sense that it respects the user's freedom.

OMNeT++ INET framework for simulating real-time Ethernet with high accuracy (Steinbach et al. 2011).

There are several INET-based model frameworks maintained by independent research groups (OpenSim 2015c):

- OverSim is an open-source overlay and peer-to-peer network simulation framework for the OMNeT++ simulation environment (KIT/TeleMatics 2010). The simulator contains several models for structured (such as Chord, Kademlia, Pastry) and unstructured (like GIA) P2P systems and overlay protocols.
- Veins is an open-source inter-vehicular communication (IVC) simulation framework composed of an event-based network simulator and a road traffic micro-simulation model (Sommer 2015).
- Others are ReaSE, HIPSim++, INET-HNRL, EPON, mCoA++, SimuLTE, TTE4INET, EBitSim, and Quagga.

Since its first release, simulation models have been developed by various individuals and research groups for several areas including wireless and ad hoc networks, sensor networks, storage area networks (SANs), optical networks, queuing networks, file systems, high-speed interconnections (InfiniBand), etc. Some of the simulation models are parts of real-life protocol implementations like the Quagga Linux routing daemon (Lamparter and Troxel 2015) or the BSD TCP/IP stack (Wikipedia 2015); others have been written directly for OMNeT++. A study on the accuracy of OMNeT++ in the WSN domain is presented in (Colesanti et al. 2007).

In addition to university research groups and nonprofit research institutions, companies like IBM, Intel, Cisco, Thales, and Broadcom are also using OMNeT++ successfully in commercial projects or for in-house research.

8.4.5.1 The Design of OMNeT++

OMNeT++ was primarily designed to support network simulation on a large scale. This objective leads to the following main design requirements:

- Enabling large-scale simulation. Simulation models need to be hierarchical and built from reusable components as much as possible.
- The simulation software should facilitate visualizing and debugging of simulation models in order to reduce debugging time that usually consumes a large portion of simulation projects.
- The simulation software should be modular and customizable and should allow embedding simulations in larger applications such as network planning software.
- Data interfaces should be open. It should be possible to generate and process input and output files with commonly available software tools.
- Providing an integrated development environment that largely facilitates model development and results analysis.

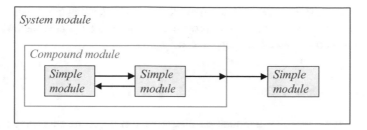

Fig. 8.5 System, compound, simple modules, and gates

8.4.5.1.1 Model Structure

An OMNeT++ model consists of modules that communicate by message passing. The active modules are termed simple modules; they are written in C++, using the simulation class library. Simple modules can be grouped into compound modules and so forth; the number of hierarchy levels is not limited.

Both simple and compound modules are instances of module types. While describing the model, the user defines module types; instances of these module types serve as components for more complex module types. Modules are implemented as C++ objects using support functions from the simulator library. The topology of module connections is specified using NED. Finally, the user creates the system module as a network module, which is a special compound module type without gates to the external world.

Modules communicate with messages that in addition to usual attributes such as timestamp may contain arbitrary data. Simple modules typically send messages via gates, but it is also possible to send them directly to their destination modules. Gates are the input and output interfaces of modules (Fig. 8.5). Modules can have parameters. Parameters are mainly used to pass configuration data to simple modules and to help define model topology. Parameters may take string, numeric, or Boolean values.

8.4.5.1.2 The NED Language

The user defines the structure of the model, i.e., the modules and their interconnection, in OMNeT++'s topology description language, NEtwork Description (NED) (OpenSim 2015). Typical ingredients of a NED description are simple module declarations, compound module definitions, and network definitions. Simple module declarations describe the interface of the module, gates, and parameters. Compound module definitions consist of the declaration of the module's external interface (gates and parameters) and the definition of submodules and their interconnection. Network definitions are compound modules that qualify as self-contained simulation models.

The NED language has been designed to scale well. However, recent growth in the amount and complexity of OMNeT++-based simulation models and model frameworks made it necessary to improve the NED language as well. Thus, in addition to a number of smaller improvements, the major features included in the coming subsections have been introduced.

8.4.5.1.3   Graphical Editor

The OMNeT++ package includes an integrated development environment, which contains a graphical editor using NED as its native file format; moreover, the editor can work with arbitrary, even hand-written NED code. The editor is a fully two-way tool, i.e., the user can edit the network topology either graphically or in the NED source view, and switch between the two views at any time. This is made possible by design decisions about the NED language itself. NED is a declarative language, and as such, it does not use an imperative programming language for defining the internal structure of a compound module. Declarative constructs, resembling loops and conditionals in imperative languages, enable parametric topologies; it is possible to create common regular topologies such as rings, grids, stars, trees, hypercubes, or random interconnections whose parameters (size, etc.) are passed in numeric-valued parameters. With parametric topologies, NED holds an advantage in many simulation scenarios both over OPNET, where only fixed model topologies can be designed, and over ns-2, where building model topology is programmed in Tcl and is often intermixed with simulation logic, so it is generally impossible to write graphical editors which could work with existing hand-written code.

8.4.5.1.4   Separation of Model and Experiments

It is a good practice to separate the different aspects of a simulation as much as possible. Model behavior is captured in C++ files as code, while model topology and the parameters defining the topology are defined by the NED files. This approach allows the user to keep the different aspects of the model in different places that in turn allows having a clearer model and better tooling support.

8.4.5.1.5   Simple Module Programming Model

Simple modules are the active elements in a model. They are atomic elements in the module hierarchy; they cannot be divided any further. Simple modules are programmed in C++, using the OMNeT++ simulation class library. OMNeT++ provides an integrated C++ development environment, so it is possible to write, run, and debug the code without leaving the OMNeT++ integrated development environment (IDE) (Techopedia 2015a).

The simulation kernel does not distinguish between messages and events; events are also represented as messages. Message sending and receiving are the most frequent tasks in simple modules. Messages can be sent either via output gates or directly to another module.

Modifying the topology of the network can be done dynamically. One can create and delete modules and rearrange connections while the simulation is executing. Even compound modules with parametric internal topology can be created on the fly.

8.4.5.1.6 Design of the Simulation Library

OMNeT++ provides a rich object library for simple module implementers. There are several distinguishing factors between this library and other general-purpose or simulation libraries. The OMNeT++ class library provides reflection functionality, which makes it possible to implement high-level debugging and tracing capability, as well as automatic animation on top of it. Memory leaks, pointer aliasing, and other memory allocation problems are common in C++ programs not written by specialists; OMNeT++ alleviates this problem by tracking object ownership and detecting bugs caused by aliased pointers and misuse of shared objects.

Recently it has become more common to do large-scale network simulations with OMNeT++, with tens of thousands of network nodes. To address this requirement, aggressive memory optimization has been implemented in the simulation kernel, based on shared objects and copy-on-write semantics.

8.4.5.1.7 Parallel Simulation Support

OMNeT++ also has support for parallel simulation execution. Very large simulations may benefit from the parallel distributed simulation (PDES) feature, either by getting speedup or by distributing memory requirements. If the simulation requires several gigabytes of memory, distributing it over a cluster may be the only way to run it. For getting speedup, the hardware or cluster should have low latency and the model should have inherent parallelism.

8.4.5.1.8 Real-Time Simulation, Network Emulation

Network emulation, together with real-time simulation and hardware-in-the-loop[4]-like functionality (Applied Dynamics International 2015), is available because the event scheduler in the simulation kernel is pluggable too. The OMNeT++ distribution contains a demo of real-time simulation and a simplistic example of network

---

[4] Hardware-in-the-Loop (HIL) simulation is a technique that is used increasingly in the development and test of complex real-time embedded systems. The purpose of HIL simulation is to provide an effective platform for developing and testing real-time embedded systems.

emulation. Interfacing OMNeT++ with other simulators (hybrid operation) is also largely a matter of implementing one's own scheduler class.

8.4.5.1.9   Animation, Tracing, and Visualizing Dynamic Behavior

An important feature of OMNeT++ is the easy debugging and traceability of simulation models. These associated features are implemented in *Tkenv*, the GUI user interface of OMNeT++. As the behavior of large and complex models is usually hard to understand because of the complex interaction between different modules, OMNeT++ helps to reduce complexity by mandating the communication between modules using predefined connections. The graphical runtime environment allows the user to follow module interactions and to animate, slow down, or single-step the simulation.

OMNeT++ 5.0b1 was released on March 6, 2015. The primary aim is to provide a new simulation environment and speed up development (OpenSim 2015a). The main highlights of this release are extended logging facilities, and the new Canvas API allows models to draw freely on the surface of a module. This beta version as available for test download should be sufficiently stable for daily use, but does not yet contain all the changes planned for the 5.0 release. More betas can be expected in the next few months, adding the missing pieces.

Simulators built on the OMNeT++ environment are presented in Sects. 8.4.17, 8.4.18, 8.4.20, 8.4.21, and 8.4.22.

8.4.6   TOSSIM

TinyOS simulator (TOSSIM) is a discrete-event simulator for TinyOS sensor networks (Levis and Lee 2003; TinyOS Wiki 2013). Instead of compiling a TinyOS application for a mote, users can compile it into the TOSSIM framework, which runs on a PC. This allows users to debug, test, and analyze algorithms in a controlled and repeatable environment. TOSSIM is a TinyOS library based on nesC language (Gay et al. 2003), an extension of C language.

A TinyOS simulator satisfies four objectives:

- Scalability. To enable handling of large networks of thousands of nodes in a wide range of configurations. The largest deployed TinyOS sensor network has approximately 850 nodes.
- Completeness. Many system interactions must be covered to accurately seize behavior at a wide range of levels.
- Fidelity. The behavior of the network at a fine grain must be captured. Catching subtle timing interactions on a mote and between motes is important both for evaluation and testing.

- Bridging. The gap between algorithm and implementation must be bridged to allow developers to test and verify the code that will run on real hardware.

TOSSIM is actually an emulator for WSNs operating under the control of TinyOS operating system; unlike simulators, emulators run actual application code. The environment simulates networks at the bit level; specifically, hundreds of simulated nodes may communicate with a number of actual nodes and create a common topology running exactly the same TinyOS applications (Levis et al. 2003). TOSSIM, as included in the TinyOS, supports the MicaZ node platform (Crossbow 2006a), emulating radio interfaces, analog-to-digital converters (ADCs), and EEPROM.

TOSSIM does not model the real environment, but it provides a probabilistic representation of transmission errors occurring between two nodes. *TinyViz* is a GUI tool that permits users to visualize, monitor, control, and debug running simulations; it may capture and inject radio messages (Levis and Lee 2003). *TinyViz* has the *AutoRun* feature, a special script allowing it to run multiple simulations, set breakpoints, and define actions taken before and after each simulation such as collecting statistics.

To enhance the thorough analysis of the WSN hardware design, the SUNSHINE project (Zhang et al. 2011) has been inaugurated; it focuses on the integration of TOSSIM with an Atmel AVR family microcontroller simulator, SimulAVR (Rivet and Klepp 2012), and a hardware simulator, GEZEL (Schaumont 2012). Tython is a scripting environment that extends TOSSIM to repeatedly execute complex simulations in different scenarios (Demmer et al. 2005). SimX is an add-on tool that complements TOSSIM with simulation speed control, topology manipulation, and variable watching (Yang et al. 2007). PowerTOSSIM is a power modeling extension to TOSSIM for energy-constrained environments (Shnayder et al. 2004).

8.4.7 ATEMU

ATEMU (ATmel EMUlator) (Blazakis et al. 2004) is an open-source tool built as a software emulator for AVR processor-based systems such as MICA2 and its peripheral devices. ATEMU picks up where TOSSIM left off. Like TOSSIM, ATEMU code is binary compatible with the MEMSIC Mica2 platform (formerly Crossbow Mica2). It emulates the processor, radio interface, timers, LEDs, and other devices, making the platform able to run TinyOS. CPU instructions are decoded and executed according to the Atmel ATmega 128L microcontroller specification. However, emulation is more fine-grained than in TOSSIM; ATEMU uses a cycle-by-cycle strategy to run application code through the emulation of the AVR CPU used by Mica2 (Crossbow 2002).

The main features of ATEMU are related to the low-level emulation of the MEMSIC Mica2 sensor platform. Nevertheless, it is customizable enough to be extended to support different hardware platforms used in heterogeneous network simulations by allowing the user to set different system parameters for each single node.

ATEMU comes with *XATDB*, a graphical debugger that allows setting breakpoints, showing values of variables, statuses of peripherals, etc.; it also provides the ability to single step through either assembly instructions or at the C instruction level. The ATEMU platform supports a configuration specification based on XML files (W3C 2015) to define hardware and software configurations along with the physical location of each node.

ATEMU offers an accurate emulation model in which each Mica2 mote can run a different application code. ATEMU accuracy over TOSSIM is attained at the expense of speed and scalability; it only runs accurately with up to 120 nodes. Besides the overhead involved in decoding instruction by instruction, ATEMU also suffers from the same overhead as object-oriented models. One radio transmission can affect every other node in the network, creating an n^2 algorithm. Despite its scalability problems, ATEMU is one of the most accurate discrete-event sensor simulators available; its simulation speed though is 30 times slower than TOSSIM (Titzer et al. 2005).

8.4.8   Avrora

The name Avrora is derived from the Latin phrase "aurora borealis," meaning "the dawn of the north." It refers to a spectacular phenomenon also known as the "Northern Lights," where charged particles streaming from the sun are caught and channeled by the Earth's magnetic field. The proper name Aurora was used by the Romans to refer to the personification of dawn as a goddess. When the Avrora project was being named, the *u* was replaced with *a v*; the first three letters became *avr*, referring to the AVR architecture from Atmel (Avrora 2004).

Avrora, a research project of the UCLA Compilers Group (UCLA 2015), is a set of simulation and analysis tools for programs written for the AVR microcontroller (Atmel 2015) produced by Atmel and for the Mica2 sensor nodes (Crossbow 2002). Avrora embodies a flexible framework for simulating and analyzing assembly programs, providing a Java API and infrastructure for experimentation, profiling, and analysis.

As simulation is a basic step in the development cycle of embedded systems, the Avrora open-source implementation is motivated by the need to acquire a more detailed inspection of the dynamic execution of microcontroller programs and an accurate diagnosis of software problems before deploying software onto the target hardware. Avrora also provides a framework for program analysis, allowing static checking of embedded software and providing an infrastructure for future program analysis. Avrora's flexibility stems from provisioning a Java API for developing analyses and removes the need to build a large support structure to investigate program analysis (Titzer et al. 2005).

Avrora main features are its accuracy and scalability for simulating the actual hardware platform on which sensor programs run; it has an almost full emulation of the Mica2 and MicaZ hardware platforms, with a nearly all-inclusive ATMega128L and CC1000 radio implementations. Avrora, as a discrete-event simulator, can also run a sensor network simulation with full-timing accuracy, allowing programs to communicate via the radio using the software stack provided in TinyOS. Avrora has also an extension point that permits users to create a new simulation type depending on the number and orientation of the nodes.

Avrora is the middle ground between TOSSIM and ATEMU. Hence, it runs code instruction by instruction and avoids synchronizing all nodes after every instruction to achieve better scalability and speed. Therefore, it conducts simulation experiments with sensor networks of up to 10,000 nodes and performs as much as 20 times faster than previous simulators with equivalent accuracy. Like ATEMU, Avrora simulates a network of motes, runs the actual microcontroller programs, rather than models of the software, and runs accurate simulations of the devices and radio communication. Avrora is implemented in Java, which boosts flexibility and portability, while TOSSIM and ATEMU are implemented in C. Avrora and ATEMU gain language and operating system independence by simulating machine code, while TOSSIM can simulate only TinyOS programs. Notably, Avrora is 50% slower than TOSSIM, while ATEMU lags behind Avrora by a factor of 20 and behind TOSSIM by a factor of 30 (Titzer et al. 2005).

A limitation of Avrora is that it does not model clock drift, a phenomenon where nodes may run at slightly different clock frequencies over time due to manufacturing tolerances, temperature, and battery performance.

8.4.9 EmStar

Current sensor networks commonly share two characteristics: the use of mote-class sensor platforms with their inherent computational and communications constraints and heterogeneous deployments consisting of both mote-class and microserver-class component systems. Characteristics of both make designing, developing, debugging, deploying, and maintaining sensor networks a tricky problem. EmStar, developed at the Center for Embedded Networked Sensing (CENS) at UCLA (CENS 2015), is a comprehensive and extensible development platform in the Linux environment that greatly reduces the costs and challenges of sensor network development (Girod et al. 2007). EmStar provides a complete mote-class simulation environment (EmTOS) and a general simulation environment (EmSim). It also supports the emulation of selected components (EmCee), including radios. Moreover, EmStar allows for full native deployment, offers visualization features (EmView), and provides a robust monitoring and restarting facility at a software component level (EmRun).

EmStar is a Linux-based software framework that addresses the difficulties in creating robust software in the sensor network domain. Broadly speaking, its contributions fall into several areas:

- EmStar's execution environments address the problem of visibility into an in situ system. It provides a spectrum of runtime platforms, a pure simulation, a true distributed deployment, and two hybrid modes that combine simulation with real wireless communication and sensors in the environment. Each of these modes runs the same code and uses the same configuration files, allowing developers to seamlessly iterate between the convenience of simulation and the reality afforded by physically situated devices.
- EmStar's programming model aims to promote software reusability while being more flexible than a strictly layered stack. EmStar's modules may be flexibly interconnected using standardized interfaces; connections can be a flow of packets, stream data, state updates, or configuration commands. EmStar permits applications' domain knowledge to affect modules that are common across applications without making application-specific changes to those modules.
- EmStar's programming model aims to be inclusive. Unlike systems such as TinyOS that are tightly coupled with a specific language, EmStar does not restrict users to use certain specific languages. In fact, in some cases, whole legacy binaries can be used unchanged. This is advantageous from a perspective of rapid development and integration. Integrated code has numerous advantages because it can leverage many features of EmStar that make it suitable for building sensor network applications. One of the languages that can be used to write EmStar modules is NesC/TinyOS. Using a wrapper library called EmTOS, a TinyOS application can be run as a single module within EmStar.

EmStar uses a very simple environmental model and network medium for two reasons: First, as the purpose is to migrate the code to a real sensor environment, simple environment and network medium abstractions are satisfactory for the developers. Secondly, the simulator will only run code for the types of nodes that it is designed to work with.

EmStar was designed to be compatible with two different types of nodes. As such, like other emulators, it can be used to develop software for Mica2 motes; it also offers support for developing software for iPAQ[5]-based microservers (Webopedia 2015a). The development cycle is the same for either hardware platform. In the development cycle, EmStar uses data collected from actual sensors in order to run its simulation; the half-simulation/half-emulation approach is adopted, similar to SensorSim's (Sect. 8.4.10), where software is running on a host machine and interfacing with the actual sensor. This allows using the actual communication channel and sensors.

---

[5] iPAQ is the name of the HP PDA. The iPAQ was initially introduced by Compaq, but after Hewlett Packard's acquisition of Compaq, the product has been marketed under the HP brand.

In its deployment, EmStar brings together a number of the strongest features of other simulators and emulators. While not as efficient and fast as other frameworks like TOSSIM, EmStar's use of the component-based model allows for fair scalability.

8.4.9.1 Experimentation

As to be presented in this section, EmStar offers several modes of operation, namely, pure simulation, testbeds, and emulation, and is further extended through EmTOS (CENS 2015; Girod et al. 2007).

8.4.9.1.1 Pure Simulation

The first EmStar execution environment, EmSim, is a pure simulation model. An important limitation of EmSim is that it can only run in real time, using real-timers and interrupts from the underlying operating system. In contrast, a discrete-event simulator such as ns-2 runs in its own virtual time and therefore can run for as long as necessary to complete the simulation without affecting the results. Discrete-event simulations can also be made completely deterministic, allowing the developer to easily reproduce an intermittent bug. Identical copies, n, of a sensor node software stack run centrally on a single machine. Models of the communication and sensor channels define the effective range of each packet and the input of sensors. As in reality, instances of the stack cannot share state directly; they are forced to communicate with each other via the simulated communication channel. EmSim allows software to be developed and debugged with the convenience of simulation.

8.4.9.1.2 Testbeds

EmStar also supports several execution environments that run all the code on a central server, making debugging easy, but use real channels for sensing and communications. Two permanent testbeds are created for such use:

- A uniform array of 54 Mica1 motes (Culler et al. 2002) on the ceiling of the laboratory. The motes are all wired for power and have a serial-port connection back to a central simulation machine.
- An array of 40 Mica2 motes (Crossbow 2002), stretched across a 200 meter × 10 meter L-shaped area. Because of the longer range of the Mica2 radio, the testbed must be physically larger in order to achieve multihop topologies. The Mica2 testbed also supports remote reprogramming, enabling the mote software to be more readily upgradable. This second testbed also includes a dozen Stargate microservers to enable rich heterogeneous topologies of motes and microservers.

8.4.9.1.3   Emulation

In emulation mode, each mote is programmed to be a wireless transceiver and a sensor interface board. EmCee, like EmSim, runs instances of each node's stack centrally and provides an interface to real low-power radio, not a simulated radio model. No channel simulator is used; instead, each simulated node is mapped to one of the motes on the laboratory ceiling. When a node sends, a packet is transmitted and received by real motes through the real channel. This mode gives developers the convenience of simulation while bringing real aspects of channel dynamics.

Using the testbeds, it is also easy to emulate heterogeneous systems, such as the Extensible Sensing System (ESS) (Guy et al. 2006). ESS is a software application that provides high-level interfaces for controlling data sampling, transformation, and collection from the sensor network that embodies a microserver basestation and sensors; it also includes lower-level tools such as energy-efficient routing algorithms and sensor interface drivers. Also, using the Mica2 (Crossbow 2002) testbed, experiments can be performed such that emulated EmStar systems interact with networks of real standalone motes. This is done by selectively reprogramming some of the testbed motes directly with the mote application code and having the remaining motes programmed for transmit and receive.

8.4.9.1.4   EmTOS

Heterogeneous systems can also be emulated using an extension of EmStar, EmTOS, which is a wrapper library that enables NesC/TinyOS code to run as an EmStar module. Using EmTOS, the mote application can be run on the central server along with other emulated EmStar nodes. This enables the mote application to be debugged in a friendly environment while being part of a complete heterogeneous system such as ESS. It is also possible to test hybrid emulation modes, where some of the motes in the system are real motes running the real application, and others are emulated motes running inside EmTOS.

8.4.10   SensorSim

SensorSim, an event-driven simulator, developed at UCLA (Park et al. 2000; Park 2001) targeted ns-2 as a base, and it was intended to extend it in several directions:

- It comprises an advanced power model that takes into account each of the hardware components that would need battery power to operate.
- It includes a sensor channel that was a forerunner to the phenomena introduced to ns-2 in 2004. Both function in approximately the same way, but SensorSim's model is slightly more complicated and includes sensing through both a geophone and a microphone.

- An interaction mechanism with external applications is provided in SensorSim targeting to interact with actual sensor node networks, which allows for real sensed events to trigger reactions within the simulated environment. To achieve such goal, each real node is given a stack in the simulation environment. The real node is then connected to the simulator via a proxy, which provides the necessary mechanism for interaction.
- The use of SensorWare, a middleware platform, makes possible the dynamic management of nodes in simulation. This gives the user the ability to provide the network with small application scripts that can be dynamically moved throughout. Thus, it is not necessary to preinstall all possible applications needed by each node, and a mechanism is provided for distributed computation.

J-Sim framework for WSNs (Sect. 8.4.12) is derived from SensorSim; however, SensorSim was unfinished and has been withdrawn due to the inability to provide the necessary support.

8.4.11 NRL SensorSim

This project, developed at Naval Research Laboratory (NRL), focused on extending ns-2 by targeting a non-tackled notion of a phenomenon such as chemical clouds or moving vehicles that could trigger nearby sensors through a channel such as air quality or ground vibrations (Downard 2004). Once a sensor detects the ping of a phenomenon in that channel, it acts according to the sensor application defined by the ns-2 user. The application determines how a sensor will react once it detects its target phenomenon. For example, a sensor may periodically report to some data collection point as long as it detects the phenomenon, or it may do more sophisticated actions, such as collaborating with neighboring sensor nodes to more accurately characterize the phenomenon before alerting any outside observer of a supposed occurrence. WSN applications accomplish phenomena detection, such as surveillance, environmental monitoring, etc.

In NRL SensorSim, sensor network simulations have phenomenon nodes that trigger sensor nodes; the traffic the sensor nodes generate once they detect phenomena depends on the function of the sensor network. For instance, sensor networks designed for energy-efficient target tracking would generate more sensor-to-sensor traffic than a sensor network designed to provide an outside observer with raw sensor data (Yang and Sikdar 2003). This function is defined by the sensor application customized according to the traffic properties associated with the sensor network being simulated.

NRL's sensor network extensions to ns-2 and NRL SensorSim are online, but are no longer under development, nor supported (Networks and Communication Systems Branch 2015).

8.4.12   J-Sim

Java-simulator (J-Sim) project was developed in 2003 with the collaborative support of Next Generation Software Program at the National Science Foundation (NSF) (The National Science Foundation 2015), Network Modeling and Simulation Program of the Defense Advanced Research Projects Agency/Information Processing Techniques Office (DARPA/IPTO) (Federal Grants 2015), the Multidisciplinary Research Program of the University Research Initiative/Air Force Office of Scientific Research (MURI/AFOSR), Cisco Systems, Inc., Ohio State University, and the University of Illinois at Urbana-Champaign.

J-Sim is a modeling, simulation, and emulation framework for WSNs; it is a real-time process-driven, open-source, application development framework, component-based compositional network simulation environment (Sobeih et al. 2006). J-Sim is built upon the autonomous component architecture (ACA) and the extensible internetworking framework (INET). Both ACA and INET have been implemented in Java, and the resulting code, along with its scripting framework and GUI interfaces, is called J-Sim. In J-Sim, essential suites of wireline and wireless network components and protocols have been implemented. Moreover, a set of classes and mechanisms to realize network emulation is also included.

J-Sim possesses noteworthy programming features, as below listed:

- The fact that J-Sim is implemented in Java, along with its autonomous component architecture, makes J-Sim a truly platform-independent, extensible, and reusable environment.
- J-Sim provides a script interface that allows its integration with different script languages such as Perl, Tcl, or Python. The latest release of J-Sim (version 1.3) has been fully integrated with a Java implementation of Tcl interpreter, called Jacl, with the Tcl/Java extension (The Tcl/Java Project 2008). Thus, similar to ns-2 (Sect. 8.4.1), J-Sim is a dual-language simulation environment in which classes are written in Java (C++ for ns-2) and "glued" together using Tcl/Java. However, unlike ns-2, classes/methods/fields in Java need not be explicitly exported in order to be accessed in the Tcl environment. Instead, all the public classes/methods/fields in Java can be accessed (naturally) in the Tcl environment.

Also, J-Sim includes a set of classes and mechanisms that realize network emulation in sensor network environments, where network emulation means that the virtual simulation environment is integrated with a small number of real hardware devices to facilitate performance evaluation of real-life devices in a large-scale, but well-controlled environment. As real-life packets have to be seamlessly transported between the two environments, the main tasks are to synchronize the virtual time used in the simulation engine with the wall time and to convert packet headers and payloads from the real-life format to that used in the simulation environment.

8.4.12.1 ACA Overview

A component-based architecture is composed of *components* that are interfaced via *ports* and bound by *contracts* (Google Sites 2003). In the coming subsections, a quick review of the component-based architecture is to be presented as a prelude to J-Sim.

8.4.12.1.1 Component

In software, a component is the basic entity in the autonomous component architecture (ACA). An application is a composition of components. The notion of components has been used in several commercial component-based software standards such as JavaBeans, CORBA[6] (TechTarget 2015), and COM/DCOM/COM+[7] (Microsoft 2015). Differently though, the components in J-Sim are loosely coupled, communicate with one another by "wiring" their ports together, and are bound to contracts.

As shown in Fig. 8.6, a clarifying analogy with the component-based architecture is the known integrated circuit (IC) architecture, where a hardware module (a software system) is assembled by connecting a set of integrated circuits (ICs) (components) through their pins (ports). When the signals arrive at the pins of an IC chip, the chip performs certain tasks in compliance with the specification in the cookbook (contract), and may send signals to some other pins. A component can be reused in other software systems with the same contract context, in much the same fashion as IC chips are used in hardware design.

The main goal of the component-based architecture is to mimic the hardware assembling architecture. Typically, a wide set of IC chips are available in the market. By selecting and connecting an appropriate set of chips, one can readily compose a hardware component with desirable functions. An important step toward such a goal is to build a set of software components that can be reused in applications of similar nature.

In its framework, J-Sim specifies the components of:

- Target, sensor, and sink nodes.
- Sensor channels and wireless communication channels.
- Physical media such as seismic channels, mobility models, and power models (both energy-producing and energy-consuming components).

---

[6]Common Object Request Broker Architecture (CORBA) is an architecture and specification for creating, distributing, and managing distributed program objects in a network.

[7]Microsoft COM (Component Object Model) technology in the Microsoft Windows-family of Operating Systems enables software components to communicate. COM is used by developers to create reusable software components, link components together to build applications, and take advantage of Windows services. The family of COM technologies includes COM+, Distributed COM (DCOM) and ActiveX® Controls.

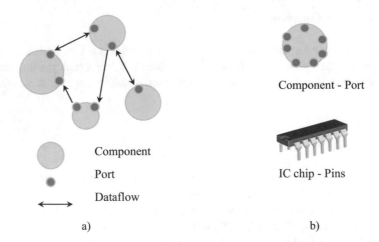

Fig. 8.6 Component model. (**a**) Component-based architecture (**b**) Component-IC analogy (Google Sites 2003)

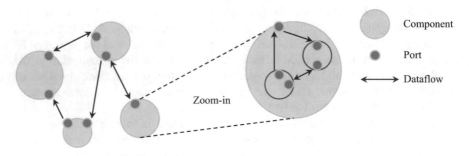

Fig. 8.7 Component hierarchy (Google Sites 2003)

New application-specific models can be defined by subclassing appropriate classes defined in the simulation framework.

8.4.12.1.1.1   Component Hierarchy

In J-Sim, a parent component may include several sub-components, called child components.

A component is uniquely identified within its parent component by its ID. Ports of a child component or the child component itself may be exposed to the outside world of its parent. Port exposure is realized by creating a port for the parent and then connecting it to the port of the child. The port of the parent component acts as a shadow port of that of the child component. Real communication occurs between the outside world and the child component's port.

Figure 8.7 illustrates the concept of hierarchy.

Child exposure is realized via a twofold procedure:

- Creating a shadow port of the parent for every port of the child when the child is included in the parent.
- Creating a port of the child when a port of the parent is created and making the parent's port a shadow of the child's port.

8.4.12.1.2 Port

A component communicates with the rest of the world via its ports. A component may own more than one port. The programming interface between a component and its port is well-defined. Since a component only interfaces with its ports, one component can be developed without the existence of other components. Also, the actual communication mechanism a component uses to communicate with the outside world is completely hidden in ports.

Ports in a component can be organized into different groups. A port group is uniquely identified within a component by its group ID. A port is uniquely identified within a port group by its assigned port ID. Therefore, a port is uniquely identified within a component by its port group and port IDs.

8.4.12.1.3 Contract

The behavior of a component is described by the port contract and the component contract. A port contract is bound to a specific port or a group of ports. It defines the communication pattern between the component that owns the port(s) and the other components that are connected to the port(s). A component is expected to work properly if all the adopted contracts are fulfilled.

Contracts specify the causality of data sent/received between components, but do not specify the components that participate in the communication. Contracts are bound at design time and components are bound at system integration time. One immediate advantage of this separation is that different components can be independently developed on different platforms and/or different programming languages, and integrated later.

When a user writes a component, he has only to follow the contracts adopted by the component, without worrying about the other components or the communication mechanism between them.

8.4.12.2 J-Sim Framework

The major objective of WSNs is to monitor and sense events of interests in a specific environment. Upon detecting an event of interest, such as change in the acoustic sound, seismic, or temperature, sensor nodes report to sink (user) nodes, either

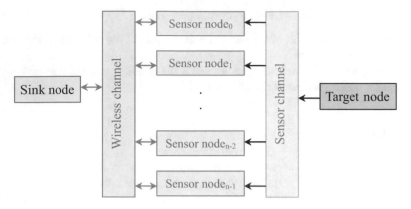

Fig. 8.8 Typical WSN environment (Sobeih et al. 2006)

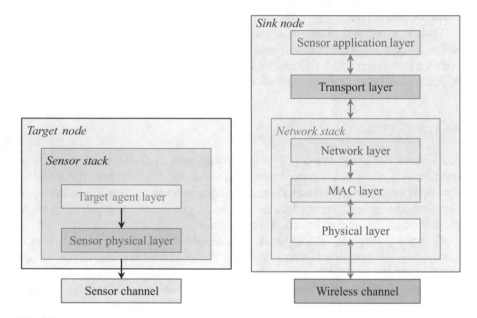

Fig. 8.9 Internal view of target and sink nodes (Sobeih et al. 2006)

periodically or on demand. Events, termed stimuli, are generated by target nodes. For instance, a moving vehicle may generate ground vibrations that can be detected by seismic sensors. From the perspective of network simulation, a WSN typically consists of three types of nodes (Fig. 8.8). Namely, sensor nodes that sense and detect the events of interest, target nodes that generate events of interest, and sink nodes that utilize and consume the sensor information. Figures 8.9 and 8.10 depict, respectively, the internal view of a target, a sink, and a sensor node as defined and implemented in the J-Sim framework.

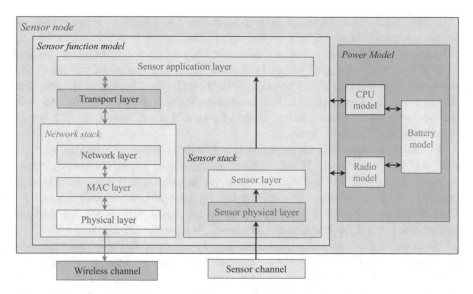

Fig. 8.10 Internal view of a sensor node (Sobeih et al. 2006)

8.4.12.2.1 Communication Model

J-Sim framework for WSNs is derived from the SensorSim framework (Sect. 8.4.10). In a nutshell, sensor nodes detect the stimuli (signals) generated by the target nodes over a sensor channel and forward the detected information to the sink nodes over a wireless channel. Figure 8.8 depicts the topmost view of the proposed simulation framework. Notably, the nature of signal propagation between target nodes and sensor nodes over the sensor channel is inherently different from that between sensor nodes and sink nodes over the wireless channel. Two different models for signal propagation are therefore included – a sensor propagation model and a wireless propagation model:

- A sensor node is equipped with a sensor protocol stack that enables it to detect signals generated by target nodes over the sensor channel and a wireless protocol stack that allows it to send reports to the other sensor nodes (and eventually to sink nodes) over the wireless channel.
- A target node has only a sensor protocol stack.
- A sink node has only a wireless protocol stack.

The operation of the J-Sim framework can be illustrated by considering a simplified event-to-sink transport protocol. A stimulus is periodically generated by a target node and propagated over the sensor channel. As shown in Fig. 8.9, a target node can "only send" data packets over the sensor channel. The neighboring sensor nodes that are within the sensing radius of the target node will then receive the stimulus over the sensor channel.

Figure 8.10 illustrates that a sensor node can "only receive" stimuli over the sensor channel. However, due to the signal attenuation in the course of being propagated over the sensor channel, a sensor node receives and detects a stimulus only if the received signal power is at least equal to a predetermined receiving threshold. The received signal power is determined by the adopted sensor propagation model (e.g., seismic or acoustic). Inside a sensor node, the coordination between the sensor protocol stack and the wireless protocol stack is done by the sensor application and transport layers.

Likewise, a sensor/sink node receives, and further processes, a data packet from the wireless channel only if the received signal power exceeds a predetermined receiving threshold. The embraced wireless propagation model determines how to calculate the received signal power. The latest release of J-Sim includes classes for three wireless propagation models, typically, the free space model, the two-ray ground model, and the irregular terrain model (Sarabandi et al. 2001).

As the sink node may not be in the vicinity of a sensor node, communication over the wireless channel is usually multihop. Specifically, to send a packet from a sensor node to a sink node, intermediate sensor nodes serve as relays (routers) to forward that packet along the route from the source sensor node to the sink node. This justifies why sensor nodes have to be able to both send and receive data packets over the wireless channel (Fig. 8.10). As sensor nodes may fail or fade out due to power depletion, the network topology of a WSN may change dynamically, and the multihop routing protocol has to adapt to the topology change through routing protocols such as ad hoc on-demand distance vector routing (AODV) (Perkins et al. 2003) or geometric routing such as greedy perimeter stateless routing (GPSR) (Karp and Kung 2000).

The information received at the sink node over the wireless channel can be further analyzed by a control server and/or a human operator. Based on the content of the information, the sink node may have to send commands/queries to the sensor nodes. This explains why, as shown in Fig. 8.9, sink nodes have to be able to both send and receive data packets over the wireless channel.

Three WSN protocols have been implemented in J-Sim and reported:

- Localization. It is how each sensor node obtains its accurate position, even in the presence of different geographic shapes of the monitoring region, different node densities, irregular radio patterns, and anisotropic terrain conditions. A distributed positioning algorithm called ad hoc positioning system/distance vector-hop (APS/DV-hop) (Niculescu and Nath 2003) has been implemented in J-Sim.
- Geographic routing. Most of the geographic routing protocols operate under the assumption that each node knows its own geographic position, and nodes can exchange their position information with their neighbors. Instead of building a routing table with the use of shortest paths and transitive reachability, geographic routing protocols make hop-by-hop routing decisions by using the geographic positions of nodes. Sensors are usually not associated with IP addresses, but instead are attributed to their geographic positions. Thus, geographic routing has been used to route data packets that contain sensed information to sink nodes. The

greedy perimeter stateless routing (GPSR) algorithm has been adopted in J-Sim (Karp and Kung 2000).

- Directed diffusion. It is a data-centric information dissemination paradigm for WSNs (Intanagonwiwat et al. 2000). In directed diffusion, a sink node periodically broadcasts to its neighbors an interest message, containing the description of a sensing task it is interested in knowing about, such as detecting a vehicle in a specific area. Interest messages are diffused throughout the network, e.g., via selective flooding, and gradients are set up within the network.

Directed diffusion consists of several elements. Data is named using attribute-value pairs. A sensing task is disseminated throughout the sensor network as an interest for named data. This dissemination sets up gradients, within the network designed to "draw" events, i.e., data matching the interest. Events start flowing toward the originators of interests along multiple paths. The sensor network reinforces one, or a small number, of these paths. Specifically, a gradient is a direction state created in each node that receives an interest message. The gradient direction is set toward the neighboring node from which the interest message is received.

After receiving an interest, a node may decide to resend the interest to some subset of its neighbors. To its neighbors, this interest appears to originate from the sending node, although it might have originated from a distant sink. In such a manner, interests diffuse throughout the network. Not all received interests are resent. A node may suppress a received interest if it recently resent a matching interest.

8.4.12.2.2 Power Model

A sensor node has also a power model that embodies the energy-producing components, such as battery, and the energy-consuming components, such as a radio and CPU (Fig. 8.10). Moreover, in order to enable simulation of mobile nodes, such as moving vehicles, a mobility model is included. The sensor function model, i.e., combination of the sensor protocol stack, the network protocol stack, and the sensor application and transport layers, is subject to the power model. The energy incurred in handling a received data packet is dictated by the CPU model, and the energy incurred in sending and/or receiving data packets is dictated by the radio model. In the J-Sim framework, both the CPU and radio models can be in one of several different operation modes that determine the amount of energy consumed. The radio model can be in idle, sleep, OFF, transmit, or receive modes. The CPU and radio models can report their operation modes to the sensor function model, and the sensor function model can change the operation modes of the CPU and radio models.

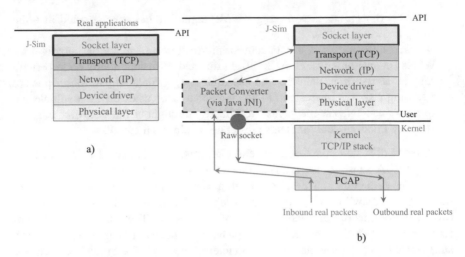

Fig. 8.11 Network emulation in J-Sim. (**a**) Top-down network emulation. (**b**) Bottom-down network emulation (Sobeih et al. 2006)

8.4.12.3   Network Emulation

Network emulation is an inexpensive approach to testing, validating, and evaluating protocols and approaches in a realistic but well-controlled network environment. The protocol or approach to be tested is usually executed in the real environment, while other interacting components are executed in the well-controlled, virtual environment.

Network emulation in J-Sim is realized in both the top-down and bottom-up hierarchy:

- In the top-down approach, a Java-compliant socket layer is developed with real applications on top. As shown in Fig. 8.11a, the socket layer essentially gives applications the illusion that they are interfacing with the operating system, rather than with a virtual network environment.
- In the bottom-up approach, real-life packets are intercepted at the device driver level and transported to the Packet Converter utility that converts packet headers and payloads from the real-life format to that of J-Sim. Packets can then be directed to different layers depending on the header information. As illustrated in Fig. 8.11b, to implement this technique, the packet capturing facility, such as PCAP[8] in Linux and Windows (Tech-FAQ 2015), is used to intercept real-life packets and redirect them to the Packet Converter. Outbound packets will be

---

[8] PCAP (Packet Capture) is a protocol for wireless Internet communication that allows a computer or device to receive incoming radio signals from another device and convert those signals into usable information. It allows a wireless device to convert information into radio signals in order to transfer them to another device.

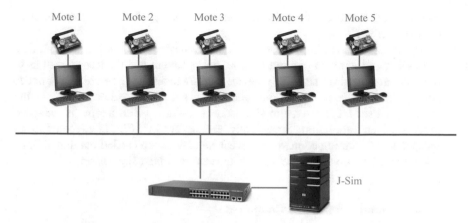

Fig. 8.12 Network emulation in J-Sim (Sobeih et al. 2006)

processed by the Packet Converter and directed via IP raw sockets to real device drivers.

WSN Network emulation, as depicted in Fig. 8.12, involves Berkeley Mica motes (Crossbow 2002). Berkeley motes, equipped with sensors and RF circuitry, are used as the real "small dust"[9] (Webopedia 2015b) devices to extract physical environment data. With the use of the TinyOS SerialForwarder program (TinyOS Wiki 2012), the generic two-way communication tool, real-life data are relayed from motes to I/O devices and vice versa. These real-life packets are then intercepted at the serial link, converted by the Packet Converter to proper formats, and then fed into one of the virtual J-Sim classes.

Several layouts of network emulation in WSNs are devised as listed below from the simplest to the most complex:

- Extracting sensor data from real devices. In this layout, motes serve as sensor hardware that provides one-way data traffic from real devices to the simulation environment. Specifically, the functionality of the sensor physical layer component is implemented in motes. All other WSN functions, such as in-network processing and information relay to the sink nodes, are simulated in J-Sim.
- Processing sensor data in real devices. Different from the previous layout, the task of processing sensor data is moved from the simulation environment to real devices. The functionalities of both the sensor physical layer and the sensor application layer components are implemented in motes, while the communication over the shared wireless channel is still simulated in J-Sim. In this form,

---

[9]Also termed "smart dust." These are millimeter-scale self-contained micro-electromechanical devices that include sensors, computational ability, bidirectional wireless communications technology and a power supply. As tiny as dust particles, smart dust motes can be spread throughout buildings or into the atmosphere to collect and monitor data. Smart dust devices have applications in everything from military to meteorological to medical fields.

packets are forwarded bidirectionally between motes and wireless physical layer components in J-Sim.

- Processing and transmitting sensor data in real devices. In this layout, both data processing and wireless communications take place in real devices as well as in the simulation environment. Real devices communicate with virtual sensor nodes and synchronize their operations and wireless communication events in the shared channel. Obviously, the simulation paradigm J-Sim adopts is real-time process-driven simulation. Specifically, each event in J-Sim is executed in an independent execution context, and event executions are carried out in real time, not at fixed time points in the discrete-time event-driven simulation.

8.4.12.4   J-Sim Performance Compared

The performance of the J-Sim simulation framework is tested and compared against ns-2 in several WSN scenarios. Of particular interest is the effect of network size on the execution time required to complete simulation, the resulting number of generated events, and the memory thus consumed. Notably, the execution time includes both the time required to set up the nodes, i.e., the time incurred in creating and configuring the network before the simulation starts, and the time required to conduct a T-second simulation run. Each data point reported was an average of 20 simulation runs.

8.4.12.4.1   Target Tracking

This simulation scenario consists of one sink node and two target nodes. The sensor nodes are evenly distributed over a 1500×1500 m^2 region. The two target nodes move according to the random waypoint model with a maximum speed equal to 10 m/s. Each target node generates a stimulus every second, and the sensing radius is 200 m. The simulation time, T, is 1000 s.

The execution time and the number of events generated versus the network size $n^2 + 2$, where n varies from 10 to 22, are evaluated for both J-Sim and ns-2. It was shown that ns-2 stops at $n = 18$ because ns-2 ran out of memory for $n > 18$. J-Sim incurs however a considerably longer execution time, even though the number of events generated is almost the same, especially in the range $10 \leq n \leq 16$. These results are due to the inherent slowness of a Java program as compared to its C/C++ counterpart. Specifically, the execution time in J-Sim is up to 41.6% higher than that in ns-2, and the number of events generated by J-Sim is up to 27.5% higher than that in ns-2.

Also evaluated was the amount of memory allocated before the start and before the end of the simulation. The memory usage before the start of the simulation represents the amount of memory allocated to set up the nodes and other components in the simulation, e.g., wireless and sensor channels. While memory usage before the end of the simulation represents the amount of memory allocated to complete the

1000-sec simulation. The rate of increase in memory usage before the start of the simulation in ns-2 is higher, causing J-Sim to outperform ns-2 for $n \geq 15$. This indicates that the data structures are used in a more scalable manner in J-Sim to represent different classes and their interactions in the WSN framework. Moreover, the memory allocated to complete the 1000 s simulation in J-Sim is at least two orders of magnitude lower than that in ns-2. This is attributed to the better garbage collection mechanism used in Java to reclaim unused memory.

8.4.12.4.2 Using GPSR Routing Protocol

The simulation scenario is identical to that in the previous subsection except that GPSR is used as the underlying routing protocol instead of AODV. Obtained results indicate that ns-2 cuts off at $n = 14$ because it ran out of memory for $n > 14$. Close execution time and the number of events values are obtained for ns-2 and J-Sim; J-Sim, though, incurs a smaller execution time to carry out the simulation.

Regarding memory usage before and after the simulation, once more, the memory usage in J-Sim is at least two orders of magnitude lower than that in ns-2. As compared with AODV, GPSR incurs much less memory as well, due to the fact that AODV incurs significant routing overhead.

From the two previous performance experiments, the ability of J-Sim to carry out simulation for $n > 18$ at the cost of a reasonable amount of memory, reveals the scalability of the WSN simulation framework in J-Sim. This is coupled with added extensibility acquired from the component architecture (ACA) (i.e., new components can be defined by subclassing appropriate base classes that are readily inserted into the framework with matched contracts).

J-Sim is, however, relatively complicated to use, no more complicated though than the much more popular ns-2. J-Sim, while more scalable than many other simulators, has its share of inefficiencies. Java, in general, is possibly less efficient than many other languages. There is also the added overhead in the intercommunication model. Moreover, IEEE 802.11 is the only supported MAC protocol that can be used, a limitation that occurs in WSN support built on top of all-purpose simulators.

J-Sim version 1.3 was released on February 2, 2004 (Table 8.1).

8.4.13 Prowler/JProwler

The Probabilistic Wireless Network Simulator (Prowler) was developed in 2003 at the Institute for Software Integrated Systems (ISIS) in Vanderbilt University (ISIS 2015). It is a free event-driven simulator that can be set to operate in either a deterministic mode to produce replicable results while testing the application or a probabilistic mode to simulate the nondeterministic nature of the communication channel and the low-level communication protocol of the motes (Simon et al. 2003).

Prowler is able to simulate all the important aspects of sensor networks built from Berkeley motes. It can incorporate an arbitrary number of motes on any possibly dynamic topology. It was designed to be easily embedded into optimization algorithms. Prowler runs under MATLAB, thus providing a fast and easy way to prototype applications; moreover, its GUI provides nice visualization capabilities. Prowler is designed to allow users, interested in algorithmic rather than implementation details, to do fast and realistic prototyping of sensor networking applications using mote-class nodes, without the need of expert knowledge for low level issues concerning the hardware platform, the operating system, the programming language, or even a special simulation environment.

Prowler models the important aspects of all levels of the communication channel and the application. A probabilistic radio channel model characterizes the nondeterministic nature of radio propagation. Also, a simplified yet accurate model is used to describe the operation of the MAC layer. Applications interact with the MAC layer through a set of events and actions.

8.4.13.1 Prowler Framework

8.4.13.1.1   Radio Propagation Models

The radio propagation model determines the strength of a transmitted signal at a particular point of space for all transmitters in the system. Based on this information, the signal reception conditions for the receivers can be evaluated and collisions can be detected. The signal strength from the transmitter to a receiver is determined by a deterministic propagation function and by random disturbances:

- The deterministic part of the propagation function models the fading of signal strength with distance, and it can be any user-supplied function, yet a frequently used model of the signal strength versus distance is given by:

$$P_{\text{rec,ideal}}(d) = P_{\text{transmit}} * \frac{1}{1 + d^{\Upsilon}} \qquad (8.1)$$

where $P_{\text{rec, ideal}}$ is the ideal reception signal strength, P_{transmit} is the transmission signal power, d is the distance between the transmitter and the receiver, and γ is a decay parameter with typical values $2 \leq \Upsilon \leq 4$.

- Random disturbances account for real signal behavior; they model the fading effect, the time-varying nature of the signal strength, and other miscellaneous transmission errors. The signal strength can be seriously impacted by distance changes. Also, the signal strength can change even if the distance between the transmitter and receiver is constant. This *fading effect* is modeled by random disturbances in the simulator. The received signal strength from node j to node i is calculated from the propagation function (Eq. 8.1) by modulating it with random functions:

$$P_{rec}(i, j) = P_{rec,ideal}\left(d_{i,j}\right) * \left[1 + \alpha\left(d_{i,j}\right)\right] * \left[1 + \beta(t)\right] \tag{8.2}$$

The random variable α depends on the distance only; thus, in the simulator, it is calculated only when the position of either the transmitter or the receiver changes; while β is time-dependent, so its value is recalculated at the beginning of every transmission. The random variables α and β have normal distributions $N(O, \sigma_\alpha)$ and $N(O, \sigma_\beta)$, respectively, with adjustable parameters σ_α and σ_β.

8.4.13.1.2 Signal Reception and Collisions

Prowler provides two models to account for signal reception and collision:

- Model 1. The signal is received if its strength is greater than a reception limit parameter. The channel is sensed idle if there is no signal that could be received. Collision occurs if two transmissions overlap in time, and both could be received.
- Model 2. Each receiver has a noise variance parameter σ_n^2. The signal to interference and noise ratio (SINR) for receiver i and transmitter j is defined by:
- The total signal strength at node i is defined by:

$$\text{SINR} = \frac{P_{rec}(i, j)}{\sigma_n^2 + \sum_{k \neq j} P_{rec}(i, k)} \tag{8.3}$$

The total signal strength at node i is given by:

$$P_{tot}(i) = \sum_k P_{rec}(i, k) \tag{8.4}$$

The signal is received if the SINR at the receiver is greater than the reception threshold during the whole transmission time. The channel is sensed idle if the total signal strength is smaller than an idle threshold, which depends on the noise variance of the receiver. There is a collision if the SINR at the receiver becomes smaller than the reception threshold at any time during the reception.

Model 1 is simple and fast, while Model 2 is more accurate; the choice of the model is a usual compromise between speed and accuracy. The radio models in the simulator are interchangeable plug-ins; thus a new model can be easily added on need.

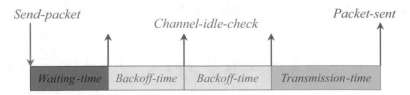

Fig. 8.13 MAC layer communication (Simon et al. 2003)

8.4.13.1.3   MAC-Layer Model

The MAC layer in Prowler simulates a CSMA MAC protocol similar to that of the Berkeley Mica motes (Crossbow 2002). A simplified event channel, illustrated in Fig. 8.13, models the MAC layer communication:

- The application emits the *Send-packet* event. After a random *waiting-time* interval, the MAC layer checks if the channel is idle.
- If not idle, it continues the idle checking until the channel is found idle; the *Backoff time* is a random interval that precedes each idle check.
- When the channel is idle, the transmission begins, and after *transmission time*, the application receives the *packet-sent* event.
- After the reception of a packet on the receiver's side, the application receives a *packet-received* or *collided-packet-received* event, depending on the success of the transmission. Notably, collision occurs if two or more stations transmit at the same time after sensing an idle channel.

The *waiting time* and *backoff time* parameters are random uniformly distributed variables in predefined intervals, while *transmission time* is constant, that is all messages have the same length.

8.4.13.1.4   The Application Layer

Similarly to the real TinyOS framework, the applications are event based. Several events can be noticed: *Init-application*, *Packet-sent*, *Packet-received*, etc. Debugging and visualization commands are also available, e.g., switch ON/OFF the LEDs on the motes, draw lines and arrows, and print messages.

8.4.13.2   Optimization Framework

In the development phase of new protocols, it is required to provide optimal performance in some metric, versus a certain set of design parameters. This optimization problem leads to the search of an error surface above a parameter space. There are multiple solution methods if the error surface is well defined. Basically, they depend on some kind of exploration of the error surface, either a gradient-based method, Monte-Carlo search, or an annealing method (Press et al. 2002). These optimization methods use "function calls" to compute the value of the cost function. The more computationally expensive the function call, the more cost-efficient it is to keep the number of function calls low.

As Prowler can be used to test protocols and algorithms, it provides metrics on the performance of the tested application. Similarly to the core of Prowler, the applications can be parameterized, so different settings can be easily tested. The adopted optimization algorithm is built around Prowler and calls it with the required parameters.

The error function can be any performance metric defined above the parameter space, such as time, energy, application-specific metrics, or a combination of them. Optimization though has its concerns:

- Due to the stochastic nature of the environment, a useful performance metric does not result from a single experiment, but rather an average value, a minimum, or a maximum. To calculate such a performance metric, several experiments must be made, i.e., several simulation runs. Thus a single "function call" of the optimizer algorithm can be very expensive.
- Some a priori knowledge would be necessary on the error surface so that the necessary precision of calculating the error surface could be determined. Generally such information is not available; thus, the necessary number of experiments is practically unknown. This can result in imprecise error surface estimations, and thus the optimization algorithm may not converge to the right minimum.

The main features of the proposed optimization algorithm are as listed:

- The search is performed over a finite set of predefined parameter values, i.e., discrete points in the parameter space.
- The function call for one point returns the outcome of one experiment only. The search method uses this "noisy" cost function value. Calls are calculated for the same point several times; thus, at certain points, the error surface becomes more and more accurate during the search process.
- The search algorithm makes steps on the discrete parameter space, after each function call, depending on the result of the last function call and the values of the averaged error function.

8.4.13.3 Prowler Performance

As illustrated, Prowler can be combined with an optimization algorithm to optimize the parameters of middleware services and applications in WSNs. It is advised to use Prowler, over other simulators, for beginners because of its user-friendly GUI, fast and easy prototyping in MATLAB, as well as easy debugging. However, it does not have sensor node energy modeling. Prowler is not widely used because of its specific use and capabilities; as well, no comparative studies with well-known simulators are provided to ascertain its performance.

Routing Modeling Application Simulation Environment (RMASE) (Zhang et al. 2006) is an application built on top of Prowler. It provides a layered routing architecture with routing scenario specifications and performance metrics for algorithm evaluations. Both Prowler and RMASE are easily extensible with new models, metrics, and components.

In Barberis et al. (2007), modifications are made to Prowler, making its MAC protocol model compliant with the Crossbow mote running on Tiny OS (Crossbow 2006a). Moreover, radio propagation models were added to the simulator, and importantly an energy consumption estimation model of the CC2420 radio chip has been implemented. WSN applications were developed with a particular emphasis on energy consumption optimization and consequently battery lifetime.

Prowler version 1.25 was released on January 28, 2004 (Table 8.1).

8.4.13.4   JProwler

The JProwler tool is a discrete event simulator for prototyping, verifying, and analyzing communication protocols of TinyOS ad hoc wireless networks (ISIS 2004). The simulator supports pluggable radio models and MAC protocols and multiple application modules. Two radio models are implemented; specifically, Gaussian and Rayleigh, and one MAC protocol for MICA2 (Crossbow 2002) with no acknowledgment. These components have the same underlying dynamic physical model as in the MATLAB Prowler. JProwler is implemented in Java and optimized for raw speed. It can run a simple network-wide broadcast protocol on a 5000-node network in real time (around 1.3 s). However, the startup time, during which the simulator creates static data structures, is 35 s for a 5000-node network and 1.5 s for a 1000-node network. The simulator can visualize the status of the network and application data.

JProwler version 1.0 was created on February 13, 2004, and is available for download (Table 8.1).

8.4.14   SENS

Sensor, Environment and Network Simulator (SENS) was developed in 2004 at the Department of Computer Science at the University of Illinois at Urbana-Champaign. SENS is a C++ written simulator for WSNs; it has a modular, layered architecture with customizable components to model an application, network communication, and the physical environment (Sundresh et al. 2004). By choosing appropriate component implementations, users may capture a variety of application-specific scenarios, with accuracy and efficiency tuned on a per-node basis. For the sake of realistic simulations, typical values from Mica2 sensors are used to represent the behavior of components' implementations. Such behavior includes sound and radio signal strength characteristics and power usage. Furthermore, SENS is layout-independent as it allows new WSN layouts with the possibility of adding and including their parameter profiles. The ability to develop portable applications is an important asset, knowing that WSN layouts constantly evolve as new sensor node implementations rise up.

Moreover, SENS introduces an innovative mechanism for modeling physical environments. WSN applications are characterized by the tight integration of computation, communication, and interaction with the physical environment. When a node drives its actuator, it may affect the environment and alter network propagation characteristics. Thus, the validity and effectiveness of simulation results depend heavily on how accurately the environment is modeled. To provide users with the flexibility of modeling the environment and its interaction with applications at different levels of detail, SENS defines an environment as a grid of interchangeable tiles.

8.4.14.1 Simulator Structure

SENS consists of several simulated sensor nodes interacting with an environment component. Each node consists of three components, namely, Application, Network, and Physical (Fig. 8.14). Each component has a virtual clock; messages can be sent with any delay past the sender's current virtual time. For instance, a node's Network component may simulate the reception of two colliding packets, and hence they are not received, while at the same virtual time, the node's Application processes some data. Clearly, components are isolated and interchangeable; a user may employ any of the implementations SENS provides, modify existing components, or write entirely new ones for custom applications, network models, sensor capabilities, or environments.

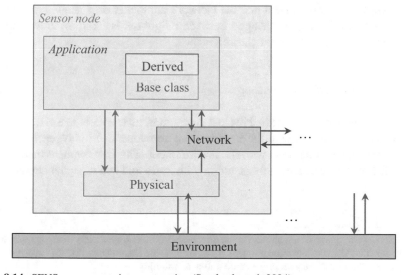

Fig. 8.14 SENS components interconnection (Sundresh et al. 2004)

8.4.14.1.1   Application Components

An Application component simulates the execution of software on a single sensor node. A node's Application component communicates with its Network component to send or receive packets and with its Physical component to read sensor values or control actuators. A C++ base class provides an interface for applications. An Application component may receive and act upon the received message; results are sent back as a message.

Users may create an Application component in two ways:

- They may derive a new class from the Application class to directly implement an application. However, such a program may not run directly on existing WSN layouts.
- Alternatively, a thin compatibility layer is developed to enable direct portability between SENS and real sensor nodes. When compiling a source code intended to run on a real sensor node, it is linked with a library that translates sensor node API calls to SENS Application APIs. This technique allows a SENS target for TinyOS (TinyOS 2012), similar to the approach used in TOSSIM (Sect. 8.4.6) and TOSSF[10] (Perrone and Nicol 2002).

SENS applications used for the simulations can be ported to real network devices, which is a noteworthy advantage.

8.4.14.1.2   Network Components

A Network component simulates the packet send and receive functions of a wireless sensor node. All such components are derived from the Network base class, which specifies the basic network interface. Each Network component is connected to a single Application component and the Network components of neighboring nodes. The format of messages exchanged between neighbors is fixed to allow multiple implementations with different characteristics.

8.4.14.1.3   Physical Components

Each simulated node includes a Physical component that models sensors, actuators, and power and interacts with the Environment component. Initially, each Physical component registers its node with the Environment. The Environment then replies with a list of the node's neighbors, along with radio and sound signal strength and

---

[10]TOSSF is a simulator, which allows for the direct execution, at source code level, of applications written for TinyOS, the operating system that executes on Smart Dust. TOSSF also provides detailed models for radio signal propagation and node mobility.

delay for each neighbor. Microphone (sensor) and speaker (actuator) devices are provided.

The Physical component also simulates a node's power usage. When the Application or Network component enters a different power mode, they notify the Physical component by actuating messages to turn ON or OFF the associated virtual hardware. When a radio message is transmitted, the Network component sends the duration of transmission to the Physical. The current is multiplied by a nominal 3 V and the power usage is accumulated over time.

8.4.14.1.4 Environment Component

The role of the Environment component is to provide a useful model of a real environment with which nodes' sensors and radios might interact. By varying the Environment component, developers can test a wide variety of settings with less effort as compared to setting up actual experiments. The Environment component yields models of various types of surfaces that influence the radio and sound propagation parameters. An environment is simulated as a 2-D grid of interchangeable square tiles to allow modular, reconfigurable scenarios. This helps in modeling sensors on the ground outdoors. Tiles use experimentally measured parameters for radio and sound wave propagation. SENS provides tiles to simulate grass, concrete sidewalks, and walls. Concrete is considered a baseline, and other tiles with greater signal attenuation or delays are called obstacles.

The Environment component models circular wave propagation through the 2-D grid of tiles. Since a tile may be anywhere on the map, propagation rules must use only local information about a wave. This information is modified by tiles as the wave propagates and passed on to neighboring tiles arranged along the propagation paths. Tiles receive and propagate the following information describing (part of) a wave:

- Source location (x_s, y_s).
- The amount of energy contained in that part of the wave.
- The delay profile along the edge through which the wave entered a tile.

An ideal 2-D circular wave propagation performed on a grid of tiles is shown in Fig. 8.15. Angle $\theta_{13} = \theta_{12} + \theta_{23}$, the angle from the source spanning tile (x, y), determines the total energy that passes through the tile. Angle θ_{23} represents the fraction that passes through tile (x, y) and the tile above $(x, y + 1)$, while θ_{12} represents the fraction through tiles (x, y) and $(x + 1, y)$

Signal strength (SS) measured by a sensor depends on energy density. Measurements are made for 2-D and 3-D propagation models. Environment data, such as sound and radio propagation parameters, is obtained using Mica2 sensor motes (Crossbow 2002).

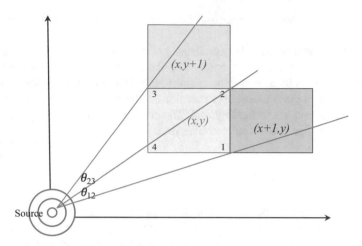

Fig. 8.15 Circular wave propagation through adjacent tiles (Sundresh et al. 2004)

8.4.14.2   Simulation Examples

Two example applications have been devised to illustrate the features of SENS. Several random environments were created, ranging from 0% to 100% obstacle tiles (grass, walls), with concrete for the remaining tiles.

8.4.14.2.1   Spanning Tree

A service which generates a partial spanning tree via flooding is initiated at time *0* by a root node in the middle of a 400 meter × 400 meter region containing 1000 nodes. The root broadcasts a single spanning tree message containing its ID. When a node receives such a message, it reads the sender's ID, stores it as its parent, broadcasts a new spanning tree message containing its own ID, and goes to sleep. It is revealed that for 25% obstacle:

- The spanning tree coverage peaks at 914 of 1000 nodes. It then drops quickly as more obstacles are added, such that some nodes become entirely cut off from the root. Tree coverage also decreases with increased obstacle density.
- Noticeably, the problem under very low obstacle density is collisions; obstacles decrease radio range and hence increase usable network capacity until the point where the network is partitioned. Also, collisions greatly outnumbered broadcast messages sent, because they were counted at each receiving node.

8.4.14.2.2   Simplified Localization

Many sensor network applications need location information to correlate measurements. A simple localization service based on acoustic ranging is studied. Simulations were performed in a 50 meter × 50 meter environment with 6 anchor nodes and 200 non-anchors. Anchors are nodes with known locations; all others are non-anchors with unknown locations. Anchors periodically broadcast their ID and location over the radio, immediately followed by a 0.1 s beep. When a non-anchor receives such a radio message, it measures the delay until it hears the beep to estimate the distance to the anchor node. Furthermore, non-anchors take the median of the past 10 measurements from an anchor node to filter out erroneous data.

For simplification, non-anchors estimate their location by averaging anchors' locations, weighted by the inverse of the approximate distance to each anchor node:

$$(x, y) = \frac{1}{r_1}(x_1, y_1) + \cdots + \frac{1}{r_n}(x_n, y_n) \tag{8.5}$$

A node is considered successfully localized if it has computed its own location to within 10 meters (the large tolerance compensates for the error in Eq. 8.5). Several findings were obtained from this test case:

- The number of successfully localized nodes varies inversely, roughly linearly with obstacle density. This linear decline indicates that anchor-based acoustic ranging has fairly predictable behavior with respect to obstacles, as opposed to the accelerating drop-off in spanning tree coverage.
- Nodes with several barriers between themselves and any anchor tend to have worse errors. Furthermore, localized positions are primarily skewed toward those anchors to which they have a more direct path. This is because direct paths appear shorter than indirect paths obstructed by walls and provide a stronger signal than those where sound signals are attenuated by grass. In general, the component structure of SENS allows users to quickly run and visualize their applications because obstacles, network models, etc. can simply be plugged-in rather than rewritten.
- The decrease in localization error for large obstacle densities may seem surprising. However, based on the low number of nodes successfully localized, it is apparent that for high obstacle densities, nodes either have relatively direct sound paths to anchor nodes, or cannot hear them at all.

8.4.14.3   SENS Performance

A simulator should offer a strong performance advantage over setting up real sensor networks. Considering the simplified localization with varying numbers of randomly positioned nodes, the performance was satisfactory. Results are encouraging, having n nodes and a $\sqrt{n} \times \sqrt{n}$ meter environment with $n/16$ anchors and 50% of the tiles assigned as obstacles. For a network of 8192 nodes, 1000 virtual seconds of the application were simulated in only 136 s of real-time on a 2.5 GHz Pentium 4 with 512MByte RAM running Linux 2.4.20; of them, 124 s were spent on environment initialization, with a mere 12 seconds dedicated to actual execution.

However, SENS is less customizable than other protocols since it does not provide the opportunity to modify the MAC protocol or any other lower-layer protocols. Although power utilization analysis is supported, phenomena detection capabilities are limited to sound. Also, SENS does not provide a GUI.

SENS latest version was available for download on January 31, 2005 (Table 8.1).

8.4.15   SENSE

Sensor Network Simulator and Emulator (SENSE) is a sensor network simulator developed in 2004 at the Department of Computer Science, Rensselaer Polytechnic Institute, New York (Chen et al. 2005). It was designed to be efficient, powerful, and easy to use while satisfying three main factors, namely, extensibility, reusability, and scalability. SENSE targets three types of users: high-level users, network builders, and component designers:

- For high-level users, the process of building a simulation merely consists of selecting appropriate models and templates and changing some parameters; there is no need for programming skills, and their main concern is scalability.
- The network builders need to create new network topologies and traffic patterns. They may not have knowledge of popular programming languages, such as C, C++, or Java. For them, models must be reusable so that they can be plugged into many simulations.
- The component designers intend to modify available models or build new ones from scratch. Their main concern is the extensibility—how easily existing models can be extended or replaced.

The most significant feature of SENSE is the balanced concern for modeling methodology and simulation efficiency. SENSE is a very fast and user-friendly simulator. Unlike object-oriented network simulators, it is based on a novel component-oriented simulation methodology that promotes extensibility and reusability to the maximum possible, without overlooking simulation efficiency and scalability.

Extensibility is achieved in SENSE by avoiding a tight coupling of objects via the introduction of the component-based model that removes the interdependency of objects often encountered in object-oriented architectures (Sect. 8.4.12.1). This is attained by the proposed "simulation component classifications." They are essentially interfaces, which allow exchanging implementations without the need to change the actual code. Reusability at the code level is a direct consequence of the component-based model.

SENSE is influenced by three other frameworks. It attempts to implement the same functionality as ns-2 (Sect. 8.4.1). However, it moves away from the object-oriented approach using J-Sim's (Sect. 8.4.12) component-based architecture. Like GloMoSim (Sect. 8.4.3), SENSE also includes support for parallelization. Through its component-based model and support for parallelization, the developers attempt to address such critical factors in simulation as extensibility, reusability, and scalability.

G-Sense is a tool that greatly improves SENSE user-friendliness, with graphical input of simulation parameters, save and load of simulation features, and management of simulation results with plot view (NetGNA 2008; Rosa et al. 2009). G-Sense uses the SENSE simulation engine in a transparent way. The user may thus be focused on the simulation itself, not on the underlying simulation tool.

8.4.15.1 Component-Based Design

SENSE is built on top of COST (Chen and Szymanski 2002), a general-purpose discrete event simulator. The design of COST was influenced by concepts of component-based software architecture and component-based simulation. In the component-based model, a component communicates with others only via inports and outports (Fig. 8.16). An inport implements certain functionality; it is similar to a function. An outport serves as an abstraction of a function pointer; it defines what functionality it expects of others. The component-based model is implemented in C++.

The existence of outports distinguishes components from objects. Outports impose constraints on the dynamic runtime interaction between components. Due to their use, the development of a component can be fully separated from the

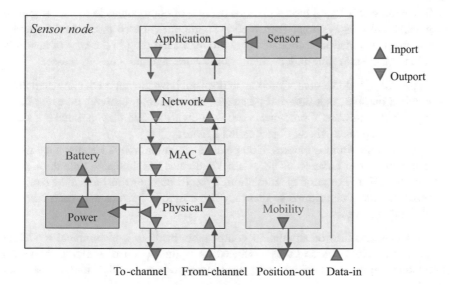

Fig. 8.16 Sensor node components (Chen et al. 2005)

application context in which the component is to be used, leading to surely reusable components. Moreover, components become more extensible, because there are fewer constraints on the component that provides the desired functionality. On the other hand, in an object-oriented environment, if an object *A* is to be replaced by an object *B*, object *B* has to be derived from *A*. In the component-based model, this constraint is no longer necessary, any component providing the required functionality can be used, regardless of its type.

8.4.15.2 Sensor Network Simulation Components

The component-based model gives the users a wide freedom in configuring sensor nodes. A sensor node is a composite component that consists of a number of smaller primitive components, each implementing a function (Fig. 8.16). Also, a sensor node has some layered network protocol components, a power component and a battery component that are related to power management, and more components such as mobility and sensors. The inports and outports of the sensor node component are directly connected to the corresponding inports and outports of internal components. This structure is modifiable; a user can freely remove or add components, as demanded by the targeted simulation. For instance, the network protocol stack can be simplified by removing the network component, or tuned up by adding a new transport layer without affecting other components. A queue component can be added between the network layer and the MAC layer to prevent packets from being dropped when the MAC layer is busy transmitting other packets. Nodes can be configured using C++, TCL (Tcl Developer Xchange 2015), or XML (W3C 2015).

The component-based model clarifies the role of components in the development of general software systems. To extend the component-based model to the simulation domain, simulation components are grouped according to the way of handling simulation time into time-independent, time-aware, and autonomous classes:

- Type I, time-independent. Component does not have the notion of simulated time, and it is passive, as it does not generate events without receiving an event first.
- Type II, time-aware. Components are time-aware in that they can make a simulated time advance via an object called a *timer*.
- Type III, autonomous classes. Components are autonomous because they maintain their own simulation clock. A clock indicates the simulated time throughout the simulation. For parallel simulation, Type III components have to be synchronized via some algorithms so that they can correctly interact with each other by exchanging events.

Such classification of simulation components leads to a hierarchical modeling process in SENSE, which gives advanced users the option of building their own simulation engines instead of using built-in engines as in other parallel network simulators.

8.4.15.3 Components Repository

As illustrated below, SENSE encompasses an extensive set of components ranging from the application layer to the physical layer, as well as energy and mobility models that are specifically targeted at sensor networks:

- IEEE 802.11. This component implements the distributed coordination function (DCF) described in the IEEE 802.11 standard (Bianchi 2000). The IEEE 802.11 implementation has the same level of details as that of ns-2 (Sect. 8.4.1); however, the source code in SENSE is twice as short as that in ns-2, due to the simplicity and effectiveness of SENSE APIs.
- AODV. The ad hoc on-demand distance vector routing (AODV) implementation in SENSE is based on the AODV Internet draft (Perkins et al. 2003). The essential components of AODV's basic operation have been implemented. This includes all the steps required to actually build routes. However, some route maintenance functions have not been included in the current simulation, such as maintaining local connectivity, processing route error packets, and implementing local repair functions.
- DSR. The dynamic source routing (DSR) (Johnson and Maltz 1996) is a widely used on-demand routing protocol for wireless networks. Like AODV, DSR provides a mechanism of route discovery if the route from the source to the destination is unknown. But unlike AODV, after the route has been discovered, the entire route is included in the packet header, and intermediate nodes will determine the next hop by looking at the routing information contained in the packet. SENSE implements an initial version of the DSR component, which imposes some restrictive assumptions within the DSR specifications. Specifically, all nodes are assumed to be bidirectional, without support for promiscuous communications, and running in a homogeneous link layer environment. Moreover, it is assumed that all communication links, once established, are not subject to damages, and hence there is no need for error handling and route recovery.
- Battery models. Two battery components have been implemented in SENSE (Chen et al. 2005):

  - In the *SimpleBattery* component, the discharge rate is always proportional to the power drawn from the battery and is independent of the current. Its capacity is a constant defined by the simulation parameter. Let E' be the previous remaining energy and P the power consumed in the time unit; the energy remaining after a consumption period of t can be expressed as:

  $$E = E' - P * t \tag{8.6}$$

  - In the more complex *RealBattery* component, the discharge rate is dependent on the current. A larger current renders the battery discharge quicker, thus resulting in less actual capacity at the end of the usage period than the smaller current would do (Park et al. 2001). A discharge rate dependence parameter, k, determines how the value of the current affects the discharge rate. More specifically, Eq. (8.6) becomes:

$$E = \frac{E'}{(1 + k^*I)} - P * t \tag{8.7}$$

– The *RealBattery* component also models relaxation (Park et al. 2001), which is the phenomenon of a battery gradually recovering some of its lost capacity if the discharge current suddenly drops to a very small level. It is assumed that relaxation only occurs if the current first sustains for a fast discharge period of at least T_R with a current larger than I_R and then suddenly drops from above I_R to 0. Let λ be the recovery rate, g the growth ratio that can be eventually reached; then, during the relaxation period the capacity is governed by:

$$E = g * E'\left(1 - e^{-\lambda t}\right) \tag{8.8}$$

• Power model. In SENSE, the power component is responsible for power management. A *SimplePower* component has been implemented, which can operate in any of five modes; typically, TRANSMIT, RECEIVE, IDLE, SLEEP, and OFF. Four parameters specify the energy consumption rate for each of the first four modes, while in the OFF mode there is no energy consumption. As a response to accepting control signals from networking components, the power component switches from one mode to another. Current is drawn from the battery depending on the operating mode.

8.4.15.4   Performance Comparison

The performance of SENSE versus ns-2 version 2.26 was tested in terms of execution speed and memory efficiency. Simulating flooding was used as the comparison benchmark. All nodes run the IEEE 802.11 protocol using only the broadcast functionality due to the nature of flooding. For the comparison, TCL and C ++ scripts were written to randomly generate traffic and topology files, while both simulators were modified to read from the same input files.

To compare the execution speeds of both simulators, a WSN composed of 60 nodes with the same random placement over a 1000 meter × 1000 meter terrain is devised. Twelve sources were randomly chosen to send 1000-Byte length packets, at fixed 10-second intervals. Simulation results revealed that SENSE is consistently twice as fast as ns-2.

The performance difference between ns-2 and SENSE is attributed to the ways they allocate and release packets. In ns-2, when a packet is being broadcast, every neighboring node will receive a copy, so the number of packet allocations is equal to the number of received packets. In SENSE, all receivers always share a packet, so the number of packet allocations is equal to the number of sent packets. In a dense WSN, a node can usually communicate with many neighbors; consequently, the number of received packets is far greater than the number of sent packets. By having all sensors use the same packet in memory, assuming that the packet should not be modified, SENSE improves scalability.

SENSE's packet sharing model improves scalability by reducing memory use, thus allowing an improvement over ns-2 and other object-oriented models. However, the model is simplistic and places some communication limitations on the user. While implementing the same basic functionality as ns-2, SENSE cannot match the ns-2 extensions. Whether because it is new or because it has not achieved the popularity of ns-2, there has not been research into adding a sensing model, eliminating physical phenomena and environmental effects.

SENSE, also, is not a match for the parallelization abilities of GloMoSim, leaving to the user a significant portion of the parallelizing. It does, however, give the user the option to optimize for parallelization or for sequential operation.

It is worth mentioning that SENSE is similar to J-Sim in that it is component based, but it is written in C++ to avoid the perceived inefficiency of Java.

While in its active development phase, SENSE version 3.0.3 was made available on April 28, 2008 (Table 8.1).

8.4.16 Shawn

Shawn (Kroller et al. 2005; Fekete et al. 2007) is an open-source discrete event simulator developed in 2004 under the SwarmNet project (SwarmNet Project 2004) funded by the German Research Foundation (DFG). For maximum performance, Shawn is written in C++ and runs under Windows and many variations of Unix/Linux. Due to its high customizability, it is extremely fast and tunable to the required accuracy. Released under the GNU General Public License (Free Software Foundation 2007),[11] it is currently in active development and successful used by different universities and companies to simulate WSNs with large numbers of nodes.

Knowingly, the utmost goal of WSN simulators is to be as "realistic" as possible by simulating physical effects, data and message encoding, wireless interference impacts, processor limitations, etc. However, high accuracy comes at the price of long simulation times. When developing sensor network algorithms at the application layer, accuracy is not really required during testing and improving an algorithm, since at such a step the developer's usual concern is results, quality, and whether termination is correct. Furthermore, the large number of factors that influence the

---

[11] The GNU General Public License is intended to guarantee the freedom to share and change all versions of a program, to make sure it remains free software for all its users.

behavior of the whole network renders it nearly impossible to isolate a specific parameter of interest.

Shawn uses the concept of discrete time to speed up simulations where node polling happens at much larger intervals, but events can be scheduled to happen in between, at any time, with any precision. Non-discrete event simulators often trigger their simulated nodes periodically, e.g., every few msec. So, even inactive nodes slow down the simulation, and the effects that happen in less than this fixed interval may go unobserved.

The central idea of Shawn is to replace low-level effects with abstract and exchangeable models; the simulation can be used for huge networks in reasonable time. With the motivation of being fast without compromising realism, Shawn was built upon design paradigms that make it different from other simulators:

- Simulating the effects. The design rationale behind Shawn is to simulate the effects of a phenomenon, not the phenomenon itself, by implementing and using abstract models. By selecting the actual granularity and behavior, a user is able to adapt the simulation to his specific needs. For instance, instead of simulating a complete MAC layer including the radio propagation model, its effects such as packet loss and corruption are modeled in Shawn. This impacts simulations in several ways; specifically, they become more predictable and meaningful. Also, there is a huge performance gain, because such a model can often be implemented very efficiently. On the other hand, this results in the inability to come up with the detail level that simulators like ns-2 provide with regard to physical layer or packet-level phenomena. For almost all aspects in Shawn, one can choose between an abstract, a simplified, or a realistic model implementation. A simplified model is fast, and it is needed if the particular aspect is not important in the current development phase. On the other hand, the realistic model is preferable when a certain simulation aspect is the focus of investigations such as radio propagation properties or lower protocol layer issues.
- Simulation of huge networks. A main benefit of the above paradigm is superior scalability. Networks consisting of millions of nodes can be studied. Successful simulations were run on standard PC equipment with more than 100,000 nodes.
- Supporting a development cycle. Shawn inherently supports the development process with a complete development cycle that begins at the initial idea and ends with a fully distributed protocol. The complete development cycle of simulations using Shawn is as listed, with each step being optional:

  - Performing a structural analysis of the problem at hand, given the first idea of an algorithm, is assumed to precede designing some protocol. To get a better understanding of the problem in this first phase, it is helpful to look at some example networks and analyze the network structure and underlying graph representation.
  - A first centralized version of the algorithm is to be implemented to achieve a rapid prototype version. A centralized algorithm has full access to all nodes and has a global, flat view of the network. This provides a simple means to obtain results and a first impression of the overall performance of the

examined algorithm. The results emerging from this process can provide optimization feedback for the algorithm design.
- The feasibility of the distributed implementation can be investigated in depth after achieving a satisfactory state of its centralized version. Only a simplified communication model between individual sensor nodes is used. Since the goal of this step is to prove that the algorithm can be transformed into a distributed implementation, the messages exchanged between the nodes are simple data structures passed in memory. This allows for efficient and fast implementation that leads to meaningful results.
- Defining the actual protocol and rules for the nodes to run the distributed algorithm ensues arriving at a fully distributed and working implementation. Messages that have been in-memory data structures and passed as references may be represented in the form of individual data packets. With the protocol and data structures in place, the performance of the distributed implementation can be evaluated. Performance issues can be explored, such as the number of messages, energy consumption, runtime, resilience to message loss, and environmental effects.

8.4.16.1 Architecture

As to be detailed below and illustrated in Fig. 8.17, Shawn consists of three major parts, namely, models, sequencer, and simulation environment.

8.4.16.1.1 Models

To attain reusability, extensibility, and flexibility, exchangeable models are used wherever possible in Shawn. A thorough distinction between models and their respective implementations supports these goals. Shawn maintains a flexible and powerful repository of model implementations that can be used to compose simulation setups simply by selecting the desired behaviors through model identifiers at runtime. Some models shape the behavior of the virtual world, while others provide more specialized data. Models that form the foundation of Shawn are the communication model, the edge model, and the transmission model:

- The communication model determines for a pair of nodes whether they can communicate. There are models that represent unit disk graphs for graph-theoretical studies, models based on radio propagation physics, and models that resort to a predefined connectivity scenario.
- The edge model uses the communication model to provide a graph representation of the network by giving access to the direct neighbors of a node. This leads to two main implications:
  - It allows for simple centralized algorithms that require information on the communication graph. In this respect, Shawn differs from ns-2 and other

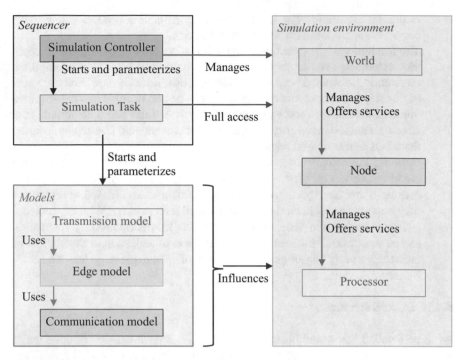

Fig. 8.17 Shawn architecture (Kroller et al. 2005)

  simulators, where the check for connectivity is based on sending test
  messages.
– Simulations of relatively small networks allow storing the complete neighborhood of each node in memory, thus providing remarkably fast replies to
  queries. However, huge networks will impose impractical demands on memory; therefore, an alternative edge model trades memory for runtime by
  recalculating the neighborhood on each request or only caches a certain
  number of neighborhoods.

• The transmission model determines the properties of an individual message
  transmission. It can arbitrarily delay, drop, or alter messages. Thus, when the
  runtime of algorithms is not in question, a simple transmission model without
  delays is sufficient. A more sophisticated model may account for contention,
  transmission time, and errors. Moreover, specialized models provide data for
  simulations:

– Random variable model. To test algorithms with different underlying random
  variables.
– Node distance estimate model. To mimic distance measurements for localization algorithms, among others.

Shawn allows running a separate application on each network node. Moreover, simulation scenarios can be saved to and loaded from XML files (W3C 2015). The runtime parameters can be typed manually or loaded from a configuration file.

8.4.16.1.2 Sequencer

The sequencer is the central coordinating unit in Shawn. It configures the simulation, executes tasks sequentially, and drives the simulation. It consists of the Simulation Controller, the Event Scheduler, and the straightforward concept of Simulation Tasks:

- The Simulation Controller is the central repository for all available model implementations; it drives the simulation by transforming the configuration input into parameterized calls of Simulation Tasks. These are arbitrary pieces of code that are configured and run from the simulation's setup files. Because they have full access to the whole simulation, they are able to perform a wide range of jobs. Instance uses include the steering of simulations, gathering data from individual nodes, or running centralized algorithms.
- The Event Scheduler triggers the execution of events that can be scheduled at arbitrary discrete points in time.

8.4.16.1.3 Simulation Environment

The simulation environment is the home for the virtual world where the simulation objects reside. All nodes of a simulation run are contained in a single-world instance. The nodes are a container for Processors, which are the workhorses of the simulations; they process incoming messages, run algorithms and emit messages.

Shawn features persistence and decoupling of the simulation environment by introducing the concept of Tags. They attach both persistent and volatile data to individual nodes and the world. They also decouple state variables from member variables, thus allowing for an easy implementation of persistence. Another benefit is that portions of a potentially complex protocol can be replaced without modifying code, because the internal state is stored in tags and not in a special node implementation.

8.4.16.2 Shawn Compared

Figure 8.18 categorizes some of the most prominent simulation frameworks according to the criteria of scalability and abstraction level. This representation does not express the maximal feasible network sizes, but rather reflects the typical application domain.

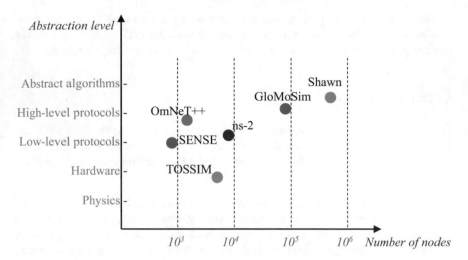

Fig. 8.18 Application levels of simulators (Kroller et al. 2005)

Conducted case studies against ns-2 revealed that the Shawn approach outperforms in terms of runtime and memory usage, especially for very large networks. To be noticed, Shawn's inability to come up with the detailed level that ns-2 provides with respect to physical layer or packet-level phenomena.

A special extension entitled JShawn is capable of interpreting simple Java-based scripts defined in the startup files (Frick 2013). The latest version of Shawn was uploaded on June 6, 2007 (Table 8.1).

8.4.17   SenSim

The design of WSNs requires the simultaneous consideration of the effects of several factors such as energy efficiency, fault tolerance, quality of service demands, synchronization, scheduling strategies, system topology, communication, and coordination protocols. SenSim is a simulator for WSNs based on the discrete event simulation framework OMNeT++; it was developed by the Sensor Network Research Group at Louisiana State University (Mallanda et al. 2005).

8.4.17.1   SenSim Design

The topology of the WSN field in SenSim is derived from the simple and compound module concept of the OMNeT++ framework (Sect. 8.4.5). The architecture of a sensor node is depicted in Fig. 8.19. Each layer of the sensor node is represented as a simple module of OMNeT++. Each layer has a reference to the Coordinator. These

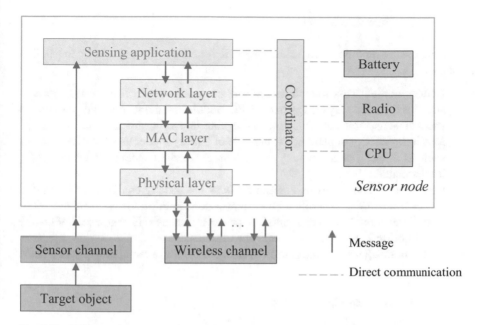

Fig. 8.19 Structure of the sensor node in SenSim. (Mallanda et al. 2005)

simple modules are connected according to the layered architecture of a sensor node. The different layers of the sensor node have gates to other layers to form the sensor node stack. A simple module with wireless channel functionality is used to communicate with the compound modules (sensor nodes) through multiple gates. The functionalities provided by each module are described below with the hardware model of a sensor node composed of the radio, CPU, and battery modules.

8.4.17.1.1   Coordinator Module

Coordinator module has the functionalities that enable coordinating the activities of the hardware and software modules of the sensor node. Through the Coordinator, any layer may access and update the properties of other layers. For example, the battery module needs to be informed of the packets transmitted or received by the physical module so as to update the energy consumption at the node. During simulation, the Coordinator class is responsible for registering the sensor node to the sensor network. Registering a sensor node indicates that it is up and functioning. On the other hand, when the available energy is completely depleted, the node is unregistered from the network.

8.4.17.1.2   Hardware Model

The hardware module encompasses several functions, as laid out in the bullets below:

- Battery model. This module is an essential component of the sensor node; it supplies the necessary energy to the CPU module, a radio module, and the sensors used to sense the environment. At regular intervals, the module updates its remaining energy depending on the type of battery model used. Various models such as linear battery model and discharge rate-dependent model are being implemented.
- CPU model. The nodes in a sensor network are usually equipped with low-end processors or microcontrollers; their power consumption for performing various operations should be very limited. A processor needs different levels of energy consumption in the idle, sleep, and active states.
- Radio model. This module is used to characterize the antenna property of a node.

8.4.17.1.3   Wireless Channel Model

The wireless channel module controls and maintains all potential connections between sensor nodes. Statistic connections are provided from all the nodes to the wireless channel module and from the module to all the nodes in the NED file. These connections enable the sensor nodes to exchange data and communicate with each other. Any message from a node is sent to all the neighbors within its transmission region with a delay d, where d is *Distance between the communicating sensor nodes/ Speed of light.*

Various radio propagation models are used to predict the received signal power of each packet. These models, as derived by the wireless channel module, affect the communication region between any two nodes:

- Free space propagation model. The free space propagation model assumes an ideal propagation condition where there is only one clear line-of-sight path between the transmitter and receiver.
- Two-ray ground reflection model. A single line-of-sight path between two mobile nodes is seldom the only means of propagation. The two-ray ground reflection model considers both the direct path and a ground-reflection path; it gives more accurate prediction at a long distance than the free space propagation model.

8.4.17.1.4   Sensor Node Stack

The simple module at the highest level of the hierarchy of the sensor node, specifically the sensor application, simulates the behavior of the application layer. New applications can be incorporated into this module.

The simple network module simulates the packets sent and received by the nodes in the network; it initially receives application layer messages and adds to them the network header. The particular features of this layer depend on the protocol implementation. Directed diffusion with geographic and energy aware routing (GEAR) protocol is implemented at the network layer. The structure of a packet sent from the network layer to the MAC layer has a field for the next hop in the route.

The MAC layer provides the interface between the physical layer and the network layer. It has the basic functionality of media access and supports the IEEE 802.11b implementation. Such a modular structure of entities simulated with OMNeT++ makes SenSim more flexible as compared to ns-2.

Based on simulations with the IEEE 802.11 MAC and directed diffusion integrated with GEAR, it was found that SenSim is at least an order of magnitude faster than ns-2 with a more efficient memory use.

Sensim 3.0 is available for free download (Table 8.1).

8.4.18 PAWiS

The power aware wireless sensors (PAWiS) simulation framework (Weber et al. 2007; Glaser et al. 2008), built at the Institute of Computer Technology, Technical University of Vienna, Austria (TU Wien 2015), helps in developing, modeling, simulating, and optimizing WSN nodes and networking protocols. Concurrent simulation of the power consumption of every sensor node is provided. Moreover, it supports detailed power reporting and modeling of wireless environments. PAWiS reduces the overall power consumption by carefully optimizing various design aspects within the context of the application, which impels new applications that would not be possible otherwise due to insufficient battery lifetime or limited energy scavenging systems. As shown in Fig. 8.20, the PAWiS simulation framework is based on OMNeT++, the object-oriented C++ based discrete event simulator (Sect. 8.4.5).

The PAWiS framework offers several capabilities:

- Enabling researchers to program models of a wide variety of abstraction
- Modeling the details of WSN nodes as well as the communication between them, with a distinction between software and hardware tasks.
- Accurately elaborating power simulation. In WSN nodes, components with different supply voltages are combined necessitating low dropout regulators (LDOs) and DC/DC converters. PAWiS allows modeling of this hierarchical supply structure as well as the efficiency factor of the converters.
- Providing powerful analysis and visualization techniques to evaluate the simulation results and proceed toward optimization.
- Supporting cross-layer design to exploit the synergy between layers.
- Modeling RF communication according to real-world wave propagation phenomena. Models include interferers, noise, and attenuation due to distance. No

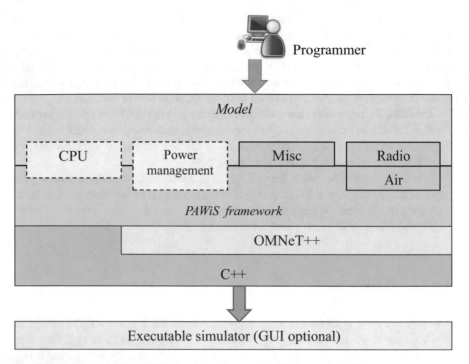

Fig. 8.20 Structure of the PAWiS simulation framework (Weber et al. 2007)

preset topology is needed as packets are transmitted to all nodes within range. The network topology originates from the link quality and the routing algorithm; thus routing protocols, especially ad hoc protocols, can be implemented. The transmission model is independent of the underlying modulation technique, enabling the simulation of any type of modulation.

- Modeling dynamic behavior, such as mobility and environmental dynamics, via an embedded scripting language.

8.4.18.1   Structure and Functions

In PAWiS, user-defined models are implemented as C++ classes (Fig. 8.20). Node composition and network layout along with environmental and setup parameters are specified in configuration and script files. The modules are compiled and linked with the simulation kernel, resulting in the simulation application. A GUI-based interface enables visual debugging of the communication processes of the model on a per-event basis at simulation runtime. An optional command line-based interface can be utilized for increased simulation performance.

The PAWiS framework is primarily focused on simulating inter- and intra-node communication (Grammarist 2014).[12] Additionally, fine-grained aspects, such as the CPU instruction set, can be easily emulated as in ATEMU (Sect. 8.4.7). However, a tradeoff between simulation details and execution performance has to be considered when the number of network nodes increases (Heidemann et al. 2001). The obtained model contains information regarding the functional description and architecture specifications, along with low-level implementation details. Simulation results comprise timing and power consumption profiles as well as event records.

Jointly processing events from OMNeT++ and SystemC (Accellera 2015)[13] is achievable by combining the OMNeT++ and the SystemC simulation kernels.

8.4.18.1.1   Modularization

A wireless sensor node is typically composed of multiple modules, specifically, CPU, timer, radio, and network layers. Every module is based on one or more tasks. PAWiS framework defines two types of tasks. The first task type models a hardware component, such as timer and ADC, whereas the second type is a software task, such as application, routing, MAC, and physical layer. Every module is implemented as a C++ class derived from a framework base class. Tasks are implemented as methods within a module class.

8.4.18.1.2   CPU

In a sensor node, the CPU executes software tasks. Notably, multiple software tasks cannot run in parallel since only one CPU is available. The CPU module in the PAWiS framework ensures that only one task's code simulation is executed at a time. To model the power consumption and timing behavior of software tasks, the PAWiS simulation framework splits the simulation into two parts. The functional part is implemented in the C++ method of the task, and the timing and power consumption part is delegated to the CPU module.

The PAWiS simulation framework supports modeling of a CPU which scales processing time and power consumption according to its individual properties and a CPU which offers special low-power modes. A high-accuracy modeling of the CPU considers the percentage of integer, floating point, memory access, and flow control operations.

---

[12] The prefix *inter-* means *between* or *among* different nodes. The prefix *intra-* means *within* a node, between CPU and memory for instance.

[13] SystemC is a set of C++ classes and macros, which provide an event-driven simulation interface, that enable a designer to simulate concurrent processes described using plain C++ syntax. SystemC processes can communicate in a simulated real-time environment using signals of all the data types offered by C++, by the SystemC library, as well as user defined.

8.4.18.1.3   Timing

Modeling time delays consider whether they occur in firmware or hardware modules. For hardware modules, the framework provides a simple wait method to suspend execution for a certain amount of time. Several distinct implementations of the wait method are available with support for fixed and conditional timeouts. On the other hand, if delays are needed in software, the corresponding module has to use a loop, or a similar construct, to wait for a certain time and therefore utilize the CPU to achieve the delay. Moreover, the CPU can be put into a low-power mode, which stops execution too, and therefore delays it until an external or timer event occurs.

8.4.18.1.4   Environment and Air

All sensor nodes are deployed at 3D positions within the environment to represent the outer world and surroundings of all nodes including the RF channel. Besides nodes, objects like walls, floors, trees, interferers, heaters, light sources, global properties, and more are defined within the environment. The environment can be configured with configuration and scripting files.

The air is a main part of the environment to handle RF channels that are defined by 3D node placement in space and obstacles between the nodes. In the PAWiS simulation framework, it is possible to model effects like an RF signal subject to wave propagation phenomena such as attenuation, interference, noise, reflection, refraction, and fading (multipath propagation).

Similar to a wired network, packet transmission is modeled without a predefined topology. Specifically, every RF message is transmitted to all other nodes; the received RF power is calculated from the transmitter power, antenna properties, and the distance and obstacles between the transmitter and the receiver. The network topology results from the reachability between nodes, which is limited by the minimum received signal quality.

8.4.18.1.5   Power Simulation

A key feature of PAWiS is simulating the power consumption of tasks. A central power meter object logs the power consumption values that are reported by all modules of all nodes. Only tasks that simulate dedicated hardware directly report the power consumption. Different electrical behaviors are proposed, that is, the current I depends in different ways on the supply voltage U. Software tasks report their CPU utilization; the CPU module calculates and reports their power consumption. Reporting power consumption is accomplished by calls to special methods offered by PAWiS framework classes.

It is worth noting that modules of a sensor node do not consume constant power throughout their lifetime. The CPU consumes less power when in sleep mode (and hence does not execute instructions); also, the power consumption of the radio differs whether in transmit or in receive, in phase-locked loop (PLL), or in idle mode. The model developer has to report new power consumption figures every time the state of the module changes.

Power simulation values are not evaluated during the simulation run; rather they are analyzed and visualized using the data post-processing tool when the simulation is over.

8.4.18.1.6   Dynamic Behavior

The PAWiS framework supports modeling dynamic behavior such as mobility and environmental dynamics via an embedded scripting language. Scripting languages are platform-independent and need a virtual machine for execution. The scripting language Lua[14] has been adopted due to its simplicity, extensibility, prevalent usage, fast execution, and maturity (PUC-Rio 2015). Simple topologies can be set up with the OMNeT++ intrinsic NED language, but more complex topologies can be easily created utilizing PAWiS scripts.

8.4.18.2   Optimization

Several strategies for the optimization of the wireless sensor system are available:

- System-level optimization. This includes the node composition and modifications of the whole system behavior like choosing different network layouts or application patterns. System-level optimization results in an adequate system architecture.
- Cross-layer optimization. More than one network layer is modified at a time. While if a single module updating itself would degrade the node performance, the interaction of all module updates leads to an improvement of the entire node performance.

In PAWiS, deciding which aspects for model optimization can be based on the pure function of the model. With the GUI, it is possible to see wrong or insufficient behavior during runtime. Additionally, the proposed data post-processing tool permits to analyze the power consumption profile of nodes and distinct modules; it can also be used to check the timing behavior and progression of certain events.

The PAWiS simulation framework version 2.0 was released on July 1, 2008 (Table 8.1).

---

[14] Lua (pronounced LOO-ah) means "Moon" in Portuguese. As such, it is neither an acronym nor an abbreviation, but a noun. Lua is designed, implemented, and maintained by a team at PUC-Rio, the Pontifical Catholic University of Rio de Janeiro in Brazil.

8.4.19 MSPsim

MSPsim was developed at the Swedish Institute of Computer Science (SICS) (Eriksson et al. 2007). MSPsim, a Java-based, extensible instruction-level emulator for the MSP430 microcontroller (Texas Instruments 2011), is intended to be a component in a larger sensor network simulation system supporting cross-level simulation (Osterlind et al. 2006). MSPsim simulates unmodified target platform firmware. Also, MSPsim is a part of the Contiki operating system (Doxygen 2012)[15] and can be used in cross-level simulations conducted with the Cooja platform (Contiki Developers 2015).[16]

MSPsim extensibility means that it is adaptable to new sensor boards while requiring no more than the implementation of a few Java classes. Its easy extensibility with peripheral devices makes it possible to simulate various types of MSP430-based sensor nodes. As an instruction-level simulator, it is easy to achieve an accurate timing simulation. MSPsim also provides a graphical representation of the sensor board in an on-screen window to verify that an application is correctly simulated.

Version 0.97 of MSPsim was made available on April 30, 2009 (Table 8.1).

8.4.20 Castalia

Castalia (Castalia 2013) was developed in C++ at the Australian National Information and Telecommunications Technology (ICT) (Australian Government 2015). The name Castalia comes from Greek mythology. Castalia was a nymph whom Apollo transformed into a fountain at Delphi (the place where all important oracles came from, in ancient Greece). All the oracle seekers stopped and washed their hair in the fountain. Metaphorically, for the designer's purposes, it can be seen as a representation of the truth.

Castalia is a simulator for WSNs, BANs, and generally networks of low-power embedded devices. It is based on the OMNeT++ platform and can be used by researchers and developers who want to test their distributed algorithms and/or protocols in realistic wireless channel and radio models, with a realistic node behavior, especially relating to radio access. Since it is highly parametric, Castalia can also be used to evaluate different platform characteristics for specific applications and can simulate a wide range of platforms. The main features of Castalia are:

---

[15] Contiki is an open source, highly portable, multitasking operating system for memory-efficient networked embedded systems and wireless sensor networks. Contiki is designed for microcontrollers with small amounts of memory.

[16] Cooja is the Contiki network simulator, it allows large and small networks of Contiki motes to be simulated.

- Advanced channel model, based on empirically measured data, that has several characteristics:

  - Defines a map of path loss, not simply connections between nodes.
  - It is a complex model for the temporal variation of path loss.
  - It fully supports the mobility of the nodes.
  - Interference is handled as received signal strength, not as a separate feature.

- Advanced radio model based on real radios for low-power communication with support for:

  - Probability of reception based on SINR, packet size, and modulation type. PSK and FSK are considered; custom modulation is allowed by defining the SNR-BER curve.
  - Multiple TX power levels with individual node variations allowed.
  - States with different power consumption and delays switching between them.
  - Flexible carrier sensing (polling-based and interrupt-based).

- Extended sensing modeling capabilities:

  - Highly flexible physical process model.
  - Sensing device noise, bias, and power consumption.
  - Measuring node clock drift and CPU power consumption.
  - Availability of MAC and routing protocols.
  - Designed for adaptation and expansion. With proper modularization, Castalia was designed right from the beginning so that users can easily implement/import their algorithms and protocols into Castalia while making use of the features the simulator provides. The modularity, reliability, and speed of Castalia are partly enabled by OMNeT++.

Notably, Castalia is not sensor platform-specific; it is meant to provide a generic reliable and realistic framework for the first-order validation of an algorithm before moving to implementation on a specific sensor platform. Castalia is not useful if one would like to test code compiled for a specific sensor node platform. For such usage, there are other simulators/emulators available, e.g., Avrora (Sect. 8.4.8).

Castalia version 3.2 was released on March 3, 2011 (Table 8.1).

8.4.21   MiXiM

OMNeT++ provides a powerful and clear simulation framework, but it lacks direct support and a concise modeling chain for wireless communication (Sect. 8.4.5). Mixed simulator (MiXiM) fills such gap by joining and extending several existing simulation frameworks developed for wireless and mobile simulations in OMNeT++ (Köpke et al. 2008; MiXiM Developers 2011). MiXiM is a merger of several OMNeT++ frameworks written to support mobile and fixed wireless networks such as WSNs, BANs, ad hoc networks, vehicular networks, etc. The built-upon previous MiXiM frameworks are the Channel Simulator (ChSim) by Universitaet Paderborn (Universitat Paderborn 2010), MAC Simulator by Technische Universiteit Delft (University of Twente/Technical University of Delft 2005), Mobility Framework (MF) by Telecommunication Networks Group at Technische Universitaet Berlin (Löbbers and Willkomm 2007), and Positif Framework (PF) by Technische Universiteit Delft (University of Twente/Technical University of Delft 2005).

MiXiM offers detailed models of radio wave propagation, interference estimation, radio transceiver power consumption, and wireless MAC protocols such as Zigbee. Also, MiXiM provides a user-friendly graphical representation of wireless and mobile networks in OMNeT++ that supports debugging and defining complex wireless scenarios. MiXiM extensive functionality and clear concept may motivate researchers to contribute to this open-source project.

Noticeably, discrete event simulators like OMNeT++ are standard tools to study protocols for wired and wireless networks. In contrast to the wired channel, the wireless channel has a complex influence on the protocol performance; it requires in-depth knowledge of the level of detail necessary to make a sound performance analysis. The basic components of MiXiM can be divided into five groups:

- Environment models. In a simulation, only relevant parts of the real world should be reflected, such as obstacles that hinder wireless communication.
- Connectivity and mobility. When nodes move, their influence on other nodes in the network varies. The simulator has to track these changes and provide an adequate graphical representation.
- Reception and collision. For wireless network simulations, the movements of objects and nodes have an influence on the reception of a message. The reception handling is responsible for modeling how a transmitted signal changes on its way to the receivers while taking transmissions of other senders into account.
- Experiment support. The experimentation support is necessary to help the researcher compare the results with an ideal state, find a suitable template for his implementation, and support different evaluation methods.
- Protocol library. A rich protocol library enables researchers to compare their ideas with already implemented ones.

Logically, MiXiM can be divided into two parts:

- The base framework that provides the general functionality needed for almost any wireless simulation, such as connection management, mobility, and wireless channel modeling.
- The protocol library complements the base framework with a rich set of standard protocols, including mobility models.

In order to have clearly defined interfaces between the base framework and the protocol library, MiXiM provides a base module for the OMNeT++ modules. Following this concept makes it easy to implement new protocols for MiXiM while facilitating reusability. An architectural overview of the modeling and the implementation of the physical layer within the MiXiM simulation framework are offered in (Wessel et al. 2009).

8.4.21.1 MiXiM Base Models

Simulating wireless communication systems requires a suitable abstraction of the environment, the radio channels, and the physical layer. This section presents the basic modeling approaches and the assumptions behind them, as well as the implementation of relevant aspects such as model abstraction level and model support for trading-off accuracy and calculation complexity.

8.4.21.1.1 Environmental Model

Simulations are usually carried out in a limited area, such as a playground, on which nodes and objects are placed. Nodes represent the wireless devices with their protocol stack and are modeled as isotropic radiators that do not have a physical dimension. An object, in contrast, is anything with a physical dimension that resides in the propagation environment and can possibly attenuate a wireless signal. Both objects and nodes may be mobile. Nodes may even be combined with objects to model a sensor node mounted on a car. The term entity is used to refer to both nodes and objects.

The mobility of objects is a time-continuous process, which raises a trade-off between accuracy and computational complexity. In MiXiM, the user can choose the level of accuracy and, thus, the computational complexity of modeling mobility. Mobility also requires handling collisions with entities and handling border crossing of the playground. Among the situations that might involve collisions and border crossing are as follows:

- A collision may cause an error.
- A new position may be randomly chosen upon collision and the entity is placed there afterwards, as long as this position does not coincide with another entity.
- The entity may be reflected in an angle that it had when it collided.

MiXiM provides the Object Manager as a central authority for managing objects in the propagation environment. Objects are characterized by dimensions, by position, optionally by angle of rotation, and by frequency-dependent attenuation factors. An object that obstructs the line of sight between any pair of interconnected nodes causes additional signal losses during transmission (Fig. 8.21). Since entities can be mobile, intersections of the line of sight of two nodes with one or more objects must be determined at runtime.

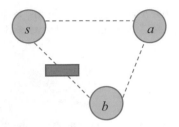

An object within the line-of-sight between two nodes s and b yields a weaker received signal than that of a non-obstructed pair s and a at the same distance.

Fig. 8.21   Signal loss (Köpke et al. 2008)

8.4.21.1.2   Connection Modeling

In contrast to wired simulations, connectivity modeling is a challenging task in wireless simulations. In wired simulations, two nodes are connected by wires, which can be easily modeled, e.g., in OMNeT++ by connections. In wireless simulations, however, the "channel" between two nodes is the air, which is a broadcast medium and cannot be easily represented by one connection. MiXiM divides the connection modeling into two parts, as described in the two coming subsections: the connectivity between nodes and the wireless channel and its attenuation property.

8.4.21.1.2.1   Nodes Connectivity

Theoretically, the signal sent out by a node affects all other nodes in the simulation, if operating in the same frequency range. However, when the signal is attenuated, the received power at nodes far away from the sending node may be as low as to be negligible. In order to reduce the computational complexity in MiXiM, nodes are connected only when they are within the maximal interference distance. The maximal interference distance is a conservative bound on the maximal distance at which a node can still possibly disturb the communication of a neighbor. A node that wants to receive a message from a communication peer also receives all interfering signals and can thus decide on the interference level and resulting bit errors.

The presence of objects in the propagation environment also impacts the maximal interference distance. As illustrated in Fig. 8.21, objects may shield two nodes from each other.

8.4.21.1.2.2   Wireless Channel Models

MiXiM's channel models support multiple parallel radio channels in frequency (e.g., in terms of OFDM subcarriers), and space (e.g., as in multi-antenna systems). For each of these channels, radio propagation effects are expressed as a time-variant factor of the linear instantaneous signal-to-noise ratio (SNR) of the received signal. Although such SNR-based models abstract the exact signal behavior, such as current phase shift, they support the separate calculation of channel effects at various timescale. This enables a trade-off between calculation complexity and model accuracy by selecting the modeled effects and timescale, specifically, reducing complexity by modeling channel variation per packet instead of per modulation symbol. Nonetheless, if more accuracy is required and increasing calculation complexity is feasible, MiXiM's modular structure enables to include models operating at a digital signal level such as modulation symbols. However, at the SNR abstraction level, MiXiM already includes widely accepted channel models for path loss, shadowing, and large- and small-scale fading (Simon and Alouini 2005).

8.4.21.1.3 Physical Layer Models

At the physical layer, essentially, the used modulation and forward error correction (FEC) coding and decoding functions define the bit error rate and throughput of a system. As for wireless channels, the effect of these functions can be modeled at an SNR level (Lichte and Valentin 2008).

MiXiM last version is 2.3. It was released on March 8, 2013 (Table 8.1).

8.4.22 NesCT

The NesCT project (OMNeT++ Wiki 2011) has been supported by the Featherlight project at the University of Twente, the Netherlands (University of Twente 2014), and the European Embedded WiseNt project (Information Society Technologies 2006) at Wireless Networks Laboratory at Yeditepe University, Turkey (Wireless Networks Laboratory 2015). NesCT is a programming language translator that uses the NesC programming language (Gay et al. 2003) as an input and produces C++ classes for OMNeT++. The primary aim is to provide a new simulation environment and speed up development.

NesCT enables to write code in NesC using TinyOS components while still making use of functionalities from OMNeT++ (Sect. 8.4.5) and the Mobility Framework (MF) by the Telecommunication Networks Group at Technische Universität Berlin (Löbbers and Willkomm 2007). Such approach grants the advantage of writing code for actual hardware. With small modifications, researchers can test their implementation on a given hardware platform. The original NesC code is translated into C++ classes.

It is worth noting that NesCT relies on other available simulation frameworks for physical and MAC layers. Thus, implementing both layers at NesCT would create needless efforts since the main focus of NesCT is to reuse TinyOS components with the available frameworks.

Currently, NesCT is compatible with OMNeT++ 4.2 and TinyOS 1.1.x releases. A NesCT release has been issued on August 4, 2011 (Table 8.1).

8.4.23   SUNSHINE

Sensor Unified Analyzer for Software and Hardware in Networked Environments (SUNSHINE), is developed at Virginia Polytechnic Institute and State University (Virginia Tech 2015). SUNSHINE is an open-source scalable hardware-software emulator for sensornet applications; it effectively supports joint evaluation and design of sensor hardware and software performance in a networked context (Zhang et al. 2011). SUNSHINE captures the performance of network protocols, software, and hardware up to cycle-level accuracy through its seamless integration of three available sensornet simulators, specifically, a network simulator TOSSIM (Sect. 8.4.6), an instruction-set simulator SimulAVR (Rivet and Klepp 2012), and a hardware simulator GEZEL (Schaumont et al. 2006). SUNSHINE handles several sensornet simulation challenges, including data exchanges and time synchronizations across different simulation domains and simulation accuracy levels. SUNSHINE also provides a hardware specification scheme for simulating flexible and customized hardware designs. Experimentation demonstrated that SUNSHINE is an efficient tool for software-hardware co-design in sensornet research.

SUNSHINE is developed to overcome the limitations of available simulators/ emulators. Sensornet simulators, such as TOSSIM (Sect. 8.4.6), ATEMU (Sect. 8.4.7), and Avrora (Sect. 8.4.8), focus on evaluating the designs of communication protocols and application software while assuming a fixed hardware platform that cannot accurately capture the impact of alternative hardware designs on the performance of network applications. As such, sensornet researchers cannot easily configure and evaluate diverse joint software-hardware designs and are bound within the constraints of available fixed sensor hardware platforms. This lack of simulator support makes it difficult to improve the sensor hardware platforms and their applications.

8.4.23.1   SUNSHINE Components

SUNSHINE buildup is based upon three sturdy constituents:

- TOSSIM. An event-based simulator for TinyOS-based WSNs (Sect. 8.4.6). TOSSIM is able to simulate a complete TinyOS-based sensor network as well as capture the network behavior and interactions. TOSSIM provides functional-level abstract implementations of both software and hardware modules for several sensor node architectures, such as the MICAz mote (Crossbow 2006a). In TOSSIM, sensor nodes' behaviors are regarded as functional-level events that are kept sequentially in TOSSIM's event queue according to the events' timestamps. These events are processed in ascending order of their timestamps.
- Even though TOSSIM is capable of capturing the sensor motes' behaviors and interactions, such as packet transmission, reception, and packet losses at a high fidelity, it does not consider the sensor motes' processors' execution time. Therefore, TOSSIM cannot capture the fine-grained timing and interrupt properties of software code.

- SimulAVR. An instruction-set simulator that supports software domain simulation for the popular Atmel AVR family of microcontrollers (Rivet and Klepp 2012). SimulAVR provides precise timing of software execution and can simulate multiple AVR microcontrollers in one simulation. SimulAVR is integrated into the hardware domain simulator in SUNSHINE to evaluate interactions between sensor hardware and software. As SimulAVR does not support simulating sleep or wakeup modes of sensor nodes, they are added to SUNSHINE to provide simulation support for energy-saving mode of sensor networks.
- GEZEL. A hardware domain simulator that includes a simulation kernel and a hardware description language (Schaumont et al. 2006). GEZEL is an open-source tool that offers stand-alone simulation, cosimulation, and code generation into synthesizable (VHDL) code. Through user-defined library-block extensions in C++, new cosimulation interfaces can be added.
- In GEZEL, a platform is defined as the combination of a microprocessor connected with one or more other hardware modules such as a coprocessor or radio chip. To simulate the operations of such a platform, there should be a combination of software simulation domain to capture software executions over the microprocessor and hardware simulation domain that captures the behaviors of hardware modules and their interaction with the microprocessor. GEZEL is capable of providing a hardware-software co-design environment that seamlessly integrates the hardware and software simulation domains at cycle-level. GEZEL models can be directly translated into a hardware implementation, which permits determining the functional correctness of that custom hardware within the actual system context and monitoring cycle-accurate performance metrics for the design.
- Among several applications, GEZEL has been used for cryptographic hashing modules (Knezzevic et al. 2008) and formal verification of security properties of hardware modules (Köpf and Basin 2007).

8.4.23.2   SUNSHINE Functioning

By integrating TOSSIM, SimulAVR, and GEZEL, SUNSHINE enables simulating sensornet in network, software, and hardware domains. A user of SUNSHINE has thus a twofold choice:

- Selecting a subset of sensor nodes to be emulated in hardware and software domains. These nodes are called cycle-level hardware-software co-simulated (*co-sim*) nodes; their cycle-level behaviors are accurately captured by SimulAVR and GEZEL.
- Simulating other nodes in the network domain by TOSSIM and capturing only the high-level functional behaviors. These nodes are named TOSSIM nodes.

SUNSHINE is capable of running multiple *co-sim* nodes with TOSSIM nodes in one simulation. The network topology on the right side of Fig. 8.22 illustrates the

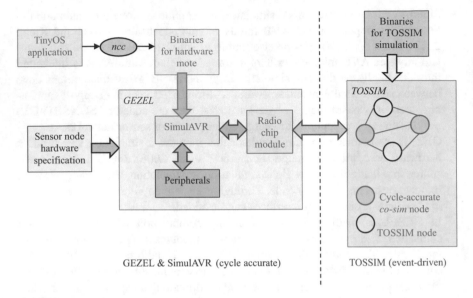

Fig. 8.22 SUNSHINE functioning (Zhang et al. 2011)

functioning of SUNSHINE. TOSSIM nodes are simulated in the network domain, while *co-sim* nodes are emulated in software and hardware domains. When running simulation, TOSSIM nodes and *co-sim* nodes interact with each other according to the network configuration and sensornet applications. Cycle-level *co-sim* nodes detail sensor nodes' behaviors, such as hardware behavior, but are relatively slower to simulate. On the other hand, TOSSIM nodes do not simulate many details of the sensor nodes but are simulated much faster. The mix of cycle-level simulation with event-based simulation guarantees that SUNSHINE can achieve the fidelity of the cycle-accurate simulation while still benefiting from the scalability of event-driven simulation.

The simulation process in SUNSHINE, as illustrated in Fig. 8.1, goes through several steps:

- For *co-sim* nodes that emulate real sensor motes, executable binaries are compiled from TinyOS applications using nesC compiler (*ncc*) and executed directly over these *co-sim* nodes that emulate a hardware platform at the cycle level.
- TinyOS executable binaries can be interpreted by SimulAVR, instruction-by-instruction. At the same time, GEZEL interprets the sensor node's hardware architecture description and simulates the AVR microcontroller's interactions with other hardware modules at every clock cycle. The GEZEL simulated radio chip module provides an interface to TOSSIM that models the wireless communication channels. Through these wireless channels, *co-sim* nodes interact with other sensor nodes, whether simulated as *co-sim* nodes by GEZEL and SimulAVR or as functional-level nodes by TOSSIM. The correct causal relationship that rules the interactions between TOSSIM nodes and *co-sim* nodes is based on timing synchronization and cross-domain data exchange techniques.

8.4.23.3 Cross-Domain Interface

To permit the simulated AVR microcontroller to exchange data with the simulated hardware modules, GEZEL creates cycle accurate hardware-software co-simulation interfaces according to the AVR microcontroller's datasheet (Crossbow 2002). With the support of GEZEL's co-simulation interfaces, SUNSHINE forms an emulator (*P-sim*) that captures the sensor nodes' hardware-software interactions and performance.

Network simulator TOSSIM and hardware-software emulator, *P-sim*, are integrated into SUNSHINE for the sake of scalability. Proper synchronization is achieved to obtain a match in simulation time between event-driven simulation and cycle-level simulation. For functionality, SUNSHINE includes a time synchronization scheme as depicted in Fig. 8.23. TOSSIM uses the Event Scheduler to handle all the network events, while *P-sim* uses the Cycle-level Simulation Engine to control at every clock cycle the simulation of the AVR microcontroller and the hardware modules. In the Event Queue, all network events are sorted according to timestamps that record their occurrence time. The Event Scheduler processes the head-of-line (HOL) event in the Event Queue only when the Cycle-level Simulation Engine has progressed to the event's timestamp. By selecting either an event or a cycle-level simulation to be simulated next, SUNSHINE maintains the correct causality between different simulation schemes in the whole network.

As shown in Fig. 8.23, SUNSHINE also provides synchronization for *co-sim* nodes in sleep mode by maintaining an Active Node List that holds the active nodes to be simulated with cycle-level accuracy. The Event Scheduler adds or removes nodes from the list upon node wakeup or sleep events. At each cycle-level simulation step, the Cycle-level Simulation Engine only processes a clock cycle for the nodes of the Active Nodes List.

For the integration of the simulators working in three different domains, SUNSHINE implements interfaces for cross-domain data exchange between SimulAVR and GEZEL, and between hardware-software emulator *P-sim* with event-based

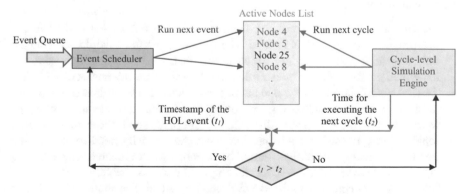

Fig. 8.23 Synchronization scheme (Zhang et al. 2011)

simulator TOSSIM. SUNSHINE experimentation integrates TinyOS version 2.1.1, SimulAVR and GEZEL version 2.5.

8.4.23.4   SUNSHINE Compared

Contending with TOSSIM, Avrora, and ATEMU, SUNSHINE provides hardware flexibility and extensibility, permits user-defined platform architecture, and allows transition between event-based and cycle-accurate simulation.

SUNSHINE is available for download (Table 8.1).

8.4.24   NetTopo

NetTopo builds a new software framework that integrates simulation and visualization functions to assist in the investigation of various WSN algorithms (Shu et al. 2011). The following two concrete scenarios especially motivate this integration:

- Researchers wantto compare the performance of running the same algorithm on both a simulator and a real testbed. The comparison can guide to improve the algorithm design and incorporate more realistic conditions. A good example is applying the face routing algorithm in greedy perimeter stateless routing (GPSR), which proved to be loop-free in theory (Karp and Kung 2000), but it is not so in realistic situations due to irregular radio coverage (Seada et al. 2007).
- Budget limitation preventing researchers from buying enough real sensor nodes for a large-scale WSN. For instance, to evaluate the performance of any sensor middleware, a large-scale sensor network is needed (Aberer et al. 2007). Hence, researchers can actually work by integrating a small number of real sensor nodes with a large number of virtual sensor nodes generated from the simulator.

The integration of simulation environment and physical testbed brings three major challenges:

- Sensor node simulation. Normally, a number of heterogeneous sensor devices can be used for building a WSN testbed. The integrated platform should not simulate only a specific sensor device, which means that, for heterogeneity's sake, the integrated platform should be flexible enough to simulate any new sensor device.
- Testbed visualization. On the one hand, sensor nodes are small in size and do not have user interfaces such as displays or keyboards, which makes the tracking of testbed communication status difficult. On the other hand, the communication topology in a testbed is invisible, but researchers usually need to look at the topology to analyze their algorithms; for example, when implementing a routing algorithm in the testbed, the actual routing path must be visible.
- Interaction between the simulated WSN and the testbed. The simulated WSN and the real testbed need to exchange information, e.g., routing packets. Their

horizontal interconnection, communication, and collaboration are all emerging challenging issues that need to be addressed.

NetTopo is an extensible integrated framework of simulation and visualization that assists in investigating WSN algorithms:

- In the simulation module, users can easily define a large number of on-demand initial parameters for sensor nodes, such as residual energy, transmission bandwidth, and radio radius. Users also can define and extend the internal processing behavior of sensor nodes, like energy consumption and bandwidth management. NetTopo allows users to simulate a large-scale heterogeneous WSN.
- For the visualization module, it works as a plug-in component to visualize testbed connection status, topology, sensed data, etc.

These two modules integrate the virtual sensor nodes and links on the same canvas. Since node attributes and internal operations are user-definable, simulated virtual nodes are guaranteed to have the same properties as those of real nodes. The sensed data captured from real sensor nodes can drive the simulation in a predeployed virtual WSN. Topology layouts and algorithms of virtual WSNs are customizable and work as user-defined plug-ins, which can easily match the corresponding topology and algorithms of a real WSN testbed.

Since NetTopo is an integrated framework of simulation and visualization for WSNs, network simulation and WSN visualization are the two main aspects of simulator comparison. Simulators can be classified into three major categories based on the level of complexity:

- Algorithm level. These simulators focus on the logic, data structure, and presentation of algorithms. They do not consider detailed communication models and, most commonly, rely on some form of graph data structure to illustrate the communication between nodes. Typical such simulators are Shawn, AlgoSensim, and Sinalgo, as respectively presented in Sect. 8.4.16 (Jacques and Marculescu 2010; Swiss Federal Institute of Technology 2013).
- Packet level. Such simulators implement the data link and physical layers in a typical OSI network stack. Hence, it is common for these types of simulators to implement 802.11b or newer MAC protocols and radio models that account for propagation, noise and wave diffraction, fading, and collision. This class of simulators includes ns-2, GloMoSim, SensorSim, and J-Sim as respectively introduced in Sects. 8.4.1, 8.4.3, 8.4.10, and 8.4.12.
- Instruction level. These simulators model the CPU execution at the level of instructions or even cycles; they are often regarded as emulators. This class contains TOSSIM, ATEMU, and Avrora, as elaborated respectively in Sects. 8.4.6, 8.4.7, and 8.4.8.

Based on the above classification, the simulation framework in NetTopo is by far algorithm-oriented. Additionally, the abovementioned simulators did not consider integrating with a real WSN testbed, except for one research on sQualNet (Sect. 8.4.3), which is an extension of Qualnet motivated by using a real testbed to simulate the network models. However, sQualNet did not consider the real testbed visualization.

With respect to visualization of real WSN testbed, there are much less related works; typically, MOTE-VIEW (Crossbow 2006b) and SpyGlass (Buschmann et al. 2005). Most visualization tools support a single type of WSNs and are highly coupled to TinyOS. However, NetTopo targets the visualization and control of a WSN testbed where heterogeneous devices are used, such as wireless cameras and Bluetooth-based body monitoring sensor devices, knowing that these devices are generally not TinyOS-based.

The process of problem-solving and planning for the NetTopo framework depends on the requirement specification. The specification is divided into functional and nonfunctional requirements.

Functional requirements include system-level functions and node-level functions that must be fulfilled:

- System-level functions to be achieved:

  - Providing a graphical user interface on which network deployment, simulation, and visualization are to be displayed.
  - Allowing the interaction and communication between simulation and visualization of WSNs.
  - Being extensible to provide device-based wrappers for WSN visualization.
  - Providing a set of high-level APIs for users to simulate their own algorithms.
  - Offering functions for debugging and logging simulation process.
  - Recording simulation results.
  - Supporting sensor data streams (XML streams) import and export.
  - Running file-related operations, including file saving, creating, and opening, to keep the deployment state and support for future reuse.
  - Supporting WSN 3D representation.

- Node-level functions that must be realized:

  - Providing basic symbols representing elements in a WSN including source node, sensor node, sink node, hole, and link. All these symbols can be used to build users' own expected network.
  - Being extensible to allow users to define expected attributes of their own virtual sensor nodes.
  - Allowing users to create their own network with maximum freedom, e.g., adding, deleting, selecting, modifying, moving, searching, distributing sensor nodes, and configuring their attributes, such as energy, expected lifetime, and transmission radius.
  - Permitting users to deploy virtual sensor nodes and holes in predefined network topologies including line, circle, tree, grid, and random or user-defined topologies.
  - Letting users disable virtual sensor nodes to make them holes.
  - Allowing users to clear all deployed virtual sensor nodes.

Nonfunctional requirements that the system must satisfy include:

- Being programmed in Java so as to integrate with a java-based WSN middleware called global sensor networks (GSNs) (Aberer et al. 2006).
- Having a local operating system look and feel.
- Warning users that they cannot reconfigure virtual sensor nodes or WSN after the simulation start, if they do so.
- The coordinates of any virtual sensor node to be created or modified must be different from those existing.

As required and constrained by the requirements specifications, the whole system environment involves five entities: explicitly, user, NetTopo framework, persistent data, sensor devices, and gateway. From the users' point of view, NetTopo, regarded as a software application, is located above an operating system in a computer. Persistent data stored in the local machine are transparent to the user. Besides the application, the user is aware of real sensor devices. Due to the heterogeneity and hardware-specific properties of sensor devices, a gateway, working as an intermediate node, is incorporated for adapting communication between NetTopo and sensor devices.

Figure 8.24 shows the NetTopo architecture. Based on this architecture, the external entities that NetTopo needs to communicate with are user, local computer, and gateway. Inside the NetTopo framework, there are four components: GUI, virtual WSN, simulator, and visualizer.

The figure shows not only the relationship between the five entities but also an overview of the inner framework of NetTopo. The user sends commands via the graphical user interface (GUI) of NetTopo and waits for response, which is a visual change on GUI. The two main components, the simulator and the visualizer, can interact with each other indirectly by sharing the same virtual WSN, which stores temporary data copied from persistent data.

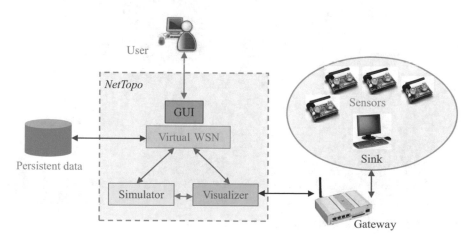

Fig. 8.24 NetTopo architecture (Shu et al. 2011)

From a network point of view, the simulator component runs locally on the machine without communicating with outside entities. However, the visualizer component should gather information from the gateway. The gateway, working as an adaptor, is responsible for communicating with various kinds of hardware devices. This information exchange could be based on various protocols, such as TCP/IP Socket and HTTP Get, and it needs to set up the corresponding driver software of devices from which NetTopo can get all kinds of sensed data. Especially, the GSN middleware can be installed in the gateway for gathering sensor data from different real sensor networks and shaping these heterogeneous data into a uniform format. By receiving the processed sensor data streams, mostly XML streams, from GSN, the NetTopo framework can be easily extended to support the visualization of multiple different real WSNs.

A screenshot of the NetTopo GUI is depicted in Fig. 8.25.

NetTopo is released as open source available from the authors.

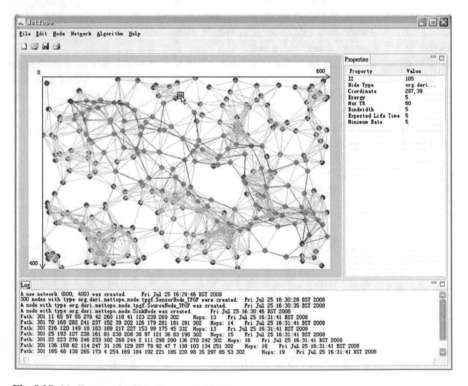

Fig. 8.25 NetTopo main GUI (Shu et al. 2011)

Table 8.1 Simulators/emulators compared

	Simulator/emulator	Features	General/wireless/WSN[a]	Programming language	Download site
ns-2 (Sect. 8.4.1)	Simulator	Discrete-event, open-source, object-oriented simulator. It provides support for wired and wireless networks	General	OTcl, C++	http://sourceforge.net/projects/nsnam/files/allinone/ns-allinone-2.35/
ns-3 (Sect. 8.4.2)	Simulator and emulator	Discrete-event, open-source, network simulator that supports research on wireless/IP networks, which involve models for Wi-Fi, WiMAX, or LTE for layers 1,2 and a variety of static or dynamic routing protocols such as OLSR and AODV for IP-based applications. Allows developing models to make it a real-time network emulator	Wireless networks	C++, Python	https://www.nsnam.org/ns-3-dev/download/
GloMoSim (Sect. 8.4.3)	Simulator	Discrete-event, open-source, object-oriented simulator. Parallel execution may be used to study large-sized IP networks	Wireless networks	C++, PARSEC	http://web.scalable-networks.com/content/landing/run-time-licensing
OPNET (Sect. 8.4.4)	Simulator	Discrete-event, open-source, object-oriented simulator. It employs a hierarchical structure for modeling. A wide range of protocols is supported at MAC, network, and transport layers, as well as several applications. Both wired and wireless networks are considered. Commercial and academic versions are available	General and some WSN	C, Java	http://www.opnet.com/university_program/itguru_academic_edition/

(continued)

Table 8.1 (continued)

	Simulator/ emulator	Features	General/ wireless/ WSN[a]	Programming language	Download site
OMNeT++ (Sect. 8.4.5)	Simulator and some emulation	Discrete-event, open-source object-oriented simulation tool for academic, educational, and research-oriented commercial institutions for the simulation of computer networks and distributed or parallel systems	General	C++, NED	http://omnetpp.org/omnetpp
TOSSIM (Sect. 8.4.6)	Simulator and emulator	Emulates at the bit level WSNs operating under TinyOS	WSN	C/C++, Python	http://www.di.unipi.it/~ste/MaD-WiSe/download-11.htm
ATEMU (Sect. 8.4.7)	Simulator and emulator	Open-source low-level emulation of the Mica2 platform and customizable to support different hardware platforms. Discrete event simulation of WSNs	WSN	C	http://sourceforge.net/projects/atemu/files/
Avrora (Sect. 8.4.8)	Simulator and emulator	Full emulation of Mica2 and MicaZ, with ATMega128L and CC1000 radio implementations. Open-source, discrete-event simulator of sensor networks with timing accuracy, allowing programs' communication via the radio using the software stack in TinyOS	WSN	Java	http://sourceforge.net/projects/avrora/files/
EmStar (Sect. 8.4.9)	Simulator and emulator	EmSim is a pure time-driven simulation model that allows software to be developed and debugged. EmCee allows emulation using Mica2 motes. EmTOS emulator enables NesC/TinyOS code to run as an EmStar module	WSN	NesC	—

	Type	Description	Category	Language	URL
SensorSim (Sect. 8.4.10)	Simulator	Focused on extending ns-2, it provides an advanced power model, sensor channel, interaction mechanism with external applications, and middleware platform to permit the dynamic management of nodes in simulation. Did not come to life	WSN	C++	Unavailable
NRL SensorSim (Sect. 8.4.11)	Simulator	Focused on extending ns-2, it tackles the notion of a phenomenon that could trigger nearby sensors. Is not maintained anymore	WSN	C++, Tcl[b]	http://downloads.pf.itd.nrl.navy.mil/archive/nrlsensorsim/
J-Sim (Sect. 8.4.12)	Simulator and emulator	An open-source, real-time, process-driven, application development framework, component-based compositional network simulation environment. Power models for CPU and radio communication are provided. It also includes a set of classes and mechanisms that realize WSN emulation based on Berkeley mica motes	General	Java	https://sites.google.com/site/jsimofficial/downloads
Prowler/JProwler (Sect. 8.4.13)	Simulator	Event-driven simulator that permits algorithms' testing in deterministic, probabilistic, and dynamically changing environments. Using it under MATLAB makes its use fast and easy, with easy debugging. Visualization makes it user-friendly. The supported application areas are communication protocols and routing, arbitrary application prototyping, and optimization and parameter tuning	WSN	MATLAB for prowler/Java for JProwler	http://www.isis.vanderbilt.edu/projects/nest#tabs%2D%2D4

(continued)

Table 8.1 (continued)

	Simulator/ emulator	Features	General/ wireless/ WSN[a]	Programming language	Download site
SENS (Sect. 8.4.14)	Simulator	It has a modular, layered architecture with customizable components that model an application, network communication, and physical environment. It also enables realistic simulation by using values from Mica2 motes to include sound, radio signal strength, and power usage. Customizability allows users to assemble application-specific environments. The same source code used on simulated nodes can be deployed on actual sensor nodes, which permits application portability. To provide users with the flexibility of modeling the environment and its interaction with applications at different levels of detail, SENS defines the environment as a grid of interchangeable tiles	WSN	C++	http://osl.cs.illinois.edu/sens/
SENSE (Sect. 8.4.15)	Simulator	It is a component-based software environment that supports parallelization. It aims at achieving extensibility, reusability, and scalability. The component-based model and autonomous nature of components support extensibility and reusability, while packet sharing improves scalability. Several components are made available, such as battery and power models, application, network, MAC, and physical layer functionalities	WSN	C++, Tcl, XML	http://www.ita.cs.rpi.edu

Shawn (Sect. 8.4.16)	Simulator	Customizable and open-source designed to support large-scale networks simulation. It simulates the effects of the phenomenon, thus it cannot provide the detailed level that simulators such as ns-2 provide with respect to physical layer or packet-level phenomena. Shawn consists of three major parts, namely, models, sequencer, and simulation environment	WSN	C++	http://sourceforge.net/projects/shawn/files/shawn/Shawn-2007-06-06_source/
SenSim (Sect. 8.4.17)	Simulator	OMNeT++-based, discrete-event, open-source object-oriented simulator. It allows researchers to develop new sensor networks protocols and investigate networking and scalability. The hardware model of a node comprises battery, radio, and CPU modules. A wireless channel module controls and establishes connections between nodes	WSN	C++, NED	http://sensim.software.informer.com/3.0/
PAWiS (Sect. 8.4.18)	Simulator and Emulator	OMNeT++-based, discrete-event, open-source object-oriented simulator. Focus is given to power efficiency and thus on capturing inefficiencies in system aspects. Aspects include all layers of the communication system, the targeted class of application, the power supply and energy management, the CPU, and the sensor-actuator interface. User can define dynamic behaviors such as mobility	WSN	C++, NED, Lua	http://sourceforge.net/projects/pawis/files/
MSPsim (Sect. 8.4.19)	Emulator	Emulator for WSN nodes based on Texas Instruments MSP430 microcontrollers	WSN	Java	http://sourceforge.net/projects/mspsim/files/

(continued)

Table 8.1 (continued)

Simulator/ emulator		Features	General/ wireless/ WSN[a]	Programming language	Download site
Castalia (Sect. 8.4.20)	Simulator	OMNeT++-based, discrete-event, open-source object-oriented application-level simulator for researchers and developers who want to test their distributed algorithms and/or protocols in realistic wireless channel and radio models, with realistic node behavior	WSN	C++, NED	https://forge.nicta.com.au/frs/?group_id=301
MiXiM (Sect. 8.4.21)	Simulator	OMNeT++-based, discrete-event, open-source object-oriented simulator. It is a cross-level platform integrating ChSim, MAC simulator, MF, and PF. Each logical network layer can be implemented by the user. The framework supports mobility, models for obstacles as well as several protocols	WSN	C++, NED	http://sourceforge.net/projects/mixim/files/mixim/
NesCT (Sect. 8.4.22)	Simulator	OMNeT++-based, discrete-event, open-source object-oriented simulator. NesCT enables to write code in NesC using TinyOS components while still making use of functionalities from OMNeT++ and the Mobility Framework (MF). It relies on other available simulation frameworks for physical and MAC layers	WSN	C++, NesC	http://sourceforge.net/projects/nesct/files/

SUNSHINE (Sect. 8.4.23)	Simulator and emulator	SUNSHINE integrates a network simulator TOSSIM, an instruction-set simulator SimulAVR, and a hardware simulator GEZEL. It provides hardware flexibility and extensibility, permits user-defined platform architecture, and allows transition between event-based and cycle-accurate simulation	WSN	C/C++ Python, VHDL	http://rijndael.ece.vt.edu/sunshine/downloads.html
NetTopo (Sect. 8.4.24)	Simulator	NetTopo is an integrated framework for simulation and visualization. Due to its algorithm-oriented design, it supports the simulation of large-scale WSNs. It is useful for rapid prototyping of an algorithm. The visualization function uncovers the real device-based WSN topology and displays sensed data. Based on modular components design and common graphical resources, visualization can drive the simulation	WSN	Java	At the authors

[a]*General*: For wired networks and for wireless networks in general. *Wireless*: For wireless networks only. *WSN*: For WSN specifically

[b]Tcl (Tool command language) is a very powerful but easy to learn dynamic programming language, suitable for a very wide range of uses, including web and desktop applications, networking, administration, testing and many more. Opensource and business-friendly, Tcl is a mature yet evolving language that is truly cross platform, easily deployed and highly extensible (Tcl Developer Xchange 2015)

8.5   Conclusion for Takeoff

Writing is intermixed with life events and moved by feelings, not just local but also all over the planet. On Monday, May 25, 2020, George Perry Floyd Jr. (October 14, 1973–May 25, 2020), an African-American man, was killed by police during an arrest in Minneapolis. Protests in response to both Floyd's death and, more broadly, police violence quickly spread the burdened screams everywhere.

In this chapter, an in-depth study of simulators and emulators has been presented, with care accorded to their features, implementation, and use. Since emulators are hardware-dependent, selecting one to use is straightforward. On the other hand, with the wide variety of simulators, the choice is rather complex, and is subject mainly to how is the simulator easy to use, and convenient for the model requirements. Remarkably, different simulators do not give similar results for the same model due to their different underlying features and implementations.

Simulation has proven to be a valued tool in many areas where analytical methods are not applicable and experimentation is not feasible. Researchers generally use simulation to analyze system performance prior to physical design or to compare multiple alternatives over a wide range of conditions. Notably, errors in simulation models or improper data analysis often produce incorrect or misleading results. Although there exists an extensive row of performance evaluation tools for WSNs, it is impractical to have an all-in-one integrated tool that simultaneously supports simulation, emulation, and testbed implementation.

In fact, there is no all-in-one stretchy simulator for WSNs. Each simulator exhibits different features and models, and each has advantages and weaknesses. Different simulators are appropriate and most effective in typical conditions, so in choosing a simulation tool from available picks, it is fruitful to elect a simulator that is best suited for the intended study and targeted application. Also, it is recommended to weigh the pros and cons of different simulators that do the same job: the level of complexity of each simulator, availability, extensibility, and scalability. Usually, WSN applications consist of a large number of sensor nodes; therefore it is recommended to settle on a simulation tool capable of simulating large-scale WSNs. Essentially, the reported simulator use and the obtained simulation results of a simulator, should not be ignored before deciding which simulator to prefer. The exercises at the end of the chapter are designed to pinpoint the simulator comparison and selection criteria suitable to the model under study.

As projected in this chapter, many simulators have been developed in the past 15 years to account for the broad range of applications entailed in the unbounded WSN realm. Each of the simulators targets a specific application domain in which it can deliver the best results. The semantics of what is actually meant by "simulation" varies heavily among researchers in their publications, depending on the goals of the simulations in question. This sometimes results in focusing on the simulation of physical phenomena such as radio signal propagation characteristics and ISO/OSI layer protocols, e.g., media access control (MAC). On the other hand, other approaches focus on algorithmic aspects, and hence they abstract the lower layers.

The first approach delivers a precise image of what happens in real networks and how the protocols interact with each other at the anticipated cost of resource-demanding simulations, leading accordingly to scalability problems. The second approach employs abstract models of the real world, instead of simulating it down to the bit level. Important questions, thus, arise about the analysis of the network structure as well as the design and evaluation of algorithms, not protocols.

When bottom-up building a simulator, many decisions need to be made. Developers must consider the pros and cons of different programming languages, whether simulation is event-based or time-based, component-based, or object-oriented architecture, the level of complexity of the simulator, features to include and to not include, use of parallel execution, ability to interact with real nodes, and other design choices that are pertinent to a typical application. Weighting decisions can thus be summarized as below listed:

- For sake of efficiency, most simulators use a discrete event engine.
- Component-based architectures scale significantly better than object-oriented architectures but are more difficult to implement in a modularized way. Defining each sensor as its own object ensures independence among the nodes. The ease of swapping in new algorithms for different protocols also appears to be easier in object-oriented designs. However, with careful programming, component-based architectures may perform more efficiently.
- With the dominant use of C++ and Java, the use of the implementation language depends on the intended application; nevertheless, there is no clear-cut answer that favors a language or makes it a default simulation language.
- Generally, the level of complexity built into the simulator has a lot to do with the goals of the developers and the time constraints imposed. It is satisfactory and time-efficient if the simulator job is achieved, using a simple MAC protocol or with less protocols.
- Other design choices are dependent on intended situation, programmer's ability, and available design time.
- A simulator is a professional work intended for researchers at large. It should encompass as many models as possible and provide a benchmark for comparison between different simulation studies. The major difference between using a professional simulator and a modeling program written for a specific research is trust. A custom-made modeling program may provide biased or incomplete comparisons between a proposed work and the literature, which blemishes the results' credibility. Focusing on one's research is a tedious task that should not be deviated by writing a modeling program that is hard by itself and cannot match the professional simulator functionalities and efficiency.
- A simulator is a research necessity when building a simulator; developers' efforts must match researchers' concerns of clarity and ease of use, as well as GUI facilities. Some researchers evade the harder task of learning how to use a professional simulator by building their own modeling programs, which results in non-truly compared work.

For researchers, choosing which simulator to use is not an easy duty. A full understanding of one's own model is however the first major step before looking into the bookshelf of simulators. Then follows a survey of the available simulators that can do the job. A major step comes after, the careful weighting of the simulator features against the model under study and the programming capabilities of the researcher. Annoyingly, a suitable simulator for the model may reveal the unfitness of the researcher on how to use and program, such a non-pleasant barrier that may restrain some researchers from wasting time learning a simulator. Learning how to use a professional simulator may cost extra months of overhead on the research work, a price to be afforded for trustworthy research. Understanding a simulator means having full awareness of questions and answers about the traffic pattern, the area size, the number of nodes, the node density, the routing protocol used, the MAC layer handling of collisions, the power model, the radio model, the mobility pattern, etc. A simulator should not be thought of as a black box that just receives inputs and produces outputs; inputs as well as the inner simulator structure should conform to the model under study. Simulator outputs should not be taken for true before adjusting and tuning the simulator to the research necessities and running several case studies that range from small to larger.

Conclusively, who drives whom? Is the researcher simulator-driven? Or it is the other way?

8.6 Exercises

Note: A technical report follows the template of a peer-reviewed journal paper.

1. What are the requirements for a credible simulation?
2. Explain the interrelationship between the different models used for simulation.
3. When is validation to be used?
4. How may simulation be incorrectly done?
5. Based on Sect. 8.3, perform a study on the latest two MobiHoc conferences.
6. Some simulators are open-source. Discuss the pros and cons of such approach.
7. Compare the simulators that are built upon object-oriented programming. As illustrated in Sect. 8.4, identify the impact of object-oriented programming on simulator performance.
8. Write a technical report on the component-based software architecture. Do not stop at what has been made available in Sect. 8.4.12.1.
9. Differentiate between object-oriented and component-based simulators.
10. Java and C++ languages are commonly used by the simulators presented in Sect. 8.4. Identify the resemblances and differences that arise when using such simulators.
11. Which of the simulators considered in Sect. 8.4 never came to life, and which are no longer maintained?

12. Which of the simulators illustrated in Sect. 8.4 are for general networking? Which are for wireless networking? And which are for WSNs?
13. Is ns-3 a replacement of ns-2? Clarify your answer in a technical report.
14. Identify and compare the simulators based on OMNeT++.
15. SENSE (Sect. 8.4.15) is built on top of COST, a discrete event simulator whose design was influenced by the concepts of component-based software architecture and component-based simulation. Write a report on COST and how it compares with the simulators presented in Sect. 8.4.
16. Compare the simulators presented in Sect. 8.4 based on their power modeling.
17. Compare the simulators presented in Sect. 8.4 based on their radio modeling.
18. Identify and compare the component-based simulators as laid out in Sect. 8.4.
19. A simulator should not be used as a black box. Discuss technically.
20. Compare the emulators presented in this chapter.

References

Aberer, K., M. Hauswirth, and A. Salehi. 2006. Global Sensor Networks. In *Communications Magazine*. IEEE, pp. 1–14.
———. 2007. Infrastructure for Data Processing in Large-Scale Interconnected Sensor Networks. In *The 8th International Conference on Mobile Data Management (MDM)*. Mannheim: IEEE, pp. 198–205.
Accellera. 2015, January 1. *SystemC*. Accellera Systems Initiative. http://accellera.org/downloads/standards/systemc. Accessed 29 June 2015.
ACM SIGMOBILE. 2000, August 11. *MobiHoc: The ACM International Symposium on Mobile Ad Hoc Networking and Computing*. ACM SIGMOBILE. http://sigmobile.org/mobihoc/. Accessed 18 Mar 2015.
Andel, T.R., and A. Yasinsac. 2006. On the Credibility of Manet Simulations. *Computer* 39 (7): 48–54.
Applied Dynamics International. 2015, January 1. *What is Hardware-In-the-Loop Simulation?* Applied Dynamics International. http://www.adi.com/technology/tech-apps/what-is-hardware-in-the-loop-simulation/. Accessed 27 May 2015.
Atmel. 2015, January 1. *Atmel AVR 8-bit and 32-bit Microcontrollers*. http://www.atmel.com/products/microcontrollers/avr/. Accessed 17 April 2015.
Australian Government. 2015, January 1. *Australian Government ICT Entry-level Programs*. Australian Government. http://www.australia.gov.au/ictentrylevel. Accessed 28 May 2015.
Avrora. 2004, December 17. *Avrora: The AVR Simulation and Analysis Framework*. http://compilers.cs.ucla.edu/avrora/. Accessed 17 April 2015.
Bagrodia, R., R. Meyer, M. Takai, Y.-A. Chen, X. Chen Zeng, J. Martin, and H.Y. Song. 1998. Parsec: A Parallel Simulation Environment for Complex Systems. *Computer* 31 (10): 77–85.
Balci, O. 1994. Validation, Verification, and Testing Techniques Throughout the Life Cycle of a Simulation Study. *Annals of Operations Research* 53 (1): 121–173.
Barberis, A., L. Barboni, and M. Valle. 2007. Evaluating Energy Consumption in Wireless Sensor Networks Applications. In *10th Euromicro Digital System Design Architectures, Methods and Tools (DSD)*, 455–462. Lubeck: IEEE.
Bianchi, G. 2000. Performance Analysis of the IEEE 802.11 Distributed Coordination Function. *IEEE Journal on Selected Areas in Communications (IEEE)* 18 (3): 535–547.
Blazakis, D., J. McGee, D. Rusk, and J.S. Baras. 2004. ATEMU: A Fine-Grained Sensor Network Simulator. In *First Annual IEEE Communications Society Conference on Sensor and Ad Hoc Communications and Networks (SECON)*, 145–152. Santa Clara: IEEE.

Brakmo, L.S., and L.L. Peterson. 1996. Experiences with Network Simulation. In *ACM SIGMETRICS International Conference on Measurement and Modeling of Computer Systems*, 80–90. Philadelphia: ACM.

Buchheim, T. 2002, July 3. *Nam: Network Animator*. ISI. http://www.isi.edu/nsnam/nam/. Accessed 26 April 2015.

Buschmann, C., D. Pfisterer, S. Fischer, S.P.P. Fekete, and A. Kröller. 2005. SpyGlass: A Wireless Sensor Network Visualizer. *SIGBED Review* 2 (1): 1–6.

Castalia. 2013, January 1. *Castalia*. https://castalia.forge.nicta.com.au/index.php/en/19-sample-data-articles/castalia/24-homepage.html. Accessed 28 May 2015.

CENS. 2015, January 1. *Generalized Software Tools: EMSTAR*. CENS. http://research.cens.ucla.edu/projects/2006/Systems_Infrastructure/Emstar/default.htm. Accessed 20 April 2015.

Chang, X. 1999. Private Network-to-Network Interface. In *The 31st Conference on Winter Simulation: Simulation a Bridge to the Future (WSC)*. Phoenix: IEEE, pp. 307–314.

Chen, X. 2002, March 31. *CONSER: Collaborative Simulation for Education and Research*. http://www.isi.edu/conser/index.html. Accessed 26 April 2015.

Chen, G., and B.K. Szymanski. 2002. COST: Component-Oriented Simulation Toolkit. In *The Winter Simulation Conference (WSC)*. San Diego: The INFORMS Computing Society (ICS).

Chen, G., J. Branch, M. Pflug, L. Zhu, and B. Szymanski. 2005. Sense: A Wireless Sensor Network Simulator. In *Advances in Pervasive Computing and Networking*, ed. B.K. Szymanski and B. Yener, 249–267. New York: Springer.

Colesanti, U.M., C. Crociani, and A. Vitaletti. 2007. On the Accuracy of OMNeT++ in the Wireless Sensor Networks Domain: Simulation vs. Testbed. In *The 4th ACM Workshop on Performance Evaluation of Wireless Ad hoc, Sensor, and Ubiquitous Networks (PE-WASUN)*, 25–31. Chania: ACM.

Contiki Developers. 2015, January 1. *Get Started with Contiki*. Contiki Developers. http://www.contiki-os.org/start.html. Accessed 8 July 2015.

Coulson, G., et al. 2012. Flexible Experimentation in Wireless Sensor Networks. *Communications of the ACM* 55 (1): 82–90.

Crossbow. 2002, January 1. *MICA2*. http://www.eol.ucar.edu/isf/facilities/isa/internal/CrossBow/DataSheets/mica2.pdf. Accessed 3 Feb 2014.

———. 2006a, January 1. *MICAz*. http://www.openautomation.net/uploadsproductos/micaz_datasheet.pdf. Accessed 5 Feb 2014.

———. 2006b, January 1. *MOTE-VIEW 1.2 User's Manual*. http://www.willow.co.uk/MOTE-VIEW_User_Manual_.pdf. Accessed 20 Jan 2014.

Culler, D., J. Hill, M. Horton, K. Pister, R. Szewczyk, and A. Woo. 2002, April 1. *MICA: The Commercialization of Microsensor Motes*. http://www.sensorsmag.com/networking-communications/mica-the-commercialization-microsensor-motes-1070. Accessed 20 April 2015.

DARPA. 2015, January 1. *Our Work*. http://www.darpa.mil/our_work/. Accessed 26 April 2015.

Demmer, M., P. Levi, A. Joki, E. Brewer, and D. Culler. 2005. *Tython: A Dynamic Simulation Environment for Sensor Networks*. Technical, Department of Electrical Engineering and Computer Science, Berkeley: University of California.

Downard, I.T. 2004. *Simulating Sensor Networks in NS-2*. Washington DC: Final, Naval Research Lab.

Doxygen. 2012, July 24. *The Contiki Operating System*. Doxygen. The Contiki Operating System. Accessed 10 Sept 2014.

Dyer, T.D., and R.V. Boppana 2001. A Comparison of TCP Performance over Three Routing Protocols for Mobile Ad Hoc Networks. In *The 2nd ACM International Symposium on Mobile Ad Hoc Networking and Computing (MobiHoc)*. Long Beach: ACM SIGMOBILE, pp. 56–66.

Egea-López, E., J. Vales-Alonso, A.S. Martínez-Sala, P. Pavón-Mariño, and J. García-Haro. 2005. Simulation Tools for Wireless Sensor Networks. In *International Symposium on Performance Evaluation of Computer and Telecommunication Systems (SPECTS)*. Philadelphia: Society for Modeling & Simulation International (SCS)/IEEE.

Eriksson, J., A. Dunkels, N. Finne, F. Österlind, and T. Voigt. 2007. MSPsim–An Extensible Simulator for MSP430-equipped Sensor Boards. In *Fourth European Conference on Wireless Sensor Networks (EWSN)*. Delft: Springer.

Federal Grants. 2015, January 1. http://www.federalgrants.com/DARPA-Information-Processing-Techniques-Office-IPTO-Broad-Agency-Announcement-24052.html. Accessed 1 Aug 2015.

Fekete, S.P., A. Kröller, S. Fischer, and D. Pfisterer. 2007. Shawn: The Fast, Highly Customizable Sensor Network Simulator. In *Fourth International Conference on Networked Sensing Systems (INSS)*. Braunschweig: Transducer Research Foundation and the IEEE Sensors Council.

Free Software Foundation. 2007, June 29. *GNU General Public License: Version 3*. Free Software Foundation, Inchttp://www.gnu.org/licenses/gpl-3.0.en.html. Accessed 23 July 2015.

Frick, M. 2013, May 28. *JShawn*. GitHub, Inc. https://github.com/itm/shawn/wiki/JShawn. Accessed 23 July 2015.

Gay, D., P. Levis, D. Culler, and E. Brewer. 2003, May 1. *nesC 1.1 Language Reference Manual*. http://nescc.sourceforge.net/papers/nesc-ref.pdf. Accessed 16 April 2015.

GCC Team. 2015, April 27. *GCC, the GNU Compiler Collection*. Free Software Foundation, Inc. https://gcc.gnu.org. Accessed 27 May 2015.

Girod, L., et al. 2004. A System for Simulation, Emulation, and Deployment of heterogeneous Sensor Networks. In *The 2nd International Conference on Embedded Networked Sensor Systems (SenSys)*. Baltimore: ACM, pp. 201–213.

Girod, L., N. Ramanathan, J. Elson, T. Stathopoulos, M. Lukac, and D. Estrin. 2007. Emstar: A Software Environment for Developing and Deploying Heterogeneous Sensor-actuator Networks. *ACM Transactions on Sensor Networks (TOSN)* 3 (3).

Glaser, J., D. Weber, S.A. Madani, and S. Mahlknecht. 2008. Power Aware Simulation Framework for Wireless Sensor Networks and Nodes. *EURASIP Journal on Embedded Systems* 2008 (3).

Goldsman, D., and G. Tokol. 2000. Output Analysis: Output Analysis Procedures for Computer Simulations. In *The 32nd Conference on Winter Simulation (WSC)*, 39–45. Orlando: INFORMS Simulation Society.

Google Sites. 2003, December 10. *Tutorial: Working With J-Sim*. https://sites.google.com/site/jsimofficial/j-sim-tutorial. Accessed 31 July 2015.

Grammarist. 2014, January 1. *Inter , intra-*. http://grammarist.com/usage/inter-intra/. Accessed 28 June 2015.

Guy, R., et al. 2006. *Experiences with the Extensible Sensing System ESS*. Technical, Center for Embedded Network Sensing, University of California. Los Angeles: UCLA.

Heidemann, J., et al. 2001. Effects of Detail in Wireless Network Simulation. In *2001 Western MultiConference-Communication Networks and Distributed Systems Modeling and Simulation Conference*. Phoenix: The Society for Modeling and Simulation International, pp. 311.

———. 2015, January 1. *OTcl and TclCL*. SourceForge. http://sourceforge.net/projects/otcl-tclcl/. Accessed 3 May 2015.

Helmy, A., and S. Kumar. 1997, October 19. *VINT: Virtual InterNetwork Testbed*. http://www.isi.edu/nsnam/vint/index.html. Accessed 26 April 2015.

Hu, Y.-C., and D.B. Johnson. 2001. Implicit Source Routes for On-Demand Ad Hoc Network Routing. In *The 2nd ACM International Symposium on Mobile Ad Hoc Networking and Computing (MobiHoc)*, 1–10. Long Beach: ACM SIGMOBILE.

Imran, M., A.M. Said, and H. Hasbullah. 2010. A Survey of Simulators, Emulators and Testbeds for Wireless Sensor Networks. In *International Symposium in Information Technology (ITSim)*. Kuala Lumpur: IEEE, pp. 897–902.

Information Society Technologies. 2006, December 31. *Embedded WiSeNts - Project FP6-004400*. Information Society Technologies. http://www.embedded-wisents.org. Accessed 11 June 2015.

Intanagonwiwat, C., R. Govindan, and D. Estrin. 2000. Directed Diffusion: A Scalable and Robust Communication Paradigm for Sensor Networks. In *The 6th Annual International Conference on Mobile Computing and Networking (MobiCom)*, 56–67. Boston: ACM SIGMOBILE.

ISI. 2011, November 5. *ns-2*. Information Sciences Institute. http://nsnam.isi.edu/nsnam/index.php/User_Information. Accessed 26 April 2015.

ISIS. 2004, February 13. *JProwler.* Institute for Software integrated Systems. http://w3.isis. vanderbilt.edu/projects/nest/jprowler/. Accessed 7 Aug 2015.

———. 2015, January 1. *The Institute for Software Integrated Systems.* Vanderbilt University - School of Engineering. http://www.isis.vanderbilt.edu. Accessed 6 Aug 2015.

Jacques, F., and A. Marculescu. 2010, January 1. *AlgoSensim.* http://tcs.unige.ch/lib/exe/ fetch.php/ code/algosensim/guide.pdf. Accessed 1 June 2020.

Jain, R. 1991. *The Art of Computer Systems Performance Analysis.* John Wiley & Sons, Inc.

Jardosh, A., E.M. Belding-Royer, K.C. Almeroth, and S. Suri. Towards Realistic Mobility Models for Mobile Ad Hoc Networks." In *The 9th Annual International Conference on Mobile Computing and Networking (MobiCom).* San Diego: ACM SIGMOBILE, 2003. pp. 217229.

Johnson, D.B. 1996, February 20. *The Monarch Project: Protocols for Adaptive Mobile and Wireless Networking.* http://www.cs.cmu.edu/~./dbj/mobile.html. Accessed 3 May 2015.

Johnson, D.B., and D.A. Maltz. 1996. Dynamic Source Routing in Ad Hoc Wireless Networks. In *Mobile Computing,* ed. T. Imielinski and H.F. Korth, vol. 353, 153–181. Springer.

Judd, G., and P. Steenkiste. 2004. Repeatable and Realistic Wireless Experimentation through Physical Emulation. *SIGCOMM Computer Communication Review* 34 (1): 63–68.

Karp, B., and H.T. Kung. 2000. GPSR: Greedy Perimeter Stateless Routing for Wireless Networks. In *The 6th Annual International Conference on Mobile Computing and Networking (MobiCom),* 243–254. Boston: ACM SIGMOBILE.

Keshav, S. 1997, August 13. *REAL 5.0 Overview.* http://www.cs.cornell.edu/skeshav/real/ overview.html. Accessed 26 April 2015.

KIT/TeleMatics. 2010, January 1. *OverSim: THE overlay Simulation Framework.* Karlsruher Institut für Technologie Institut für Telematik. http://www.oversim.org. Accessed 4 June 2015.

Knezzevic, M., K. Sakiyama, Y.K. Lee, and I. Verbauwhede. 2008. On the High-Throughput Implementation of RIPEMD-160 Hash Algorithm. In *International Conference on Application-Specific Systems, Architectures and Processors (ASAP),* 85–90. Leuven: IEEE.

Köpf, B., and D. Basin. 2007. An Information-Theoretic Model for Adaptive Side-Channel Attacks. In *14th ACM Conference on Computer and Communications Security (CCS),* 286–296. Alexandria: ACM.

Köpke, A., et al. 2008. Simulating Wireless and Mobile Networks in OMNeT++ the MiXiM Vision. In *First International Conference on Simulation Tools and Techniques for Communications, Networks and Systems (SIMUTools).* Marseille: The Institute for Computer Sciences, Social Informatics and Telecommunications Engineering (ICST).

Kroller, A., D. Pfisterer, C. Buschmann, S.P. Fekete, and S. Fischer. "Shawn: A New Approach to Simulating Wireless Sensor Networks." In *Design, Analysis, and Simulation of Distributed Systems (DASD).* San Diego: The Institute for Operations Research and the Management Sciences (INFORMS), 2005. pp. 117–124.

Krop, T., M. Bredel, M. Hollick, and R. Steinmetz. 2007. JiST/MobNet: Combined Simulation, Emulation, and Real-World Testbed for Ad Hoc Networks. In *The Second ACM International Workshop on Wireless Network Testbeds, Experimental Evaluation and Characterization (WinTECH),* 27–34. San Juan: ACM SIGMOBILE and USENIX.

Kurkowski, S., T. Camp, and M. Colagrosso. 2005. MANET Simulation Studies: The Incredibles. *ACM SIGMOBILE Mobile Computing and Communications Review - Special Issue on Medium Access and Call Admission Control Algorithms for Next Generation Wireless Networks (ACM)* 9 (4): 50–61.

L'Ecuyer, P., and R. Simard. 2001. On the Performance of Birthday Spacings Tests with Certain Families of Random Number Generators. *Mathematics and Computers in Simulation* 55 (1–3): 131–137.

Lamparter, D., and G. Troxel. 2015, January 1. *Quagga Routing Suite.* Slashdot Media. http:// sourceforge.net/projects/quagga/. Accessed 27 May 2015.

Lan, K.-C. 2001, July 12. *SAMAN: Simulation Augmented by Measurement and Analysis for Networks.* http://www.isi.edu/saman/index.html. Accessed 26 April 2015.

Law, A.M., and W.D. Kelton. 2000. *Simulation Modeling and Analysis.* McGraw-Hill.

LBNL. 2009, August 1. *LBNL's Network Research Group*. Lawrence Berkeley National Laboratory. http://ee.lbl.gov. Accessed 26 April 2015.

Lee, S., and C. Kim. 2000. Neighbor Supporting Ad Hoc Multicast Routing Protocol. In *First Annual Workshop on Mobile and Ad Hoc Networking and Computing (MobiHOC)*. Boston: ACM/IEEE, pp. 37–44.

Levis, P., and N. Lee. 2003, September 17. *TOSSIM: A Simulator for TinyOS Networks*. http://www.tinyos.net/tinyos-1.x/doc/nido.pdf. Accessed 14 April 2015.

Levis, P., N. Lee, M. Welsh, and D. Culler. 2003. TOSSIM: Accurate and Scalable Simulation of Entire TinyOS Applications. In *The 1st International Conference on Embedded Networked Sensor Systems (SenSys)*, 126–137. Los Angeles: ACM.

Lichte, H.S., and S. Valentin. 2008. Implementing MAC Protocols for Cooperative Relaying: A Compiler-Assisted Approach. In *The 1st International Conference on Simulation Tools and Techniques for Communications, Networks and Systems (SIMUTools)*. Brussels: Institute for Computer Sciences, Social-Informatics and Telecommunications Engineering (ICST).

Löbbers, M., and D. Willkomm. 2007, January 12. *A Mobility Framework for OMNeT++ User Manual*. Telecommunication Networks Group-Technische Universitaet Berlin. http://mobility-fw.sourceforge.net/manual/index.html. Accessed 5 June 2015.

Mallanda, C., et al. 2005, January 24. *Simulating Wireless Sensor Networks with OMNeT++*. Department of Computer Science, Louisiana State University. Sensor Network Research Group. http://citeseerx.ist.psu.edu/viewdoc/download?doi=10.1.1.111.8475&rep=rep1&type=pdf. Accessed 11 June 2015.

Maria, A. 1997. Introduction to Modeling and Simulation. In *The 29th Conference on Winter Simulation (WSC)*. Atlanta: IEEE, pp. 7–13.

McNickle, D., K. Pawlikowski, and G. Ewing. 2010. Akaroa2: A Controller of Discrete-Event Simulation Which Exploits the Distributed Computing Resources of Networks. In *24th European Conference on Modelling and Simulation (ECMS)*. European Counsil for Modelling and Simulation (ECMS).

Melodia, T., D. Pompili, V.C. Gungor, and I.F. Akyildiz. 2005. A Distributed Coordination Framework for Wireless Sensor and Actor Networks. In *The 6th ACM International Symposium on Mobile Ad hoc Networking and Computing (MobiHoc)*. Kuala Lumpur: ACM SIGMOBILE, pp. 99–110.

Microsoft. 2015, January 1. *What is COM?* Microsoft. https://www.microsoft.com/com/default.mspx. Accessed 31 July 2015.

MiXiM Developers. 2011, January 1. *MiXiM*. MiXiM Developers. http://mixim.sourceforge.net/index.html. Accessed 5 June 2015.

Musznicki, B., and P. Zwierzykowski. 2012. Survey of Simulators for Wireless Sensor Networks. *International Journal of Grid and Distributed Computing* 5(3): 23–49.

NetGNA. 2008, January 1. *G-SENSE*. Next Generation Networks and Applications Group (NetGNA). http://www.ita.cs.rpi.edu. Accessed 15 July 2015.

Networks and Communication Systems Branch. 2015, January 1. *NRL's Sensor Network Extension to NS-2*. U.S. Naval Research Lab. http://www.nrl.navy.mil/itd/ncs/products/sensorsim. Accessed 23 April 2015.

Niculescu, D., and B. Nath. 2003. Ad hoc Positioning System (APS) using AOA. In *Twenty-Second Annual Joint Conference of the IEEE Computer and Communications (INFOCOM)*. San Francisco: IEEE, pp. 1734–1743.

ns-3 Consortium. 2015, January 1. *About*. https://www.nsnam.org/consortium/about/. Accessed 10 May 2015.

Nuevo, J. 2004, March 4. *A Comprehensible GloMoSim Tutorial*. INRS-Quebec University. http://www.ccs.neu.edu/course/csg250/Glomosim/glomoman.pdf. Accessed 3 May 2015.

OMNeT++ Wiki. 2011. *NesCT*. https://omnetpp.org/pmwiki/index.php?n=Main.NesCT. Accessed 11 June 2015.

Open Source Initiative. 2015, January 1. *Academic Free License ("AFL") v. 3.0*. Open Source Initiative. http://opensource.org/licenses/AFL-3.0. Accessed 27 May 2015.

OpenSim. 2015a, March 11. *OMNeT++ 5.0b1 released.* OpenSim Ltd. https://omnetpp.org/9-articles/software/3726-omnet-5-0b1-released. Accessed 11 June 2015.

———. 2015b, January 1. *OMNeT++ version 4.6.* OpenSim Ltd. https://omnetpp.org/doc/omnetpp/manual/usman.html. Accessed 28 May 2015.

———. 2015c, January 1. *What is OMNeT++?* OpenSim Ltd. https://omnetpp.org/intro. Accessed 27 May 2015.

Oracle. 2015, January 1. *Oracle and Sun Microsystems.* Oracle. http://www.oracle.com/us/sun/index.html. Accessed 3 May 2015.

Osterlind, F., A. Dunkels, J. Eriksson, N. Finne, and T. Voigt. 2006. Cross-Level Sensor Network Simulation with COOJA. In *31st IEEE Conference on Local Computer Networks*, 641–648. Tampa: IEEE.

Park, S. 2001, October 22. *SensorSim: A Simulation Framework for Sensor Networks.* http://nesl.ee.ucla.edu/projects/sensorsim/. Accessed 23 April 2015.

Park, S., A. Savvides, and M.B. Srivastava. 2000. SensorSim: A Simulation Framework for Sensor Networks. In *The 3rd ACM International Workshop on Modeling Analysis and Simulation of Wireless and Mobile Systems (MSWiM)*. Huntington Beach: ACM, pp. 104–111.

Park, S., A. Savvides, and M. Srivastava. 2001. Battery Capacity Measurement and Analysis Using Lithium Coin Cell Battery. In *The 2001 International Symposium on Low power Electronics and Design (ISLPED)*. Boston: ACM/IEEE, pp. 382–387.

Pawlikowski, K., H.-D.J. Jeong, and J.-S.R. Lee. 2002. On Credibility of Simulation Studies of Telecommunication Networks. *Communications Magazine* 40 (1): 132–139.

Perkins, C., E. Belding-Royer, and S. Das. 2003, July 1. *RFC 3561-Ad Hoc on-Demand Distance Vector (AODV) Routing.* Network Working Group. http://www.rfc-editor.org/rfc/pdfrfc/rfc3561.txt.pdf. Accessed 15 July 2015.

Perrone, L.F. 2003. Modeling and Simulation Best Practices for Wireless Ad Hoc Networks. In *The Winter Simulation Conference*, 685–693. New Orleans: IEEE.

Perrone, L.F., and D.M. Nicol. 2002. A Scalable Simulator for TinyOS Applications. In *The 2002 Winter Simulation Conference (WSC)*, 679–687. San Diego: INFORMS.

Pidd, M. 1998. *Computer Simulation in Management Science.* 4th ed. Wiley & Sons, Ltd: Chichester.

Press, W.H., S.A. Teukolsky, W.T. Vetterling, and B.P. Flannery. 2002. *Numerical Recipes in C++: The Art of scientifque Computing.* Cambridge: Cambridge University Press.

PUC-Rio. 2015, June 17. *Lua.* The Pontifical Catholic University of Rio de Janeiro, Brazil. http://www.lua.org/about.html. Accessed 27 June 2015.

Rahman, M.A., A. Pakštas, and F.Z. Wang. 2009. Network Modelling and Simulation Tools. *Simulation Modelling Practice and Theory* 17(6): 1011–1031.

Riverbed Technology. 2015, January 1. *Riverbed Application and Network Performance Management Solutions.* Riverbed Technology. http://www.riverbed.com/products/performance-management-control/opnet.html. Accessed 17 May 2015.

Rivet, B., and T. Klepp. 2012, February 14. *SimulAVR.* http://www.nongnu.org/simulavr/. Accessed 16 April 2015.

Roache, P.J. 1998. *Verification and Validation in Computational Science and Engineering.* Albuquerque: Hermosa.

Robinson, S. 2004. *Simulation: The Practice of Model Development and Use.* Chichester: John Wiley & Sons, Ltd.

Rosa, P.M., P.A. Neves, B. Vaidya, and J.J. Rodrigues. 2009. G-SENSE: A Graphical Interface for Sense Simulator. In *The First International Conference on Advances in System Simulation (SIMUL)*, 88–93. Porto: IARIA/IEEE.

Sadagopan, N., F. Bai, B. Krishnamachari, and A. Helmy. 2003. PATHS: Analysis of PATH Duration Statistics and Their Impact on Reactive MANET Routing Protocols. In *The 4th ACM International Symposium on Mobile Ad Hoc Networking and Computing (MobiHoc)*, 245–256. Annapolis: ACM SIGMOBILE.

Sarabandi, K., I. Koh, G. Liang, and H. Bertoni. 2001. Propagation Modeling for FCS. In *Military Communications Conference (Milcom)*. IEEE: Washington, DC.

Sargent, R.G. 2005. Verification and Validation of Simulation Models. In *The 37th Conference on Winter Simulation (WSC)*. Orlando: ACM, pp. 130–143.

SCALABLE Network Technologies. 2014, January 1. *Qualnet*. SCALABLE Network Technologies, Inc. http://web.scalable-networks.com/content/qualnet. Accessed 8 May 2015.

Schaumont, P. 2012, November 23. *GEZEL Manual*. http://rijndael.ece.vt.edu/gezel2/manual.html. Accessed 16 April 2015.

Schaumont, P., D. Ching, and I. Verbauwhede. 2006. An Interactive Codesign Environment for Domain-specific Coprocessors. *ACM Transactions on Design Automation of Electronic Systems (TODAES)* 11 (1): 70–87.

Schlesinger, S., et al. 1979. Terminology for model credibility. *Simulation* 32: 103–104.

Schruben, L.W. 1982. Detecting Initialization Bias in Simulation Output. *Operations Research* 30 (3): 569–590.

Seada, K., A. Helmy, and R. Govindan. 2007, August. Modeling and Analyzing the Correctness of Geographic Face Routing Under Realistic Conditions. *Ad Hoc Networks* 5(6): 855–871.

Shnayder, V., M. Hempstead, B. R. Chen, G.W. Allen, and M. Welsh. 2004. Simulating the Power Consumption of Large-Scale Sensor Network Applications. In *The 2nd International Conference on Embedded Networked Sensor Systems (SenSys)*. ACM, pp. 188–200.

Shu, L., M. Hauswirth, H.-C. Chao, M. Chen, and Y. Zhang. 2011, July. NetTopo: A Framework of Simulation and Visualization for Wireless Sensor Networks. In *Ad Hoc Networks* (Elsevier, B. V.) 9(5): 799–820.

Simon, M.K., and M.S. Alouini. 2005. *Digital Communication over Fading Channels*. Edited by J.C. Proakis. John Wiley & Sons, Inc.

Simon, G., P. Völgyesi, M. Maróti, and A. Lédeczi. 2003. Simulation-based Optimization of Communication Protocols for Large scale Wireless Sensor Networks. In *IEEE Aerospace Conference*. Big Sky: IEEE.

Singh, C.P., O.P. Vyas, and M.K. Tiwari. 2008. A Survey of Simulation in Sensor Networks. In *International Conference on Computational Intelligence for Modelling Control & Automation*. IEEE, pp. 867–872.

Sobeih, A., et al. 2006. J-Sim: A Simulation and Emulation Environment for Wireless Sensor Networks. *Wireless Communications* 13 (4): 104–119.

Sommer, C. 2015, February 25. *Veins: The Open Source Vehicular Network Simulation Framework*. http://veins.car2x.org. Accessed 4 June 2015.

Steinbach, T., H.D. Kenfack, F. Korf, and T.C. Schmidt. 2011. An Extension of the OMNeT++ INET Framework for Simulating Real-Time Ethernet with High Accuracy. In *The 4th International ICST Conference on Simulation Tools and Techniques (SIMUTools)*, 375–382. Barcelona: Institute for Computer Sciences, Social-Informatics and Telecommunications Engineering (ICST).

Stojmenovic, I. 2008, December. Simulations in Wireless Sensor and Ad Hoc Networks: Matching and Advancing Models, Metrics, and Solutions. *Communications Magazine* 46(12): 102–107.

Sundresh, S., W. Kim, and G. Agha. 2004. SENS: A Sensor, Environment and Network Simulator. In *The 37th Annual Symposium on Simulation (ANSS)*. Arlington: IEEE.

SwarmNet Project. 2004, January 1. *SwarmNet*. SwarmNet Project. http://www.swarmnet.de. Accessed 23 July 2015.

Swiss Federal Institute of Technology. 2013, January 1. *Sinalgo*. https://sourceforge.net/projects/sinalgo/. Accessed 1 June 2020.

Takai, M., L. Bajaj, R. Ahuja, R. Bagrodia, and M. Gerla. 1999. *GloMoSim: A Scalablenetwork Simulation Environment*. Technical, Computer Science Department, UCLA, Los Angeles: UCLA.

Tcl Developer Xchange. 2015, January 1. *Welcome to the Tcl Developer Xchange!* http://www.tcl.tk. Accessed 23 April 2015.

Tech-FAQ. 2015, January 1. *What is PCAP?* Independent Media. http://www.tech-faq.com/pcap.
    html. Accessed 1 Aug 2015.
Techopedia. 2015a, January 1. *Integrated Development Environment (IDE).* Techopedia. http://
    www.techopedia.com/definition/26860/integrated-development-environment-ide.   Accessed
    27 May 2015.
———. 2015b, January 1. *Proof of Concept (POC).* Techopedia. http://www.techopedia.com/
    definition/4066/proof-of-concept-poc. Accessed 12 April 2015.
TechTarget. 2015, January 1. *Corba (Common Object Request Broker Architecture) Definition.*
    TechTarget. http://searchsqlserver.techtarget.com/definition/CORBA. Accessed 31 July 2015.
Texas Instruments. 2011, January 1. *MSP430C11x1, MSP430F11x1A Mixed Signal Microcontrol-
    ler.* Texas Instruments. http://www.ti.com/lit/ds/symlink/msp430f1611.pdf. Accessed
    1 Feb 2014.
Thacker, B.H., S.W. Doebling, F.M. Hemez, M.C. Anderson, J.E. Pepin, and E.A. Rodriguez.
    2004. *Concepts of Model Verification and Validation.* Technical, Los Alamos National Labo-
    ratory, University of California. Los Alamos: University of California, pp. 1–27.
The ICSI Networking and Security Group. 2015, April 19. *Overview.* The ICSI Networking and
    Security Group. http://www.aciri.org. Accessed 26 April 2015.
The National Science Foundation. 2015, January 1. *Home.* The National Science Foundation. http://
    www.nsf.gov. Accessed 2 Aug 2015.
The Tcl/Java Project. 2008, April 4. *Latest Tcl/Java News.* SourceForge. http://tcljava.sourceforge.
    net/docs/website/index.html. Accessed 1 Aug 2015.
TinyOS. 2012, August 20. *TinyOS.* http://www.tinyos.net. Accessed 9 Oct 2013.
TinyOS Wiki. 2012, October 21. *Mote-PC serial communication and Serial Forwarder.* TinyOS
    Wiki. http://tinyos.stanford.edu/tinyos-wiki/index.php/Mote-PC_serial_communication_and_
    SerialForwarder_(TOS_2.1.1_and_later). Accessed 7 Aug 2014.
———. 2013, May 10. TOSSIM. TinyOS Wiki. http://tinyos.stanford.edu/tinyos-wiki/index.php/
    TOSSIM. Accessed 1 Sept 2014.
Titzer, B.L., D.K. Lee, and J. Palsberg. 2005. Avrora: Scalable Sensor Network Simulation with
    Precise Timing. In *The 4th International Symposium on Information Processing in Sensor
    Networks (IPSN).* Los Angeles: ACM/IEEE.
TU Wien. 2015, May 6. *About TU Wien.* TU Wien. https://www.tuwien.ac.at/en/about_us/.
    Accessed 17 June 2015.
Ubuntu. 2010, January 1. *Ubuntu Manuals.* Ubuntu. http://manpages.ubuntu.com/manpages/
    utopic/man1/protoc.1.html. Accessed 17 May 2015.
UCB. 2015, January 1. *University of California Berkeley.* University of California Berkeley. http://
    www.berkeley.edu. Accessed 26 April 2015.
UCLA. 2015, January 1. *UCLA Compilers Group.* http://compilers.cs.ucla.edu. Accessed
    17 April 2015.
Universitat Paderborn. 2010, October 16. *Chsim: A Wireless Channel Simulator for OMNeT++.*
    Universitat Paderborn. http://www.cs.uni-paderborn.de/en/fachgebiete/research-group-com
    puter-networks/projects/chsim.html. Accessed 5 June 2015.
University of Twente. 2014, February 5. *Home.* University of Twente. http://dies.ewi.utwente.nl.
    Accessed 11 June 2015.
University of Twente/Technical University of Delft. 2005, August 25. *Software.* University of
    Twente/Technical University of Delft. http://www.consensus.tudelft.nl/software.html.
    Accessed 5 June 2015.
USC/ISI. 2015, January 1. *Information Sciences Institute.* University of Southern California. http://
    www.isi.edu/home. Accessed 26 April 2015.
Varga, A., and R. Hornig. 2008. An Overview of the OMNeT++ Simulation Environment. In *The
    1st International Conference on Simulation Tools and Techniques for Communications, Net-
    works and Systems & Workshops (SIMUTools).* Marseille: Institute for Computer Sciences,
    Social-Informatics and Telecommunications Engineering (ICST), Brussels.

Virginia Tech. 2015, January 1. Department of Electrical and Computer Engineering. Virginia Tech. http://www.ece.vt.edu. Accessed 8 July 2015.

W3C. 2015, May 19. *Extensible Markup Language (XML)*. W3C. http://www.w3.org/XML/. Accessed 15 July 2015.

Weber, D., J. Glaser, and S. Mahlknecht. 2007. Discrete Event Simulation Framework for Power Aware Wireless Sensor Networks. In *5th IEEE International Conference on Industrial Informatics*, 335–340. Vienna: IEEE.

Webopedia. 2015a, January 1. *iPAQ*. Webopedia. http://www.webopedia.com/TERM/I/iPAQ. html. Accessed 20 April 2015.

———. 2015b, January 1. *Smart Dust*. QuinStreet Inc. http://www.webopedia.com/TERM/S/ smart_dust.html. Accessed 6 August 6, 2015).

Wessel, K., M. Swigulski, A. Köpke, and D. Willkomm. 2009. MiXiM–The Physical Layer an Architecture Overview. In *The 2nd International Workshop on OMNeT++*. Rome: ICST/ACM.

Wikipedia. 2015, May 26. *Berkeley Software Distribution*. Wikipedia. http://en.wikipedia.org/wiki/ Berkeley_Software_Distribution. Accessed 27 May 2015.

Wireless Networks Laboratory. 2015, January 1. *Projects*. Wireless Networks Laboratory. http:// cse.yedltepe.edu.tr/wnl/projects/. Accessed 11 June 2015.

Wu, H., Q. Luo, P. Zheng, B. He, and L.M. Ni. 2004. Accurate Emulation of Wireless Sensor Networks. In *IFIP International Conference on Network and Parallel Computing (NPC)*, 576–583. Berlin: Springer.

Xerox PARC. 2015, January 1. *PARC*. Xerox. http://www.parc.com. Accessed 26 April 2015.

Yang, H., and B. Sikdar. 2003. A Protocol for Tracking Mobile Targets Using Sensor Networks. In *The First IEEE International Workshop on Sensor Network Protocols and Applications*, 71–81. Anchorage: IEEE.

Yang, X., M. Xu, P. Stickney, and W.-Z. Song. 2007. SimX: An Integrated Sensor Network Simulation and Evaluation Environment. In *IEEE International Parallel and Distributed Processing Symposium (IPDPS)*, 1–6. Long Beach: IEEE.

Zhang, Y., G. Simon, and G. Balogh. 2006. High-Level Sensor Network Simulations for Routing Performance Evaluations. In *Third International Conference on Networked Sensing Systems*. Salt Lake City: UbiComp.

Zhang, J., Y. Tang, S. Hirve, S. Iyer, P. Schaumont, and Y. Yang. 2011. A Software-Hardware Emulator for Sensor Networks. In *8th Annual IEEE Communications Society Conference on Sensor, Mesh and Ad Hoc Communications and Networks (SECON)*, 440–448. Chicago: IEEE.

Zheng, R., J.C. Hou, and L. Sha. 2003. Asynchronous Wakeup for Ad Hoc Networks. In *The 4th ACM International Symposium on Mobile Ad hoc Networking and Computing (MobiHoc)*. Annapolis: ACM SIGMOBILE, pp. 35–45.

Part IV
WSNs Manufacturers and Datasheets

Chapter 9
WSN Manufacturers

Manufacturers and products drive each other.

9.1 Adaptive Wireless Solutions (Adaptive Wireless Solutions 2015)

Adaptive Wireless Solutions Ltd., located in the United Kingdom, specializes in industrial and commercial monitoring and control solutions using wireless and other remote telemetry systems. It offers a full range of products and services from individual system elements to complete solutions tailored to customer requirements.

9.2 AlertMe (AlertMe 2014) and British Gas (British Gas 2015)

AlertMe was a leader in connected homes with more than five years' expertise in building a platform for scale deployment. Its vision was to make the connected home accessible to the mass market, simple, useful, and affordable for all, with services that make consumers' lives easier and safer and transform the productivity of the businesses that provide them. For consumers, AlertMe provided a full range of applications in energy, home monitoring, and home automation in one ecosystem where things work together simply and intelligently. This allowed the user to connect as many devices as they like, customize notifications and alerts, and trigger actions automatically.

On March 17, 2015, British Gas acquired AlertMe to create the United Kingdom's leading connected home provider. This move brings together British Gas' ability to innovate for customers with AlertMe's next-generation Internet-of-Things technology and expertise. The acquisition has created a highly experienced and fully integrated team which will accelerate the development of new connected home services in the United Kingdom and worldwide.

H. M. A. Fahmy, *Concepts, Applications, Experimentation and Analysis of Wireless Sensor Networks*, Signals and Communication Technology, https://doi.org/10.1007/978-3-031-20709-9_9

9.3   ANT Wireless Division of Dynastream (Dynastream Innovations 2014)

The ANT+ Alliance is an open special interest group of companies that have adopted the ANT+ promise of interoperability. The Alliance ensures standardized communications through optimized brand value and partnerships with other top-tier products. ANT, ANT+, and the ANT+ Alliance are all managed by the ANT Wireless division of Dynastream Innovations, Inc. Established in 1998, Dynastream introduced the first accelerometer-based speed and distance monitor for runners in 2000. In 2003, the wireless protocol ANT was launched, and in 2004, the first ultra-low power wireless standard, ANT+, was created. In 2005, integrated PAN solutions were elaborated with Nordic Semiconductor (Nordic Semiconductor 2004). In December 2006, Garmin Ltd., once a valued Dynastream customer, purchased the company.

Today, hundreds of companies are members of the ANT+ Alliance, building products across a range of personal area network (PAN) applications, including sports, wellness, and home health monitoring.

9.4   Atmel (Atmel 2015)

Founded in 1984, Atmel's corporate headquarters are located in San Jose, California. Atmel microcontrollers deliver a rich blend of efficient integrated designs, proven technology, and groundbreaking innovation that is ideal for today's smart, connected products. In this era of the Internet of Things (IoT), microcontrollers comprise a key technology that fuels machine-to-machine (M2M) communications.

Building on decades of experience and industry leadership, Atmel offers proven architectures that are optimized for low power, high-speed connectivity, optimal data bandwidth, and rich interface support. By using a wide variety of configuration options, developers can devise complete system solutions for all kinds of applications.

Atmel microcontrollers can also support the seamless integration of capacitive touch technology to implement buttons, sliders, and wheels (BSW). In addition, Atmel microcontrollers (MCUs) deliver wireless and security support. Atmel offers a compelling solution that is tailored to customer needs today and tomorrow. Applications for Atmel microcontrollers include automotive, building automation, home appliances and entertainment, industrial automation, lighting, smart energy, mobile electronics, PC peripherals, and Internet of Things.

9.5   Cisco (Cisco 2015)

Cisco, whose headquarter is located in San Jose, California, is a communications giant that provides communication devices that cover almost everything. Among many things, at Cisco, they deliver solutions for Internet of Things and for wireless networking.

9.6   Coalesenses (Coalesenses 2014)

Coalesenses is a young company providing solutions for massively distributed systems with a focus on wireless sensor networks (WSNs). Coalesenses originates from a university background in this new application area and holds on to the concept of cooperation with public research facilities. Hence, at Coalesenses, they have under their command a state-of-the-art research knowledge. Coalesenses employs PhDs, engineers, and students to incorporate the latest research results into their projects.

Coalesenses line of products includes WSN solutions, WSN software and protocol stacks, WSN devices, and WSN modules.

9.7   Crossbow Technologies (Aol 2015)

Crossbow Technology, Inc. manufactured and supplied WSNs and inertial sensor systems. It offered accelerometers, angular rate sensors (gyros), magnetometers, GPS, and air-data sensors for instrumentation, navigation, and control applications in land, marine, and airborne environments. Furthermore, it provided inertial systems, accelerometers, tilt sensors, and magnetometers for general aviation, automotive testing, antenna stabilization, unmanned aerial vehicles, agriculture and construction, vibration monitoring, and towed sonar arrays. The company also delivered WSN development kits, wireless modules, sensor boards, gateways, and wireless solutions blogs for industrial, environmental monitoring, building automation, academic programs, and asset management applications. In addition, it offered MoteWorks, an OEM software platform that offers developers the comprehensive benefits of wireless technology in a given sensor application.

Crossbow Technology, Inc. was acquired by Moog (Sect. 9.18) on June 3rd, 2011.

9.8   Dust Networks (Dust Networks 2015)

Dust Networks, a pioneer in the field of WSN, is defining the way to connect smart devices. Dust Networks delivers reliable, resilient, and scalable wireless embedded products with advanced network management and comprehensive security features. Dust Networks products are built on breakthrough Eterna 802.15.4 System-on-Chip (SoC) technology, delivering ultra-low power consumption for wire-free operation on batteries or energy harvesting.

Dust's portfolio of standards-based products includes (Chap. 1 of this book) the following:

- SmartMesh IP, which is built for IP compatibility and is based on the 6LoWPAN and 802.15.4e standards. The SmartMesh IP solution is widely applicable and cost-effective and enables low power consumption even in harsh, dynamically changing RF environments.
- SmartMesh WirelessHART products are designed for the harshest industrial environments, where low power, reliability, resilience, and scalability are essential, making them well-suited for general industrial applications as well as WirelessHART-specific designs. SmartMesh WirelessHART complies with the WirelessHART (IEC 62591) standard. It offers the lowest power consumption in its class and is the most widely used WirelessHART product available.

9.9   EasySen (EasySen 2015)

EasySen located in South Bend, Indiana, is a small company specializing in state-of-the-art wireless sensing solutions. It offers customized hardware designs, algorithms, and other consulting services for designing commercial and research applications of wireless sensor networks. EasySen expertise includes autonomous mobile sensor and actuator platforms, ultra-low complexity swarm systems, energy harvesting solutions for sensor networks, and software algorithms for sensor signal processing and navigation. EasySen takes pride in excellent customer service that assists in the exploration of new research frontiers in sensor network applications.

9.10   EcoLogicSense (EcoLogicSense 2015)

EcoLogicSense, located in Rousset, France, specializes in the design of communicating products for real-time monitoring of air quality. EcoLogicSense is the leader in producing sensors for measuring air quality via innovative technology.

EcoLogicSense main objective is the development and design of molecular and particle sensors to meet real-time measurements of air quality in controlled environments (clean rooms), indoor and outdoor environments. Based on its expertise in atmospheric chemistry for over 10 years, the EcoLogicSense technical team has combined its experience in the field of chemistry, electronics, and computing to meet the collection, treatment, and diffusion of environmental data.

9.11 EpiSensor (EpiSensor 2015)

EpiSensor whose headquarters are located in Limerick, on the west coast of Ireland, is one of the world's leading suppliers of easy to deploy, secure, and reliable wireless sensors. EpiSensor was founded in 2007 at the intersection of three technology waves; specifically, wireless sensor networks, cloud computing, and mobile communication. This new technology stack, known as the "Internet of Things," is applied to the world's energy and efficiency problems.

All EpiSensor products are designed by EpiSensor in Ireland. A "full chain" of technology from the sensor to the server has been developed, which means that EpiSensor can be extremely responsive and flexible. EpiSensor Internet-of-Things platform can dramatically increase efficiency, reduce costs, and improve sustainability. Data produced by EpiSensor systems can transform the efficiency of an organization by providing insight into areas of waste that could not be achieved using traditional monitoring, control, and automation systems. EpiSensor's products are trusted by some of the world's largest and most secure organizations.

At the research level, EpiSensor has contributed to many local and international research projects; it is the only small and medium-sized enterprise (SME) core member of the CLARITY Centre for Sensor Web Technologies. CLARITY is a partnership between University College Dublin, Dublin City University, and Tyndall National Institute in Ireland; it focuses on the intersection between two important research areas: adaptive sensing and information discovery.

9.12 ERS (ERS 2015)

Embedded Research Solutions (ERS) is privately held and located in Annapolis, Maryland. ERS creates products and enabling technologies for pervasive computing applications. The technology base includes software and hardware that allows multihop communication, ad-hoc networking, mesh networking, power management, location awareness, real-time responsiveness, and dynamic reconfiguration.

ERS has created the first truly deployable solution for pervasive applications. The software architecture is uniquely scalable. It has been shrunk to fit on the smallest and least expensive embedded processors, yet maintains a rich set of functions. In addition, the software can run as middleware on standard operating systems, including Linux and Windows. Other attributes of ERS solutions include low-cost, guaranteed real-time performance, dynamic configurability, modularity, ease of use, and low maintenance.

ERS has also developed a platform software known as ZEE that consists of a configurable, modular, hardware-independent, framework uniquely suited for highly distributed systems and networks required to deliver data in real time.

9.13   GainSpan (GainSpan 2015)

GainSpan was founded in 2006, its headquarters are located in San Jose, California. Since 2006, GainSpan has designed and marketed Wi-Fi chips, modules, and solutions to connect traditionally non-connected devices to smartphones or to the Internet. More recently, a combo Wi-Fi/Thread/6LoWPAN chip and modules were added. It is planned to expand the GainSpan portfolio to offer the most suitable wireless solutions to continue to connect "things" to the Internet and people to "things."

In the near future, GainSpan anticipates seeing commercial buildings where Wi-Fi-equipped sensors will detect temperatures and initiate heating or cooling responses wirelessly. In the not-too-distant future, GainSpan-embedded Wi-Fi could be used to control the lights at home, monitor elderly parents' health, or turn OFF the air conditioner during periods of peak energy use when no one is home.

9.14   Infineon (Infineon 2015)

Infineon has headquarters for production, R&D, and sales in El Segundo, California, and headquarters for R&D and sales in Neubiberg, Germany, and in Reigate, United Kingdom. For industrial applications, Infineon develops a wide variety of sensors to account for renewable energy, industrial automation, and e-mobility. Offerings include products such as magnetic position and speed sensors, as well as integrated pressure sensors and current sensors. For electric drives, a key area in industrial applications, the portfolio comprises a full range of energy-saving sensors for electric commutated drives. In the increasingly important solar sector, Infineon sensors help customers achieve optimum system efficiency and meet country-specific regulations.

For automotive applications, Infineon has a noticeable record of perfection. Over three billion integrated magnetic sensors are installed in cars all over the world, delivering reliable results in safety-relevant applications, such as ABS, and in harsh environments, such as engines and transmissions.

9.15   Libelium (Libelium 2015)

Libelium located in Zaragoza, Spain, delivers a powerful, modular, easy-to-program open source sensor platform for the Internet of Things, enabling system integrators to implement reliable smart cities and machine-to-machine (M2M) solutions with minimum time to market. The Libelium versatile platform allows the implementation of any WSN, from smart parking to smart irrigation solutions.

9.16   MEMSIC (MEMSIC 2015)

MEMSIC delivers powerful sensing solutions to enhance life. With sight and sound, touch and smell, communication with the world around us is established. Just as eyes sense light and ears sense sound and relay that information to the brain enabling to sense the environment, sensors comprehend the surrounding world and relay these electrical signals back through intricate integrated circuitry (IC) and electronic systems.

MEMSIC (MEMS + IC) enables intelligent powerful sensing solutions by combining all the essential elements for the application needs. By integrating IC and electronic system functionality, with manufactured solid-state low-cost sensors, MEMSIC solutions are launched to optimize life. As such, gaming systems have taken the player experience to new levels by sensing actions and motion, cars are intrinsically safer by automatically sensing and controlling the movement of the vehicle, industrial equipment and machines perform their functions without human intervention, and mobile phones have built-in intelligence so they can respond to gestures and position with a simple user interface and location-aware services. The surrounding environment responds to humans when instrumented with wireless sensors. Avionics equipment has been retrofitted with high-performance yet lower-cost systems. The possibilities are endless when the solutions are effective. MEMSIC technology drives the advancement of sensors and sensing solutions to create a better life for all.

9.17   Millennial Net (Millennial Net 2012)

Millennial Net, Inc. is a privately held company headquartered in Chelmsford, Massachusetts. Millennial Net develops wireless sensor networking software, systems, and services that enable original equipment manufacturers (OEMs) and systems integrators to quickly and cost-effectively implement WSNs. WSNs enable the remote monitoring and management of critical devices while providing data to enable more informed decision-making, better control, and increased revenue opportunities. Millennial Net is an industry leader in real-world deployments with networks installed across commercial buildings and industrial environments.

MeshScape wireless sensor networking products reveal innovative Millennial Net solutions.

9.18   Moog Crossbow (Moog Crossbow 2014)

Moog Crossbow specializes in connecting the physical world to the digital world within the Moog Aircraft Group. Founded in 1995, the company is a leading supplier of low-cost, smart-sensor technology to military programs and high-value, asset-tracking operations. Moog Crossbow has shipped more than half a million sensors to customers including Raytheon, Lockheed Martin, Airbus, US DOD, DRS, and Israel Aerospace Industries, as well as leading global logistics companies. Moog Crossbow is headquartered in Milpitas, California.

Over sixty years ago, Moog Crossbow started as a designer and supplier of aircraft and missile components. Today, its motion control technology enhances performance in a variety of markets and applications, from commercial aircraft cockpits and power-generation turbines to Formula One racing and medical infusion systems.

History begins with the founder, William C. Moog, an inventor, entrepreneur, and visionary. In 1951, Bill Moog developed the electro-hydraulic servo-valve, a device that translates tiny electrical impulses into a precise and powerful movement. In July of 1951, Bill, his brother Arthur, and Lou Geyer rented a corner of the abandoned Proner Airport in East Aurora and formed the Moog Valve Company.

9.19   Moteiv (Sensors Online 2007)

Founded in 2003, Moteiv Corp Moteiv Corp. is a leading provider of wireless sensor network solutions. Moteiv makes wireless sensor network technology accessible through innovative hardware platforms, robust open-source software, and whole-solution development services. Headquartered in San Francisco, Moteiv's products are used in a wide variety of applications, including climate monitoring, asset management, homeland security, and industrial control. Moteiv's mission is to broaden the adoption of this remarkable technology by making it approachable, affordable, and intimately familiar to those accustomed to traditional IT infrastructures.

9.20   National Instruments (National Instruments 2015)

National Instruments (NI) was founded in 1976. Its headquarters are located in Austin, Texas, with offices distributed over nearly 50 countries. With the NI WSN platform, it is easy to monitor assets or environments with reliable, battery-powered measurement nodes that offer industrial ratings and local analysis and control capabilities. Each wireless network can scale from tens to hundreds of nodes and seamlessly integrate with existing wired measurement and control systems.

WSN architectures combine different types of nodes and gateways to meet the unique needs of customer applications. Using such architectures, it is possible to create a simple, PC-based WSN monitoring system with the NI WSN-9791 Ethernet gateway; also possible is the formation of a headless, embedded monitoring system with the NI 9792 programmable gateway, which can run deployed NI LabVIEW real-time applications. For applications that require the combination of high-speed I/O (or control) and distributed wireless monitoring, the NI 9795 C Series WSN gateway may be used.

9.21   OmniVision Technologies (OmniVision Technologies 2011)

Founded in 1995 and headquartered in Santa Clara, California, OmniVision currently houses 19 offices in 12 different countries worldwide, including a design center and state-of-the-art testing facility in Shanghai, China. OmniVision Technologies (NASDAQ: OVTI) is a leading developer of advanced digital imaging solutions. OmniVision's workforce is 2200 worldwide. It has shipped over 4,3 billion CMOS image sensors. Their award-winning CMOS imaging technology enables superior image quality in many of today's consumer and commercial applications, including mobile phones, notebooks, netbooks and webcams, security and surveillance, entertainment, digital still and video cameras, and automotive and medical imaging systems.

9.22   Sensirion (Sensirion 2015)

Sensirion is the leading manufacturer of high-quality sensors and sensor solutions for the measurement and control of humidity, as well as of gas and liquid flows. Founded in 1998 as a spin-off from the Swiss Federal Institute of Technology (ETH) Zurich, the company is based in Stäfa near Zurich, Switzerland, and employs people in countries such as the USA, South Korea, Japan, China, Taiwan, and Germany. The headquarters in Switzerland are responsible for research, development, and production.

Together with the capacitive humidity sensor, the product range includes liquid flow sensors, mass flow meters, mass flow controllers, and differential pressure sensors. Using Sensirion microsensor solutions, OEM customers benefit from the proven CMOSens Technology and excellent technical support. Among a large variety of applications, flow and humidity sensors are successfully used in the automotive and medical industries.

9.23   Shimmer (Shimmer 2015)

Headquartered in Dublin, Ireland, with an R&D center located in Boston, USA, Shimmer has been a leading provider of wearable wireless sensor products and solutions since its foundation in 2008. For academic, applied, and clinical researchers integrating wearable sensing technologies into a wide range of applications, Shimmer offers a flexible wireless sensor platform, scientifically reliable data, and complete control of data capture, interpretation, and analysis.

Also, Shimmer delivers mature, robust, and reliable market-ready technology that eliminates up to 80% of the development time and expense for wearable wireless sensing applications.

9.24   Silicon Labs (Sillicon Labs 2015)

Silicon Labs, headquartered in Austin, Texas, is a leading supplier of mixed-signal intelligent sensor solutions that are characterized by high reliability, compact size, high levels of integration, and ease of use for a variety of applications. Diverse sensor product portfolio includes optical sensors, digital relative I2C humidity and temperature sensor ICs, and capacitive touch sense microcontroller devices. Also, Silicon Labs offers integrated, robust, reliable, and easy-to-use wireless and RF IC solutions. By using mixed-signal ICs designed in standard CMOS from Silicon Labs, designers are able to eliminate many discrete components and use fewer external components. Customers can focus on value-added features and speed time with ZigBee, Bluetooth, Wi-Fi, ISM band, and Wireless MCUs from Silicon Labs.

9.25   SOWNet Technologies (SOWNet Technologies 2014)

SOWNet Technologies is a company specializing in wireless sensor network solutions, founded in 2006 as a spinout of the Dutch research institute TNO (TNO 2014). SOWNet Technologies is dedicated to providing cutting-edge, high-quality sensor network solutions to several different markets. After successfully completing projects in a variety of sectors, including logistics, public transport, and precision agriculture, there is a focus on developing the GuArtNet art security system with the partner Automatic Signal (Automatic Signal 2014).

9.26   SPI (SPI 2015)

Sensor Products Inc. (SPI) is a world leader in the niche field of tactile surface pressure and force sensors. The privately held company was founded in 1990; it is headquartered in Madison, New Jersey, with offices in Toronto, Canada, and Guadalajara, Mexico. The line of products includes sensors for cars, industry, homes, and daily life.

9.27   Terabee (Terabee 2015)

Terabee is a French company that aims at innovating the world of measurements and sensing by shifting analog toward digital, manual towards automated, 2D toward 3D, local toward global, and grounded toward airborne. With excellence in sensors, Terabee has strong international partnerships with academic and industrial institutions. Terabee studies and uses:

- Video and still cameras.
- Multispectral and hyper-spectral devices (for spotting what the human eye cannot see).
- Laser scanners (i.e., Lidar).
- Radioactivity meters.

At Terabee, they have developed the fastest, smallest, and lightest distance sensors for advanced robotics in challenging environments. TeraRanger distance sensors are born from a fruitful collaboration with the European Centre for Nuclear Research (CERN) while developing flying indoor inspection systems. A technology is established for the fast acquisition of distance data and having it packed in an only 8-gram distance sensor chip (Terabee 2017).

Also, flying and drones are accorded professional focus. With experience in aviation, the advent of automated unmanned robots/vehicles to cover basic tasks is foreseen. Terabee uses proven technology when available (i.e., from the military) and develops solutions in-house when nonavailable. Each application requires a dedicated system:

- Unmanned fixed wing (i.e., small planes) for scanning large areas.
- Unmanned vertical takeoff and landing (VTOL) (i.e., multi-copters) for proximity tasks.
- Aerial manned operations.
- Ground vehicles (mainly wheeled).

Terabee is application-driven, not technology-driven, by:

- Studying the market.
- Immersing into the application to fully understand its needs.
- Mapping pros and cons—and risks—and then providing a real solution that makes both technical and business sense. This could include a modern sensor, a specific existing or to-be-developed drone, or none of them.

9.28   Texas Instruments (TI 2015)

Texas Instruments (TI) headquartered in Dallas, Texas, is a technology giant that has products that cover almost everything. TIers are differentiators; TI is a global semiconductor company operating in 35 countries. From the TIer who unveiled the first working integrated circuit in 1958 to the more than 30,000 TIers around the world today who design, manufacture and sell analog and embedded processing chips; they are problem-solvers collaborating to change the world through technology. For WSN, a line of innovations includes tools and software, microcontrollers, ARM, and digital signal processors (DSP).

9.29   Valarm (Valarm 2015)

Valarm was founded in the spring of 2012; it is based in the new "Silicon Beach" of Los Angeles, California. Valarm products cover monitoring anything, anywhere.

From oil and gas to agriculture and viticulture, to fleets of vehicles with precious cargo and remote tanks of liquids, to environmental factors (such as air and water quality, water usage, flood alerts, and liquid level detection) in distant facilities.

In industrial applications, Valarm sensor solutions (water levels, air quality, temperature, humidity, switches, GPS, tanks, light, water usage/flow, and others) with powerful connectivity (any mobile network carrier, Wi-Fi, Ethernet) perform remote environmental monitoring and telemetry wherever, whenever needed.

Cloud-based web tools provide powerful real-time ad hoc mobile sensor networks due to Valarm's open platform (with easy-to-use APIs) and integration of a variety of sensors, such as water, temperature, humidity, CO_2, volatile organic compounds (VOCs), switches, 0–10 V, PWM, electrical resistance, location, liquids, and 4–20 mA.

9.30   WhizNets (WhizNets 2015)

WhizNets is located in San Ramon, California; its solutions address the challenges of a continuously evolving connected ecosystem through innovation and collaboration. The portfolio of solutions and services includes a broad range of embedded Wi-Fi modules, embedded wireless solutions, and cloud and mobile solutions.

WhizNets' innovative low-cost, low–footprint, highly integrated Wi-Fi module solutions allow making microcontroller-based products Wi-Fi ready at the lowest cost. Wi-Fi starter kits are available for many processor technologies based on Cortex-M3, Cortex-M4, Cortex-M0+, AVR32, PIC32, ARM7, and ARM9 CPUs with SPI and SDIO interfaces.

WhizNets IoT cloud solution is a complete solution offering the IoT devices, cloud platform, infrastructure, and applications that enable people and enterprises to be mobile, interactive, and available, thus increasing the overall productivity, effectiveness, and efficiency. The solution serves the application requirements across many verticals:

- Energy conservation.
- Security and surveillance.
- Transportation.
- Smart buildings.
- Transportation.
- Healthcare.

WhizNets service and product offerings enable the delivery of intelligent wireless solutions that simplify the connected world. For many years WhizNets provides easy integration of:

- Wireless connectivity (Wi-Fi, WIMAX, LTE).
- Sensors for machine-to-machine (M2M) and IoT.
- Cloud Internet of Things/Internet of Everything (IoT/IoE) platform integration.
- Android platform and applications.
- Customized system development.
- Testing and Inter-op.

9.31   Willow Technologies (Willow Technologies 2012)

Established in 1989, Willow Technologies is located in Copthorne, West Sussex, United Kingdom. They provide electronic solutions to customers by designing, manufacturing, and supplying components and systems globally to the electrical and electronic marketplace. Willow Technologies are specialists in switching, sensing, resistive and hermetic seal solutions and have a wide portfolio of sensing technologies. Their in-house engineering capability and rapid prototyping facility for custom parts enable them to develop products to match specific application requirements (ISO9001:2000 registered).

9.32   Xandem (Xandem 2015)

Xandem, located in Salt Lake City, Utah, has a line of products that cover security, elderly care, automation, and customizing applications. Xandem's technology is brilliant in security sensing; it covers large areas, remains totally hidden, without being blocked or fooled, and smartly detects motion through walls and obstructions. Elderly care systems based on Xandem products monitor motion over the entire home. The system remains completely hidden, and the patient does not need to wear a device. As for automation, knowledge about where people are is essential for automation in smart buildings and homes. Xandem products provide this knowledge to systems that control lights, appliances, and heating, ventilating, and air conditioning (HVAC) units.

Developers and integrators can use Xandem products to build exciting and innovative applications.

References

Adaptive Wireless Solutions. 2015, January 1. *Wireless Sensors*. (Adaptive Wireless Solutions) Retrieved August 24, 2015, from http://adaptive-wireless.co.uk/wireless-sensors/

AlertMe. 2014, January 1. *About Us*. (AlertMe) Retrieved August 13, 2014, from http://www.alertme.com/about-us/

Aol. 2015, January 1. *Crossbow Technologies*. (Aol Inc.) Retrieved August 24, 2015, from https://www.crunchbase.com/organization/crossbow-technologies

Atmel. 2015, January 1. *Fact Sheet*. (Atmel Corporation) Retrieved August 26, 2015, from http://www.atmel.com/about/corporate/factsheet.aspx

Automatic Signal. 2014, January 1. *Profiel*. (Automatic Signal) Retrieved July 19, 2014, from Automatic Signal : http://www.automaticsignal.nl/Profiel_nl.html

British Gas. 2015, January 1. *Products & Services*. (British Gas) Retrieved August 26, 2015, from https://www.britishgas.co.uk/products-and-services/

Cisco. 2015, January 1. *Internet of Things (IoT)*. (Cisco) Retrieved August 25, 2015, from http://www.cisco.com/web/solutions/trends/iot/overview.html

Coalesenses. 2014, January 1. *Portrait*. (Coalesenses) Retrieved July 17, 2014, from Coalesenses : http://www.coalesenses.com/index.php/company/

Dust Networks. 2015, January 1. *Wireless Sensor Networks – Dust Networks*. (Linear Technology) Retrieved August 24, 2015, from http://www.linear.com/products/wireless_sensor_networks_-_dust_networks

Dynastream Innovations. 2014, January 1. *About Us*. (Dynastream Innovations) Retrieved August 13, 2014, from http://www.thisisant.com/company/d1/history/

EasySen. 2015, January 1. *About EasySen*. (EasySen) Retrieved August 26, 2015, from http://www.zoominfo.com/s/#!search/profile/company?companyId=88116085&targetid=profile

EcoLogicSense. 2015, January 1. *Communicating Products for Real-time Monitoring of Air Quality*. (EcoLogicSense) Retrieved August 25, 2015, from https://sites.google.com/site/ecologicsenseen/

EpiSensor. 2015, Janaury 1. *About Us*. (EpiSensor) Retrieved August 24, 2015, from http://episensor.com/about-us/

ERS. 2015, January 1. *Company – Overview*. (Embedded Research Solutions, Inc.) Retrieved August 25, 2015, from http://www.embedded-zone.com/Company/Overview

GainSpan. 2015, January 1. *Markets: Wi-Fi Anywhere and Everywhere*. (GainSpan) Retrieved August 24, 2015, from http://www.gainspan.com/application/market

Infineon. 2015, January 1. *Products*. (Infineon Technologies AG) Retrieved August 26, 2015, from http://www.infineon.com/cms/en/product/products.html

Libelium. 2015, January 1. *Company*. (Libelium Comunicaciones Distribuidas S.L) Retrieved August 25, 2015, from http://www.libelium.com/company/

MEMSIC. 2015, January 1. *About MEMSIC*. (MEMSIC, Inc) Retrieved August 24, 2015, from http://www.memsic.com/about-memsic/index.cfm

Millennial Net. 2012, January 1. *MeshScape Wireless Sensor Networking System*. (Millennial Net) Retrieved August 25, 2015, from http://www.millennialnet.com/Technology.aspx

Moog Crossbow. 2014, January 1. *About Us*. (Moog Crossbow) Retrieved March 27, 2014, from Moog Crossbow: http://www.xbow.com/about-us/

National Instruments. 2015, January 1. *What Is A Wireless Sensor Network?* (National Instruments Corporation) Retrieved August 24, 2015, from http://www.ni.com/wsn/whatis/

Nordic Semiconductor. 2004, June 1. *nRF24E1: 2.4GHz RF transceiver with embedded 8051 compatible micro-controller and 9 input, 10 bit ADC*. (Nordic Semiconductor) Retrieved March 21, 2014, from http://www.datasheetarchive.com/dlmain/Datasheets-23/DSA-442985.pdf

OmniVision Technologies. 2011, January 1. *About Us*. (OmniVision Technologies) Retrieved March 27, 2014, from OmniVision Technologies: http://www.ovt.com/aboutus/

Sensirion. 2015, January 1. *Home*. (Sensirion) Retrieved January 4, 2015, from http://www.sensirion.com/en/about-us/company/

Sensors Online. 2007, May 11. *Wireless Applications Moteiv Launches Wireless Mote with Interface to Mobile Devices*. (Sensors Online) Retrieved August 2014, 2014, from http://www.sensorsmag.com/wireless-applications/news/moteiv-launches-wireless-mote-with-interface-mobile-devices-2523

Shimmer. 2015, January 1. *Home*. (Shimmer) Retrieved August 25, 2015, from http://www.shimmersensing.com

Sillicon Labs. 2015, January 1. *About Us*. (Sillicon Labs) Retrieved August 24, 2015, from http://www.silabs.com/about/pages/default.aspx

SOWNet Technologies. 2014, January 1. *Company Profile*. (SOWNet Technologies) Retrieved July 19, 2014, from SOWNet Technologies: http://www.sownet.nl/index.php/company/company-profile

SPI. 2015, January 1. *Applications*. (Sensor Products Inc.) Retrieved August 26, 2015, from http://www.sensorprod.com/index.php

Terabee. 2015, January 1. *About Us*. (Terabee) Retrieved February 1, 2017, from http://www.terabee.com/about/

———. 2017, January 1. *The Smallest, Lightest and Fastest Distance Sensors!* Retrieved February 1, 2017, from http://www.teraranger.com

TI. 2015, Janaury 1. *Tools & Software - Microcontrollers, ARM, and DSP*. (Texas Instruments Inc.) Retrieved August 26, 2015, from http://www.ti.com/lsds/ti/tools-software/sw_portal.page

TNO. 2014, January 1. *About Us*. (TNO) Retrieved July 19, 2014, from https://www.tno.nl/home.cfm?context=overtno&content=overtno&laag1=overtno

Valarm. 2015, January 1. *Home*. (Valarm) Retrieved August 25, 2015, from http://www.valarm.net
WhizNets. 2015, January 1. *WhiZnets Inc. a Connected Wireless Solutions Company*. (WhizNets Inc.) Retrieved August 25, 2015, from http://www.whiznets.com
Willow Technologies. 2012, January 1. *Facts*. (Willow Technologies) Retrieved July 16, 2014, from Willow Technologies: http://www.willow.co.uk/html/facts.html
Xandem. 2015, January 1. *Xandem*. (Xandem Technology) Retrieved August 25, 2015, from http://www.xandem.com

Chapter 10
Datasheets

There should be a datasheet for everybody.

10.1   Agilent ADCM-1670 CIF Resolution CMOS Camera Module (Agilent Technologies 2003a)

Agilent ADCM-1670 CIF Resolution CMOS Camera Module, UART Output

Product Overview

Description

The Agilent ADCM-1670 ultra compact CMOS camera module is an advanced, low-power CIF resolution camera component for embedded applications. The camera module combines an CMOS image sensor and image processing pipeline with a high-quality lens to deliver images in JPEG or video format, ready for storage or transmission. Output data is transmitted using a serial port.

The ADCM-1670 camera module features a quality, integral lens in a tightly integrated sensor and image processing design. The camera module is optimized for use in a variety of embedded applications from cell phones to handheld wireless devices to image-enabled appliances and automotive design.

Incorporating JPEG, YCbCr, RGB or grayscale outputs, the ADCM-1670 supports industry-leading data resolutions as well as subsampling.

The ADCM-1670 camera module also supports a range of programmable modes, which extend design flexibility.

Features

- 352 x 288 CIF resolution
- Bayer color filters – blue, red and green
- Frame rate – 15 frames per second at CIF resolution
- Flexible orientation
- Programmable to many image formats:
  - CIF (352 x 288)
  - QVGA (320 x 240)
  - QCIF (176 x 144)
  - QQVGA (160 x 120)
  - QQCIF (88 x 72)
  - Any other format 352 x 352 or smaller
- Panning – window can be placed anywhere in the 352 x 352 array
- Low power – 100 mW typical at 13 MHz input clock
- High intrinsic sensitivity for enhanced low light performance
- Fully configurable image processing
- Single 2.8V power supply with internal voltage regulation

- High quality F/2.6 lens
- Programmable gamma correction and color balancing
- Direct JPEG, YCbCr, RGB or grayscale output
- Internal buffer for slow readout of compressed images
- Horizontal/vertical mirroring and subsampling
- Programmable VSYNC-HSYNC setup and hold, vertical line blanking and external clock polarity
- Optimized temperature performance
- Excellent image quality – JPEG based compression with selectable quantization tables
- Automatic gathering of frame statistics including histograms for each color channel
- Automatic adjustment of compression rate for constant image file sizes
- Image resizer
- Auto exposure and auto white balance
- Integrated IR filter
- Compact size – 12.5 x 10.5 x 5.2 mm

Applications

- Mobile phones
- Video phones
- Personal Digital Assistants
- Image-enabled appliances
- Digital still mini cameras
- Embedded automotive
- Monitoring equipment

Agilent Technologies

10.2   Agilent ADCM-1700-0000 CMOS Camera Module (Agilent Technologies 2003b)

Agilent ADCM-1700-0000
Landscape CIF Resolution
CMOS Camera Module

Data Sheet

Description

The ADCM-1700-0000 ultra compact CMOS camera module is an advanced, low-power CIF resolution camera component for embedded applications. The camera module combines an Agilent CMOS image sensor and image processing design with a high quality lens to deliver images in formats that are ready for storage or transmission. Output data can be transmitted using a serial or parallel port.

The ADCM-1700-0000 camera module features a quality, integral lens in a tightly integrated sensor and image processing design. The camera module is optimized for use in a variety of embedded applications from cell phones and handheld wireless devices to image-enabled appliances and automotive design.

Incorporating a CCIR 656-compatible 8-bit parallel interface, or an RGB or YCbCr interface (serial or parallel), the ADCM-1700-0000 supports industry-leading data resolutions as well as subsampling.

The ADCM-1700-0000 camera module also supports a range of programmable modes, including support for embedded or external synchronization capabilities, extending design flexibility.

Features

- 352 x 288 landscape CIF resolution
- 24 bit color depth (16 million colors)
- Bayer color filters – blue, red and green
- Frame rate – 15 frames per second at CIF resolution
- Programmable to many image formats:
  - CIF (352 x 288) landscape only
  - QVGA (320 x 240) landscape only
  - QCIF (176 x 144)
  - QQVGA (160 x 120)
  - QQCIF (88 x 72)
  - Any other format 352 x 288 or smaller
- Flexible orientation
- Panning – window can be placed anywhere in the 352 x 288 array
- Low power – 42 mW typical at 13 MHz input clock
- Single 2.8V power supply with internal voltage regulation

- High quality F/2.8 lens
- Fully configurable image processing
- Direct RGB or YCbCr 8-bit parallel output port (CCIR 656-compatible)
- Embedded synchronization capability – CCIR 656
- Horizontal/vertical mirroring and subsampling
- Excellent image quality
- Noise adaptive processing
- Statistics gathering – automatic gathering of frame statistics including histograms for each color channel
- Image resizer
- Auto exposure and auto white balance
- High intrinsic sensitivity for enhanced low light performance
- Integrated IR filter
- Compact size – 8.0 x 7.0 x 5.4 mm

Applications

- Mobile phones
- Video phones
- Personal Digital Assistants
- Digital still mini cameras
- Image-enabled appliances
- Embedded automotive
- Monitoring equipment

Agilent Technologies

General Specifications

Feature	Value
Output format	8-bit parallel YCbCr CCIR 656-compliant 8-bit parallel YCbCr or RGB
Maximum frame rates	15 fps at 352 x 288 (CIF)
Image modes	Grayscale and full color
YCbCr (YUV) formats	4:4:4 YCbCr 4:2:2 $Y_1Cb_{12}Y_2Cr_{12}$ 4:2:2 $Cb_{12}Y_1Cr_{12}Y_2$ 4:2:2 $Y_1Cr_{12}Y_2Cb_{12}$ 4:2:2 $Cr_{12}Y_1Cb_{12}Y_2$
Gamma correction	33 value programmable interpolated table
Data synchronization	End_of_Line, End_of_Frame, Data_Clock
Video synchronization	HSYNC, VSYNC, VCLK
Serial control identification	0x51
Supply voltage requirements	2.65 to 3.1 V
External clock frequency	4 to 32 MHz
Power consumption	42 mW typical, CIF output, 13 MHz clock
Scene illumination (minimum)	5 lux

Optical Specifications

Function	Description
Pixel count	352 x 288 (CIF landscape mode)
Pixel size	5.6 μm x 5.6 μm
Effective fill factor	~ 80%
IR filter	Integrated
Lens type	Plastic singlet aspheric
Focal length	2.10 mm
F/#	2.8
Focus	Fixed focus
Depth of focus	100 mm to infinity
Field of view	52° full angle (horizontal)
Distortion	≤ 4%

10.3   Agilent ADCM-2650 CMOS Camera Module (Agilent Technologies 2003c)

Agilent ADCM-2650 Portrait VGA Resolution CMOS Camera Module

Product Overview

Description

The ADCM-2650 ultra compact CMOS camera module is an advanced, low-power VGA resolution camera component for embedded applications. The camera module combines an Agilent CMOS image sensor and image processing pipeline with a high-quality lens to deliver images in JPEG or video format, ready for storage or transmission. Output data is transmitted using a parallel port.

The ADCM-2650 camera module features a quality, integral lens in a tightly integrated sensor and image processing design. The camera module is optimized for use in a variety of embedded applications from cell phones to handheld wireless devices to image-enabled appliances and automotive design.

Incorporating a CCIR 656-compatible 8-bit parallel interface, or a JPEG or YCbCr interface, the ADCM-2650 supports industry-leading data resolutions as well as sub-sampling.

The ADCM-2650 camera module also supports a range of programmable modes, including support for embedded or external sync capabilities, which extend design flexibility.

Features

* Portrait VGA resolution
* Bayer color filters: blue, red and green
* Frame rate: 15 frames per second @ VGA resolution
* Flexible orientation
* Programmable to many image formats, portrait or landscape:
  - VGA portrait only (480 x 640)
  - CIF (352 x 288)
  - QVGA (320 x 240)
  - QCIF (176 x 144)
  - QQVGA (160 x 120)
  - QQCIF (88 x 72)
  - Any other format 480 x 640 or smaller
* Panning - window can be placed anywhere in the 480 x 640 array
* Low power: 70 mW typical at 13 MHz input clock
* High intrinsic sensitivity for enhanced low light performance

* Single power supply with internal voltage regulation
* High quality F/2.8 lens
* Fully configurable image processing
* Direct JPEG, YCbCr 8-bit parallel output port (CCIR 656-compatible)
* Embedded sync capability - CCIR 656
* Horizontal/vertical mirroring and sub-sampling
* Optimized temperature performance
* Excellent image quality; JPEG based compression with selectable quantization tables
* Automatic gathering of frame statistics including histograms for each color channel
* Adaptive quantization in JPEG
* Image resizer
* Auto exposure and auto white balance
* Integrated IR filter
* Compact size: 12.5 x 10.5 x 6.8 mm (without cover glass)

Applications

* Mobile phones
* Video phones
* Personal Digital Assistants
* Image-enabled appliances
* Digital still mini cameras
* Embedded automotive
* Monitoring equipment

10.4   Agilent ADNS-3060 Optical Mouse Sensor (Agilent Technologies 2004)

Agilent ADNS-3060
High-performance
Optical Mouse Sensor
Data Sheet

Description

The ADNS-3060 is a high performance addition to Agilent's popular ADNS family of optical mouse sensors.

The ADNS-3060 is based on a new, faster architecture with improved navigation. The sensor is capable of sensing high speed mouse motion - up to 40 inches per second and acceleration up to 15g – for increased user precision and smoothness.

The ADNS-3060 along with the ADNS-2120 (or ADNS-2120-001) lens, ADNS-2220 (or ADNS-2220-001) assembly clip and HLMP-ED80-XX000 form a complete, compact optical mouse tracking system. There are no moving parts, which means high reliability and less maintenance for the end user. In addition, precision optical alignment is not required, facilitating high volume assembly.

The sensor is programmed via registers through a four-wire serial port. It is packaged in a 20-pin staggered dual inline package (DIP).

Theory of Operation

The ADNS-3060 is based on Optical Navigation Technology, which measures changes in position by optically acquiring sequential surface images (frames) and mathematically determining the direction and magnitude of movement.

It contains an Image Acquisition System (IAS), a Digital Signal Processor (DSP), and a four-wire serial port.

The IAS acquires microscopic surface images via the lens and illumination system. These images are processed by the DSP to determine the direction and distance of motion. The DSP calculates the Δx and Δy relative displacement values.

An external microcontroller reads the Δx and Δy information from the sensor serial port. The microcontroller then translates the data into PS2 or USB signals before sending them to the host PC or game console.

Features

- High speed motion detection – up to 40 ips and 15g
- New architecture for greatly improved optical navigation technology
- Programmable frame rate over 6400 frames per second
- SmartSpeed self-adjusting frame rate for optimum performance
- Serial port burst mode for fast data transfer
- 400 or 800 cpi selectable resolution
- Single 3.3 volt power supply
- Four-wire serial port along with Chip Select, Power Down, and Reset pins

Applications

- Mice for game consoles and computer games
- Mice for desktop PC's, Workstations, and portable PC's
- Trackballs
- Integrated input devices

Agilent Technologies

10.5   AL440B High Speed FIFO Field Memory (AverLogic Technologies 2002)

   **AL440B**

4M-Bit High Speed FIFO Field Memory

Applications

- Multimedia systems
- Video capture or editing systems for NTSC/PAL or SVGA resolution
- Security systems
- Scan rate converter
- PIP(Picture In Picture) video display
- TBC(Time Base Correction)
- Frame Synchronizer
- Digital Video Camera
- Hard Disk cache memory
- Buffer for Communication System

* 80MHz High-Speed Version

- DTV/HDTV video stream buffer

Description

The AL440B is a high-performance FIFO (First-In-First-Out) field memory chip designed to buffer audio/video/graphic digital data for a wide range of applications.

Features

- 4Mbits (512k x 8 bits) organization FIFO
- Independent 8-bit data I/O port operations
- Available in 2 speed grades: 80 and 40Mhz
- Input Enable control (write mask)
- Output Enable control (data skipping)
- Supports Input ready/Output ready flags
- Selectable control signal polarity
- Programmable window mode data access with mirroring function support
- Self-refresh
- 5V signals input tolerance
- 3.3V±10% power supply
- Standard 44-pin TSOP (II) package

Ordering Information

Part number	Package	Power supply
AL440B-24 (40MHz)	44-pin plastic TSOP(II)	+3.3V±10%
AL440B-12 (80MHz)	44-pin plastic TSOP(II)	+3.3V±10%

10.6   Atmel AT29BV040A Flash Memory (Atmel 2003a)

Features
- Single Supply Voltage, Range 2.7V to 3.6V
- Single Supply for Read and Write
- Software Protected Programming
- Fast Read Access Time – 200 ns
- Low Power Dissipation
  - 15 mA Active Current
  - 40 µA CMOS Standby Current
- Sector Program Operation
  - Single Cycle Reprogram (Erase and Program)
  - 2048 Sectors (256 Bytes/Sector)
  - Internal Address and Data Latches for 256 Bytes
- Two 16K Bytes Boot Blocks with Lockout
- Fast Sector Program Cycle Time – 20 ms Max.
- Internal Program Control and Timer
- $\overline{\text{DATA}}$ Polling for End of Program Detection
- Minimum Endurance 10,000 Cycles
- CMOS and TTL Compatible Inputs and Outputs
- Commercial and Industrial Temperature Ranges

**4-megabit
(512K x 8)
Single 2.7-volt
Battery-Voltage™
Flash Memory**

AT29BV040A

Description
The AT29BV040A is a 3-volt-only in-system Flash Programmable and Erasable Read Only Memory (PEROM). Its 4 megabits of memory is organized as 524,288 words by 8 bits. Manufactured with Atmel's advanced nonvolatile CMOS EEPROM technology, the device offers access times to 200 ns, and a low 54 mW power dissipation. When the device is deselected, the CMOS standby current is less than 40 µA. The device

Pin Configurations

Pin Name	Function
A0 - A18	Addresses
$\overline{\text{CE}}$	Chip Enable
$\overline{\text{OE}}$	Output Enable
$\overline{\text{WE}}$	Write Enable
I/O0 - I/O7	Data Inputs/Outputs
NC	No Connect

**TSOP Top View
Type 1**

Rev. 0383G–FLASH–5/03

10.7   Atmel AT91 ARM Thumb-Based Microcontrollers (Atmel 2008)

Features
* Incorporates the ARM7TDMI® ARM® Thumb® Processor
  - High-performance 32-bit RISC Architecture
  - High-density 16-bit Instruction Set
  - Leader in MIPS/Watt
  - EmbeddedICE™ In-circuit Emulation, Debug Communication Channel Support
* Internal High-speed Flash
  - 512 Kbytes (AT91SAM7S512) Organized in Two Contiguous Banks of 1024 Pages of 256 Bytes (Dual Plane)
  - 256 Kbytes (AT91SAM7S256) Organized in 1024 Pages of 256 Bytes (Single Plane)
  - 128 Kbytes (AT91SAM7S128) Organized in 512 Pages of 256 Bytes (Single Plane)
  - 64 Kbytes (AT91SAM7S64) Organized in 512 Pages of 128 Bytes (Single Plane)
  - 32 Kbytes (AT91SAM7S321/32) Organized in 256 Pages of 128 Bytes (Single Plane)
  - 16 Kbytes (AT91SAM7S161/16) Organized in 256 Pages of 64 Bytes (Single Plane)
  - Single Cycle Access at Up to 30 MHz in Worst Case Conditions
  - Prefetch Buffer Optimizing Thumb Instruction Execution at Maximum Speed
  - Page Programming Time: 6 ms, Including Page Auto-erase, Full Erase Time: 15 ms
  - 10,000 Write Cycles, 10-year Data Retention Capability, Sector Lock Capabilities, Flash Security Bit
  - Fast Flash Programming Interface for High Volume Production
* Internal High-speed SRAM, Single-cycle Access at Maximum Speed
  - 64 Kbytes (AT91SAM7S512/256)
  - 32 Kbytes (AT91SAM7S128)
  - 16 Kbytes (AT91SAM7S64)
  - 8 Kbytes (AT91SAM7S321/32)
  - 4 Kbytes (AT91SAM7S161/16)
* Memory Controller (MC)
  - Embedded Flash Controller, Abort Status and Misalignment Detection
* Reset Controller (RSTC)
  - Based on Power-on Reset and Low-power Factory-calibrated Brown-out Detector
  - Provides External Reset Signal Shaping and Reset Source Status
* Clock Generator (CKGR)
  - Low-power RC Oscillator, 3 to 20 MHz On-chip Oscillator and one PLL
* Power Management Controller (PMC)
  - Software Power Optimization Capabilities, Including Slow Clock Mode (Down to 500 Hz) and Idle Mode
  - Three Programmable External Clock Signals
* Advanced Interrupt Controller (AIC)
  - Individually Maskable, Eight-level Priority, Vectored Interrupt Sources
  - Two (AT91SAM7S512/256/128/64/321/161) or One (AT91SAM7S32/16) External Interrupt Source(s) and One Fast Interrupt Source, Spurious Interrupt Protected
* Debug Unit (DBGU)
  - 2-wire UART and Support for Debug Communication Channel interrupt, Programmable ICE Access Prevention
  - Mode for General Purpose 2-wire UART Serial Communication
* Periodic Interval Timer (PIT)
  - 20-bit Programmable Counter plus 12-bit Interval Counter
* Windowed Watchdog (WDT)
  - 12-bit key-protected Programmable Counter
  - Provides Reset or Interrupt Signals to the System

AT91 ARM Thumb-based Microcontrollers

AT91SAM7S512
AT91SAM7S256
AT91SAM7S128
AT91SAM7S64
AT91SAM7S321
AT91SAM7S32
AT91SAM7S161
AT91SAM7S16
Summary

NOTE: This is a summary document. The complete document is available on the Atmel website at www.atmel.com.

6175HS-ATARM-07-Dec-08

- – Counter May Be Stopped While the Processor is in Debug State or in Idle Mode
- • Real-time Timer (RTT)
  - – 32-bit Free-running Counter with Alarm
  - – Runs Off the Internal RC Oscillator
- • One Parallel Input/Output Controller (PIOA)
  - – Thirty-two (AT91SAM7S512/256/128/64/321/161) or twenty-one (AT91SAM7S32/16) Programmable I/O Lines Multiplexed with up to Two Peripheral I/Os
  - – Input Change Interrupt Capability on Each I/O Line
  - – Individually Programmable Open-drain, Pull-up resistor and Synchronous Output
- • Eleven (AT91SAM7S512/256/128/64/321/161) or Nine (AT91SAM7S32/16) Peripheral DMA Controller (PDC) Channels
- • One USB 2.0 Full Speed (12 Mbits per Second) Device Port (Except for the AT91SAM7S32/16).
  - – On-chip Transceiver, 328-byte Configurable Integrated FIFOs
- • One Synchronous Serial Controller (SSC)
  - – Independent Clock and Frame Sync Signals for Each Receiver and Transmitter
  - – I²S Analog Interface Support, Time Division Multiplex Support
  - – High-speed Continuous Data Stream Capabilities with 32-bit Data Transfer
- • Two (AT91SAM7S512/256/128/64/321/161) or One (AT91SAM7S32/16) Universal Synchronous/Asynchronous Receiver Transmitters (USART)
  - – Individual Baud Rate Generator, IrDA® Infrared Modulation/Demodulation
  - – Support for ISO7816 T0/T1 Smart Card, Hardware Handshaking, RS485 Support
  - – Full Modem Line Support on USART1 (AT91SAM7S512/256/128/64/321/161)
- • One Master/Slave Serial Peripheral Interface (SPI)
  - – 8- to 16-bit Programmable Data Length, Four External Peripheral Chip Selects
- • One Three-channel 16-bit Timer/Counter (TC)
  - – Three External Clock Input and Two Multi-purpose I/O Pins per Channel (AT91SAM7S512/256/128/64/321/161)
  - – One External Clock Input and Two Multi-purpose I/O Pins for the first Two Channels Only (AT91SAM7S32/16)
  - – Double PWM Generation, Capture/Waveform Mode, Up/Down Capability
- • One Four-channel 16-bit PWM Controller (PWMC)
- • One Two-wire Interface (TWI)
  - – Master Mode Support Only, All Two-wire Atmel EEPROMs and I²C Compatible Devices Supported (AT91SAM7S512/256/128/64/321/32)
  - – Master, Multi-Master and Slave Mode Support, All Two-wire Atmel EEPROMs and I²C Compatible Devices Supported (AT91SAM7S161/16)
- • One 8-channel 10-bit Analog-to-Digital Converter, Four Channels Multiplexed with Digital I/Os
- • SAM-BA™ Boot Assistant
  - – Default Boot program
  - – Interface with SAM-BA Graphic User Interface
- • IEEE® 1149.1 JTAG Boundary Scan on All Digital Pins
- • 5V-tolerant I/Os, Including Four High-current Drive I/O lines, Up to 16 mA Each (AT91SAM7S161/16 I/Os Not 5V-tolerant)
- • Power Supplies
  - – Embedded 1.8V Regulator, Drawing up to 100 mA for the Core and External Components
  - – 3.3V or 1.8V VDDIO I/O Lines Power Supply, Independent 3.3V VDDFLASH Flash Power Supply
  - – 1.8V VDDCORE Core Power Supply with Brown-out Detector
- • Fully Static Operation: Up to 55 MHz at 1.65V and 85·C Worst Case Conditions
- • Available in 64-lead LQFP Green or 64-pad QFN Green Package (AT91SAM7S512/256/128/64/321/161) and 48-lead LQFP Green or 48-pad QFN Green Package (AT91SAM7S32/16)

2   **AT91SAM7S Series Summary** ━━━━━━━━━━━━━

10.8   Atmel AT91SAM ARM-Based Embedded MPU (Atmel 2011c)

**AT91SAM
ARM-based
Embedded MPU**

AT91SAM9261

Features

- Incorporates the ARM926EJ-S™ ARM® Thumb® Processor
  - DSP Instruction Extensions
  - ARM Jazelle® Technology for Java® Acceleration
  - 16 Kbyte Data Cache, 16 Kbyte Instruction Cache, Write Buffer
  - 210 MIPS at 190 MHz
  - Memory Management Unit
  - EmbeddedICE™, Debug Communication Channel Support
  - Mid-level Implementation Embedded Trace Macrocell™
- Additional Embedded Memories
  - 32 Kbytes of Internal ROM, Single-cycle Access at Maximum Bus Speed
  - 160 Kbytes of Internal SRAM, Single-cycle Access at Maximum Processor or Bus Speed
- External Bus Interface (EBI)
  - Supports SDRAM, Static Memory, NAND Flash and CompactFlash®
- LCD Controller
  - Supports Passive or Active Displays
  - Up to 16-bits per Pixel in STN Color Mode
  - Up to 16M Colors in TFT Mode (24-bit per Pixel), Resolution up to 2048 x 2048
- USB
  - USB 2.0 Full Speed (12 Mbits per second) Host Double Port
    - Dual On-chip Transceivers
    - Integrated FIFOs and Dedicated DMA Channels
  - USB 2.0 Full Speed (12 Mbits per second) Device Port
    - On-chip Transceiver, 2 Kbyte Configurable Integrated FIFOs
- Bus Matrix
  - Handles Five Masters and Five Slaves
  - Boot Mode Select Option
  - Remap Command
- Fully Featured System Controller (SYSC) for Efficient System Management, including
  - Reset Controller, Shutdown Controller, Four 32-bit Battery Backup Registers for a Total of 16 Bytes
  - Clock Generator and Power Management Controller
  - Advanced Interrupt Controller and Debug Unit
  - Periodic Interval Timer, Watchdog Timer and Real-time Timer
  - Three 32-bit PIO Controllers
- Reset Controller (RSTC)
  - Based on Power-on Reset Cells, Reset Source Identification and Reset Output Control
- Shutdown Controller (SHDWC)
  - Programmable Shutdown Pin Control and Wake-up Circuitry
- Clock Generator (CKGR)
  - 32,768 Hz Low-power Oscillator on Battery Backup Power Supply, Providing a Permanent Slow Clock
  - 3 to 20 MHz On-chip Oscillator and two PLLs
- Power Management Controller (PMC)
  - Very Slow Clock Operating Mode, Software Programmable Power Optimization Capabilities
  - Four Programmable External Clock Signals

6062N–ATARM–3-Oct-11

- Advanced Interrupt Controller (AIC)
  - Individually Maskable, Eight-level Priority, Vectored Interrupt Sources
  - Three External Interrupt Sources and One Fast Interrupt Source, Spurious Interrupt Protected
- Debug Unit (DBGU)
  - 2-wire USART and support for Debug Communication Channel, Programmable ICE Access Prevention
  - Mode for General Purpose Two-wire UART Serial Communication
- Periodic Interval Timer (PIT)
  - 20-bit Interval Timer plus 12-bit Interval Counter
- Watchdog Timer (WDT)
  - Key Protected, Programmable Only Once, Windowed 12-bit Counter, Running at Slow Clock
- Real-Time Timer (RTT)
  - 32-bit Free-running Backup Counter Running at Slow Clock
- Three 32-bit Parallel Input/Output Controllers (PIO) PIOA, PIOB and PIOC
  - 96 Programmable I/O Lines Multiplexed with up to Two Peripheral I/Os
  - Input Change Interrupt Capability on Each I/O Line
  - Individually Programmable Open-drain, Pull-up Resistor and Synchronous Output
- Nineteen Peripheral DMA (PDC) Channels
- Multimedia Card Interface (MCI)
  - SDCard and MultiMediaCard™ Compliant
  - Automatic Protocol Control and Fast Automatic Data Transfers with PDC, MMC and SDCard Compliant
- Three Synchronous Serial Controllers (SSC)
  - Independent Clock and Frame Sync Signals for Each Receiver and Transmitter
  - I²S Analog Interface Support, Time Division Multiplex Support
  - High-speed Continuous Data Stream Capabilities with 32-bit Data Transfer
- Three Universal Synchronous/Asynchronous Receiver Transmitters (USART)
  - Individual Baud Rate Generator, IrDA® Infrared Modulation/Demodulation
  - Support for ISO7816 T0/T1 Smart Card, Hardware and Software Handshaking, RS485 Support
- Two Master/Slave Serial Peripheral Interface (SPI)
  - 8- to 16-bit Programmable Data Length, Four External Peripheral Chip Selects
- One Three-channel 16-bit Timer/Counters (TC)
  - Three External Clock Inputs, Two multi-purpose I/O Pins per Channel
  - Double PWM Generation, Capture/Waveform Mode, Up/Down Capability
- Two-wire Interface (TWI)
  - Master Mode Support, All Two-wire Atmel EEPROMs Supported
- IEEE® 1149.1 JTAG Boundary Scan on All Digital Pins
- Required Power Supplies:
  - 1.06V to 1.32V for VDDCORE and VDDBU
  - 3.0V to 3.6V for VDDOSC and for VDDPLL
  - 2.7V to 3.6V for VDDIOP (Peripheral I/Os)
  - 1.65V to 1.95V and 3.0V to 3.6V for VDDIOM (Memory I/Os)
- Available in a 217-ball LFBGA RoHS-compliant Package

10.9   Atmel Microcontroller with 4/8/16 K Bytes in-System Programmable Flash (Atmel 2011b)

Features

* High performance, low power Atmel® AVR® 8-bit microcontroller
* Advanced RISC architecture
  - 131 powerful instructions – most single clock cycle execution
  - 32 × 8 general purpose working registers
  - Fully static operation
  - Up to 20 MIPS throughput at 20MHz
  - On-chip 2-cycle multiplier
* High endurance non-volatile memory segments
  - 4/8/16 Kbytes of in-system self-programmable flash program memory
  - 256/512/512 bytes EEPROM
  - 512/1K/1Kbytes internal SRAM
  - Write/erase cyles: 10,000 flash/100,000 EEPROM
  - Data retention: 20 years at 85°C/100 years at 25°C[1]
  - Optional boot code section with independent lock bits
    In-system programming by on-chip boot program
    True read-while-write operation
  - Programming lock for software security
* QTouch® library support
  - Capacitive touch buttons, sliders and wheels
  - QTouch and QMatrix acquisition
  - Up to 64 sense channels
* Peripheral features
  - Two 8-bit timer/counters with separate prescaler and compare mode
  - One 16-bit timer/counter with separate prescaler, compare mode, and capture mode
  - Real time counter with separate oscillator
  - Six PWM channels
  - 8-channel 10-bit ADC in TQFP and QFN/MLF package
  - 6-channel 10-bit ADC in PDIP Package
  - Programmable serial USART
  - Master/slave SPI serial interface
  - Byte-oriented 2-wire serial interface (Philips I²C compatible)
  - Programmable watchdog timer with separate on-chip oscillator
  - On-chip analog comparator
  - Interrupt and wake-up on pin change
* Special microcontroller features
  - DebugWIRE on-chip debug system
  - Power-on reset and programmable brown-out detection
  - Internal calibrated oscillator
  - External and internal interrupt sources
  - Five sleep modes: Idle, ADC noise reduction, power-save, power-down, and standby
* I/O and packages
  - 23 programmable I/O lines
  - 28-pin PDIP, 32-lead TQFP, 28-pad QFN/MLF and 32-pad QFN/MLF
* Operating voltage:
  - 1.8V - 5.5V for Atmel ATmega48V/88V/168V
  - 2.7V - 5.5V for Atmel ATmega48/88/168
* Temperature range:
  - -40°C to 85°C
* Speed grade:
  - ATmega48V/88V/168V: 0 - 4MHz @ 1.8V - 5.5V, 0 - 10MHz @ 2.7V - 5.5V
  - ATmega48/88/168: 0 - 10MHz @ 2.7V - 5.5V, 0 - 20MHz @ 4.5V - 5.5V
* Low power consumption
  - Active mode:
    250µA at 1MHz, 1.8V
    15µA at 32kHz, 1.8V (including oscillator)
  - Power-down mode:
    0.1µA at 1.8V

Note: 1.   See "Data Retention" on page 7 for details.

8-bit Atmel Microcontroller with 4/8/16K Bytes In-System Programmable Flash

ATmega48/V
ATmega88/V
ATmega168/V

Summary

Rev. 2545TS-AVR–05/11

10.10   Atmel Microcontroller with 128KBytes in-System Programmable Flash (Atmel 2011a)

Features
- High-performance, Low-power Atmel®AVR®8-bit Microcontroller
- Advanced RISC Architecture
  - 133 Powerful Instructions – Most Single Clock Cycle Execution
  - 32 x 8 General Purpose Working Registers + Peripheral Control Registers
  - Fully Static Operation
  - Up to 16MIPS Throughput at 16MHz
  - On-chip 2-cycle Multiplier
- High Endurance Non-volatile Memory segments
  - 128Kbytes of In-System Self-programmable Flash program memory
  - 4Kbytes EEPROM
  - 4Kbytes Internal SRAM
  - Write/Erase cycles: 10,000 Flash/100,000 EEPROM
  - Data retention: 20 years at 85°C/100 years at 25°C[1]
  - Optional Boot Code Section with Independent Lock Bits
    In-System Programming by On-chip Boot Program
    True Read-While-Write Operation
  - Up to 64Kbytes Optional External Memory Space
  - Programming Lock for Software Security
  - SPI Interface for In-System Programming
- QTouch® library support
  - Capacitive touch buttons, sliders and wheels
  - QTouch and QMatrix acquisition
  - Up to 64 sense channels
- JTAG (IEEE std. 1149.1 Compliant) Interface
  - Boundary-scan Capabilities According to the JTAG Standard
  - Extensive On-chip Debug Support
  - Programming of Flash, EEPROM, Fuses and Lock Bits through the JTAG Interface
- Peripheral Features
  - Two 8-bit Timer/Counters with Separate Prescalers and Compare Modes
  - Two Expanded 16-bit Timer/Counters with Separate Prescaler, Compare Mode and Capture Mode
  - Real Time Counter with Separate Oscillator
  - Two 8-bit PWM Channels
  - 6 PWM Channels with Programmable Resolution from 2 to 16 Bits
  - Output Compare Modulator
  - 8-channel, 10-bit ADC
    8 Single-ended Channels
    7 Differential Channels
    2 Differential Channels with Programmable Gain at 1x, 10x, or 200x
  - Byte-oriented Two-wire Serial Interface
  - Dual Programmable Serial USARTs
  - Master/Slave SPI Serial Interface
  - Programmable Watchdog Timer with On-chip Oscillator
  - On-chip Analog Comparator
- Special Microcontroller Features
  - Power-on Reset and Programmable Brown-out Detection
  - Internal Calibrated RC Oscillator
  - External and Internal Interrupt Sources
  - Six Sleep Modes: Idle, ADC Noise Reduction, Power-save, Power-down, Standby, and Extended Standby
  - Software Selectable Clock Frequency
  - ATmega103 Compatibility Mode Selected by a Fuse
  - Global Pull-up Disable
- I/O and Packages
  - 53 Programmable I/O Lines
  - 64-lead TQFP and 64-pad QFN/MLF
- Operating Voltages
  - 2.7 - 5.5V ATmega128L
  - 4.5 - 5.5V ATmega128
- Speed Grades
  - 0 - 8MHz ATmega128L
  - 0 - 16MHz ATmega128

8-bit Atmel Microcontroller with 128KBytes In-System Programmable Flash

ATmega128
ATmega128L

Rev. 2467X–AVR–06/11

10.11   Atmel FPSLIC (Atmel 2002)

Features

- Monolithic Field Programmable System Level Integrated Circuit (FPSLIC™)
  - AT40K SRAM-based FPGA with Embedded High-performance RISC AVR® Core, Extensive Data and Instruction SRAM and JTAG ICE
- 5,000 to 40,000 Gates of Patented SRAM-based AT40K FPGA with FreeRAM™
  - 2 - 18.4 Kbits of Distributed Single/Dual Port FPGA User SRAM
  - High-performance DSP Optimized FPGA Core Cell
  - Dynamically Reconfigurable In-System – FPGA Configuration Access Available On-chip from AVR Microcontroller Core to Support Cache Logic® Designs
  - Very Low Static and Dynamic Power Consumption – Ideal for Portable and Handheld Applications
- Patented AVR Enhanced RISC Architecture
  - 120+ Powerful Instructions – Most Single Clock Cycle Execution
  - High-performance Hardware Multiplier for DSP-based Systems
  - Approaching 1 MIPS per MHz Performance
  - C Code Optimized Architecture with 32 x 8 General-purpose Internal Registers
  - Low-power Idle, Power-save and Power-down Modes
  - 100 µA Standby and Typical 2-3 mA per MHz Active
- Up to 36 Kbytes of Dynamically Allocated Instruction and Data SRAM
  - Up to 16 Kbytes x 16 Internal 15 ns Instructions SRAM
  - Up to 16 Kbytes x 8 Internal 15 ns Data SRAM
- JTAG (IEEE std. 1149.1 Compliant) Interface
  - Extensive On-chip Debug Support
  - Limited Boundary-scan Capabilities According to the JTAG Standard (AVR Ports)
- AVR Fixed Peripherals
  - Industry-standard 2-wire Serial Interface
  - Two Programmable Serial UARTs
  - Two 8-bit Timer/Counters with Separate Prescaler and PWM
  - One 16-bit Timer/Counter with Separate Prescaler, Compare, Capture Modes and Dual 8-, 9- or 10-bit PWM
- Support for FPGA Custom Peripherals
  - AVR Peripheral Control – 16 Decoded AVR Address Lines Directly Accessible to FPGA
  - FPGA Macro Library of Custom Peripherals
- 16 FPGA Supplied Internal Interrupts to AVR
- Up to Four External Interrupts to AVR
- 8 Global FPGA Clocks
  - Two FPGA Clocks Driven from AVR Logic
  - FPGA Global Clock Access Available from FPGA Core
- Multiple Oscillator Circuits
  - Programmable Watchdog Timer with On-chip Oscillator
  - Oscillator to AVR Internal Clock Circuit
  - Software-selectable Clock Frequency
  - Oscillator to Timer/Counter for Real-time Clock
- V_{CC}: 3.0V – 3.6V
- 3.3V 33 MHz PCI-compliant FPGA I/O
  - 20 mA Sink/Source High-performance I/O Structures
  - All FPGA I/O Individually Programmable
- High-performance, Low-power 0.35µ CMOS Five-layer Metal Process
- State-of-the-art Integrated PC-based Software Suite Including Co-verification
- 5V I/O Tolerant

FP$SLIC$™

5K - 40K Gates of AT40K FPGA with 8-bit AVR® Microcontroller, up to 36K Bytes of SRAM and On-chip JTAG ICE

AT94KAL Series Field Programmable System Level Integrated Circuit

Rev. 1138G-FPSLI-11/03

10.12   Bluegiga WT12 (Bluegiga Technologies 2007)

 **WT12 Bluetooth ® module**

DESCRIPTION

WT12 is a next-generation, class 2, Bluetooth® 2.0+EDR (Enhanced Data Rates) module. It introduces three times faster data rates compared to existing Bluetooth® 1.2 modules even with lower power consumption! WT12 is a highly integrated and sophisticated Bluetooth® module, containing all the necessary elements from Bluetooth® radio to antenna and a fully implemented protocol stack. Therefore WT12 provides an ideal solution for developers who want to integrate Bluetooth® wireless technology into their design with limited knowledge of Bluetooth® and RF technologies.

By default WT12 module is equipped with powerful and easy-to-use iWRAP firmware. iWRAP enables users to access Bluetooth® functionality with simple ASCII commands delivered to the module over serial interface - it's just like a Bluetooth® modem.

FEATURES:

- Fully Qualified Bluetooth system v2.0 + EDR, CE and FCC
- Integrated chip antenna
- Industrial temperature range from -40°C to +85°C
- Enhanced Data Rate (EDR) compliant with v2.0.E.2 of specification for both 2Mbps and 3Mbps modulation modes
- RoHS Compliant
- Full Speed Bluetooth Operation with Full Piconet
- Scatternet Support
- USB version 2.0 compatible
- UART with bypass mode
- Support for 802.11 Coexistence
- 8Mbits of Flash Memory

APPLICATIONS:

- Hand held terminals
- Industrial devices
- Point-of-Sale systems
- PCs
- Personal Digital Assistants (PDAs)
- Computer Accessories
- Access Points
- Automotive Diagnostics Units

10.13   C8051F121 Mixed-Signal MCU (Silicon Laboratories 2004)

C8051F121
100 MIPS, 128 kB Flash, 12-Bit ADC, 64-Pin Mixed-Signal MCU

Analog Peripherals
12-Bit ADC
- ±1 LSB INL; no missing codes
- Programmable throughput up to 100 ksps
- 8 external inputs; programmable as single-ended or differential
- Programmable amplifier gain: 16, 8, 4, 2, 1, 0.5
- Data-dependent windowed interrupt generator
- Built-in temperature sensor (±3 °C)

8-Bit ADC
- ±1 LSB INL; no missing codes
- Programmable throughput up to 500 ksps
- 8 external inputs
- Programmable amplifier gain: 4, 2, 1, 0.5

Two 12-Bit DACs
- Can synchronize outputs to timers for jitter-free waveform generation

Two Comparators

Internal Voltage Reference

V$_{DD}$ Monitor/Brown-out Detector

On-Chip JTAG Debug & Boundary Scan
- On-chip debug circuitry facilitates full speed, non-intrusive in-system debug (no emulator required)
- Provides breakpoints, single stepping, watchpoints, stack monitor
- Inspect/modify memory and registers
- Superior performance to emulation systems using ICE-chips, target pods, and sockets
- IEEE1149.1 compliant boundary scan

High-Speed 8051 µC Core
- Pipelined instruction architecture; executes 70% of instructions in 1 or 2 system clocks
- Up to 100 MIPS throughput with 100 MHz system clock
- 16 x 16 multiply/accumulate engine (2-cycle)

Memory
- 8448 bytes data RAM
- 128 kB Flash; in-system programmable in 1024-byte sectors (1024 bytes are reserved)
- External parallel data memory interface

Digital Peripherals
- 32 port I/O; all are 5 V tolerant
- Hardware SMBus™ (I2C™ Compatible), SPI™, and two UART serial ports available concurrently
- Programmable 16-bit counter/timer array with six capture/compare modules
- 5 general-purpose 16-bit counter/timers
- Dedicated watchdog timer; bidirectional reset
- Real-time clock mode using Timer 3 or PCA

Clock Sources
- Internal oscillator: 24.5 MHz, 2% accuracy supports UART operation
- On-chip programmable PLL: up to 100 MHz
- External oscillator: Crystal, RC, C, or Clock

Supply Voltage: 3.0 to 3.6 V
- Typical operating current: 50 mA at 100 MHz
- Typical stop mode current: 0.4 uA

64-Pin TQFP

Temperature Range: −40 to +85 °C

Selected Electrical Specifications
(T$_A$ = −40 to +85 C°, V$_{DD}$ = 3.0 V unless otherwise specified)

PARAMETER	CONDITIONS	MIN	TYP	MAX	UNITS
GLOBAL CHARACTERISTICS					
Supply Voltage		3.0		3.6	V
Supply Current (CPU active)	Clock = 100 MHz		50		mA
	Clock = 1 MHz		0.6		mA
	Clock = 32 kHz		16		µA
Supply Current (shutdown)	Oscillator off; V$_{DD}$ Monitor Enabled		10		µA
	Oscillator off; V$_{DD}$ Monitor Disabled		0.4		µA
Clock Frequency Range		DC		100	MHz
INTERNAL CLOCKS					
Oscillator Frequency		24.0	24.5	25.0	MHz
PLL Frequency		96	98	100	MHz
A/D CONVERTER					
Resolution			12		bits
Integral Nonlinearity				±1	LSB
Differential Nonlinearity	Guaranteed Monotonic			±1	LSB
Signal-to-Noise Plus Distortion		66	69		dB
Throughput Rate				100	ksps
D/A CONVERTERS					
Resolution			12		bits
Differential Nonlinearity	Guaranteed Monotonic			±1	LSB
Output Settling Time			10		µS

10.14   CC1000 (Texas Instruments 2007a)

CC1000
Single Chip Very Low Power RF Transceiver

Applications
- Very low power UHF wireless data transmitters and receivers
- 315 / 433 / 868 and 915 MHz ISM/SRD band systems
- RKE – Two-way Remote Keyless Entry

- Home automation
- Wireless alarm and security systems
- AMR – Automatic Meter Reading
- Low power telemetry
- Game Controllers and advanced toys

Product Description

CC1000 is a true single-chip UHF transceiver designed for very low power and very low voltage wireless applications. The circuit is mainly intended for the ISM (Industrial, Scientific and Medical) and SRD (Short Range Device) frequency bands at 315, 433, 868 and 915 MHz, but can easily be programmed for operation at other frequencies in the 300-1000 MHz range.

The main operating parameters of CC1000 can be programmed via a serial bus, thus making CC1000 a very flexible and easy to use transceiver. In a typical system CC1000 will be used together with a microcontroller and a few external passive components.

CC1000 is based on Chipcon's SmartRF® technology in 0.35 µm CMOS.

Features

- True single chip UHF RF transceiver
- Very low current consumption
- Frequency range 300 – 1000 MHz
- Integrated bit synchroniser
- High sensitivity (typical -110 dBm at 2.4 kBaud)
- Programmable output power –20 to 10 dBm
- Small size (TSSOP-28 or UltraCSP™ package)
- Low supply voltage (2.1 V to 3.6 V)
- Very few external components required
- No external RF switch / IF filter required

- RSSI output
- Single port antenna connection
- FSK data rate up to 76.8 kBaud
- Complies with EN 300 220 and FCC CFR47 part 15
- Programmable frequency in 250 Hz steps makes crystal temperature drift compensation possible without TCXO
- Suitable for frequency hopping protocols
- Development kit available
- Easy-to-use software for generating the CC1000 configuration data

10.15   CC1020 (Texas Instruments 2014a)

CC1020

CC1020
Low-Power RF Transceiver for Narrowband Systems

Applications

- Narrowband low power UHF wireless data transmitters and receivers with channel spacing as low as 12.5 and 25 kHz
- 402 / 424 / 426 / 429 / 433 / 447 / 449 / 469 / 868 / 915 / 960 MHz ISM/SRD band systems

- AMR - Automatic Meter Reading
- Wireless alarm and security systems
- Home automation
- Low power telemetry

Product Description

CC1020 is a true single-chip UHF transceiver designed for very low power and very low voltage wireless applications. The circuit is mainly intended for the ISM (Industrial, Scientific and Medical) and SRD (Short Range Device) frequency bands at 402, 424, 426, 429, 433, 447, 449, 469, 868, 915, and 960 MHz, but can easily be programmed for multi-channel operation at other frequencies in the 402 - 470 and 804 - 960 MHz range.

The CC1020 is especially suited for narrowband systems with channel spacing of 12.5 or 25 kHz complying with ARIB STD-T67 and EN 300 220.

The CC1020 main operating parameters can be programmed via a serial bus, thus making CC1020 a very flexible and easy to use transceiver.

In a typical system CC1020 will be used together with a microcontroller and a few external passive components.

Features

- True single chip UHF RF transceiver
- Frequency range 402 MHz - 470 MHz and 804 MHz - 960 MHz
- High sensitivity (up to -118 dBm for a 12.5 kHz channel)
- Programmable output power
- Low current consumption (RX: 19.9 mA)
- Low supply voltage (2.3 V to 3.6 V)
- No external IF filter needed
- Low-IF receiver
- Very few external components required
- Small size (QFN 32 package)
- Pb-free package
- Digital RSSI and carrier sense indicator
- Data rate up to 153.6 kBaud

- OOK, FSK and GFSK data modulation
- Integrated bit synchronizer
- Image rejection mixer
- Programmable frequency and AFC make crystal temperature drift compensation possible without TCXO
- Suitable for frequency hopping systems
- Suited for systems targeting compliance with EN 300 220, FCC CFR47 part 15, ARIB STD-T67, and ARIB STD-T96
- Development kit available
- Easy-to-use software for generating the CC1020 configuration data

---

TI recommends using the latest RF performance line device CC1120 as successor of CC1020: www.ti.com/rfperformanceline*

10.16   CC1100 (Texas Instruments 2005a)

 **Chipcon Products from Texas Instruments**

CC1100

CC1100
Low-Power Sub- 1 GHz RF Transceiver

Applications

- *Ultra low-power wireless applications operating in the 315/433/868/915 MHz ISM/SRD bands*
- *Wireless alarm and security systems*
- *Industrial monitoring and control*

- *Wireless sensor networks*
- *AMR – Automatic Meter Reading*
- *Home and building automation*

Product Description

The *CC1100* is a low-cost sub- 1 GHz transceiver designed for very low-power wireless applications. The circuit is mainly intended for the ISM (Industrial, Scientific and Medical) and SRD (Short Range Device) frequency bands at 315, 433, 868, and 915 MHz, but can easily be programmed for operation at other frequencies in the 300-348 MHz, 400-464 MHz and 800-928 MHz bands.

The RF transceiver is integrated with a highly configurable baseband modem. The modem supports various modulation formats and has a configurable data up to 500 kBaud.

CC1100 provides extensive hardware support for packet handling, data buffering, burst transmissions, clear channel assessment, link quality indication, and wake-on-radio.

The main operating parameters and the 64-byte transmit/receive FIFOs of *CC1100* can be controlled via an SPI interface. In a typical system, the *CC1100* will be used together with a microcontroller and a few additional passive components.

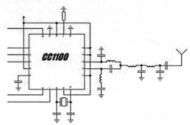

**Chipcon Products
from Texas Instruments**

Key Features

RF Performance

- High sensitivity (−111 dBm at 1.2 kBaud, 868 MHz, 1% packet error rate)
- Low current consumption (14.4 mA in RX, 1.2 kBaud, 868 MHz)
- Programmable output power up to +10 dBm for all supported frequencies
- Excellent receiver selectivity and blocking performance
- Programmable data rate from 1.2 to 500 kBaud
- Frequency bands: 300-348 MHz, 400-464 MHz and 800-928 MHz

Analog Features

- 2-FSK, GFSK, and MSK supported as well as OOK and flexible ASK shaping
- Suitable for frequency hopping systems due to a fast settling frequency synthesizer: 90us settling time
- Automatic Frequency Compensation (AFC) can be used to align the frequency synthesizer to the received centre frequency
- Integrated analog temperature sensor

Digital Features

- Flexible support for packet oriented systems: On-chip support for sync word detection, address check, flexible packet length, and automatic CRC handling
- Efficient SPI interface: All registers can be programmed with one "burst" transfer
- Digital RSSI output
- Programmable channel filter bandwidth
- Programmable Carrier Sense (CS) indicator

- Programmable Preamble Quality Indicator (PQI) for improved protection against false sync word detection in random noise
- Support for automatic Clear Channel Assessment (CCA) before transmitting (for listen-before-talk systems)
- Support for per-package Link Quality Indication (LQI)
- Optional automatic whitening and de-whitening of data

Low-Power Features

- 400nA SLEEP mode current consumption
- Fast startup time: 240us from sleep to RX or TX mode (measured on EM reference design [5] and [6])
- Wake-on-radio functionality for automatic low-power RX polling
- Separate 64-byte RX and TX data FIFOs (enables burst mode data transmission)

General

- Few external components: Completely on-chip frequency synthesizer, no external filters or RF switch needed
- Green package: RoHS compliant and no antimony or bromine
- Small size (QLP 4x4 mm package, 20 pins)
- Suited for systems targeting compliance with EN 300 220 (Europe) and FCC CFR Part 15 (US).
- Support for asynchronous and synchronous serial receive/transmit mode for backwards compatibility with existing radio communication protocols

10.17   CC1101 (Texas Instruments 2014b)

Low-Power Sub-1 GHz RF Transceiver

Applications

- Ultra low-power wireless applications operating in the 315/433/868/915 MHz ISM/SRD bands
- Wireless alarm and security systems
- Industrial monitoring and control

- Wireless sensor networks
- AMR – Automatic Meter Reading
- Home and building automation
- Wireless MBUS

Product Description

CC1101 is a low-cost sub-1 GHz transceiver designed for very low-power wireless applications. The circuit is mainly intended for the ISM (Industrial, Scientific and Medical) and SRD (Short Range Device) frequency bands at 315, 433, 868, and 915 MHz, but can easily be programmed for operation at other frequencies in the 300-348 MHz, 387-464 MHz and 779-928 MHz bands.

The RF transceiver is integrated with a highly configurable baseband modem. The modem supports various modulation formats and has a configurable data rate up to 600 kbps.

CC1101 provides extensive hardware support for packet handling, data buffering, burst transmissions, clear channel assessment, link quality indication, and wake-on-radio.

The main operating parameters and the 64-byte transmit/receive FIFOs of *CC1101* can be controlled via an SPI interface. In a typical system, the *CC1101* will be used together with a microcontroller and a few additional passive components.

The *CC1190* 850-950 MHz range extender [21] can be used with *CC1101* in long range applications for improved sensitivity and higher output power.

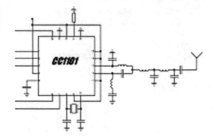

CC1101

Key Features

RF Performance

- High sensitivity
  - -116 dBm at 0.6 kBaud, 433 MHz, 1% packet error rate
  - -112 dBm at 1.2 kBaud, 868 MHz, 1% packet error rate
- Low current consumption (14.7 mA in RX, 1.2 kBaud, 868 MHz)
- Programmable output power up to +12 dBm for all supported frequencies
- Excellent receiver selectivity and blocking performance
- Programmable data rate from 0.6 to 600 kbps
- Frequency bands: 300-348 MHz, 387-464 MHz and 779-928 MHz

Analog Features

- 2-FSK, 4-FSK, GFSK, and MSK supported as well as OOK and flexible ASK shaping
- Suitable for frequency hopping systems due to a fast settling frequency synthesizer; 75 µs settling time
- Automatic Frequency Compensation (AFC) can be used to align the frequency synthesizer to the received signal centre frequency
- Integrated analog temperature sensor

Digital Features

- Flexible support for packet oriented systems; On-chip support for sync word detection, address check, flexible packet length, and automatic CRC handling
- Efficient SPI interface; All registers can be programmed with one "burst" transfer
- Digital RSSI output
- Programmable channel filter bandwidth
- Programmable Carrier Sense (CS) indicator
- Programmable Preamble Quality Indicator (PQI) for improved protection against false sync word detection in random noise
- Support for automatic Clear Channel Assessment (CCA) before transmitting (for listen-before-talk systems)
- Support for per-package Link Quality Indication (LQI)
- Optional automatic whitening and de-whitening of data

Low-Power Features

- 200 nA sleep mode current consumption
- Fast startup time; 240 µs from sleep to RX or TX mode (measured on EM reference design [1] and [2])
- Wake-on-radio functionality for automatic low-power RX polling
- Separate 64-byte RX and TX data FIFOs (enables burst mode data transmission)

General

- Few external components; Completely on-chip frequency synthesizer, no external filters or RF switch needed
- Green package: RoHS compliant and no antimony or bromine
- Small size (QLP 4x4 mm package, 20 pins)
- Suited for systems targeting compliance with EN 300 220 (Europe) and FCC CFR Part 15 (US)
- Suited for systems targeting compliance with the Wireless MBUS standard EN 13757-4:2005
- Support for asynchronous and synchronous serial receive/transmit mode for backwards compatibility with existing radio communication protocols

Improved Range using CC1190

- The *CC1190* [21] is a range extender for 850-950 MHz and is an ideal fit for *CC1101* to enhance RF performance
- High sensitivity
  - -118 dBm at 1.2 kBaud, 868 MHz, 1% packet error rate
  - -120 dBm at 1.2 kBaud, 915 MHz, 1% packet error rate
- +20 dBm output power at 868 MHz
- +27 dBm output power at 915 MHz
- Refer to AN094 [22] and AN096 [23] for more performance figures of the *CC1101* + *CC1190* combination

10.18   C2420 (Texas Instruments 2005b)

**Chipcon Products
from Texas Instruments**

CC2420

2.4 GHz IEEE 802.15.4 / ZigBee-ready RF Transceiver

Applications

- *2.4 GHz IEEE 802.15.4 systems*
- *ZigBee systems*
- *Home/building automation*
- *Industrial Control*

- *Wireless sensor networks*
- *PC peripherals*
- *Consumer Electronics*

Product Description

The **CC2420** is a true single-chip 2.4 GHz IEEE 802.15.4 compliant RF transceiver designed for low power and low voltage wireless applications. **CC2420** includes a digital direct sequence spread spectrum baseband modem providing a spreading gain of 9 dB and an effective data rate of 250 kbps.

The **CC2420** is a low-cost, highly integrated solution for robust wireless communication in the 2.4 GHz unlicensed ISM band. It complies with worldwide regulations covered by ETSI EN 300 328 and EN 300 440 class 2 (Europe), FCC CFR47 Part 15 (US) and ARIB STD-T66 (Japan).

The **CC2420** provides extensive hardware support for packet handling, data buffering, burst transmissions, data encryption, data authentication, clear channel assessment, link quality indication and packet timing information. These

features reduce the load on the host controller and allow **CC2420** to interface low-cost microcontrollers.

The configuration interface and transmit / receive FIFOs of **CC2420** are accessed via an SPI interface. In a typical application **CC2420** will be used together with a microcontroller and a few external passive components.

CC2420 is based on Chipcon's SmartRF®- 03 technology in 0.18 μm CMOS.

Key Features

- True single-chip 2.4 GHz IEEE 802.15.4 compliant RF transceiver with baseband modem and MAC support
- DSSS baseband modem with 2 MChips/s and 250 kbps effective data rate.
- Suitable for both RFD and FFD operation
- Low current consumption (RX: 18.8 mA, TX: 17.4 mA)
- Low supply voltage (2.1 – 3.6 V) with integrated voltage regulator
- Low supply voltage (1.6 – 2.0 V) with external voltage regulator

- Programmable output power
- No external RF switch / filter needed
- I/Q low-IF receiver
- I/Q direct upconversion transmitter
- Very few external components
- 128(RX) + 128(TX) byte data buffering
- Digital RSSI / LQI support
- Hardware MAC encryption (AES-128)
- Battery monitor
- QLP-48 package, 7x7 mm
- Complies with ETSI EN 300 328, EN 300 440 class 2, FCC CFR-47 part 15 and ARIB STD-T66
- Powerful and flexible development tools available

10.19   CC2430 (Texas Instruments 2006)

Not Recommended for New Designs

Chipcon Products from Texas Instruments

CC2430

A True System-on-Chip solution for 2.4 GHz IEEE 802.15.4 / ZigBee®

Applications

- *2.4 GHz IEEE 802.15.4 systems*
- *ZigBee® systems*
- *Home/building automation*
- *Industrial Control and Monitoring*

- *Low power wireless sensor networks*
- *PC peripherals*
- *Set-top boxes and remote controls*
- *Consumer Electronics*

Product Description

The **CC2430** comes in three different flash versions: CC2430F32/64/128, with 32/64/128 KB of flash memory respectively. The **CC2430** is a true System-on-Chip (SoC) solution specifically tailored for IEEE 802.15.4 and ZigBee® applications. It enables ZigBee® nodes to be built with very low total bill-of-material costs. The **CC2430** combines the excellent performance of the leading **CC2420** RF transceiver with an industry-standard enhanced 8051 MCU, 32/64/128 KB flash memory, 8 KB RAM and many other powerful features. Combined with the industry leading ZigBee® protocol stack (Z-Stack™) from Texas Instruments, the **CC2430** provides the market's most competitive ZigBee® solution.

The **CC2430** is highly suited for systems where ultra low power consumption is required. This is ensured by various operating modes. Short transition times between operating modes further ensure low power consumption.

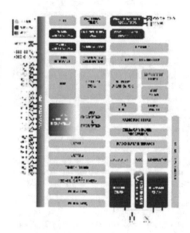

Key Features

- **RF/Layout**
  - 2.4 GHz IEEE 802.15.4 compliant RF transceiver (industry leading CC2420 radio core)
  - Excellent receiver sensitivity and robustness to interferers
  - Very few external components
  - Only a single crystal needed for mesh network systems
  - RoHS compliant 7x7mm QLP48 package

- **Low Power**
  - Low current consumption (RX: 27 mA, TX: 27 mA, microcontroller running at 32 MHz)
  - Only 0.5 µA current consumption in powerdown mode, where external interrupts or the RTC can wake up the system
  - 0.3 µA current consumption in stand-by mode, where external interrupts can wake up the system
  - Very fast transition times from low-power modes to active mode enables ultra low average power consumption in low dutycycle systems
  - Wide supply voltage range (2.0V - 3.6V)

- **Microcontroller**
  - High performance and low power 8051 microcontroller core
  - 32, 64 or 128 KB in-system programmable flash
  - 8 KB RAM, 4 KB with data retention in all power modes
  - Powerful DMA functionality
  - Watchdog timer
  - One IEEE 802.15.4 MAC timer, one general 16-bit timer and two 8-bit timers
  - Hardware debug support

- **Peripherals**
  - CSMA/CA hardware support.
  - Digital RSSI / LQI support
  - Battery monitor and temperature sensor
  - 12-bit ADC with up to eight inputs and configurable resolution
  - AES security coprocessor
  - Two powerful USARTs with support for several serial protocols
  - 21 general I/O pins, two with 20mA sink/source capability

- **Development tools**
  - Powerful and flexible development tools available

10.20 CC2431 (Texas Instruments 2005c)

Not Recommended for New Designs

**Chipcon Products
from Texas Instruments**

CC2431

System-on-Chip for 2.4 GHz ZigBee® / IEEE802.15.4 with Location Engine

Applications

- ZigBee® systems
- 2.4 GHz IEEE 802.15.4 systems
- Home/building automation
- Industrial Control and Monitoring
- Low power wireless sensor networks
- Access Control

- PC peripherals
- Set-top boxes and remote controls
- Consumer Electronics
- Container/Vehicle Tracking
- Active RFID
- Inventory Control

Product Description

The *CC2431* is a true System-On-Chip (SOC) for wireless sensor networking ZigBee®/IEEE 802.15.4 solutions. The chip includes a location detection hardware module that can be used in so-called blind nodes (i.e. nodes with unknown location) to receive signals from nodes with known location's. Based on this the location engine calculates an estimate of a blind node's position. The *CC2431* enables ZigBee® nodes to be built with very low total bill-of-material costs. The *CC2431* combines the excellent performance of the leading *CC2420* RF transceiver with an industry-standard enhanced 8051 MCU, 128 KB flash memory, 8 KB RAM and many other powerful features. Combined with the industry leading ZigBee® protocol stack (Z-Stack™) from Texas Instruments, the *CC2431* provides the market's most competitive ZigBee® solution.

The *CC2431* is highly suited for systems where ultra low power consumption is required. This is achieved by various operating modes. Short transition times between these modes further ensure low power consumption.

Key Features

- Location Engine calculates the location of a node in a network
- High performance and low power 8051 microcontroller core.
- 2.4 GHz IEEE 802.15.4 compliant RF transceiver (industry leading *CC2420* radio core).
- ZigBee® protocol stack (Z-Stack™) from Texas Instruments includes support for *CC2431* 's location engine.
- Excellent receiver sensitivity and robustness to interferers
- 128 KB in-system programmable flash
- 8 KB RAM, 4 KB with data retention in all power modes
- Powerful DMA functionality
- Very few external components
- Only a single crystal needed for mesh network systems

- Low current consumption (RX: 27 mA, TX: 27 mA, microcontroller running at 32 MHz)
- Only 0.5µA current consumption in power-down mode, where external interrupts or the RTC can wake up the system
- 0.3 µA current consumption in power-down mode, where external interrupts can wake up the system
- Very fast transition times from low-power modes to active mode enables ultra low average power consumption in low duty-cycle systems
- CSMA/CA hardware support
- Wide supply voltage range (2.0 V – 3.6 V)
- Digital RSSI/ LQI support
- Battery monitor and temperature sensor
- ADC with up to eight inputs and configurable resolution
- 128-bit AES security coprocessor

Not Recommended for New Designs

 **Chipcon Products from Texas Instruments**

CC2431

Key Features (continued)

- Two powerful USARTs with support for several serial protocols.
- Hardware debug support
- Watchdog timer
- One IEEE 802.15.4 MAC Timer, one general 16-bit timer and two 8-bit timers

- RoHS compliant 7x7 mm QLP48 package
- 21 general I/O pins, two with 20 mA sink/source capability
- Powerful and flexible development tools available

Note:

The CC2431 and the CC2430 are pin compatible, and the MCU and RF parts of the CC2430-F128 are identical to the CC2431 except the Location Engine. This data sheet complements the CC2430 data sheet with a description of the Location Engine. For complete information about the CC2431, please refer to the CC2430 data sheet in addition to this data sheet. The CC2430 data sheet can be found here:

http://focus.ti.com/lit/ds/symlink/cc2430.pdf

10.21   CC2530 (Texas Instruments 2011a)

TEXAS
INSTRUMENTS
www.ti.com

CC2530F32, CC2530F64
CC2530F128, CC2530F256

SWRS081B – APRIL 2009 – REVISED FEBRUARY 2011

A True System-on-Chip Solution for 2.4-GHz IEEE 802.15.4 and ZigBee Applications

Check for Samples: CC2530F32, CC2530F64, CC2530F128, CC2530F256

FEATURES

- **RF/Layout**
  - 2.4-GHz IEEE 802.15.4 Compliant RF Transceiver
  - Excellent Receiver Sensitivity and Robustness to Interference
  - Programmable Output Power Up to 4.5 dBm
  - Very Few External Components
  - Only a Single Crystal Needed for Asynchronous Networks
  - 6-mm × 6-mm QFN40 Package
  - Suitable for Systems Targeting Compliance With Worldwide Radio-Frequency Regulations: ETSI EN 300 328 and EN 300 440 (Europe), FCC CFR47 Part 15 (US) and ARIB STD-T-66 (Japan)
- **Low Power**
  - Active-Mode RX (CPU Idle): 24 mA
  - Active Mode TX at 1 dBm (CPU Idle): 29 mA
  - Power Mode 1 (4 µs Wake-Up): 0.2 mA
  - Power Mode 2 (Sleep Timer Running): 1 µA
  - Power Mode 3 (External Interrupts): 0.4 µA
  - Wide Supply-Voltage Range (2 V–3.6 V)
- **Microcontroller**
  - High-Performance and Low-Power 8051 Microcontroller Core With Code Prefetch
  - 32-, 64-, 128-, or 256-KB In-System-Programmable Flash
  - 8-KB RAM With Retention in All Power Modes
  - Hardware Debug Support

- **Peripherals**
  - Powerful Five-Channel DMA
  - Integrated High-Performance Op-Amp and Ultralow-Power Comparator

- IEEE 802.15.4 MAC Timer, General-Purpose Timers (One 16-Bit, Two 8-Bit)
- IR Generation Circuitry
- 32-kHz Sleep Timer With Capture
- CSMA/CA Hardware Support
- Accurate Digital RSSI/LQI Support
- Battery Monitor and Temperature Sensor
- 12-Bit ADC With Eight Channels and Configurable Resolution
- AES Security Coprocessor
- Two Powerful USARTs With Support for Several Serial Protocols
- 21 General-Purpose I/O Pins (19 × 4 mA, 2 × 20 mA)
- Watchdog Timer
- **Development Tools**
  - CC2530 Development Kit
  - CC2530 ZigBee® Development Kit
  - CC2530 RemoTI™ Development Kit for RF4CE
  - SmartRF™ Software
  - Packet Sniffer
  - IAR Embedded Workbench™ Available

APPLICATIONS

- 2.4-GHz IEEE 802.15.4 Systems
- RF4CE Remote Control Systems (64-KB Flash and Higher)
- ZigBee Systems (256-KB Flash)
- Home/Building Automation
- Lighting Systems
- Industrial Control and Monitoring
- Low-Power Wireless Sensor Networks
- Consumer Electronics
- Health Care

CC2530F32, CC2530F64
CC2530F128, CC2530F256

SWRS081B – APRIL 2009 – REVISED FEBRUARY 2011

TEXAS
INSTRUMENTS

www.ti.com

DESCRIPTION

The CC2530 is a true system-on-chip (SoC) solution for IEEE 802.15.4, Zigbee and RF4CE applications. It enables robust network nodes to be built with very low total bill-of-material costs. The CC2530 combines the excellent performance of a leading RF transceiver with an industry-standard enhanced 8051 MCU, in-system programmable flash memory, 8-KB RAM, and many other powerful features. The CC2530 comes in four different flash versions: CC2530F32/64/128/256, with 32/64/128/256 KB of flash memory, respectively. The CC2530 has various operating modes, making it highly suited for systems where ultralow power consumption is required. Short transition times between operating modes further ensure low energy consumption.

Combined with the industry-leading and golden-unit-status ZigBee protocol stack (Z-Stack™) from Texas Instruments, the CC2530F256 provides a robust and complete ZigBee solution.

Combined with the golden-unit-status RemoTI stack from Texas Instruments, the CC2530F64 and higher provide a robust and complete ZigBee RF4CE remote-control solution.

10.22   CP2102/9 Single-Chip USB to UART Bridge (Silicon Laboratories 2013)

CP2102/9

SINGLE-CHIP USB TO UART BRIDGE

Single-Chip USB to UART Data Transfer
- Integrated USB transceiver; no external resistors required
- Integrated clock; no external crystal required
- Internal 1024-byte programmable ROM for vendor ID, product ID, serial number, power descriptor, release number, and product description strings
  - EEPROM (CP2102)
  - EPROM (One-time programmable) (CP2109)
- On-chip power-on reset circuit
- On-chip voltage regulator
  - 3.3 V output (CP2102)
  - 3.45 V output (CP2109)
- 100% pin and software compatible with CP2101

USB Function Controller
- USB Specification 2.0 compliant; full-speed (12 Mbps)
- USB suspend states supported via SUSPEND pins

Asynchronous Serial Data BUS (UART)
- All handshaking and modem interface signals
- Data formats supported:
  - Data bits: 5, 6, 7, and 8
  - Stop bits: 1, 1.5, and 2
  - Parity: odd, even, mark, space, no parity
- Baud rates: 300 bps to 1 Mbps
- 576 Byte receive buffer; 640 byte transmit buffer
- Hardware or X-On/X-Off handshaking supported
- Event character support
- Line break transmission

Virtual COM Port Device Drivers
- Works with existing COM port PC Applications
- Royalty-free distribution license
- Windows 8/7/Vista/Server 2003/XP/2000
- Mac OS-X/OS-9
- Linux

USBXpress™ Direct Driver Support
- Royalty-Free Distribution License
- Windows 7/Vista/XP/Server 2003/2000
- Windows CE

Example Applications
- Upgrade of RS-232 legacy devices to USB
- Cellular phone USB interface cable
- USB interface cable
- USB to RS-232 serial adapter

Supply Voltage
- Self-powered: 3.0 to 3.6 V
- USB bus powered: 4.0 to 5.25 V

Package
- RoHS-compliant 28-pin QFN (5x5 mm)

Ordering Part Numbers
- CP2102-GM
- CP2100-A01-GM

Temperature Range: –40 to +85 °C

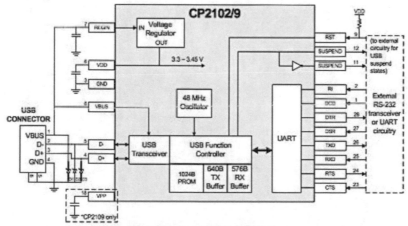

Figure 1. Example System Diagram

10.23   Digital Compass Solutions HMR3300 (Honeywell 2012)

The Honeywell HMR3300 is a digital compass solution for use in precision heading applications. Honeywell's magnetoresistive sensor is utilized to provide the reliability and accuracy of these small, solid-state compass designs. This compass solution is designed for generic precision compass integration into customer systems using a 5-voltage logic level serial data interface with commands in ASCII format. The HMR3300 includes a MEMS accelerometer for a horizontal three-axis, tilt-compensated precision compass for performance up to a $\pm$ 60° tilt range. Table 10.1 lists HMR3300 specifications.

Table 10.1 HMR3300 specifications

Characteristics	Conditions	Min	Typ	Max	Units
Heading					
Accuracy	Level 0° to ±30° (HMR3300 only) ±30° to ±60° (HMR3300 only)		1.0 3.0 4.0		deg RMS
Hysteresis	HMR3300		0.2	0.4	deg
Repeatability	HMR3300		0.2	0.4	deg
Pitch and Roll					
Range	Roll and Pitch Range			± 60	deg
Accuracy	0° to ± 30° ± 30° to ± 60°		0.4 1.0	0.5 1.2	deg
Null Accuracy*	Level -20° to +70°C Thermal Hysterisis -40° to +85°C Thermal Hysterisis		0.4 1.0 5.0		deg
Resolution			0.1		deg
Hysteresis			0.2		deg
Repeatability			0.2		deg
Magnetic Field					
Range	Maximum Magnetic Flux Density		± 2		gauss
Resolution			0.1	0.5	milli-gauss
Electrical					
Input Voltage	Unregulated Regulated	6 4.75	- -	15 5.25	volts DC
Current	HMR3300		22	24	mA
Digital Interface					
UART	ASCII (1 Start, 8 Data, 1 Stop, 0 Parity) User Selectable Baud Rate	2400	-	19200	Baud
SPI	CKE = 0, CKP = 0 Psuedo Master				
Update	Continuous/Strobed/Averaged HMR3300		8		Hz
Connector	In-Line 8-Pin Block (0.1" spacing)				
Physical					
Dimensions	Circuit Board Assembly	25.4 x 36.8 x 11			mm
Weight	HMR3300		7.50		grams
Environment					
Temperature	Operating Storage	-20 -55	-	+70 +125	°C

* Null zeroing prior to use of the HMR3300 and upon exposure to temperature excursions beyond the Operating Temperature limits is required to achieve highest performance.

10.24　DS18B20 Programmable Resolution 1-Wire Digital Thermometer (Maxim Integrated 2008)

maxim integrated™

DS18B20
Programmable Resolution
1-Wire Digital Thermometer

DESCRIPTION

The DS18B20 digital thermometer provides 9-bit to 12-bit Celsius temperature measurements and has an alarm function with nonvolatile user-programmable upper and lower trigger points. The DS18B20 communicates over a 1-Wire bus that by definition requires only one data line (and ground) for communication with a central microprocessor. It has an operating temperature range of -55°C to +125°C and is accurate to ±0.5°C over the range of -10°C to +85°C. In addition, the DS18B20 can derive power directly from the data line ("parasite power"), eliminating the need for an external power supply.

Each DS18B20 has a unique 64-bit serial code, which allows multiple DS18B20s to function on the same 1-Wire bus. Thus, it is simple to use one microprocessor to control many DS18B20s distributed over a large area. Applications that can benefit from this feature include HVAC environmental controls, temperature monitoring systems inside buildings, equipment, or machinery, and process monitoring and control systems.

FEATURES

- Unique 1-Wire® Interface Requires Only One Port Pin for Communication
- Each Device has a Unique 64-Bit Serial Code Stored in an On-Board ROM
- Multidrop Capability Simplifies Distributed Temperature-Sensing Applications
- Requires No External Components
- Can Be Powered from Data Line; Power Supply Range is 3.0V to 5.5V
- Measures Temperatures from -55°C to +125°C (-67°F to +257°F)
- ±0.5°C Accuracy from -10°C to +85°C
- Thermometer Resolution is User Selectable from 9 to 12 Bits
- Converts Temperature to 12-Bit Digital Word in 750ms (Max)

- User-Definable Nonvolatile (NV) Alarm Settings
- Alarm Search Command Identifies and Addresses Devices Whose Temperature is Outside Programmed Limits (Temperature Alarm Condition)
- Available in 8-Pin SO (150 mils), 8-Pin μSOP, and 3-Pin TO-92 Packages
- Software Compatible with the DS1822
- Applications Include Thermostatic Controls, Industrial Systems, Consumer Products, Thermometers, or Any Thermally Sensitive System

PIN CONFIGURATIONS

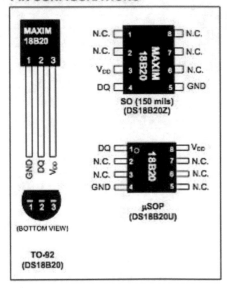

1-Wire is a registered trademark of Maxim Integrated Products, Inc.

10.25 DS18S20 High-Precision 1-Wire Digital Thermometer (Maxim Integrated 2010)

 maxim integrated™

DS18S20
High-Precision 1-Wire Digital Thermometer

FEATURES

- Unique 1-Wire® Interface Requires Only One Port Pin for Communication
- Each Device has a Unique 64-Bit Serial Code Stored in an On-Board ROM
- Multidrop Capability Simplifies Distributed Temperature Sensing Applications
- Requires No External Components
- Can Be Powered from Data Line. Power Supply Range is 3.0V to 5.5V
- Measures Temperatures from -55°C to +125°C (-67°F to +257°F)
- ±0.5°C Accuracy from -10°C to +85°C
- 9-Bit Thermometer Resolution
- Converts Temperature in 750ms (max)
- User-Definable Nonvolatile (NV) Alarm Settings
- Alarm Search Command Identifies and Addresses Devices Whose Temperature is Outside Programmed Limits (Temperature Alarm Condition)
- Applications Include Thermostatic Controls, Industrial Systems, Consumer Products, Thermometers, or Any Thermally Sensitive System

PIN CONFIGURATIONS

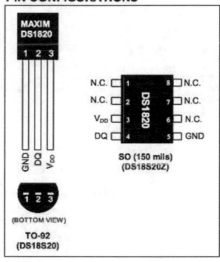

DESCRIPTION

The DS18S20 digital thermometer provides 9-bit Celsius temperature measurements and has an alarm function with nonvolatile user-programmable upper and lower trigger points. The DS18S20 communicates over a 1-Wire bus that by definition requires only one data line (and ground) for communication with a central microprocessor. It has an operating temperature range of -55°C to +125°C and is accurate to ±0.5°C over the range of -10°C to +85°C. In addition, the DS18S20 can derive power directly from the data line ("parasite power"), eliminating the need for an external power supply.

Each DS18S20 has a unique 64-bit serial code, which allows multiple DS18S20s to function on the same 1-Wire bus. Thus, it is simple to use one microprocessor to control many DS18S20s distributed over a large area. Applications that can benefit from this feature include HVAC environmental controls, temperature monitoring systems inside buildings, equipment, or machinery, and process monitoring and control systems.

1-Wire is a registered trademark of Maxim Integrated Products, Inc.

10.26   G-Node G301 (SOWNet Technologies 2014)

SOWNet technologies

GUARTNET

Home   Company   Technology   **Products**   Contact

GuArtNet   Testbed   **G-Node**   Order Online

You are here: Home >> Products >> G-Node   English (UK)   search...   Search

Introducing the G-Node

The SOWNet G-Node G301 Wireless Sensor Node, or G-Node in short, is a new Wireless Sensor Network development platform based on L-Nodes technology developed by SOWNet Technologies in collaboration with the Technical University of Delft.

Top features

- USB reprogramming
- Supports TinyOS 2.1
- Getting started DVD kit with all required tools in one package

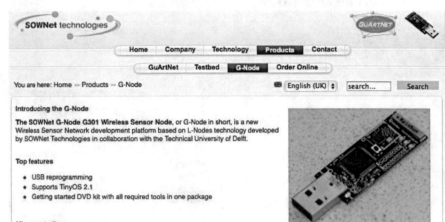

Microcontroller

The Texas Instruments MSP430F2418 microcontroller supplies all the processing power and memory you need for your Wireless Sensor Network applications.

- 116 KB ROM
- 8 KB RAM
- 8 Mbit external flash memory
- 16 MHz top speed

Radio

A versatile CC1101 radio chip from Texas Instruments allows the G-Node to shine in dense multi-hop deployments. This makes it possible to do large scale experiments in limited physical space.

- Packet radio with receiver sensitivity down to -112dBm
- Variable data rates from 0.8kbps up to 500kbps supported
- Variable transmission power from -30dBm up to +10dBm
- Antenna with decent range on-board

Interfacing

Finally, it is easy to interface the G-Node with existing sensors and other equipment having several different interfacing options available.

- 8 I/O pins for external connectivity
- SPI/I2C, UART, interrupt and ADC functions available
- 3V power supply up to 70mA strong can power most sensors

Links and contact information

The G-Node G301 is used in our Testbed T301 product: T301 product page
For more details on the G-Node G301 Wireless Sensor Node: download G-Node G301 whitepaper
To order online please visit our order page
Or contact us for more information and other purchasing options.

10.27   GS-1 Low Frequency Seismometer (Geospace Technologies 2014b)

GS-1 Low Frequency Seismometer

The GS-1 is a high-sensitivity, self-generating velocity detector with extremely low natural frequencies. It is an excellent choice for detecting seismic activity for structural analysis, geologic hazards, vibration isolation, etc.

GS-1 seismometers are available in 1.0 and 2.0 Hz natural frequencies and in vertically or horizontally oriented models. Sensitivities range from 3.0 to 15.0 V/in/sec depending on coil configurations. Optional weather-resistant cover (w/level bubble), calibration coil, and adjustable leg bases are available. A Vernier spring adjustment is included.

Spec Sheet
GS-1 Specifications

Natural frequency	1.0 Hz ± 10%
Orientation angle	Vertical ±7.5° Horizontal ±0.5%
DC resistance	450 ohms ±5% 4550 ohms ±5% 17,400 ohms ±5%
Sensitivity (V/in/s)	3.0 V/in/sec ± 10% 7.0 V/in/sec ± 10% 15.0 V/in/sec ± 10%
Open circuit damping	0.54 ± 20%
Moving mass	700 grams ±5%
Coil excursion	> 0.25 in. P-P
Operating temperature	−40° to 100 ° C
Dimensions (basic unit)	Height: 6.45 in. Diameter: 3.0 in. Weight: 69 oz.
Shock	50 G

Notes:

1. Specification temperature is 25 ° C.
2. Two Hz model is also available.

10.28   GS-11D Geophone (Geospace Technologies 2014a)

Geophones GS-11D
Rotating Coil Geophone

- Field-proven design
- Shock-resistant, rotating dual coil construction
- Gold-plated contacts for positive electrical connection

- Precision springs, computer designed and matched
- Full one-year warranty

The GS-11D is a high-output, rotating coil geophone designed and built to withstand the shocks of rough handling. The precision springs of this field-proven geophone are computer designed and matched to optimize performance specifications even under the most extreme conditions.

Gold-plated contacts assure positive electrical connections. The Geo Space manufacturing process includes checking all geophone operating parameters with the ATS, an automated computerized test system.

Natural frequencies are 4.5, 8, 10, and 14 Hz, with standard coil resistance of 380 ohms. The PC-21 Land Case is used with the GS-11D geophone.

Cases Available PC-21 Land Case
Spec Sheet:
GS-11D Specifications

Natural Frequency	4.5 ± 0.75 Hz	8 ± 0.75 Hz	10 ± 0.75 Hz	14 ± 0.75 Hz
Coil resistance at 25 °C ± 5%	380 ohms			
Intrinsic voltage sensitivity with 380 ohm coil ±10%	0.81 V/in/sec (0.32 V/cm/sec)			
Normalized transduction constant (V/in/sec)	042 (sq. root of Rc)			
Open circuit damping	0.34 ± 20%	0.39 ± 10%	0.32 ± 10%	0.23 ± 10%
Damping constant with 380 ohm coil	762	602	482	344
Optional coil resistances ±5%	4000 Ohms			
Moving mass ± 5%	23.6 g	16.8 g	16.8 g	16.8 g
Typical case to coil motion P-P	0.07 in (0.18 cm)	0.07 in (0.18 cm)	0.07 in (0.18 cm)	0.07 in (0.18 cm)
Harmonic distortion with a driving velocity of 0.7 in/sec (1.8 cm/sec) P-P	N/S	0.2% or less @ 12 Hz	@ 12 Hz	@ 12 Hz
Dimensions				
Height (less terminals[a])	1.32 in (3.35 cm)			
Diameter	1.25 in (3.18 cm)			
Weight	3.9 oz. (111 g)			

[a]Terminal height is 0.135 inches

10.29   Imote2 (Crossbow 2005)

Imote2
HIGH-PERFORMANCE WIRELESS SENSOR NETWORK NODE

- Intel PXA271 XScale® Processor at 13 – 416MHz

- Intel Wireless MMX DSP Coprocessor

- 256kB SRAM, 32MB FLASH, 32MB SDRAM

- Integrated 802.15.4 Radio

- Integrated 2.4GHz Antenna, Optional External SMA Connector

- Multi-color Status Indicator LED

- USB Client With On-board mini-B Connector and Host Adapters

- Rich Set of Standard I/O: 3xUART, 2xSPI, I²C, SDIO, GPIOs

- Application Specific I/O: I²S, AC97, Camera Chip Interface, JTAG

- Compact Size: 36mm x 48mm x 9mm

Applications

- Digital Image Processing

- Condition Based Maintenance

- Industrial Monitoring and Analysis

- Seismic and Vibration Monitoring

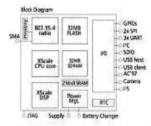

Block Diagram

Imote2

The Imote2 is an advanced wireless sensor node platform. It is built around the low power PXA271 XScale CPU and also integrates an 802.15.4 compliant radio. The design is modular and stackable with interface connectors for expansion boards on both the top and bottom sides. The top connectors provide a standard set of I/O signals for basic expansion boards. The bottom connectors provide additional high-speed interfaces for application specific I/O. A battery board supplying system power can be connected to either side.

Processor

The Imote2 contains the Intel PXA271 CPU. This processor can operate in a low voltage (0.85V), low frequency (13MHz) mode, hence enabling very low power operation. The frequency can be scaled from 13MHz to 416MHz with Dynamic Voltage Scaling. The processor has a number of different low power modes such as sleep and deep sleep. The PXA271 is a multi-chip module that includes three chips in a single package, the CPU with 256kB SRAM, 32MB SDRAM and 32MB of FLASH memory. It integrates many I/O options making it extremely flexible in supporting different sensors, A/Ds, radios, etc. These I/O features include I²C, 2 Synchronous Serial Ports (SPI) one of which is dedicated to the radio, 3 high speed UARTs, GPIOs, SDIO, USB client and host, AC97 and I²S audio codec interfaces, a fast infrared port, PWM, a Camera Interface and a high speed bus (Mobile Scaleable Link). The processor also

supports numerous timers as well as a real time clock. The PXA271 includes a wireless MMX coprocessor to accelerate multimedia operations. It adds 30 new media processor (DSP) instructions, support for alignment and video operations and compatibility with Intel MMX and SSE integer instructions. For more information on the PXA271, please refer to the Intel/Marvell datasheet.

Radio & Antenna

The Imote2 uses the CC2420 IEEE 802.15.4 radio transceiver from Texas Instruments. The CC2420 supports a 250kb/s data rate with 16 channels in the 2.4GHz band.

The Imote2 platform integrates a 2.4GHz surface mount antenna which provides a nominal range of about 30 meters. For longer range a SMA connector can be soldered directly to the board to connect to an external antenna.

Power Supply

The Imote2 can be powered by various means:

Primary Battery: This is typically accomplished by attaching a Crossbow Imote2 Battery Board to either the basic or advanced connectors.

Crossb✷w

Rechargeable Battery: This requires a specially con-figured battery board attached to either the basic or advanced connectors. The Imote2 has a built-in charger for Li-Ion or Li-Poly batteries.

USB: The Imote2 can be powered via the on-board mini-B USB connector. This mode can also be used to charge an attached battery.

Battery Pads: A suitable primary battery or other power source can be connected via a dedicated set of solder pads on the Imote2 board.

Processor/Radio Board	IPR2400	Remarks
CPU		
Processor	Intel PXA271	
SRAM Memory	256 kB	
SDRAM Memory	32MB	
FLASH Memory	32MB	
POWER CONSUMPTION		
Current Draw in Deep Sleep Mode	390 µA	
Current Draw in Active Mode	31 mA	13MHz, radio off
Current Draw in Active Mode	44 mA	13MHz, radio Tx/Rx
Current Draw in Active Mode	66 mA	104MHz, radio Tx/Rx
Radio		
Transceiver	TI CC2420	
Frequency Band (ISM)	2400.0 – 2483.5 MHz	
Data Rate	250 kb/s	
Tx Power	-24 – 0 dBm	
Rx Sensitivity	-94 dBm	
Range (line of sight)	~30 m	With integrated antenna
I/O		
USB Client (mini-B), USB Host		
UART 3x, GPIOs, I²C, SDIO, SPI 2x, I²S, AC97, Camera		
Power		
Battery Board	3x AAA	
USB Voltage	5.0 V	
Battery Voltage	3.2 – 4.5 V	
Li-Ion Battery Charger		
Mechanical		
Dimensions Imote2 Board	36mm x 48mm x 9mm	
Weight	12g	

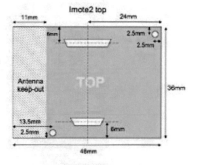

Imote2 top

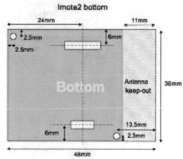

Imote2 bottom

Imote2 design licensed from Intel® Corporation.

Document Part Number: 6020-0117-02 Rev A

10.30　Intel PXA270 Processor (Intel 2005a)

 Intel® PXA270 Processor

Electrical, Mechanical, and Thermal Specification

Data Sheet

- High-performance processor:
  - Intel XScale® microarchitecture with Intel® Wireless MMX™ Technology
  - 7 Stage pipeline
  - 32 KB instruction cache
  - 32 KB data cache
  - 2 KB "mini" data cache
  - Extensive data buffering
- 256 Kbytes of internal SRAM for high speed code or data storage preserved during low-power states
- High-speed baseband processor interface (Mobile Scalable Link)
- Rich serial peripheral set:
  - AC'97 audio port
  - I²S audio port
  - USB Client controller
  - USB Host controller
  - USB On-The-Go controller
  - Three high-speed UARTs (two with hardware flow control)
  - FIR and SIR infrared communications port
- Hardware debug features — IEEE JTAG interface with boundary scan
- Hardware performance-monitoring features with on-chip trace buffer
- Real-time clock
- Operating-system timers
- LCD Controller
- Universal Subscriber Identity Module interface

- Low power:
  - Wireless Intel Speedstep® Technology
  - Less than 500 mW typical internal dissipation
  - Supply voltage may be reduced to 0.85 V
  - Four low-power modes
  - Dynamic voltage and frequency management
- High-performance memory controller:
  - Four banks of SDRAM: up to 104 MHz @ 2.5V, 3.0V, and 3.3V I/O interface
  - Six static chip selects
  - Support for PCMCIA and Compact Flash
  - Companion chip interface
- Flexible clocking:
  - CPU clock from 104 to 624 MHz
  - Flexible memory clock ratios
  - Frequency changes
  - Functional clock gating
- Additional peripherals for system connectivity:
  - SD Card / MMC Controller (with SPI mode support)
  - Memory Stick card controller
  - Three SSP controllers
  - Two I²C controllers
  - Four pulse-width modulators (PWMs)
  - Keypad interface with both direct and matrix keys support
  - Most peripheral pins double as GPIOs

10.31   Intel StrataFlash Embedded Memory (Intel 2005b)

 Intel StrataFlash® Embedded Memory (P30)

1-Gbit P30 Family

Datasheet

Product Features

- **High performance**
  - 85 ns initial access
  - 40 MHz with zero wait states, 20 ns clock-to-data output synchronous-burst read mode
  - 25 ns asynchronous-page read mode
  - 4-, 8-, 16-, and continuous-word burst mode
  - Buffered Enhanced Factory Programming (BEFP) at 5 μs/byte (Typ)
  - 1.8 V buffered programming at 7 μs/byte (Typ)
- **Architecture**
  - Multi-Level Cell Technology: Highest Density at Lowest Cost
  - Asymmetrically-blocked architecture
  - Four 32-KByte parameter blocks: top or bottom configuration
  - 128-KByte main blocks
- **Voltage and Power**
  - V_{CC} (core) voltage: 1.7 V – 2.0 V
  - V_{CCQ} (I/O) voltage: 1.7 V – 3.6 V
  - Standby current: 55 μA (Typ) for 256-Mbit
  - 4-Word synchronous read current: 13 mA (Typ) at 40 MHz
- **Quality and Reliability**
  - Operating temperature: –40 °C to +85 °C
  - Minimum 100,000 erase cycles per block
  - ETOX™ VIII process technology (130 nm)

- **Security**
  - One-Time Programmable Registers:
    - 64 unique factory device identifier bits
    - 64 user-programmable OTP bits
    - Additional 2048 user-programmable OTP bits
  - Selectable OTP Space in Main Array:
    - Four pre-defined 128-KByte blocks (top or bottom configuration)
  - Absolute write protection: $V_{PP} = V_{SS}$
  - Power-transition erase/program lockout
  - Individual zero-latency block locking
  - Individual block lock-down
- **Software**
  - 20 μs (Typ) program suspend
  - 20 μs (Typ) erase suspend
  - Intel® Flash Data Integrator optimized
  - Basic Command Set and Extended Command Set compatible
  - Common Flash Interface capable
- **Density and Packaging**
  - 64/128/256-Mbit densities in 56 Lead TSOP package
  - 64/128/256/512-Mbit densities in 64-Ball Intel® Easy BGA package
  - 64/128/256-Mbit densities in Intel® QUAD+ SCSP
  - 16-bit wide data bus

The Intel StrataFlash® Embedded Memory (P30) product is the latest generation of Intel StrataFlash® memory devices. Offered in 64-Mbit up through 1-Gbit densities, the P30 device brings reliable, two-bit-per-cell storage technology to the embedded flash market segment. Benefits include more density in less space, high-speed interface, lowest cost-per-bit NOR device, and support for code and data storage. Features include high-performance synchronous-burst read mode, fast asynchronous access times, low power, flexible security options, and three industry standard package choices. The P30 product family is manufactured using Intel® 130 nm ETOX™ VIII process technology.

Order Number: 306666, Revision: 004
01-Nov-2005

10.32 Intel StrongARM* SA-1110 (Intel 2000)

Intel® StrongARM* SA-1110 Microprocessor

Brief Datasheet

Product Features

The Intel®StrongARM SA-1110 Microprocessor (SA-1110) is a device optimized for meeting portable and embedded application requirements. The SA-1110 incorporates a 32-bit StrongARM RISC processor capable of running at up to 206 MHz. The SA-1110 has a large instruction and data cache, memory-management unit (MMU), and read/write buffers. The SA-1110 memory bus interfaces to many device types including synchronous DRAM (SDRAM), synchronous mask ROM (SMROM), and SRAM-like variable latency I/O devices with a shared data ready signal. In addition, the SA-1110 provides system support logic, multiple serial communication channels, a color/gray scale LCD controller, PCMCIA support for up to two sockets, and general-purpose I/O ports.

- High performance
  - 150 Dhrystone 2.1 MIPS @ 133 MHz
  - 235 Dhrystone 2.1 MIPS @ 206 MHz

- Low power (normal mode) [†]
  - <240 mW @1.55 V/133 MHz
  - <400 mW @1.75 V/206 MHz
- Integrated clock generation
  - Internal phase-locked loop (PLL)
  - 3.686-MHz oscillator
  - 32.768-kHz oscillator
- Power-management features
  - Normal (full-on) mode
  - Idle (power-down) mode
  - Sleep (power-down) mode
- Big and little endian operating modes

- 3.3-V I/O interface

- Memory bus
  - Interfaces to ROM, synchronous mask ROM (SMROM), Flash, SRAM, SRAM-like variable latency I/O, DRAM, and synchronous DRAM (SDRAM)
  - Supports two PCMCIA sockets
- 32-way set-associative caches
  - 16 Kbyte instruction cache
  - 8 Kbyte write-back data cache
- 32-entry MMUs
  - Maps 4 Kbyte, 8 Kbyte, or 1 Mbyte

- Write buffer
  - 8-entry, between 1 and 16 bytes each

- Read buffer
  - 4-entry, 1, 4, or 8 words
- 256 mini-ball grid array (mBGA)

[†] Power dissipation, particularly in idle mode, is strongly dependent on the details of the system design

Order Number: 278241-005
April 2000

10.33   iSense Security Sensor Module (Coalesenses 2014)

iSense Security Sensor Module
Preliminary product brief

Product

The iSense Security Sensor Module series features a passive infrared (PIR) sensor and/or a 3-axis accelerometer. In addition, a camera module can be attached.

The PIR Sensor can be used to detect moving objects that feature a temperature different from the environment (such as humans) in distances of up to 10 meters. The sensor offers a wide range of 110° for comprehensive monitoring.

The 3-axis accelerometer can be configured to cover accelerations of ±2g or ±6g. In addition to delivering acceleration values via a digital interface, it can generate interrupts on movement, direction change or free fall.

In addition, a camera module that can take color pictures with a mega pixel resolution can be attached. The images are preprocessed, so they can be scaled down to lower resolutions and compressed according to the JPEG standard.

Applications

- Building monitoring and security
- Automated lighting control
- Structure monitoring
- Valuable goods monitoring

Accelerometer	
Range	±2g or ±6g
Frequency	40Hz or 640 Hz
Current draw operation	~650µA
Current draw sleep mode	~1µA
Passive Infrared Sensor	
Range	~10m
Angle (hor./vert.)	93°/110°
Current draw operation	~300µA
Current draw sleep mode	0µA
Electromechanical	
Supply voltage	3.3 V
Dimensions	35mm x 30mm
Weight	8g
Temperature range	-20 to 70°C

This product brief shows the specification of a product in planning or in development. The functionality and electrical performance specifications are target values and may be used as a guide to the final specification.

10.34   MICA2 Mote (Crossbow 2002a)

MICA2

WIRELESS MEASUREMENT SYSTEM

▼ 3rd Generation, Tiny, Wireless
  Smart Sensors

▼ TinyOS - Unprecedented
  Communications and Processing

▼ > 1Yr Battery Life on AA Batteries (Using Sleep Modes)

▼ Wireless Communications with
  Every Node as Router Capability

▼ 315, 433 or 868/916 MHz Multi-
  Channel Radio Transceiver

▼ Light, Temperature, RH, Baromet-
  ric Pressure, Acceleration/Seismic,
  Acoustic, Magnetic, and other
  Sensors available

Applications

▼ Wireless Sensor Networks

▼ Security, Surveillance, and
  Force Protection

▼ Environmental Monitoring

▼ Large Scale Wireless Networks
  (1000+ points)

▼ Distributed Computing Platform

MICA2

The MICA2 Mote is a third
generation mote module used for
enabling low-power, wireless,
sensor networks. The MICA2
Mote features several new
improvements over the original
MICA Mote. The following
features make the MICA2 better
suited to commercial deployment:

• 868/916MHz, 433 or 315MHz
  multi-channel transceiver with
  extended range
• TinyOS (TOS) Distributed
  Software Operating System v1.0
  with improved networking stack
  and improved debugging
  features
• Support for wireless remote
  reprogramming
• Wide range of sensor boards
  and data acquisition add-on
  boards
• Compatible with MICA2DOT
  (MPR500) quarter-sized Mote

TinyOS 1.0 is a small, open-
source, energy efficient, software
operating system developed by
UC Berkeley which supports large
scale, self-configuring sensor
networks. The source code and
software development tools are
publicly available at:

http://webs.cs.berkeley.edu/tos

**Processor and Radio Platform
(MPR400CB):**
The MPR400CB is based on the
Atmel ATmega 128L. The
ATmega 128L is a low-power
microcontroller which runs TOS
from its internal flash memory.
Using TOS, a single processor
board (MPR400CB) can be
configured to run your sensor
application/processing and the
network/radio communications
stack simultaneously. The
MICA2 51-pin expansion
connector supports Analog Inputs,
Digital I/O, I2C, SPI, and UART
interfaces. These interfaces
make it easy to connect to a wide
variety of external peripherals.

Sensor Boards:
Various sensor and data
acquisition boards are available
from Crossbow . These boards
connect to the MICA2 through
a surface mount 51-pin connector
Crossbow supplies the following
sensor boards:

• MTS101CA Photocell/
  Thermistor/Proto and
  Experiment Board
• MTS300CA/MTS310CA Photo-
  cell, Thermistor, Microphone,
  Sounder, Magnetic (310 only),
  Acceleration (310 only)
• Contact Crossbow for
  information on other boards

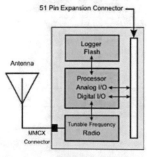

MPR400CB Block Diagram

Crossbøw

Processor/Radio Board	MPR400CB	MPR410CB	MPR420CB	Remarks
Processor Performance				
Program Flash Memory	128K bytes	128K bytes	128K bytes	
Measurement (Serial) Flash	512K bytes	512K bytes	512K bytes	>100,000 Measurements
Configuration EEPROM	4 K bytes	4 K bytes	4 K bytes	
Serial Communications	UART	UART	UART	0-3V transmission levels
Analog to Digital Converter	10 bit ADC	10 bit ADC	10 bit ADC	8 channel, 0-3Vin
Other Interfaces	DIO,I2C,SPI	DIO,I2C,SPI	DIO,I2C,SPI	
Current Draw	8 mA	8 mA	8 mA	active mode
	< 15uA	< 15 uA	< 15 uA	sleep mode
Multi-Channel Radio				
Center Frequency	868/916 MHz	433 MHz	315 MHz	ISM bands
Number of Channels	> 4, > 50	> 4	> 4	programmable, country specific
Data Rate	38.4 Kbaud	38.4 Kbaud	38.4 Kbaud	manchester encoded
RF Power	-20 to +5 dBm	-20 to +10 dBm	-20 to +10 dBm	programmable, typical
Receive Sensitivity	-98 dBm	-101 dBm	-101 dBm	typical, analog RSSI at AD Ch. 0
Outdoor Range	500 ft	1000 ft	1000 ft	1/4 Wave dipole, line of sight
Current Draw	27 mA	25 mA	25 mA	transmit with maximum power
	10 mA	8 mA	8 mA	receive
	< 1 uA	< 1 uA	< 1 uA	sleep
Electromechanical				
Battery	2X AA batteries	2X AA batteries	2X AA batteries	attached pack
External Power	2.7 - 3.3 V	2.7 - 3.3 V	2.7 - 3.3 V	connector provided
User Interface	3 LEDs	3 LEDs	3 LEDs	user programmable
Size (in)	2.25 x 1.25 x 0.25	2.25 x 1.25 x 0.25	2.25 x 1.25 x 0.25	excl. battery pack
(mm)	58 x 32 x 7	58 x 32 x 7	58 x 32 x 7	excl. battery pack
Weight (oz)	0.7	0.7	0.7	excl. batteries
(grams)	18	18	18	excl. batteries
Expansion Connector	51 pin	51 pin	51 pin	all major I/O signals

Base Stations:

A base station allows the aggregation of sensor network data onto a PC or other computer platform. Any MICA2 node (MPR400CB) can function as a base station by plugging the MPR400CB processor/radio board into a standard PC interface board, known as the Mote Interface Board (MIB510CA). The MIB510CA provides a serial interface for RS-232 as well as a parallel port programming interface for the Motes.

Crossbow also offers a stand-alone gateway solution, the MICA-WEB for both TCP/IP-based Ethernet networks and serial networks.

▼ MIB510CA Mote Interface Board

Table 10.6 compares the family of Berkeley motes up to Telos (Sect. 10.59).

10.35   MICA2DOT (Crossbow 2002b)

MICA2DOT

WIRELESS MICROSENSOR MOTE

▼ 3rd Generation, Quarter-Sized
(25mm), Wireless Smart Sensor

▼ TinyOS - Unprecedented
Communications and Processing

▼ Battery-Powered, Low-Mass

▼ Fits Anywhere, Wireless
Reprogrammable

▼ Wireless Communications with
Every Node as Router Capability

▼ 868/916 MHz, 433 MHz or
315 MHz Multi-channel Radio
Transceiver (MICA2 Compatible)

Applications

▼ Wireless Sensor Networks

▼ Temperature and Environmental
Monitoring

▼ Remote Data Logging

▼ Smart Badges, Wearable Computing

▼ Active 2-Way "Smart" Tags

MICA2DOT

The MICA2DOT Mote is a third generation mote module used for enabling low-power, wireless, sensor networks. The MICA2DOT is similar to the MICA2, except for its quarter-sized (25mm) form factor and reduced input/output channels. The following features make the MICA2DOT better suited for commercial deployment;

• 868/916MHz, 433MHz or 315MHz multi-channel transceiver with extended range
• TinyOS (TOS) Distributed Software Operating System v1.0 with improved networking stack and improved debugging features
• Support for wireless remote reprogramming
• Compatible with MICA2 (MPR400) Mote
• On Board Temperature Sensor, Battery Monitor, and LED

TinyOS 1.0 is a small, open-source, energy efficient, software operating system developed by UC Berkeley which supports large scale, self-configuring sensor networks. The source code and software development tools are publicly available at:

http://webs.cs.berkeley.edu/tos

Processor and Radio Platform (MPR500CA):

The MPR500CA is based on the Atmel ATmega 128L. The ATmega 128L is a low-power microcontroller which runs TOS from its internal flash memory. Using TOS, a single processor board (MPR500CA) can be configured to run your sensor application/processing and the network/radio communications stack simultaneously. The MICA2DOT features 18 solderless expansion pins for connecting 6 Analog Inputs, Digital I/O, and a serial communication or UART interface. These interfaces make it easy to connect to a wide variety of external peripherals.

Sensor Boards:

Various sensor boards and data acquisition boards are available from Crossbow . These boards connect onto the MICA2DOT through a ring of 18 solderless expansion pins. These pins allow boards to be stacked both above and below the MICA2DOT processor radio board. Crossbow supplies the following expansion boards:

• MDA500CA: Protoboard
• Contact Crossbow for information on other boards

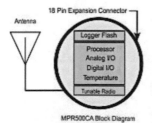

MPR500CA Block Diagram

Crossbøw

Processor/Radio Board	MPR500CA	MPR510CA	MPR520CA	Remarks
Processor Performance				
Program Flash Memory	128K bytes	128K bytes	128K bytes	
Measurement (Serial) Flash	512K bytes	512K bytes	512K bytes	>100,000 Measurements
Configuration EEPROM	4 K bytes	4 K bytes	4 K bytes	
Serial Communications	UART	UART	UART	0-3V transmission levels
Analog to Digital Converter	10 bit ADC	10 bit ADC	10 bit ADC	6 channels, 0-3Vin
Other Interfaces	DIO	DIO	DIO	9 channels
Current Draw	8 mA	8 mA	8 mA	active mode
	< 15 uA	< 15 uA	< 15 uA	sleep mode
Multi-Channel Radio				
Center Frequency	868/916 MHz	433 MHz	315MHz	ISM bands
Number of Channels	> 8, > 100	> 8	> 8	programmable, country specific
Data Rate	38.4 Kbaud	38.4 Kbaud	38.4 Kbaud	manchester encoded
RF Power	-20 - +5 dBm	-20 - +10 dBm	-20 - +10 dBm	programmable, typical
Receive Sensitivity	-98 dBm	-101 dBm	-101 dBm	typical, analog RSSI at A0 Ch. 0
Outdoor Range	500 ft	1000 ft	1000 ft	1/4 Wave dipole, line of sight
Current Draw	27 mA	25 mA	25 mA	transmit with maximum power
	10 mA	8 mA	8 mA	receive
	< 1 uA	< 1 uA	< 1 uA	sleep
Electromechanical				
Battery	3V Coin Cell	3V Coin Cell	3V Coin Cell	
External Power	2.7 - 3.3 V	2.7 - 3.3 V	2.7 - 3.3 V	connector provided
User Interface	1 LED	1 LED	1 LED	user programmable
Size (in)	1.0 x 0.25	1.0 x 0.25	1.0 x 0.25	excl. battery pack
(mm)	25 x 6	25 x 6	25 x 6	excl. battery pack
Weight (oz)	0.11	0.11	0.11	excl. batteries
(grams)	3	3	3	excl. batteries
Expansion Connector	18 pins	18 pins	18 pins	all major I/O signals

Base Stations:
The MICA2DOT communicates with base stations that use the MICA2 radio module. These include a standard MICA2 (MPR400CB) mated to a Mote Interface Board (MIB510CA), as well as the MICA-WEB Gateway.

Packaging:
The MICA2DOT is presently distributed as a stand-alone subassembly without packaging In future, a small plastic housing will be available.

Developers Kits:
Crossbow offers a variety of development kits for the MICA2 and MICA2DOT Motes.

▼ MIB510CA Mote Interface Board

Table 10.6 compares the family of Berkeley motes up to Telos (Sect. 10.59).

10.36   MICAz Mote (Crossbow 2006a)

MICAz
WIRELESS MEASUREMENT SYSTEM

- 2.4 GHz IEEE 802.15.4, Tiny Wireless Measurement System

- Designed Specifically for Deeply Embedded Sensor Networks

- 250 kbps, High Data Rate Radio

- Wireless Communications with Every Node as Router Capability

- Expansion Connector for Light, Temperature, RH, Barometric Pressure, Acceleration/Seismic, Acoustic, Magnetic and other Crossbow Sensor Boards

Applications

- Indoor Building Monitoring and Security

- Acoustic, Video, Vibration and Other High Speed Sensor Data

- Large Scale Sensor Networks (1000+ Points)

MICAz

The MICAz is a 2.4 GHz Mote module used for enabling low-power, wireless sensor networks.

Product features include:

- IEEE 802.15.4 compliant RF transceiver
- 2.4 to 2.48 GHz, a globally compatible ISM band
- Direct sequence spread spectrum radio which is resistant to RF interference and provides inherent data security
- 250 kbps data rate
- Supported by MoteWorks™ wireless sensor network platform for reliable, ad-hoc mesh networking
- Plug and play with Crossbow's sensor boards, data acquisition boards, gateways, and software

MoteWorks™ enables the development of custom sensor applications and is specifically optimized for low-power, battery-operated networks. MoteWorks is based on the open-source TinyOS operating system and provides reliable, ad-hoc mesh networking, over-the-air-programming capabilities, cross development tools, server middleware for enterprise network integration and client user interface for analysis and a configuration.

Processor & Radio Platform (MPR2400CA)

The MPR2400 is based on the Atmel ATmega128L. The ATmega128L is a low-power microcontroller which runs MoteWorks from its internal flash memory. A single processor board (MPR2400) can be configured to run your sensor application/ processing and the network/radio communications stack simultaneously. The 51-pin expansion connector supports Analog Inputs, Digital I/O, I2C, SPI and UART interfaces. These interfaces make it easy to connect to a wide variety of external peripherals. The MICAz (MPR2400) IEEE 802.15.4 radio offers both high speed (250 kbps) and hardware security (AES-128).

Sensor Boards

Crossbow offers a variety of sensor and data acquisition boards for the MICAz Mote. All of these boards connect to the MICAz via the standard 51-pin expansion connector. Custom sensor and data acquisition boards are also available. Please contact Crossbow for additional information.

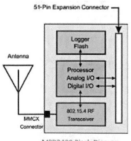

MPR2400 Block Diagram

Crossbow

Processor/Radio Board	MPR2400CA	Remarks
Processor Performance		
Program Flash Memory	128K bytes	
Measurement (Serial) Flash	512K bytes	> 100,000 Measurements
Configuration EEPROM	4K bytes	
Serial Communications	UART	0-3V transmission levels
Analog to Digital Converter	10 bit ADC	8 channel, 0-3V input
Other Interfaces	Digital I/O,I2C,SPI	
Current Draw	8 mA	Active mode
	< 15 µA	Sleep mode
RF Transceiver		
Frequency band'	2400 MHz to 2483.5 MHz	ISM band, programmable in 1 MHz steps
Transmit (TX) data rate	250 kbps	
RF power	-24 dBm to 0 dBm	
Receive Sensitivity	-90 dBm (min), -94 dBm (typ)	
Adjacent channel rejection	47 dB	+ 5 MHz channel spacing
	38 dB	- 5 MHz channel spacing
Outdoor Range	75 m to 100 m	1/2 wave dipole antenna, LOS
Indoor Range	20 m to 30 m	1/2 wave dipole antenna
Current Draw	19.7 mA	Receive mode
	11 mA	TX, -10 dBm
	14 mA	TX, -5 dBm
	17.4 mA	TX, 0 dBm
	20 µA	Idle mode, voltage regular on
	1 µA	Sleep mode, voltage regulator off
Electromechanical		
Battery	2X AA batteries	Attached pack
External Power	2.7 V - 3.3 V	Molex connector provided
User Interface	3 LEDs	Red, green and yellow
Size (in)	2.25 x 1.25 x 0.25	Excluding battery pack
(mm)	58 x 32 x 7	Excluding battery pack
Weight (oz)	0.7	Excluding batteries
(grams)	18	Excluding batteries
Expansion Connector	51-pin	All major I/O signals

Notes
'5 MHz steps for compliance with IEEE 802.15.4/D18-2002.
Specifications subject to change without notice

MIB520CB Mote Interface Board

Base Stations

A base station allows the aggregation of sensor network data onto a PC or other computer platform. Any MICAz Mote can function as a base station when it is connected to a standard PC interface or gateway board. The MIB510 or MIB520 provides a serial/USB interface for both programming and data communications. Crossbow also offers a stand-alone gateway solution, the MIB600 for TCP/IP-based Ethernet networks.

10.37   ML675K Series (Oki Semiconductor 2004)

Oki Semiconductor

ad*v*antage™
microcontrollers

ML675K Series
ML675001/ML67Q5002/ML67Q5003
32-Bit ARM®-Based General Purpose Microcontrollers

Description

The Oki ML675001/ML67Q5002/ML67Q5003 family of microcontrollers (MCUs) are the newest members of an extensive and growing family of 32-bit ARM®-based standard products for general-purpose applications that require 32-bit CPU performance and low cost afforded by MCU integrated features.

The ML675001, ML67Q5002 and ML67Q5003 devices each provide 8 Kbytes of unified cache memory, 32 Kbytes of built-in SRAM, 4 Kbytes of built-in boot ROM, and a host of other useful peripherals such as auto-reload timers, a watchdog timer (WDT), two pulse-width modulators (PWM), A/D converters, multiple UARTs, synchronous serial port, I2C serial interface, GPIOs, DMA controller, external memory controller, and boundary scan capability. In addition, the ML67Q5002 and ML67Q5003 devices offer 256 Kbytes and 512 Kbytes of built-in Flash memory respectively. The ML675001, ML67Q5002 and ML67Q5003 devices are pin-to-pin compatible with each other, and are pin-to-pin compatible with the Oki ML674001/Q4002/Q4003 family of microcontrollers for easy performance updates.

The ARM7TDMI® Advantage

The ML675001/ML67Q5002/ML67Q5003 family of low-cost ARM-based MCUs offers system designers a bridge from 8- and 16-bit proprietary MCU architectures to ARM's higher-performance, affordable, widely-accepted industry standard architecture and its industry-wide support infrastructure. The ARM industry infrastructure offers the system developers many advantages including software compatibility, many ready-to-use software applications, large choices among hardware and software development tools. These ARM-based advantages allow Oki's customers to better leverage engineering resources, lower development costs, minimize project risks, and reduce their product time to market. In addition, migration of a design with an Oki standard MCU to an Oki custom solution is easily facilitated with its award-winning µPLAT™ product development architecture.

Features

- ARM7TDMI 32-bit RISC CPU
  - 16-bit Thumb™ instruction set for power efficiency applications
- 32-bit mode (ARM) and/or 16-bit mode (Thumb)
- Built-in external memory controller supports glueless connectivity to memory (including SDRAM and EDO DRAM) and I/O
- Built in Flash ROM
  - 256 KB (ML67Q5002)
  - 512 KB (ML67Q5003)
- 32-KBytes built in zero-wait-state SRAM
- 28 interrupt sources

- DMA: 2 channels with external access
- Timers: 7 16-bit timers
- Watch-Dog Timer: dual stage 16 bit
- PWM: Two 16-bit channels
- Serial Interfaces: SIO, UART, USART, I2C
- GPIO: 42 bits
- A/D Converter: Four 10-bit channels
- Built-in boot ROM accommodates in-circuit Flash ROM re-programming and field-updates
- Package
  - 144-pin plastic LQFP
  - 144-pin plastic LFBGA

ML675001/Q5002/Q5003 MCUs

Part Number	Clock Frequency	Built-in Flash Size	Packages
ML675001	60 MHz	n/a	144-pin plastic LQFP (ML675001TC) 144-pin plastic LFBGA (ML675001LA)
ML67Q5002	60 MHz	256 KB	144-pin plastic LQFP (ML67Q5002TC) 144-pin plastic LFBGA (ML67Q5002LA)
ML67Q5003	60 MHz	512 KB	144-pin plastic LQFP (ML67Q5003TC) 144-pin plastic LFBGA (ML67Q5003LA)

10.38   MOTE-VIEW 1.2 (Crossbow 2006b)

MOTE-VIEW is designed to be an interface (client layer) between a user and a deployed network of wireless sensors. MOTE-VIEW provides users with the tools to simplify deployment and monitoring. It also makes it easy to connect to a database, to analyze, and to graph sensor readings. Figure 10.1 depicts a three-part framework for deploying a sensor network system. The left column represents the wireless sensor network itself. The server layer aggregates the data and allows for a connection to another network or terminal. The client layer software is for viewing and manipulating sensor network data. MOTE-VIEW is a free software tool and is available at www.xbow.com.

All of Crossbow's sensor and data acquisition boards are also supported by MOTE-VIEW (Table 10.2 Sensor (MTS series) and data acquisition boards supported by MOTE-VIEW). MOTE-VIEW supports the MICA-series platforms of wireless sensor network hardware, including the MICA2, MICA2DOT, and MICAz Motes (Table 10.3). In addition, sensor-integrated platforms such as the security/intrusion detection system based on the MSP motes and the environmental monitoring system (based on the MEP Motes) can be deployed and monitored (Table 10.4).

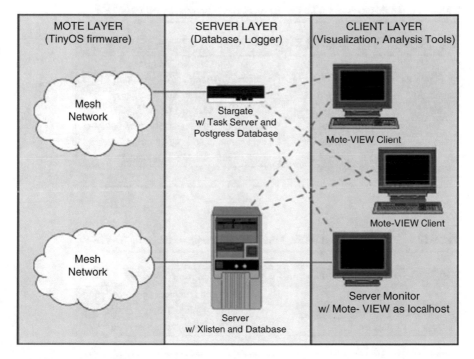

Fig. 10.1 Software framework for a WSN

Table 10.2 Sensor (MTS series) and data acquisition boards supported by MOTE-VIEW and their plug-and-play compatible mote platforms

	Mote Platforms		
Sensor and Data Acquisition Boards	**MICAz**	**MICA2**	**MICA2DOT**
MTS101	✓	✓	
MTS300/310	✓	✓	
MTS400/MTS420	✓	✓	
MTS410	✓	✓	
MTS510			✓
MDA100	✓	✓	
XBW-DA100	✓	✓	
MDA300	✓	✓	
MDA320	✓	✓	
XBW-DA325	✓	✓	
MDA500			✓

Table 10.3 Mote processor/radio (MPR) platforms supported by MOTE-VIEW

Mote Platforms	Model Number(s)	RF Frequency Band(s)
MICAz	MPR2400	2400 MHz to 2483.5 MHz
MICA2	MPR400	868 MHz to 870 MHz; 903 MHz to 928 MHz
	MPR410	433.05 to 434.8 MHz
	MPR420	315 MHz (for Japan only)
MICA2DOT	MPR510	868 MHz to 870 MHz; 903 MHz to 928 MHz
	MPR510	433.05 to 434.8 MHz
	MPR520	315 MHz (for Japan only)

Table 10.4 Sensor-integrated (MEP, MSP) platforms supported by MOTE-VIEW

Sensor Integrated Mote Platforms	Description of Usage
MEP410	Microclimate and ambient light monitoring
MEP510	Temperature and humidity monitoring
MSP410	Physical security and intrusion detection

10.39   MSB-A2 Platform (Baar et al. 2008)

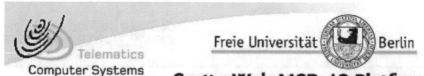

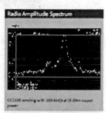

The latest ScatterWeb research platform

Platform Concept
- Versatile, extendable module
- Sophisticated hard- and software integration
- Enabling demanding real-world applications

Core Features
- 32-bit ARM7TDMI-S based microcontroller (512 KiB Flash-ROM, 98 KiB RAM, up to 72 MHz)
- 868 MHz ISM-band TI Chipcon CC1100 radio with wake-on-radio support
- Serial (USB) programming

Energy
- Good scalability of power consumption for computational power and peripherals
- Low overall power consumption < 1 μA ... 150 mA

Extensive Connectivity Options
- microSD-Card storage
- Primary USB for programming
- Secondary native USB port
- Native Ethernet and CAN bus support

High-Potential Challenging Applications
- Low-energy sensor networks
- Autonomous indoor localisation
- Resource demanding networked applications

Robust Software Platform
- Based on Contiki 2.2 core (SICS)
- Object-oriented Hardware Abstraction Layer
- Micromesh routing protocol
- ScatterWeb middleware services

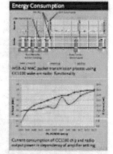

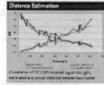

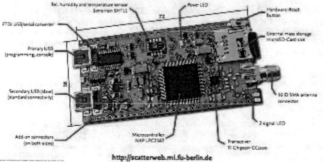

http://scatterweb.mi.fu-berlin.de

10.40   MSP430F1611 Microcontroller (Texas Instruments 2011b)

MSP430F15x, MSP430F16x, MSP430F161x
MIXED SIGNAL MICROCONTROLLER

SLAS368G – OCTOBER 2002 – REVISED MARCH 2011

- Low Supply-Voltage Range: 1.8 V to 3.6 V
- Ultralow Power Consumption
  - Active Mode: 330 µA at 1 MHz, 2.2 V
  - Standby Mode: 1.1 µA
  - Off Mode (RAM Retention): 0.2 µA
- Five Power-Saving Modes
- Wake-Up From Standby Mode in Less Than 6 µs
- 16-Bit RISC Architecture, 125-ns Instruction Cycle Time
- Three-Channel Internal DMA
- 12-Bit Analog-to-Digital (A/D) Converter With Internal Reference, Sample-and-Hold, and Autoscan Feature
- Dual 12-Bit Digital-to-Analog (D/A) Converters With Synchronization
- 16-Bit Timer_A With Three Capture/Compare Registers
- 16-Bit Timer_B With Three or Seven Capture/Compare-With-Shadow Registers
- On-Chip Comparator
- Serial Communication Interface (USART0), Functions as Asynchronous UART or Synchronous SPI or I²C™ Interface
- Serial Communication Interface (USART1), Functions as Asynchronous UART or Synchronous SPI Interface
- Supply Voltage Supervisor/Monitor With Programmable Level Detection
- Brownout Detector
- Bootstrap Loader

I²C is a registered trademark of Philips Incorporated.

- Serial Onboard Programming, No External Programming Voltage Needed, Programmable Code Protection by Security Fuse
- Family Members Include
  - MSP430F155
      16KB+256B Flash Memory
      512B RAM
  - MSP430F156
      24KB+256B Flash Memory
      1KB RAM
  - MSP430F157
      32KB+256B Flash Memory,
      1KB RAM
  - MSP430F167
      32KB+256B Flash Memory,
      1KB RAM
  - MSP430F168
      48KB+256B Flash Memory,
      2KB RAM
  - MSP430F169
      60KB+256B Flash Memory,
      2KB RAM
  - MSP430F1610
      32KB+256B Flash Memory
      5KB RAM
  - MSP430F1611
      48KB+256B Flash Memory
      10KB RAM
  - MSP430F1612
      55KB+256B Flash Memory
      5KB RAM
- Available in 64-Pin QFP Package (PM) and 64-Pin QFN Package (RTD)
- For Complete Module Descriptions, See the MSP430x1xx Family User's Guide, Literature Number SLAU049

description

The Texas Instruments MSP430 family of ultralow power microcontrollers consist of several devices featuring different sets of peripherals targeted for various applications. The architecture, combined with five low power modes is optimized to achieve extended battery life in portable measurement applications. The device features a powerful 16-bit RISC CPU, 16-bit registers, and constant generators that contribute to maximum code efficiency. The digitally controlled oscillator (DCO) allows wake-up from low-power modes to active mode in less than 6 µs.

 This integrated circuit can be damaged by ESD. Texas Instruments recommends that all integrated circuits be handled with appropriate precautions. Failure to observe proper handling and installation procedures can cause damage. ESD damage can range from subtle performance degradation to complete device failure. Precision integrated circuits may be more susceptible to damage because very small parametric changes could cause the device not to meet its published specifications. These devices have limited built-in ESD protection.

 Please be aware that an important notice concerning availability, standard warranty, and use in critical applications of Texas Instruments semiconductor products and disclaimers thereto appears at the end of this data sheet.

POST OFFICE BOX 655303 • DALLAS, TEXAS 75265

MSP430F15x, MSP430F16x, MSP430F161x
MIXED SIGNAL MICROCONTROLLER

SLAS368G – OCTOBER 2002 – REVISED MARCH 2011

description (continued)

The MSP430F15x/16x/161x series are microcontroller configurations with two built-in 16-bit timers, a fast 12-bit A/D converter, dual 12-bit D/A converter, one or two universal serial synchronous/asynchronous communication interfaces (USART), I²C, DMA, and 48 I/O pins. In addition, the MSP430F161x series offers extended RAM addressing for memory-intensive applications and large C-stack requirements.

Typical applications include sensor systems, industrial control applications, hand-held meters, etc.

AVAILABLE OPTIONS

T$_A$	PACKAGED DEVICES	
	PLASTIC 64-PIN QFP (PM)	PLASTIC 64-PIN QFN (RTD)
–40°C to 85°C	MSP430F155IPM	MSP430F155IRTD
	MSP430F156IPM	MSP430F156IRTD
	MSP430F157IPM	MSP430F157IRTD
	MSP430F167IPM	MSP430F167IRTD
	MSP430F168IPM	MSP430F168IRTD
	MSP430F169IPM	MSP430F169IRTD
	MSP430F1610IPM	MSP430F1610IRTD
	MSP430F1611IPM	MSP430F1611IRTD
	MSP430F1612IPM	MSP430F1612IRTD

† For the most current package and ordering information, see the Package Option Addendum at the end of this document, or see the TI web site at www.ti.com.
‡ Package drawings, thermal data, and symbolization are available at www.ti.com/packaging.

DEVELOPMENT TOOL SUPPORT

All MSP430 microcontrollers include an Embedded Emulation Module (EEM) allowing advanced debugging and programming through easy to use development tools. Recommended hardware options include the following:

- Debugging and Programming Interface
  - MSP-FET430UIF (USB)
  - MSP-FET430PIF (Parallel Port)
- Debugging and Programming Interface with Target Board
  - MSP-FET430U64 (PM package)
- Standalone Target Board
  - MSP-TS430PMG4 (PM package)
- Production Programmer
  - MSP-GANG430

10.41   MSP430F2416 Microcontroller (Texas Instruments 2007b)

MSP430x241x, MSP430x261x
MIXED SIGNAL MICROCONTROLLER

SLAS541A - JUNE 2007 - REVISED OCTOBER 2007

- Low Supply-Voltage Range, 1.8 V to 3.6 V
- Ultralow-Power Consumption:
  - Active Mode: 365 µA at 1 MHz, 2.2 V
  - Standby Mode (VLO): 0.5 µA
  - Off Mode (RAM Retention): 0.1 µA
- Wake-Up From Standby Mode in Less Than 1 µs
- 16-Bit RISC Architecture, 62.5-ns Instruction Cycle Time
- Three-Channel Internal DMA
- 12-Bit Analog-to-Digital (A/D) Converter With Internal Reference, Sample-and-Hold, and Autoscan Feature
- Dual 12-Bit Digital-to-Analog (D/A) Converters With Synchronization
- 16-Bit Timer_A With Three Capture/Compare Registers
- 16-Bit Timer_B With Seven Capture/Compare-With-Shadow Registers
- On-Chip Comparator
- Four Universal Serial Communication Interfaces (USCIs)
  - USCI_A0 and USCI_A1
    - Enhanced UART Supporting Auto-Baudrate Detection
    - IrDA Encoder and Decoder
    - Synchronous SPI
  - USCI_B0 and USCI_B1
    - I²C™
    - Synchronous SPI

- Supply Voltage Supervisor/Monitor With Programmable Level Detection
- Brownout Detector
- Bootstrap Loader
- Serial Onboard Programming, No External Programming Voltage Needed Programmable Code Protection by Security Fuse
- Family Members Include:
  - MSP430F2416:
    92KB+256B Flash Memory, 4KB RAM
  - MSP430F2417:
    92KB+256B Flash Memory, 8KB RAM
  - MSP430F2418:
    116KB+256B Flash Memory, 8KB RAM
  - MSP430F2419:
    120KB+256B Flash Memory, 4KB RAM
  - MSP430F2616:
    92KB+256B Flash Memory, 4KB RAM
  - MSP430F2617:
    92KB+256B Flash Memory, 8KB RAM
  - MSP430F2618:
    116KB+256B Flash Memory, 8KB RAM
  - MSP430F2619:
    120KB+256B Flash Memory, 4KB RAM
- Available in 80-Pin Quad Flat Pack (QFP) and 64-Pin QFP (See Available Options)
- For Complete Module Descriptions, See the *MSP430x2xx Family User's Guide*, Literature Number SLAU144

† The MSP430F241x devices are identical to the MSP430F261x devices, with the exception that the DAC12 modules and the DMA controller are not implemented.

description

The Texas Instruments MSP430 family of ultralow-power microcontrollers consists of several devices featuring different sets of peripherals targeted for various applications. The architecture, combined with five low-power modes is optimized to achieve extended battery life in portable measurement applications. The device features a powerful 16-bit RISC CPU, 16-bit registers, and constant generators that contribute to maximum code efficiency. The calibrated digitally controlled oscillator (DCO) allows wake-up from low-power modes to active mode in less than 1 µs.

The MSP430F261x/241x series are microcontroller configurations with two built-in 16-bit timers, a fast 12-bit A/D converter, a comparator, dual 12-bit D/A converters, four universal serial communication interface (USCI) modules, DMA, and up to 64 I/O pins. The MSP430F241x devices are identical to the MSP430F261x devices, with the exception that the DAC12 and the DMA modules are not implemented.

Typical applications include sensor systems, industrial control applications, hand-held meters, etc.

⚠ Please be aware that an important notice concerning availability, standard warranty, and use in critical applications of Texas Instruments semiconductor products and disclaimers thereto appears at the end of this data sheet.

I²C is a registered trademark of Philips Incorporated.

POST OFFICE BOX 655303 • DALLAS, TEXAS 75265

MSP430x241x, MSP430x261x
MIXED SIGNAL MICROCONTROLLER

SLAS541A - JUNE 2007 - REVISED OCTOBER 2007

AVAILABLE OPTIONS		
T_A	PACKAGED DEVICES	
	PLASTIC 80-PIN LQFP (PN)	PLASTIC 64-PIN LQFP (PM)
-40°C to 105°C	MSP430F2416TPN	MSP430F2416TPM
	MSP430F2417TPN	MSP430F2417TPM
	MSP430F2418TPN	MSP430F2418TPM
	MSP430F2419TPN	MSP430F2419TPM
	MSP430F2616TPN	MSP430F2616TPM
	MSP430F2617TPN	MSP430F2617TPM
	MSP430F2618TPN	MSP430F2618TPM
	MSP430F2619TPN	MSP430F2619TPM

TEXAS
INSTRUMENTS
POST OFFICE BOX 655303 • DALLAS, TEXAS 75265

10.42   MSX-01F Solar Panel (BP Solar 2014)

BP SOLAR - MSX-01F - SOLAR PANEL, 1.2W

Product Information

SOLAR PANEL, 1.2W
- **Power Rating:** 1.2W
- **Power Voltage Max:** 7.5V
- **Current at P Max:** 150mA
- **Open Circuit Voltage:** 10.3V
- **Short Circuit Current:** 160mA
- **Length:** 127mm
- **Width:** 127mm
- **Height:** 3mm
- **DC Power:** 1.2W
- **External Depth:** 10mm
- **External Length / Height:** 161mm
- **External Width:** 139mm
- **Lead Length:** 750mm
- **Nom Voltage:** 6V
- **Output Current Max:** 0.15A
- **Output Voltage Max:** 8.4V
- **Weight (kg):** 0.34

10.43   MTS/MDA (Crossbow 2007a)

The MTS series of sensor boards and the MDA series of sensor/data acquisition boards are designed to interface with Crossbow's MICA, MICA2, and MICA2DOT families of wireless Motes. There are a variety of sensor boards available, and the sensor boards are specific to the MICA, MICA2 board, or MICA2DOT form factor. The sensor boards allow for a range of different sensing modalities as well as interfaces to external sensors via prototyping areas or screw terminals. Table 10.5 lists the currently available sensor boards for each Mote family.

Table 10.5 Crossbow's sensor and data acquisition boards

Crossbow Part Name	Motes Supported	Sensors and Features
MTS101CA	MICAz, MICA2, MICA	Light, temperature, prototyping area
MTS300CA MTS300CB	IRIS, MICAz, MICA2, MICA	Light, temperature, microphone, and buzzer
MTS310CA MTS310CB	IRIS, MICAz, MICA2, MICA	Light, temperature, microphone, buzzer, 2-axis accelerometer, and 2-axis magnetometer
MTS400CA MTS400CB MTS400CC	IRIS, MICAz, MICA2	Ambient light, relative humidity, temperature, 2-axis accelerometer, and barometric pressure
MTS420CA MTS420CB MT3420CC	IRIS, MICAz, MICA2	Same as MTS400CA plus a GPS module
MTS510CA	MICA2DOT	Light, microphone, and 2-axis accelerometer
MDA100CA MDA100CB	IRIS, MICAz, MICA2	Light, temperature, prototyping area
MDA300CA	IRIS, MICAz, MICA2	Light, relative humidity, general purpose interface for external sensors
MDA320CA	IRIS, MICAz, MICA2	General purpose interface for external sensors
MDA500CA	MICA2DOT	Prototyping area

MTS/MDA

SENSOR, DATA ACQUISITION BOARDS

- Selection of 3 Standard Sensor/DAQ Boards
- MoteWorks™ Drivers Support Sensor Readings
- Supports IRIS, MICAz and MICA2 Motes
- Individual Power Control for Each Sensor

Applications

- Vibration and Magnetic Anomaly Detection
- External Sensor Connection
- Localization and Acoustic Tracking
- Robotics
- Wireless Sensor Networking

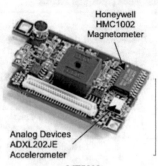

Honeywell
HMC1002
Magnetometer

Analog Devices
ADXL202JE
Accelerometer

MTS310

MTS300

MDA100

MTS310

The MTS310 is a flexible sensor board with a variety of sensing modalities. These modalities include a Dual-Axis Accelerometer, Dual-Axis Magnetometer, Light, Temperature, Acoustic and Sounder. The MTS310 is for use with the IRIS, MICAz and MICA2 Motes.

MTS300

The MTS300CB is a flexible sensor board with a variety of sensing modalities. These modalities include Light, Temperature, Acoustic and Sounder. The MTS300CB is for use with the IRIS, MICAz and MICA2 Motes.

MDA100

The MDA100CB sensor and data acquisition board has a precision thermistor, a light sensor/photocell and general prototyping area. Designed for use with the IRIS, MICAz and MICA2 Motes, the prototyping area supports connection to all 51 pins on the expansion connector, and provides an additional 42 unconnected solder points for breadboarding.

10.44   Omron Subminiature Basis Switch (Omron 2014)

SS
Subminiature Basic Switch

Subminiature Basic Switch Offers High Reliability and Security

- The OMRON's best-selling micro switches of a wide variety from 0.1A to 10.1A.
- A variety of models are available, with operating force ranging from low to high.
- Two split springs ensure a high stability and durability of 30,000,000 operations.

RoHS Compliant

Model Number Legend

SS-①②③④⑤⑥

1. Ratings
- 10 : 250 VAC 10.1A
- 5 : 125 VAC 5 A
- 01 : 30 VDC 0.1A

2. Actuator
- None : Pin plunger
- GL : Hinge lever
- GL111 : Long hinge lever
- GL13 : Simulated roller lever
- GL2 : Hinge roller lever
- GL02 : Hinge roller lever
  (Roller material: Stainless) heat-resistant

3. Maximum Operating Force (OF)
- None : 1.47 N {150 gf}
- -F : 0.49 N {50 gf} (0.1 A, 5 A)
- -F : 0.25 N {25 gf} (0.1 A)

4. Contact form
- None : SPDT
- -2 : SPST-NC
- -3 : SPST-NO

5. Terminals
- None : Solder terminals
- T : Quick-connect terminals (#110)
- D : PCB terminals

6. Heat resistance
- None : Standard (85°C)
- -T : Heat-resistant (120°C)

Note. These values are for the pin plunger models.

10.45   OV528 Serial Bus Camera System (OmniVision Technologies 2002)

DATASHEET

Architecture

General Description
The OV528 Serial Bus Camera System performs as a video camera or a JPEG compressed still camera and can be attached to a wireless or PDA host. When it performs as a video camera, the TFT-LCD panel of the host operates as a viewfinder. Users can send out a snapshot command from the host in order to capture a full resolution single-frame still picture. The picture is then compressed by the JPEG engine and transferred to the host.

Functional Description
OV528, the Single Chip Camera-to-Serial Bridge, is a low-cost, single-chip & low-powered solution for high-resolution serial bus PDA/cellular phone camera accessory applications. Along with OV76x0/OV66x0 CMOS VGA/CIF color digital CameraChips, OV528 comprises a low-cost, highly integrated serial camera system. There is no additional DRAM required.

The OV528 system, as shown in Figure 1, consists of a CameraChip, Program Memory and OV528 Serial Bridge.

Camera Sensors
The OV528 supports OmniVision OV76x0/ and OV66x0 CameraChips with an 8-bit YC_bC_r interface.

Program Memory
A program memory is required for the embedded MC to respond to host commands correctly, as well as to store all necessary parameters for adjusting image/compression qualities. A serial type program memory is required for OV528, while both serial and parallel types of program memory can be adapted to OV528-100.

The contents of the program memory can be updated on the fly.

OV528 Serial Bridge
The OV528 Serial Bridge is a controller chip that can transfer image data from CameraChips to wireless/PDA hosts.

The OV528 takes 8-bit YC_bC_r 422 progressive video data from an OV76x0/OV66x0 CameraChip. The camera interface synchronizes with input video data and performs down-sampling, clamping and windowing functions with desired resolution, as well as color conversion that is requested by the user through serial bus host commands.

The JPEG CODEC with variable quality settings can achieve higher compression ratio & better image quality for various image resolutions.

The Serial Camera Control Bus is used to achieve greater flexibility in camera interface.

Figure 1.   **System Block Diagram**

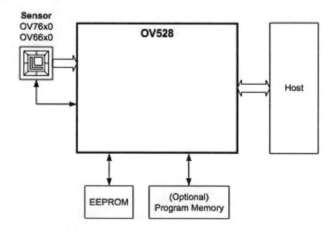

10.46   OV6620/OV6120 Single-Chip CMOS Digital Camera (OmniVision Technologies 1999)

Advanced Information
Preliminary

OV6620/OV6120

OV6620 SINGLE-CHIP CMOS CIF COLOR DIGITAL CAMERA
OV6120 SINGLE-CHIP CMOS CIF B&W DIGITAL CAMERA

Features

- 101,376 pixels, 1/4" lens, CIF/QCIF format
- Progressive scan read out
- Data format - YCrCb 4:2:2, GRB 4:2:2, RGB Raw Data
- 8/16 bit video data: CCIR601, CCIR656, ZV port
- Wide dynamic range, anti-blooming, zero smearing
- Electronic exposure / Gain / white balance control
- Image enhancement - brightness, contrast, gamma, saturation, sharpness, window, etc.
- Internal/external synchronization

- Frame exposure/line exposure option
- 5-Volt operation, low power dissipation
  - < 80 mW active power
  - < 10 μA in power-save mode
- Gamma correction (0.45/0.55/1.00)
- I²C programmable (400 kb/s):
  - color saturation, brightness, contrast, white balance, exposure time, gain

General Description

The OV6620 (color) and OV6120 (black and white) CMOS Image sensors are single-chip video/imaging camera devices designed to provide a high level of functionality in a single, small-footprint package. Both devices incorporate a 352 x 288 image array capable of operating up to 60 frames per second image capture. Proprietary sensor technology utilizes advanced algorithms to cancel Fixed Pattern Noise (FPN), eliminate smearing, and drastically reduce blooming. All needed

camera functions including exposure control, gamma, gain, white balance, color matrix, windowing, and more, are programmable through an I²C interface. Both devices can be programmed to provide image output in either 4-, 8- or 16-bit digital formats.

Applications include: Video Conferencing, Video Phone, Video Mail, Still Image, and PC Multimedia.

Array Elements (CIF) (QCIF)	356 x 292 (176 x 144)
Pixel Size	9.0 x 8.2 μm
Image Area	3.1 x 2.5 mm
Max Frames/Sec	Up to 60 FPS
Electronic Exposure	Up to 500 : 1 (for selected FPS)
Scan Mode	progressive
Gamma Correction	0.45/.55/1.0
Min. Illumination (3000K)	OV6620 - < 3 lux @ f1.2 OV6120 - < 0.5 lux @ f1.2
S/N Ratio (Digital Camera Out)	> 48 dB (AGC = Off, Gamma = 1)
FPN	< 0.03% Vp.p
Dark Current	< 0.2 nA/cm²
Dynamic Range	> 72 dB
Power Supply	5VDC, ±5% (Anal.) 5VDC or 3.3VDC (DIO)
Power Requirements	< 80mW Active < 30μW Standby
Package	48 pin LCC

Note: Outputs UV0-UV7 are not available on the OV6120. The inputs associated with these respective pins are still functional.

OV6620/OV6120 PIN ASSIGNMENTS

OmniVision Technologies, Inc. 930 Thompson Place Sunnyvale, CA 94086 U.S.A.
Tel: (408) 733-3030 Fax: (408) 733-3061
e-mail: info@ovt.com
Website: http://www.ovt.com

Version 1.2, 2 June 1999

10.47   OV7640/OV7140 CMOS VGA CAMERACHIPS
(OmniVision Technologies 2003)

Advanced Information
Preliminary Datasheet

OV7640 Color CMOS VGA (640 x 480) CAMERACHIP™
OV7140 B&W CMOS VGA (640 x 480) CAMERACHIP™

General Description

The OV7640 (color) and OV7140 (black and white) CAMERACHIPS™ are low voltage CMOS image sensors that provide the full functionality of a single-chip VGA (640 x 480) camera and image processor in a small footprint package. The OV7640/OV7140 provides full-frame, sub-sampled or windowed 8-bit images in a wide range of formats, controlled through OmniVision's Serial Camera Control Bus (SCCB) interface.

This product family has an image array capable of operating at up to 30 frames per second (fps) with complete user control over image quality, formatting and output data transfer. All required image processing functions, including exposure control, gamma, white balance, color saturation, hue control and more, are also programmable through the SCCB interface. In addition, OmniVision CAMERACHIPS use proprietary sensor technology to improve image quality by reducing or eliminating common lighting/electrical sources of image contamination such as fixed pattern noise, smearing, blooming, etc. to produce a clean, fully stable color image.

Features

- High sensitivity for low-light operation
- 2.5V operating voltage for embedded portable applications
- Standard Serial Camera Control Bus (SCCB) interface
- VGA, QVGA (sub-sampled) and Windowed outputs with Raw RGB, RGB (GRB 4:2:2), YUV (4:2:2) and YCbCr (4:2:2) formats
- Automatic image control functions including: Automatic Exposure Control (AEC), Automatic Gain Control (AGC), Automatic White Balance (AWB), Automatic Brightness Control (ABC), Automatic Band Filter (ABF) for 60Hz noise and Automatic Black-Level Calibration (ABLC)
- Image quality controls including color saturation, hue, gamma, sharpness (edge enhancement), anti-blooming and zero smearing

Ordering Information

Product	Package
OV7640 (Color)	PLCC-28
OV7140 (B&W)	PLCC-28

Applications

- Cellular and Picture Phones
- Toys
- PC Multimedia

Key Specifications

	Array Size	640 x 480 (VGA)
	Core	2.5VDC ± 10%
Power Supply	Analog	2.5VDC ± 4%
	I/O	2.25V to 3.3V
Power Requirements	Active	40 mW (30 fps, including I/O power)
	Standby	30 µW
Temperature Range	Operation	-10°C to 70°C
	Stable Image	0°C to 50°C
Output Formats (8-bit)		• YUV/YCbCr 4:2:2 • RGB 4:2:2 • Raw RGB Data
	Lens Size	1/4"
Maximum Image Transfer Rate	VGA	30 fps
	QVGA	60 fps
Sensitivity	B&W	2.20 V/Lux-sec
	Color	1.12 V/Lux-sec
	S/N Ratio	46 dB
	Dynamic Range	62 dB
	Scan Mode	Progressive/Interlaced
Maximum Exposure Interval		523 x t$_{ROW}$
Gamma Correction		0.45
	Pixel Size	5.6 µm x 5.6 µm
	Dark Current	30 mV/s
	Well Capacity	60 Ke
Fixed Pattern Noise		< 0.03% of V$_{PEAK-TO-PEAK}$
	Image Area	3.6 mm x 2.7 mm
Package Dimensions		11.43 mm x 11.43 mm

Figure 1   OV7640/OV7140 Pin Diagram

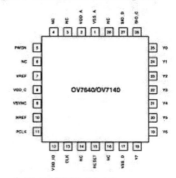

10.48   OV9655/OV9155 (OmniVision Technologies 2006)

Advanced Information
Preliminary Datasheet

OV9655/OV9155 CMOS SXGA (1.3 MegaPixel) CAMERACHIP™ Sensor with OmniPixel® Technology

General Description

The OV9655 (color) and OV9155 (B&W) CAMERACHIP™ image sensors are low voltage CMOS devices that provide the full functionality of a single-chip SXGA (1280x1024) camera and image processor in a small footprint package. The OV9655/OV9155 provides full-frame, sub-sampled, scaled or windowed 8-bit/10-bit images in a wide range of formats, controlled through the Serial Camera Control Bus (SCCB) interface.

This product has an image array capable of operating at up to 15 frames per second (fps) in SXGA resolution with complete user control over image quality, formatting and output data transfer. All required image processing functions, including exposure control, gamma, white balance, color saturation, hue control, white pixel canceling, noise canceling, and more, are also programmable through the SCCB interface. In addition, OmniVision CAMERACHIP sensors use proprietary sensor technology to improve image quality by reducing or eliminating common lighting/electrical sources of image contamination, such as fixed pattern noise, smearing, etc., to produce a clean, fully stable color image.

 Note: The OV9655/OV9155 uses a lead-free package.

Features

- High sensitivity for low-light operation
- Low operating voltage for embedded portable apps
- Standard SCCB interface
- Output support for Raw RGB, RCB (GRB 4:2:2, RGB565/555), YUV (4:2:2) and YCbCr (4:2:2) formats
- Supports image sizes: SXGA, VGA, CIF, and any size scaling down from CIF to 40x30
- VarioPixel® method for sub-sampling
- Automatic image control functions including Automatic Exposure Control (AEC), Automatic Gain Control (AGC), Automatic White Balance (AWB), Automatic Band Filter (ABF), and Automatic Black-Level Calibration (ABLC)
- Image quality controls including color saturation, gamma, sharpness (edge enhancement), lens correction, white pixel canceling, noise canceling, and 50/60 Hz luminance detection
- Supports LED and flash strobe mode
- Supports scaling

Ordering Information

Product	Package
OV09655-VL1A (Color, lead-free)	28-pin CSP2
OV09155-VL1A (B&W, lead-free)	28-pin CSP2

Applications

- Cellular and Picture Phones
- Toys
- PC Multimedia
- Digital Still Cameras

Key Specifications

	Active Array Size	1280 x 1024
Power Supply	Core	1.8VDC ± 10%
	Analog	2.45 to 3.0VDC
	I/O	1.7V to 3.3V[a]
Power Requirements	Active	90 mW typical (15fps SXGA YUV format)
	Standby	<20 μA
Temperature Range	Operation	-30°C to 70°C
	Stable Image	0°C to 50°C
Output Formats (8-bit)		• YUV/YCbCr 4:2:2 • RGB565/555 • GRB 4:2:2 • Raw RGB Data
	Lens Size	1/4"
	Chief Ray Angle	25°
Maximum Image Transfer Rate	SXGA	15 fps
	VGA, CIF and down scaling	30 fps
	Sensitivity	1.1 V/(Lux - sec)
	S/N Ratio	42 dB
	Dynamic Range	50 dB
	Scan Mode	Progressive
Maximum Exposure Interval		1050 x t_ROW
Gamma Correction		Programmable
	Pixel Size	3.18 μm x 3.18 μm
	Dark Current	15 mV/s at 60°C
	Well Capacity	10 K e
	Fixed Pattern Noise	<0.03% of V_PEAK-TO-PEAK
	Image Area	4.17 mm x 3.29 mm
	Package Dimensions	5145 μm x 6145 μm

a. I/O power should be 2.45V or higher when using the internal regulator for Core (1.8V); otherwise, it is necessary to provide an external 1.8V for the Core power supply.

Figure 1   OV9655/OV9155 Pin Diagram (Top View)

10.49   PCF50606/605 Single-Chip Power Management Unit + (Philips 2002)

PCF50606/605

Each Philips PMU+ is a complete, highly integrated power management unit designed for PDAs, smartphones, webpads and other handheld products. The PMU+ incorporates on-chip battery management, a touchscreen interface and more.

Features

> Complete system power management on a single chip

> Integrates functions to reduce board space by up to 50%

> Fully compatible with Intel® XScale™ and other processors

> Software-programmable power supplies for design flexibility: up to three integrated DC/DC converters and five LDO voltage regulators with parallel, low-current LDOs

> Software-controlled real-time power management can dramatically reduce power consumption and extend battery life

> Battery management system with battery voltage monitor, backup battery support, charger control with multiple charge modes, and 10-bit ADC

> 10-bit touchscreen interface

> Optional 32-kHz Xtal oscillator output for generating real-time clock

> Up to six programmable GPOs

> Space-saving, 56-pin leadless HVQFN package (8 x 8 x 1 mm)

> Complete evaluation kit with PC control software

Single-chip Power Management Unit +

Semiconductors

The latest additions to a growing family of highly integrated mixed-signal companion chips, the Philips PCF50606 and PCF50605 power management unit 'plus' (PMU+) ICs dramatically reduce board space and significantly lower overall cost in handheld devices such as PDAs, smartphones, and webpads. Each PMU+ also supports real-time, software-controlled power management, a feature that can reduce overall system power consumption by up to 70%. The full-featured PCF50606 PMU+ combines an efficient set of programmable power supplies, a flexible battery management system, a 10-bit analog-to-digital converter (ADC) and a touchscreen interface. The PCF50605 PMU+ offers the same power management features without an on-chip touchscreen interface, ADC, or battery management.

Each PMU+ is fully compatible with the Intel XScale architecture and other processors designed for mobile use. It operates from a three-cell NiCd/NiMH or a one-cell Li-Ion/Li-Polymer battery pack and is available in a space-saving, 56-pin leadless HVQFN package measuring only 8 x 8 x 1 mm.

Real-time Power Management

As battery-powered systems take on more functionality and incorporate more advanced processors and DSPs, power management becomes increasingly complex. To address this issue, each PMU+ supports real-time power management, a feature that allows various settings to be adjusted dynamically. Full software control is provided through an I²C interface, so power management can be programmed for different options and can adapt to different operating conditions.

PHILIPS

10.50   PIC18 Microcontroller Family (Microchip 2000)

PIC18 Microcontroller Family

The PIC18 microcontroller family provides PICmicro® devices in 18- to 80-pin packages, that are both socket and software upwardly compatible to the PIC16 family. The PIC18 family includes all the popular peripherals, such as MSSP, ESCI, CCP, flexible 8- and 16-bit timers, PSP, 10-bit ADC, WDT, POR and CAN 2.0B Active for the maximum flexible solution. Most PIC18 devices will provide FLASH program memory in sizes from 8 to 128 Kbytes and data RAM from 256 to 4 Kbytes; operating from 2.0 to 5.5 volts, at speeds from DC to 40 MHz. Optimized for high-level languages like ANSI C, the PIC18 family offers a highly flexible solution for complex embedded applications.

High Performance RISC CPU:

* 77 instructions
* C-Language friendly architecture
* PIC16 source code compatible
* Linear program memory addressing to 2 Mbyte
* Linear data memory addressing up to 4 Kbytes
* Up to 10 MIPs operation:
  – DC - 40 MHz osc/clock input
  – 4 MHz - 10 MHz clock with PLL active
* 16-bit wide instructions, 8-bit wide data path
* Priority levels for interrupts
* 8 x 8 Single Cycle Hardware Multiplier

Peripheral Features:

* High current sink/source 25 mA/25 mA
* Up to four external interrupt pins
* Up to three 16-bit timer/counters
* Up to two 8-bit timer/counters with 8-bit period register (time-base for PWM)
* Secondary LP oscillator clock option - Timer1
* Up to five Capture/Compare/PWM (CCP) modules
  CCP pins can be configured as:
  – Capture input: 16-bit, resolution 6.25 ns (Tcy/16)
  – Compare: 16-bit, max. resolution 100 ns (Tcy)
  – PWM output: PWM resolution is 1- to 10-bit
    Max. PWM frequency @: 8-bit resolution = 156 kHz
                          10-bit resolution = 39 kHz
* Master Synchronous Serial Port (MSSP) module
  Two modes of operation:
  – 3-wire SPI™ (supports all 4 SPI modes)
  – I²C™ Master and Slave mode
* Up to 2 Addressable USART modules (ESCI)
  – Supports interrupt on Address bit
* Parallel Slave Port (PSP) module

Analog Features:

* 10-bit Analog-to-Digital Converter module (A/D) with:
  – Fast sampling rate
  – Up to 16 channels input multiplexor
  – Conversion available during SLEEP
  – DNL = ±1 LSb, INL = ±1 LSb

Analog Features (Continued):

* Programmable Low Voltage Detection (LVD) module
  – Supports interrupt-on-low voltage detection
* Programmable Brown-out Reset (BOR)
* Comparators

Special Microcontroller Features:

* Power-on Reset (POR), Power-up Timer (PWRT) and Oscillator Start-up Timer (OST)
* Watchdog Timer (WDT) with its own on-chip RC oscillator for reliable operation
* Programmable code protection
* In-Circuit Serial Programming™ (ICSP™) via two pins

CMOS Technology:

* Fully static design
* Wide operating voltage range (2.0V to 5.5V)
* Industrial and Extended temperature ranges

Power Managed Features:

* Dynamically switch to secondary LP oscillator
* Internal RC oscillator for ADC operation during SLEEP
* SLEEP mode (IPD < 1 μA typ.)
  – up to 23 individually selectable wake-up events
    3 edge selectable wake-up inputs
  – 4 state change wake-up inputs
* Internal RC oscillator for WDT (period wake-up)
* RAM retention mode (VDD as low as 1.5V)
* Up to 6 more Power Managed modes available on selected models (PIC18F1320/2320/4320 and PIC18F1220/2220/4220)

MICROCHIP
PICmicro® Microcontrollers

Microchip Technology Inc. · *The Embedded Control Solutions Company®*

Additional Information:

- Microchip's web site: www.microchip.com
- Microchip's *PICmicro 18C MCU Reference Manual*, Order No. DS39500
- Microchip's CD-ROMs available:
  - *Technical Library*, Order No. DS00161
- Microchip's Data Sheets available:
  - *PIC18CXX2*, Order No. DS39026
  - *PIC18CXX8*, Order No. DS30475
  - *PIC18C601/801*, Order No. DS39541
- Application Notes are available in:
  - *Embedded Control Handbook*, Order No. DS00092
  - *Embedded Control Handbook, Volume 2, Math Library*, Order No. DS00167
  - *Embedded Control Handbook Update 2000*, Order No. DS00711

- Microchip's *Quality Systems and Customer Interface System*, Order No. DS00169
- Demo Boards Available:
  - PICDEM™ 2 Demonstration Board
  - ROMless
  - CAN/LIN bus
- Third Party Tools Available:
  - C Compilers
    HI-TECH - PICC™, www.htsoft.com
    IAR - EWB-PIC, www.iar.com
    CCS PIC18 C Compiler, www.ccsinfo.com

	PIC18 Microcontroller Family												
	Program Memory		Data Memory							CCP/	Timers		
Product	Type	Bytes	RAM Bytes	EEPROM Bytes	I/O Ports	ADC 10-bit	MSSP	USART	Other	PWM	8/16-bit	Packages	Pins
PIC18F1220	FLASH	4K	256	256	16	7	—	1	6x PMM	1	1/3	DIP SOIC, SSOP QFN	18
PIC18F1320	FLASH	8K	256	256	16	7	—	1	6x PMM	1	1/3	DIP SOIC, SSOP QFN	18
PIC18F2220	FLASH	4K	512	256	23	10	I²C/SPI	1	6x PMM	2	1/3	DIP SOIC	28
PIC18F2320	FLASH	8K	512	256	23	10	I²C/SPI	1	6x PMM	2	1/3	DIP SOIC	28
PIC18C242	OTP	16K	512	—	23	5	I²C/SPI	1	—	2	1/3	DIP SOIC	28
PIC18C252	OTP	32K	1536	—	23	5	I²C/SPI	1	—	2	1/3	DIP SOIC	28
PIC18F242	FLASH	16K	512	256	23	5	I²C/SPI	1	—	2	1/3	DIP SOIC, SSOP	28
PIC18F252	FLASH	32K	1536	256	23	5	I²C/SPI	1	—	2	1/3	DIP SOIC, SSOP	28
PIC18F258	FLASH	32K	1536	256	22	5	I²C/SPI	1	CAN 2.0B	1	1/3	DIP SOIC	28
PIC18F4220	FLASH	4K	512	256	34	13	I²C/SPI	1	6x PMM	2	1/3	DIP TQFP QFN	40/44
PIC18F4320	FLASH	8K	512	256	34	13	I²C/SPI	1	6x PMM	2	1/3	DIP TQFP QFN	40/44
PIC18C442	OTP	16K	512	—	34	8	I²C/SPI	1	—	2	1/3	DIP PLCC, TQFP	40/44
PIC18C452	OTP	32K	1536	—	34	8	I²C/SPI	1	—	2	1/3	DIP PLCC, TQFP	40/44
PIC18F442	FLASH	16K	512	256	34	8	I²C/SPI	1	—	2	1/3	DIP PLCC, TQFP	40/44
PIC18F452	FLASH	32K	1536	256	34	8	I²C/SPI	1	—	2	1/3	DIP PLCC, TQFP	40/44
PIC18F458	FLASH	32K	1536	256	33	5	I²C/SPI	1	CAN 2.0B	1	1/3	DIP PLCC, TQFP	40/44
PIC18C601	ROMless		1536	—	31	8	I²C/SPI	1	—	2	1/3	PLCC, TQFP	64/68
PIC18C658	OTP	32K	1536	—	52	12	I²C/SPI	1	CAN 2.0B	2	1/3	PLCC, TQFP	64/68
PIC18F6520	FLASH	32K	2048	1024	52	12	I²C/SPI	2	—	5	2/3	TQFP	64
PIC18F6620	FLASH	64K	3840	1024	52	12	I²C/SPI	2	—	5	2/3	TQFP	64
PIC18F6720	FLASH	128K	3840	1024	52	12	I²C/SPI	2	—	5	2/3	TQFP	64
PIC18C801	—	ROMless	1536	—	42	12	I²C/SPI	1	—	2	1/3	PLCC, TQFP	80/84
PIC18C858	OTP	32K	1536	—	68	16	I²C/SPI	1	CAN 2.0B	2	1/3	PLCC, TQFP	80/84
PIC18F8520	FLASH	32K	2048	1024	68	16	I²C/SPI	2	EMA	5	2/3	TQFP	80
PIC18F8620	FLASH	64K	3840	1024	68	16	I²C/SPI	2	EMA	5	2/3	TQFP	80
PIC18F8720	FLASH	128K	3840	1024	68	16	I²C/SPI	2	EMA	5	2/3	TQFP	80

Abbreviation: ADC = Analog-to-Digital Converter   CCP = Capture/Compare/PWM   I²C = Inter-Integrated Circuit Bus   PMM = Power Managed Mode
PWM = Pulse Width Modulation   SPI = Serial Peripheral Interface   USART = Universal Synchronous/Asynchronous Receiver/Transmitter

10.51   Qimonda HYB18L512160BF-7.5 (Qimonda AG 2006)

Data Sheet.
www.DataSheet4U.com

HY[B/E]18L512160BF-7.5
512-Mbit Mobile-RAM

1   Overview

1.1   Features

- 4 banks × 8 Mbit × 16 organization
- Fully synchronous to positive clock edge
- Four internal banks for concurrent operation
- Programmable CAS latency: 2, 3
- Programmable burst length: 1, 2, 4, 8 or full page
- Programmable wrap sequence: sequential or interleaved
- Programmable drive strength
- Auto refresh and self refresh modes
- 8192 refresh cycles / 64 ms
- Auto precharge
- Commercial (0°C to +70°C) and Extended (-25°C to +85°C) operating temperature range
- Dual-Die 54-ball PG-TFBGA package (12.0 × 8.0 × 1.2 mm)
- RoHS Compliant Products[1]

Power Saving Features

- Low supply voltages: V_{DD} = 1.70 V to 1.95 V, V_{DDQ} = 1.70 V to 1.95 V
- Optimized self refresh (I_{DD6}) and standby currents (I_{DD2} / I_{DD3})
- Programmable Partial Array Self Refresh (PASR)
- Temperature Compensated Self Refresh (TCSR), controlled by on-chip temperature sensor
- Power-Down and Deep Power Down modes

Part Number Speed Code		- 7.5	Unit	Performance
Speed Grade		133	MHz	
Access Time (t_{ACmax})	CL = 3	6.0	ns	
	CL = 2	7.0	ns	
Clock Cycle Time (t_{CKmin})	CL = 3	7.5	ns	
	CL = 2	9.5	ns	

---

1) RoHS Compliant Product: Restriction of the use of certain hazardous substances (RoHS) in electrical and electronic equipment as defined
     in the directive 2002/95/EC issued by the European Parliament and of the Council of 27 January 2003. These substances include mercury,
     lead, cadmium, hexavalent chromium, polybrominated biphenyls and polybrominated biphenyl ethers.

10.52   SBT30EDU Sensor and Prototyping Board (EasySen LLC 2008a)

SBT30EDU

Sensor and prototyping board

Plug-in sensor and prototyping board for IEEE 802.15.4/Zigbee compliant TelosB wireless motes for education and prototyping purposes

Three Sensor channels including Visual Light, Infrared and Low Pass Filtered Acoustic Signal Envelope; Prototyping Area for Up to Three User-Selected Channels; Experimental Board for University Lab/Classroom Use

Product Description

The SBT30EDU is a low-power, multi-channel sensor and prototyping board designed for use with IEEE 802.15.4 compliant TelosB[1] wireless sensor network platforms. The SBT30EDU features visual light, infrared and a low pass filtered acoustic signal envelope which allows the user to sample at a much lower speed to detect "activities" on the acoustic channel. Its prototyping area allows up to three user selected additional channels, as desired in experiments and customized application development. It can be directly connected to the expansion connectors of the TelosB platform using an IDC header.

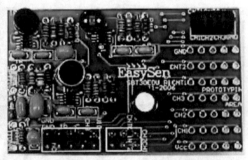

The SBT30EDU is accompanied by tested TinyOS and Java code that can be used to gather data across all associated sensor channels, display sensor readings on a computer, and to switch the SBT30EDU into a power saving "sleep" mode. The SBT30EDU has a smaller form-factor than TelosB platforms. By leveraging industry standard wireless sensor platforms and specially designed prototyping areas, the SBT30EDU enables a wide variety of customized sensor applications and experiments for labs and classrooms.

Key Features

- Prototyping area and external connector with 3 open analog channels for soldering custom sensors or connecting external signals
- Visual light, infrared and acoustic sensors on board
- Rectified acoustic envelope output with adjustable time constant and adjustable low-pass filter for customized applications
- Can be directly plugged in to the external connector of IEEE 802.15.4 / Zigbee compatible TelosB motes
- Smaller form factor than TelosB motes
- Power saving mode

---

[1] CrossBow Inc, TelosB (TPR2400CA) Product Datasheet, http://www.xbow.com/Products/Product_pdf_files/Wireless_pdf/TelosB_Datasheet.pdf

10.53   SBT80 Multi-Modality Sensor Board for TelosB Wireless Motes (EasySen LLC 2008a)

SBT80

Multi-Modality Sensor Board for TelosB wireless motes

Plug-in multi-sensor board for IEEE 802.15.4 compliant TelosB wireless motes

Eight Sensor channels including Visual Light, Infrared, Acoustic, Temperature, Magnetometer (X and Y axis), Accelerometer (X and Y axis)

Product Description

The SBT80 is a low-power, multi-channel sensor board designed for use with IEEE 802.15.4 compliant and TelosB™ wireless sensor network platforms. The SBT80 features visual light, infrared, acoustic, temperature, dual-axis magnetometer and dual-axis accelerometer sensors. It can be directly connected to the expansion connectors of the TelosB platform using an IDC header.

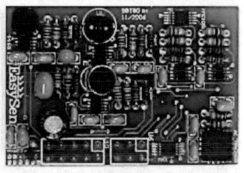

The SBT80 is accompanied by tested TinyOS and Java code that can be used to gather data over 8 different sensor channels, display sensor readings on a computer, and to switch the SBT80 into a power saving "sleep" mode.

The SBT80 has a smaller form-factor than TelosB platforms. By leveraging industry standard wireless sensor platforms, the SBT80 enables a wide variety of multi-modal sensor applications for security, surveillance using wireless sensor networks.

Key Features

- Can be directly plugged in to the external connector of IEEE 802.15.4 compliant TelosB motes

- Eight Sensor channels with high sensitivity sensors

- Smaller form factor than TelosB motes

- TinyOS and Java code support available for sampling sensor channels, display readings on PC and switching the sensor board to a power saving mode

- Can be used for multi-modal surveillance and monitoring applications using sensor networks

10.54   Spartan-3 FPGA (XILINX 2013)

Σ XILINX.

**Spartan-3 FPGA Family
Data Sheet**

DS099 June 27, 2013   **Product Specification**

**Module 1:
Introduction and Ordering Information**

DS099 (v3.1) June 27, 2013

- Introduction
- Features
- Architectural Overview
- Array Sizes and Resources
- User I/O Chart
- Ordering Information

Module 2: Functional Description

DS099 (v3.1) June 27, 2013

- Input/Output Blocks (IOBs)
  - IOB Overview
  - SelectIO™ Interface I/O Standards
- Configurable Logic Blocks (CLBs)
- Block RAM
- Dedicated Multipliers
- Digital Clock Manager (DCM)
- Clock Network
- Configuration

**Module 3:
DC and Switching Characteristics**

DS099 (v3.1) June 27, 2013

- DC Electrical Characteristics
  - Absolute Maximum Ratings
  - Supply Voltage Specifications
  - Recommended Operating Conditions
  - DC Characteristics
- Switching Characteristics
  - I/O Timing
  - Internal Logic Timing
  - DCM Timing
  - Configuration and JTAG Timing

Module 4: Pinout Descriptions

DS099 (v3.1) June 27, 2013

- Pin Descriptions
  - Pin Behavior During Configuration
- Package Overview
- Pinout Tables
  - Footprints

DS099 (v3.1) June 27, 2013

**Spartan-3 FPGA Family:
Introduction and Ordering Information**

Product Specification

Introduction

The Spartan®-3 family of Field-Programmable Gate Arrays is specifically designed to meet the needs of high volume, cost-sensitive consumer electronic applications. The eight-member family offers densities ranging from 50,000 to 5,000,000 system gates, as shown

The Spartan-3 family builds on the success of the earlier Spartan-IIE family by increasing the amount of logic resources, the capacity of internal RAM, the total number of I/Os, and the overall level of performance as well as by improving clock management functions. Numerous enhancements derive from the Virtex®-II platform technology. These Spartan-3 FPGA enhancements, combined with advanced process technology, deliver more functionality and bandwidth per dollar than was previously possible, setting new standards in the programmable logic industry.

Because of their exceptionally low cost, Spartan-3 FPGAs are ideally suited to a wide range of consumer electronics applications, including broadband access, home networking, display/projection and digital television equipment.

The Spartan-3 family is a superior alternative to mask programmed ASICs. FPGAs avoid the high initial cost, the lengthy development cycles, and the inherent inflexibility of conventional ASICs. Also, FPGA programmability permits design upgrades in the field with no hardware replacement necessary, an impossibility with ASICs.

Features

- Low-cost, high-performance logic solution for high-volume, consumer-oriented applications
  - Densities up to 74,880 logic cells
- SelectIO™ interface signaling
  - Up to 633 I/O pins
  - 622+ Mb/s data transfer rate per I/O
  - 18 single-ended signal standards
  - 8 differential I/O standards including LVDS, RSDS
  - Termination by Digitally Controlled Impedance
  - Signal swing ranging from 1.14V to 3.465V
  - Double Data Rate (DDR) support
  - DDR, DDR2 SDRAM support up to 333 Mb/s
- Logic resources
  - Abundant logic cells with shift register capability
  - Wide, fast multiplexers
  - Fast look-ahead carry logic
  - Dedicated 18 x 18 multipliers
  - JTAG logic compatible with IEEE 1149.1/1532
- SelectRAM™ hierarchical memory
  - Up to 1,872 Kbits of total block RAM
  - Up to 520 Kbits of total distributed RAM
- Digital Clock Manager (up to four DCMs)
  - Clock skew elimination
  - Frequency synthesis
  - High resolution phase shifting
- Eight global clock lines and abundant routing
- Fully supported by Xilinx ISE® and WebPACK™ software development systems
- MicroBlaze™ and PicoBlaze™ processor, PCK®, PCI Express® PIPE Endpoint, and other IP cores
- Pb-free packaging options
- Automotive Spartan-3 XA Family variant

Summary of Spartan-3 FPGA Attributes

Device	System Gates	Equivalent Logic Cells(1)	CLB Array (One CLB = Four Slices)			Distributed RAM Bits (K=1024)	Block RAM Bits (K=1024)	Dedicated Multipliers	DCMs	Max. User I/O	Maximum Differential I/O Pairs
			Rows	Columns	Total CLBs						
XC3S50(2)	50K	1,728	16	12	192	12K	72K	4	2	124	56
XC3S200(2)	200K	4,320	24	20	480	30K	216K	12	4	173	76
XC3S400(2)	400K	8,064	32	28	896	56K	288K	16	4	264	116
XC3S1000	1M	17,280	48	40	1,920	120K	432K	24	4	391	175
XC3S1500	1.5M	29,952	64	52	3,328	208K	576K	32	4	487	221
XC3S2000	2M	46,080	80	64	5,120	320K	720K	40	4	565	270
XC3S4000	4M	62,208	96	72	6,912	432K	1,728K	96	4	633	300
XC3S5000	5M	74,880	104	80	8,320	520K	1,872K	104	4	633	300

Notes:
1. Logic Cell = 4-input Look-Up Table (LUT) plus a 'D' flip-flop. "Equivalent Logic Cells" equals "Total CLBs" x 8 Logic Cells/CLB x 1.125 effectiveness.
2. These devices are available in Xilinx Automotive versions as described in DS314: Spartan-3 Automotive XA FPGA Family.

10.55   Stargate (Crossbow 2004)

STARGATE
X-SCALE, PROCESSOR PLATFORM

▼ 400MHz, Intel PXA255 Processor

▼ Low Power Consumption
   <500 mA

▼ Embedded Linux BSP Package,
   Source Code Shipped with Kit

▼ Small, 3.5" x 2.5" Form Factor

▼ PCMCIA and Compact Flash
   Connector

▼ 51-pin Expansion Connector
   for MICA2 Motes and other
   Peripherals

▼ Ethernet, Serial, JTAG, USB
   Connectors via 51-pin
   Daughter Card Interface

▼ Li-Ion Battery Option

Applications

▼ Single-Board Embedded Linux
   Computer

▼ Sensor Network Gateway

▼ Custom 802.11a/b Gateway

▼ Robotics Controller Card

▼ Distributed Computing Platform

STARGATE

The Stargate is a high-performance processing platform designed for sensor, signal processing, control, and wireless sensor networking applications. The Stargate is based on Intel's Xscale® processor, and it is the same processor found in today's most powerful handheld computers including the Compaq IPAQ® and the Dell Axim®.

The Stargate processor board is the result of the combined design efforts of several different Ubiquitous Computing research groups within Intel. The completed design is licensed to Crossbow Technology for commercial production. The Stargate processor board is preloaded with a Linux distribution and basic drivers. A variety of useful applications and development tools is also provided.

The Stargate processor module is compatible with Crossbow's MICA2 family of wireless sensor networking products (using the embedded TinyOS platform) and the public domain software from Intel's Open-Source Robotics initiative. The Stargate processor module is also an ideal solution for standalone Linux-based Single Board Computer (SBC) applications.

With it's strong communications capability and Crossbow's ongoing commitment to its open-source architecture, the Stargate platform offers tremendous flexibility. The SPB400CA Processor Board has both Compact Flash and PCMCIA connectors as well as an optional installable header for I2C and 4 general-purpose I/O (GPIO) lines. The SDC400CA Daughter Card supports a variety of additional interfaces, including:

- RS-232 Serial
- 10/100 Ethernet
- USB Host
- JTAG

Finally, the standard MICA2 Mote connector on the SPB400CA Processor Board provides support for synchronous serial port, (SSP), UART, and other GPIO connections.

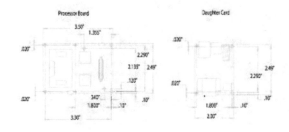

Specifications	Remarks
STARGATE Processor Board	
Intel PXA255, Xscale®	400 MHz, RISC Processor
Intel SA1111, StrongARM®	Multiple I/O Companion Chip
Memory	
64 MB SDRAM	
32 MB FLASH	Linux Software < 10 MBytes
Communications	
PCMCIA Slot	
Compact Flash Slot	Type II
51-pin MICA2/GPIO	UART, SSP via MICA2 Connector
Optional I2C Port	Installable Header
General	
Li-Ion Battery Option	
Watch Dog Timer (WDT)	Configurable up to 60 seconds
Battery Gas Gauge	
LED and User Application Switch	
Power Switch	
STARGATE Daughter Card	
Communications	
10/100 Base-T Ethernet Port	RJ-45 Connector
RS-232 Serial Port	DB-9 Connector
JTAG Debug Port	
USB Host Port	Version 1.1
General	
A/C Power Adaptor	5-6 VDC, 1 Amp
Reset Button	
Real-Time Clock	
Physical	
Processor Board	
(in) 3.50 x 2.49 x 0.73	Excluding Daughter Card
(cm) 9.53 x 6.33 x 1.86	Excluding Daughter Card
Weight (oz) 1.68	
(g) 47.47	
Daughter Card	
(in) 2.49 x 2.00 x 0.60	
(cm) 6.33 x 5.08 x 1.52	
Weight (oz) 1.42	
(g) 40.16	
Environmental	
0 to +70 (°C)	Operating Temperature

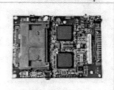

▼ Processor Board - Top

▼ Processor Board - Bottom

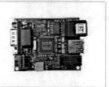

▼ Daughter Card

Stargate Basic Kit Contents
Stargate Processor Board
Stargate Daughter Card
Power Supply
Null Modem Cable
Developer's Guide
CD-ROM
Advanced Kit Contents
All Basic Kit Contents
JTAG Pod with Cable
WiFi Compact Flash Card
CD-ROM Contents
Linux - Kernel & Driver Sources
GNU Cross Platform Dev. Tools
Bootloader with Source Code
Flash Programming Utility
Shareware & Test Applications
Developer's Guide - PDF Format

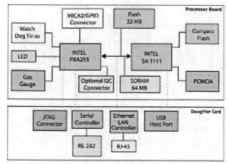

Stargate Block Diagram

10.56   Stargate NetBridge (Crossbow 2007b)

Stargate NetBridge
EMBEDDED SENSOR NETWORK GATEWAY

The Stargate NetBridge is an embedded Sensor Network gateway device. Its purpose is to connect Crossbow Sensor Nodes to an existing Ethernet network. It is based on the Intel IXP420 XScale processor running at 266MHz. It features one wired Ethernet and two USB 2.0 ports. The device is further equipped with 8MB of program FLASH, 32MB of RAM and a 2GB USB 2.0 system disk.

- Access Point and Network Coordinator
- Web Interface for Real-time Viewing Sensor Data and Network Health
- Ethernet Interface Backhaul for Remote Access and Viewing
- Plug-and-Play Capability Using Crossbow's BU Series Wireless Sensor Network Base Station

Stargate NetBridge runs the Debian Linux operating system. This is a full fledged standard Linux distribution for the ARM architecture. It comes preloaded with Crossbow's Sensor Network management and data visualization software packages, XServe and MoteExplorer. Those programs are started automatically at boot time of the Stargate NetBridge. In order to set up a Sensor Network gateway configuration, a base station (BU900, BU2400, BU2110 or MIB520, available separately) should be plugged into the secondary USB port of the Stargate NetBridge.

Stargate NetBridge contains a built-in Web server (MoteExplorer) and Sensor Network management tool (XServe). The latter can automatically identify what types of sensor boards are plugged into the nodes of the wireless sensor network and will instruct MoteExplorer to display the data accordingly. Stargate NetBridge's hosted Sensor Network is truly plug-and-play with minimal overhead for configuration and administration.

User Interface
- 4x status LEDs
- 1x power LED
- 1x soft power button
- 1x reset button
- 1x beeper

Preloaded Software
- Debian Linux (2.6.18 kernel)
- Crossbow MoteExplorer
- Crossbow XServe

Crossbow Accessories
- BU900 – MICA2 base station
- BU2400 – MICAz base station
- BU2110 – IRIS base station
- MIB520 – USB interface board

Note: Crossbow Telos and Imote2 Motes can be connected via the USB port, however, Crossbow does not provide software support for such configurations.

10.57 T-Node (SOWNet 2014)

T-Node product sheet

SOWNet's T-Node

General specifications of the Wireless sensor network OEM module:

- 868 MHz FSK Transceiver with frequency hopping
- Range: Up to 120 meter line of sight in free space and 40 meter indoor
- Power Consumption: active TX 18 mA - RX 13 mA, sleep mode: 20 µA at 3V
- Self-organizing ultra low power multi-hop protocols
- Ultra-compact design diameter Ø 25mm. *(Figure 1)*
- Two 10 pin expansion connectors with:
  - Analog inputs
  - 10-bit ADC
  - Interrupts
  - Digital I/O
  - I²C
  - SPI
  - UART interfaces
- Wide variety of sensor boards based on the T-Nodes OEM module available.

Applications

T-Nodes can be used for and implemented in:

- Building management systems
- Micro-climate monitoring like Relative Humidity monitoring
- Dynamic evacuation systems (DES)
- Seat monitoring in the public railway sector
- Fire systems
- Intruder systems
- Men guarding
- Access Control systems
- Temperature monitoring (HVAC systems)

Figure 1: T-Node

Schematic diagram T-Node

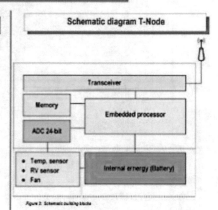

Figure 2: Schematic building blocks

Technology

The T-Node allows OEM partners to build their own wireless solution for the market using the 868 MHz. radios together with ultra low power multi-hop network protocols like Mesh-, Star-, and Single hop topologies.

The SOWNet™ protocol provides an automatic self-organizing capability for easy deployment and recovery.

T-Node use an 868 MHz radio to achieve up to 120 meter communication distance outdoors line of sight free space, while consuming as little as 20µA in sleep mode.
The T-Node support analog inputs, 10-bit ADC, interrupt, digital I/O, I2C, SPI and UART interfaces. These interfaces make it easy to connect to a wide variety of sensors and actuators.

The ADC interface has single-ended and differential channels with programmable gain. For high resolution measurements a 24-bit ADC can be interfaced with the T-Nodes. *(Figure 2)*

SOWNet Technologies BV offers a wide variety of sensor and data-acquisition systems based on the T-Node OEM modules. Casing, battery and antenna can be tailored to customer specifications.
Custom sensor and data acquisition systems are also available.

Please contact SOWNet Technologies BV for additional information.

Demonstrator kit

Graphical User Interface

BNV-Node

Wireless Passive Infrared

Dynamic Evacuation System

Outdoor Gateway

GreenNode

T-Node impression

Technical specifications

Specifications	T-Node	Comments
Micro-controller		
Program flash memory	128 k bytes	
SRAM	4 k bytes	
Data flash memory	512 k bytes	Optional on request
Serial communications	Dual UART	
Other interfaces	I2C, SPI, Digital I/O	
AD converter	8 channel 10 bit	Single ended and differential with 1, 10 and 200 gain
Radio		
Center frequency	868 MHz	ISM band
Number of channels	50	Programmable
Data rate	52.2 kbps	Simplex mode
	38.6 kbps	Duplex mode
RF power	-20 to +5 dBm	Programmable
Receiver sensitivity	-105 dBm	at 10⁻³ BER
Range	120 mtr.*	Line of sight free space
	40 mtr.*	Indoor
Modulation	FSK	Manchester encoded
Power consumption		
Transmit mode	25 mA	Tx is +5 dBm
Receive mode	12 mA	
Sleep mode	20 µA	
Operating conditions		
Supply voltage	2.4-3.6V	Easily extendable up to 12V by DC-DC converter
Size	Ø 25 x 5 mm.	
Operating temperature range	-10 °C to 75 °C	
Interfaced sensors		
Alcohol	Resitive	Temperature compensated range 50 – 1,200 ppm
Temperature	Bandgap	Range -40°C to 120 °C – accuracy @ 25°C ± 0.3°C
Passive Infrared	Optes	
Relative Humidity	Capacitive	Range 0 % to 100 % - accuracy @ 25°C ± 1.8%
Heartbeat		
Light	LDR	
Magnetic	Inductive	Range ± 1100 µT - 15 nT resolution
Additional interfaces in progress		

Note: * 'Actual' performance depending on specific environmental condi-

Interfaces

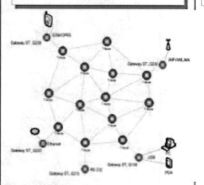

Figure 3 Currently available interfaces

SOWNet Gateway

- The SOWNet Gateway allows processing and data collection of the sensor network to a PC or PDA via an USB connection.
- Housing in accordance with IP54 standards - dust and splash water proof.
- Connections to a wide variety of other technologies as: GSM/GPRS, Wifi, Ethernet etc. (Figure 3)
- Windows, Linux and MacOS based Gateway Graphical User Interface (GUI) software available for connection to a PC.
- Pocket PC/PDA's supported with Windows Mobile 2005 and newer

Demonstrator kit

10.58   TC55VCM208ASTN40,55 CMOS Static RAM (Toshiba 2002)

TOSHIBA   TC55VCM208ASTN40,55

TENTATIVE   TOSHIBA MOS DIGITAL INTEGRATED CIRCUIT SILICON GATE CMOS

524,288-WORD BY 8-BIT FULL CMOS STATIC RAM

DESCRIPTION

The TC55VCM208ASTN is a 4,194,304-bit static random access memory (SRAM) organized as 524,288 words by 8 bits. Fabricated using Toshiba's CMOS Silicon gate process technology, this device operates from a single 2.3 to 3.6 V power supply. Advanced circuit technology provides both high speed and low power at an operating current of 3 mA/MHz and a minimum cycle time of 40 ns. It is automatically placed in low-power mode at 0.7 μA standby current (at V_{DD} - 3 V, Ta - 25°C, typical) when chip enable ($\overline{CE1}$) is asserted high or (CE2) is asserted low. There are three control inputs. $\overline{CE1}$ and CE2 are used to select the device and for data retention control, and output enable ($\overline{OE}$) provides fast memory access. This device is well suited to various microprocessor system applications where high speed, low power and battery backup are required. And, with a guaranteed operating extreme temperature range of -40° to 85°C, the TC55VCM208ASTN can be used in environments exhibiting extreme temperature conditions. The TC55VCM208ASTN is available in a plastic 40-pin thin-small outline package (TSOP).

FEATURES

- Low-power dissipation
  Operating: 9 mW/MHz (typical)
- Single power supply voltage of 2.3 to 3.6 V
- Power down features using $\overline{CE1}$ and CE2
- Data retention supply voltage of 1.5 to 3.6 V
- Direct TTL compatibility for all inputs and outputs
- Wide operating temperature range of -40° to 85°C
- Standby Current (maximum):

3.6 V	10 μA
3.0 V	5 μA

- Access Times:

	TC55VCM208ASTN	
	40	55
Access Time	40 ns	55 ns
$\overline{CE1}$ Access Time	40 ns	55 ns
CE2 Access Time	40 ns	55 ns
$\overline{OE}$ Access Time	25 ns	30 ns

- Package:
  TSOP I 40-P-1014-0.50   (Weight 0.30 g typ)

PIN ASSIGNMENT (TOP VIEW)

40 PIN TSOP

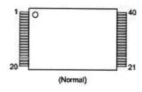

(Normal)

PIN NAMES

A0~A18	Address Inputs
$\overline{CE1}$, CE2	Chip Enable
R/W	Read/Write Control
$\overline{OE}$	Output Enable
$\overline{LB}$, $\overline{UB}$	Data Byte Control
I/O1~I/O16	Data Inputs/Outputs
V_{DD}	Power
GND	Ground
NC	No Connection
OP*	Option

*: OP pin must be open or connected to GND.

Pin No.	1	2	3	4	5	6	7	8	9	10	11	12	13	14	15	16	17	18	19	20
Pin Name	A16	A15	A14	A13	A12	A11	A9	A8	R/W	CE2	OP	NC	A18	A7	A6	A5	A4	A3	A2	A1
Pin No.	21	22	23	24	25	26	27	28	29	30	31	32	33	34	35	36	37	38	39	40
Pin Name	A0	$\overline{CE1}$	GND	$\overline{OE}$	I/O1	I/O2	I/O3	I/O4	NC	V_{DD}	V_{DD}	I/O5	I/O6	I/O7	I/O8	A10	NC	NC	GND	A17

10.59   Telos (Moteiv 2004)

Telos
Rev B (Low Power Wireless Sensor Module)

Telos
Ultra low power IEEE 802.15.4 compliant wireless sensor module
Revision B : Humidity, Light, and Temperature sensors with USB

Product Description

Telos is an ultra low power wireless module
for use in sensor networks, monitoring
applications, and rapid application
prototyping. Telos leverages industry
standards like USB and IEEE 802.15.4 to
interoperate seamlessly with other devices.
By using industry standards, integrating
humidity, temperature, and light sensors,
and providing flexible interconnection with
peripherals, Telos enables a wide range of
mesh network applications. Telos Revision B is a drop-in replacement for Moteiv's successful
Revision A design. Revision B includes increased performance, functionality, and expansion.
With TinyOS support out-of-the-box, Telos leverages emerging wireless protocols and the open
source software movement. Telos is part of a line of modules featuring on-board sensors to
increase robustness while decreasing cost and package size.

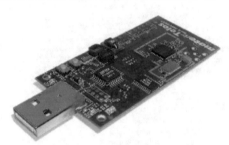

Key Features

- 250kbps 2.4GHz IEEE 802.15.4 Chipcon Wireless Transceiver
- Interoperability with other IEEE 802.15.4 devices
- 8MHz Texas Instruments MSP430 microcontroller (10k RAM, 48k Flash)
- Integrated ADC, DAC, Supply Voltage Supervisor, and DMA Controller
- Integrated onboard antenna with 50m range indoors / 125m range outdoors
- Integrated Humidity, Temperature, and Light sensors
- Ultra low current consumption
- Fast wakeup from sleep (<6µs)
- Hardware link-layer encryption and authentication
- Programming and data collection via USB
- 16-pin expansion support and optional SMA antenna connector
- TinyOS support : mesh networking and communication implementation

Table 10.6 compares the family of Berkeley motes up to Telos.

Table 10.6 Berkeley family of motes up to Telos (Polastre et al. 2005)

Mote type / Year	WeC 1998	René' 1999	René'2 2000	Dot 2000	Mica 2001	Mica2Dot 2002	Mica 2 2002	Telos 2004
Microcontroller								
Type	AT90LS8535		ATmega163		ATmega128			TI MSP430
Program memory (KB)	8		16		128			48
RAM (KB)	0.5		1		4			10
Active power (mW)	15		15		8		3	3
Sleep power (µW)	45		45		75		15	15
Wakeup time (µs)	1000		36		180		6	6
Nonvolatile storage								
Chip	24LC256				AT45DB041B			ST M25P80
Connection type	I²C				SPI			SPI
Size (KB)	32				512			1024
Communication								
Radio	TR1000				TR1000	CC1000		CC2420
Data rate (kbps)	10				40	38.4		250
Modulation type	OOK				ASK	FSK		O-QPSK
Receive power (mW)	9				12	29		38
Transmit power at 0 dBm (mW)	36				36	42		35
Power consumption								
Minimum operation (V)	2.7		2.7		2.7		·	1.8
Total active power (mW)	24				27	44	89	41
Programming and sensor interface								
Expansion	None	51-pin	51-pin	None	51-pin	19-pin	51-pin	16-pin
Communication	IEEE 1284 (programming) and RS232 (requires additional hardware							USB
Integrated sensors	No	No	No	Yes	No	No	No	Yes

10.60   TinyNode (Dubois-Ferrière et al. 2006)

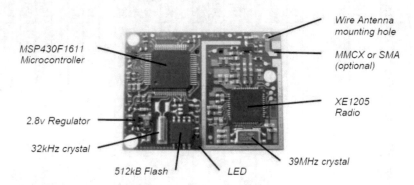

Fig. 10.2   TinyNode core module (Dubois-Ferrière et al. 2006)

10.61   Tmote Connect (Moteiv 2006a)

Wireless Gateway Appliance Software
Access Tmote wireless sensor modules through Ethernet
with Moteiv's Tmote Connect gateway software.

Product Description

Tmote Connect software allows a Linksys
NSLU2 Network Attached Storage device to
function as a gateway appliance, connecting
Tmote wireless sensor modules to a wired local
area network. Each Tmote wireless module
connected to a gateway appliance can be
remotely administered through a concise web-
based graphical user interface. Tmote Connect
integrates quickly and conveniently with TinyOS
and provides control over remotely connected
Tmote wireless sensor modules.

Key Features

Tmote Connect Software includes:
- Bridging between Tmote wireless
  networks and Ethernet infrastructure.
- Support for up to 2 Tmote wireless
  modules per Tmote Connect gateway.
- Bi-directional connectivity for data
  transfers to and from Tmote wireless
  modules over TCP/IP sockets.

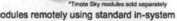

*Tmote Sky modules sold separately

- Flash reprogramming of Tmote wireless modules remotely using standard in-system
  programming protocols.
- Integration with TinyOS development system and tools (Both TinyOS 1.x and 2.x).
- Web-based status interface – mote identification, reset, and performance counters.
- Operates in networks with and without DHCP support.
- Field upgradeable for new software features from Moteiv.

10.62   Tmote Sky (Moteiv 2006b)

Low Power Wireless Sensor Module

**Ultra low power IEEE 802.15.4 compliant
wireless sensor module**
Humidity, Light, and Temperature sensors with USB

Product Description

Tmote Sky is an ultra low power wireless
module for use in sensor networks,
monitoring applications, and rapid
application prototyping. Tmote Sky
leverages industry standards like USB and
IEEE 802.15.4 to interoperate seamlessly
with other devices. By using industry
standards, integrating humidity,
temperature, and light sensors, and
providing flexible interconnection with
peripherals, Tmote Sky enables a wide
range of mesh network applications.

Tmote Sky is a drop-in replacement for Moteiv's successful Telos design. Tmote Sky includes
increased performance, functionality, and expansion. With TinyOS support out-of-the-box,
Tmote leverages emerging wireless protocols and the open source software movement. Tmote
Sky is part of a line of modules featuring on-board sensors to increase robustness while
decreasing cost and package size.

Key Features

- 250kbps 2.4GHz IEEE 802.15.4 Chipcon Wireless Transceiver
- Interoperability with other IEEE 802.15.4 devices
- 8MHz Texas Instruments MSP430 microcontroller (10k RAM, 48k Flash)
- Integrated ADC, DAC, Supply Voltage Supervisor, and DMA Controller
- Integrated onboard antenna with 50m range indoors / 125m range outdoors
- Integrated Humidity, Temperature, and Light sensors
- Ultra low current consumption
- Fast wakeup from sleep (<6μs)
- Hardware link-layer encryption and authentication
- Programming and data collection via USB
- 16-pin expansion support and optional SMA antenna connector
- TinyOS support : mesh networking and communication implementation
- Complies with FCC Part 15 and Industry Canada regulations

Low Power Wireless Sensor Module

Module Description

The Tmote Sky module is a low power "mote" with integrated sensors, radio, antenna, microcontroller, and programming capabilities.

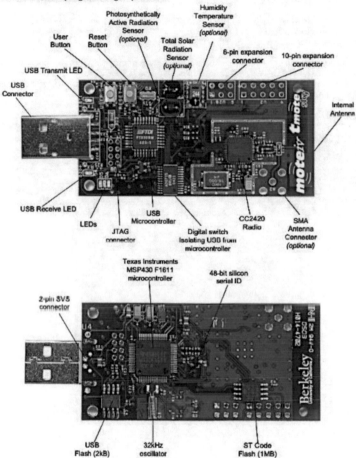

Front and Back of the Tmote Sky module

Low Power Wireless Sensor Module

Power

Tmote Sky is powered by two AA batteries. The module was designed to fit the two AA battery form factor. AA cells may be used in the operating range of 2.1 to 3.6V DC, however the voltage must be at least 2.7V when programming the microcontroller flash or external flash.

If the Tmote Sky module is plugged into the USB port for programming or communication, it will receive power from the host computer. The mote operating voltage when attached to USB is 3V. If Tmote will always be attached to a USB port, no battery pack is necessary.

The 16-pin expansion connector (described in the Section on page 17) can provide power to the module. Any of the battery terminal connections may also provide power to the module. At no point should the input voltage exceed 3.6V—doing so may damage the microcontroller, radio, or other components.

Typical Operating Conditions

	MIN	NOM	MAX	UNIT
Supply voltage	2.1		3.6	V
Supply voltage during flash memory programming	2.7		3.6	V
Operating free air temperature	-40		85	°C
Current Consumption: MCU on, Radio RX		21.8	23	mA
Current Consumption: MCU on, Radio TX		19.5	21	mA
Current Consumption: MCU on, Radio off		1800	2400	µA
Current Consumption: MCU idle, Radio off		54.5	1200	µA
Current Consumption: MCU standby		5.1	21.0	µA

Caution! ESD sensitive device. Precaution should be used when handling the device in order to prevent permanent damage.

10.63   TSL250R, TSL251R, TSL252R Light-to-Voltage Optical Sensors (TAOS 2001)

TSL250R, TSL251R, TSL252R
LIGHT-TO-VOLTAGE OPTICAL SENSORS

TAOS028A – MAY 2001

- Monolithic Silicon IC Containing Photodiode, Operational Amplifier, and Feedback Components
- Converts Light Intensity to a Voltage
- High Irradiance Responsivity, Typically 137 mV/(μW/cm^2) at λ_p = 635 nm (TSL250R)
- Compact 3-Lead Clear Plastic Package
- Single Voltage Supply Operation
- Low Dark (Offset) Voltage....10mV Max
- Low Supply Current......1.1 mA Typical
- Wide Supply-Voltage Range.... 2.7 V to 5.5 V
- Replacements for TSL250, TSL251, and TSL252

PACKAGE
(FRONT VIEW)

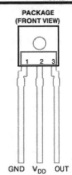

GND   V$_{DD}$ OUT

Description

The TSL250R, TSL251R, and TSL252R are light-to-voltage optical sensors, each combining a photodiode and a transimpedance amplifier (feedback resistor = 16 MΩ, 8 MΩ, and 2.8 MΩ respectively) on a single monolithic IC. Output voltage is directly proportional to the light intensity (irradiance) on the photodiode. These devices have improved amplifier offset-voltage stability and low power consumption and are supplied in a 3-lead clear plastic sidelooker package with an integral lens

Functional Block Diagram

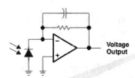

Voltage
Output

Terminal Functions

TERMINAL NAME	NO.	DESCRIPTION
GND	1	Ground (substrate). All voltages are referenced to GND.
OUT	3	Output voltage
V$_{DD}$	2	Supply voltage

Texas Advanced Optoelectronic Solutions Inc.
800 Jupiter Road, Suite 205 • Plano, TX 75074 • (972) 673-0759

10.64   WiEye Sensor Board for Wireless Surveillance and Security Applications (EasySen LLC 2008b)

 **WiEye®**
Sensor board for wireless surveillance and security applications

Plug-in sensor board for IEEE 802.15.4/Zigbee compliant TelosB wireless motes

Four Sensor channels including Long-Range Passive Infrared (PIR), Visual Light, Low Pass Filtered Acoustic Envelope and Unprocessed Acoustic Channels

Product Description

EasySen WiEye is a low-power, multi-channel sensor board designed for use with IEEE 802.15.4/Zigbee compliant TelosB[1] wireless sensor network platforms in security and surveillance applications. WiEye features passive infrared (PIR), visual light, low pass filtered acoustic envelope and unprocessed acoustic channels. It can be directly connected to the expansion connectors of the TelosB platforms using an IDC header.

WiEye has a smaller form-factor than TelosB platforms and enables a wide variety of multi-modal sensor applications for security, surveillance and monitoring using wireless sensor networks. It is particularly suited to support perimeter security applications and can detect movement of individuals from a distance of 30 feet and the movement of vehicles from distances up to 150 feet.

Key Features

*   Long-range passive infrared (PIR) sensor well suited for detecting human presence or vehicles (90-100° wide detection cone, 20-30 feet detection range for human presence, 50-150 feet detection range for vehicles depending on the size of vehicle)

*   Visual light and acoustic sensors complementing the PIR sensor for improved detection

*   Adjustable time constant in PIR sensor output for customized applications

*   Rectified acoustic envelope output with adjustable time constant and adjustable low-pass filter for customized applications

*   Additional acoustic channel allowing high frequency sampling of voice and other noise signals

*   Can be directly plugged into the external connector of IEEE 802.15.4/Zigbee compliant TelosB motes

*   Smaller form factor than TelosB motes

*   TinyOS and Java code support available for sampling sensor channels, display readings on a PC

*   Power saving mode

---

[1] CrossBow Inc, TelosB (TPR2400CA) Product Datasheet, http://www.xbow.com/Products/Product_pdf_files/Wireless_pdf/TelosB_Datasheet.pdf

10.65   WM8950 (Wolfson Microelectronics 2011)

WM8950

ADC with Microphone Input and Programmable Digital Filters

DESCRIPTION

The WM8950 is a low power, high quality mono ADC designed for portable applications such as Digital Still Camera, Digital Voice Recorder or games console accessories.

The device integrates support for a differential or single ended mic. External component requirements are reduced as no separate microphone amplifiers are required.

Advanced Sigma Delta Converters are used along with digital decimation filters to give high quality audio at sample rates from 8 to 48ks/s. Additional digital filtering options are available, to cater for application filtering such as wind noise reduction, noise rejection, plus an advanced mixed signal ALC function with noise gate is provided.

An on-chip PLL is provided to generate the required Master Clock from an external reference clock. The PLL clock can also be output if required elsewhere in the system.

The WM8950 operates at supply voltages from 2.5 to 3.6V, although the digital supplies can operate at voltages down to 1.71V to save power. Different sections of the chip can also be powered down under software control by way of the selectable two or three wire control interface.

WM8950 is supplied in a very small 4x4mm QFN package, offering high levels of functionality in minimum board area, with high thermal performance.

FEATURES

Mono ADC:
- Audio sample rates: 8, 11.025, 16, 22.05, 24, 32, 44.1, 48kHz
- SNR 94dB, THD -83dB ('A'-weighted @ 8 – 48ks/s)
- Multiple auxiliary analogue inputs

Mic Preamps:
- Differential or single end Microphone Interface
  - Programmable preamp gain
  - Pseudo differential inputs with common mode rejection
  - Programmable ALC / Noise Gate in ADC path
- Low-noise bias supplied for electret microphones

OTHER FEATURES
- 5 band EQ
- Programmable High Pass Filter (wind noise reduction)
- Fully Programmable IIR Filter (notch filter)
- On-chip PLL
- Low power, low voltage
  - 2.5V to 3.6V (digital: 1.71V to 3.6V)
  - power consumption 10mA all-on 48ks/s mode
- 4x4x0.9mm 24 lead QFN package

APPLICATIONS
- Digital Still Camera
- General Purpose low power audio ADC
- Games console accessories
- Voice recorders

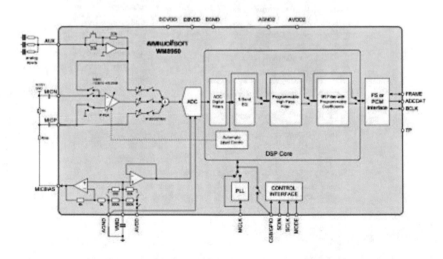

10.66   Xbee/Xbee-PRO OEM RF Modules (MaxStream 2007)

XBee/XBee-PRO OEM RF Modules

The XBee and XBee-PRO OEM RF Modules were engineered to meet IEEE 802.15.4 standards and support the unique needs of low-cost, low-power wireless sensor networks. The modules require minimal power and provide reliable delivery of data between devices.

The modules operate within the ISM 2.4 GHz frequency band and are pin-for-pin compatible with each other.

Key Features

Long Range Data Integrity

XBee

- Indoor/Urban: up to 100' (30 m)
- Outdoor line-of-sight: up to 300' (100 m)
- Transmit Power: 1 mW (0 dBm)
- Receiver Sensitivity: -92 dBm

XBee-PRO

- Indoor/Urban: up to 300' (100 m)
- Outdoor line-of-sight: up to 1 mile (1500 m)
- Transmit Power: 100 mW (20 dBm) EIRP
- Receiver Sensitivity: -100 dBm

RF Data Rate: 250,000 bps

Advanced Networking & Security

Retries and Acknowledgements

DSSS (Direct Sequence Spread Spectrum)

Each direct sequence channels has over 65,000 unique network addresses available

Source/Destination Addressing

Unicast & Broadcast Communications

Point-to-point, point-to-multipoint and peer-to-peer topologies supported

Coordinator/End Device operations

Low Power

XBee

- TX Current: 45 mA (@3.3 V)
- RX Current: 50 mA (@3.3 V)
- Power-down Current: < 10 µA

XBee-PRO

- TX Current: 215 mA (@3.3 V)
- RX Current: 55 mA (@3.3 V)
- Power-down Current: < 10 µA

ADC and I/O line support

Analog-to-digital conversion, Digital I/O

I/O Line Passing

Easy-to-Use

No configuration necessary for out-of box RF communications

Free X-CTU Software
(Testing and configuration software)

AT and API Command Modes for configuring module parameters

Extensive command set

Small form factor

Free & Unlimited RF-XPert Support

Worldwide Acceptance

FCC Approval (USA) Refer to Appendix A [p59] for FCC Requirements.
Systems that contain XBee/XBee-PRO RF Modules inherit MaxStream Certifications.

ISM (Industrial, Scientific & Medical) **2.4 GHz frequency band**

Manufactured under **ISO 9001:2000** registered standards

XBee/XBee-PRO RF Modules are optimized for use in the **United States, Canada, Australia, Israel and Europe**. Contact MaxStream for complete list of government agency approvals.

FC   **C €**

MaxStream © 2007 MaxStream, Inc.

10.67   XC2C256 CoolRunner-II CPLD (XILINX 2007)

XC2C256 CoolRunner-II CPLD

DS094 (v3.2) March 8, 2007   **Product Specification**

Features

- Optimized for 1.8V systems
  - As fast as 5.7 ns pin-to-pin delays
  - As low as 13 µA quiescent current
- Industry's best 0.18 micron CMOS CPLD
  - Optimized architecture for effective logic synthesis. Refer to the CoolRunner™-II family data sheet for architecture description.
  - Multi-voltage I/O operation — 1.5V to 3.3V
- Available in multiple package options
  - 100-pin VQFP with 80 user I/O
  - 144-pin TQFP with 118 user I/O
  - 132-ball CP (0.5mm) BGA with 106 user I/O
  - 200-pin PQFP with 173 user I/O
  - 256-ball FT (1.0mm) BGA with 184 user I/O
  - Pb-free available for all packages
- Advanced system features
  - Fastest in system programming
    - 1.8V ISP using IEEE 1532 (JTAG) interface
  - IEEE1149.1 JTAG Boundary Scan Test
  - Optional Schmitt-trigger input (per pin)
  - Unsurpassed low power management
    - DataGATE enable (DGE) signal control
  - Two separate I/O banks
  - RealDigital 100% CMOS product term generation
  - Flexible clocking modes
    - Optional DualEDGE triggered registers
    - Clock divider (divide by 2,4,6,8,10,12,14,16)
    - CoolCLOCK
  - Global signal options with macrocell control
    - Multiple global clocks with phase selection per macrocell
    - Multiple global output enables
    - Global set/reset
  - Advanced design security
  - PLA architecture
    - Superior pinout retention
    - 100% product term routability across function block
  - Open-drain output option for Wired-OR and LED drive
  - Optional bus-hold, 3-state or weak pull-up on selected I/O pins
  - Optional configurable grounds on unused I/Os
  - Mixed I/O voltages compatible with 1.5V, 1.8V, 2.5V, and 3.3V logic levels
    - SSTL2-1, SSTL3-1, and HSTL-1 I/O compatibility
  - Hot pluggable

Description

The CoolRunner™-II 256-macrocell device is designed for both high performance and low power applications. This lends power savings to high-end communication equipment and high speed to battery operated devices. Due to the low power stand-by and dynamic operation, overall system reliability is improved

This device consists of sixteen Function Blocks inter-connected by a low power Advanced Interconnect Matrix (AIM). The AIM feeds 40 true and complement inputs to each Function Block. The Function Blocks consist of a 40 by 56 P-term PLA and 16 macrocells which contain numerous configuration bits that allow for combinational or registered modes of operation.

Additionally, these registers can be globally reset or preset and configured as a D or T flip-flop or as a D latch. There are also multiple clock signals, both global and local product term types, configured on a per macrocell basis. Output pin configurations include slew rate limit, bus hold, pull-up, open drain and programmable grounds. A Schmitt-trigger input is available on a per input pin basis. In addition to storing macrocell output states, the macrocell registers may be configured as "direct input" registers to store signals directly from input pins.

Clocking is available on a global or Function Block basis. Three global clocks are available for all Function Blocks as a synchronous clock source. Macrocell registers can be individually configured to power up to the zero or one state. A global set/reset control line is also available to asynchronously set or reset selected registers during operation. Additional local clock, synchronous clock-enable, asynchronous set/reset and output enable signals can be formed using product terms on a per-macrocell or per-Function Block basis.

A DualEDGE flip-flop feature is also available on a per macrocell basis. This feature allows high performance synchronous operation based on lower frequency clocking to help reduce the total power consumption of the device.

Circuitry has also been included to divide one externally supplied global clock (GCK2) by eight different selections. This yields divide by even and odd clock frequencies.

The use of the clock divide (division by 2) and DualEDGE flip-flop gives the resultant CoolCLOCK feature.

DataGATE is a method to selectively disable inputs of the CPLD that are not of interest during certain points in time.

10.68   XE1205I Integrated UHF Transceiver (Semtech 2008)

XE1205

XE1205

180 MHz – 1GHz

Low-Power, High Link Budget Integrated UHF Transceiver

GENERAL DESCRIPTION

The XE1205 is an integrated transceiver operating in the 433, 868 and 915 MHz license-free ISM (Industrial, Scientific and Medical) frequency bands; it can also address other frequency bands in the 180-1000 MHz range. Its highly integrated architecture allows for minimum external components while maintaining design flexibility. All major RF communication parameters are programmable and most of them can be dynamically set. The XE1205 offers the unique advantage of narrow-band and wide-band communication, this without the need to modify the number or parameters of the external components. The XE1205 is optimized for low power consumption while offering high RF output power and channelized operation suited for both the European (ETSI EN 300-220-1) and the North American (FCC part 15) regulatory standards. TrueRF™ technology enables a low-cost external component count (elimination of the SAW filter) whilst still satisfying ETSI and FCC regulations.

APPLICATIONS

- Narrow-band and wide-band security systems
- Voice and data over an RF link
- Process and building control
- Access control
- Home automation
- Home appliances interconnection

KEY PRODUCT FEATURES

- Programmable RF output power: up to +15 dBm
- High Rx sensitivity: down to -121 dBm at 1.2 kbit/s, -116 dBm at 4.8 kbits.
- Low power: RX=14 mA; TX = 62 mA @ 15 dBm
- Can accommodate 300-1000 MHz frequency range
- Wide band operation: up to 304.7 kbit/s, NRZ coding
- Narrow band operation: 25 kHz channels for data rates up to 4.8 kbit/s, NRZ coding; optional transmitter pre-filtering to enable adjacent channel power below -37 dBm at 25 kHz
- On-chip frequency synthesizer with minimum frequency resolution of 500 Hz
- Continuous phase 2-level FSK modulation
- Incoming data pattern recognition
- Built-in Bit-Synchronizer for incoming data and clock synchronization and recovery
- FEI (Frequency Error Indicator) with built-in AFC
- RSSI (Received Signal Strength Indicator)
- 16-byte FIFO for transmit / receive data buffering and transfer via SPI bus

ORDERING INFORMATION

Part number	Temperature range	Package
XE1205I074TRLF[1]	-40 °C to +85 °C	VQFN48

[1]   TR refers to tape & reel.
LF refers to Lead Free package.
This device is WEEE and RoHS compliant

Rev 10 December 2008   www.semtech.com

References

Agilent Technologies. 2003a. *Agilent ADCM-1670 CIF Resolution CMOS Camera Module, UART Output*. Agilent Technologies. May 13, 2003. http://www.digchip.com/datasheets/parts/datasheet/021/ADCM-1670.php. Accessed 10 Apr 2014.

———. 2003b. *Agilent ADCM-1700-0000: Landscape CIF Resolution CMOS Camera Module*. Agilent Technologies. November 7, 2003. http://www.zhopper.narod.ru/mobile/adcm1700.pdf. Accessed 23 Mar 2014.

———. 2003c. *Agilent ADCM-2650 Portrait VGA Resolution CMOS Camera Module*. Agilent Technologies. February 19, 2003. http://centerforartificialvision.com/pdf/c4avdcm/ADCM2650 PB.pdf. Accessed 30 Mar 2014.

———. 2004. *Agilent ADNS-3060 High-performance Optical Mouse Sensor*. Agilent Technologies. October 20, 2004. http://datasheet.eeworld.com.cn/pdf/HP/48542_ADNS-3060.pdf. Accessed 30 Mar 2014.

Atmel. 2003. *4-megabit (512K x 8) Single 2.7-Volt Battery-Voltage Flash Memory*. Atmel. 2003. http://pdf.datasheetcatalog.com/datasheet/atmel/doc0383.pdf. Accessed 14 Apr 2014.

———. 2011a. *8-bit Atmel Microcontroller with 128KBytes In-System Programmable Flash*. Atmel. January 1, 2011. http://www.atmel.com/Images/doc2467.pdf. Accessed 21 Mar 2014.

———. 2011b. *8-bit Atmel Microcontroller with 4/8/16K Bytes In-System Programmable Flash*. Atmel. January 1, 2011. http://www.atmel.com/Images/2545s.pdf. Accessed 9 Mar 2014.

———. 2008. *AT91 ARM Thumb-based Microcontrollers*. Atmel. December 7, 2008. http://www.tme.eu/en/Document/84227da6b1aeb2914051d2f6e285168f/at91sam7sxxx.pdf. Accessed 30 Mar 2014.

———. 2011c. *AT91SAM ARM-based Embedded MPU*. Atmel. October 3, 2011. http://www.atmel.com/Images/doc6062.pdf. Accessed 5 Apr 2014.

———. 2002. *Atmel FPSLIC 5K - 40K Gates of AT40K FPGA with 8-bit AVR*. Atmel. January 1, 2002. http://media.digikey.com/pdf/Data%20Sheets/Atmel%20PDFs/AT94K05_10_40AL%20Complete.pdf. Accessed 31 Mar 2014.

AverLogic Technologies. 2002. *AL440B: 4M-Bit High Speed FIFO Field Memory*. AverLogic Technologies. November 11, 2002. http://www.averlogic.com/pdf/AL440B_Flyer.pdf. Accessed 29 Mar 2014.

Baar, M., et al. 2008. *The ScatterWeb MSB-A2 Platform for Wireless Sensor Networks*. Technical, Department of Mathematics and Computer Scienc, Freie Universität Berlin, 1–3. Berlin: Freie Universität Berlin.

Bluegiga Technologies. 2007. *WT12 Data Sheet - Version 2.4*. January 11, 2007. http://www.kelag.ch/industriesensoren/bluegiga/WT12_DS.pdf. Accessed 21 Jan 2014.

BP Solar. 2014. *MSX-01F*. January 1, 2014. http://au.element14.com/bp-solar/msx-01f/solar-panel-1-2w/dp/654012. Accessed 2 Feb 2014.

Coalesenses. 2014. *iSense Security Sensor Module*. Coalesenses. January 1, 2014. http://www.coalesenses.com/download/product_briefs/ProductBriefSecurityModule.pdf. Accessed 17 July 2014.

Crossbow. 2005. *Imote2: High-performance Wireless Sensor Network Node*. Crossbow. January 1, 2005. http://web.univ-pau.fr/~cpham/ENSEIGNEMENT/PAU-UPPA/RESA-M2/DOC/Imote2_Datasheet.pdf. Accessed 19 Mar 2014.

———. 2002a. *MICA2*. January 1, 2002. http://www.eol.ucar.edu/isf/facilities/isa/internal/CrossBow/DataSheets/mica2.pdf. Accessed 3 Feb 2014.

———. 2002b. *MICA2DOT*. January 1, 2002. http://www.eol.ucar.edu/isf/facilities/isa/internal/CrossBow/DataSheets/mica2dot.pdf. Accessed 5 Feb 2014.

———. 2006a. *MICAz*. January 1, 2006. http://www.openautomation.net/uploadsproductos/micaz_datasheet.pdf. Accessed 5 Feb 2014.

———. 2006b. *MOTE-VIEW 1.2 User's Manual*. January 1, 2006. http://www.willow.co.uk/MOTE-VIEW_User_Manual_.pdf. Accessed 20 Jan 2014.

————. 2007a. *MTS/MDA Sensor Board Users Manual.* August 1, 2007. http://www. investigacion.frc.utn.edu.ar/sensores/Equipamiento/Wireless/MTS-MDA_Series_Users_Man ual.pdf. Accessed 7 Jan 2014.

————. 2007b. *Stargate NetBridge: Embedded Sensor Network Gateway.* Crossbow. January 1, 2007. http://www.openautomation.net/uploadsproductos/stargate_netbridge_datasheet.pdf. Accessed 12 Aug 2014.

————. 2004. *Stargate: X-Scale, Processor Platform.* Crossbow. January 1, 2004. http://www.eol. ucar.edu/isf/facilities/isa/internal/CrossBow/DataSheets/stargate.pdf. Accessed 19 Mar 2014.

Dubois-Ferrière, H., L. Fabre, R. Meier, and P. Metrailler. 2006. TinyNode: A Comprehensive Platform for Wireless Sensor Network Applications. In *The 5th International Conference on Information Processing in Sensor Networks (IPSN)*, 358–365. Nashville: ACM.

EasySen LLC. 2008a. *SBT30EDU: Sensor and Prototyping Board.* EasySen LLC. January 29, 2008. http://www.easysen.com/support/SBT30EDU/DatasheetSBT30EDU.pdf. Accessed 12 Aug 2014.

————. 2008b. *WiEye: Sensor board for wireless surveillance and security applications.* EasySen LLC. January 29, 2008. http://www.easysen.com/support/WiEye/DatasheetWiEye.pdf. Accessed 12 Aug 2014.

Geospace Technologies. 2014a. *Geophones GS-11D.* January 1, 2014. http://www.geospace.com/ geophones-gs-11d/. Accessed 1 Feb 2014.

————. 2014b. *GS-1 Low Frequency Seismometer.* January 1, 2014. http://www.geospace.com/ tag/gs-1-low-frequency-seismometer/. Accessed 1 Feb 2014.

Honeywell. 2012. *Digital Compass Solutions HMR3300.* 2012. http://www51.honeywell.com/aero/ common/documents/myaerospacecatalog-documents/MissilesMunitions-documents/ HMR3300_Datasheet.pdf. Accessed 19 Dec 2013.

Intel. 2005a. *Intel PXA270 Processor.* Intel. January 1, 2005. http://www.armkits.com/download/ PXA270datasheet.pdf. Accessed 2 Apr 2014.

————. 2005b. *Intel StrataFlash Embedded Memory (P30).* Intel. November 1, 2005. http://www. xilinx.com/products/boards/ml505/datasheets/30666604.pdf. Accessed 3 Apr 2014.

————. 2000. *Intel StrongARM* SA-1110 Microprocessor.* Intel. April 1, 2000. http://access.ee. ntu.edu.tw/course/SoC_Lab_961/reference/StrongARM%20Datasheet.pdf.   Accessed 23 Mar 2014.

Maxim Integrated. 2008. *DS18B20 Programmable Resolution 1-Wire Digital Thermometer.* Maxim Integrated. Janaury 1, 2008. http://datasheets.maximintegrated.com/en/ds/DS18B20. pdf. Accessed 6 Apr 2014.

————. 2010. *DS18S20 High-Precision 1-Wire Digital Thermometer.* Maxim Integrated. January 1, 2010. http://datasheets.maximintegrated.com/en/ds/DS18S20.pdf. Accessed 11 Mar 2014.

MaxStream. 2007. *XBeeTM/XBee-PROTM OEM RF Modules.* MaxStream. January 1, 2007. https://www.sparkfun.com/datasheets/Wireless/Zigbee/XBee-Manual.pdf.   Accessed 9 Mar 2014.

Microchip. 2000. *PIC18 Microcontroller Family.* January 1, 2000. http://ww1.microchip.com/ downloads/en/DeviceDoc/30327b.pdf. Accessed 26 Jan 2014.

Moteiv. 2004. *Telos: Ultra low power IEEE 802.15.4 compliant wireless sensor module.* Moteiv. May 12, 2004. http://www2.ece.ohio-state.edu/~bibyk/ee582/telosMote.pdf.   Accessed 27 Mar 2014.

————. 2006a. *Tmote Connect: Wireless Gateway Appliance Software.* Moteiv. December 12, 2006. http://automatica.dei.unipd.it/public/Schenato/PSC/2010_2011/gruppo4-Building_ termo_identification/Bibliografia%20Casuale/tmote-connect-datasheet.pdf.   Accessed 25 Jan 2015.

————. 2006b. *Tmote sky Data Sheet: Ultra low power IEEE 802.15.4 compliant wireless sensor module.* June 2, 2006. http://www.eecs.harvard.edu/~konrad/projects/shimmer/references/ tmote-sky-datasheet.pdf. Accessed 3 Jan 2014.

Oki Semiconductor. 2004. *ML675K Series.* Oki Semiconductor. February 1, 2004. http://www.keil. com/dd/docs/datashts/oki/ml675xxx_ds.pdf. Accessed 21 Mar 2014.

OmniVision Technologies. 1999. *OV6620/OV6120*. OmniVision Technologies. June 1, 1999. http://coecsl.ece.illinois.edu/ge423/datasheets/DS-OV6620-1.2.pdf. Accessed 29 Mar 2014.

———. 2003. *OV7640 Color CMOS VGA (640 x 480) CAMERACHIP, OV7140 B&W CMOS VGA (640 x 480) CAMERACHIP*. OmniVision Technologies. January 15, 2003. http://www.datasheet4u.com/datasheet/O/V/7/OV7640_OmniVision.pdf.html. Accessed 27 Mar 2014.

———. 2006. *OV9655/OV9155 CMOS SXGA (1.3 MegaPixel) CameraChip Sensor with OmniPixel Technology*. OmniVision Technologies. November 21, 2006. http://www.surveyor.com/blackfin/OV9655-datasheet.pdf. Accessed 2 Apr 2014.

———. 2002. *OVT OV528: Single Chip Camera-to-Serial Bridge* . OmniVision Technologies. October 10, 2002. http://www.datasheet-pdf.com/datasheet-html/O/V/5/OV528-OmniVision.pdf.html. Accessed 29 Mar 2014.

Omron. 2014. *Omron Subminiature Basic Switch*. Omron. January 1, 2014. http://datasheet.octopart.com/SS-5-Omron-datasheet-17932117.pdf. Accessed 10 Mar 2014.

Philips. 2002. *PCF50606/605: Single-chip Power Management Unit+*. Philips. May 1, 2002. http://www.datasheetarchive.com/dl/Datasheet-03/DSA0051772.pdf. Accessed 3 Apr 2014.

Polastre, J., R. Szewczyk, and D. Culler. 2005. Telos: Enabling Ultra-low Power Wireless Research. In *Fourth International Symposium on Information Processing in Sensor Networks (IPSN)*, 364–369. Los Angeles: ACM/IEEE.

Qimonda AG. 2006. *HYB18L512160BF-7.5 DRAMs for Mobile Applications 512-Mbit Mobile-RAM* . Qimonda AG. December 1, 2006. http://www.datasheet4u.com/datasheet/H/Y/B/HYB18L512160BF-75_QimondaAG.pdf.html. Accessed 2 Apr 2014.

Semtech. 2008. *XE1205*. December 10, 2008. http://www.semtech.com/images/datasheet/xe1205.pdf. Accessed 2 Feb 2014.

Silicon Laboratories. 2004. *C8051F121: 100 MIPS, 128 kB Flash, 12-Bit ADC, 64-Pin Mixed-Signal MCU*. Silicon Laboratories. June 15, 2004. https://www.silabs.com/Support%20Documents/TechnicalDocs/C8051F121-Short.pdf. Accessed 28 Feb 2014.

———. 2013. *CP2102/9: Single-Chip USB to UART Data*. Silicon Laboratories. December 1, 2013. http://www.silabs.com/Support%20Documents/TechnicalDocs/CP2102-9.pdf. Accessed 3 Apr 2014.

SOWNet. 2014. *SOWNet's T-Node*. SOWNet. January 1, 2014. http://www.sownet.nl/download/T-Node_product_sheet.pdf. Accessed 18 July 2014.

SOWNet Technologies. 2014. *Introducing the G-Node*. SOWNet Technologies. January 1, 2014. http://www.sownet.nl/index.php/en/products/gnode. Accessed 16 July 2014.

TAOS. 2001. *TSL250R, TSL251R, TSL252R Light to Voltage Optical Sensors*. Texas Advanced Optoelectronic Solutions. January 1, 2001. http://www.goblack.de/desy/digitalt/sensoren/tsl-250/tsl250r.pdf. Accessed 10 Mar 2014.

Texas Instruments. 2011a. *CC2530F32, CC2530F64, CC2530F128, CC2530F256: A True System-on-Chip Solution for 2.4-GHz IEEE802.15.4 and ZigBee Applications*. Texas Instruments Inc. February 1, 2011. http://www.ti.com/lit/ds/symlink/cc2530.pdf. Accessed 30 Jan 2017.

———. 2007a. *CC1000: Single Chip Very Low Power RF Transceiver*. Texas Instruments. 2007. http://www.ti.com/lit/ds/symlink/cc1000.pdf. Accessed 23 Dec 2013.

———. 2014a. *CC1020: Low-Power RF Transceiver for Narrowband Systems*. Texas Instruments. January 1, 2014. http://www.ti.com/lit/ds/symlink/cc1020.pdf. Accessed 3 Jan 2015.

———. 2005a. *CC1100: Low-Cost Low-Power Sub-1 GHz RF Transceiver*. Texas Instruments. January 1, 2005. http://datasheet.octopart.com/CC1100RTKR-Texas-Instruments-datasheet-10422951.pdf. Accessed 3 Feb 2014.

———. 2014b. *CC1101: Low-Power Sub-1 GHz RF Transceiver*. Texas Instruments. January 1, 2014. http://www.ti.com/lit/ds/symlink/cc1101.pdf. Accessed 10 Sept 2014.

———. 2005b. *CC2420: First Single-chip 2.4 GHZ Ieee 802.15.4 Compliant And Zigbeetm Ready RF Transceiver*. January 1, 2005. http://www.ti.com/lit/ds/symlink/cc2420.pdf. Accessed 3 Jan 2014.

————. 2006. *CC2430: A True System-on-Chip solution for 2.4 GHz IEEE 802.15.4 / ZigBee.* Texas Instruments. January 1, 2006. http://www.ti.com.cn/cn/lit/ds/symlink/cc2430.pdf. Accessed 6 Apr 2014.

————. 2005c. *CC2431: System-on-Chip for 2.4 GHz ZigBee®/ IEEE 802.15.4 with Location Engine.* January 1, 2005. http://www.ti.com/lit/ds/symlink/cc2431.pdf. Accessed 3 Jan 2014.

————. 2011b. *MSP430C11x1, MSP430F11x1A Mixed Signal Microcontroller.* January 1, 2011. http://www.ti.com/lit/ds/symlink/msp430f1611.pdf. Accessed 1 Feb 2014.

————. 2007b. *MSP430x241x, MSP430x261x Mixed Signal Micontroller.* Texas Instruments. October 1, 2007. http://datasheet.octopart.com/MSP430F2618TPM-Texas-Instruments-datasheet-148968.pdf. Accessed 10 Sept 2014.

Toshiba. 2002. *TC55VCM208ASTN40,55: 524,288-Word by 8-bit Full CMOS Static RAM.* Toshiba. July 4, 2002. http://pdf.datasheetcatalog.com/datasheet2/1/03ace2kfgo489cs3i32 xflsi1xcy.pdf. Accessed 4 Apr 2014.

Wolfson Microelectronics. 2011. *WM8950: ADC with Microphone Input and Programmable Digital Filters.* Wolfson Microelectronics. November 1, 2011. http://www.wolfsonmicro.com/ documents/uploads/data_sheets/en/WM8950_1.pdf. Accessed 3 Apr 2014.

Xilinx. 2013. *Spartan-3 FPGA Family Data Sheet.* Xilinx. June 27, 2013. http://www.xilinx.com/ support/documentation/data_sheets/ds099.pdf. Accessed 19 Dec 2013.

————. 2007. *XC2C256 CoolRunner-II CPLD.* XILINX. March 8, 2007. http://www.xilinx.com/ support/documentation/data_sheets/ds094.pdf. Accessed 23 Mar 2014.

Part V
Ignition

Chapter 11
Third Takeoff

Is takeoff possible with unfastened seatbelts?

Throughout this book, all aspects related to WSNs were presented in full detail as made available in the literature. References were checked and double-checked for accuracy. Authors were contacted for missing or unclear information, some did thankfully reply.

The start, in Chap. 1, was by laying out the main features and distinguished characteristics of WSNs, followed in Chap. 2 by the standards that arose with the awakening of their endless applications. In Chap. 3, the applications of WSNs are updated and given ample space to stress their wide use in military and civil domains. Starting from military applications, WSNs have penetrated all daily life applications, from health to mining, industry, environment, agriculture, traffic, car parking, robotics, etc.

The network layer is added to this edition in Chap. 4. It focuses on the energy and lifetime-aware routing protocols designed to maintain sustained WSN functionality. The transport layer in WSNs is the main topic in this book. As detailed in Chap. 5, it differs widely from that of other wireless and wired networks due to the limited processing, storage, and energy. In Chap. 6, cross-layering is presented, as an evolving design approach for WSN protocols to tackle the efficient energy utilization requirements and the lack of resources.

Modeling and analyzing WSNs are major domains for study, research, and industry. To test and check WSN deployment and protocols, testbeds are required before typical in-field deployment. Chapter 7 provides an in-depth coverage of the testbeds obtainable for public use or those bound by restricted use in several research and industrial institutions for a variety of applications. Testbeds acquire particular importance in WSNs due to the risk of deploying a network with large numbers of nodes before full awareness of the underlying protocols from the physical layer up to the application layer.

Complementary to testbeds, simulators and emulators for WSNs are profoundly exposed in Chap. 8. Emulators are software tools to check hardware compatibility with the protocols to be used and the surrounding environment; they are hardware-

© The Author(s), under exclusive license to Springer Nature Switzerland AG 2023
H. M. A. Fahmy, *Concepts, Applications, Experimentation and Analysis of Wireless Sensor Networks*, Signals and Communication Technology,
https://doi.org/10.1007/978-3-031-20709-9_11

dependent and thus are easy to pick up. On the contrary, simulators are numerous, for general networking, for wireless, or for WSNs. Selecting a typical simulator is not an easy decision; it depends on the model under study, on the simulator convenience, and importantly on the researcher's skills at the programming level. A full analysis of main emulators and simulators is covered in this book to help whoever are interested understand and make the appropriate choice.

WSNs are not just theories a broad spectrum of WSN industry products and technologies are available, as well as a wide diversity of manufacturers engage in the market with astounding innovations. Chapters 9 and 10 discuss the leading manufacturers and the full range of WSN industry products. They are not to be left over; they must be accessed whenever a product or a manufacturer is cited in the text.

This book is a helper and mentor at more than one level. It is for senior undergraduates willing to understand WSNs and build their graduation projects. Also, it is intended for graduate students making a thesis and in need for specific knowledge on WSNs protocols and the related simulators and emulators. Moreover, it targets practitioners interested in the features and applications of WSNs and the available testbeds.

This edition is the third takeoff, and there is one more hop forward.

Index

H. M. A. Fahmy, *Concepts, Applications, Experimentation and Analysis of Wireless Sensor Networks*, Signals and Communication Technology,
https://doi.org/10.1007/978-3-031-20709-9

Printed in the United States
by Baker & Taylor Publisher Services